ADDITIONAL STUDENT PROBLEM SET WITH SOLUTIONS

TO ACCOMPANY

# ENGINEERING CIRCUIT ANALYSIS

FIFTH EDITION

WILLIAM H. HAYT, Jr.
Professor of Electrical Engineering, Emeritus
Purdue University

JACK E. KEMMERLY
Professor of Engineering, Emeritus
California State University, Fullerton

**McGraw-Hill, Inc.**
New York St. Louis San Francisco Auckland Bogotá
Caracas Lisbon London Madrid Mexico City Milan Montreal
New Delhi San Juan Singapore Sydney Tokyo Toronto

This book is printed on recycled, acid-free paper containing a minimum of 50% recycled de-inked fiber.

Additional Student Problem Set
With Solutions to Accompany
ENGINEERING CIRCUIT ANALYSIS
Fifth Edition



ISBN 0-07-027412-6

2 3 4 5 6 7 8 9 0 WHT WHT 9 0 9 8 7 6 5 4 3

# CONTENTS

## PREFACE

Although this manual is intended mainly for use by students who are studying basic circuit analysis from the textbook *Engineering Circuit Analysis, Fifth Edition,* by William H. Hayt, Jr. and Jack E. Kemmerly, it can be used with considerable benefit by students who are using other comparable, calculus-based, introductory circuits textbooks. The manual contains more than 600 problem statements, plus detailed solutions to those problems. Most of the problems were taken from *Engineering Circuit Analysis, Fourth Edition*; but there are also approximately forty additional problems which deal with computer-aided analysis, using SPICE™, and with state-variable analysis, topics that were not included in the Fourth Edition.

The problem statements pertaining to each chapter are immediately followed by the solutions to the problems in that chapter. Moreover, within each chapter grouping of problem statements, sub-groups of problems are identified (by section number and title) with specific sections within the chapter. For example, students who are troubled by the use of Thévenin's and Norton's theorems can refer to Section 2-6 in *Engineering Circuit Analysis* for a theoretical coverage of those theorems; then they should scrutinize the corresponding problem statements and problem solutions in Section 2-6 in this manual in which there are nine problems that deal with this important subject. Thus, students who are troubled by some particular analysis technique may easily locate the group of problems that will hopefully eradicate their difficulties.

Many students who study circuit analysis often plead with their instructors for more "worked examples" from which to study. Their pleas are met with mixed reactions. More of such examples cannot be included in the textbook itself; their inclusion would make the text too big! Furthermore, many collections of example problems that *are* available use non-standard symbology that is different from that used in *Engineering Circuit Analysis*; this can lead to confusion. Finally, many instructors feel that students are lulled into a false sense of accomplishment when their studies become too concentrated on simply reading problem solutions that they did not produce themselves. Hence, this manual must be used with caution. The following analogy may be appropriate.

If we sit in front of our TV sets and watch a world-class weight lifter hoist a 500-pound barbell over his head, we may very well understand how he did it, simply by watching. But without extensive exercise and muscle development, we cannot do it ourselves. So it is with circuit analysis. It is relatively easy for us to look at, and understand, someone else's solutions to a particular problem; but solving that same problem, or a similar one, ourselves is another matter entirely. We must exercise and develop our intellectual muscles by attempting to work the problem ourselves, *before* we rely too heavily on studying the work of others.

So, in summary, students should use the solutions in this manual by first reading a problem statement carefully; then they should make an honest effort to solve the problem; and, finally, they should refer to the detailed solution if necessary. The subject of circuit analysis is extremely problem-oriented. Unfortunately there are no short cuts, no royal roads to knowledge. The "secret" to mastering the subject is to work lots of problems — the more the better.

William H. Hayt, Jr.
Jack E. Kemmerly
1992

# CHAPTER 1: UNITS, DEFINITIONS, EXPERIMENTAL LAWS, AND SIMPLE CIRCUITS

## Section 1-2 Systems of Units

1 A famous mild-mannered reporter has a mass of 80 kg, can leap a tall (250-m) building at a single bound, and is as fast as a speeding (600 m/s) bullet. (*a*) What is his top speed in miles per hour? (*b*) What energy in joules must he impart to his leap to just clear the tall building? (*c*) For how many days would this energy power an electronic calculator requiring 100 mW? (*d*) What is the name of the reporter's girl friend?

2 A 70-kg student burns up energy at an average rate of 120 W over a 24-h day. This metabolic rate drops to 75 W while sleeping, rises to 230 W while walking and about 1000 W while running 10 mi/h. (*a*) What horsepower is expended running at 10 mi/h? (*b*) How many cans of light beer (100 kcal/can) should be consumed daily, just to keep the old body going?

3 When a certain electric clothes dryer is in full operation, it draws 5.2 kW. (*a*) How many "average horses" would have to be used to generate this power? (*b*) Assuming that fuel oil has a density of 50 lbm/ft$^3$ and yields 18 500 Btu per pound-mass, how many liters (1 liter = 1000 cubic centimeters) of fuel oil would be used to operate the dryer for one hour?

## Section 1-3 Charge, current, voltage and power

4 The net charge that has moved to the right in a conductor at point $x$ is given as a function of time by: $t \leq -4$ s, $q = 0$; $-4 \leq t \leq -2$, $q = 4t + 16$; $-2 \leq t \leq 1$, $q = 2t^2$; $1 \leq t \leq 6$, $q = 5\sqrt{t+3} - 8$; $t \geq 6$, $q = 7$ C. (*a*) Plot $q$ versus $t$. (*b*) Calculate $i$ at $t = -4.5, -3.5, -2.5, \ldots, 5.5, 6.5$ s and plot $i$ versus $t$.

5 Given the plot of $i$ vs. $t$ of Fig. 1-1, calculate the total charge that has moved past the reference point in the time interval, $-2 < t < 5$ s.

6 The wave form shown in Fig. 1-2 has a period of 5 ms. (*a*) What is the average value of the current over one period? (*b*) How much charge is transferred in the interval $1 \leq t \leq 4$ ms? (*c*) If $q(0) = 0$, sketch $q(t)$, $0 < t < 12$ ms.

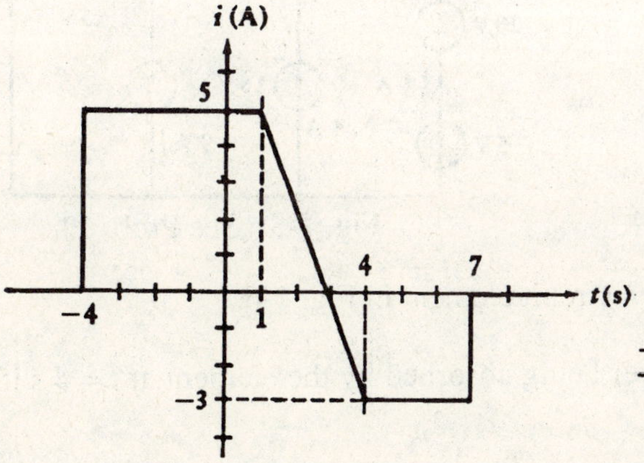


**Fig. 1-1:** See Prob. 5.

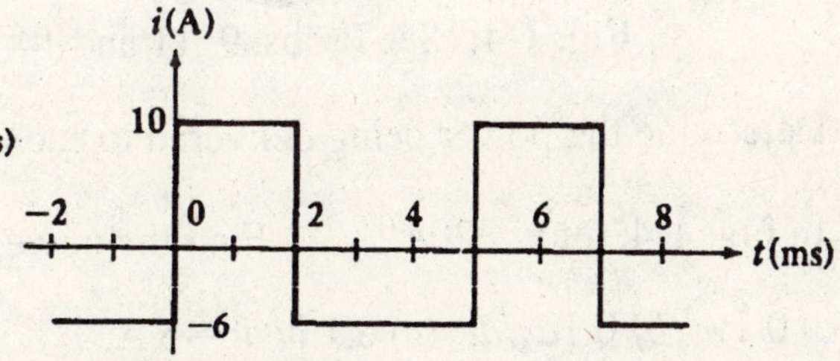


**Fig. 1-2:** See Probs. 6 and 10.

**7** The total charge that has flowed to the right past point $A$ in a certain conductor between $t = 0$ and $t$ is identified as $q_A(t) = 100e^{-200t} \cos 500t$ mC. *(a)* How much charge flows past $A$ to the right between $t = 1$ ms and $t = 2$ ms? *(b)* What is the current to the right at $A$ at $t = 1$ ms? *(c)* Now let the current directed to the right at $A$ be $i_A(t) = 2(e^{-5000t} - e^{-8000t})$ A and find the amount of charge flowing to the right between $t = 10$ μs and $t = 80$ μs.

**8** Determine the power being absorbed by each of the circuit elements shown in Fig. 1-3.

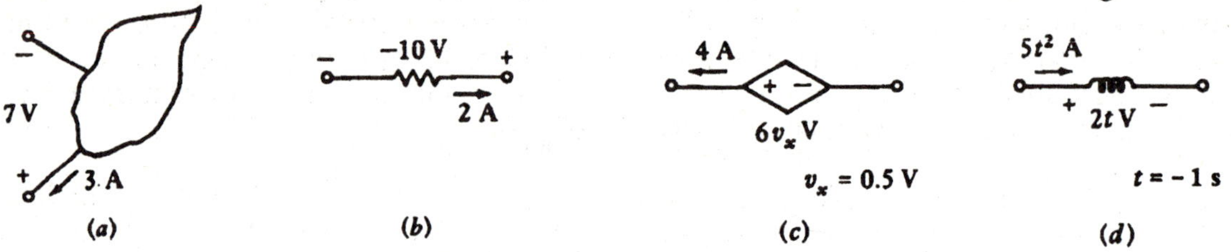


**Fig. 1-3:** See Prob. 8.

**9** Let $i = 4e^{-50t}$ A and $v = 20 - 30e^{-50t}$ V for the circuit element of Fig. 1-4. *(a)* How much power is being absorbed by the device at $t = 10$ ms? (b) How much energy is delivered to the device in the interval, $0 < t < \infty$?

**10** The current $i$ shown in Fig. 1-2 enters the terminal with the (+) sign on it for a circuit element with a voltage $v$ across it, where $v = 20 \sin 400\pi t$ V. *(a)* What is the maximum power absorbed by the element, and at what time does it occur? *(b)* What is the maximum power supplied *by* the element, and when does it occur? *(c)* How much energy is delivered to the circuit element in the time interval $0 < t < 5$ ms? *(d)* What average power is being absorbed by the element over this time interval?

**11** In Fig. 1-4, let $v = 2t^2 - 8t + 6$ V. The power being absorbed by the circuit element is $p = 4(t^3 - 6t^2 + 11t - 6)$ W. *(a)* How much energy is delivered to the element from $t = 1$ to $t = 3$ s? *(b)* How much charge is delivered to the element in the time interval $1 < t < 3$ s?

**12** Determine which of the five sources in Fig. 1-5 are being charged (absorbing positive power), and show that the algebraic sum of the five absorbed power values is zero.

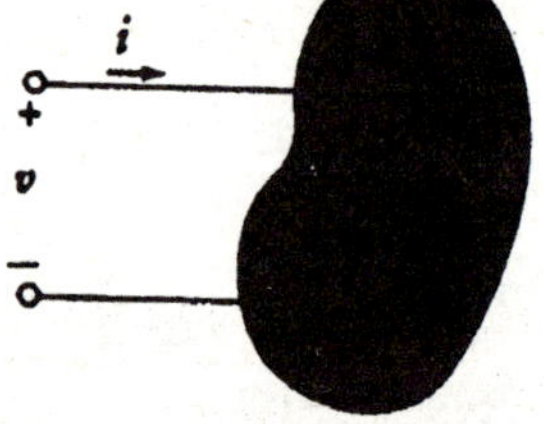


**Fig. 1-4:** See Probs. 9, 11 and 14.

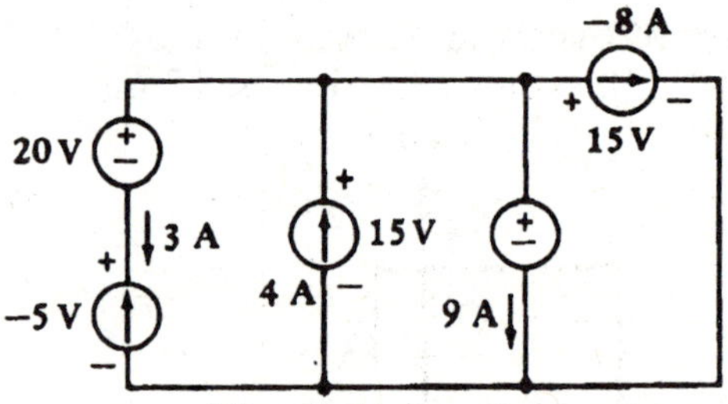


**Fig. 1-5:** See Prob. 12.

**13** Determine the power being delivered to each circuit element in Fig. 1-6.

**14** In Fig. 1-4, let $v = 20e^{-0.2t}$ V. Find the power being absorbed by the element at $t = 4$ s if $i =$:

*(a)* $0.1v$; *(b)* $0.1dv/dt$; *(c)* $0.1 \int_0^t v\, dt + 5$ A.

## Section 1-5 Ohm's law

**15** Find the resistance of a resistor if: (*a*) a constant current of 5 C/s delivers 10 kJ to the element in 4 min; (*b*) it dissipates 1 kW when connected across a 12-V battery. (*c*) What average power does the source $v_s = 80\cos 500\pi t$ V deliver to a 5-$\Omega$ resistor over the time interval $0 \le t \le 12$ ms?

**16** The current $i(t)$ in Fig. 1-7 is 4 A in the interval $0 < t < 1$ s, -4 A for $1 < t < 2$ s, and continues in this periodic manner. If $R = 100\ \Omega$, sketch $i(t)$, $v(t)$, and $p(t)$ for $0 < t < 4$s.

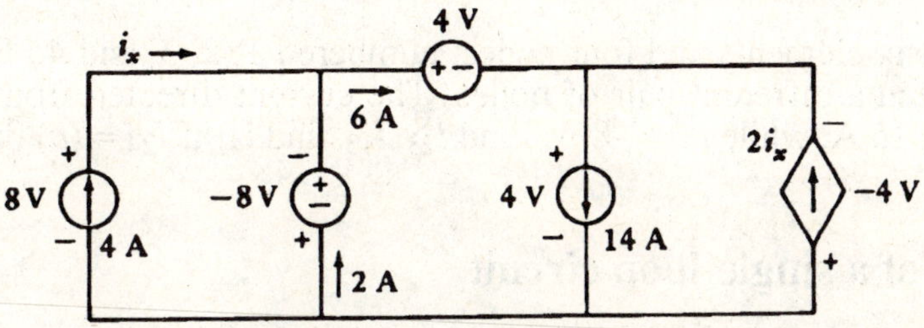


Fig. 1-6: See Prob. 13.

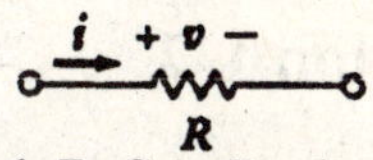


Fig. 1-7: See Prob. 16.

**17** The dc resistance of a conductor having a length $l$ and a uniform cross-sectional area $A$ is given by $R = \rho l/A = l/\sigma A$, where $\rho$ (rho) is the electrical resistivity and $\sigma$ (sigma) is the electrical conductivity. If $\sigma = 5.8 \times 10^7$ S/m for copper: (*a*) what length (in feet) of #18 copper wire (diameter = 1.024 mm) has a resistance of 1 $\Omega$? (*b*) If a circuit board has a copper-foil conducting ribbon 0.0013 in thick and 0.02 in wide that can carry 3 A safely at 60°C, what power does this current deliver to a conductor 6 in long?

## Section 1-6 Kirchhoff's laws

**18** In the circuit of Fig. 1-8, find $R$ and $G$ if the 5-A source is supplying 125 W.

**19** Use Ohm's law and Kirchhoff's laws on the circuit of Fig. 1-9 to find: (*a*) $v_{in}$; (*b*) the power provided by the dependent source; (*c*) $V_s$.

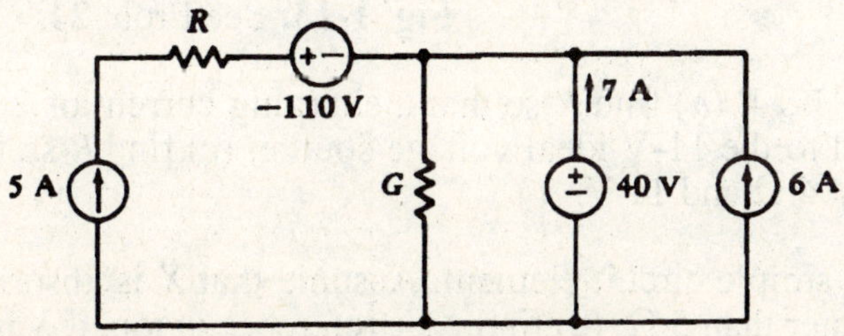


Fig. 1-8: See Prob. 18.

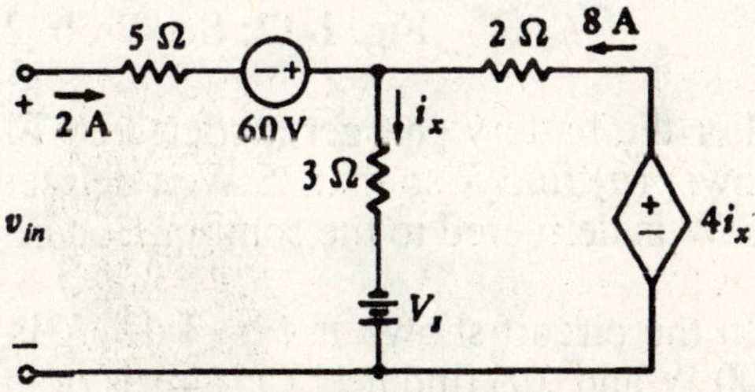


Fig. 1-9: See Prob. 19.

**20** (*a*) Use Ohm's and Kirchhoff's laws in a step-by-step procedure to determine all the voltages and currents in the circuit of Fig. 1-10. (*b*) Compute the power absorbed by every element and show that the sum is zero.

**21** In the circuit shown in Fig. 1-11, find the power absorbed by the: (*a*) 4-A source; (*b*) 2-$\Omega$ resistor; (*c*) 12-$\Omega$ resistor; (*d*) 6-$\Omega$ resistor; (*e*) voltage source $V_s$.

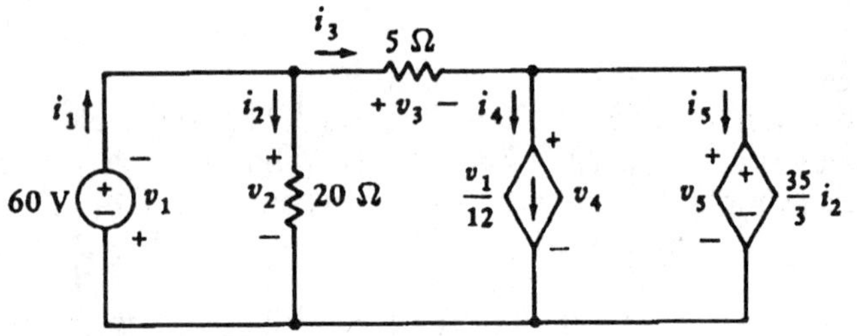


Fig. 1-10: See Prob. 20.

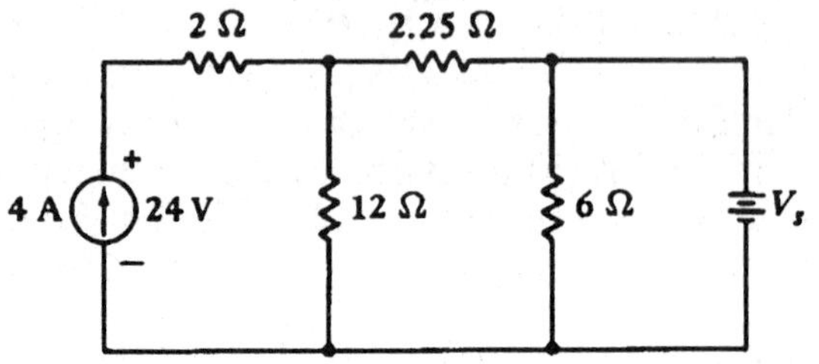


Fig. 1-11: See Prob. 21.

**22** A certain circuit contains six elements and four nodes, numbered 1, 2, 3, and 4. Each circuit element is connected between a different pair of nodes. The current directed from node 1 to node 2 in that branch is $i_{12} = 15$ A, while $i_{34} = -8$ A. Find $i_{23}$, $i_{13}$, and $i_{41}$ if $i_{24} =$ (*a*) 0; (*b*) 18 A; (*c*) -18 A.

## Section 1-7 Analysis of a single-loop circuit

**23** Find the power being absorbed by element *X* in Fig. 1-12 if it is a: (*a*) 70-Ω resistor; (*b*) 2-V independent voltage source, + reference on left; (*c*) dependent voltage source, + reference on left, labeled $19i_x$; (*d*) 52-mA independent current source, arrow directed to right.

**24** (*a*) Find *i* in the circuit shown in Fig. 1-13. (*b*) Let $v_4$ now extend across both the 4-Ω and 50-V elements, + reference on the left, and again find *i*.

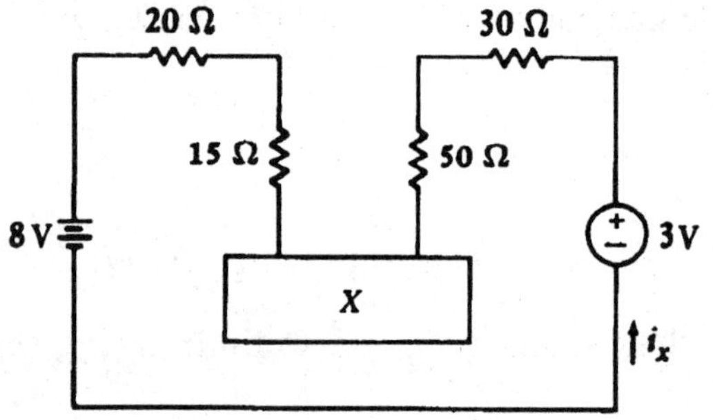


Fig. 1-12: See Prob. 23.

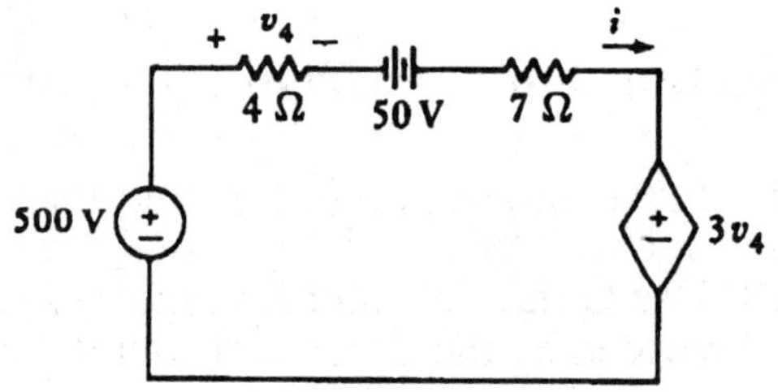


Fig. 1-13: See Prob. 24.

**25** For the battery charger modeled by Fig. 1-14: (*a*) find *R* so that a charging current of 2.5 A flows; (*b*) find *R* so that 25 W is delivered to the 11-V ideal voltage source; (*c*) find *R* so that 25 W is delivered to the combination of 0.04 Ω and 11 V.

**26** In the circuit shown in Fig. 1-15, *X* is a simple circuit element. Assume that *X* is absorbing 100 W and: (*a*) find *R* if *X* is a resistor greater than 5 Ω; (b) find $v_s$, + reference at top, if *X* is an independent voltage source, $v_s < 5$ V.

**27** Let the voltage across the 0.2-Ω resistor in Fig. 1-15 be 4 V, + reference at the left. Assume that *X* represents an unknown circuit element. (*a*) Find the absorbed power for each element in the circuit. (*b*) If *X* is a resistor, give its resistance *R* (which might be negative); if it is an independent voltage source, give $v_s$, including polarity; and if it is an independent current source, give $i_s$, including reference direction. (*c*) Repeat parts *a* and *b* if the + reference for the 4 V is at the right.

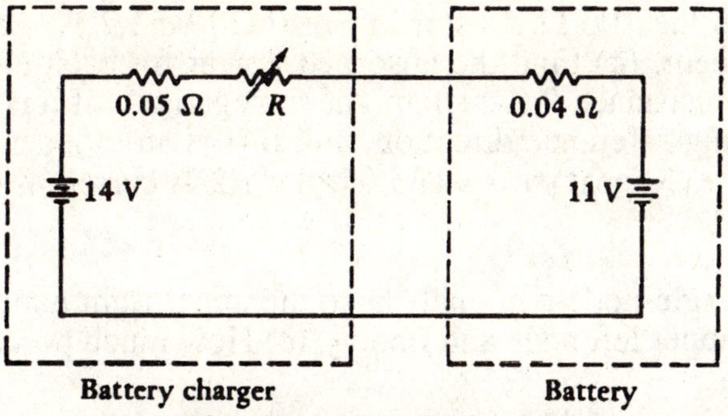


Fig. 1-14: See Prob. 25.

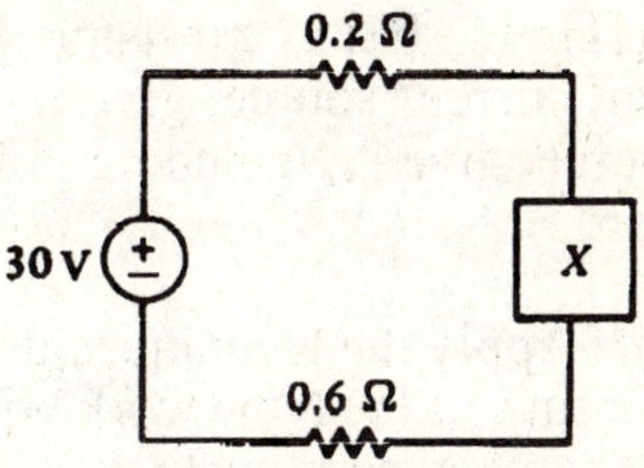


Fig. 1-15: See Probs. 26 and 27.

**28** Find the power absorbed by each of the six elements in the circuit in Fig. 1-16.

## Section 1-8 The single-node-pair circuit

**29** Find the power absorbed by each element in the circuit illustrated in Fig. 1-17.

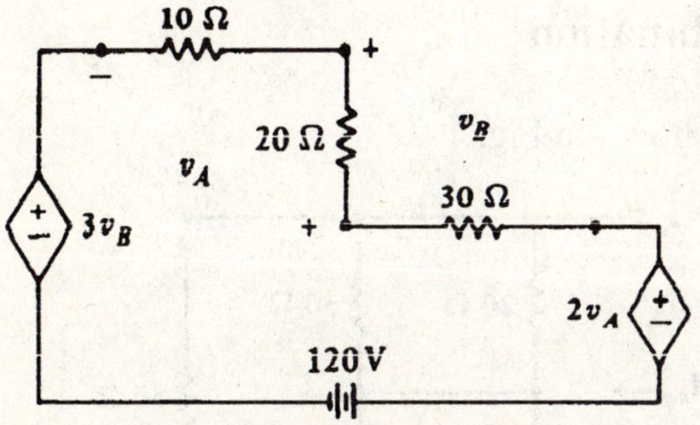


Fig. 1-16: See Prob. 28.

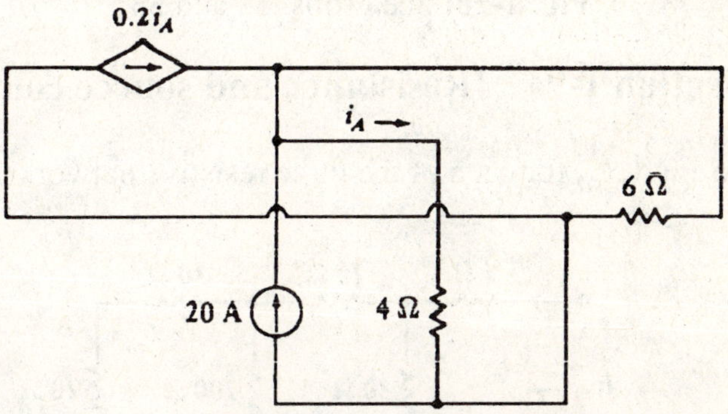


Fig. 1-17: See Prob. 29.

**30** In Fig. 1-18, find the power supplied by each of the sources if the dependent source is labeled: (*a*) $3i_A$; (*b*) $3i_B$.

**31** Find the power absorbed by each circuit element in the single-node-pair circuit of Fig. 1-19.

**32** In the circuit of Fig. 1-20, $X$ is a simple circuit element. Assume that $X$ is absorbing 100 W and: (*a*) find $R$ if $X$ is a resistor larger than 50 Ω; (*b*) find $i_s$, reference arrow directed down, if $X$ is an independent current source, $i_s > 2$ A.

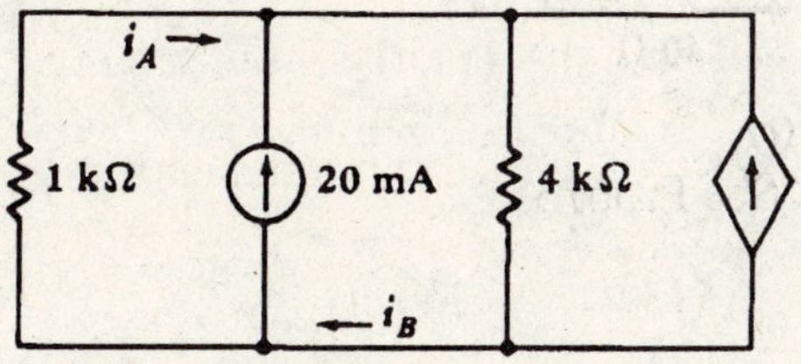


Fig. 1-18: See Prob. 30.

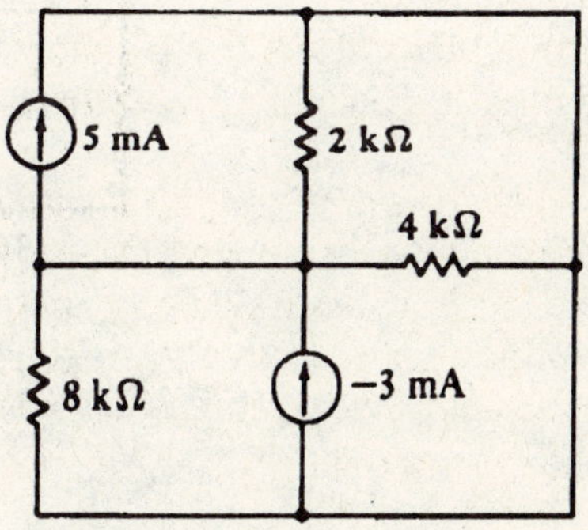


Fig. 1-19: See Prob. 31.

**33** Let the downward current through the 100-Ω resistor in Fig. 1-20 be 1.2 A. Assume that $X$ represents an unknown circuit element. (*a*) Find the absorbed power for each element in the circuit. (*b*) If $X$ is a resistor, give its resistance $R$ (which might be negative) ; if it is an independent current source, give $i_s$, including reference direction; and if it is an independent voltage source, give $v_s$, including polarity. (*c*) Repeat parts *a* and *b* if the 1.2-A current is directed upward.

**34** (*a*) Apply the techniques used in single-node-pair analysis to the upper right node in Fig. 1-21 and find $i_x$. (*b*) Now work with the upper left node and find $v_5$. (*c*) How much power is the 16-A source generating?

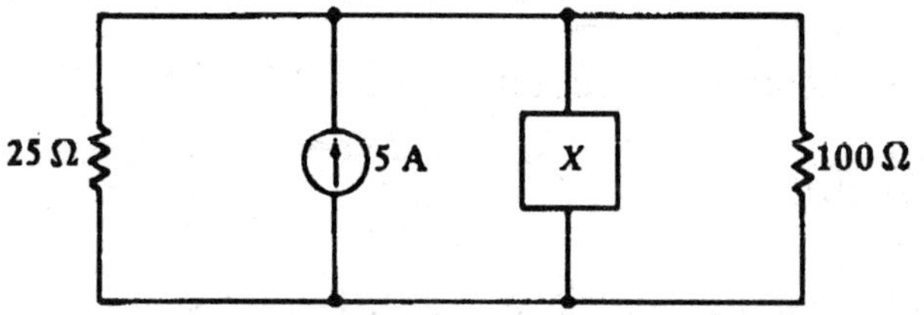


**Fig. 1-20:** See Probs. 32 and 33.

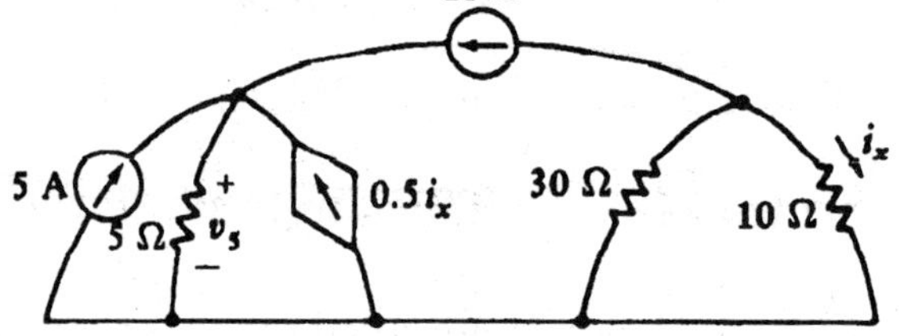


**Fig. 1-21:** See Prob. 34.

## Section 1-9 Resistance and source combination

**35** Find $R_{eq}$ for each of the three resistive networks shown in Fig. 1-22.

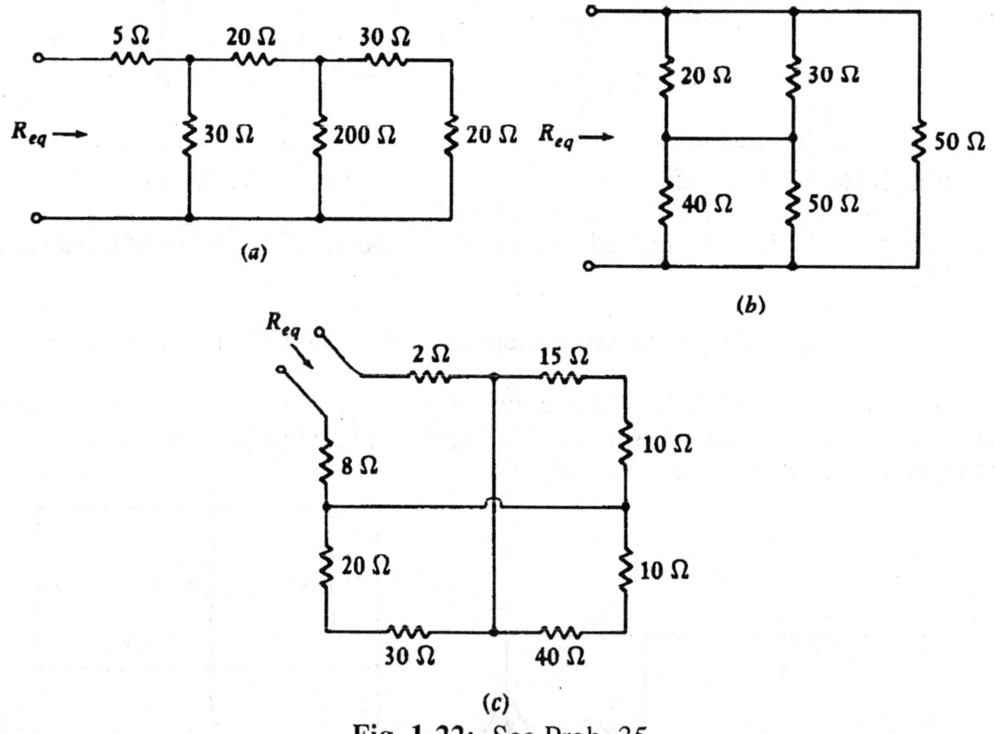


**Fig. 1-22:** See Prob. 35.

**36** For the network of Fig. 1-23: (*a*) find $R_{eq}$ if $R = 14$ Ω; (*b*) find $R$ if $R_{eq} = 14$ Ω.

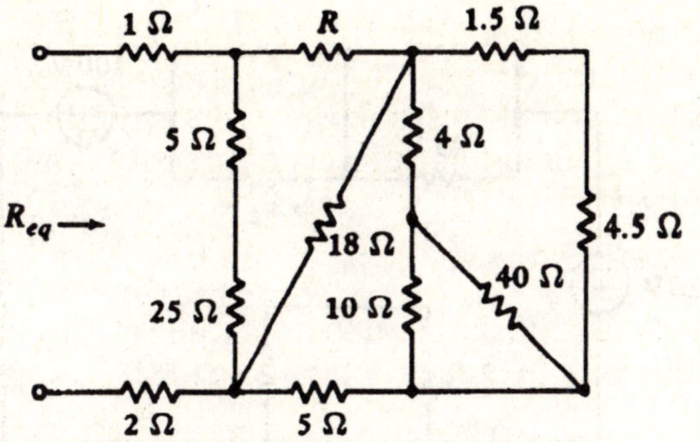


Fig. 1-23: See Prob. 36.

**37** Find $R_{eq}$ for each network shown in Fig. 1-24.

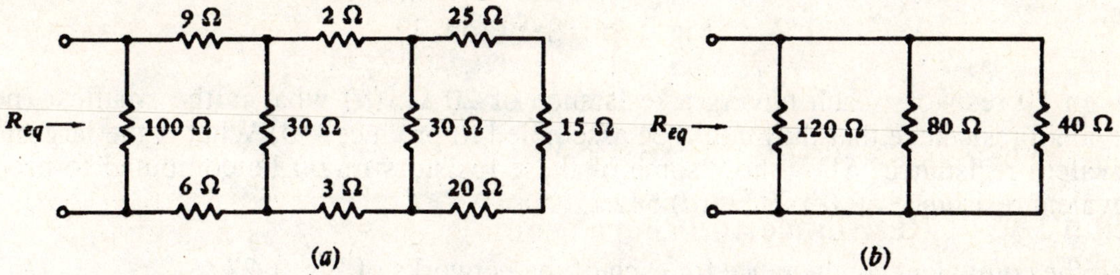

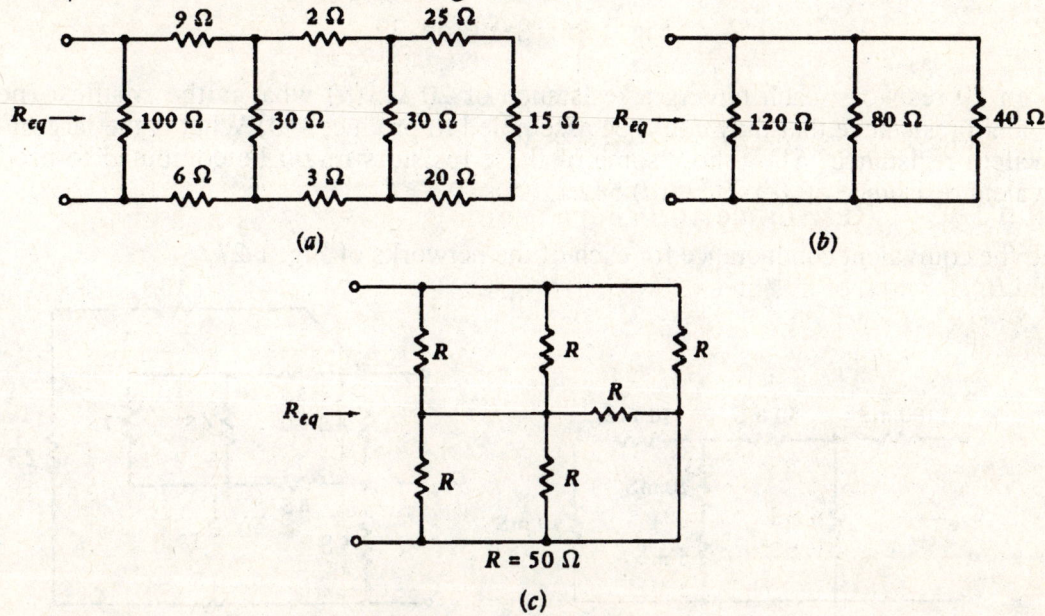

Fig. 1-24: See Prob. 37.

**38** By combining sources and resistances in the circuit of Fig. 1-25, find $v_x$ and $i_x$.

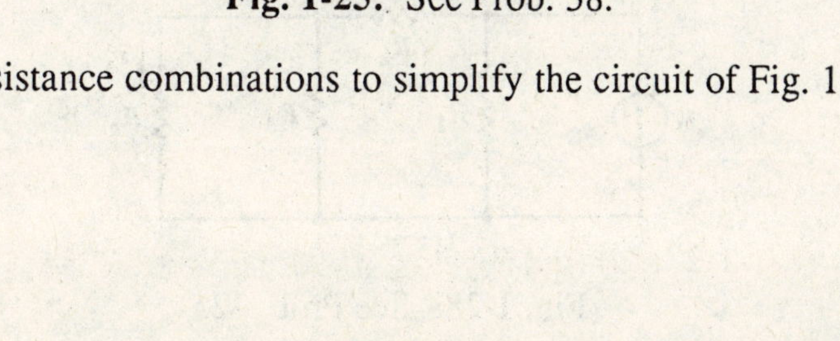

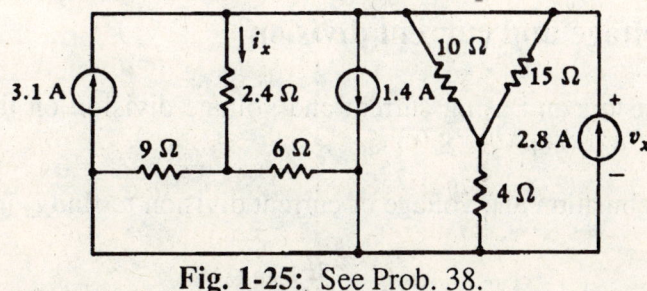

Fig. 1-25: See Prob. 38.

**39** Use source and resistance combinations to simplify the circuit of Fig. 1-26 and then evaluate $i_1$ and $v_2$.

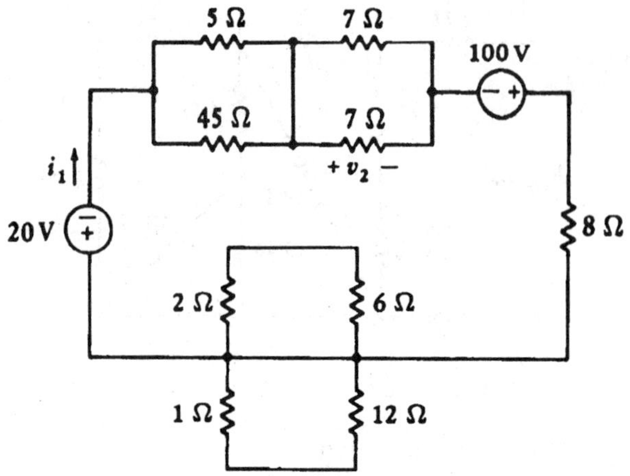


**Fig. 1-26:** See Prob. 39.

**40** Given 10 resistors, each having a resistance of 20 Ω: (*a*) what is the smallest (nonzero) equivalent resistance that they could be assembled to produce? (*b*) What is the largest (finite) equivalent resistance? Show how some of these resistors might be combined to produce an equivalent resistance of: (*c*) 29 Ω; (*d*) 6 Ω.

**41** Find the equivalent conductance for each of the networks of Fig. 1-27.

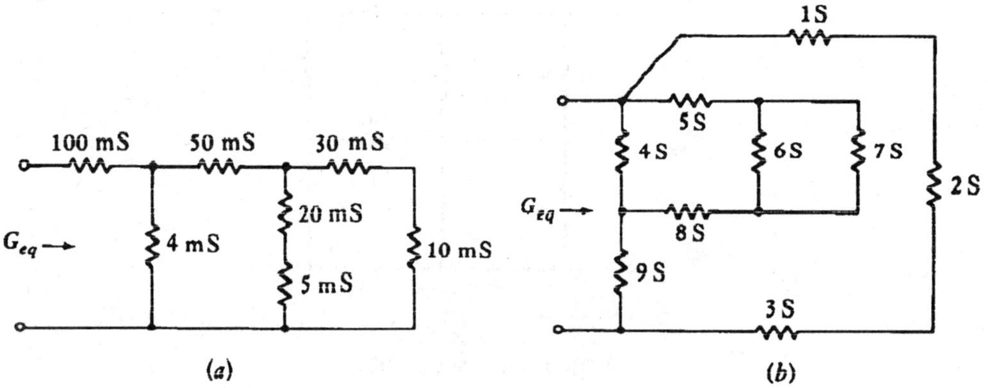


**Fig. 1-27:** See Prob. 41.

## Section 1-10 Voltage and current division

**42** By combining resistors and using current and voltage division on the circuit of Fig. 1-28, write an expression for: (*a*) $i_4$; (*b*) $v_2$; (*c*) $i_o$.

**43** Use resistance combination and voltage or current division to find $i_x$ in each circuit shown in Fig. 1-29.

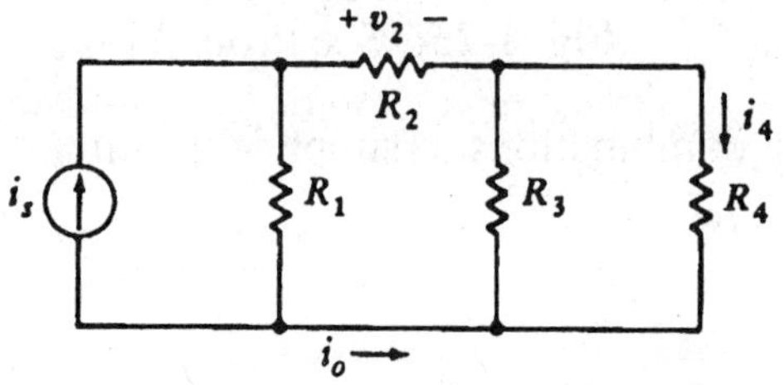


**Fig. 1-28:** See Prob. 42.

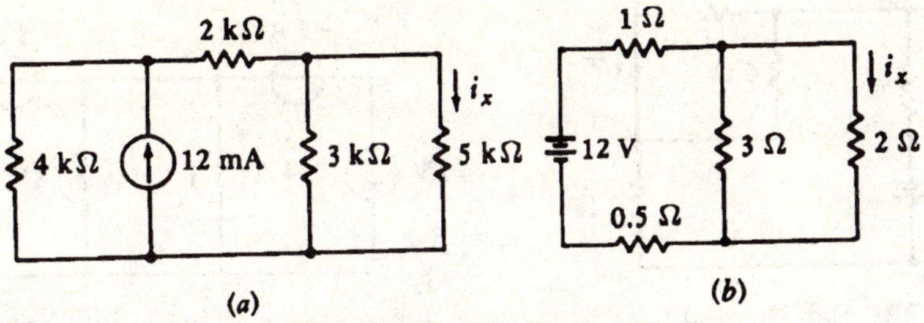


**Fig. 1-29:** See Prob. 43.

**44** (*a*) Write a one-line expression, using resistance combination and voltage division, to find $v_{57}$ in the circuit of Fig. 1-30*a*. (*b*) Write a one-line expression, using resistance combination and current division, to find $i_{57}$ for the circuit of Fig. 1-30*b*.

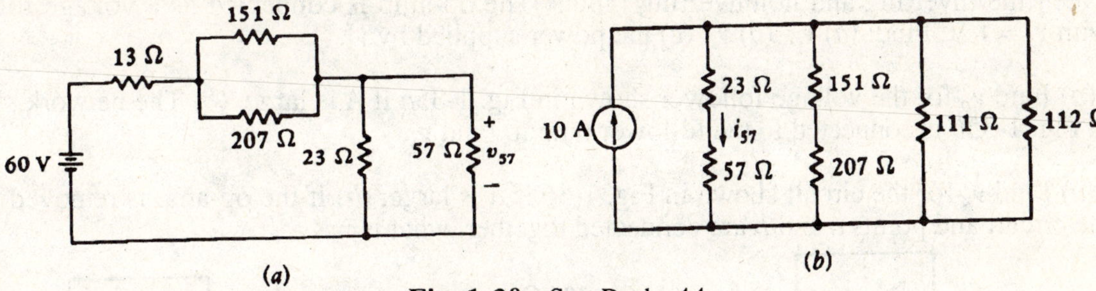


**Fig. 1-30:** See Prob. 44.

**45** The circuit shown in Fig. 1-31 contains a three-section resistive ladder network. (*a*) Assume $i_o = 1$ A and work one step at a time from the right side of the network toward the source and find $V_s$. (*b*) What is $V_s$ if $i_o = 0.4$ A? (*c*) If $V_s = 100$ V, find $i_o$.

**46** Find the value of $R$ in the circuit of Fig. 1-32.

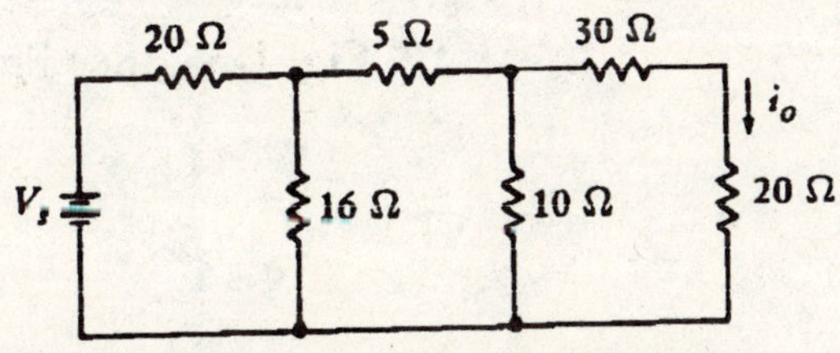


**Fig. 1-31:** See Prob. 45.

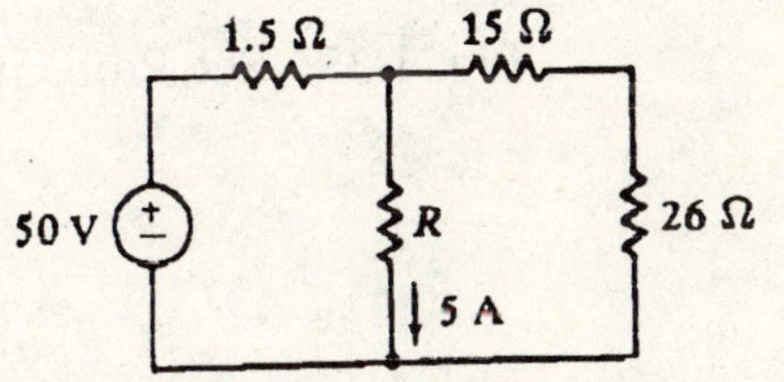


**Fig. 1-32:** See Prob. 46.

**47** (*a*) Use voltage division, current division, and resistance combination to find $i_y$ in the circuit of Fig. 1-33. (*b*) To what value should the 3-Ω resistance be changed to make $i_y = 1$ A?

**48** The circuit shown in Fig. 1-34 shows several examples of independent current and voltage sources in series and parallel. (*a*) Find the power absorbed by each source. (*b*) To what value should the 4-V source be changed to reduce the power supplied by the -5-A source to zero?

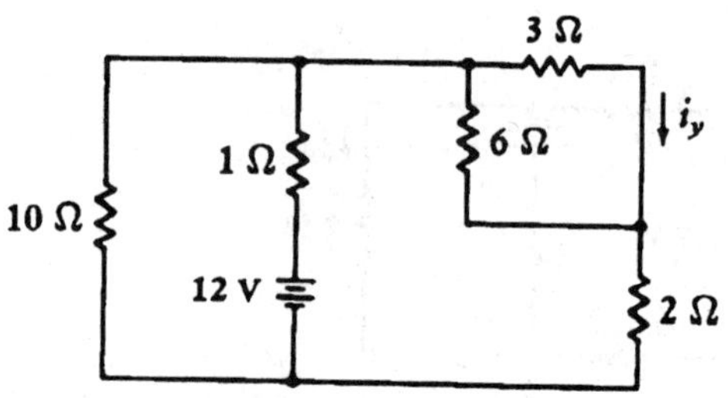


**Fig. 1-33:** See Prob. 47.

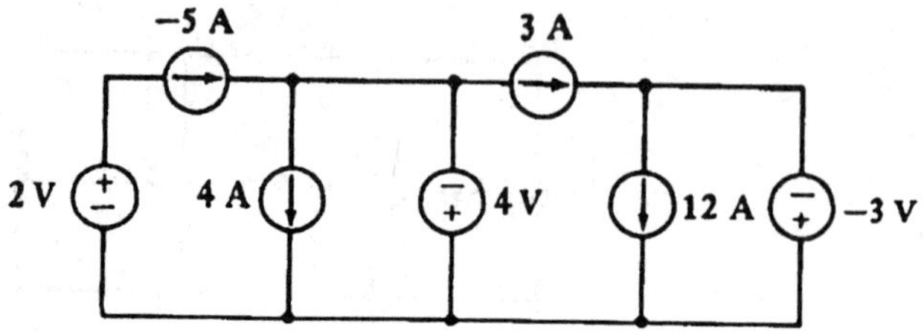


**Fig. 1-34:** See Prob. 48.

## Section 1-11 A practical example: the operational amplifier

**49** A certain op-amp has a gain $A$ of 20 000 and a resistance, $R_i = 50$ kΩ, inside the op-amp between the inverting and noninverting inputs. The op-amp is connected as a voltage follower with $v_s = 1$ V. Find: (*a*) $v_o$; (*b*) $v_i$; (*c*) the power supplied by $v_s$.

**50** (*a*) Find $v_o$ for the voltage follower shown in Fig. 1-35*a* if $A$ is large. (*b*) The network shown in Fig. 1-35*b* is connected to the follower output. Find $v_x$.

**51** (*a*) Find $v_x$ for the circuit shown in Fig. 1-36 if $A$ is large. (*b*) If the op-amp is removed from the circuit and points $a$ and $b$ are connected together, what is $v_x$?

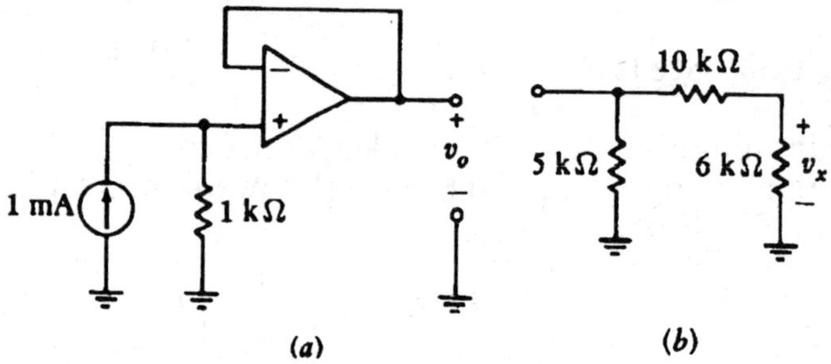


**Fig. 1-35:** See Prob. 50.

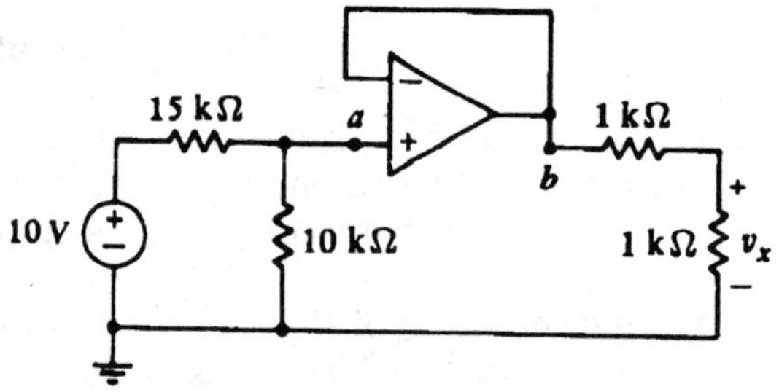


**Fig. 1-36:** See Prob. 51.

# Chapter 1

**1-1** (a) $\frac{600\,m}{1\,s} \times \frac{100\,cm}{1\,m} \times \frac{1\,in}{2.54\,cm} \times \frac{1\,ft}{12\,in} \times \frac{1\,mi}{5280\,ft} \times \frac{3600\,s}{1\,hr} = \underline{1342.2\ mph}$

(b) $g = 9.80665\,\frac{m}{s^2}$ $\therefore$ $P.E = mgh = 80 \times 9.80665 \times 250 = \underline{196{,}133\,J}$

(c) $t = \frac{196{,}133\,J}{0.1\,J/s} \times \frac{1\,hr}{3600\,s} \times \frac{1\,day}{24\,hr} = \underline{22.70\ days}$

(d) Lois Lane

**1-2** (a) $1000\,W \times \frac{1\,hp}{745.7\,W} = \underline{1.3410\,hp}$

(b) $120\,\frac{J}{s} \times \frac{3600\,s}{1\,hr} \times \frac{24\,hr}{day} \times \frac{1\,kcal}{4186.8\,J} \times \frac{1\,can}{100\,kcal} = \underline{24.76\ cans}$

**1-3** (a) $5200\,W \times \frac{1\,hp}{745.7\,W} = \underline{6.973\ horses}$

(b) $\frac{5200\,J}{1\,s} \times \frac{3600\,s}{1\,hr} \times \frac{1\,Btu}{1055.1\,J} \times \frac{1\,lbm}{18{,}500\,Btu} \times \frac{1\,ft^3}{50\,lbm} \times \frac{1728\,in^3}{1\,ft^3} \times \frac{(2.54)^3\,cm^3}{1\,in^3} \times \frac{1\,liter}{1000\,cm^3} = \underline{0.5431\ liter}$

**1-4** (a)

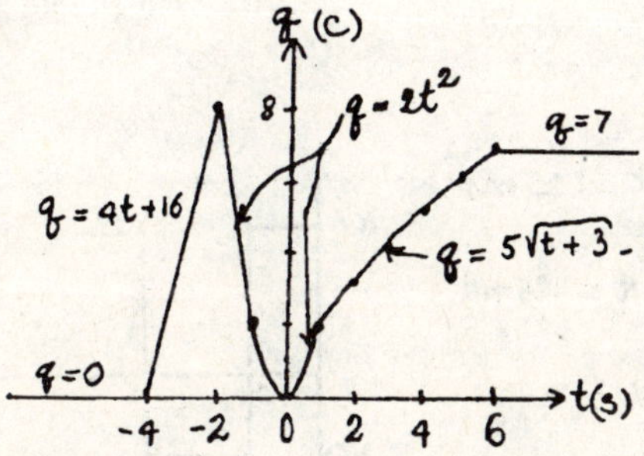


(b)

| t (s) | i (A) |
|---|---|
| -4.5 | 0 |
| -3.5 | 4 |
| -2.5 | 4 |
| -1.5 | -6 |
| -0.5 | -2 |
| 0.5 | 2 |
| 1.5 | 1.179 |
| 2.5 | 1.066 |
| 3.5 | 0.981 |
| 4.5 | 0.913 |
| 5.5 | 0.857 |

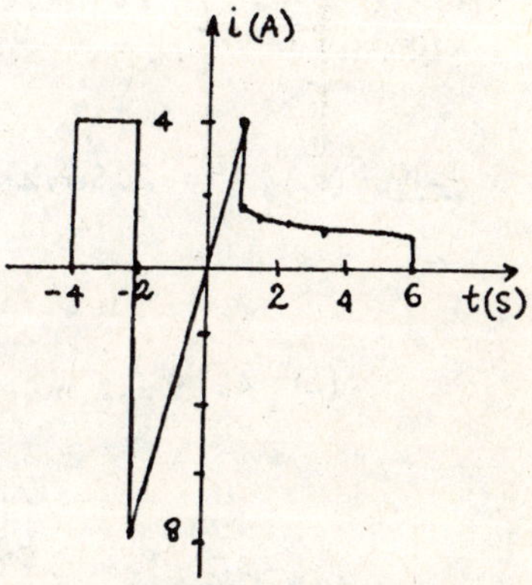


**1-5** $q = \int_{-2}^{1} 5\,dt + \int_{1}^{4}\left(-\frac{8}{3}t + \frac{23}{3}\right)dt + \int_{4}^{5} -3\,dt = 15 + \left(-\frac{4}{3}t^2 + \frac{23}{3}t\right)_1^4 - 3 = 12 - \frac{4}{3}(16-1) + \frac{23}{3} \times 3$

$q = 12 - 20 + 23 = \underline{15C}$

**1-6** (a) $i_{av} = \frac{2 \times 10 - 3 \times 6}{5} = \underline{0.4A}$

(b) $q = \int_{1ms}^{4ms} i\,dt = \int_{1ms}^{2ms} 10\,dt - \int_{2ms}^{4ms} 6\,dt = (10-12)10^{-3} = \underline{-2mC}$

(c) $0\text{-}2:\ q = 10t$ ; $2\text{-}5:\ q = 20 \times 10^{-3} - 6(t-2)10^{-3}$ ;

$5\text{-}7:\ q = 2 \times 10^{-3} + 10(t-5)10^{-3}$ ;

$7\text{-}10:\ q = 22 \times 10^{-3} - 6(t-7)10^{-3}$ ; $10\text{-}12:\ q = 4 \times 10^{-3} + 10(t-10)10^{-3}$

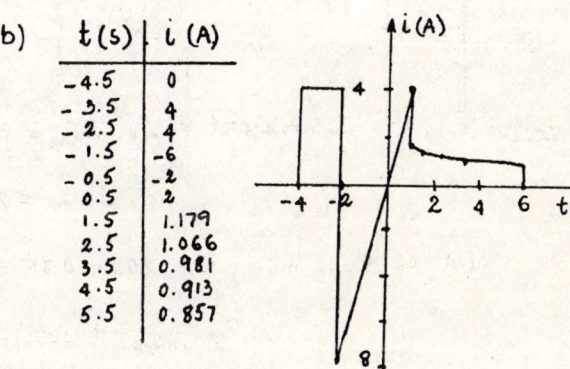

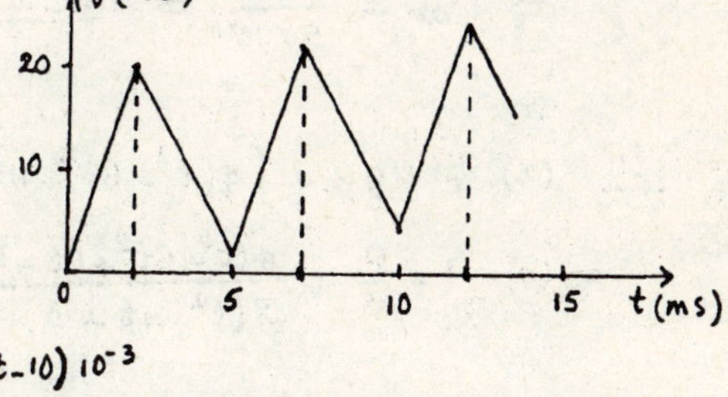

1-7 (a) $q_A = 0.1e^{-200t}\cos 500t$ C $\therefore q(2ms) - q(1ms) = 0.1(e^{-0.4}\cos 1 - e^{-0.2}\cos 0.5) = \underline{-35.63 mC}$

(b) $i_A = q_A' = -e^{-200t}(50\sin 500t + 20\cos 500t)$

$\therefore i_A(1ms) = -e^{-0.2}(50\sin 0.5 + 20\cos 0.5) = \underline{-34.00 mA}$

(c) $i_A = 2(e^{-5000t} - e^{-8000t})$ A $\therefore q_A = \int_{10\mu s}^{80\mu s}(2e^{-5000t} - 2e^{-8000t})\,dt$

$q_A = \frac{-2}{5000}(e^{-0.4} - e^{-0.05}) + \frac{2}{8000}(e^{-0.64} - e^{-0.08}) = \underline{13.408 \mu C}$

1-8 (a)

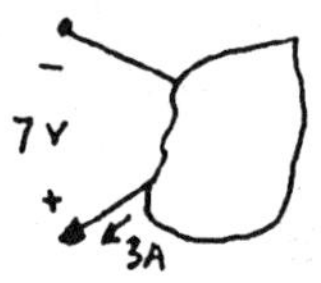


(b)

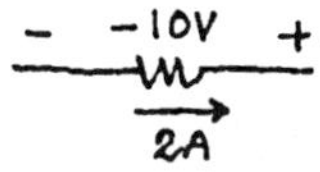


(c)

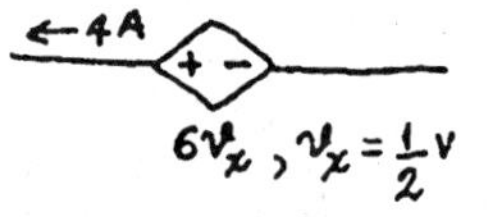


(d)

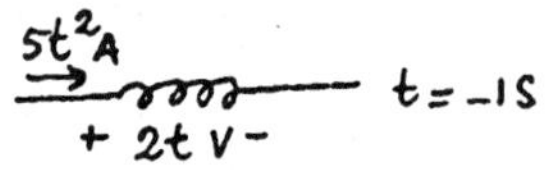


$P = 7(-3) = \underline{-21W}$ $\quad P = (-10)(-2) = \underline{20W}$ $\quad P = 6(\frac{1}{2})(-4) = \underline{-12W}$ $\quad P = 2(-1)5(-1)^2 = \underline{-10W}$

1-9 (a) $P(10ms) = vi = (20 - 30e^{-0.5})(4e^{-0.5}) = \underline{4.377 W}$

(b) Energy $= \int_0^\infty p\,dt = \int_0^\infty (80e^{-50t} - 120e^{-100t})\,dt = 1.6 - 1.2 = \underline{0.4 J}$

1-10 (a) $v = 20\sin 400\pi t$ V $\therefore v_{max} = 20V$ at $t = 1.25ms$

$\therefore P_{max} = 200W$ at $t = 1.25ms$

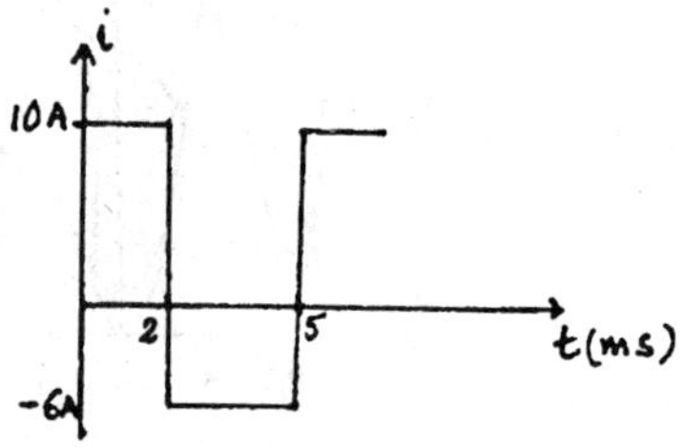


(b) At $t = 2^+ms$ ; $v = 20\sin 0.8\pi = 11.756V$

$\therefore P_{gen,max} = 11.756(+6) = \underline{70.53W(gen)}$

(c) Energy $= \int_0^{5ms} p\,dt = \int_0^{2ms} 200\sin 400\pi t\,dt - \int_{2ms}^{5ms} 120\sin 400\pi t\,dt$

$= -\frac{1}{2\pi}(\cos 0.8\pi - 1) + \frac{0.3}{\pi}(\cos 2\pi - \cos 0.8\pi) = 0.2879 + 0.1727 = \underline{0.4607 J}$

(d) $P_{av} = \frac{0.4607}{5 \times 10^{-3}} = \underline{92.13W}$

1-11 (a) Energy $= \int_1^3 4(t^3 - 6t^2 + 11t - 6)\,dt = 4(\frac{1}{4} \times 80 - 2 \times 26 + \frac{11}{2} \times 8 - 12) = \underline{0}$

(b) $i = \frac{P}{v} = \frac{4(t^3 - 6t^2 + 11t - 6)}{2(t^2 - 4t + 3)} = 2t - 4 \therefore q = \int_1^3 (2t - 4)\,dt = (t^2 - 4t)\Big]_1^3 = \underline{0}$

**1-12** 20V : $p = 20\times3 = \underline{60W}$ ; 3A : $p = -5(3) = \underline{-15W}$ ; 4A : $p = 15(-4) = \underline{-60W}$

15V±: $p = 15\times9 = \underline{135W}$ ; 8A : $p = 15(-8) = \underline{-120W}$ $\Sigma P = 0$ ✓

Upper left and 2nd-from-right source are being charged.

**1-13** $P_{4A} = 8(-4) = \underline{-32W}$ ; $P_{8V} = -8(2) = \underline{-16W}$ ; $P_{4V} = 4(6) = \underline{24W}$

$P_{14A} = 4(14) = \underline{56W}$ ; $P_{\diamond} = -4(2\times4) = \underline{-32W}$ ; $\Sigma P = 0$ ✓

**1-14** (a) $P = vi = (20e^{-0.2t})(2e^{-0.2t}) = 40e^{-0.4t}$ , $P_{4s} = 40e^{-1.6} = \underline{8.076W}$

(b) $i = 0.1v' = -0.4e^{-0.2t}$ $\therefore P_{4s} = (20e^{-0.8})(-0.4e^{-0.8}) = \underline{-1.6152W}$

(c) $i = 0.1\int_0^t 20e^{-0.2t}\,dt + 5 = -10e^{-0.2t} + 10 + 5$ $\therefore P_{4s} = (20e^{-0.8})(-10e^{-0.8} + 15) = \underline{94.42W}$

**1-15** (a) $i = 5A$ ; $P = \dfrac{10^4}{4\times60} = i^2R$ $\therefore R = \dfrac{10^4}{(4\times60)(5)^2} = \underline{1.6667\,\Omega}$

(b) $P = \dfrac{v^2}{R}$ $\therefore R = \dfrac{v^2}{P} = \dfrac{12^2}{1000} = \underline{0.1440\,\Omega}$

(c) $P = \dfrac{6400}{5}\cos^2 500\pi t$ ; $P_{av} = \dfrac{1000}{12}\int_0^{0.012}(1280\cos^2 500\pi t)\,dt = \dfrac{128}{12}\times10^4\int_0^{0.012}(0.5 + 0.5\cos10^3\pi t)\,dt$

$P_{av} = \dfrac{128}{12}\times10^4\left(0.5\times0.012 + \dfrac{0.5}{10^3\pi}\sin12\pi\right) = \underline{640W}$ $\left(P_{av} = \dfrac{1}{2}\times\dfrac{80^2}{5}\right)$

**1-16**

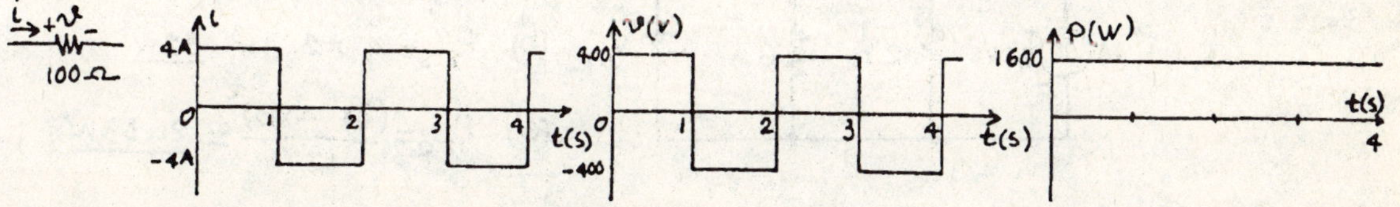


**1-17** $R = \dfrac{\rho l}{A} = \dfrac{l}{\sigma A}$ ; $\sigma_{cu} = 5.8\times10^7$ S/m

(a) $l = R\sigma A = (1)(5.8\times10^7)\left(\dfrac{\pi\times(1.024)^2\times10^{-6}}{4}\right) = 47.77m = \underline{156.71ft}$

(b) $R = \dfrac{6\times2.54\times0.01\times10^4}{5.8\times10^7\times0.013\times0.02\times2.54^2} = 0.15664\,\Omega$ , $P = i^2R = 3^2(0.15664) = \underline{1.4098W}$

1-18

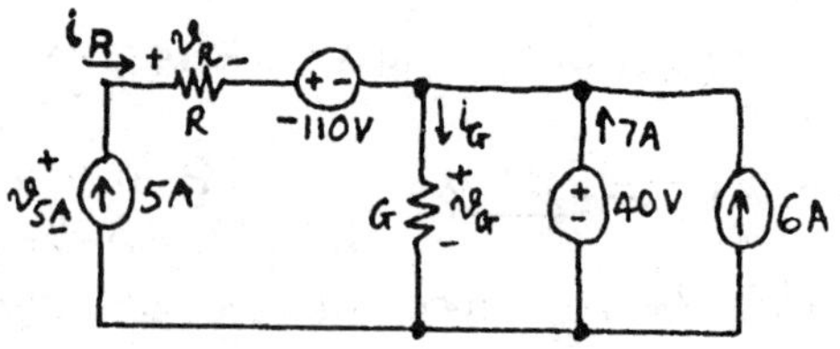

$v_G = 40V$ ; $i_G = 7+6+5 = 18A$ ∴ $G = \frac{18}{40} = 0.45\ S$

$P_{5A} = 125W = 5v_{5A}$ ∴ $v_{5A} = 25V$ ; $i_R = 5A$

$v_R = v_{5A} - 40 + 110 = 25 + 70 = 95V$

∴ $R = \frac{95}{5} = 19\Omega$

1-19

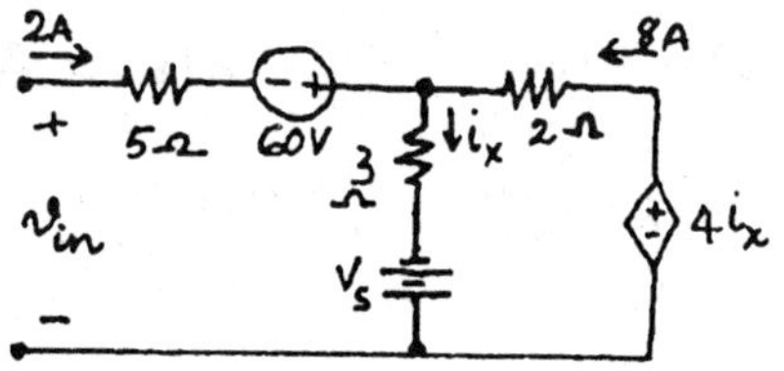

$i_x = 2 + 8 = 10A$ , $4i_x = 40V$

(a) ∴ $v_{in} = 2\times5 - 60 - 2\times8 + 40 = -26V$

(b) $P_{\diamond} = (4i_x)8 = 320W$ (gen)

(c) $3i_x - V_s - 4i_x + 2\times8 = 0$ ∴ $V_s = 30 - 40 + 16 = 6V$

1-20

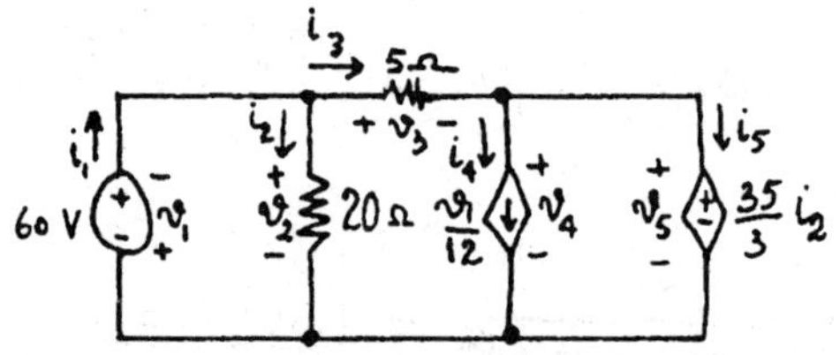

(a) $v_1 = -60V$ ; $v_2 = 60V$ ; $i_2 = \frac{60}{20} = 3A$

∴ $v_5 = \frac{35}{3}\times3 = 35V$ ; $v_4 = 35V$ ; $v_3 = 60 - 35 = 25V$

$i_3 = \frac{25}{5} = 5A$ ; $i_4 = \frac{-60}{12} = -5A$

$i_5 = 5 - (-5) = 10A$ ; $i_1 = 3 + 5 = 8A$

(b) $P_1 = 60(-8) = -480W$ ; $P_2 = 3^2\times20 = 180W$ ; $P_3 = 5^2\times5 = 125W$ ; $P_4 = 35(-5) = -175W$

$P_5 = 35\times10 = 350W$ ; $\Sigma P = -480 + 180 + 125 - 175 + 350 = 0$

1-21

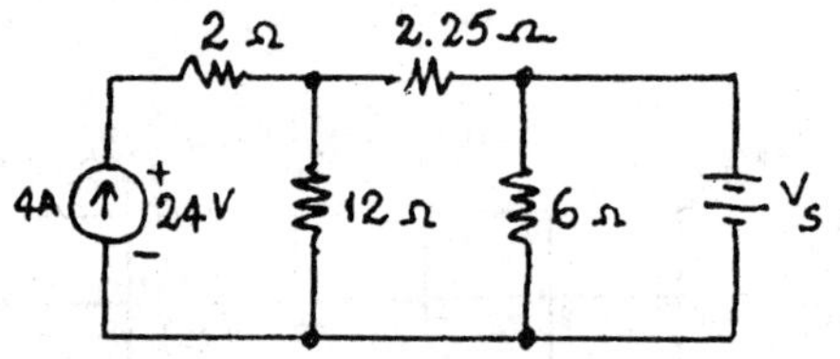

(a) $P_{4A} = 24(-4) = -96W$

(b) $P_{2\Omega} = 4^2\times2 = 32W$

(c) $P_{12\Omega} = \frac{(24 - 4\times2)^2}{12} = 21.33W$ ; $i_{12}\downarrow = \frac{4}{3}A$

(d) $\overrightarrow{i_{2.25}} = 4 - \frac{4}{3} = \frac{8}{3}A$ ∴ $v_{2.25}^{+-} = \frac{8}{3}\times2.25 = 6V$ ∴ $v_6^{+-} = V_s = 16 - 6 = 10V$ ∴ $P_6 = \frac{10^2}{6} = 16.667W$

(e) $i_6\downarrow = \frac{10}{6}$ ∴ $I_s\downarrow = \frac{8}{3} - \frac{5}{3} = 1A$ ∴ $P_s = 10\times1 = 10W$

**1-22**

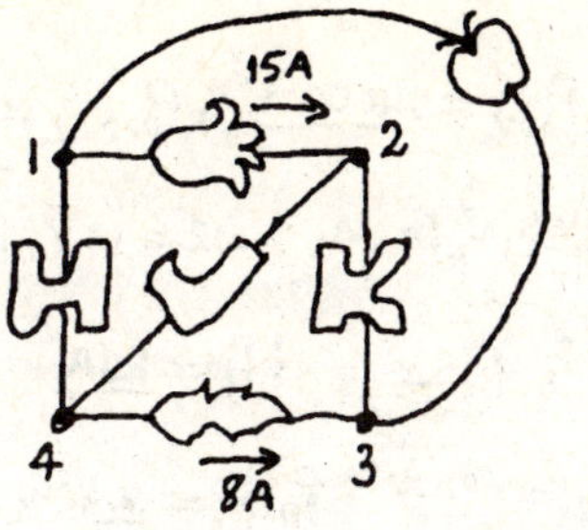


(a) $i_{24}=0 \quad \therefore i_{23}=\underline{15A}$ ; $i_{13}=-15-8=\underline{-23A}$ ; $i_{41}=-\underline{8A}$

(b) $i_{24}=18A \therefore i_{23}=15-18=\underline{-3A}$ ; $i_{13}=3-8=\underline{-5A}$ ; $i_{41}=18-8=\underline{10A}$

(c) $i_{24}=-18A \therefore i_{23}=15+18=\underline{33A}$ ; $i_{13}=-33-8=\underline{-41A}$ ; $i_{41}=-18-8=\underline{-26A}$

**1-23**

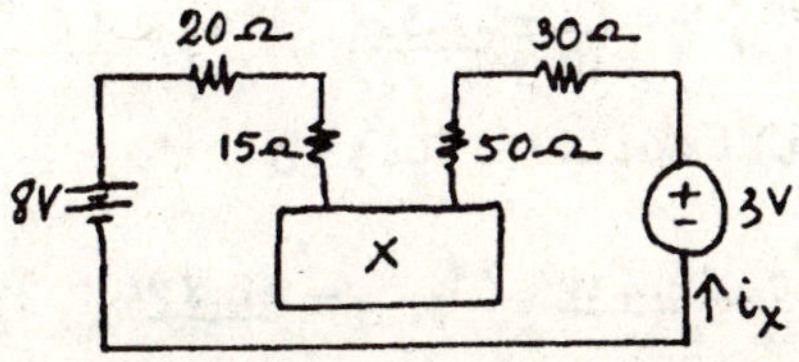


(a) $X \to 70\Omega \quad \therefore i_x=\dfrac{3-8}{20+15+70+50+30}=-\dfrac{1}{37}A$

$\therefore P_x=\dfrac{70}{37^2}=\underline{51.13\text{ mW}}$

(b) $X \to$ (+ −) 2V $\quad \therefore i_x=\dfrac{3+2-8}{35+80}=-\dfrac{3}{115}A$

$\therefore P_x=(2)\left(\dfrac{3}{115}\right)=\underline{52.17\text{ mW}}$

(c) $19i_x$ (+ −) $\quad \therefore -3+115i_x+8-19i_x=0 \quad \therefore i_x=-\dfrac{5}{96}A$ ; $P_x=-\dfrac{5}{96}\times 19\times\dfrac{5}{96}=\underline{-51.54\text{ mW}}$

(d) 52mA (→), $-\ v_x\ +$ $\quad i_x=-0.052A \quad \therefore v_x=3-8+0.052\times115=0.98V$ ; $P_x=-0.98\times0.052=\underline{-50.96\text{ mW}}$

**1-24**

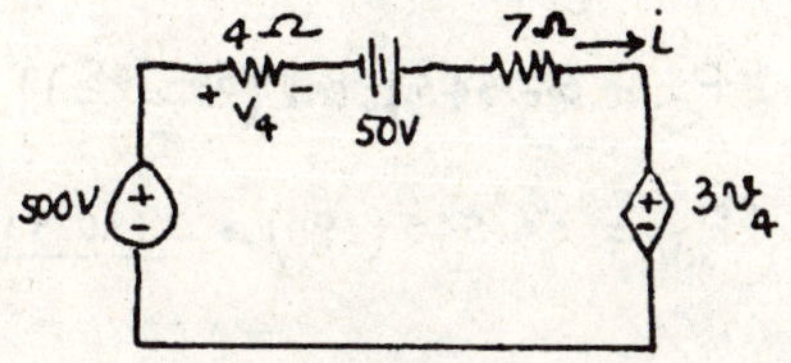


(a) $-500-50+11i+3(4i)=0 \quad \therefore i=\dfrac{550}{23}=\underline{23.91A}$

(b) $+\ v_4\ -$ $\quad \therefore -500-50+11i+3(4i-50)=0$

$\therefore i=\dfrac{700}{23}=\underline{30.43A}$

**1-25**

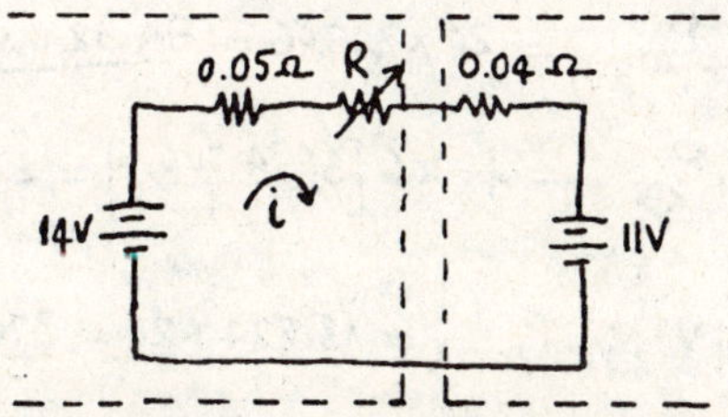


(a) $i=2.5=\dfrac{14-11}{0.05+R+0.04}= \quad \therefore \underline{R=1.1100\,\Omega}$

(b) $P_{11V}=25 \quad \therefore i=\dfrac{25}{11}=\dfrac{3}{0.09+R} \quad \therefore \underline{R=1.2300\,\Omega}$

(c) $25=0.04i^2+11i \quad \therefore i=\dfrac{-11\pm\sqrt{121+4}}{0.08}=2.2542A$

$\therefore 2.2542=\dfrac{3}{0.09+R} \quad \therefore R=\underline{1.2408\,\Omega}$

**1-26** (a)

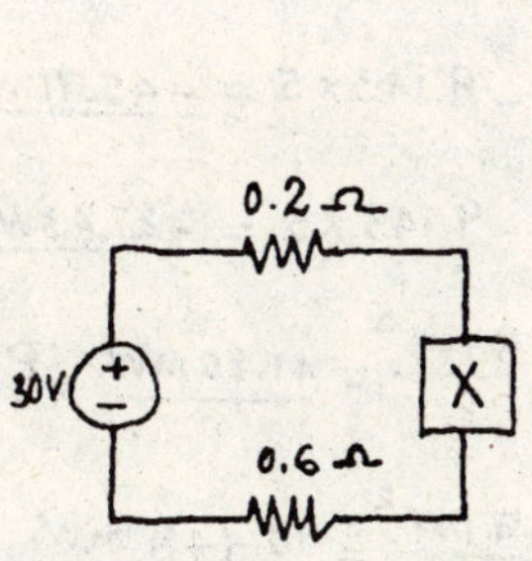


(b)

$P_x=100=i^2R$ ; $30=0.8i+Ri=(0.8+R)\sqrt{\dfrac{100}{R}}$ $\quad (R>5\Omega)$

$\therefore \quad 900R=100(R^2+1.6R+0.64)$ ; $R^2-7.4R+0.64=0 \quad \therefore R=\underline{7.3125\,\Omega}$

$v_s$ $(v_s<5V)$ ; $100=v_s i \quad \therefore i=\dfrac{100}{v_s}$ ; $30=0.8i+v_s=\dfrac{80}{v_s}+v_s$

$\therefore v_s^2-30v_s+80=0$ ; $v_s=\dfrac{30\pm\sqrt{900-320}}{2}=\underline{2.9584V}$, ~~27.0416V~~

**1-27** (a)

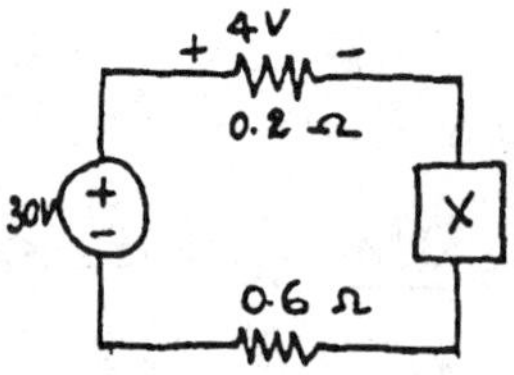


$\curvearrowright i = \frac{4}{0.2} = 20\text{A} \;\therefore\; P_{30V} = \underline{-600\text{W}}$, $P_{0.2\Omega} = 4\times 20 = \underline{80\text{W}}$

$P_{0.6\Omega} = 20^2 \times 0.6 = \underline{240\text{W}}$; $v_x^{+-} = 30 - 4 - 12 = 14\text{V}$, $P_x = \underline{280\text{W}}$, $\Sigma P = 0$ ✓

(b) $R = \frac{14}{20} = \underline{0.7\Omega}$; $\pm v_s = \underline{14\text{V}}$, $\downarrow i_s = \underline{20\text{A}}$

(c)

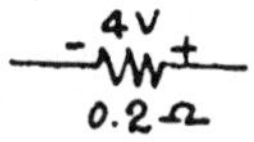


$\curvearrowleft i = 20\text{A} \;\therefore\; P_{30V} = \underline{600\text{W}}$; $P_{0.2} = \underline{80\text{W}}$; $P_{0.6} = \underline{240\text{W}}$; $v_x^{+-} = 30 + 4 + 12 = 46\text{V}$

$P_x = -46 \times 20 = \underline{-920\text{W}}$; $\Sigma P = 0$; $R = \frac{46}{-20} = \underline{-2.3\Omega}$, $\pm v_s = \underline{46\text{V}}$; $\uparrow i_s = \underline{20\text{A}}$

**1-28**

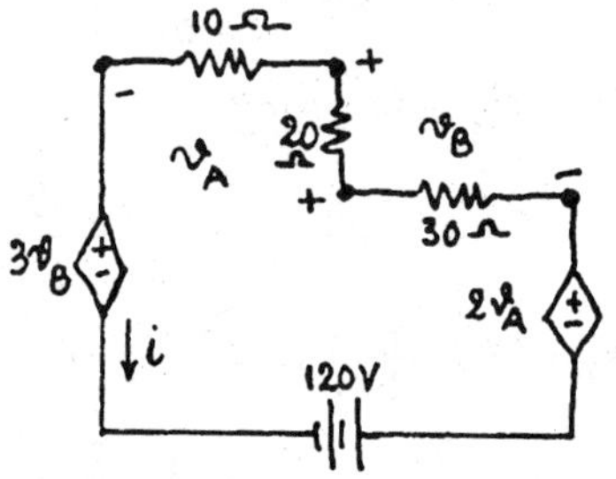


$-120 - 2(30i) + 60i + 3(-50i) = 0 \quad \therefore\; i = -0.8\text{A}$

$P_{10\Omega} = 0.8^2 \times 10 = \underline{6.4\text{W}}$; $P_{20\Omega} = \underline{12.8\text{W}}$; $P_{30\Omega} = \underline{19.2\text{W}}$

$P_{120V} = +120 \times 0.8 = \underline{96\text{W}}$; $P_{2v_A} = (2)(30i)(-i) = -60(.64) = \underline{-38.4\text{W}}$

$P_{3v_B} = 3(-50i)i = -150i^2 = \underline{-96.0\text{W}}$; $\Sigma P = 0$

**1-29**

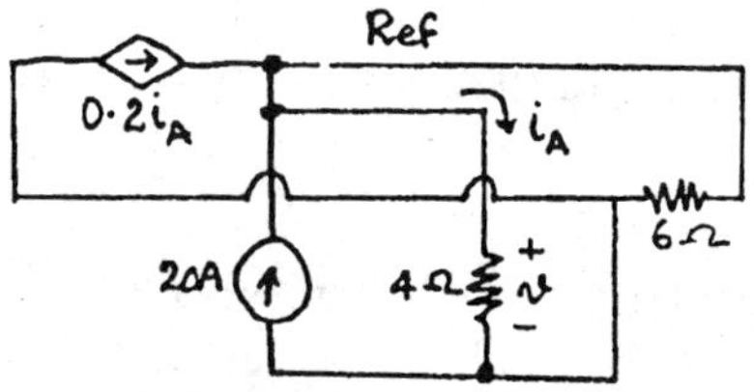


$-0.2\left(\frac{v}{4}\right) - 20 + \frac{v}{4} + \frac{v}{6} = 0 \quad \therefore\; v = 54.545\text{V}$

$\therefore P_6 = \frac{v^2}{6} = \underline{495.9\text{W}}$; $P_4 = \frac{v^2}{4} = \underline{743.8\text{W}}$

$P_{\diamond} = 54.545\left[-0.2\left(\frac{54.545}{4}\right)\right] = \underline{-148.76\text{W}}$

$P_{20A} = 54.545(-20) = \underline{-1090.9\text{W}}$; $\Sigma P = 0$

**1-30**

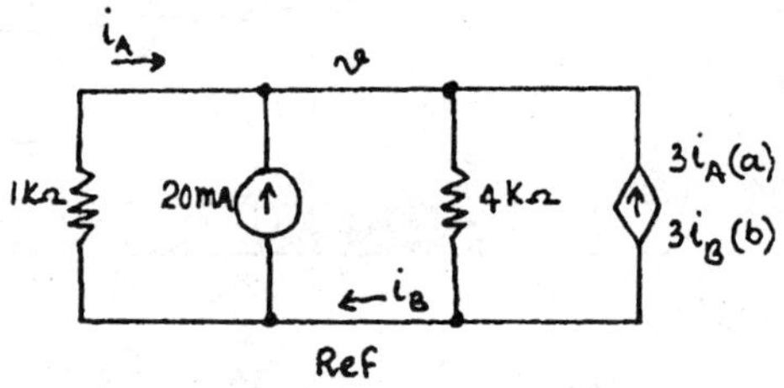


(a) $-20 - 3\left(-\frac{v}{1}\right) + \frac{v}{1} + \frac{v}{4} = 0 \quad \therefore\; v = 4.706\text{V}$

$P_{20mA} = 20 \times 4.706 = \underline{94.12\text{mW}}$ (gen)

$P_{\diamond} = 4.706\left[3\left(-\frac{4.706}{1}\right)\right] = \underline{-66.44\text{mW}}$

(b) $-20 - 3\left(20 - \frac{v}{1}\right) + 1.25v = 0 \quad \therefore\; v = 18.824\text{V}$; $P_{20mA} = 18.824 \times 20 = \underline{376.5\text{mW}}$

$P_{\diamond} = 18.824[3(20 - 18.824)] = \underline{66.44\text{mW}}$

**1-31**

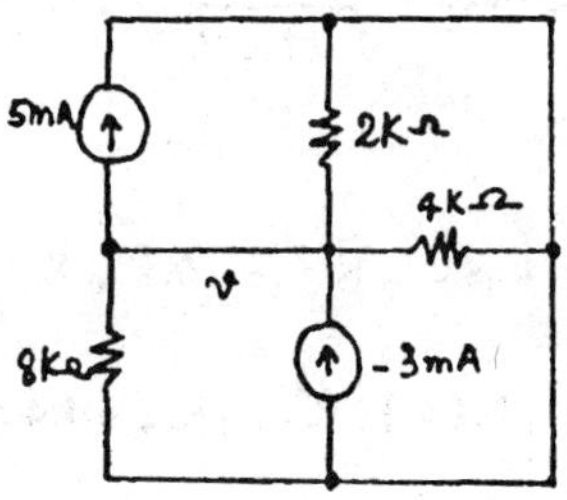


$5 + 3 + \frac{v}{2} + \frac{v}{4} + \frac{v}{8} = 0 \quad \therefore\; v = -9.143\text{V}$

$P_{5mA} = -9.143 \times 5 = \underline{-45.71\text{W}}$

$P_{-3mA} = -9.143 \times 3 = \underline{-27.43\text{W}}$

$P_{2k\Omega} = \frac{9.143^2}{2} = \underline{41.80\text{mW}}$; $P_{4k\Omega} = \frac{9.143^2}{4} = \underline{20.90\text{mW}}$

$P_{8k\Omega} = \frac{9.143^2}{8} = \underline{10.449\text{mW}}$; $\Sigma P = 0$

**1-32**

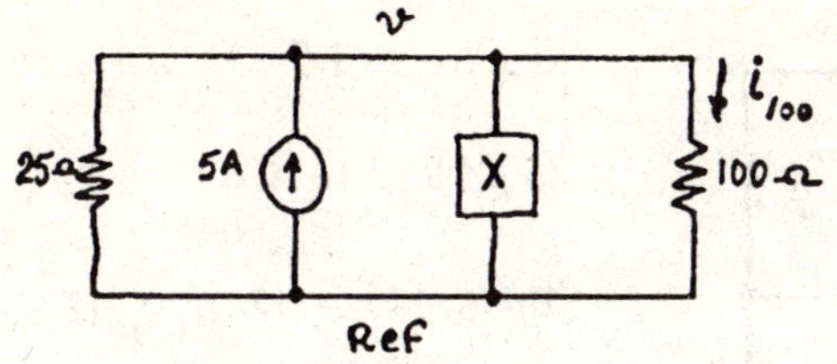


(a) $P_x = 100W$ ; $R > 50\,\Omega$ $\therefore \frac{v}{25} + \frac{v}{100} + \frac{v}{R} = 5$

$P = 100 = \frac{v^2}{R}$ $\therefore \frac{v}{20} + \frac{100}{v} - 5 = 0$

$v^2 - 100v + 2000 = 0$ ; $v = \frac{100 \pm \sqrt{10{,}000 - 8{,}000}}{2}$

$v = \cancel{27.64}$ , $\underline{72.36\,V}$ $\therefore R = \underline{52.36}$ , $\cancel{7.639}\,\Omega$

(b) $\downarrow i_s$ ; $i_s > 2$ $\therefore 5 - i_s = \frac{v}{20}$ ; $v i_s = 100$ ; $5 - i_s - \frac{5}{i_s} = 0$

$i_s^2 - 5i_s + 5 = 0$ ; $i_s = \frac{5 \pm \sqrt{25-20}}{2} = \cancel{1.382}$ , $\underline{3.618A}$

**1-33**

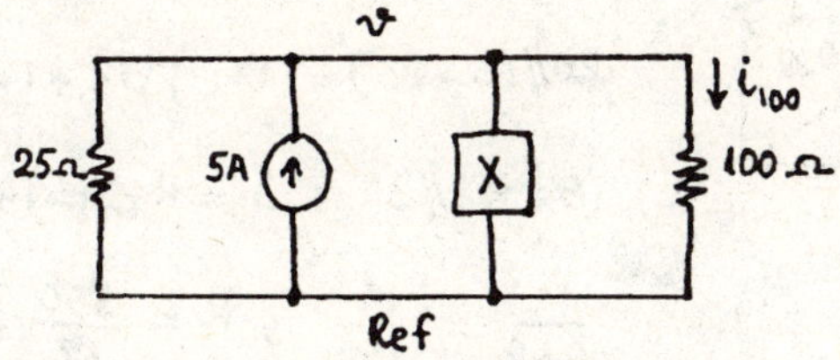


(a) $i_{100} = 1.2A$ $\therefore v = 120V$ ; $P_{100\Omega} = \frac{120^2}{100} = \underline{144W}$

$P_{25\Omega} = \frac{120^2}{25} = \underline{576W}$ ; $P_{5A} = 5(-120) = \underline{-600W}$

$i_x\downarrow = 5 - 1.2 - \frac{120}{25} = -1$ $\therefore P_x = \underline{-120W}$ ; $\Sigma P = 0$

(b) $R = \frac{120}{-1} = \underline{-120\,\Omega}$ ; $i_s\uparrow = \underline{1A}$ ; $v_s = v\pm = \underline{120V}$

(c) $i_{100} = -1.2A$ $\therefore v = \underline{-120V}$ ; $P_{100\Omega} = \frac{120^2}{100} = \underline{144W}$ ; $P_{25\Omega} = \frac{120^2}{25} = \underline{576W}$

$P_{5A} = -120(-5) = \underline{600W}$ ; $i_x\downarrow = 5 + 1.2 + \frac{120}{25} = 11A$ $\therefore P_x = -120 \times 11 = \underline{-1320W}$ ; $\Sigma P = 0$

$i_s\downarrow = \underline{11A}$ ; $R = \frac{-120}{11} = \underline{-10.909\,\Omega}$ ; $v_s = v\mp = \underline{120V}$

**1-34**

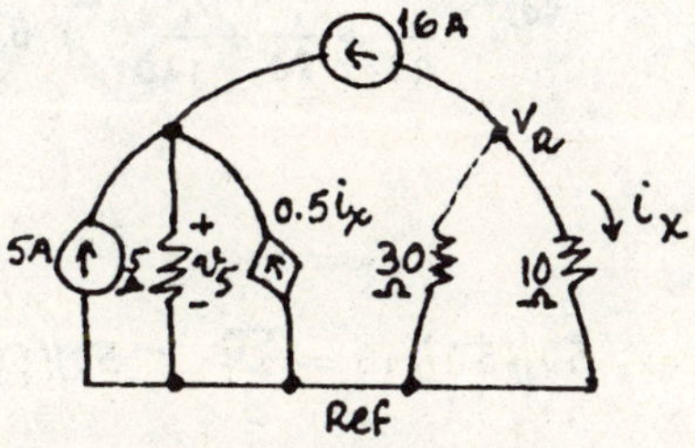


(a) $\frac{v_R}{30} + \frac{v_R}{10} + 16 = 0$ $\therefore v_R = -120V$ $\therefore i_x = \underline{-12A}$

(b) $-5 + \frac{v_5}{5} - 0.5(-12) - 16 = 0$ $\therefore v_5 = \underline{75V}$

(c) $P_{16A} = [75 - (-120)]16 = \underline{3120W}$

**1-35** (a)

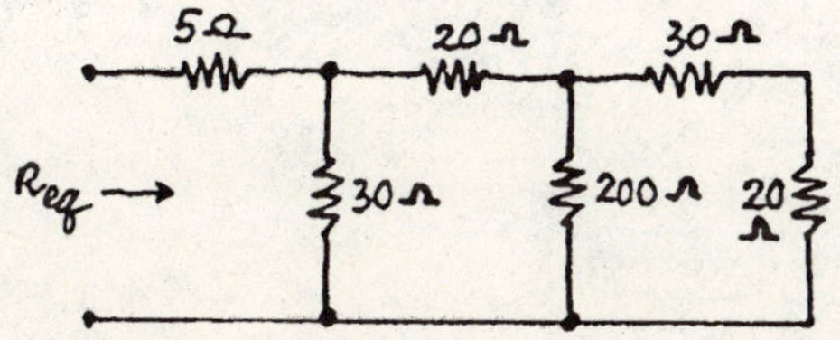


$50 \| 200 = 40$ ; $40 + 20 = 60$ ; $60 \| 30 = 20$

$R_{eq} = 20 + 5 = \underline{25\,\Omega}$

(b)

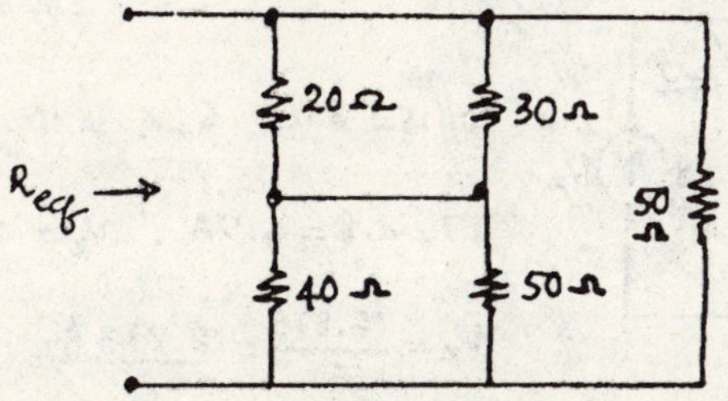


$20 \| 30 = 12$ ; $40 \| 50 = \frac{200}{9}$ ; $12 + \frac{200}{9} = \frac{308}{9}$

$\frac{308}{9} \| 50 = 20.32\,\Omega$  $R_{eq} = \underline{20.32\,\Omega}$

1-35 (c)

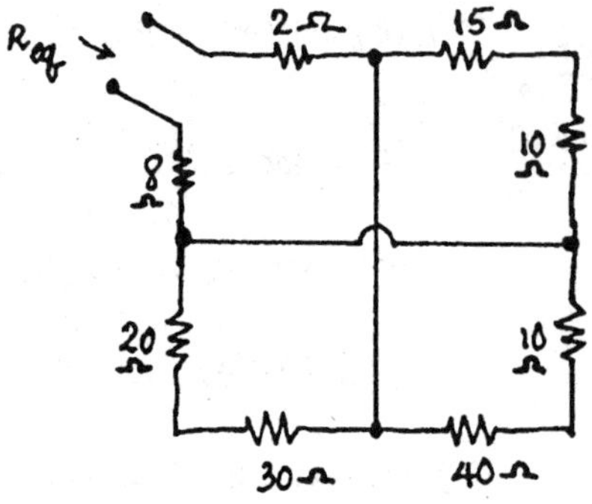


$15 + 10 = 25$ ; $40 + 10 = 50$

$25 \| 50 = 16.667$ ; $20 + 30 = 50$

$50 \| 16.667 = 12.5$ ; $12.5 + 10 = 22.5\ \Omega$

$R_{eq} = \underline{22.5\ \Omega}$

1-36

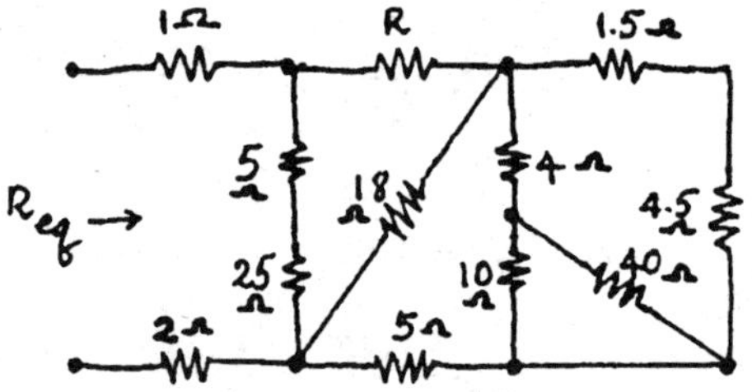


(a) $10 \| 40 = 8$ ; $8 + 4 = 12$ ; $12 \| (1.5 + 4.5) = 4$

$4 + 5 = 9$ ; $9 \| 18 = 6$ ; $6 + R = 6 + 14 = 20$

$20 \| (5 + 25) = 12$ ; $12 + 1 + 2 = \underline{15\ \Omega} = R_{eq}$

(b) $(6 + R) \| 30 + 3 = R_{in} = 14$ ; $\frac{1}{6+R} + \frac{1}{30} = \frac{1}{11}$

$\frac{1}{6+R} = \frac{19}{330}$ , $R = \frac{330}{19} - 6 = \underline{11.368\ \Omega} = R_{eq}$

1-37 (a)

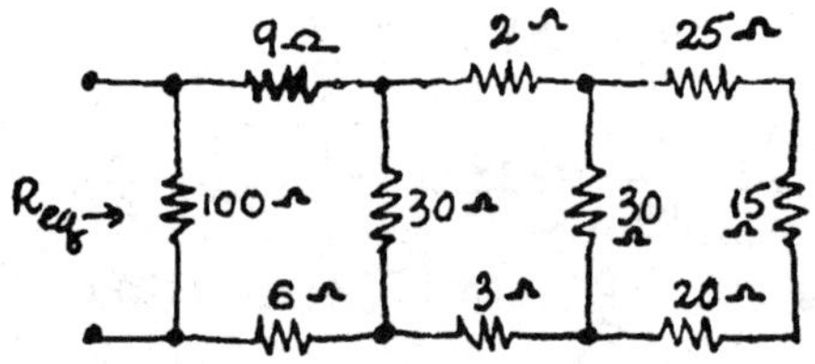


$25 + 15 + 20 = 60$ ; $60 \| 30 = 20$ ; $20 + 2 + 3 = 25$

$25 \| 30 = 13.636$ ; $13.636 + 9 + 6 = 28.636$

$28.636 \| 100 = \underline{22.26\ \Omega} = R_{eq}$

(b)

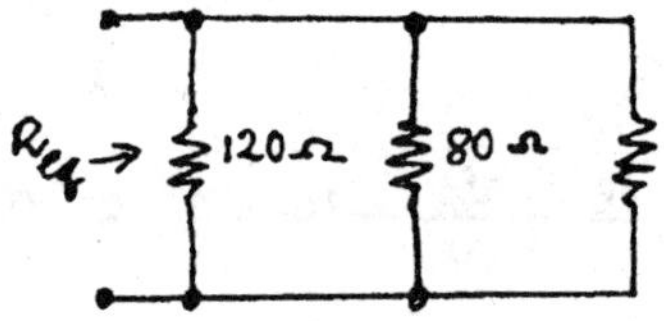


$$R_{eq} = \frac{1}{\frac{1}{40} + \frac{1}{80} + \frac{1}{120}} = \frac{240}{6 + 3 + 2} = \underline{21.82\ \Omega}$$

(c)

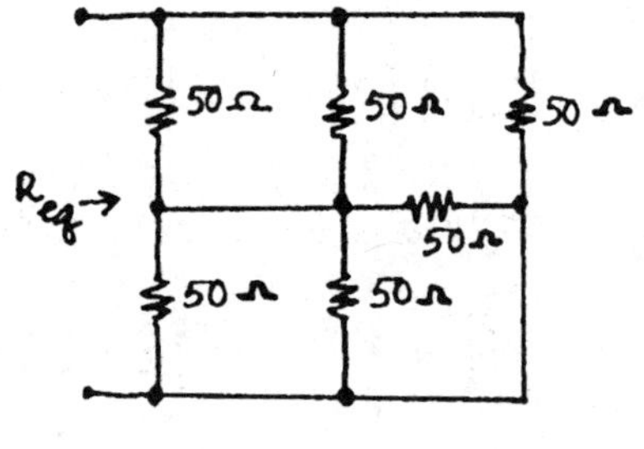


$50 \| 50 \| 50 = \frac{50}{3}$ ; $50 \| 50 = 25$

$\frac{50}{3} + 25 = \frac{125}{3}$ ; $\frac{125}{3} \| 50 = \underline{22.73\ \Omega} = R_{eq}$

1-38

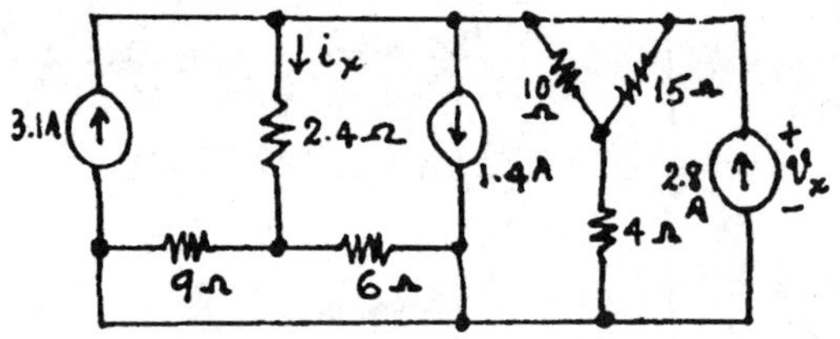


$9 \| 6 = 3.6$ ; $3.6 + 2.4 = 6\ \Omega$ ; $3.1 - 1.4 = 1.7\text{A}$

$10 \| 15 = 6\ \Omega$ ; $6 + 4 = 10\ \Omega$ ; $10 \| 6 = 3.75\ \Omega$

$1.7 + 2.8 = 4.5\text{A}$ $\therefore v_x = 4.5 \times 3.75 = \underline{16.875\text{ V}}$

$i_x = \frac{16.875}{6} = \underline{2.813\text{ A}}$

1-39

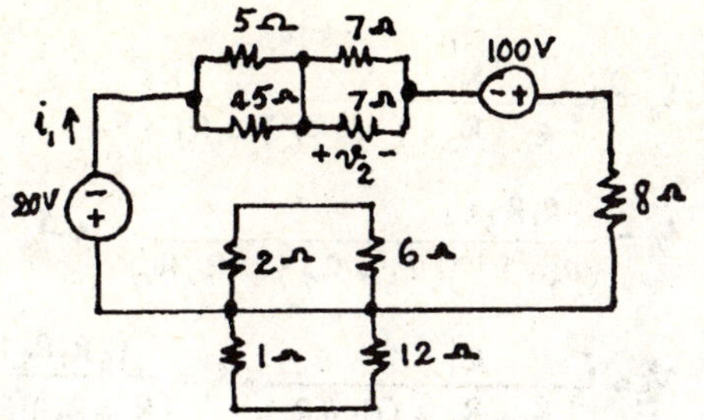


Bottom $R_{eq} = 0$

$5 \| 4.5 = 4.5\,\Omega$ ; $7\|7 = 3.5\,\Omega$

$\therefore i_1 = \dfrac{100-20}{4.5+3.5+8} = \underline{5A}$ $\therefore v_2 = 5 \times 3.5 = \underline{17.5V}$

1-40 (a) $R_{eq(min)} = \dfrac{20}{10} = \underline{2\,\Omega}$ (b) $R_{eq(max)} = 20 \times 10 = \underline{200\,\Omega}$

(c)

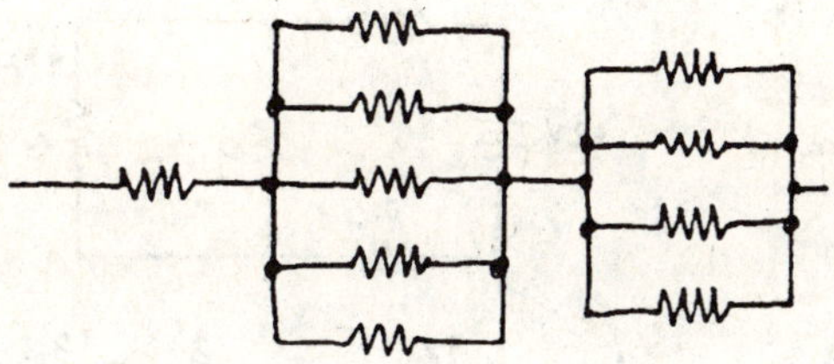

$R_{eq} = 20 + 4 + 5 = 29\,\Omega$

(d)

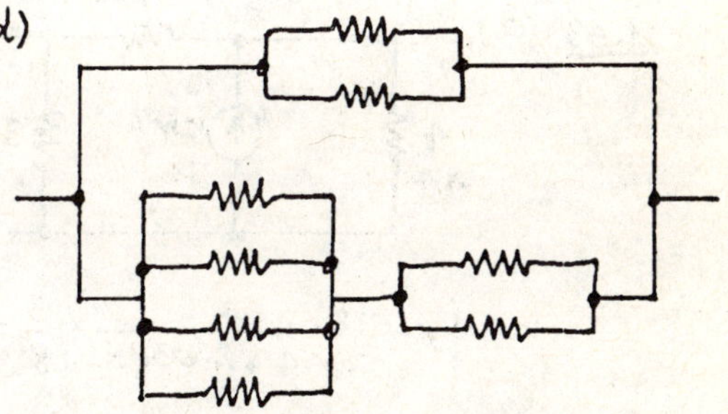

$R_{eq} = 10 \| (5+10) = 6\,\Omega$

1-41 (a)

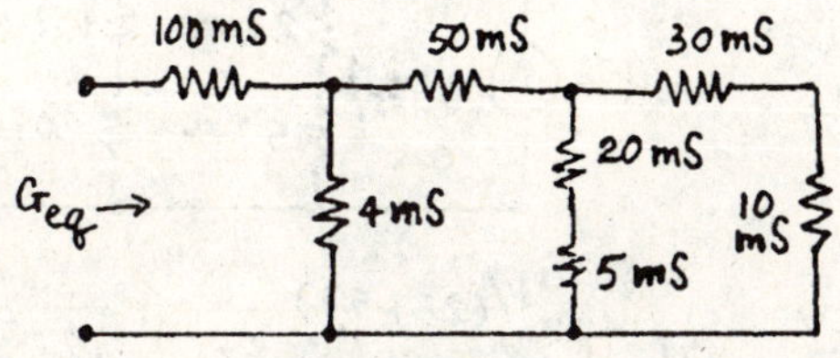


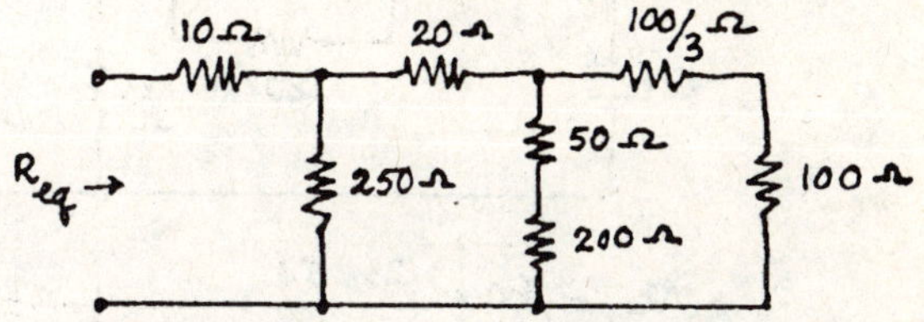


$\dfrac{100}{3} + 100 = \dfrac{400}{3}$ ; $\dfrac{400}{3} \| 250 = 86.96\,\Omega$ ; $86.96 + 20 = 106.96$ ; $106.96 \| 250 = 74.91$

$R_{eq} = 74.91 + 10 = 84.91\,\Omega$ $\therefore G_{eq} = \underline{11.777\text{ mS}}$

(b)

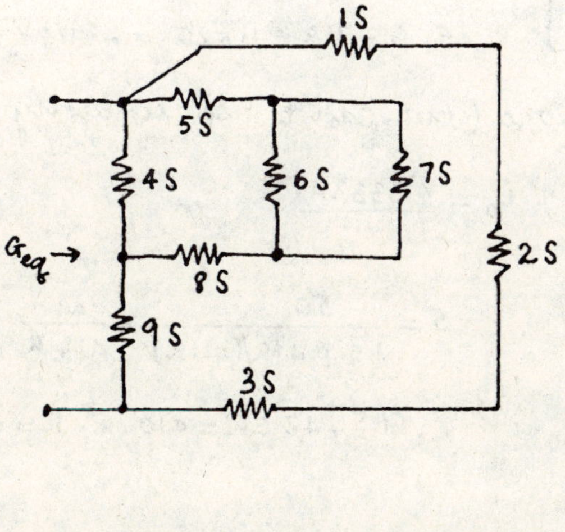


$1 + \dfrac{1}{2} + \dfrac{1}{3} = \dfrac{11}{6}\,\Omega$ ; $\dfrac{1}{6+7} + \dfrac{1}{5} + \dfrac{1}{8} = 0.4019\,\Omega$

$4 + \dfrac{1}{0.4019} = 6.488\text{ S}$ ; $\dfrac{1}{6.488} + \dfrac{1}{9} = 0.2652\,\Omega$

$0.2652 \| \dfrac{11}{6} = 0.2317\,\Omega$

$G_{eq} = \dfrac{1}{0.2317} = \underline{4.316\text{ S}}$

**1-42**

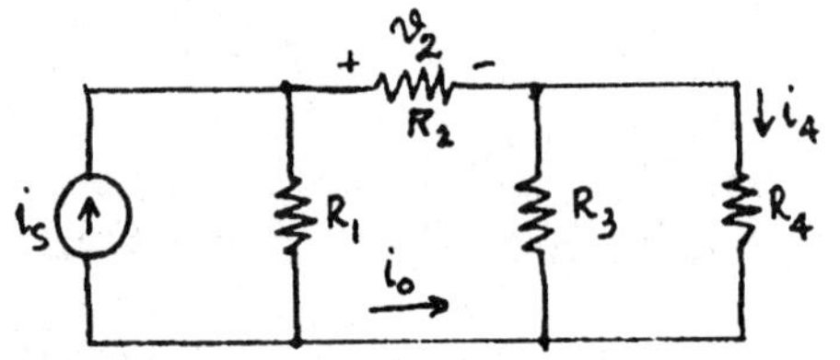


(a) $i_4 = i_s \dfrac{R_1}{R_1+R_2+(R_3\|R_4)} \times \dfrac{R_3}{R_3+R_4}$

$i_4 = \dfrac{R_1R_3i_s}{(R_1+R_2)(R_3+R_4)+R_3R_4}$ ←

(b) $v_2 = i_s \dfrac{R_1}{R_1+R_2+(R_3\|R_4)} \times R_2 = \dfrac{i_sR_1R_2(R_3+R_4)}{(R_1+R_2)(R_3+R_4)+R_3R_4}$ ←

(c) $i_o = \dfrac{-i_sR_1}{R_1+R_2+(R_3\|R_4)} = \dfrac{-i_sR_1(R_3+R_4)}{(R_1+R_2)(R_3+R_4)+R_3R_4}$ ←

**1-43** (a)

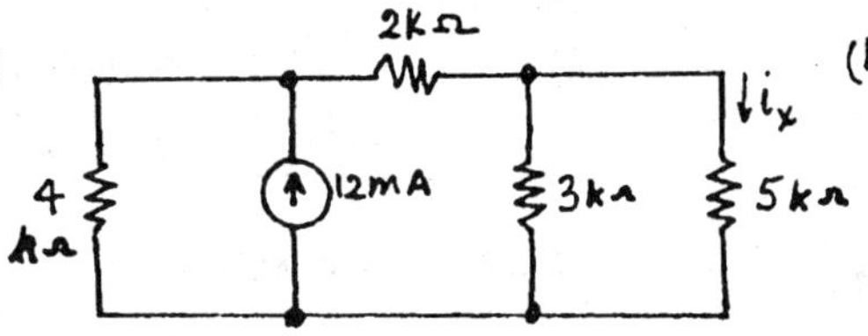


(b)

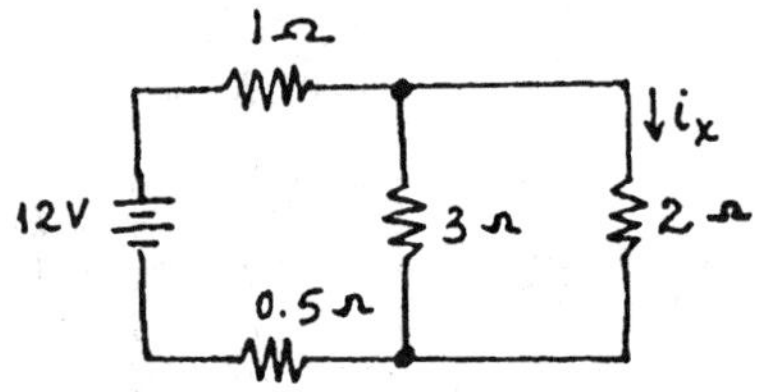


$i_x = 12 \times \dfrac{4}{4+2+3\|5} \times \dfrac{3}{8} = \underline{2.286\,mA}$

$i_x = \dfrac{12}{1+0.5+3\|2} \times \dfrac{3}{5} = \dfrac{36}{7.5+6} = \underline{2.667\,A}$

**1-44** (a)

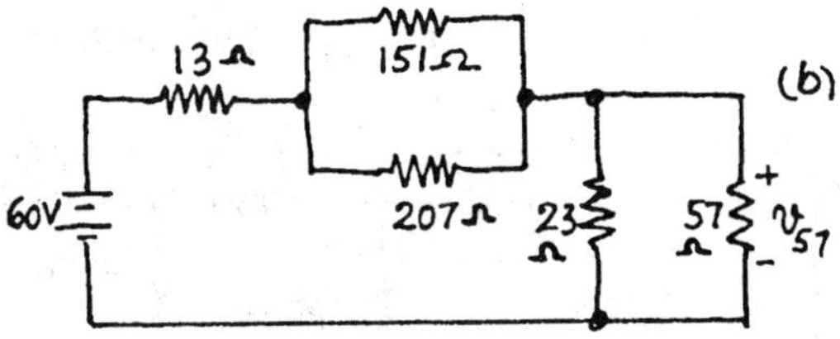


(b)

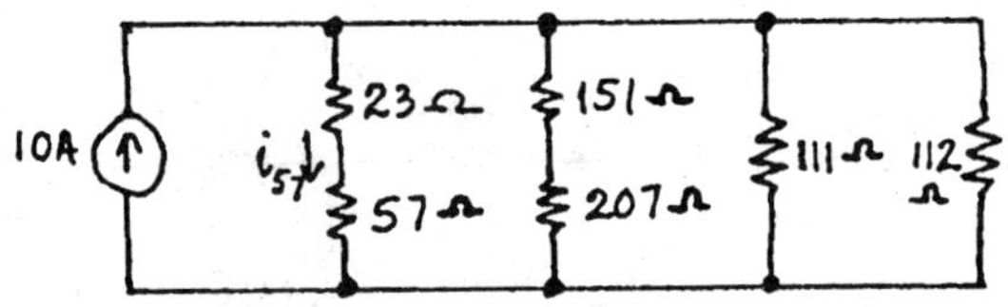


$v_{57} = 60 \times \dfrac{23\|57}{23\|57 + 151\|207 + 13} = \underline{8.426V}$

$i_{57} = 10 \times \dfrac{1/(23+57)}{1/80 + 1/358 + 1/111 + 1/112} = \underline{3.762A}$

**1-45**

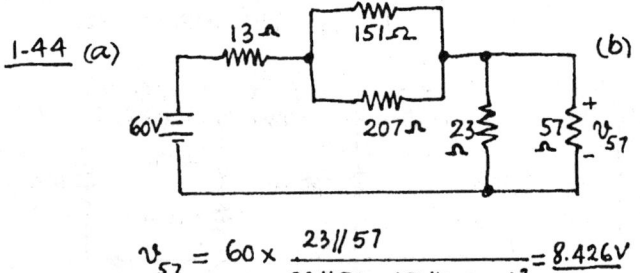

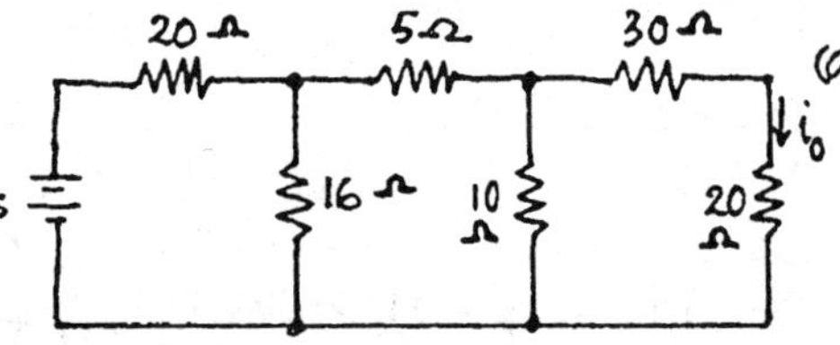


(a) $i_o = 1A$ ; $1(30+20) = 50V$ ; $\frac{50}{10} = 5A$ ; $5+1 = 6A$

$6 \times 5 = 30V$ ; $30 + 50 = 80V$ ; $\frac{80}{16} = 5A$

$5+6 = 11A$ ; $11 \times 20 = 220V$; $V_s = 220 + 80 = \underline{300V}$

(b) Every value is 0.4×previous one (but can't use linearity yet). ∴ $V_s = \underline{120V}$

(c) $V_s = 100V$ ∴ M by $\frac{100}{300} = \frac{1}{3}$ ; $i_o = \underline{0.3333A}$

**1-46**

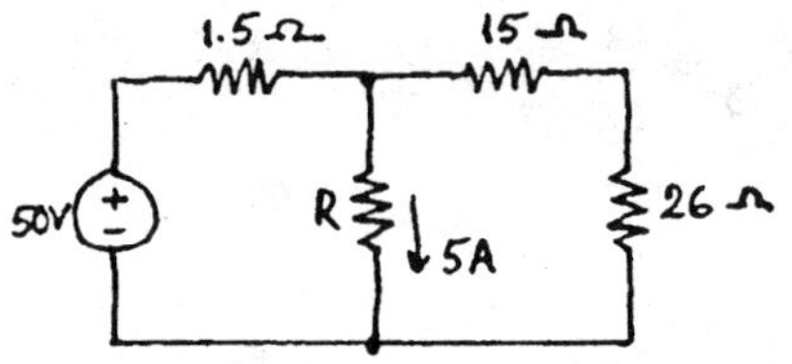


$5 = \dfrac{50}{1.5 + 41R/(41+R)} \times \dfrac{41}{41+R} = \dfrac{2050}{61.5+42.5R}$

∴ $61.5 + 42.5R = 410$ ∴ $R = \dfrac{348.5}{42.5} = \underline{8.2\Omega}$

1-47

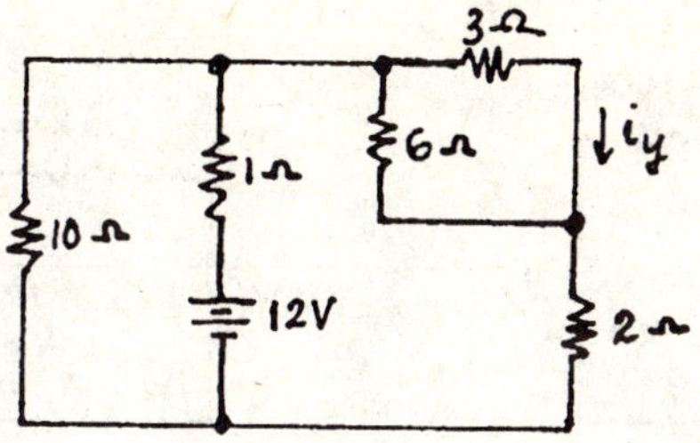


(a) $i_y = \frac{12}{1+10\|4} \times \frac{10}{14} \times \frac{2}{3} = \frac{80}{54} = \underline{1.4815A}$

(b) $1 = \frac{12}{1+10\|[2+6R/(6+R)]} \times \frac{10}{12+6R/(6+R)} \times \frac{6}{6+R}$

$1 = \frac{12}{1+10\|[2+6R\|(R+6)]} \times \frac{10}{12+3R}$

$1 + \frac{20+60R/(R+6)}{12+6R/(R+6)} = \frac{120}{12+3R}$ $\therefore R = \underline{5.388\Omega}$

1-48 (a)

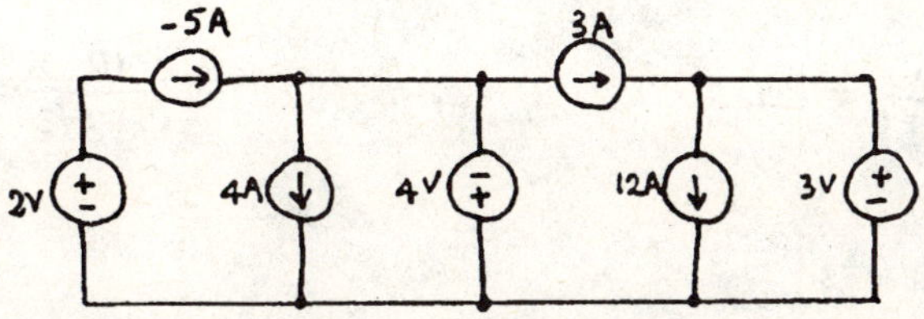


$P_{2V} = 2\times5 = \underline{10W}$ ; $P_{5A} = (4+2)(-5) = \underline{-30W}$

$P_{4A} = 4(-4) = \underline{-16W}$ ; $P_{4V} = 4(3+4+5) = \underline{48W}$

$P_{3A} = (4+3)(-3) = \underline{-21W}$ ; $P_{12A} = 3\times12 = \underline{36W}$

$P_{3V} = 3(3-12) = \underline{-27W}$ ; $\Sigma P = 0$

(b) $P_{5A} \rightarrow 0W$ $\therefore$ $4V \rightarrow \underline{-2V}$

1-49

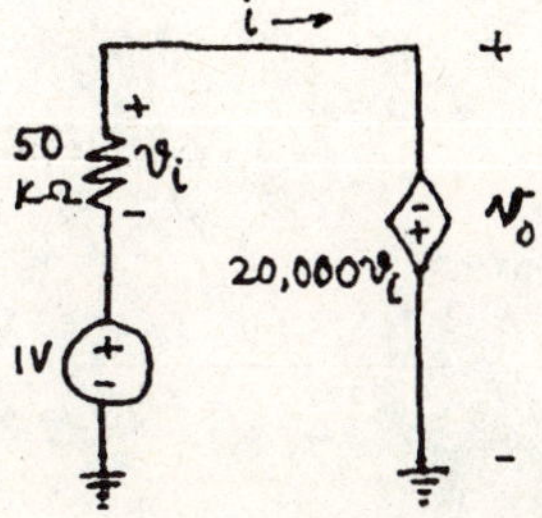


(a) $-1 + 50{,}000i - 20{,}000v_i = 0$
$v_i = -50{,}000i = -\frac{v_o}{20{,}000}$
$\Rightarrow$ $-1 + \frac{v_o}{20000} + v_o = 0$
$v_o = \frac{1}{1+1/20{,}000} = \underline{0.999950V}$

(b) $v_i = -\frac{v_o}{20{,}000} = \underline{-50.00\mu V}$

(c) $i = \frac{v_o}{(50{,}000)(20{,}000)} = 10^{-9}v_o = 0.999950nA$ $\therefore$ $p = \underline{0.999950nW}$

1-50 (a)

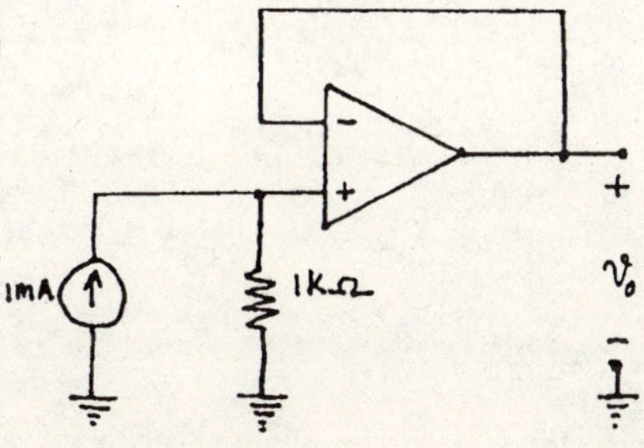


$v_o = \underline{1V}$

(b)

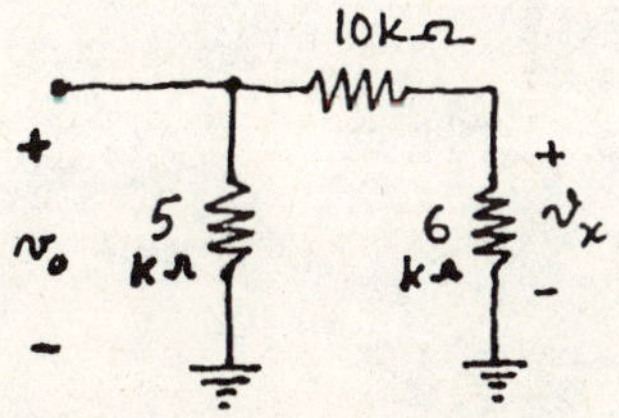


$v_x = \frac{6}{16} = \underline{0.375V}$

1-51 (a)

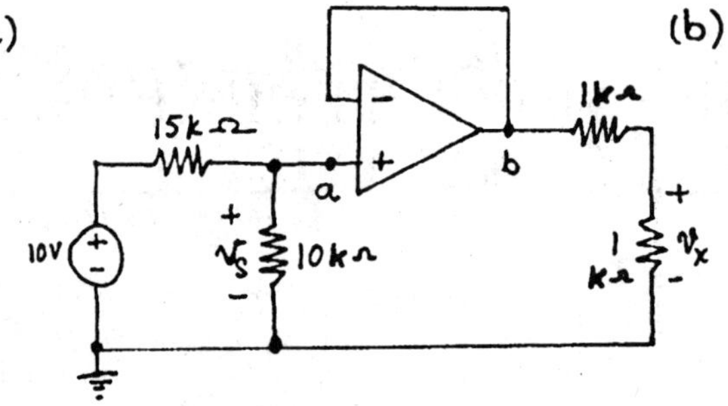


(b)

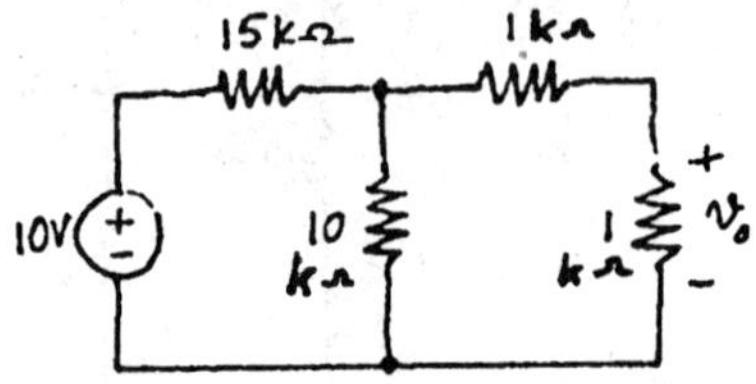


$$v_s = 10 \times \frac{10}{10+15} = 4\,\text{V}$$

$$\therefore\ v_x = v_o = 4 \times \frac{1}{1+1} = \underline{2\text{V}}$$

$$v_o = v_x = \frac{10}{15 + 10/\!/2} \times \frac{10}{10+2} \times 1$$

$$= \frac{10}{15 + 5/3} \times \frac{5}{6} = \underline{0.5\text{V}}$$

# CHAPTER 2: SOME USEFUL TECHNIQUES OF CIRCUIT ANALYSIS

## Section 2-2 Nodal Analysis

**1** (*a*) Find $v_2$ if $v_1 + 2v_2 + 3v_3 = 20$, $v_1 - 7v_2 - 5v_3 = -5$, and $v_3 + 3v_2 + 4v_1 - 10 = 0$. (*b*) Evaluate the determinant:

$$\begin{vmatrix} 1 & 2 & 3 & 4 \\ 2 & 3 & 4 & 1 \\ 3 & 4 & 1 & -2 \\ 4 & -1 & 2 & 3 \end{vmatrix}$$

**2** Use nodal analysis to find $v_P$ in the circuit shown in Fig. 2-1.

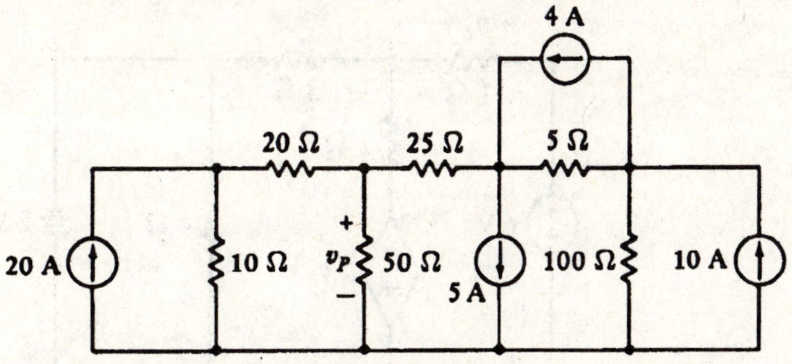


**Fig. 2-1:** See Prob. 2.

**3** Use nodal analysis on the circuit given in Fig. 2-2 to find: (*a*) $v_3$; (*b*) the power being supplied by the 5-A source.

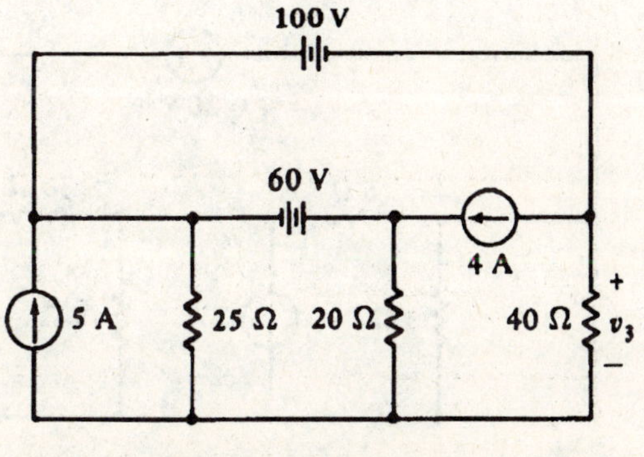


**Fig. 2-2:** See Prob. 3.

**4** Make use of nodal analysis to find $v_x$ and the power delivered to the 50-Ω resistor in the circuit of Fig. 2-3.

**5** Set up nodal equations for the circuit illustrated in Fig. 2-4, and then find the power supplied by the 5-V source.

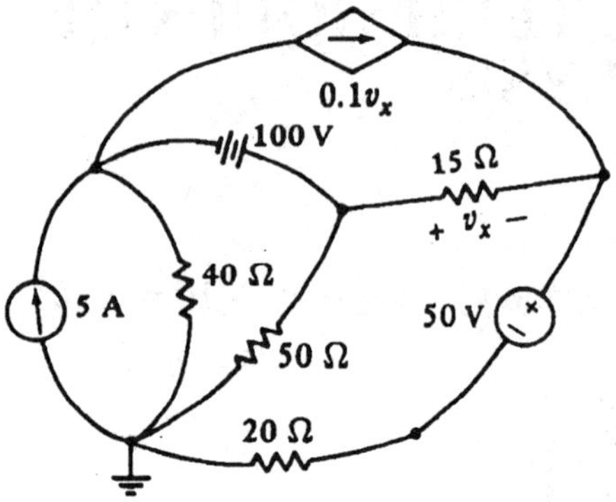


**Fig. 2-3:** See Prob. 4.

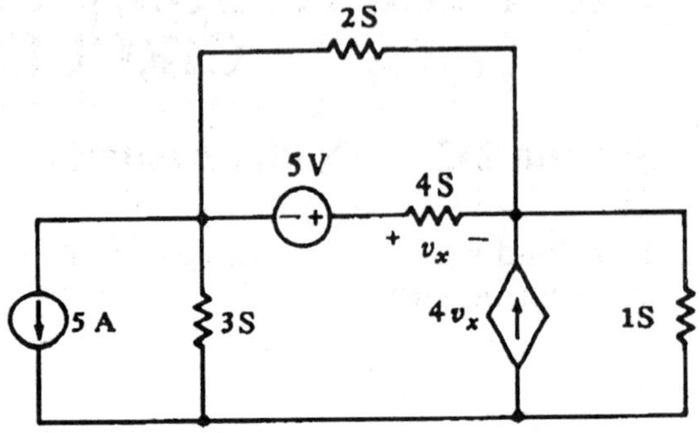


**Fig. 2-4:** See Prob. 5.

**6** Use nodal analysis to find $v_x$ in the circuit of Fig. 2-5.

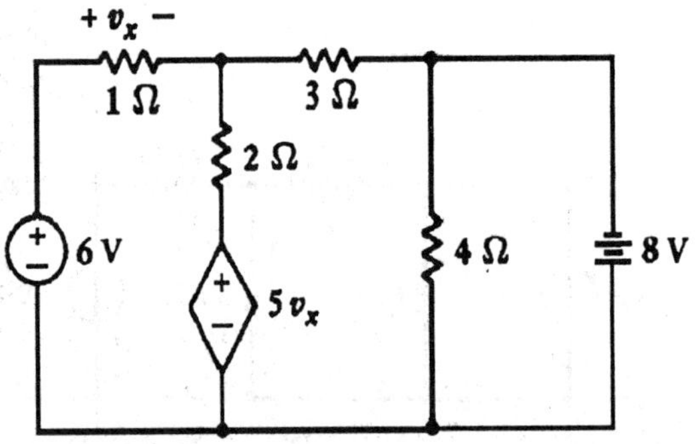


**Fig. 2-5:** See Probs. 6, 12, and 28.

**7** Analyze the circuit of Fig. 2-6 using node voltages and find the power being supplied by the 6-A source.

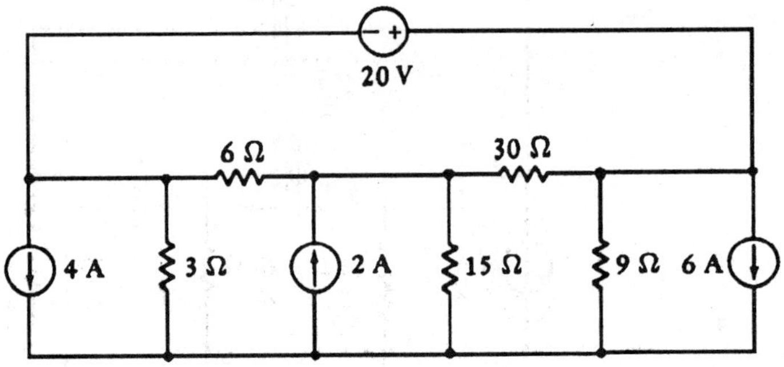


**Fig. 2-6:** See Prob. 7.

**8** In Fig. 2-7, find $v_2$ through the use of nodal analysis.

## Section 2-3 Mesh Analysis

**9** In the circuit of Fig. 2-8, use mesh analysis to: (*a*) find the power delivered to the 4-Ω resistor. (*b*) To what voltage should the 100-V battery be changed so that no power is delivered to the 4-Ω resistor?

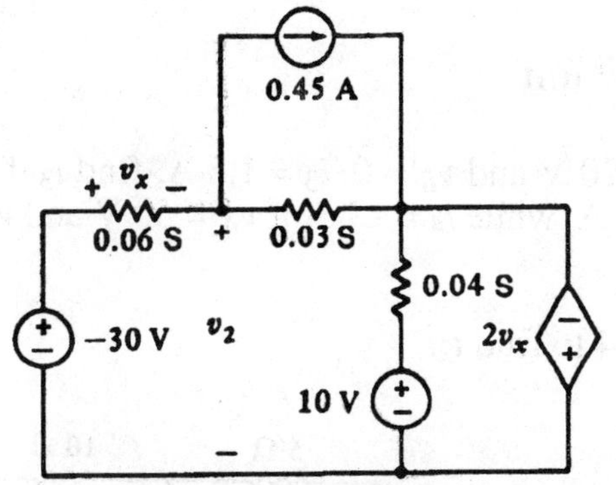


Fig. 2-7: See Prob. 8.

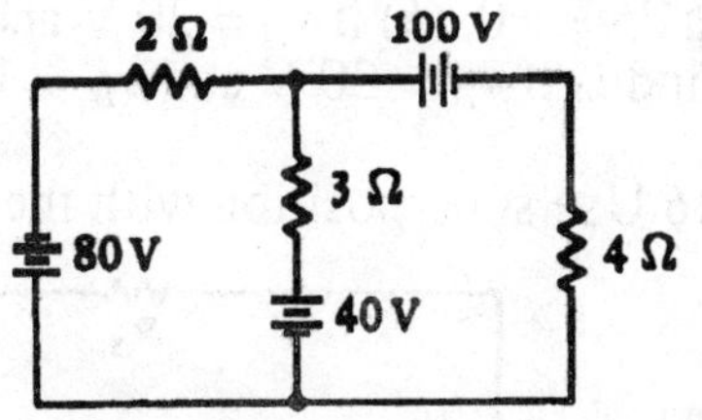


Fig. 2-8: See Prob. 9.

**10** Use mesh analysis on the circuit shown in Fig. 2-9 to find the power supplied by the 4-V battery.

**11** In Fig. 2-10, every resistance is 6 Ω and every battery voltage is 12 V. Find $i_A$.

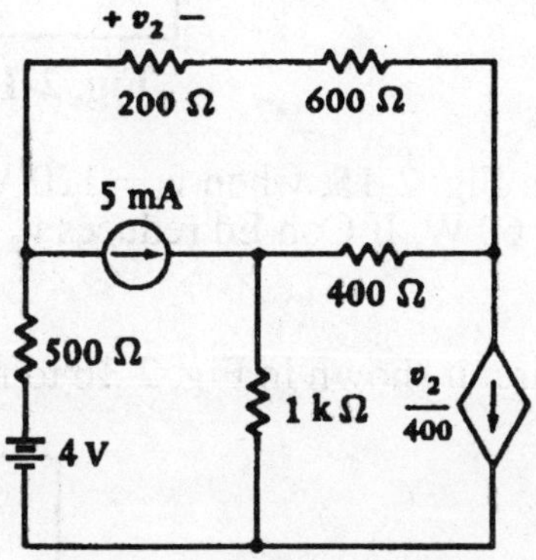


Fig. 2-9: See Prob. 10.

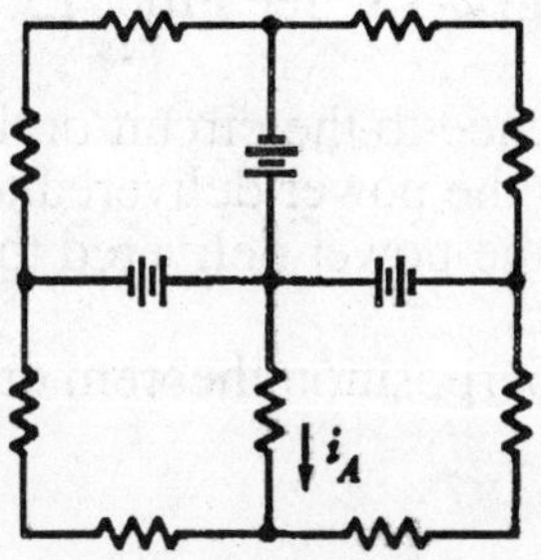


Fig. 2-10: See Prob. 11.

**12** In the circuit of Fig. 2-5, change the right-hand element to an 8-A independent current source, arrow directed upward, and use mesh analysis to find the power absorbed by the 3-Ω resistor.

**13** Use mesh analysis on the circuit of Fig. 2-11 to find the values of all the mesh currents.

**14** In the circuit of Fig. 2-12, use mesh analysis to find $i_o$, the current flowing downward in $R_L$ if $v_2 = 1.234\ 321$ V.

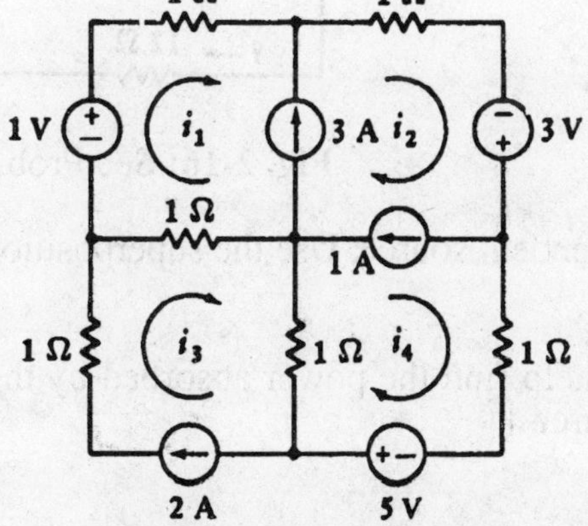


Fig. 2-11: See Prob. 13.

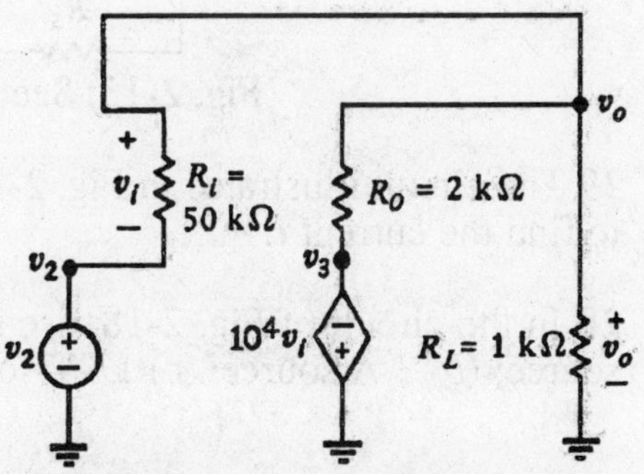


Fig. 2-12: See Prob. 14.

## Section 2-4 Linearity and Superposition

**15** In the circuit shown in Fig. 2-13: (*a*) if $v_A = 20$ V and $v_B = 0$, $i_3 = 1.5$ A; find $i_3$ if $v_A = 50$ V and $v_B = 0$; (*b*) if $v_A = 20$ V and $v_B = 50$ V, $i_4 = 2$ A, while $i_4 = -1$ A if $v_A = 50$ V and $v_B = 20$ V; find $i_4$ if $v_A = 30$ V and $v_B = 100$ V.

**16** Use superposition with the circuit of Fig. 2-14 to find $i_x$.

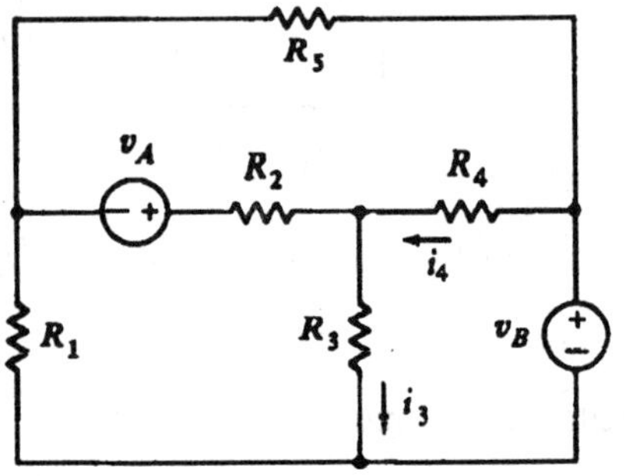


**Fig. 2-13:** See Prob. 15.

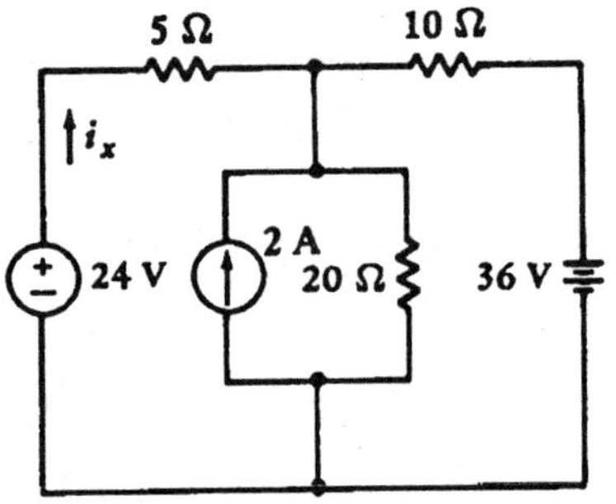


**Fig. 2-14:** See Prob. 16.

**17** With reference to the circuit outlined in Fig. 2-15, when $v_s = 120$ V, it is found that $i_1 = 3$ A, $v_2 = 50$ V, and the power delivered to $R_3$ is 60 W. If Con Ed reduces $v_s$ to 105 V, find new values for $i_1$, $v_2$, and the power delivered to $R_3$.

**18** Use the superposition theorem on the circuit shown in Fig. 2-16 to find $i$.

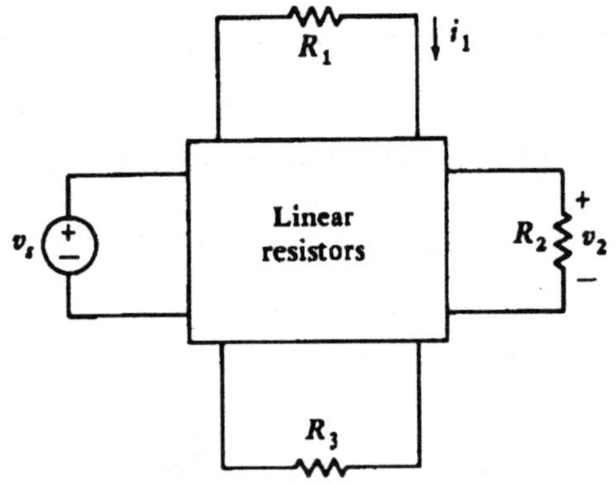


**Fig. 2-15:** See Prob. 17.

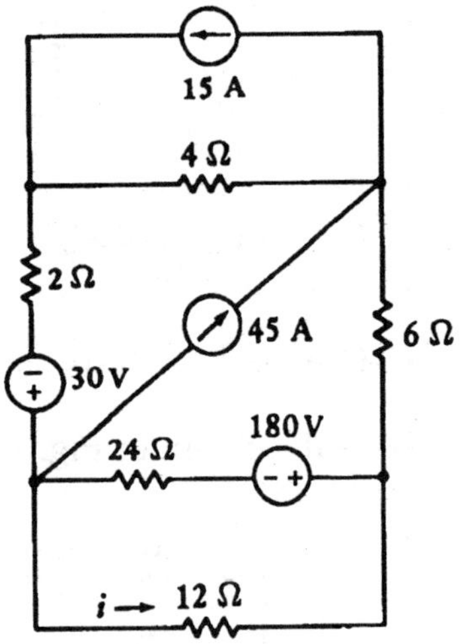


**Fig. 2-16:** See Prob. 18.

**19** The circuit illustrated in Fig. 2-17 contains a dependent source. Use the superposition theorem to find the current $I$.

**20** In the circuit of Fig. 2-18, use superposition to help find the power absorbed by the: (*a*) 6-V source; (*b*) 3-A source; (*c*) 12-V source; (*d*) 2-A source.

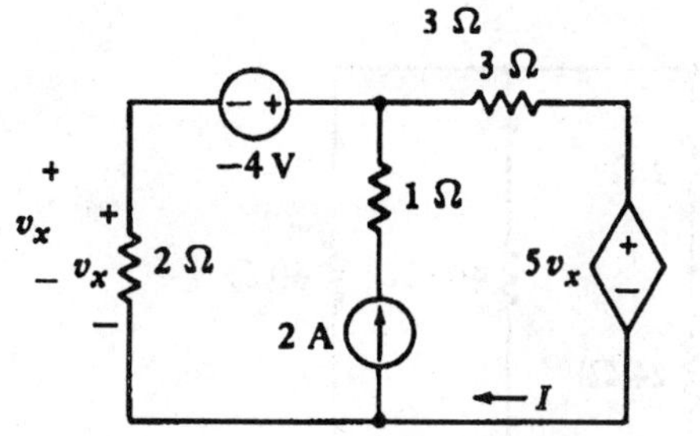


Fig. 2-17: See Probs. 19 and 29.

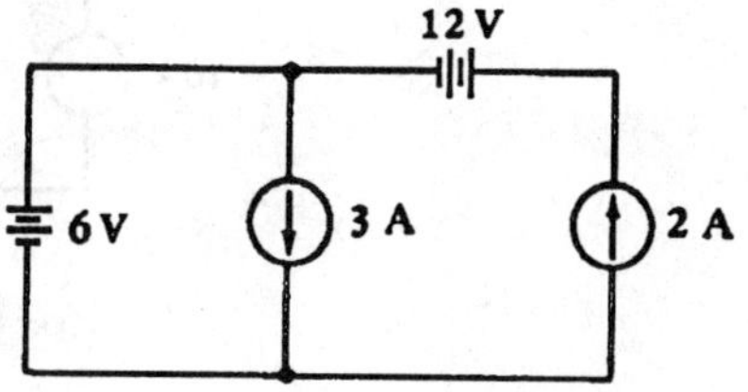


Fig. 2-18: See Prob. 20.

## Section 2-5 Source Transformations

**21** A Radio Shack type 222 flashlight bulb is intended for use with two 1.55-V penlight (AA) batteries in series. The bulb, however, is marked "2.25 V, 0.25 A." Assuming all markings are correct, what practical voltage source might be used to model one of the batteries?

**22** By making repeated source transformations and resistance combinations for each network in Fig. 2-19, replace the network to the left of terminals *a*-*b* with the series combination of a single independent voltage source and a single resistor.

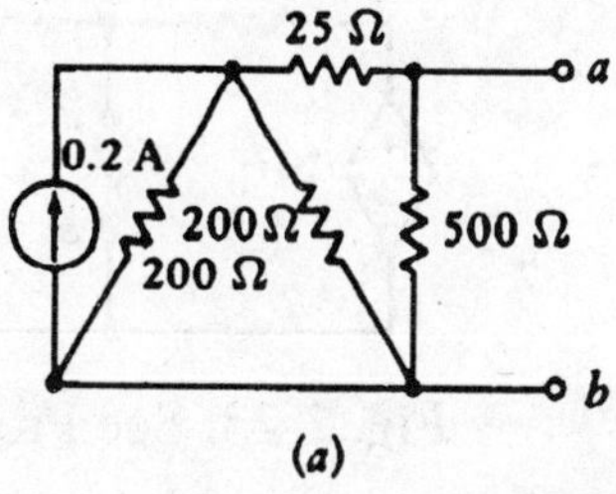


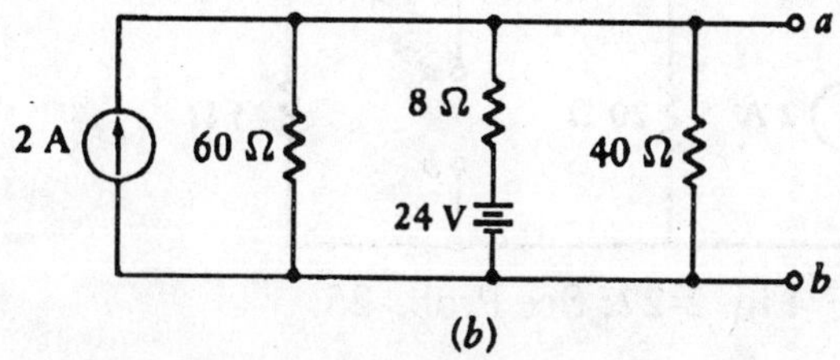


Fig. 2-19: See Prob. 22.

**23** In the circuit of Fig. 2-20, what value of $R_L$: (*a*) will absorb a maximum power from this network and what is $p_{L,\max}$? (*b*) will have the maximum voltage across it and what is $v_{L,\max}$? (*c*) will have a maximum current through it and what is $i_{L,\max}$?

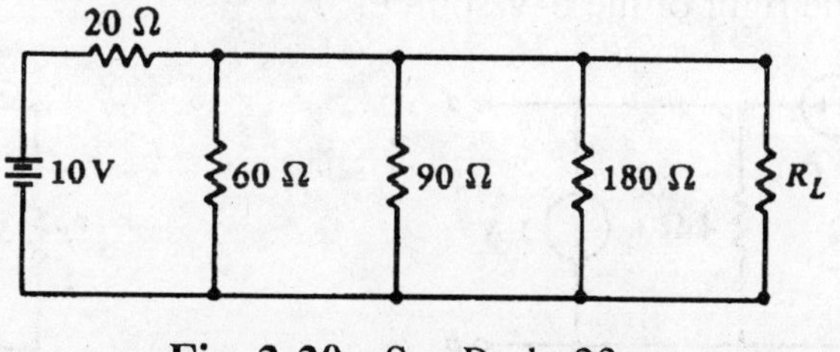


Fig. 2-20: See Prob. 23.

**24** Consider the practical voltage source to the left of terminals *a*-*b* in Fig. 2-21. (*a*) What must be the value of $R_L$ to cause the maximum possible power to be drawn from the *practical source*? (*b*) What is the value of this maximum power?

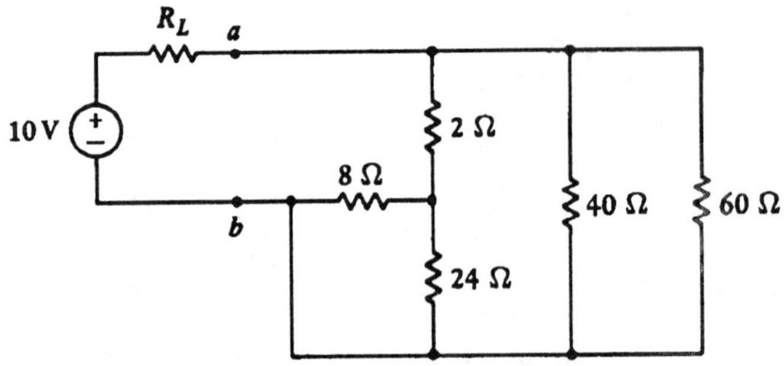


**Fig. 2-21:** See Prob. 24.

## Section 2-6 Thévenin's and Norton's Theorems

**25** (*a*) Use three separate analyses to find $v_{oc}$, $i_{sc}$, and $R_{th}$ with respect to terminals *a-b* for the circuit shown in Fig. 2-22. (*b*) Draw the Thévenin and Norton equivalents as seen at *a-b*.

**26** Find the Thévenin equivalent circuit with respect to terminals *a-b* for the circuit shown in Fig. 2-23.

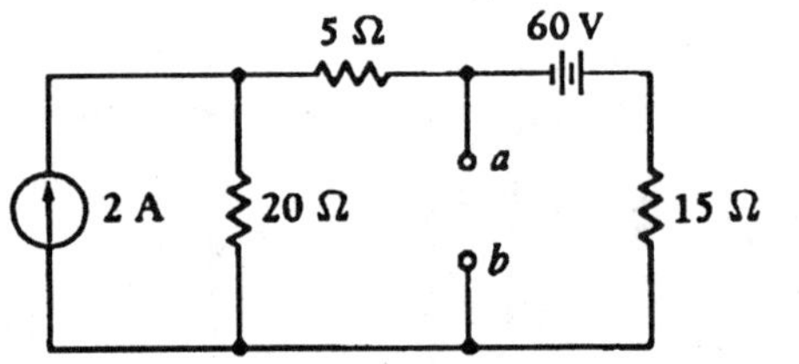


**Fig. 2-22:** See Prob. 25.

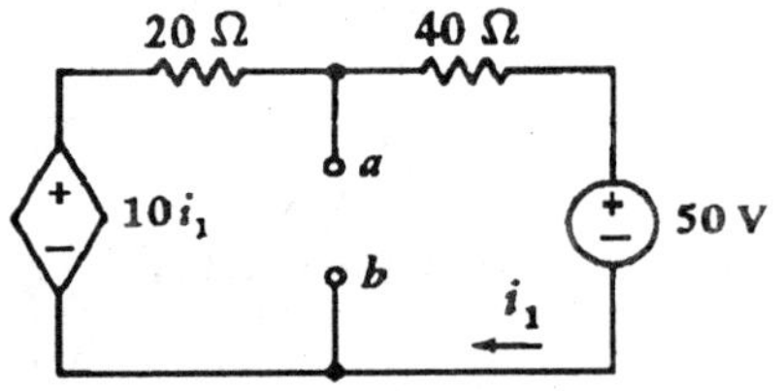


**Fig. 2-23:** See Probs. 26 and 30.

**27** (*a*) Determine the Thévenin and Norton equivalent circuits as seen at terminals *a-b* for the network of Fig. 2-24. (*b*) Replace the 5-A source with a dependent voltage source labeled $5i_x$ (+ reference at the right) and again find the Thévenin and Norton equivalents.

**28** In the circuit of Fig. 2-5, find the Thévenin equivalent of the network: (*a*) to the left of the 8-V source; (*b*) to the right of the 6-V source.

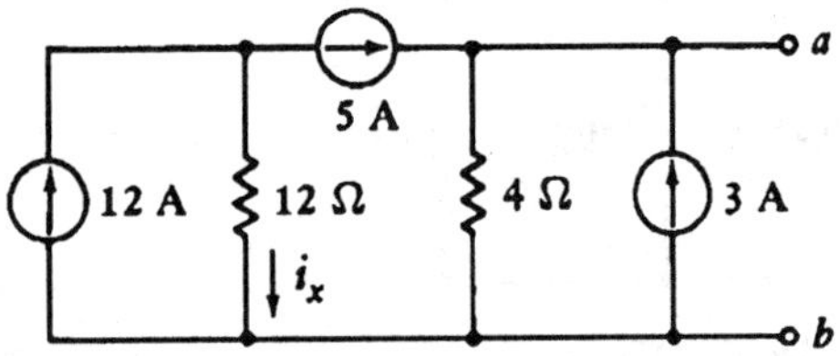


**Fig. 2-24:** See Prob. 27.

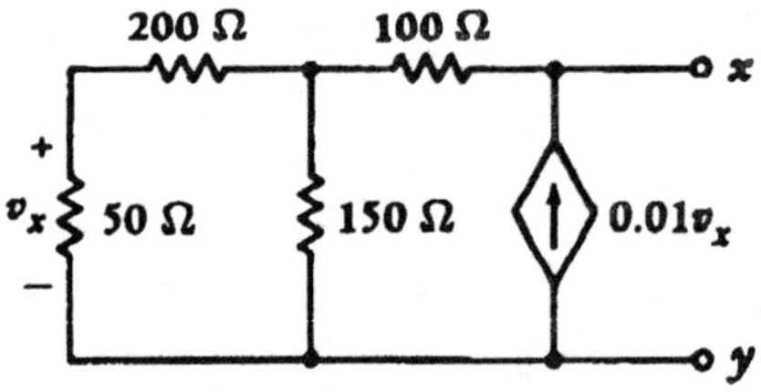


**Fig. 2-25:** See Prob. 31.

**29** (*a*) Refer to the circuit shown in Fig. 2-17 and find the Thévenin equivalent circuit faced by the 3-Ω resistor. (*b*) Find *I*. (*c*) Change the 3-Ω resistor to 13 Ω and again find *I*.

**30** If the voltage $v_L$ across or the current $i_L = v_L/R_L$ through a general load resistance $R_L$ is known, then the Thévenin or Norton equivalent may be determined easily since $v_{oc} = \lim_{R_L \to \infty} v_L$ and $i_{sc} = \lim_{R_L \to 0} i_L$. (*a*) Find $v_L$ (+ sign at terminal "*a*") for the circuit of Fig. 2-23 if a resistance $R_L$ is connected between *a* and *b*. (*b*) Use the expressions above to determine $v_{oc}$ and $i_{sc}$.

**31** Find the Thévenin equivalent circuit for the network shown in Fig. 2-25.

**32** Determine the Thévenin equivalent circuit of the network shown in Fig. 2-26.

**33** The voltage follower shown in Fig. 2-27 is modified by inserting a finite $R_i = 10$ kΩ between the terminals across which $v_i$ is defined. Find the new Thévenin equivalent.

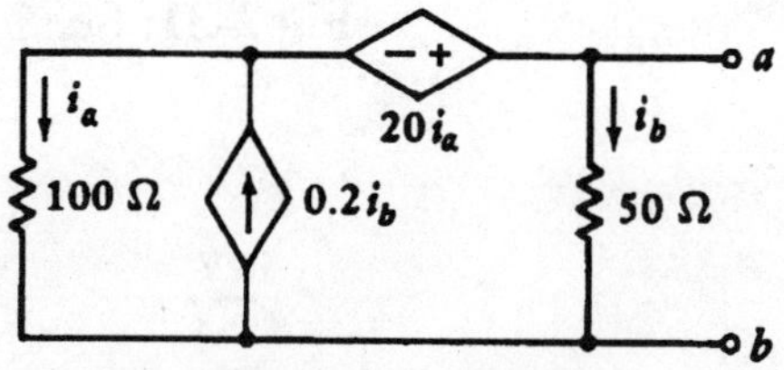


**Fig. 2-26:** See Prob. 32.

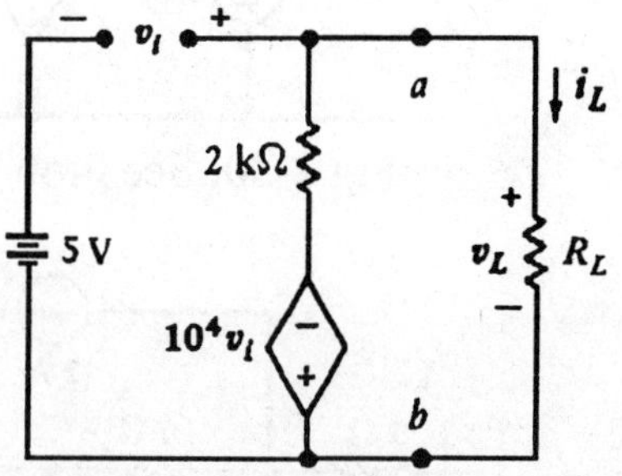


**Fig. 2-27:** See Prob. 33.

## Section 2-7 Trees and General Nodal Analysis

**34** (*a*) Construct all possible trees for the linear graph shown in Fig. 2-28. (*b*) If the top two branches are voltage sources and the left branch is a current source, show all possible trees.

**35** For the circuit shown in Fig. 2-29, construct a tree in which $v_1$ and $v_2$ are tree-branch voltages, write nodal equations, and solve for $v_1$.

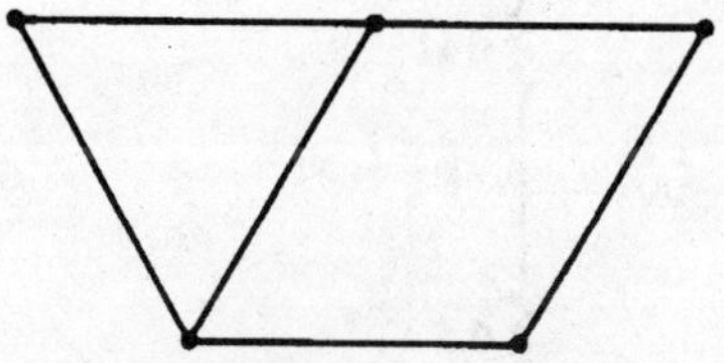

**Fig. 2-28:** See Prob. 34.

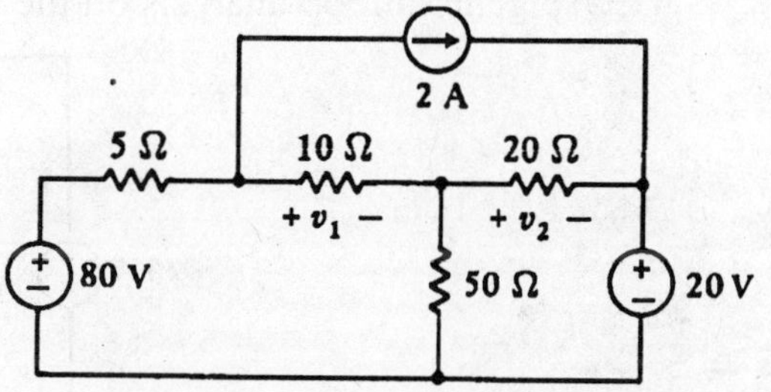


**Fig. 2-29:** See Prob. 35.

**36** Construct a suitable tree for the circuit of Fig. 2-30, assign tree-branch voltages, write KCL and control equations, and find $i_2$.

**37** Use nodal analysis with tree-branch voltages on the circuit of Fig. 2-31 to determine what value of $V_2$ will cause $v_z = 0$.

## Section 2-8 Links and Loop Analysis

**38** For the circuit shown in Fig. 2-32, construct a tree in which $i_1$ and $i_2$ are link currents, write loop equations, and evaluate $i_1$ and $i_2$.

**39** (*a*) Construct a suitable tree for the circuit of Fig. 2-33 and write the single equation necessary to find $i_1$. (*b*) Find $i_1$.

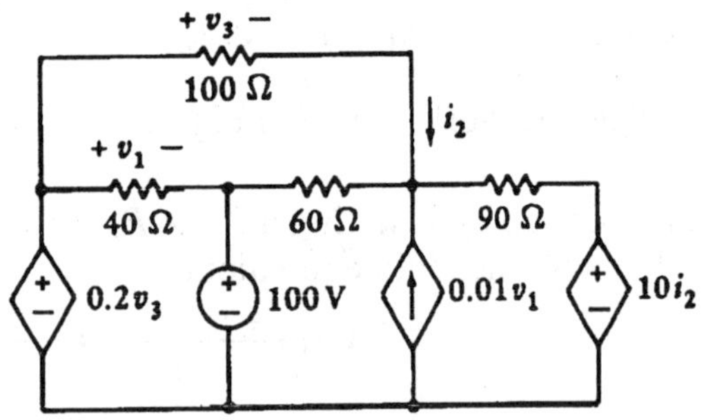


Fig 2-30: See Prob. 36.

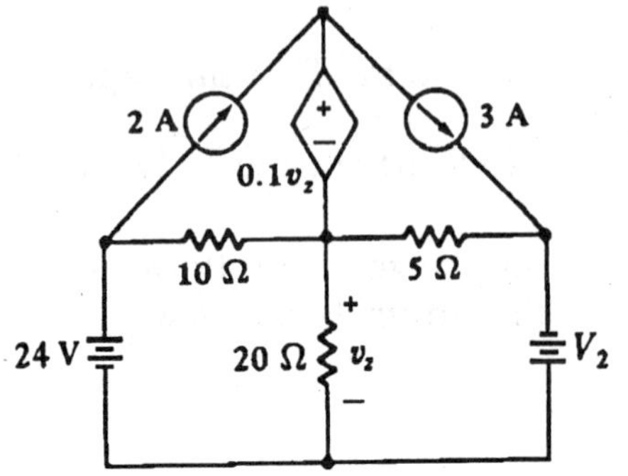


Fig 2-31: See Prob. 37.

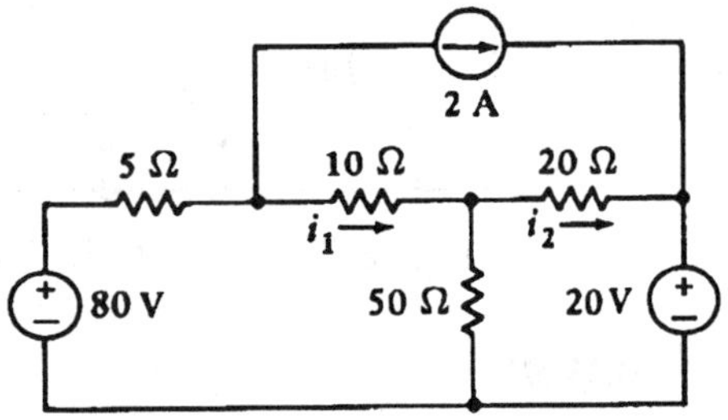


Fig 2-32: See Prob. 38.

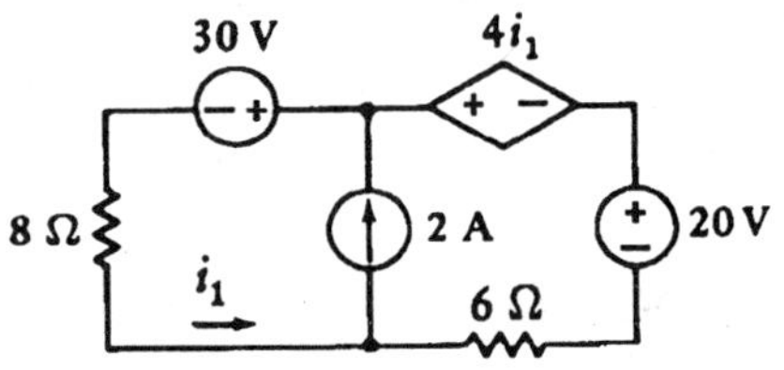


Fig 2-33: See Prob. 39.

**40** Devise a tree for the circuit shown in Fig. 2-34 for which all loop currents flow through the 1-Ω resistor. Find $i$.

**41** Use general loop analysis on the nonplanar circuit of Fig. 2-35 to find $i_x$.

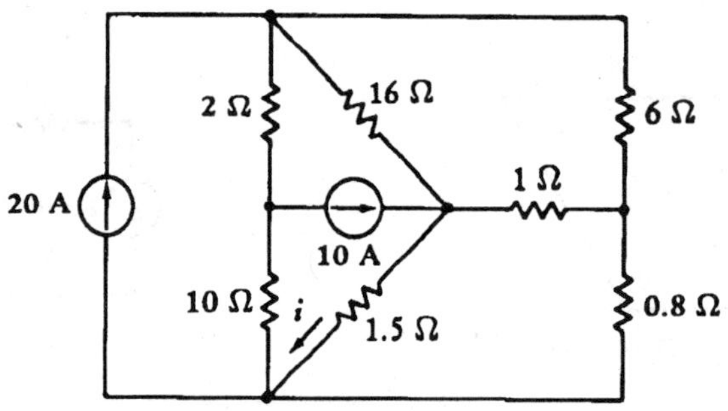


Fig. 2-34 :See Prob. 40.

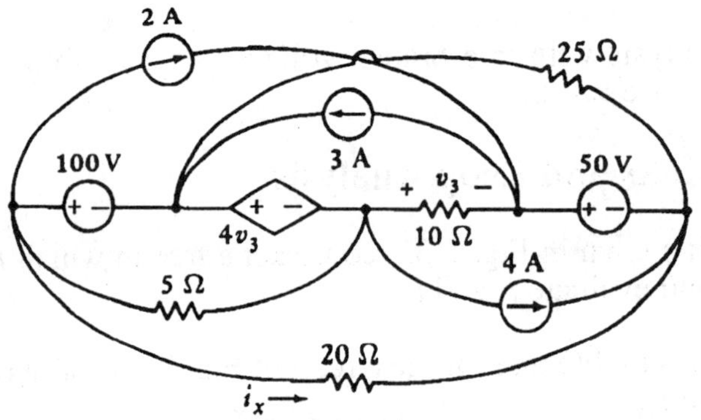


Fig. 2-35: See Prob. 41.

**42** Figure 2-36 shows one form of the equivalent circuit for a transistor amplifier. Determine the open-circuit value of $v_2$ and the output resistance ($R_{th}$) of the amplifier.

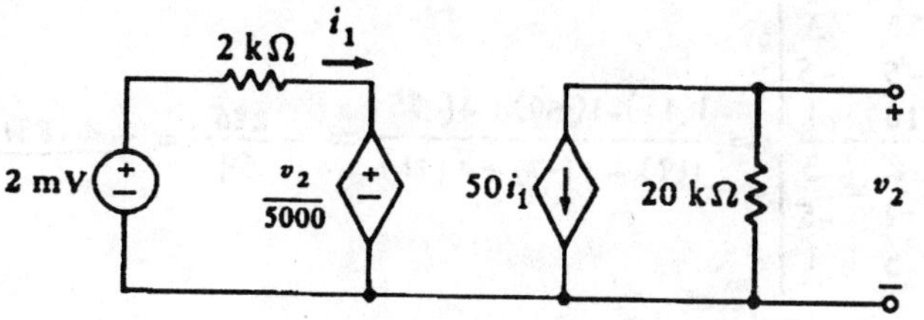


Fig. 2-36: See Prob. 42.

**43** (SPICE) Use SPICE to find $v_4$ in the circuit of Fig. 2-37.

**44** (SPICE) Use SPICE to find $i_{10}$ in the circuit of Fig. 2-38.

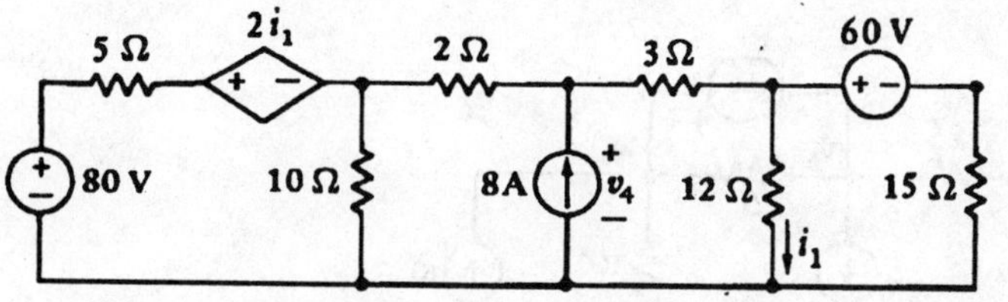


Fig. 2-37: See Prob. 43.

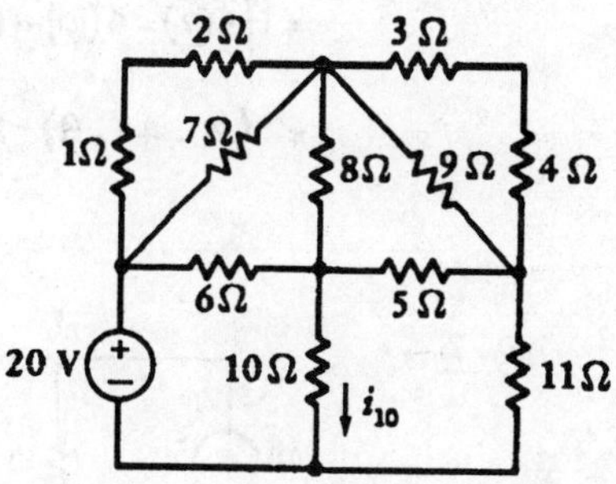


Fig. 2-38: See Prob. 44.

**45** (SPICE) Use SPICE to find $v_4$ in the circuit shown in Fig. 2-39.

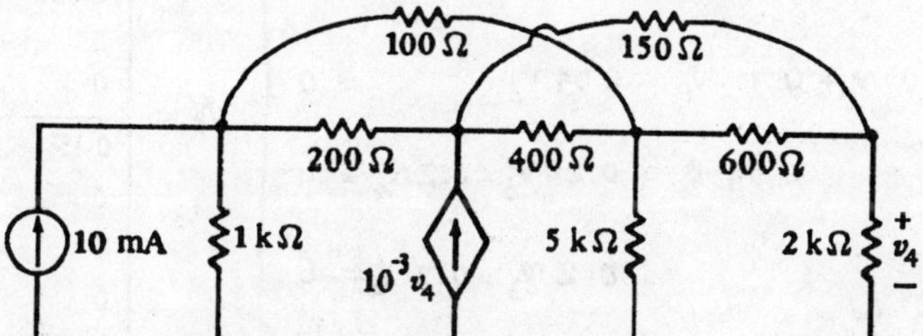


Fig. 2-39: See Prob. 45.

# Chapter 2

**2-1** (a)

$$v_2 = \frac{\begin{vmatrix} 1 & 20 & 3 \\ 1 & -5 & -5 \\ 4 & 10 & 1 \end{vmatrix}}{\begin{vmatrix} 1 & 2 & 3 \\ 1 & -7 & -5 \\ 4 & 3 & 1 \end{vmatrix}} = \frac{1(45)-1(-10)+4(-85)}{1(8)-1(-7)+4(11)} = -\frac{285}{59} = \underline{-4.831}$$

(b)

$$\begin{vmatrix} 1 & 2 & 3 & 4 \\ 2 & 3 & 4 & 1 \\ 3 & 4 & 1 & -2 \\ 4 & -1 & 2 & 3 \end{vmatrix} = 1\begin{vmatrix} 3 & 4 & 1 \\ 4 & 1 & -2 \\ -1 & 2 & 3 \end{vmatrix} - 2\begin{vmatrix} 2 & 3 & 4 \\ 4 & 1 & -2 \\ -1 & 2 & 3 \end{vmatrix} + 3\begin{vmatrix} 2 & 3 & 4 \\ 3 & 4 & 1 \\ -1 & 2 & 3 \end{vmatrix} - 4\begin{vmatrix} 2 & 3 & 4 \\ 3 & 4 & 1 \\ 4 & 1 & -2 \end{vmatrix}$$

$$= 1\left[3(7)-4(10)-1(-9)\right] - 2\left[2(7)-4(1)-1(-10)\right] + 3\left[2(10)-3(1)-1(-13)\right] - 4\left[2(-9)-3(-10)+4(-13)\right]$$

$$= 1(21-40+9) - 2(14-4+10) + 3(20-3+13) - 4(-18+30-52) = -10-40+90+160 = \underline{200}$$

**2-2**

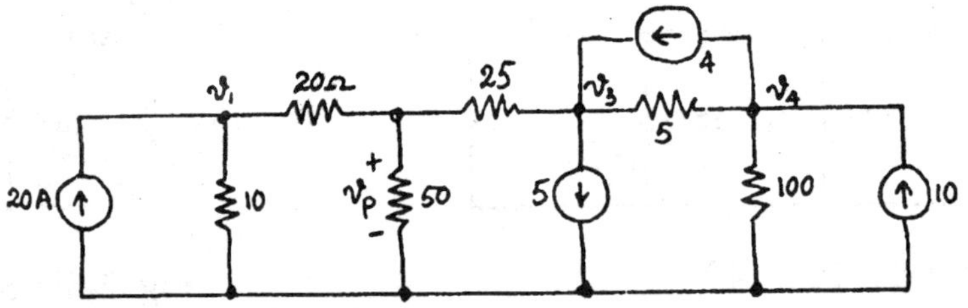


$$\left.\begin{aligned} 0.15v_1 - 0.05v_p &= 20 \\ -0.05v_1 + 0.11v_p - 0.04v_3 &= 0 \\ -0.04v_p + 0.24v_3 - 0.2v_4 &= -1 \\ -0.2v_3 + 0.21v_4 &= 6 \end{aligned}\right\} \quad v_p = \frac{\begin{vmatrix} 0.15 & 20 & 0 & 0 \\ -0.05 & 0 & -0.04 & 0 \\ 0 & -1 & 0.24 & -0.2 \\ 0 & 6 & -0.2 & 0.21 \end{vmatrix}}{\begin{vmatrix} 0.15 & -0.05 & 0 & 0 \\ -0.05 & 0.11 & -0.04 & 0 \\ 0 & -0.04 & 0.24 & -0.2 \\ 0 & 0 & -0.2 & 0.21 \end{vmatrix}} = \underline{171.639 \text{ V}}$$

**2-3**

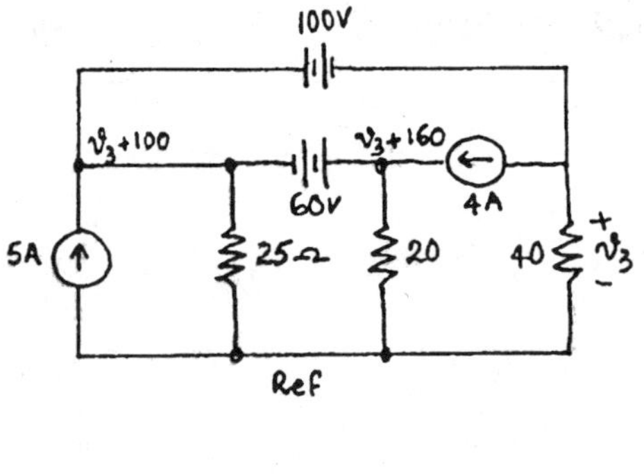


(a) Super: $-5 + \dfrac{v_3+100}{25} + \dfrac{v_3+160}{20} + \dfrac{v_3}{40} = 0$

$$-1000 + 8v_3 + 800 + 10v_3 + 1600 + 5v_3 = 0$$

$$23v_3 = -1400 \; ; \; v_3 = \underline{-60.87 \text{ V}}$$

(b) $P_{5A} = 5(v_3+100) = 5(100-60.87) = \underline{195.65^{+} \text{ W}}$

2-4

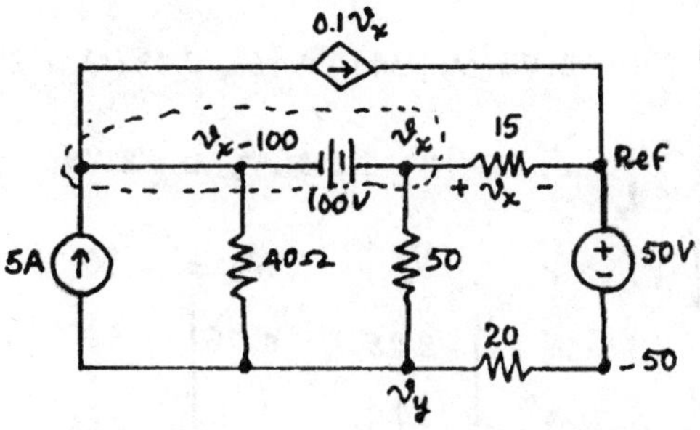


(a) $0.1v_x - 5 + \frac{v_x - 100 - v_y}{40} + \frac{v_x - v_y}{50} + \frac{v_x}{15} = 0$

$\therefore\ 127v_x - 27v_y = 4500$

$5 + \frac{v_y - v_x + 100}{40} + \frac{v_y - v_x}{50} + \frac{v_y + 50}{20} = 0$

$\therefore\ -9v_x + 19v_y = -2000;\ v_x = \frac{\begin{vmatrix} 4500 & -27 \\ -2000 & 19 \end{vmatrix}}{\begin{vmatrix} 127 & -27 \\ -9 & 19 \end{vmatrix}} = \underline{14.516\text{ V}}$

(b) $v_y = \frac{-2000 + 9v_x}{19} = -98.39\text{ V} \therefore P_{50\Omega} = \frac{(v_x - v_y)^2}{50} = \underline{254.9\ \ \text{W}}$

2-5

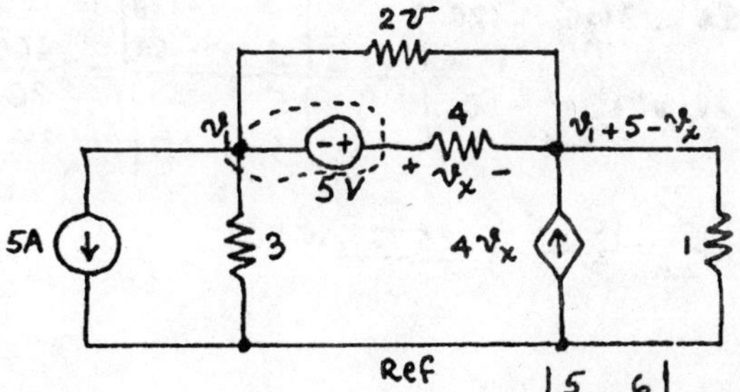


Super: $6 + 3v_1 + 2(v_1 - v_1 - 5 + v_x) + 4v_x = 0$

$\therefore\ 3v_1 + 6v_x = 5$

Rt: $-4v_x + 2(5 - v_x) - 4v_x + 1(v_1 + 5 - v_x) = 0$

$\therefore\ v_1 - 11v_x = -15$

$v_1 = \frac{\begin{vmatrix} 5 & 6 \\ -15 & -11 \end{vmatrix}}{\begin{vmatrix} 3 & 6 \\ 1 & -11 \end{vmatrix}} = \frac{35}{-39};\ v_x = \frac{\begin{vmatrix} 3 & 5 \\ 1 & -15 \end{vmatrix}}{-39} = \frac{-50}{-39};\ P_{5V} = 5(4v_x) = \underline{25.64\ \text{W}}$

2-6

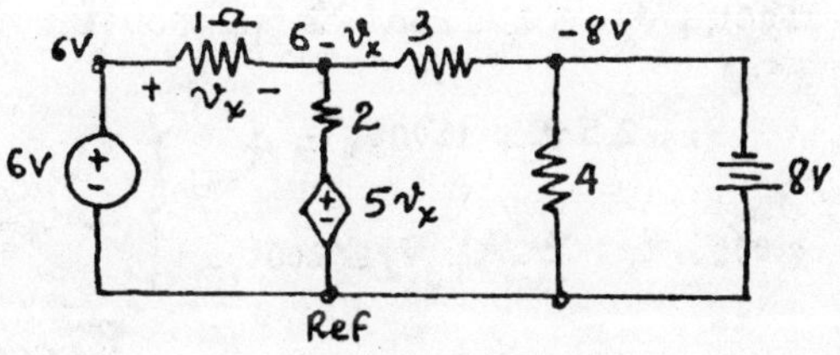


$-\frac{v_x}{1} + \frac{6 - v_x - 5v_x}{2} + \frac{6 - v_x + 8}{3} = 0$

$-6v_x + 18 - 18v_x + 28 - 2v_x = 0;\ -26v_x = -46$

$v_x = \frac{23}{13} = \underline{1.7692\ \text{V}}$

2-7

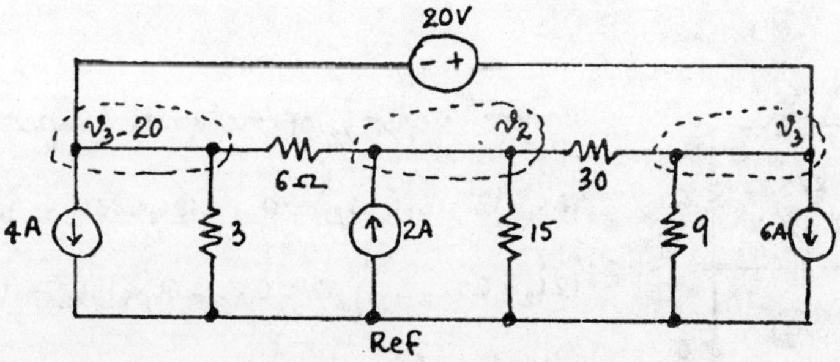


Center:

$\frac{v_2 - v_3 + 20}{6} - 2 + \frac{v_2}{15} + \frac{v_2 - v_3}{30} = 0$

$5v_2 - 5v_3 + 100 - 60 + 2v_2 + v_2 - v_3 = 0$

$\therefore\ 8v_2 - 6v_3 = -40$

Super: $4 + \frac{v_3 - 20}{3} + \frac{v_3 - 20 - v_2}{6} + \frac{v_3 - v_2}{30} + \frac{v_3}{9} + 6 = 0 \therefore -18v_2 + 58v_3 = 0$

$\therefore\ v_3 = \frac{\begin{vmatrix} 8 & -40 \\ -18 & 0 \end{vmatrix}}{\begin{vmatrix} 8 & -6 \\ -18 & 58 \end{vmatrix}} = \frac{-720}{356} = -\underline{2.022}\text{ V};\ P_{6A,gen} = -6v_3 = 6 \times 2.022 = \underline{12.135\ \ \text{W}}$

2-8

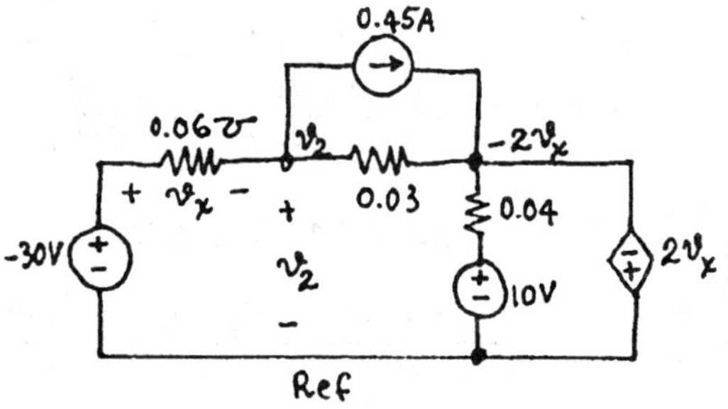


$$0.06(v_2+30)+0.45+0.03(v_2+2v_x)=0$$

$$\therefore\ 0.09v_2+0.06v_x=-2.25 \quad ; \quad v_x=-30-v_2$$

$$\therefore\ v_2+v_x=-30$$

$$v_2=\frac{\begin{vmatrix}-2.25 & 0.06\\ -30 & 1\end{vmatrix}}{\begin{vmatrix}0.09 & 0.06\\ 1 & 1\end{vmatrix}}=\underline{-15\ \text{V}}$$

2-9 (a)

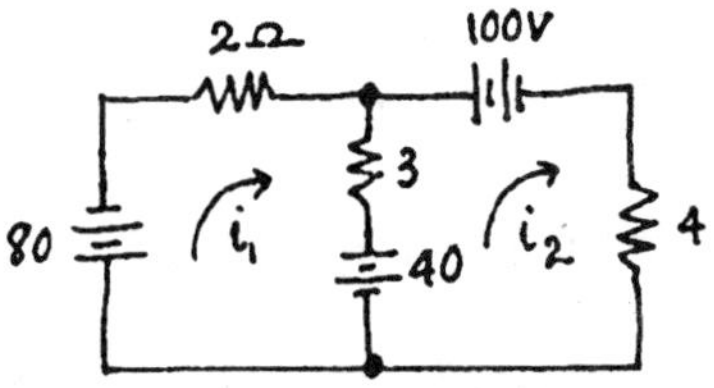


$$\left.\begin{aligned}5i_1-3i_2&=-120\\ -3i_1+7i_2&=-60\end{aligned}\right\}\ i_2=\frac{\begin{vmatrix}5 & -120\\ -3 & -60\end{vmatrix}}{\begin{vmatrix}5 & -3\\ -3 & 7\end{vmatrix}}=-\frac{660}{26}\ \text{A}$$

$$\therefore\ P_{4\Omega}=i_2^2\times 4=\underline{2578}\ \text{W}$$

(b)

$$i_2=\frac{\begin{vmatrix}5 & -120\\ -3 & 40-E\end{vmatrix}}{26}=\frac{-160-5E}{26}\quad \therefore\ E=\frac{-160}{5}=\underline{-32\ \text{V}}$$

2-10

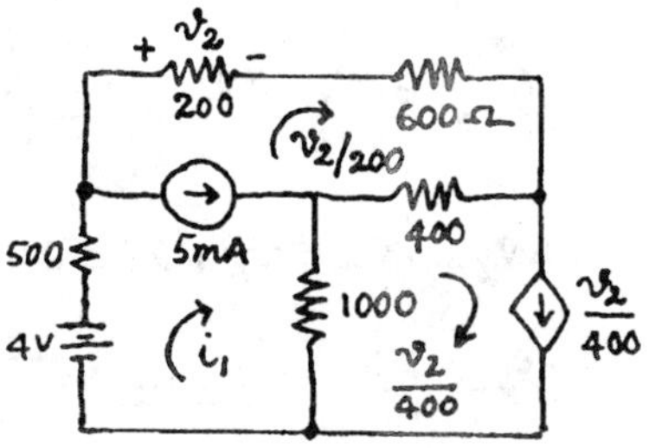


$$500i_1+v_2+3v_2+400\left(\frac{v_2}{400}\right)+1000\left(i_1-\frac{v_2}{400}\right)=4$$

$$\left.\begin{aligned}&\therefore\ 2.5v_2+1500i_1=4\\ &0.005=i_1-\frac{v_2}{200}\quad \therefore\ v_2=200i_1-1\end{aligned}\right\}\ \therefore\ i_1=3.25\ \text{mA}$$

$$P_{4V,gen}=4i_1=\underline{13\ \text{mW}}$$

2-11

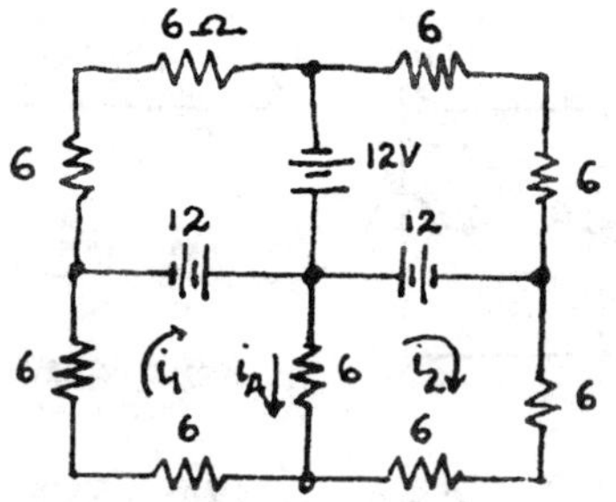


Ignore upper ½ of network (superfluous)

$$12i_1-12+6(i_1-i_2)=0\ \therefore\ 18i_1-6i_2=12$$

$$12i_2+6(i_2-i_1)+12=0\ \therefore\ -6i_1+18i_2=-12$$

$$i_1=\frac{\begin{vmatrix}12 & -6\\ -12 & 18\end{vmatrix}}{\begin{vmatrix}18 & -6\\ -6 & 18\end{vmatrix}}=\frac{144}{288}=0.5\ ;\ i_2=\frac{\begin{vmatrix}18 & 12\\ -6 & -12\end{vmatrix}}{288}=-\frac{144}{288}=-0.5$$

$$\therefore\ i_A=i_1-i_2=\underline{1\ \text{A}}$$

**2-12**

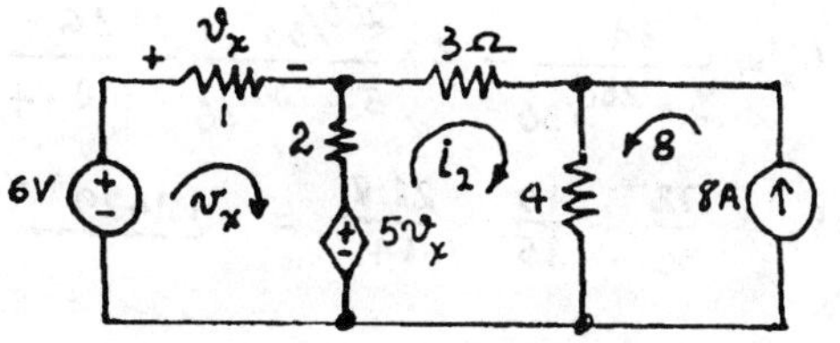


$$-6+v_x+2v_x+5v_x-2i_2=0 \quad \therefore 8v_x-2i_2=6$$

$$-5v_x-2v_x+2i_2+3i_2+4i_2=-32 \quad \therefore -7v_x+9i_2=-32$$

$$i_2=\frac{\begin{vmatrix}8 & 6\\ -7 & -32\end{vmatrix}}{\begin{vmatrix}8 & -2\\ -7 & 9\end{vmatrix}}=\frac{-214}{58}=-3.690 \quad \therefore P_{3\Omega}=3i_2^2=\underline{40.84\ \text{W}}$$

**2-13**

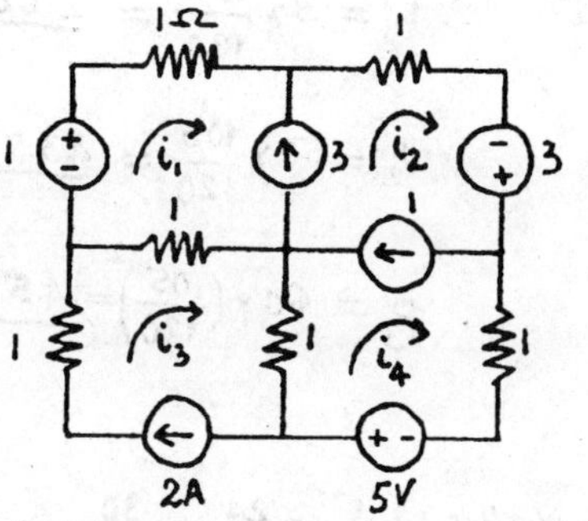


$i_3 = \underline{2\ \text{A}}$ ; $i_2 - i_1 = 3$ or $-i_1 + i_2 = 3$

$i_2 - i_4 = 1$

Around ①, ②, ④:

$$-1+i_1+i_2-3+i_4-5+(i_4-2)+(i_1-2)=0$$

$$\therefore 2i_1+i_2+2i_4=13$$

$$\therefore i_1=\frac{\begin{vmatrix}3 & 1 & 0\\ 1 & 1 & -1\\ 13 & 1 & 2\end{vmatrix}}{\begin{vmatrix}-1 & 1 & 0\\ 0 & 1 & -1\\ 2 & 1 & 2\end{vmatrix}}=\frac{9-2-13}{-3-2}=\underline{1.2\ \text{A}} \quad \therefore i_2=\underline{4.2\ \text{A}}$$

$$i_4=\underline{3.2\ \text{A}}$$

**2-14**

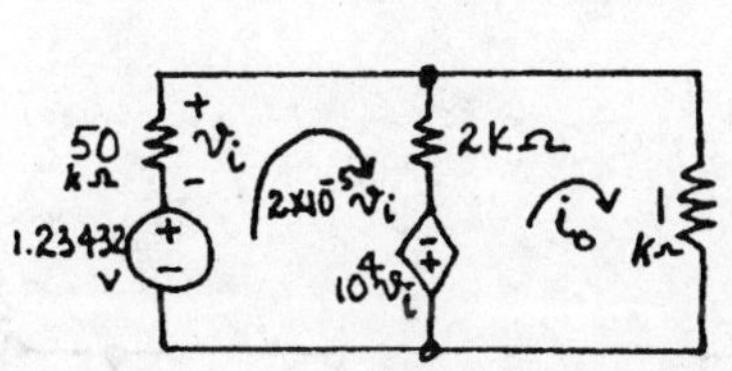


$$-v_s-v_i+2000(2\times10^{-5}v_i-i_o)-10^4v_i=0 \quad \therefore -10000.96v_i-2000i_o=v_s$$

$$10^4v_i+2000(i_o-2\times10^{-5}v_i)+1000i_o=0 \quad \therefore 9999.96v_i+3000i_o=0$$

$$i_o=\frac{\begin{vmatrix}-10{,}000.96 & v_s\\ 9{,}999.96 & 0\end{vmatrix}}{\begin{vmatrix}-10{,}000.96 & -2000\\ 9999.96 & +3000\end{vmatrix}}=\frac{-9999.96\,v_s}{-10{,}002{,}960}=0.999700\,(1.234321)$$

$$i_o=\underline{1.233951\ \text{mA}}$$

**2-15**

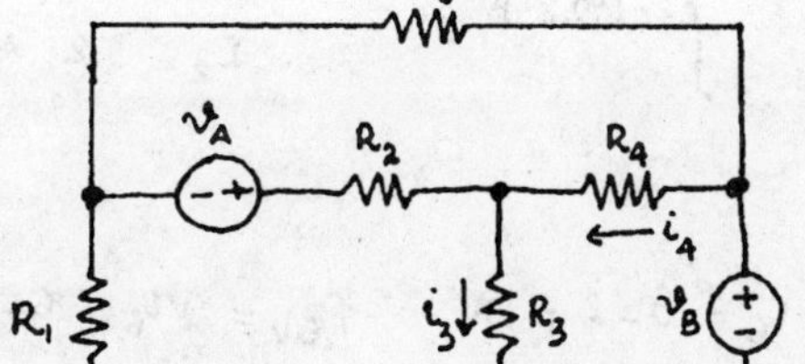

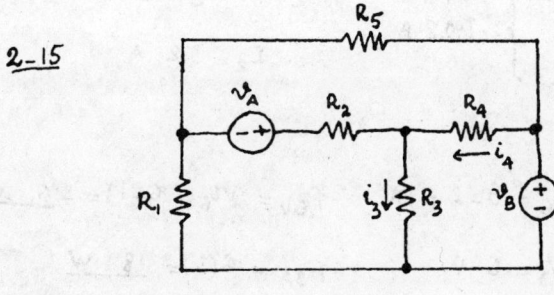

(a) $v_A=20,\ v_B=0 \rightarrow i_3=1.5$

$v_A=50,\ v_B=0 \quad \therefore i_3=\frac{50}{20}\times1.5=\underline{3.75\ \text{A}}$

(b) $i_4=Av_A+Bv_B$

$v_A=20,\ v_B=50 \rightarrow i_4=2 \quad \therefore 2=20A+50B$

$v_A=50,\ v_B=20 \rightarrow i_4=-1 \quad \therefore -1=50A+20B$

$$\therefore A=-\frac{4.5}{105}\ ;\ B=\frac{1}{20}\left(-1+\frac{225}{105}\right)=\frac{6}{105} \quad \therefore i_4=-\frac{4.5}{105}\times30+\frac{6}{105}\times100=\underline{4.429\ \text{A}}$$

**2-16**

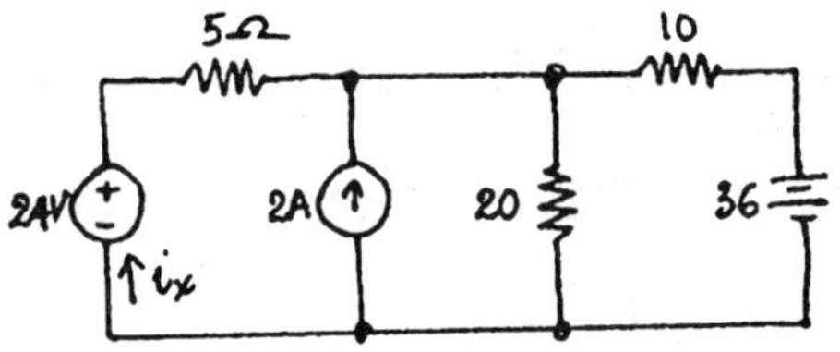


$$i_x = \frac{24}{5+200/30} - 2\times\frac{200/30}{5+200/30} - \frac{36}{10+4}\times\frac{20}{25}$$

$$i_x = \frac{72}{35} - \frac{40}{35} - \frac{28.8}{14} = \underline{-1.1429\ \text{A}}$$

**2-17**

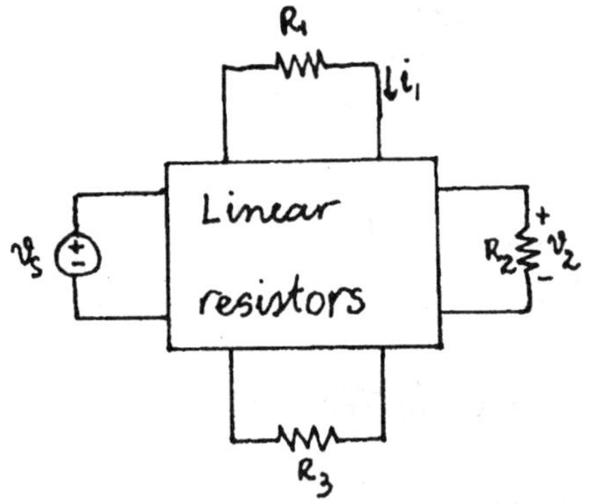


$v_s = 120 \rightarrow i_1 = 3\ ,\ v_2 = 50\ ,\ P_3 = 60$

$$v_s = 105 \qquad \therefore i_1 = 3\times\frac{105}{120} = \underline{2.625\ \text{A}}$$

$$v_2 = 50\times\frac{105}{120} = \underline{43.75\ \text{V}}$$

$$P_3 = 60\times\left(\frac{105}{120}\right)^2 = \underline{45.94\ \text{W}}$$

**2-18**

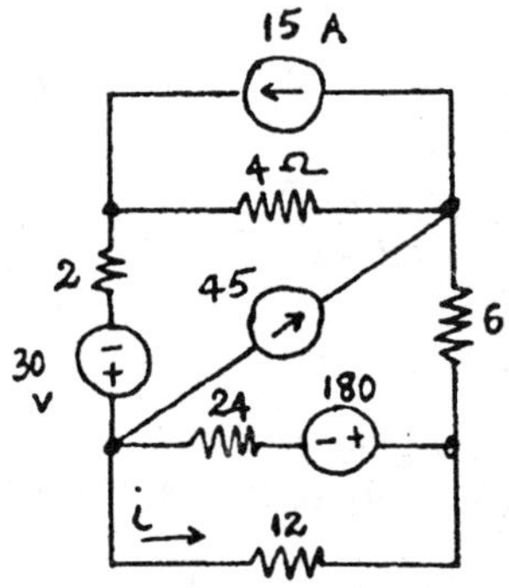


$$i = 15\times(4\|16)\times\frac{1}{16}\times\frac{24}{36} + \frac{30}{12+8}\times\frac{24}{36} - \frac{180}{24+12\|12}\times\frac{12}{24} - 45\times(14\|6)\times\frac{1}{14}\times\frac{24}{36}$$

$$i = 2+1-3-9 = \underline{-9\ \text{A}}$$

**2-19**

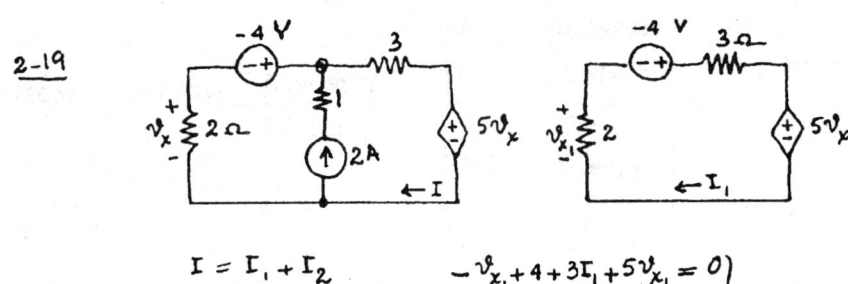

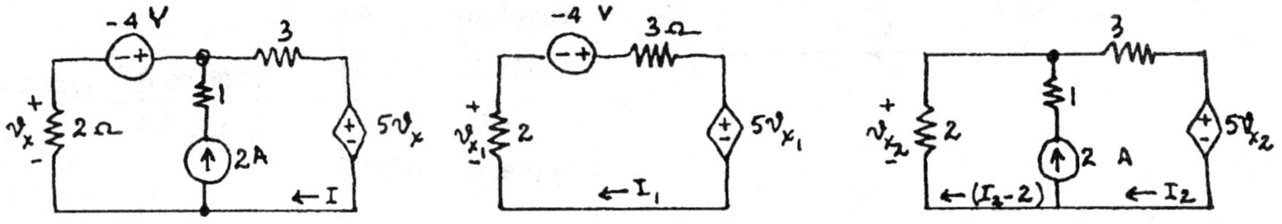


$I = I_1 + I_2$

$$\left.\begin{aligned} -v_{x_1} + 4 + 3I_1 + 5v_{x_1} &= 0 \\ \text{and } v_{x_1} &= -2I_1 \end{aligned}\right\} \therefore I_1 = 0.8\ \text{A}$$

$2(I_2-2) + 3I_2 + 5(2)(2-I_2) = 0$

$\therefore I_2 = 3.2$ A

$\therefore I = 0.8 + 3.2 = \underline{4\ \text{A}}$

**2-20**

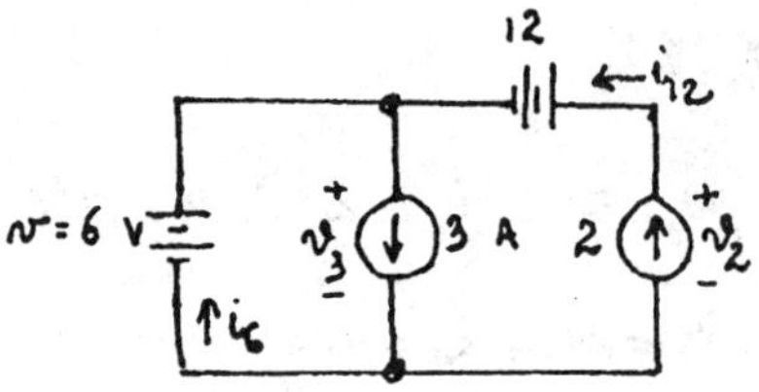


(a) $i_6 = 3-2 = 1\text{A} \quad \therefore P_{6V} = v i_6 = 6(-1) = \underline{-6\ \text{W}}$

(b) $v_3 = 6$ V $\quad \therefore P_{3A} = 6(3) = \underline{18\ \text{W}}$

(c) $i_{12} = 2$ A $\quad \therefore P_{12V} = 12(2) = \underline{24\ \text{W}}$

(d) $v_2 = 18$ V $\quad \therefore P_2 = 18(-2) = \underline{-36\ \text{W}};\ \Sigma = 0$

**2-21**

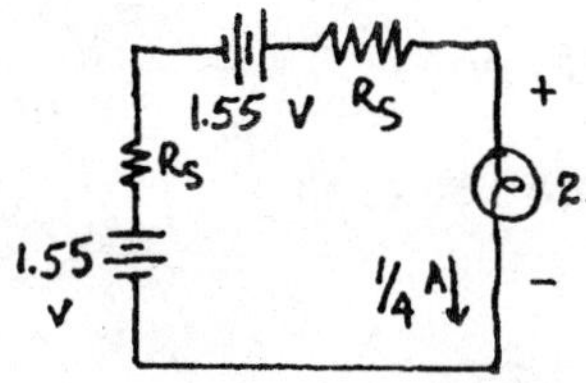

$$-1.55 + 0.25R_S - 1.55 + 0.25R_S + 2.25 = 0$$

$$\therefore R_S = \frac{3.10 - 2.55}{0.5} = 1.7\ \Omega$$

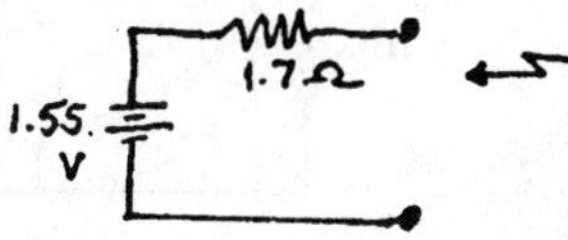

**2-22** (a)

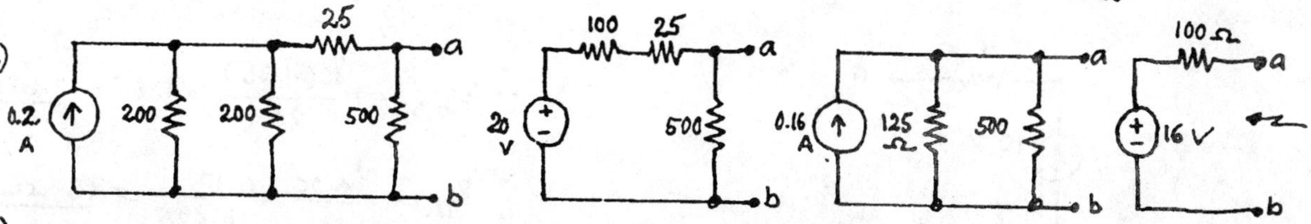

(b)

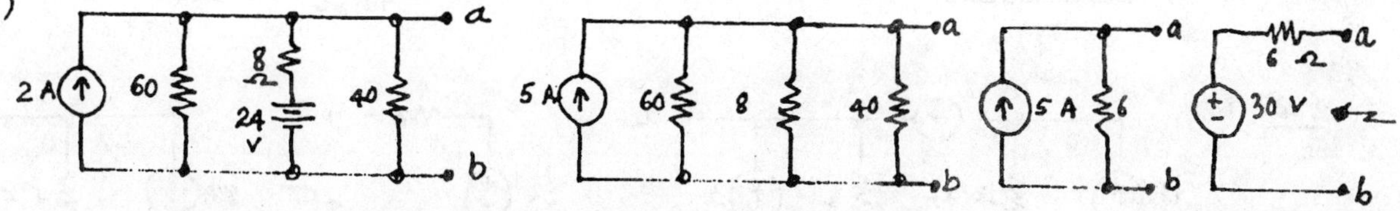

**2-23**

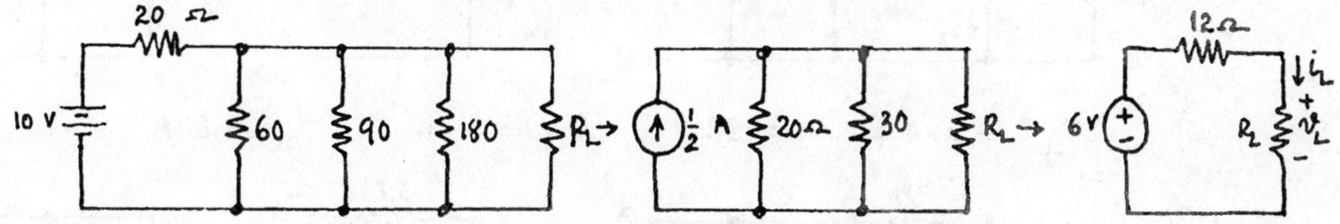

(a) $R_L = \underline{12\ \Omega}$ for $p_{L,max}$ ; $p_{L,max} = \left(\frac{6}{24}\right)^2 12 = \underline{0.75\ \text{W}}$

(b) $R_L = \underline{\infty}$ for $v_{L,max}$ ; $v_{L,max} = \underline{6\ \text{V}}$

(c) $R_L = \underline{0}$ for $i_{L,max}$ ; $i_{L,max} = \underline{0.5\ \text{A}}$

**2-24**

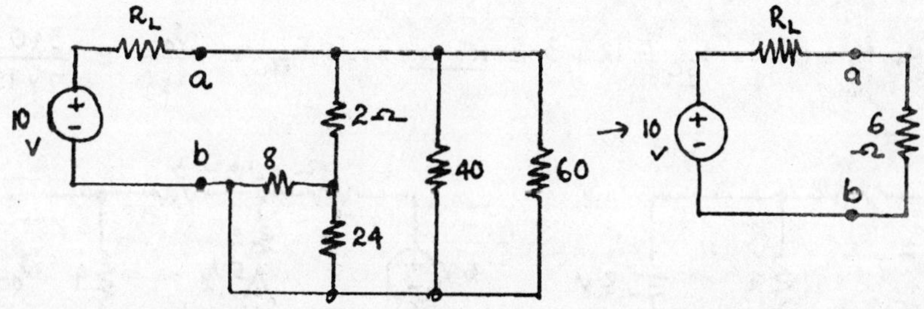

(a) $R_L = \underline{6\ \Omega}$

(b) $P_{L,max} = \left(\frac{10}{12}\right)^2 \times 6$

$P_{L,max} = \underline{4.167\text{W}}$

**2-25** (a)

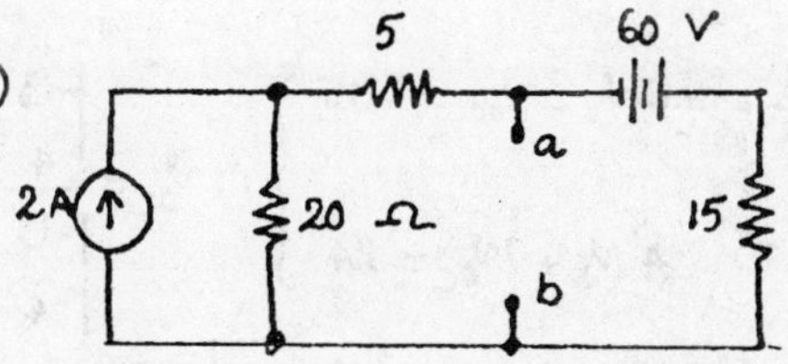

$$v_{oc}\overset{+}{} = 60 \times \frac{-25}{40} + 2\,(20 // 20)\,\frac{15}{20} = -37.5 + 15 = \underline{-22.5\ \text{V}}$$

$$i_{sc}\downarrow = \frac{-60}{15} + 2 \times \frac{20}{25} = -4 + 1.6 = \underline{-2.4\ \text{A}}$$

$$R_{Th} = (5+20) // 15 = \underline{9.375\ \Omega}$$

(b) Thev:

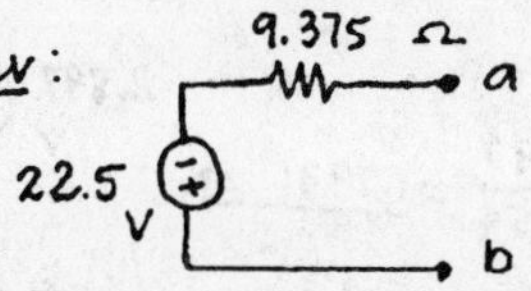

Nort:

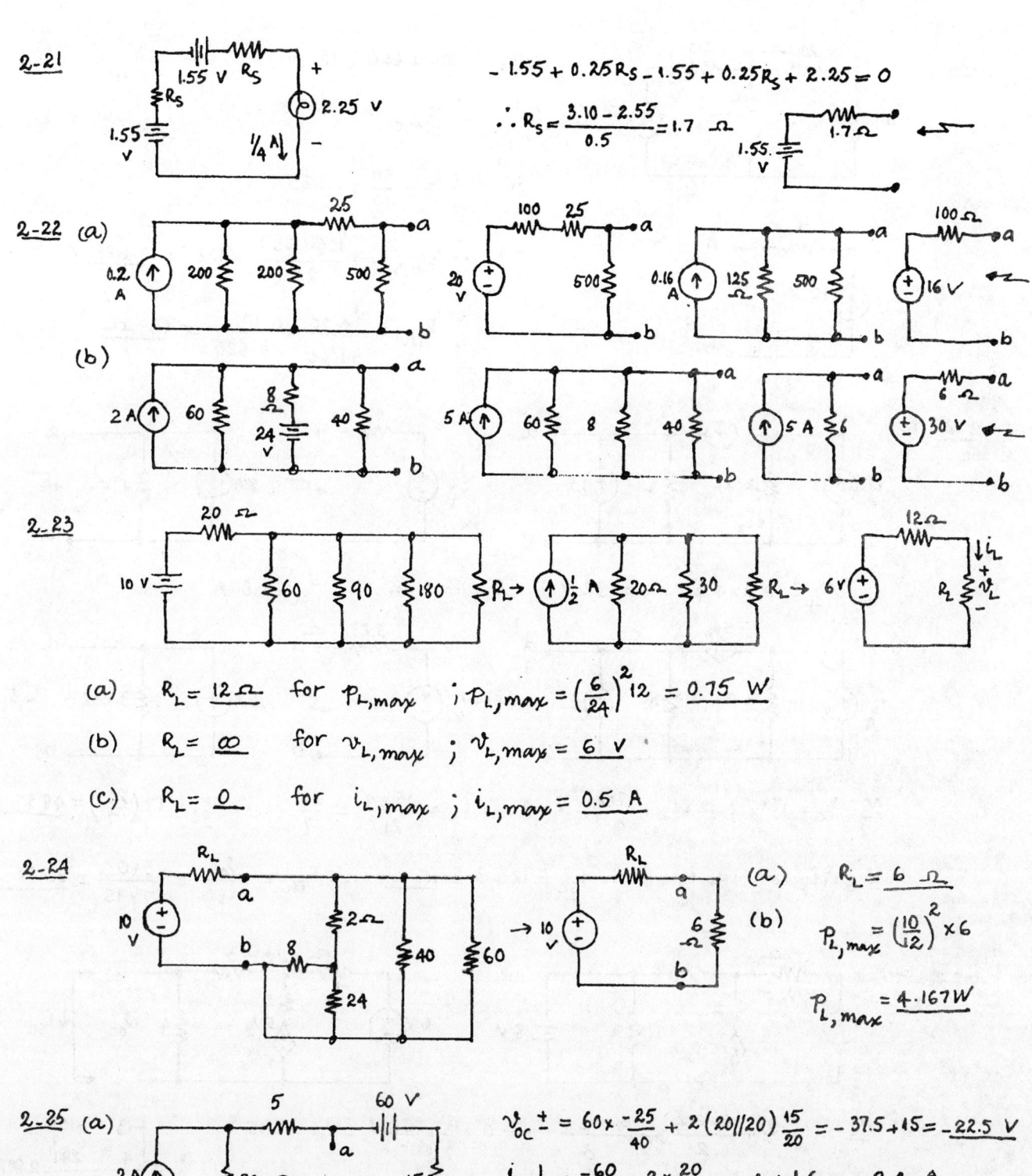

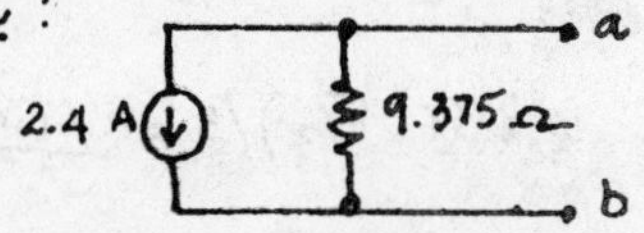

**2-26**

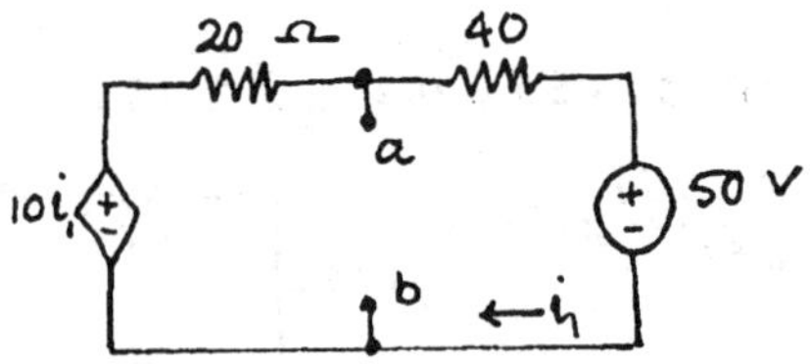


OC: $-10i_1 + 60i_1 + 50 = 0 \quad \therefore i_1 = -1$

$\therefore v_{ab,oc} = -40 + 50 = \underline{10\ V}$

SC: $i_1 = -\dfrac{50}{40} = -1.25$

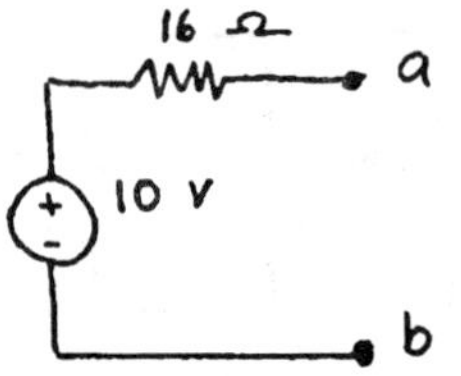


$\therefore i_{ab,sc} = \dfrac{10(-1.25)}{20} + 1.25 = \underline{0.625\ A}$

$R_{Th} = \dfrac{v_{ab,oc}}{i_{ab,sc}} = \dfrac{10}{0.625} = \underline{16\ \Omega}$

**2-27** (a)

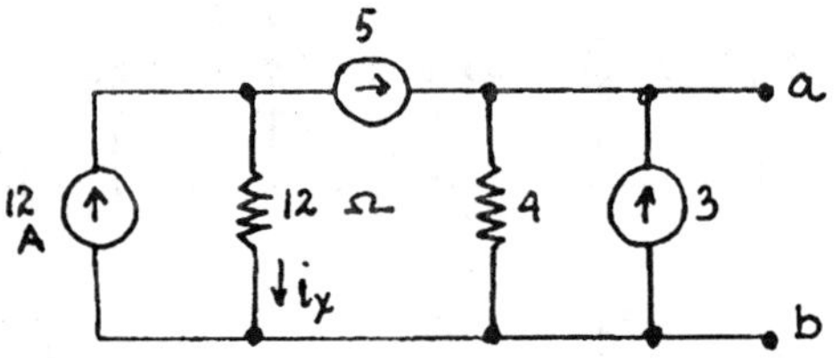


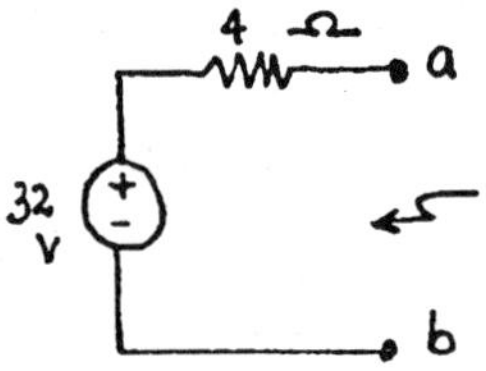


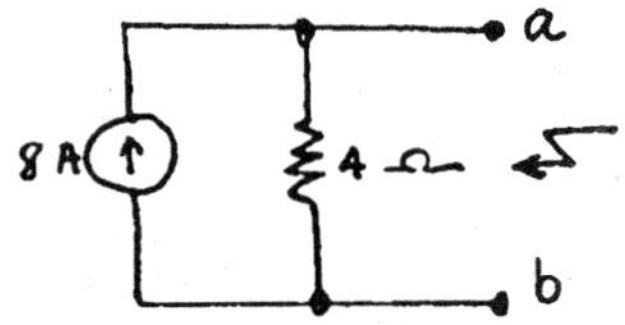


$v_{oc} = 3\times4 + 5\times4 = 32\ V$; $\quad R_{Th} = 4\ \Omega \quad \therefore i_{sc} = 8\ A$

(b)

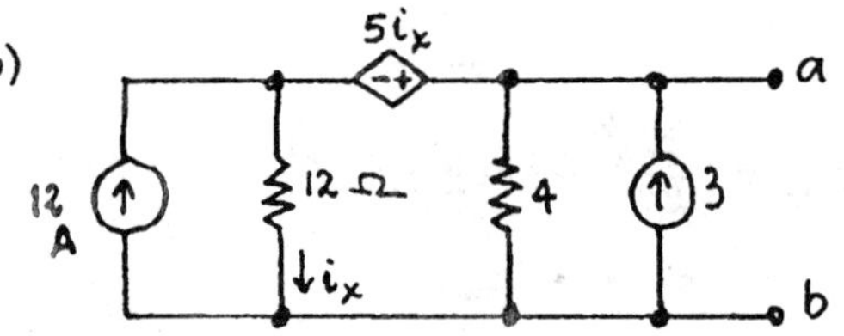


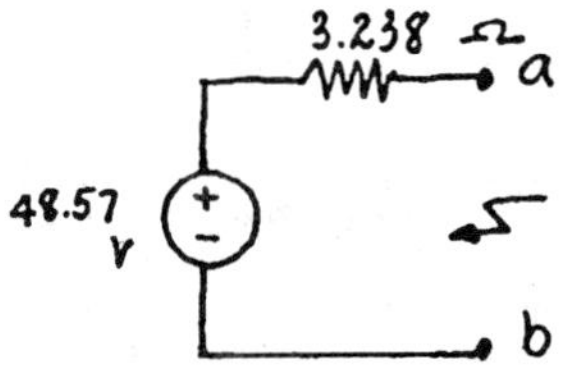


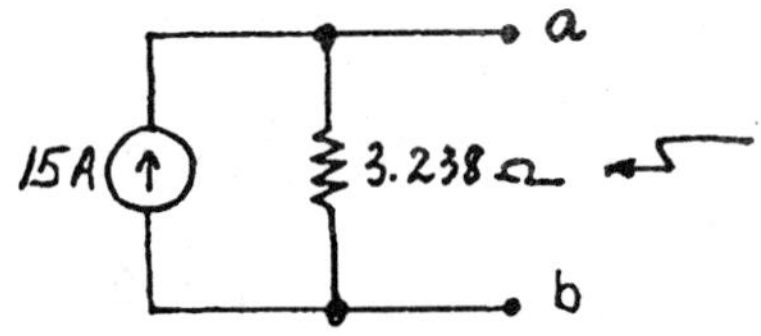


OC: $v_{oc} = 17i_x$; $-12 + i_x + \dfrac{17i_x}{4} - 3 = 0 \quad \therefore i_x = \dfrac{15\times4}{21} = \dfrac{20}{7} \quad \therefore v_{oc} = 17\left(\dfrac{20}{7}\right) = \underline{48.57\ V}$

SC: $17i_x = 0 \quad \therefore i_x = 0 \quad \therefore i_{sc} = 12 + 3 = \underline{15\ A} \quad \therefore R_{Th} = \dfrac{v_{oc}}{i_{sc}} = \dfrac{340}{7\times15} = \underline{3.238\ \Omega}$

**2-28** (a)

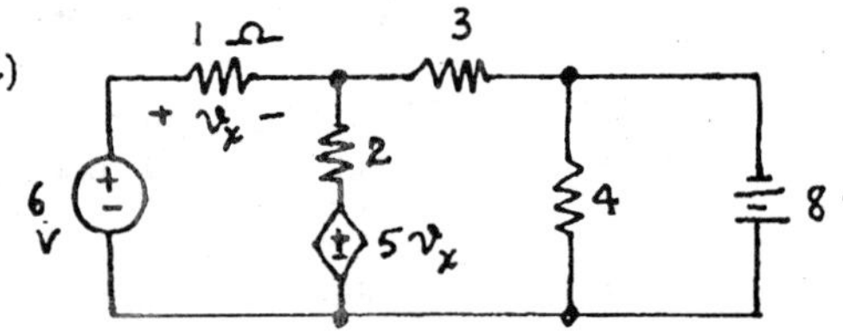


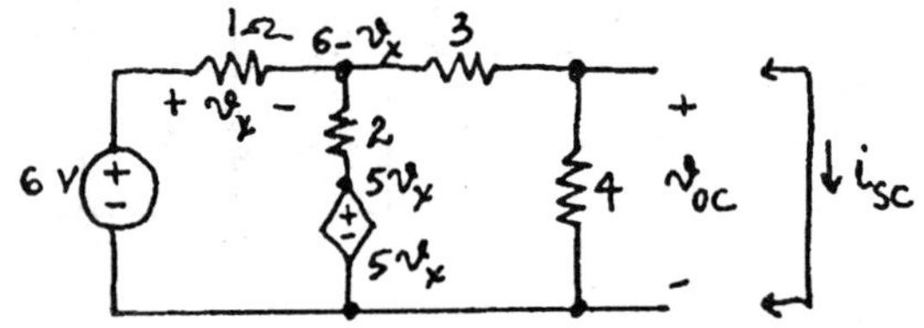


OC: $-v_x + \dfrac{6-6v_x}{2} + \dfrac{6-v_x-v_{oc}}{3} = 0 \quad \therefore -13v_x - v_{oc} = -15$

$\dfrac{v_{oc} - 6 + v_x}{3} + \dfrac{v_{oc}}{4} = 0 \quad \therefore 4v_x + 7v_{oc} = 24$

$\therefore v_{oc} = \dfrac{\begin{vmatrix} -13 & -15 \\ 4 & 24 \end{vmatrix}}{\begin{vmatrix} -13 & -1 \\ 4 & 7 \end{vmatrix}} = \underline{2.897\ V}$

SC: $-v_x + \dfrac{6-6v_x}{2} + \dfrac{6-v_x}{3} = 0 \quad \therefore v_x = \dfrac{15}{13}\ V$

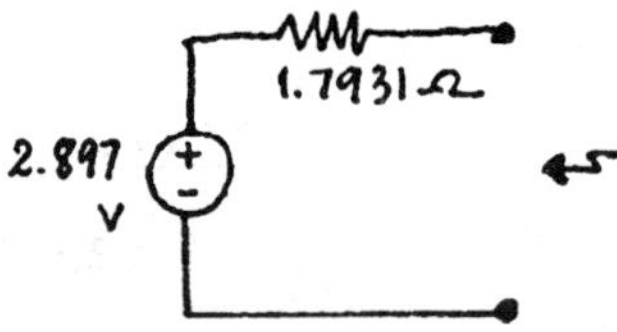


$\therefore i_{sc} = (6 - 15/13)/3 = \underline{1.6154\ A}$; $\quad R_{Th} = \dfrac{2.897}{1.6154} = \underline{1.7931\ \Omega}$

**2-28** (b)

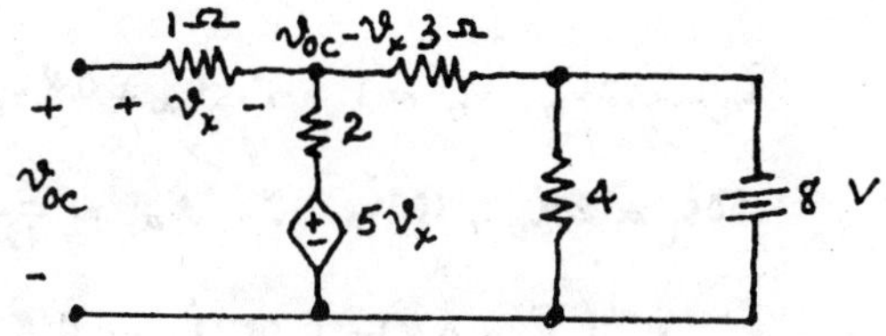


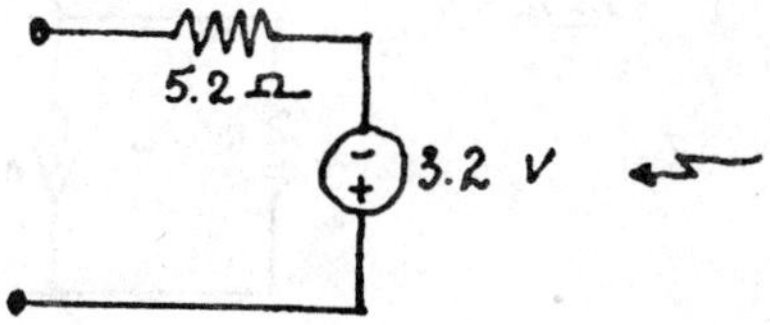


OC: $v_x = 0$ (O.C.) ∴ $v_{oc} = -8 \times \frac{2}{5} = \underline{-3.2\ V}$

SC: $-\frac{v_x}{1} + \frac{0 - v_x - 5v_x}{2} + \frac{0 - v_x + 8}{3} = 0$ ∴ $v_x = \frac{8}{13}$ V ; $i_{sc}\!\downarrow = -v_x = \underline{-\frac{8}{13}\ A}$

$R_{Th} = \frac{-3.2 \times 13}{-8} = \underline{5.2\ \Omega}$

**2-29** (a)

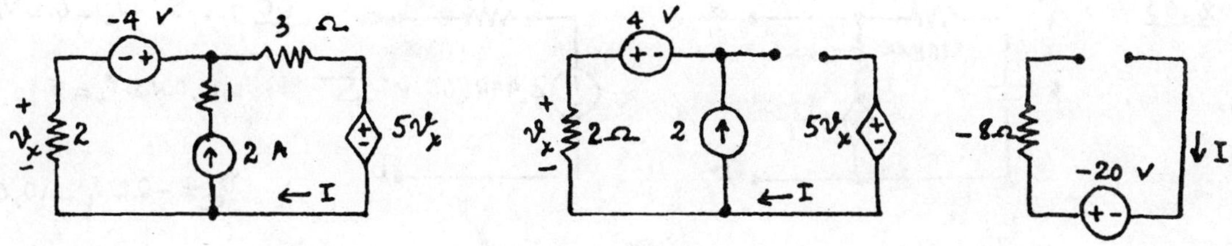


OC: $v_x = 4$ ; $\overset{+\ -}{v_{oc}} = v_x - 4 - 5v_x = -20$ V

SC: $-v_x + 4 + 5v_x = 0$ ∴ $v_x = -1$V ∴ $\vec{i}_{sc} = -\frac{v_x}{2} + 2 = 2.5$ A ; $R_{Th} = \frac{-20}{2.5} = -8\ \Omega$

(b) $I = \frac{-20}{-8+3} = \underline{4\ A}$

(c) 13Ω ∴ $I = \frac{-20}{-8+13} = \underline{-4\ A}$

**2-30** (a)

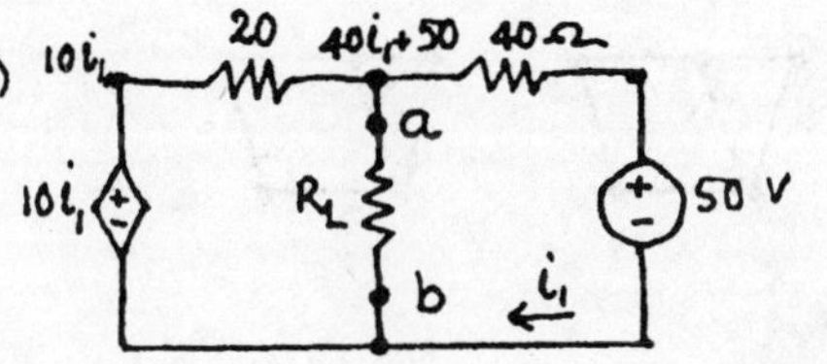


$i_1 + \frac{40i_1 + 50}{R_L} + \frac{40i_1 + 50 - 10i_1}{20} = 0$ ∴ $i_1 = -\frac{2.5R_L + 50}{2.5R_L + 40}$

$v_L^{\pm} = 40i_1 + 50$ and $i_1 = -\frac{R_L + 20}{R_L + 16}$

∴ $v_L = -\frac{40R_L + 800}{R_L + 16} + \frac{50R_L + 800}{R_L + 16} = \frac{10R_L}{R_L + 16}$

(b) $v_{oc} = \lim_{R_L \to \infty} v_L = \lim_{R_L \to \infty} \frac{10R_L}{R_L + 16} = \underline{10\ V}$ ; $i_{sc} = \lim_{R_L \to 0} \frac{v_L}{R_L} = \lim_{R_L \to 0} \frac{10}{R_L + 16} = \underline{0.625\ A}$

**2-31**

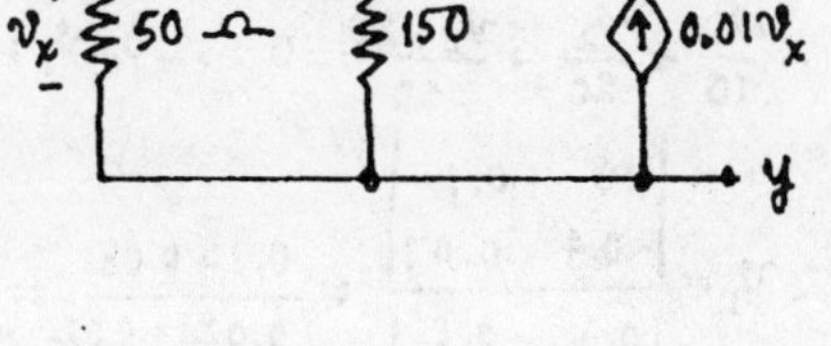


$\frac{5v_x}{250} + \frac{5v_x}{150} + \frac{5v_x - v_{in}}{100} = 0$ ∴ $v_x = 0.09677 v_{in}$

and $\frac{v_{in} - 5v_x}{100} - 0.01v_x = 1$

∴ $0.01v_{in} - 0.06(0.09677 v_{in}) = 1$ ∴ $v_{in} = 238.5$ V

$R_{Th} = \underline{238.5\ \Omega}$

**2-32**

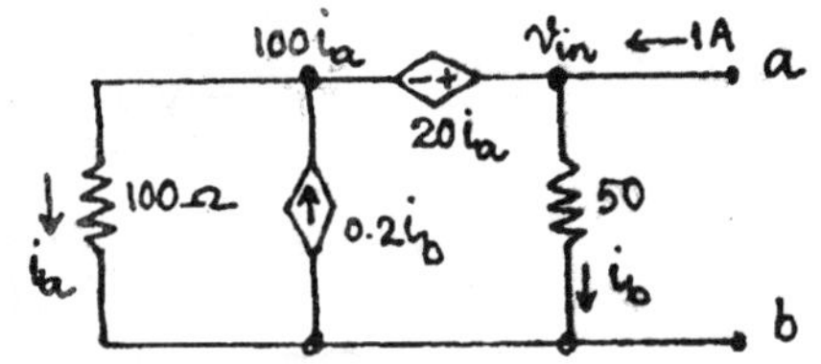


$i_a - 0.2i_b + i_b = 1 \quad \therefore\ i_a + 0.8i_b = 1$

$50i_b = 20i_a + 100i_a \quad \therefore\ i_a = \frac{5}{12}i_b$

$\therefore\ i_b\left(\frac{5}{12} + 0.8\right) = 1$ or $i_b = 0.8219$ A

$v_{in} = 50i_b = 50(0.8219) = \underline{41.10\text{ V}}$

$R_{Th} = \underline{41.10\ \Omega}$

**2-33**

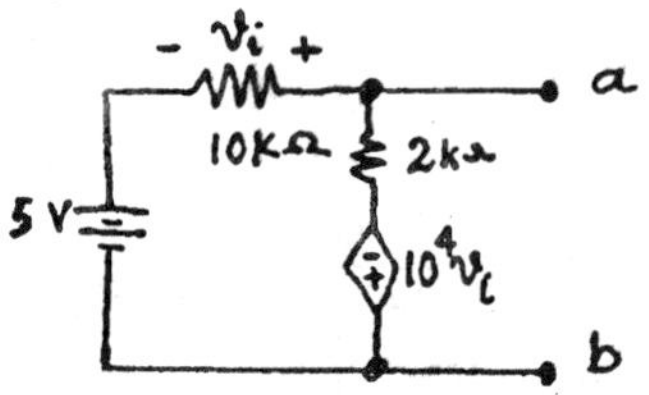


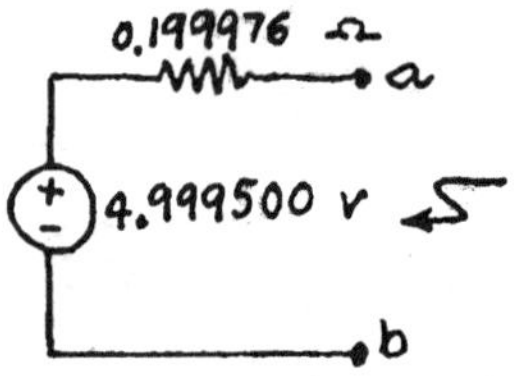


OC : $-5 - v_i - 0.2v_i - 10^4 v_i = 0$

$-10{,}001.2v_i = 5;\quad v_i = -\frac{5}{10{,}001.2}$

$v_{oc} = -0.2v_i - 10{,}000v_i = \underline{4.999500\text{ V}}$

SC : $v_i = -5$ ; $i_{sc}\downarrow = -\frac{v_i}{10{,}000} - \frac{10^4 v_i}{2000} = 25.0005$ A $\therefore R_{Th} = \frac{50{,}001}{10{,}001.2}\ \frac{1}{25.0005} = \underline{0.199976\ \Omega}$

**2-34** (a)

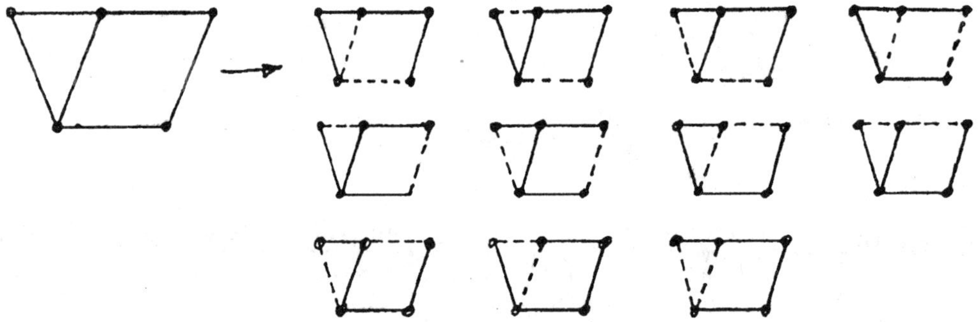

(b)

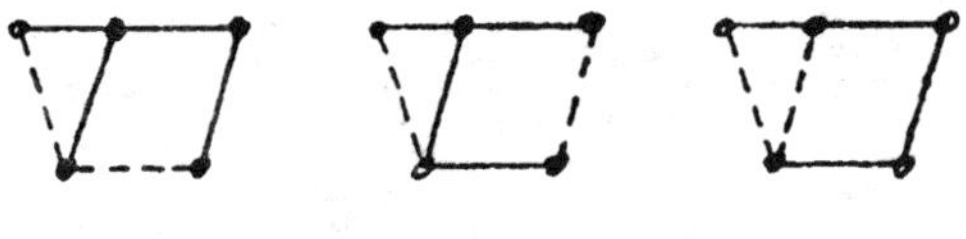

**2-35**

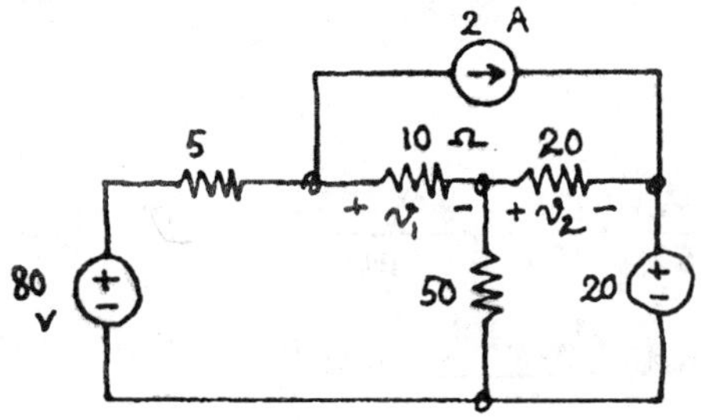


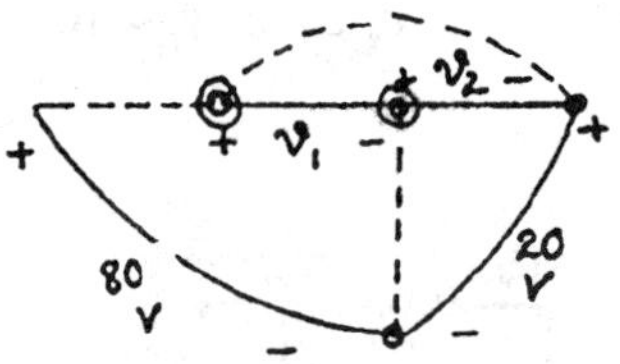


$\frac{v_1 + v_2 + 20 - 80}{5} + \frac{v_1}{10} + 2 = 0$

$\therefore\ 0.3v_1 + 0.2v_2 = 10$

$-\frac{v_1}{10} + \frac{v_2}{20} + \frac{v_2 + 20}{50} = 0 \quad \therefore\ -0.1v_1 + 0.07v_2 = -0.4$

$$\therefore v_1 = \frac{\begin{vmatrix} 10 & 0.2 \\ -0.4 & 0.07 \end{vmatrix}}{\begin{vmatrix} 0.3 & 0.2 \\ -0.1 & 0.07 \end{vmatrix}} = \frac{0.7 + 0.08}{0.021 + 0.02} = \underline{19.024\text{ V}}$$

**2-36**

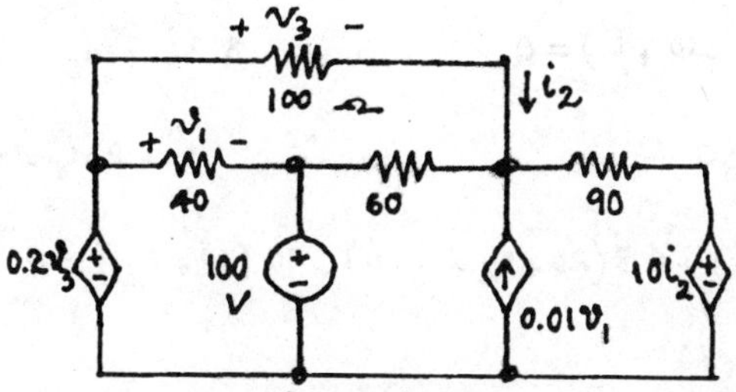


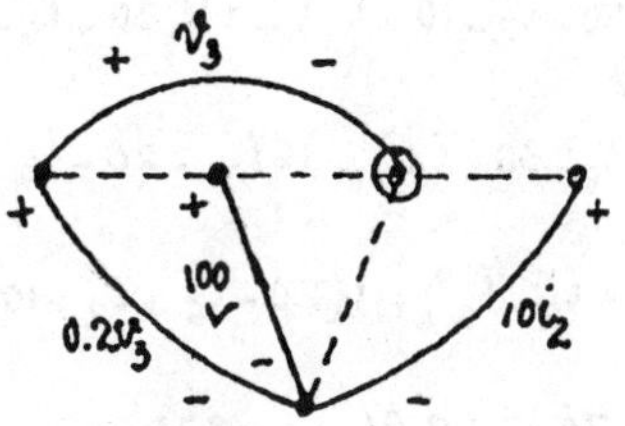


$V_3 = 100\,i_2$

$$\therefore i_2 = \frac{\begin{vmatrix} 300 & -1.8 \\ 100 & -1 \end{vmatrix}}{\begin{vmatrix} -600 & -1.8 \\ 20 & -1 \end{vmatrix}}$$

$$-i_2 + \frac{-100i_2 + 20i_2 - 100}{60} - 0.01v_1 + \frac{-100i_2 + 20i_2 - 10i_2}{90} = 0 \quad \therefore -600i_2 - 1.8v_1 = 300$$

$$\text{and} \quad v_1 = 20i_2 - 100 \quad \therefore \quad 20i_2 - v_1 = 100$$

$\therefore i_2 = \underline{-0.18868 \text{ A}}$

**2-37**

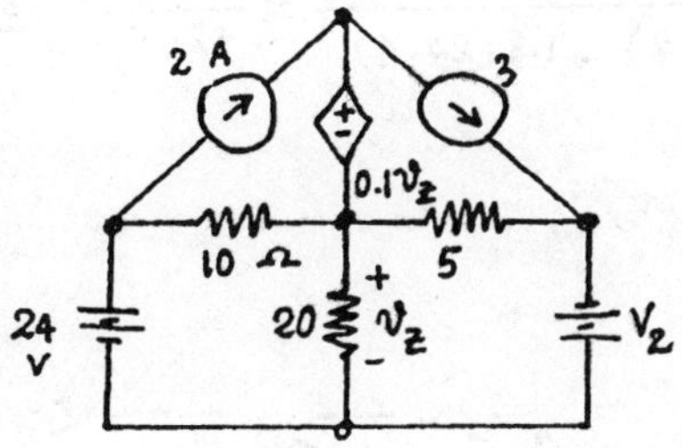


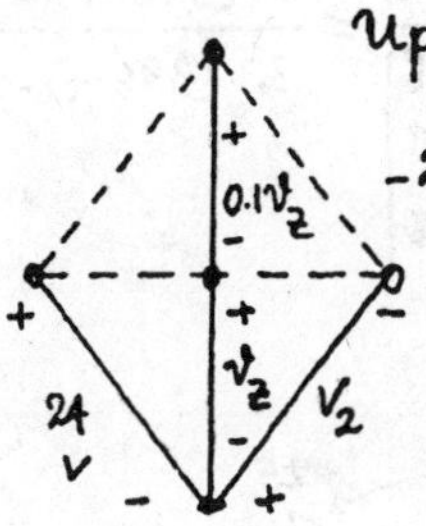


Upper supernode:

$$-2 + 3 + 0.05v_z + 0.1(v_z - 24) + 0.2(v_z + V_2) = 0$$

$$\text{If } v_z = 0\ ; \ +1 - 2.4 + 0.2V_2 = 0$$

$$V_2 = \underline{7 \text{ V}}$$

**2-38**

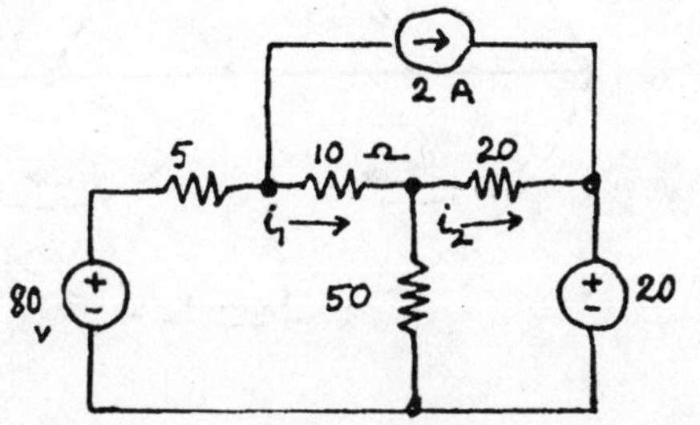


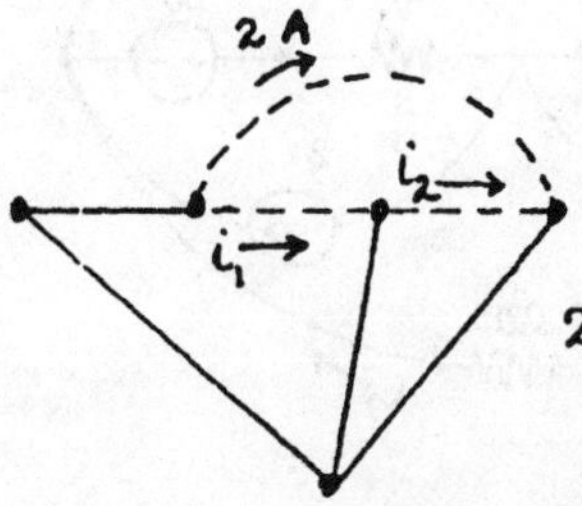


$$10i_1 + 50(i_1 - i_2) - 80 + 5(i_1 + 2) = 0$$

$$\therefore 65i_1 - 50i_2 = 70$$

$$20i_2 + 20 + 50(i_2 - i_1) = 0$$

$$\therefore -50i_1 + 70i_2 = -20$$

$$\therefore i_1 = \frac{\begin{vmatrix} 70 & -50 \\ -20 & 70 \end{vmatrix}}{\begin{vmatrix} 65 & -50 \\ -50 & 70 \end{vmatrix}} = \frac{3900}{2050} = \underline{1.9024 \text{ A}} \qquad i_2 = \frac{\begin{vmatrix} 65 & 70 \\ -50 & -20 \end{vmatrix}}{2050} = \frac{2200}{2050} = \underline{1.0732 \text{ A}}$$

**2-39**

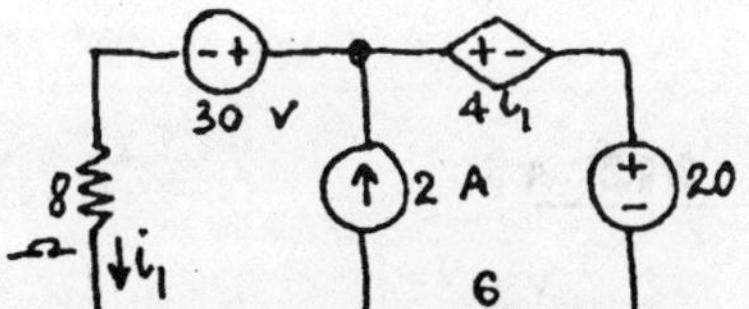


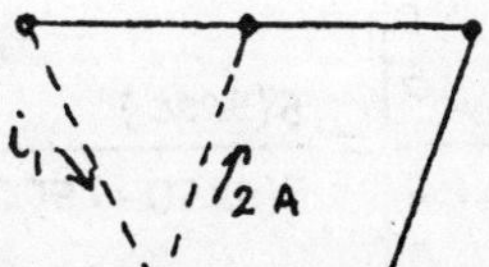


(a) $8i_1 + 6(i_1 - 2) - 20 - 4i_1 + 30 = 0$ or $10i_1 = 2$

(b) $i_1 = \underline{0.2 \text{ A}}$

**2-40**

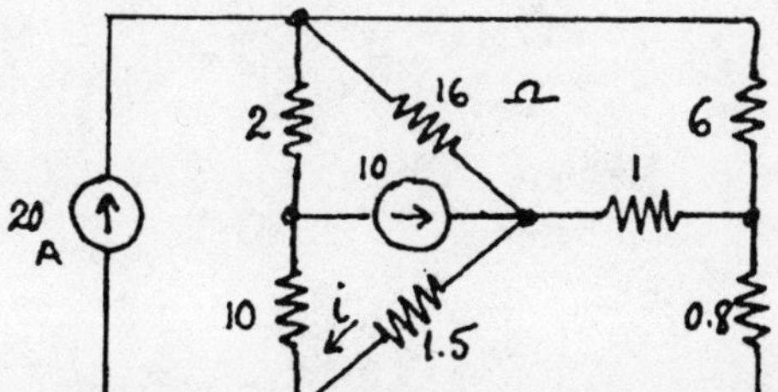


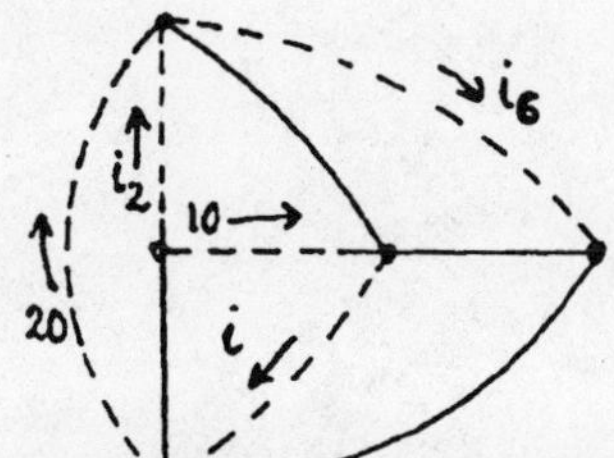


all loops include 1 Ω

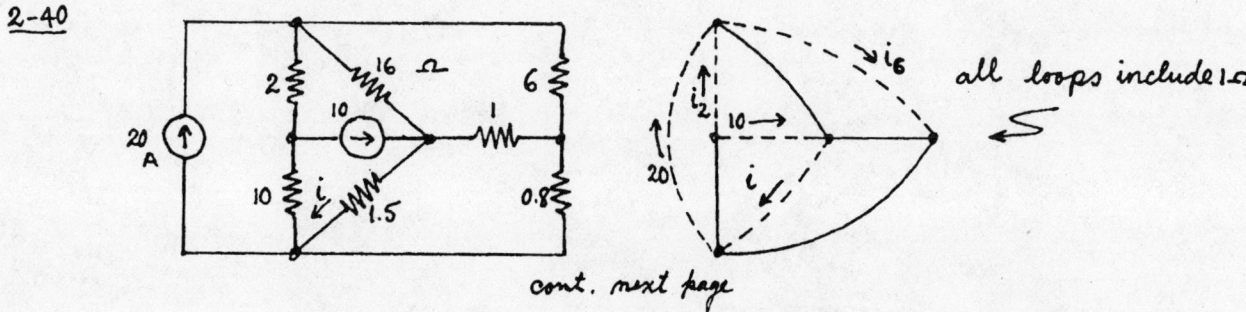

cont. next page

2-40 (cont) $1.5i + 0.8(-20 - i_2 - 10 + i) + 1(-20 + i_6 - i_2 - 10 + i) = 0$ $\therefore\ 3.3i + i_6 - 1.8i_2 = 54$

$6i_6 + 1(i + i_6 - 30 - i_2) + 16(i_6 - 20 - i_2) = 0$ $\therefore\ i + 23i_6 - 17i_2 = 350$

$2i_2 + 16(20 + i_2 - i_6) + 1(20 - i_6 + i_2 + 10 - i) + 0.8(20 + i_2 + 10 - i) + 10(10 + i_2) = 0$

$\therefore\ -1.8i - 17i_6 + 29.8i_2 = -474$

$$i = \frac{\begin{vmatrix} 54 & 1 & -1.8 \\ 350 & 23 & -17 \\ -474 & -17 & 29.8 \end{vmatrix}}{\begin{vmatrix} 3.3 & 1 & -1.8 \\ 1 & 23 & -17 \\ -1.8 & -17 & 29.8 \end{vmatrix}} = \frac{54(396.4) - 350(-0.8) - 474(24.4)}{3.3(396.4) - 1(-0.8) - 1.8(24.4)} = \frac{10.120}{1265} = \underline{8\ \text{A}}$$

2-41

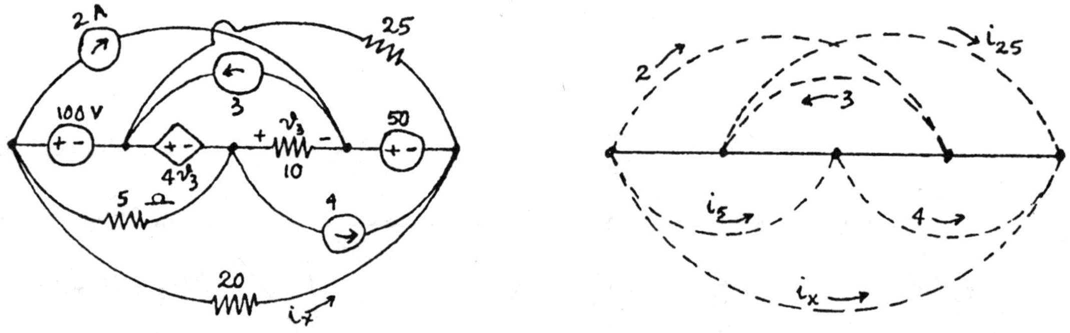


$i_x$ : $20i_x - 50 + 10(i_x + 4 - 3 + 2 + i_{25}) - 4(10)(-i_x - 4 + 3 - 2 - i_{25}) - 100 = 0$ $\therefore\ 70i_x + 50i_{25} = 0$

$i_{25}$: $25i_{25} - 50 + 10(i_x + 4 - 3 + 2 + i_{25}) - 4(10)(-i_x - 4 + 3 - 2 - i_{25}) = 0$ $\therefore\ 50i_x + 75i_{25} = -100$

$i_5$ : $5i_5 - 4(10)(-i_x - 4 + 3 - 2 - i_{25}) - 100 = 0$ $\therefore\ 40i_x + 40i_{25} + 5i_5 = -20$

$$i_x = \frac{\begin{vmatrix} 0 & 50 & 0 \\ -100 & 75 & 0 \\ -20 & 40 & 5 \end{vmatrix}}{\begin{vmatrix} 70 & 50 & 0 \\ 50 & 75 & 0 \\ 40 & 40 & 5 \end{vmatrix}} = \frac{5(5000)}{5(5250 - 2500)} = \underline{1.8182\ \text{A}}$$

2-42

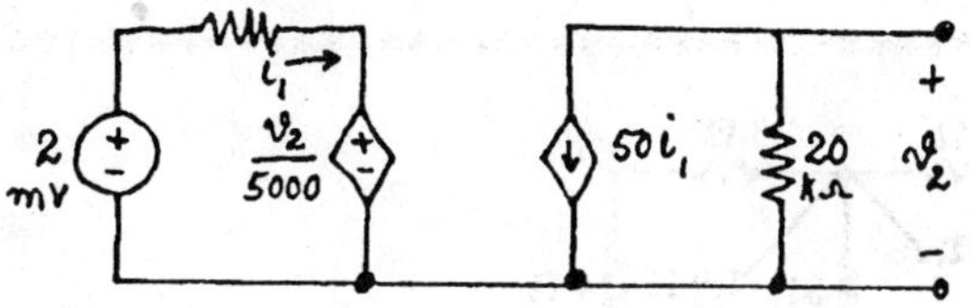


OC: $0.002 = 2000 i_1 + 0.0002 v_2$
$v_2 = -20{,}000 \times 50 i_1$
$\therefore v_{2,OC} = \underline{-1.111\ V}$

SC: $v_2 = 0 \quad \therefore i_1 = \frac{0.002}{2000} = 10^{-6}$ A

$i_{sc}\downarrow = -50 i_1 = -5 \times 10^{-5}$ A

$R_{Th} = \frac{v_{2,oc}}{i_{sc}} = \frac{-1.111}{-5 \times 10^{-5}} = \underline{22.22\ k\Omega}$

2-43

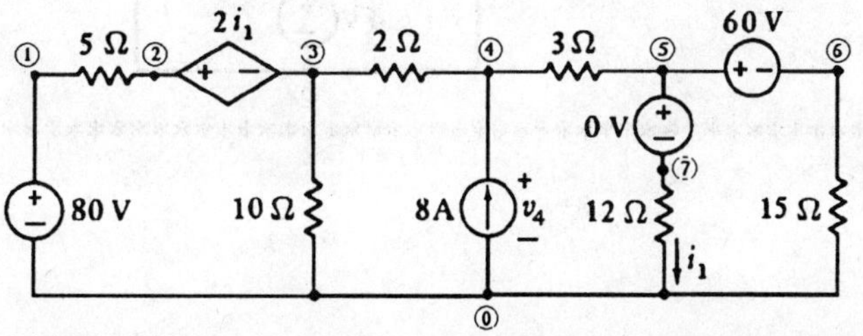


*************************************************************************

**PROGRAM:**

```
V10 1 0 80
R12  1  2  5
V57  5  7
H23  2  3  V57  2
R30  3  0  10
R34  3  4  2
IO4  0  4  DC  8
R45  4  5  3
R70  7  0  12
V56  5  6  60
R60  6  0  15
.PRINT  DC  V(4)
.END
```

*************************************************************************

**ANSWERS: Node voltages**

| NODE | NODE | NODE | NODE |
|---|---|---|---|
| (1) 80.0000 | (2) 68.9880 | (3) 59.8590 | **(4) 67.4260** |
| (5) 54.7760 | (6) -5.2235 | (7) 54.7760 | |

**ANSWERS: Source currents**

| NAME | CURRENT |
|---|---|
| V10 | -2.202E+00 |
| V57 | 4.565E+00 |
| V56 | -3.482E-01 |

*************************************************************************

## 2-44

**********************************************************************

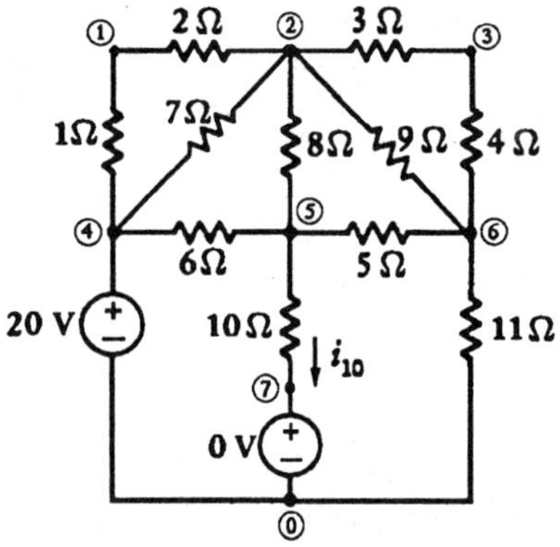


**********************************************************************

**PROGRAM:**

```
R14 1 4 1
R12 1 2 2
R24 2 4 7
R23 2 3 3
R36 3 6 4
R25 2 5 8
R26 2 6 9
R45 4 5 6
R56 5 6 5
R57 5 7 10
V40 4 0 20
V70 7 0
R60 6 0 11
.PRINT DC I(V70)
.END
```

**********************************************************************

**ANSWERS: Node Voltages**

| NODE | NODE | NODE | NODE | NODE |
|---|---|---|---|---|
| (1) 18.9810 | (2) 16.9440 | (3) 15.2010 | (4) 20.0000 | (5) 13.5670 |
| (6) 12.8770 | (7) 0.0000 | | | |

**ANSWERS: Source currents**

| NAME | CURRENT |
|---|---|
| V40 | -2.527E+00 |
| **V70** | **1.357E+00** = $i_{10}$ |

*2-45*

**************************************************************************

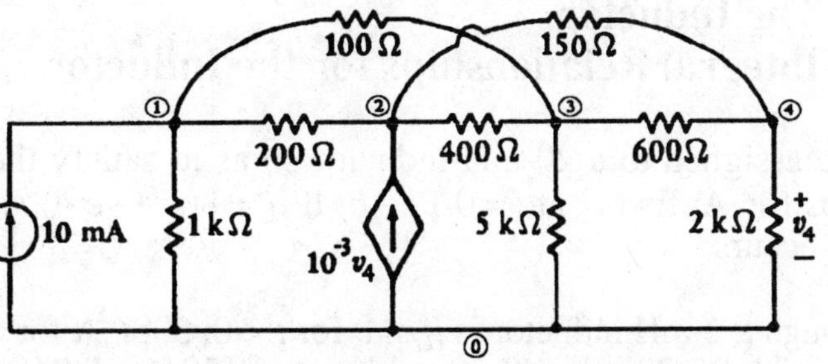


**************************************************************************

**PROGRAM:**

```
IO1 0 1 10M
R10 1 0 1K
R12 1 2 200
R13 1 3 100
GO2 0 2 4 0 1M
R23 2 3 400
R24 2 4 150
R30 3 0 5K
R34 3 4 600
R40 4 0 2K
.PRINT DC V(4)
.END
```

**************************************************************************

**ANSWERS: Node Voltages**

| NODE | NODE | NODE | NODE |
|---|---|---|---|
| (1) 14.2240 | (2) 15.1620 | (3) 14.1770 | (4) **14.1180 V** = $v_4$ |

# CHAPTER 3: INDUCTANCE AND CAPACITANCE

## Section 3-2 The Inductor
## Section 3-3 Integral Relationships for the Inductor

**1** Let $v_L$ and $i_L$ be assigned to a 20-mH inductor so as to satisfy the passive sign convention. (*a*) If $i_L = 12\, e^{-5t}\cos 10t$ A, find $v_L$ at $t = 0.1$ s. (*b*) If $i_L = 8(e^{-6t} - e^{-2t})$ A, find $v_{L,\max}$ and the value of time at which it occurs.

**2** The current through a 2-mH inductor is $i_L = 0$ for $t < 0$, $500t$ A for $0 < t < 10$ ms, $10 - 500t$ A for $10 < t < 30$ ms, -5 A for $30 < t < 40$ ms, and $-5\cos^2[50\pi(t - 0.04)]$ A for $40 < t < 50$ ms. For the interval, $0 < t < 50$ ms: (*a*) sketch $i_L$ versus $t$; (*b*) sketch $v_L$ versus $t$, assuming the passive sign convention.

**3** In Fig. 3-1, let $v_s(t) = 20t$ V for $0 < t < 3$s, and $v_s = 0$ for $t < 0$ and $t > 3$. Find: (*a*) the value of $i$ at $t = 4$ s; (*b*) the energy stored in the inductor at $t = 2$ s; (*c*) the power entering the inductor at $t = 2$ s.

**4** The current in a 0.4-H inductor is zero for $t < 0$ and $3te^{-0.1t}$ A for $t > 0$. (*a*) At what instant is maximum power being delivered to the inductor? (*b*) At what instant is the energy stored in the inductor a maximum?

**5** In the circuit of Fig. 3-2, let $v_s = 100\cos 500t$ V for $t > -0.5$ s, and let $i_L(0) = -1$ A. (*a*) Find $i_s(t)$. (*b*) Find $w_L(t)$ at $t = 1$ ms.

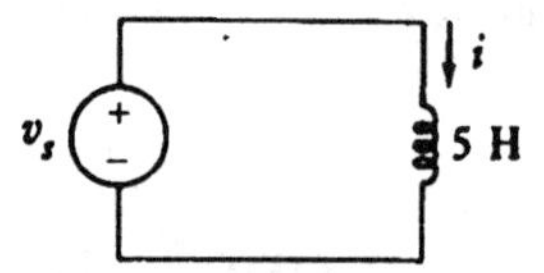


**Fig. 3-1:** See Prob. 3.

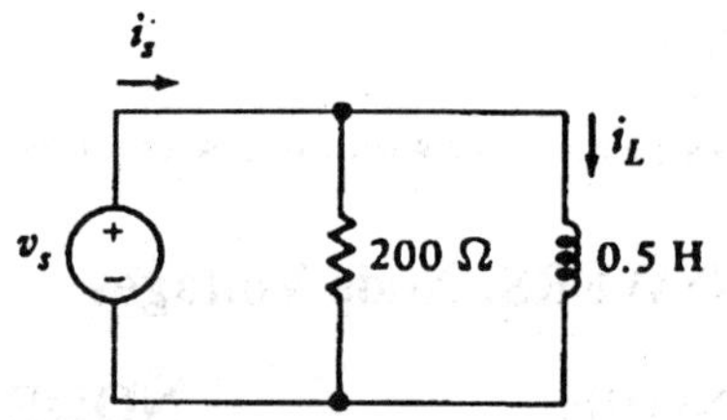


**Fig. 3-2:** See Prob. 5

## Section 3-4 The Capacitor

**6** (*a*) If $v_C(t)$ is given by the waveform shown in Fig. 3-3, sketch $i_C(t)$ for $-0.1 < t < 0.2$ s. (*b*) Sketch the power entering the capacitor over the same time interval.

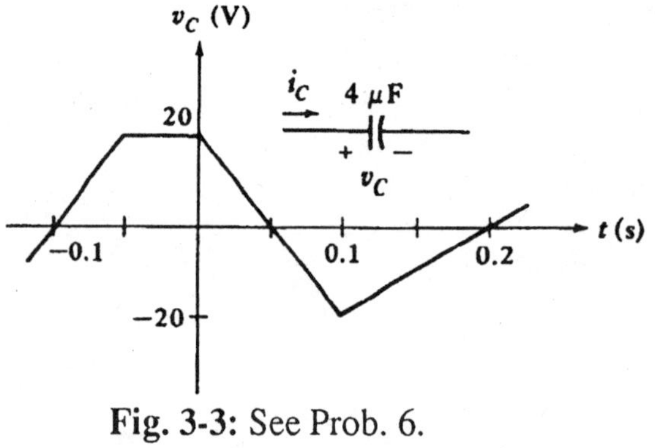


**Fig. 3-3:** See Prob. 6.

**7** The current through a 0.2-$\mu$F capacitor is 60cos $(10^4 t + 36°)$ mA for all time. The average voltage across the capacitor is zero. (*a*) What is the maximum value of energy stored in the capacitor? (*b*) What is the first nonnegative value of $t$ at which maximum energy is stored?

**8** The energy stored in a 50-$\mu$F capacitor is given as $w_C(t) = 2e^{-50t}$ J for $t \geq 0$. Find the capacitor voltage, current, and absorbed power at $t = 30$ ms.

**9** In the circuit of Fig. 3-4*a*, let $R = 1$ MΩ, $C = 1\mu$F, $R_i = \infty$, and $R_o = 0$. Suppose that we wish the output to be $v_o(t) = e^{-10t} - 1$ V. Differentiate Eq. (16) in Sec. 3-4 of your text to obtain the necessary $v_s(t)$ if: (*a*) $A = 1000$; (*b*) $A$ is infinite.

**10** Interchange the location of $R$ and $C$ in the circuit of Fig. 3-4*a* and assume that $R_i = \infty$, $R_o = 0$ and $A = \infty$ for the op-amp. (*a*) Find $v_o(t)$ as a function of $v_s(t)$. (*b*) Obtain an equation relating $v_o(t)$ and $v_s(t)$ if $A$ is not assumed to be infinite.

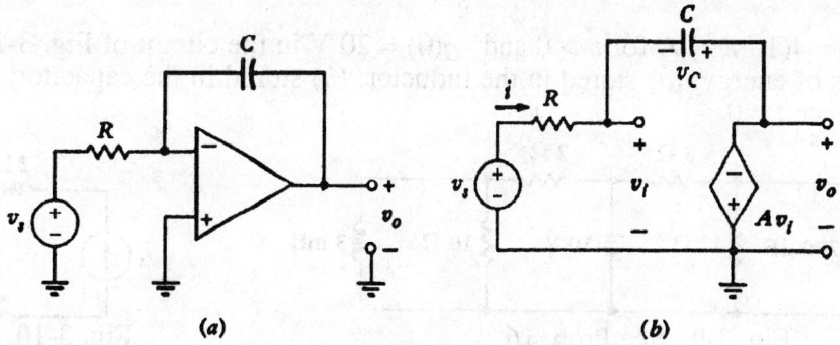


**Fig. 3-4:** See Probs. 9 and 10.

**11** In the circuit shown in Fig. 3-5, $v_C(t) = 4te^{-2t}$ V. At $t = 0.5$ s, find the value of: (*a*) the energy stored in the capacitor; (*b*) the energy stored in the inductor; (*c*) $v_s$.

## Section 3-5 Inductance and Capacitance Combinations

**12** (*a*) If each inductance in the network of Fig. 3-6 is 1 H, find the equivalent inductance at *a-b*. (*b*) Replace each inductor by a 1-F capacitor and find $C_{eq}$.

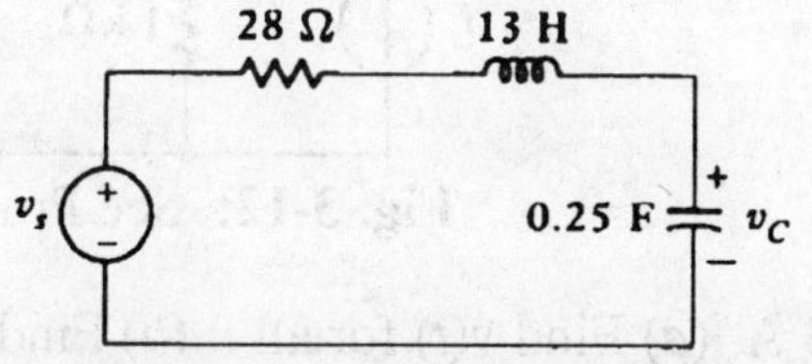


**Fig. 3-5:** See Prob. 11.

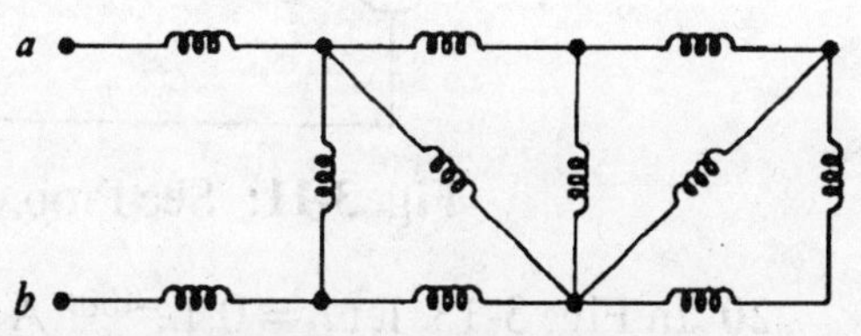


**Fig. 3-6:** See Prob. 12.

**13** Find the equivalent inductance offered at terminals *a-b* in Fig. 3-7 if terminals *x-x'* are: (*a*) open-circuited; (*b*) short-circuited.

**14** Each capacitor in Fig. 3-8 is 1 μF. Find $C_{eq}$ at *a-b* if: (*a*) 1-2 and 1-3 are both open-circuited; (*b*) 1-2 and 1-3 are both short-circuited; (*c*) 1-2 is open-circuited and 1-3 is short-circuited; (*d*) 1-2 is short-circuited and 1-3 is open-circuited.

**15** Given a bucketful of 1-nF capacitors, show how an equivalent capacitance of 0.7 nF might be obtained. Use as few capacitors as possible.

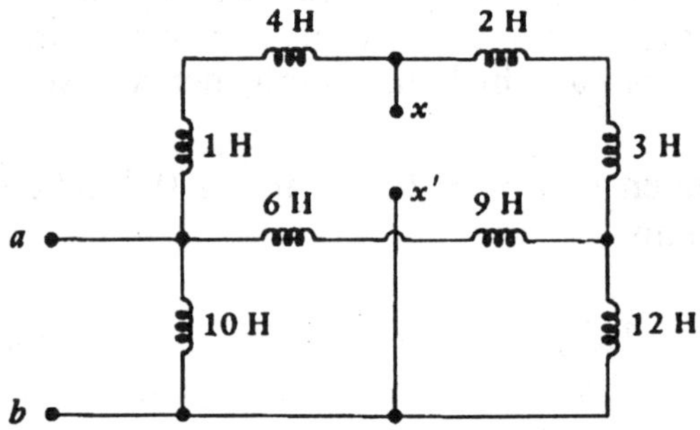


**Fig. 3-7:** See Prob. 13

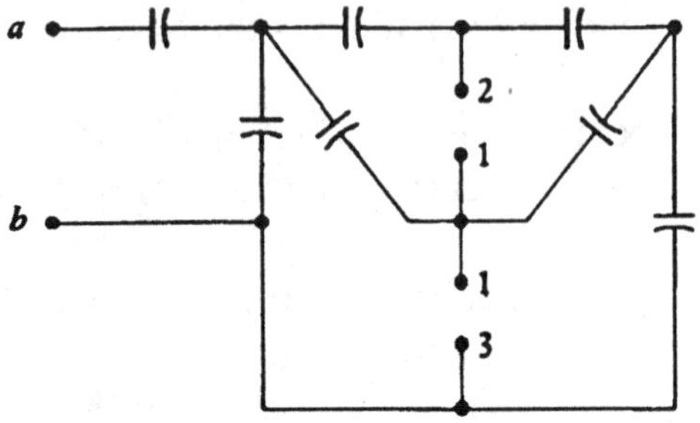


**Fig. 3-8:** See Prob. 14

**16** For the circuit of Fig. 3-9, find: (*a*) $w_C$; (*b*) $w_L$; (*c*) the current in each circuit element; (*d*) the voltage across each circuit element.

**17** Let $i_s = 4(1 - e^{-3t})$ A for $t > 0$ and $v_C(0) = 20$ V in the circuit of Fig. 3-10. At $t = 0.5$ s, find the values of energy: (*a*) stored in the inductor; (*b*) stored in the capacitor; (*c*) dissipated by the resistor since $t = 0$.

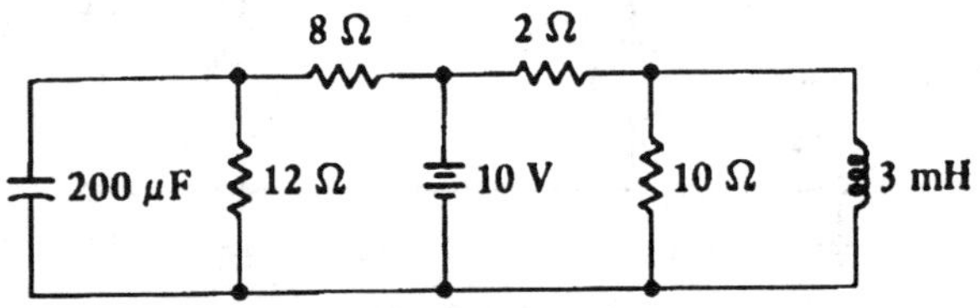


**Fig. 3-9:** See Prob. 16.

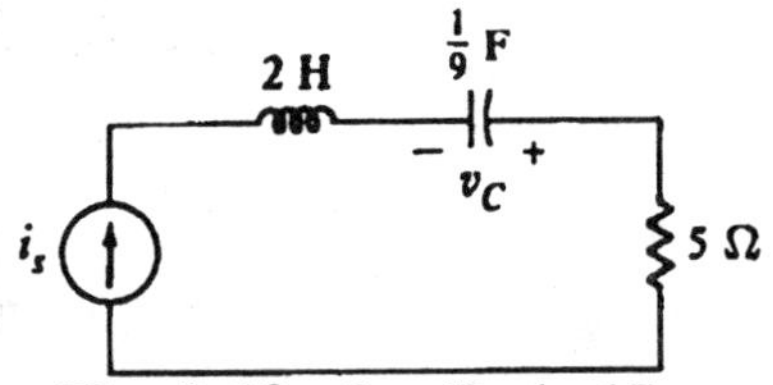


**Fig. 3-10:** See Prob. 17.

**18** Let $v(0) = 10$ V in the circuit shown in Fig. 3-11. (*a*) Find $i(t)$ for all $t$. (*b*) Find $v(t)$ for $t \geq 0$.

**19** (*a*) Write nodal equations for the circuit of Fig. 3-12. (*b*) Write mesh equations for the same circuit.

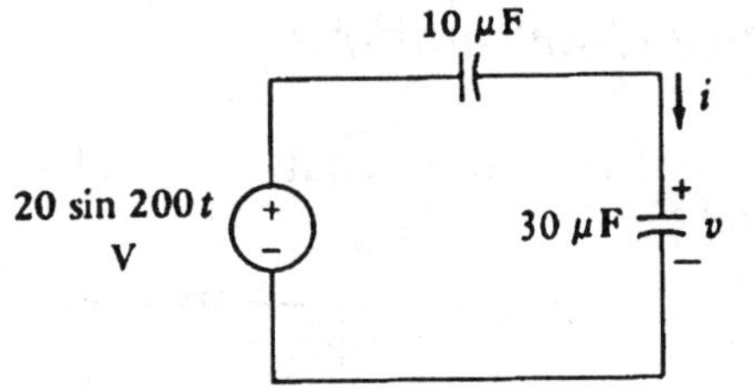

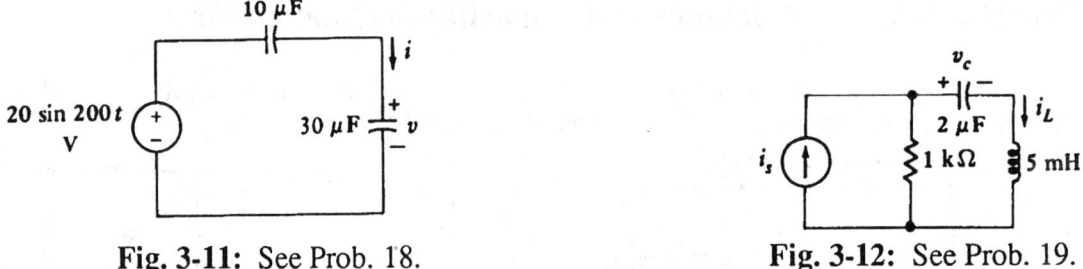

**Fig. 3-11:** See Prob. 18.

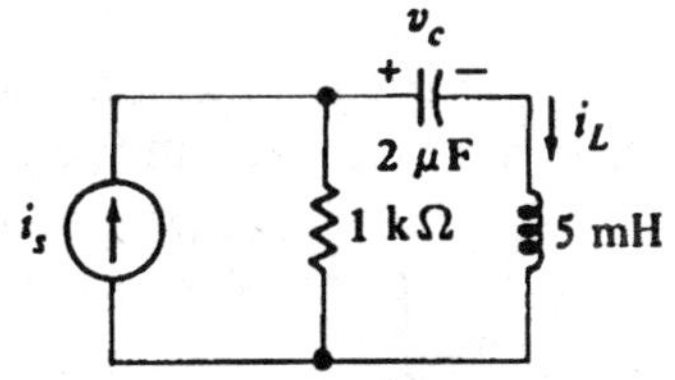


**Fig. 3-12:** See Prob. 19.

**20** In Fig. 3-13, let $i_s = 0.1e^{-400t}$ A with $i_2(0) = 0.03$ A. (*a*) Find $v(t)$ for all $t$. (*b*) Find $i_2$(t) for $t \geq 0$. (*c*) Find $i_1(t)$ for $t \geq 0$.

**21** Draw a tree for the circuit shown in Fig. 3-14 that satisfies not only the criteria enumerated in Secs. 2-7 and 2-8, but also places all capacitors in the tree and all inductors in the cotree. (*a*) Assign tree-branch voltages and write a set of nodal equations. Assume that there is no energy storage at $t = 0$. (*b*) Assign link currents and write a set of loop equations, again assuming no energy storage at $t = 0$.

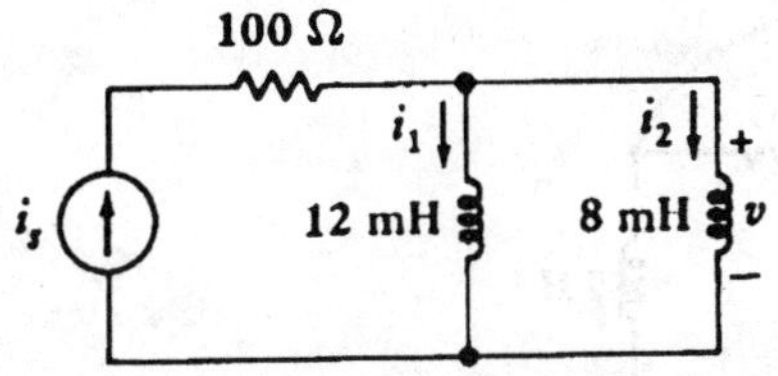


Fig. 3-13: See Prob. 20.

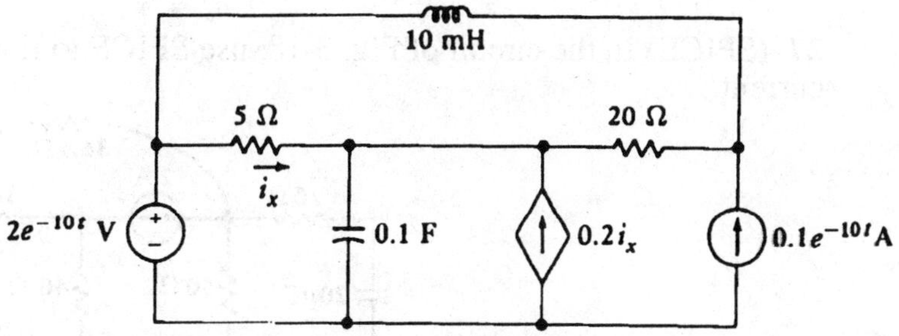


Fig. 3-14: See Probs. 21 and 24.

22 In Fig. 3-15, let $v_{C1}(0) = 5$ V, $v_{C2}(0) = 6$ V, $v_{C3}(0) = 7$ V, $i_{L1}(0) = 8$ A, and $i_{L2}(0) = -4$ A. (*a*) Write a set of nodal equations in the variables $v_{C1}$, $v_{C2}$, and $v_{C3}$. (*b*) Write a set of mesh equations in the variables $i_s$, $i_{L1}$, and $i_{L2}$.

## Section 3-6 Duality

23 Construct the exact dual of the circuit shown in Fig. 3-15 and described in Prob. 22. Label all inductor currents and capacitor voltages and give their initial values.

24 Draw the exact dual of the circuit of Fig. 3-14.

25 If $i_L(0) = 10$ A and $v_C(0) = 4$ V in the circuit of Fig. 3-16, write a single nodal equation and (*a*) show that $v_C = -10 \sin 5000t + 4 \cos 5000t$ V satisfies your equation. (*b*) Obtain the exact dual circuit and find an expression for the capacitor voltage. (*c*) How do the original and the dual circuits compare topologically?

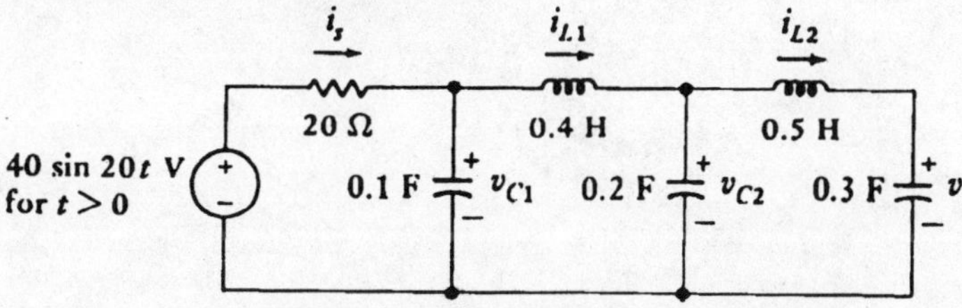

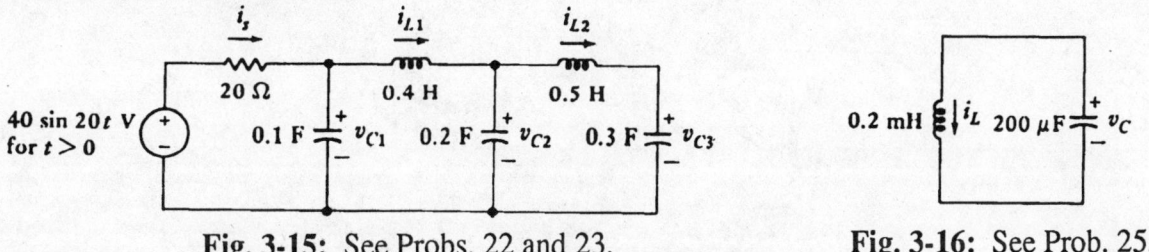

Fig. 3-15: See Probs. 22 and 23.

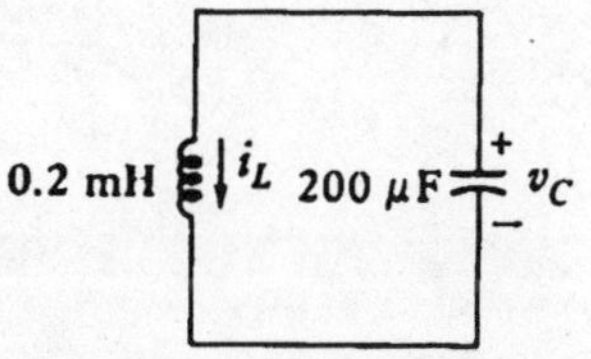


Fig. 3-16: See Prob. 25.

## Section 3-7 Linearity and Its Consequences Again

26 If all four sources in the circuit of Fig. 3-17 have been applied forever and ever, use the principle of superposition to find: (*a*) $i_C(t)$; (*b*) $i_L(t)$.

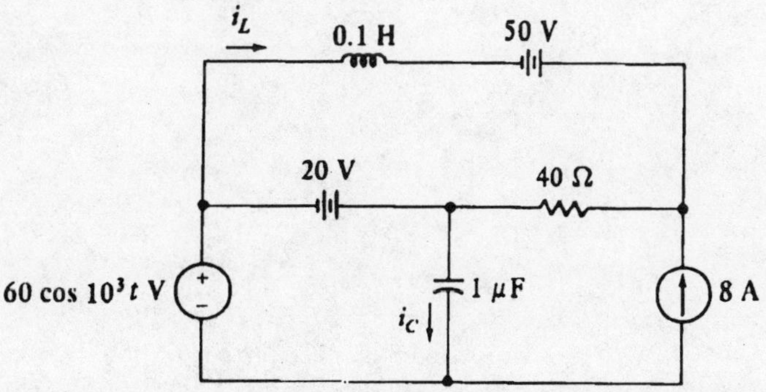


Fig. 3-17: See Prob. 26.

**27** (SPICE) In the circuit of Fig. 3-18, use SPICE to find the capacitor voltage and the inductor current.

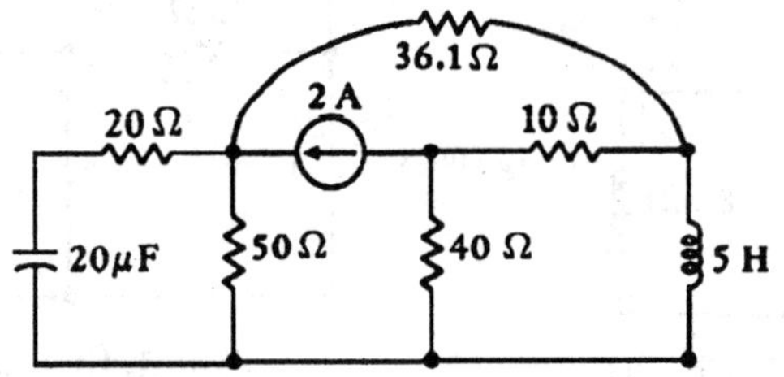


**Fig. 3-18:** See Prob. 27.

# Chapter 3

3-1 (a) 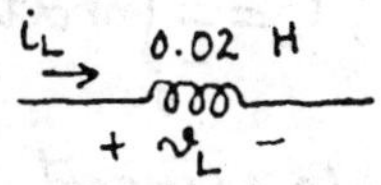


$$v_L = Li_L' = 0.02\frac{d}{dt}(12e^{-5t}\cos 10t) = 12(0.02)e^{-5t}(-10\sin 10t - 5\cos 10t)$$

$$v_L = 0.24e^{-0.5}(-10\sin 1^{rad} - 5\cos 1^{rad}) = \underline{-1.6182\ V}$$

(b) $i_L = 8(e^{-6t} - e^{-2t})$ $\quad\therefore v_L = 0.02 \times 8(2e^{-2t} - 6e^{-6t}) = 0.32(e^{-2t} - 3e^{-6t})$

$$\frac{dv_L}{dt} = 0 \quad \therefore -2e^{-2t_m} + 18e^{-6t_m} = 0 \quad \therefore 9e^{-4t_m} = 1,\ t_m = \underline{0.5493\ s} \quad \therefore v_{L,max} = \underline{0.07111\ V}$$

3-2 (a) 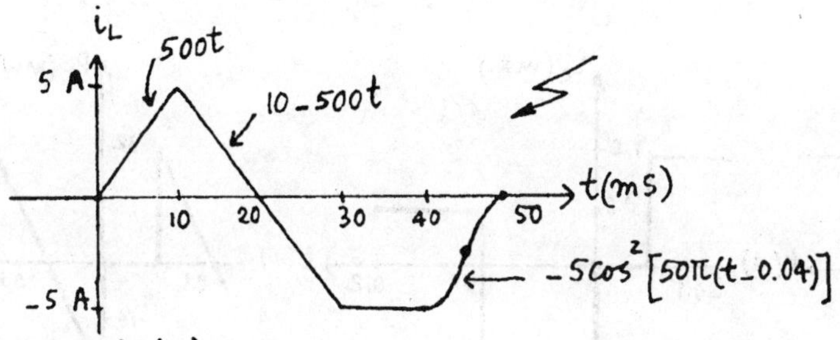


(b) 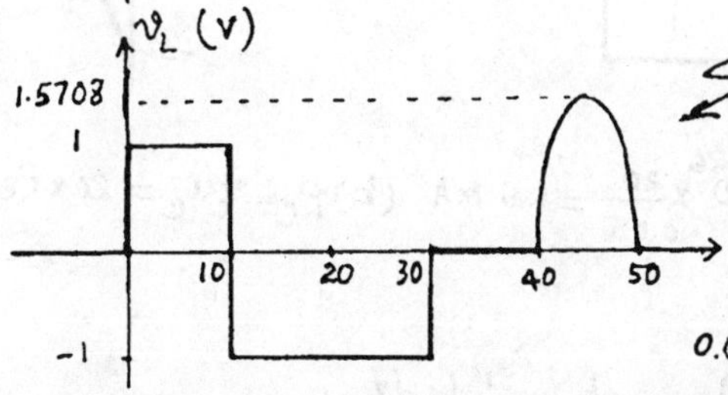


$$v_L = Li_L' = 2\times 10^{-3} i_L'$$

$0 < t < 0.01$ : $v_L = 2\times 10^{-3} \times 500 = 1\ V$

$0.01 < t < 0.03$ : $v_L = 2\times 10^{-3}(-500) = -1\ V$

$0.04 < t < 0.05$ : $v_L = \pi \sin[50\pi(t-0.04)]\cos[50\pi(t-0.04)]\ V$

3-3 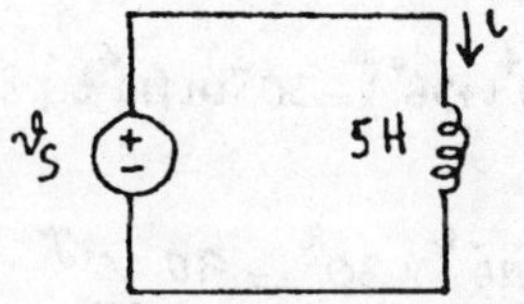


$v_S = 20t$ V, $0 < t < 3$ ; $v_S = 0$ elsewhere

(a) $i = \frac{1}{5}\int_0^{t<3} v_S\,dt \quad \therefore i(4) = \frac{1}{5}\int_0^3 20t\,dt = 2t^2\Big]_0^3 = \underline{18\ A}$

(b) $i(2) = \frac{1}{5}\int_0^2 20t\,dt = 8\ A$ ; $W_L = \frac{1}{2}Li^2$ ; $W_L(2) = \frac{1}{2}\times 5 \times 8^2 = \underline{160\ J}$

(c) $p = vi$ ; $p(2) = v(2)\,i(2) = 20(2)(8) = \underline{320\ W}$

3-4 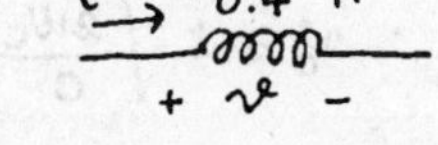


$i = 0,\ t < 0$ ; $i = 3te^{-0.1t}$ A, $t > 0$

$v = Li' = 0.4 \times 3e^{-0.1t}(-0.1t + 1)$ ; $P = vi = 3.6e^{-0.2t}(t - 0.1t^2)$

(a) $\frac{dp}{dt} = 0 \quad \therefore e^{-0.2t_m}(1 - 0.2t_m) + (t_m - 0.1t_m^2)(-0.2e^{-0.2t_m}) = 0$

$$\therefore 0.02t_m^2 - 0.4t_m + 1 = 0 \quad \therefore t_m = \underline{2.929\ s}$$

(b) $W_L = \frac{1}{2}Li^2 = \frac{1}{2} \times 0.4 \times 9t^2 e^{-0.2t}$

$\frac{dW_L}{dt} = 0 \quad \therefore t_m^2(-0.2)e^{-0.2t_m} + 2t_m e^{-0.2t_m} = 0$ or $t_m(-0.2) + 2 = 0 \quad \therefore t_m = \underline{10\ s}$

**3-5** (a)

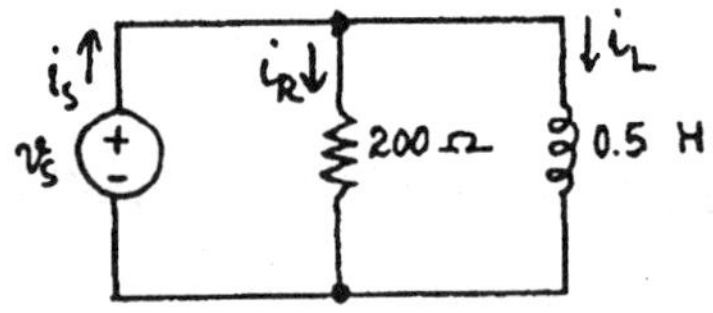


$v_s = 100\cos 500t$ V, $t > -0.5$ s ; $i_L(0) = -1$ A

$i_R = \frac{v_s}{R} = 0.5\cos 500t$ A ; $i_L = \frac{1}{L}\int_0^t v\,dt + i_L(0)$

$i_L = 2\int_0^t 100\cos 500t\,dt - 1 = (0.4\sin 500t - 1)$ A

$i_s = i_R + i_L = \underline{-1 + 0.5\cos 500t + 0.4\sin 500t \text{ A}}$

(b) $i_L(0.001) = -1 + 0.4\sin 0.5 = -0.8082$ ; $W_L = \frac{1}{2}Li^2$ ; $W_L(0.001) = \frac{1}{2}\times 0.5(-0.8082)^2 = \underline{0.16331 \text{ J}}$

**3-6**

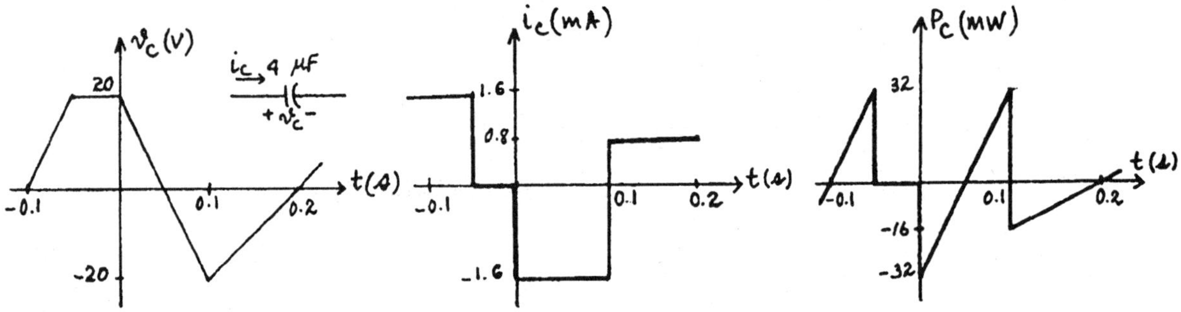


(a) $i_c = C\frac{dv_c}{dt} = 4\times10^{-6}\times\frac{20}{0.05} = 1.6$ mA (b) $P_c = v_c i_c = 20\times1.6\times10^{-3} = 32$ mW

**3-7**

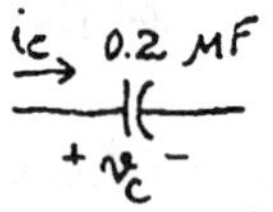


$i_c = 60\times10^{-3}\cos(10^4t + 36°)$ ; $v_c = \frac{1}{C}\int i_c dt$

(a) $v_c = \frac{10^6}{0.2}\int 60\times10^{-3}\cos(10^4t + 36°)\,dt = \frac{60\times10^{-3}}{0.2\times10^4}\sin(10^4t+36°) = 30\sin(10^4t + 36°)$ V

$\therefore v_{c,max} = 30$ V; $W_c = \frac{1}{2}Cv^2$ $\therefore W_{c,max} = \frac{1}{2}\times0.2\times10^{-6}\times30^2 = \underline{90 \text{ μJ}}$

(b) $\sin(10^4t + 36°) = 1 = \sin\frac{\pi}{2}$ $\therefore 10^4t_m + 36°\times\frac{\pi}{180°} = \frac{\pi}{2}$ $\therefore t_m = \underline{94.25 \text{ μs}}$

**3-8**

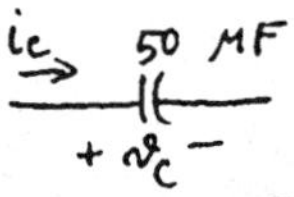


$W_c = 2e^{-50t}$ J, $t > 0$ ; $W_c = \frac{1}{2}Cv_c^2$ $\therefore v_c = \pm\sqrt{\frac{2W_c}{C}}$

$v_c = \pm\sqrt{\frac{2\times2e^{-50t}}{50\times10^{-6}}} = \pm 282.8e^{-25t}$ V $\therefore v_c(0.03) = \underline{\pm 133.61 \text{ V}}$

$i_c = 50\times10^{-6}\frac{d}{dt}v_c = \pm 50\times10^{-6}\times282.8(-25)e^{-25t}$ $\therefore i_c(0.03) = \underline{\mp 0.16701 \text{ A}}$

$P_c = v_c(0.03)\,i_c(0.03) = \underline{-22.31 \text{ W}}$

**3-9**

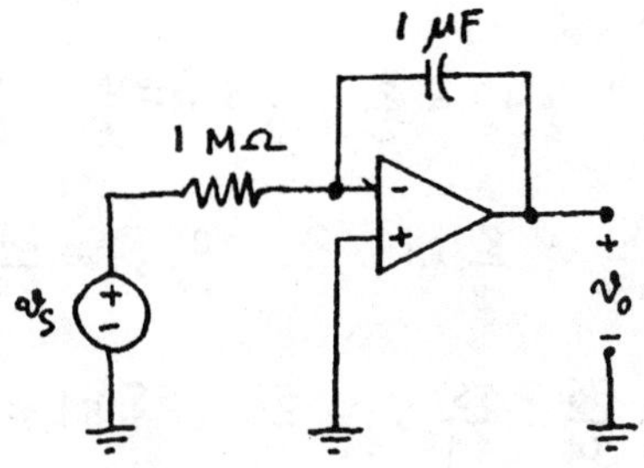


$v_o = e^{-10t} - 1$ V ; $\left(1+\frac{1}{A}\right)v_o = -\frac{1}{RC}\int_0^t \left(v_s + \frac{v_o}{A}\right)dt - v_c(0)$ (RC = 1)

$$\therefore \left(1+\frac{1}{A}\right)v_o' = -\left(v_s+\frac{v_o}{A}\right) = \left(1+\frac{1}{A}\right)\left(-10e^{-10t}\right)$$

$$v_s = -\frac{v_o}{A} + \left(1+\frac{1}{A}\right)\left(-10e^{-10t}\right)$$

(a) $A = 1000$ $\therefore v_s = -0.001e^{-10t} + 0.001 + 10.01e^{-10t} = \underline{10.009e^{-10t} + 0.001}$ V

(b) $A \to \infty$ $\therefore v_s = \underline{10e^{-10t}}$ V

**3-10** (a)

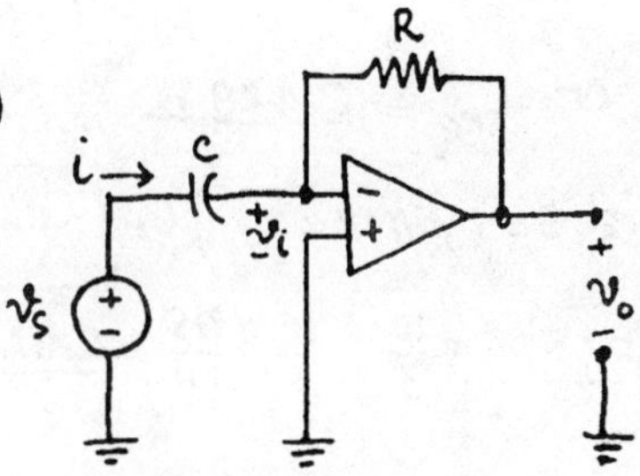


(b)

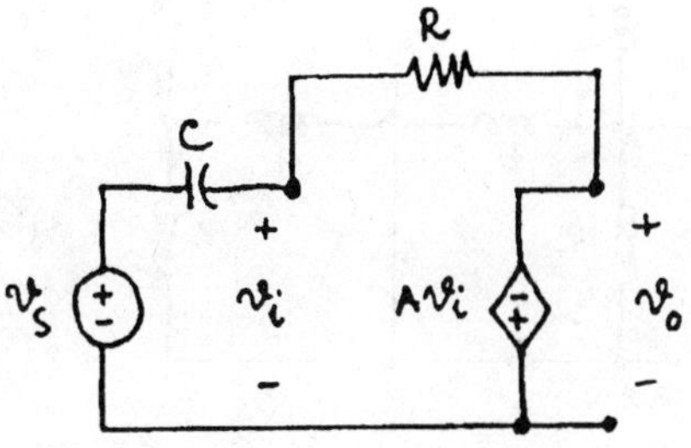


(a) $A = \infty$ , $v_i = 0$

$i = Cv_s' = -\frac{v_o}{R}$

$\therefore v_o = -RC\frac{dv_s}{dt}$ ←

(b) $C(v_s' - v_i') = \frac{v_i - v_o}{R}$ ; $v_i = -\frac{v_o}{A}$ $\Big\}$ $\therefore RC\left(v_s' + \frac{v_o'}{A}\right) = -v_o - \frac{v_o}{A}$

or $v_o\left(1+\frac{1}{A}\right) + \frac{RC}{A}v_o' = -RCv_s'$ ←

**3-11**

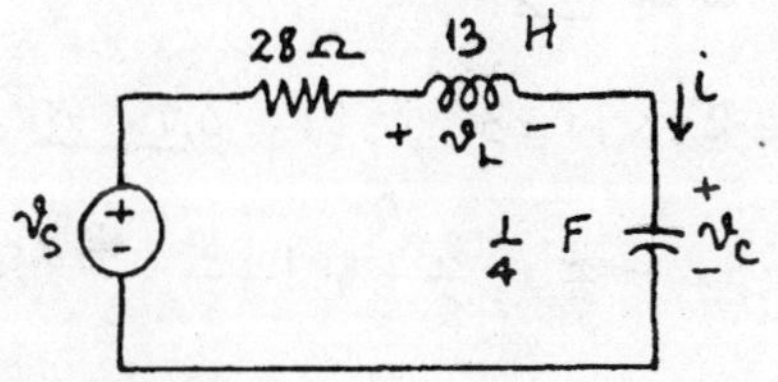


(a) $v_c(t) = 4te^{-2t}$ V

$W_c(0.5) = \frac{1}{2}\left(\frac{1}{4}\right)v_c^2(0.5) = 0.125\times 4e^{-2} = \underline{0.06767 \text{ J}}$

(b) $i = C\frac{dv}{dt} = 0.25\times 4e^{-2t}(1-2t) = e^{-2t}(1-2t)$ A

$W_L(0.5) = \frac{1}{2}\times 13\, i^2(0.5) = 6.5e^{-2}(1-1)^2 = \underline{0}$

(c) $v_L = Li' = 13e^{-2t}[-2-2(1-2t)] = 13e^{-2t}(4t-4)$ $\therefore v_L(0.5) = 13e^{-1}(-2) = -9.565$ V

$v_s(0.5) = 28i(0.5) + v_L(0.5) + v_c(0.5) = 28e^{-1}(0) - 9.565 + 2e^{-1} = \underline{-8.829 \text{ V}}$

**3-12** (a)

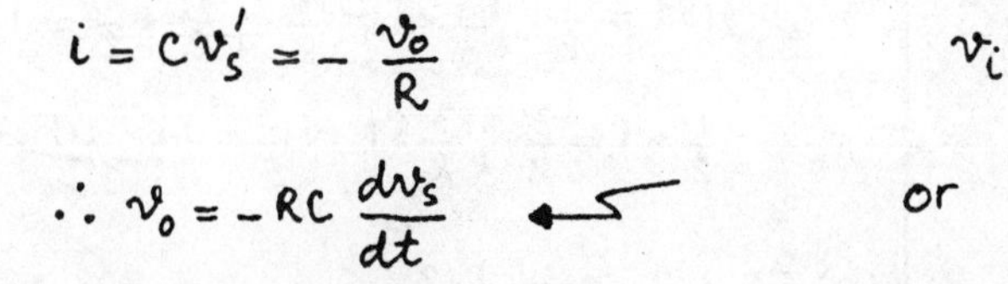


$1+1 = 2$ ; $2\|1 = 2/3$ ; $1 + 2/3 = 5/3$

$5/3\|1 = 5/8$ ; $5/8 + 1 = 13/8$ ; $13/8\|1 = 13/21$

$13/21 + 1 = 34/21$ ; $34/21\|1 = 34/55$ ; $\frac{34}{55} + 2 = \underline{2.618 \text{ H}}$

**3-12** (b)

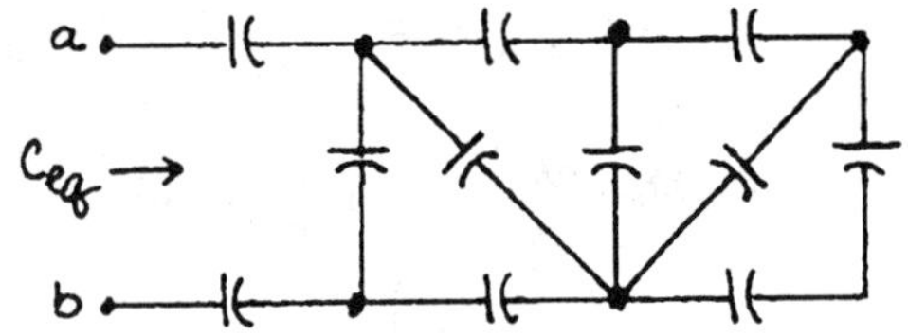


Series: $1 S 1 = \frac{1}{2}$ ; $\frac{1}{2}+1 = \frac{3}{2}$ ; $\frac{3}{2} S 1 = \frac{3}{5}$ ; $\frac{3}{5}+1 = \frac{8}{5}$

$\frac{8}{5} S 1 = \frac{8}{13}$ ; $\frac{8}{13}+1 = \frac{21}{13}$ ; $\frac{21}{13} S 1 = \frac{21}{34}$

$\frac{21}{34}+1 = \frac{55}{34}$ ; $\frac{55}{34} S 1 = \frac{55}{89}$ ; $\frac{55}{89} S 1 = \underline{0.3819\ F}$

**3-13**

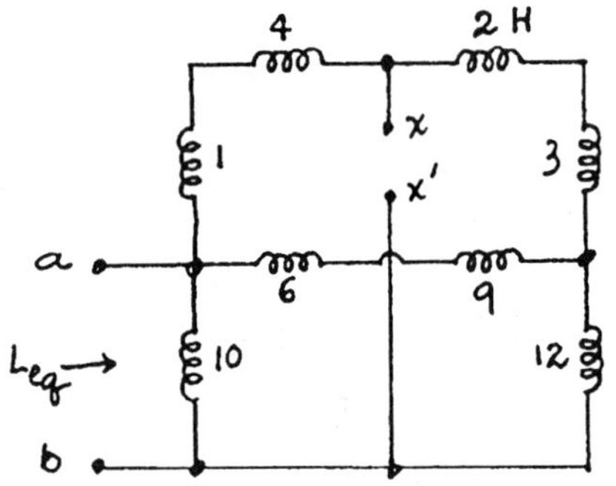


(a) $x - x'$ OC : $1+4+2+3 = 10$ ; $6+9 = 15$

$10 // 15 = 6$ ; $6+12 = 18$ ; $18 // 10 = \frac{180}{28}$

or $L_{eq} = \underline{6.429\ H}$

(b) $x - x'$ SC : $2+3 = 5$ ; $5 // 12 = \frac{60}{17}$ ; $6+9 = 15$

$15 + \frac{60}{17} = \frac{315}{17}$ ; $5 // 10 = \frac{10}{3}$ ; $\frac{10}{3} // \frac{315}{17} = \underline{2.825^{+}\ H}$

**3-14**

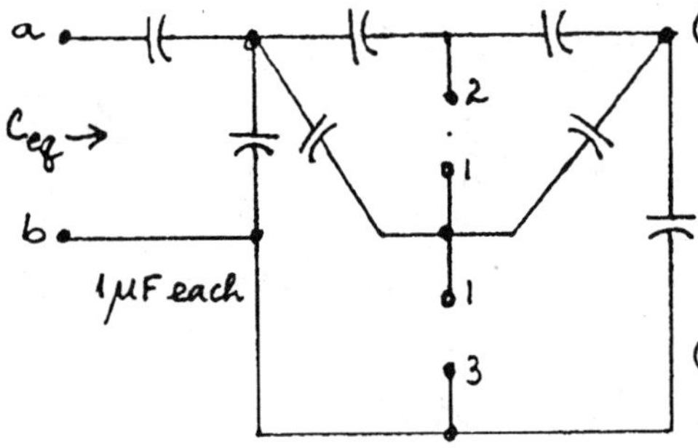


(a) OC 1-2 and 1-3 ; |S|: series

$1 S 1 = \frac{1}{2}$ ; $1 S 1 = \frac{1}{2}$ ; $\frac{1}{2}+\frac{1}{2} = 1$ ; $1 S 1 = \frac{1}{2}$

$\frac{1}{2}+1 = \frac{3}{2}$ ; $\frac{3}{2} S 1 = \frac{3}{5} = \underline{0.6\ \mu F} = C_{eq}$

(b) SC 1-2 and 1-3

$1+1 = 2$ ; $2+1 = 3$ ; $3 S 1 = \underline{0.75\ \mu F} = C_{eq}$

(c) OC 1-2 and SC 1-3. $1+1 = 2$ ; $1 S 1 = \frac{1}{2}$ ; $\frac{1}{2} S 2 = \frac{2}{5}$ ; $\frac{2}{5}+1+1 = \frac{12}{5}$ ; $\frac{12}{5} S 1 = \underline{0.7059\ \mu F}$

(d) SC 1-2 and OC 1-3. $1+1 = 2$ ; $1+1 = 2$ ; $2 S 2 = 1$ ; $1 S 1 = \frac{1}{2}$ ; $\frac{1}{2}+1 = \frac{3}{2}$ ; $\frac{3}{2} S 1 = \underline{0.6\ \mu F} = C_{eq}$

**3-15**

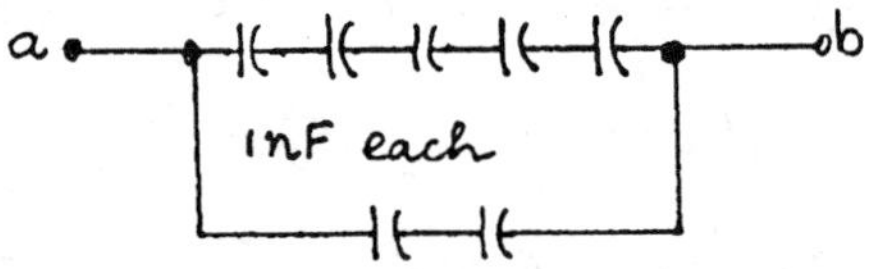


$C_{ab} = \frac{1}{5} + \frac{1}{2} = \underline{0.7\ nF}$

**3-16**

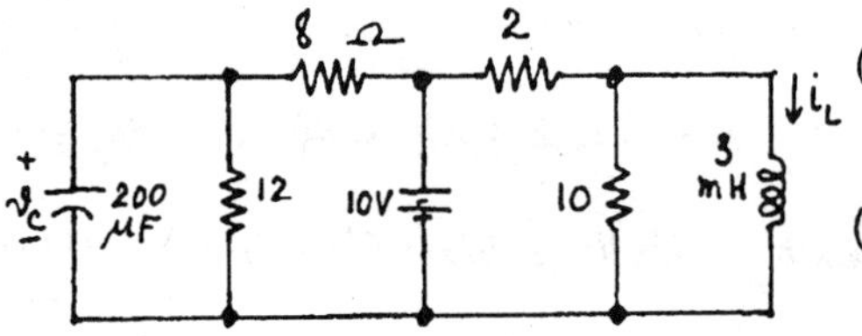


(a) $v_C = 6$ V ; $W_C = \frac{1}{2} C v_C^2 = \frac{1}{2} \times 2 \times 10^{-4} \times 36 = \underline{3.6\ mJ}$

(b) $i_L = 5$ A ; $W_L = \frac{1}{2} L i_L^2 = \frac{1}{2} \times 3 \times 10^{-3} \times 25 = \underline{37.5\ mJ}$

**3-16** (c)

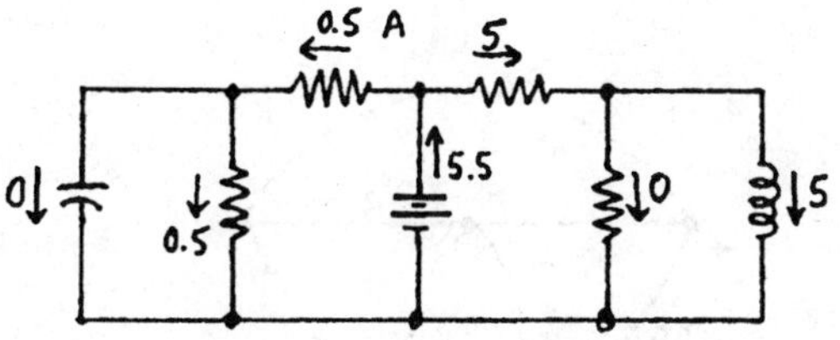


(d)

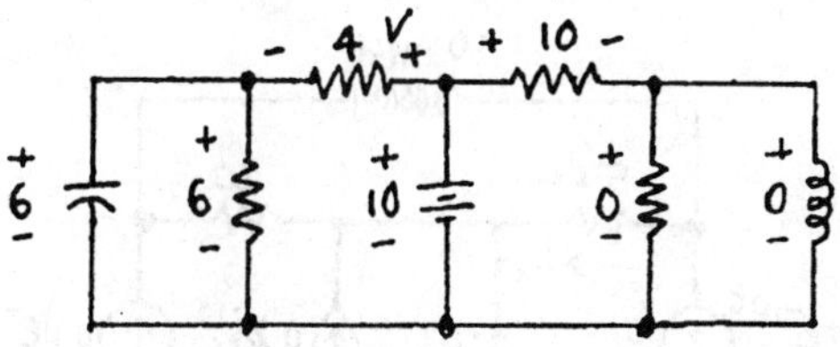


**3-17** (a)

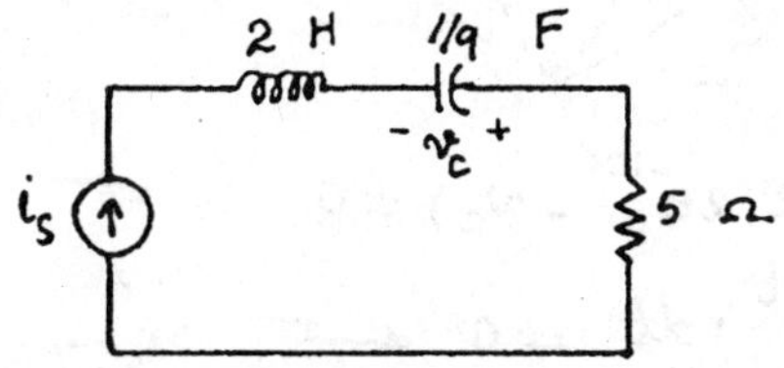


$i_s = 4(1 - e^{-3t})$ A, $t > 0$ ; $v_c(0) = 20$ V

$i_s(0.5) = 4(1 - e^{-1.5}) = 3.107$ A ; $W_L(0.5) = \frac{1}{2}L[i_s(0.5)]^2$

$W_L(0.5) = \frac{1}{2} \times 2\,(3.107)^2 = \underline{9.656\ \text{J}}$

(b) $v_c = \frac{1}{C}\int_0^t -i_s\,dt + v_c(0)$ ; $v_c(.5) = -9\int_0^{0.5} 4(1 - e^{-3t})\,dt + 20 = -18 - 12(e^{-1.5} - 1) + 20 = 11.322$ V

$W_c(0.5) = \frac{1}{2}C[v_c(0.5)]^2 = \frac{1}{2} \times \frac{1}{9}(11.322)^2 = \underline{7.122\ \text{J}}$

(c) $W_R = \int_0^t P_R\,dt = \int_0^t R i_s^2\,dt$ ; $W_R(0.5) = \int_0^{0.5} 5 \times 4^2(1 - e^{-3t})^2\,dt = 80\int_0^{0.5}(1 - 2e^{-3t} + e^{-6t})\,dt$

$W_R(0.5) = 40 + \frac{160}{3}(e^{-1.5} - 1) - \frac{80}{6}(e^{-3} - 1) = \underline{11.236\ \text{J}}$

**3-18**

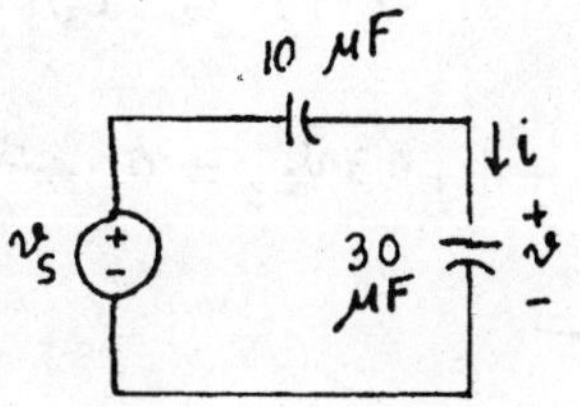


$v_s = 20\sin 200t$ V ; $v(0) = 10$ V ; $C_{eq} = 7.5\ \mu F$

(a) $i = C_{eq} v_s' = 7.5 \times 10^{-6} \times 20 \times 200\cos 200t = \underline{0.03\cos 200t\ \text{A}}$

(b) $v = \frac{1}{C}\int_0^t i\,dt = \frac{10^6}{30}\int_0^t 0.03\cos 200t\,dt + 10 = \underline{5\sin 200t + 10\ \ \text{V}}$

**3-19**

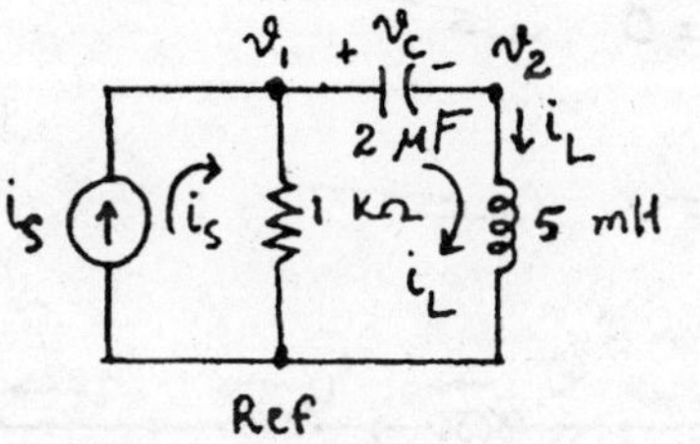


$i_L(0) = 3$ A ; $v_c(0) = 8$ V

(a) $10^{-3}v_1 + 2\times10^{-6}(v_1' - v_2') = i_s$

$2\times10^{-6}(v_2' - v_1') + 200\int_0^t v_2\,dt + 3 = 0$ ; $v_1(0) - v_2(0) = 8$

(b) $1000(i_L - i_s) + 5\times10^5\int_0^t i_L\,dt + 8 + 0.005 i_L' = 0$ ; $i_L(0) = 3$ A

**3-20**

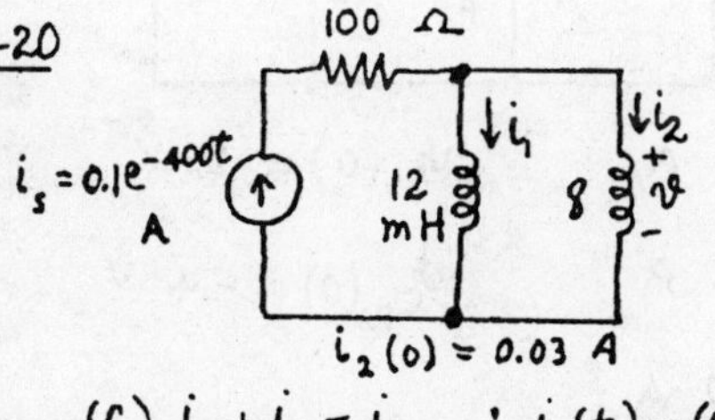


(a) $L_{eq} = 12 \| 8 = 4.8$ mH ; $v = Li_s' = 4.8\times10^{-3}(-40)e^{-400t} = \underline{-0.192e^{-400t}\ \text{V}}$

(b) $i_2(t) = i_2(0) + \frac{1}{L}\int_0^t v\,dt = 0.03 + 125\int_0^t -0.192e^{-400t}\,dt$

$\therefore\ i_2(t) = \underline{-0.03 + 0.06e^{-400t}\ \ \text{A}}$

(c) $i_1 + i_2 = i_s$ $\therefore\ i_1(t) = (0.1 - 0.06)e^{-400t} + 0.03 = \underline{0.03 + 0.04e^{-400t}\ \ \text{A}}$

**3-21**

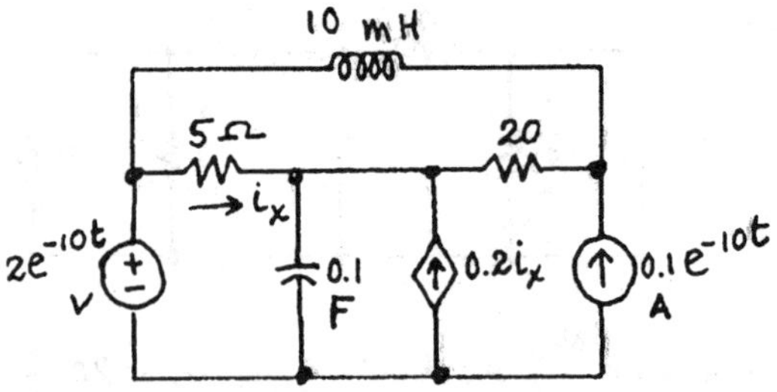


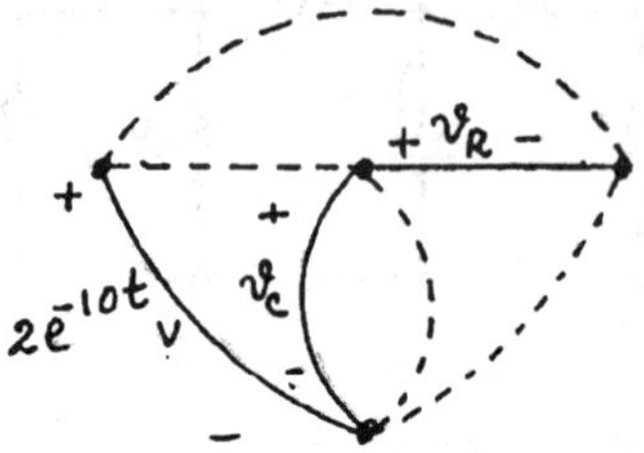


(a) $0.2(v_C - 2e^{-10t}) + 0.1v_C' + 0.05v_R - 0.2\times 0.2(2e^{-10t} - v_C) = 0$

$-0.05v_R - 0.1e^{-10t} + 100\int_0^t (-v_R + v_C - 2e^{-10t})\,dt = 0$

(b) $5i_x + 10\int_0^t (i_x + 0.2i_x + i_L + 0.1e^{-10t})\,dt - 2e^{-10t} = 0$

$0.01i_L' + 20(i_L + 0.1e^{-10t}) + 10\int_0^t (1.2i_x + i_L + 0.1e^{-10t})\,dt - 2e^{-10t} = 0$

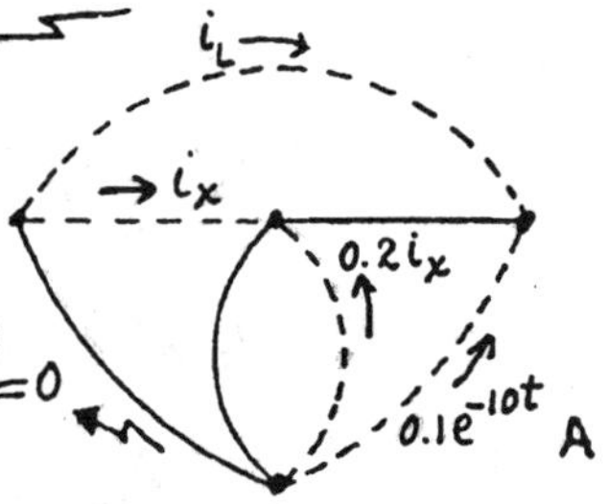


**3-22** (a)

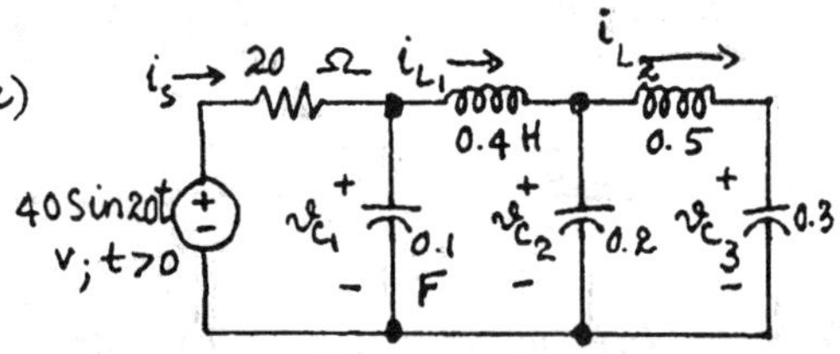


$0.05(v_{C_1} - 40\sin 20t) + 0.1v_{C_1}' + 2.5\int_0^t (v_{C_1} - v_{C_2})\,dt + 8 = 0$

$2.5\int_0^t (v_{C_2} - v_{C_1})\,dt - 8 + 0.2v_{C_2}' + 2\int_0^t (v_{C_2} - v_{C_3})\,dt - 4 = 0$

$2\int_0^t (v_{C_3} - v_{C_2})\,dt + 4 + 0.3v_{C_3}' = 0$

(b) $-40\sin 20t + 20i_s + 10\int_0^t (i_s - i_{L_1})\,dt + 5 = 0$

$10\int_0^t (i_{L_1} - i_s)\,dt - 5 + 0.4i_{L_1}' + 5\int_0^t (i_{L_1} - i_{L_2})\,dt + 6 = 0$

$5\int_0^t (i_{L_2} - i_{L_1})\,dt - 6 + 0.5i_{L_2}' + \frac{10}{3}\int_0^t i_{L_2}\,dt + 7 = 0$

**3-23**

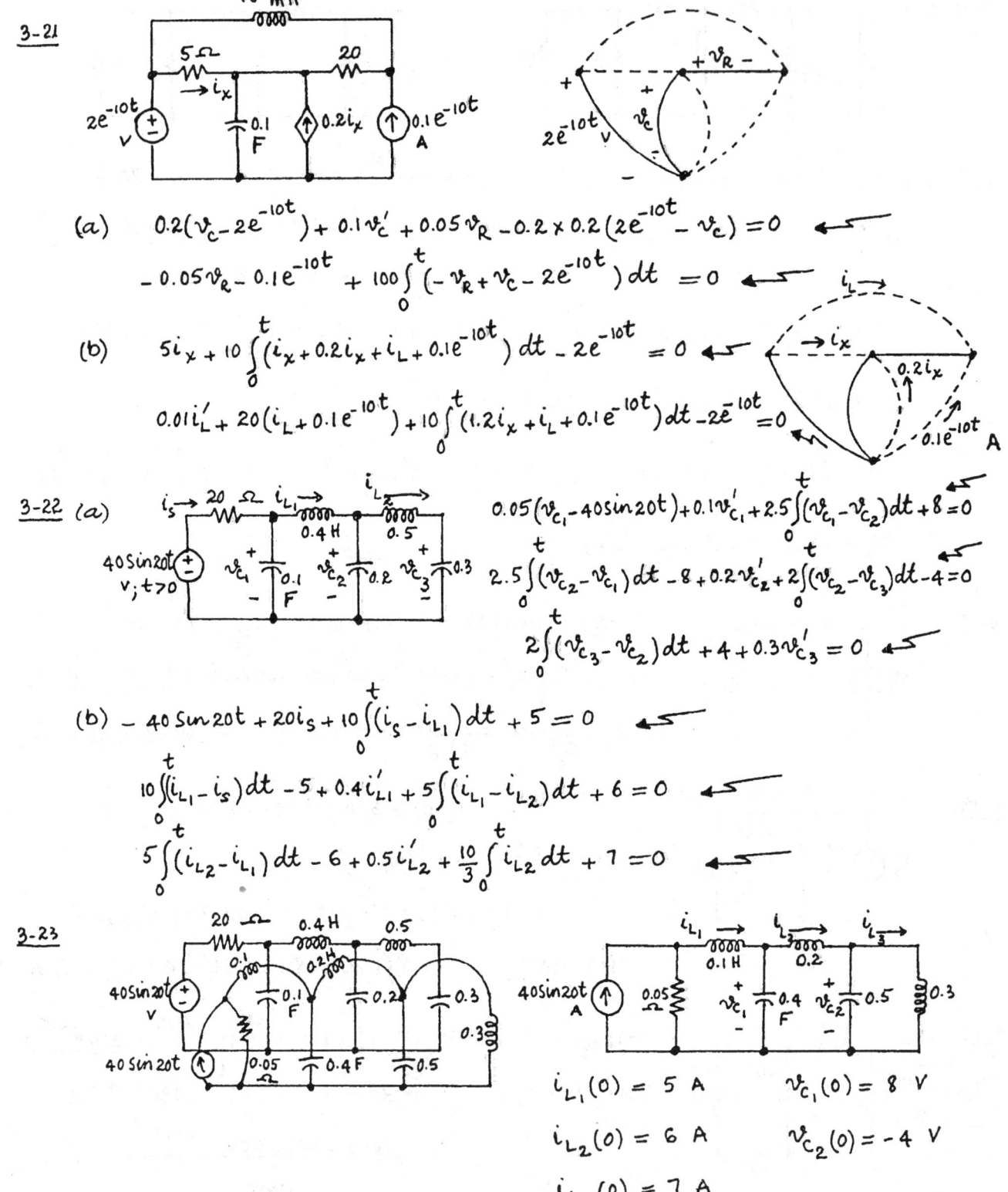

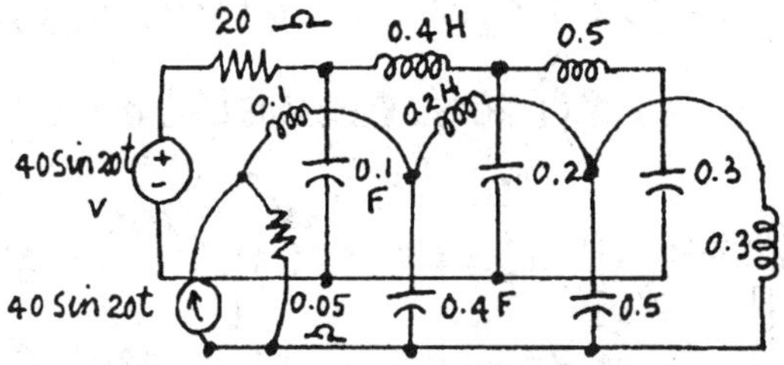


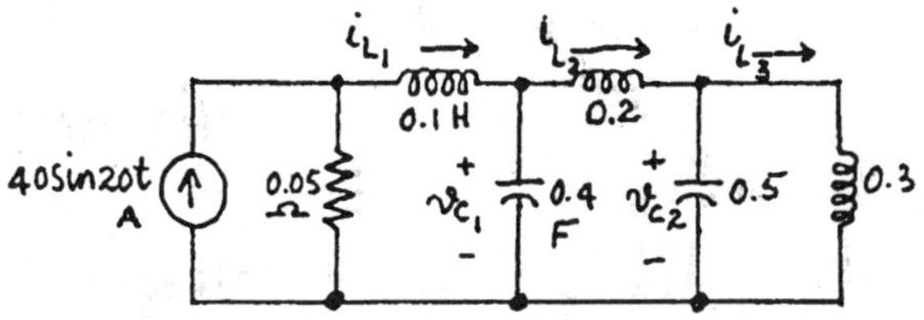


$i_{L_1}(0) = 5$ A $\qquad v_{C_1}(0) = 8$ V

$i_{L_2}(0) = 6$ A $\qquad v_{C_2}(0) = -4$ V

$i_{L_3}(0) = 7$ A

3-24

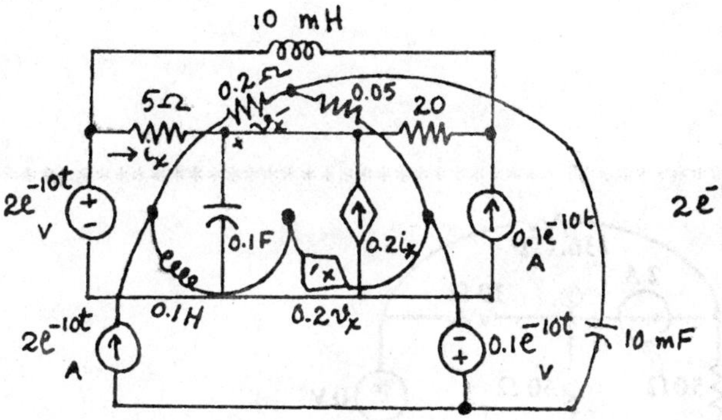


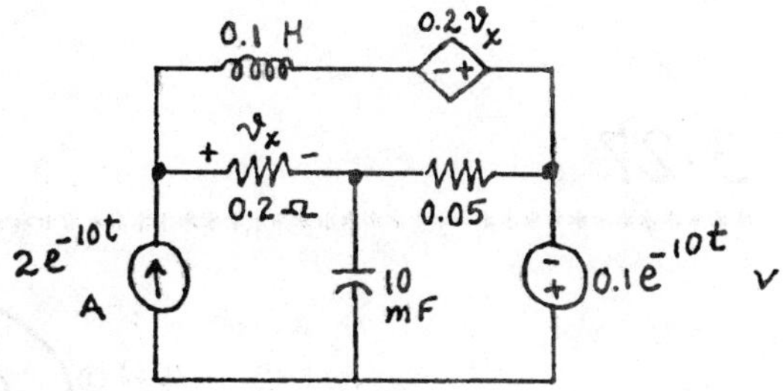


3-25 (a)

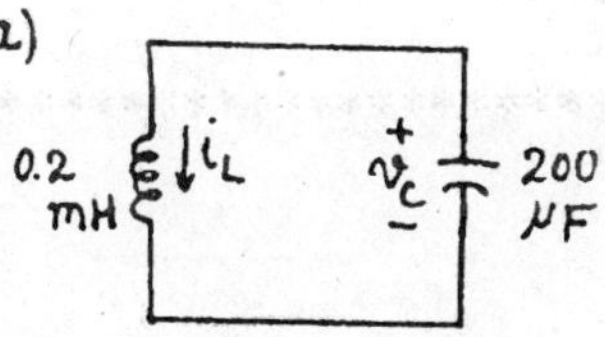


$i_L(0) = 10$ A; $v_C(0) = 4$ V

$v_C = -10\sin 5000t + 4\cos 5000t$

$v_C' = -5\times10^4\cos 5000t - 2\times10^4\sin 5000t$

$2\times10^{-4}v_C' + \frac{10^3}{0.2}\int_0^t v_C\,dt + 10 = 0$

also, $\int_0^t v_C\,dt = \frac{10}{5000}\cos 5000t + \frac{4}{5000}\sin 5000t - \frac{1}{500}$

$\therefore -10\cos 5000t - 4\sin 5000t + 10\cos 5000t + 4\sin 5000t - 10 + 10 = 0$ ✓

(b)

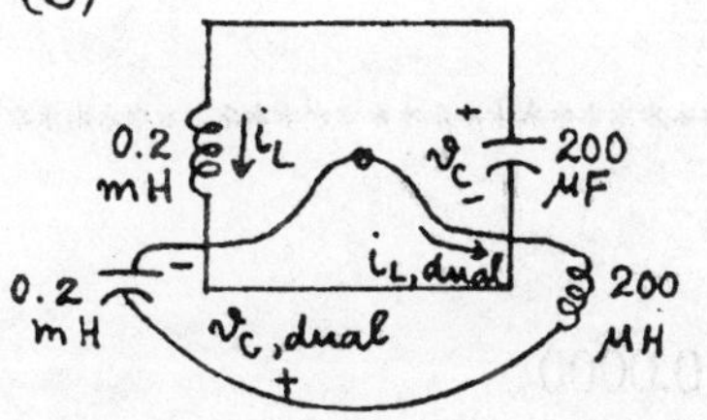


$i_{L,dual} = \underline{-10\sin 5000t + 4\cos 5000t}$ A $(= v_C)$

$v_{C,dual} = Li_{L,dual}' = -200\times10^{-6}\,i_{L,dual}'$

$= -2\times10^{-4}(-5\times10^4\cos 5000t - 2\times10^4\sin 5000t)$

$v_{C,dual} = \underline{10\cos 5000t + 4\sin 5000t}$ V $(= i_L)$

3-26

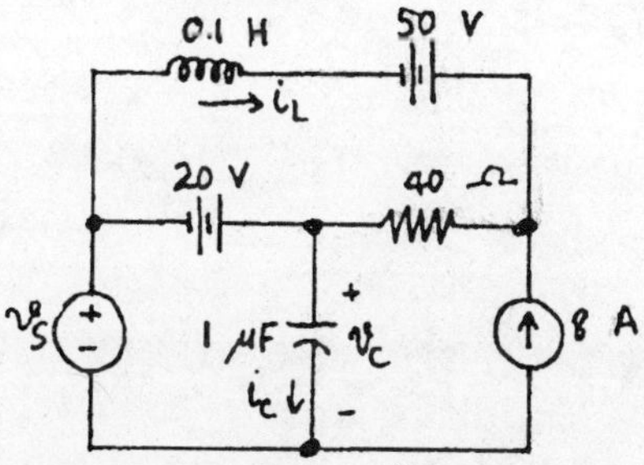


$v_S = 60\cos 10^3 t$ V

(a) $v_C(t) = 20 + 60\cos 1000t$

$i_C(t) = Cv_C'(t) = 10^{-6}(60)(-1000)\sin 1000t$

$i_C(t) = \underline{-0.06\sin 1000t\ \text{A}}$

(b) $i_{L,50V} = 1.25$ A ; $i_{L,20V} = -\frac{20}{40} = -0.5$ A

$i_{L,8A} = -8$ A ; $i_{L,ac} = 0$

$i_L(t) = i_{L,50V} + i_{L,20V} + i_{L,8A} + i_{L,ac} = \underline{-7.25\ \text{A}}$

## 3-27

************************************************************************

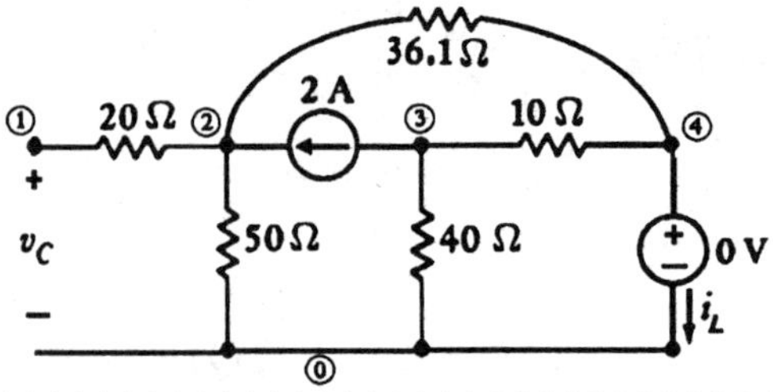


************************************************************************

**PROGRAM:**

```
*Inductor replaced by sc
*Capacitor replaced by oc
I32 3 2 2
R20 2 0 50
R30 3 0 40
R34 3 4 10
R24 2 4 36.1
V40 4 0
*V(2) = V(1) = vC
.PRINT DC V(2) I(V40)
.END
```

************************************************************************

**ANSWERS: Node voltages**

| NODE | | NODE | | NODE | |
|---|---|---|---|---|---|
| (2) | **41.9280 V** = $v_C$ | (3) | -16.0000 | (4) | 0.0000 |

**ANSWERS: Source currents**

| NAME | CURRENT |
|---|---|
| V40 | **-4.386E-01 A** = $i_L$ |

# CHAPTER 4: SOURCE-FREE *RL* AND *RC* CIRCUITS

## Section 4-2 The simple *RL* circuit

**1** After being closed for a very long time, the switch in Fig. 4-1 is opened at $t = 0$. (*a*) Find $i_L(0^+)$ and $w_L(0^+)$. (*b*) Find $i_L(t)$. (*c*) Find $v_{10}(t)$.

**2** In Fig. 4-2, let $i_s = 12$ A for $t < 0$ and zero for $t > 0$. (*a*) Find $i_L(t)$ and sketch it for $-0.01 < t < 0.05$ s. (*b*) Find $v_L(t)$ and sketch it for the same time interval.

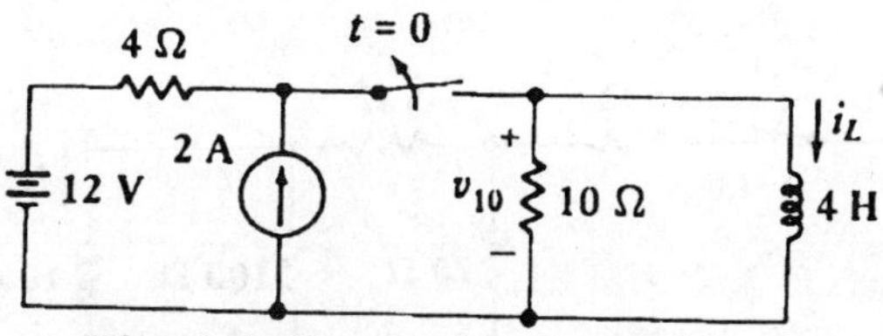

**Fig. 4-1:** See Probs. 1 and 18.

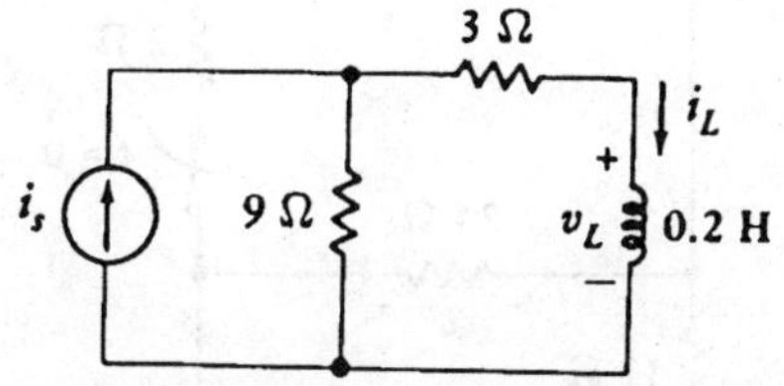

**Fig. 4-2:** See Prob. 2.

**3** (*a*) A 0.2-H inductor is in parallel with a 25-Ω resistor. If $i_L(0) = 5$ A, find $i_L(t)$ and the value of $t$, $t = t_1$, at which $i_L = 1$ A. (*b*) At this instant, a 5-Ω resistor is connected in parallel with the 25-Ω resistor. Write $i_L(t)$ for $t > t_1$ and find the value of $t$, $t = t_2$ at which $i_L = 1$ mA.

**4** For the circuit of Fig. 4-3, find $i_L(t)$ and sketch it for $t > 0$.

**5** Determine the energy stored in the inductor of Fig. 4-4 twenty microseconds after the switch is thrown.

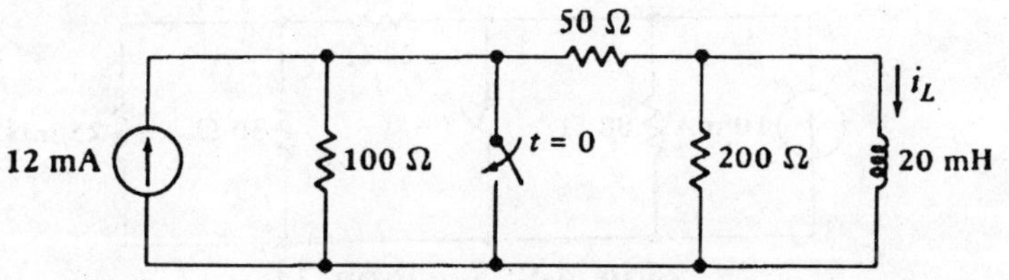

**Fig. 4-3:** See Prob. 4.

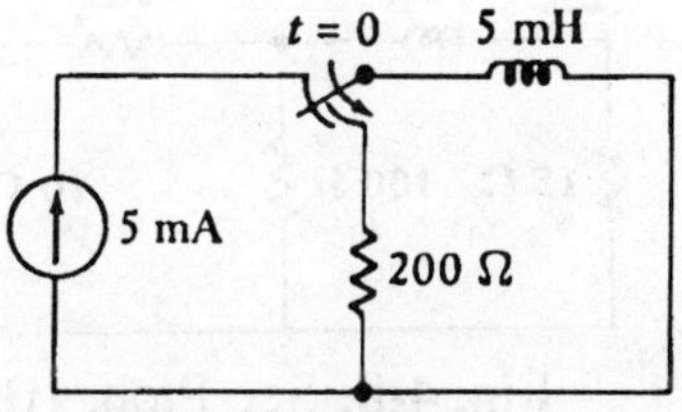

**Fig. 4-4:** See Prob. 5.

**6** (*a*) In Fig. 4-5, find the inductor current as a function of time, and sketch curves of $v_1$, $v_2$, and $v_3$ versus $t$, $-1 < t < 1$ s. (*b*) Sketch $v_3(t)$ versus $t$, $-1 < t < 1$ s, if another switch places a second 10-Ω resistor in parallel with the original 10-Ω resistor at $t = 0.25$ s.

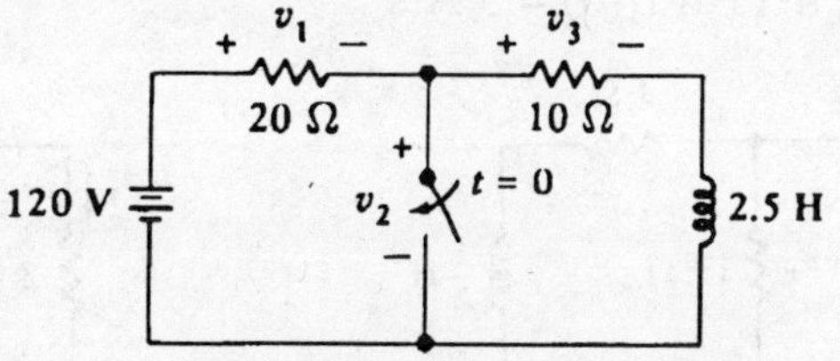

**Fig. 4-5:** See Prob. 6.

**7** A source-free *RL* circuit contains a 50-H inductor discharging through a 20-Ω resistor. The inductor current amplitude is 18 A at $t = 2$ s. (*a*) When was the energy stored in the inductor twice as great as it is at $t = 2$ s? (*b*) At what instant will the power dissipated in the resistor be 1 kW?

**8** The switch in the circuit shown in Fig. 4-6 opens at $t = 0$. (*a*) Find $v(t)$ for $t > 0$. (*b*) Sketch $v(t)$ for $-1 < t < 1$ s.

## Section 4-3 Properties of the exponential response
## Section 4-4 A more general *RL* circuit

**9** The battery in Fig. 4-7 is disconnected at $t = 0$. (*a*) Find $i_L(0^-)$. (*b*) Find $i_L(0^+)$. (*c*) Find the (Thévenin) equivalent resistance seen by the inductor for $t > 0$. (*d*) Find $\tau$. (*e*) Find $i_L(t)$, $t > 0$.

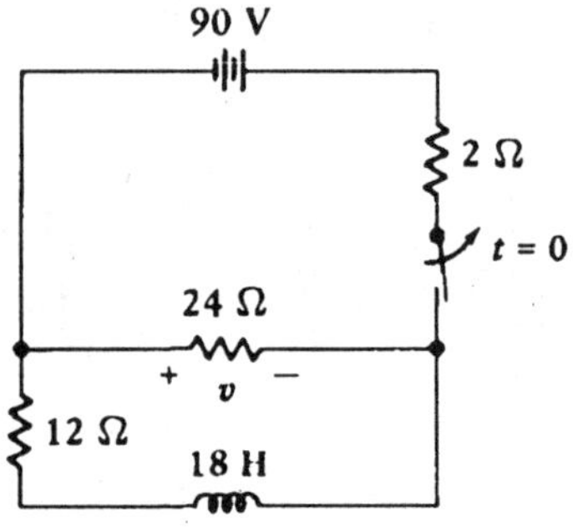


**Fig. 4-6:** See Prob. 8.

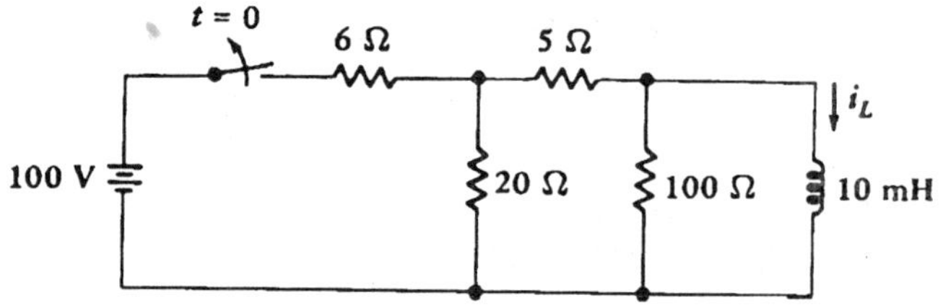


**Fig. 4-7:** See Prob. 9.

**10** In Fig. 4-8, let $i_L(0) = 10$ A. Find $v_x(t)$ for $t > 0$ and sketch it over a three-time-constant interval.

**11** Refer to Fig. 4-9 and find $i_x(t)$ for all time. Give its numerical value at $t = -2$, $0^-$, $0^+$, 2, and 4 ms.

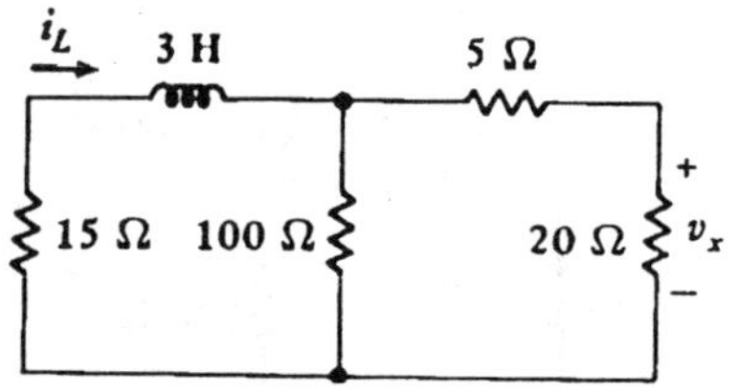


**Fig. 4-8:** See Prob. 10.

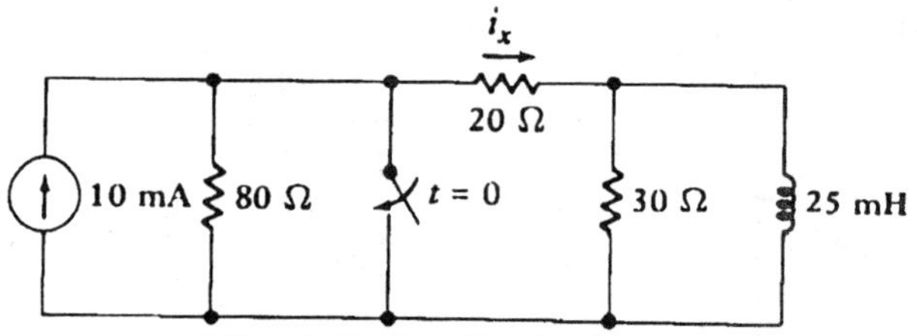


**Fig. 4-9:** See Prob. 11.

**12** The switch in Fig. 4-10 opens at $t = 0$ after being long-closed. Calculate $v_x$ at $t = -10$, $0^-$, $0^+$, 10, and 20 ms.

**13** Find $i_1(t)$ for $t > 0$ in Fig. 4-11 if $i_1(0) = 5$ A.

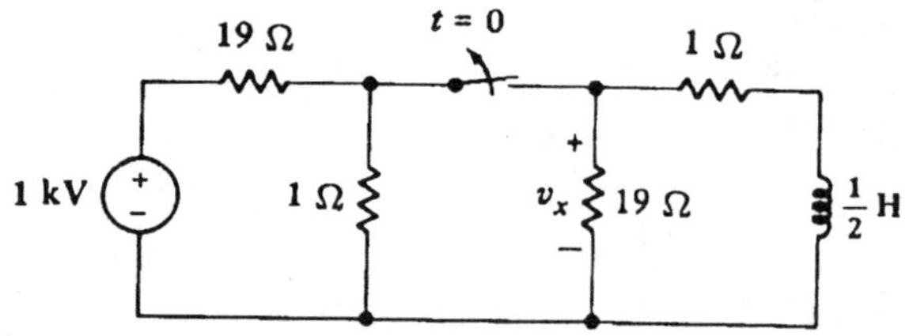


**Fig. 4-10:** See Prob. 12.

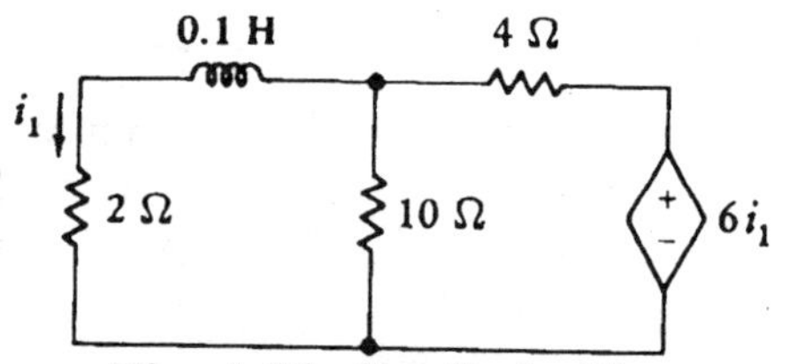


**Fig. 4-11:** See Prob. 13.

**14** The circuit shown in Fig. 4-12 contains two inductors in parallel, thus providing the opportunity of a trapped current circulating around the inductive loop. Let $i_1(0^-) = 20$ A and $i_2(0^-) = 10$ A. (*a*) Find $i_1(0^+)$, $i_2(0^+)$, and $i(0^+)$. (*b*) Determine the time constant $\tau$ for $i(t)$. (*c*) Find $i(t)$, $t > 0$. (*d*) Find $v(t)$. (*e*) Find $i_1(t)$ and $i_2(t)$ from $v(t)$ and the initial values. (*f*) Show that the stored energy at $t = 0$ is equal to the sum of the energy dissipated in the resistive network between $t = 0$ and $t = \infty$, plus the energy stored in the inductors at $t = \infty$.

**15** Calculate $i_L$ and $i_1$ in the circuit of Fig. 4-13 at $t$ = -10, 0$^-$, 0$^+$, 10, and 20 ms.

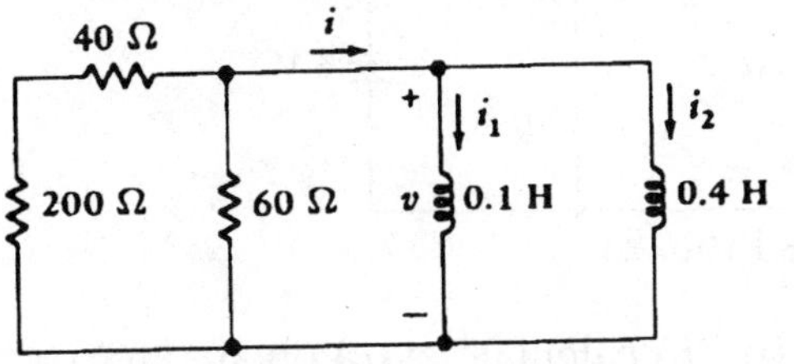


Fig. 4-12: See Prob. 14.

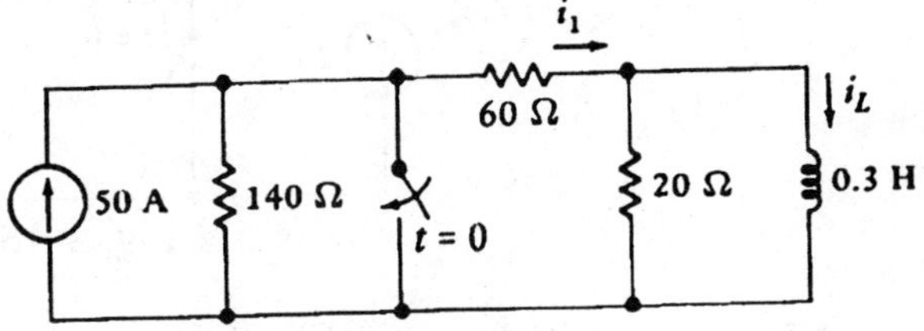


**Fig. 4-13:** See Prob. 15.

## Section 4-5 The simple *RC* circuit

**16** The switch in Fig. 4-14 is opened at $t$ = 0 after having been closed for a long time. Plot $v_C(t)$ and $v_x(t)$ on the same time axis, -0.5 < $t$ < 1 ms.

**17** Refer to the circuit shown in Fig. 4-15 and determine values at $t$ = 1 s for: (*a*) $v_C$; (*b*) $v_R$; (*c*) $v_{SW}$.

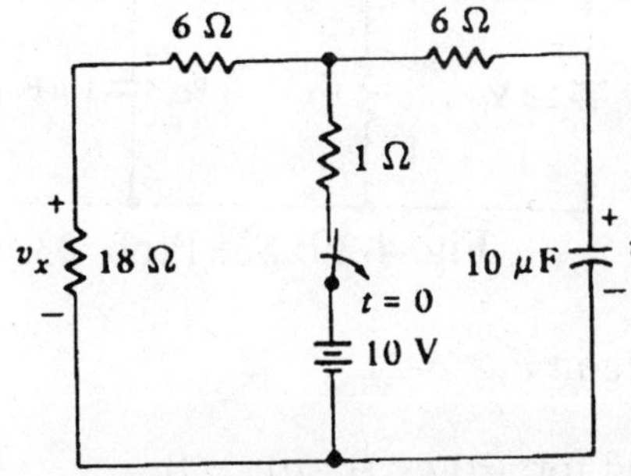


**Fig. 4-14:** See Prob. 16.

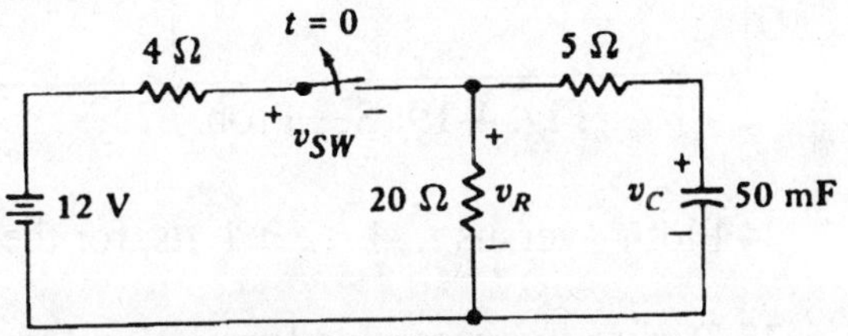


**Fig. 4-15:** See Prob. 17.

**18** Construct the exact dual of the circuit in Prob. 1, Fig. 4-1, write the exact dual of the problem statement, and then solve your new problem.

**19** At $t$ = 0, the left-hand switch in the circuit shown in Fig. 4-16 is thrown down and the right-hand switch is simultaneously opened. Find the total charge that leaves the lower terminal of the capacitor during the time interval, 0.3 < $t$ < 1 s.

## Section 4-6 A more general *RC* circuit

**20** In the circuit shown in Fig. 4-17, the switch moves from *A* to *B* at $t$ = 0. Find expressions for and sketch $i(t)$ in the interval -2 < $t$ < 2 ms.

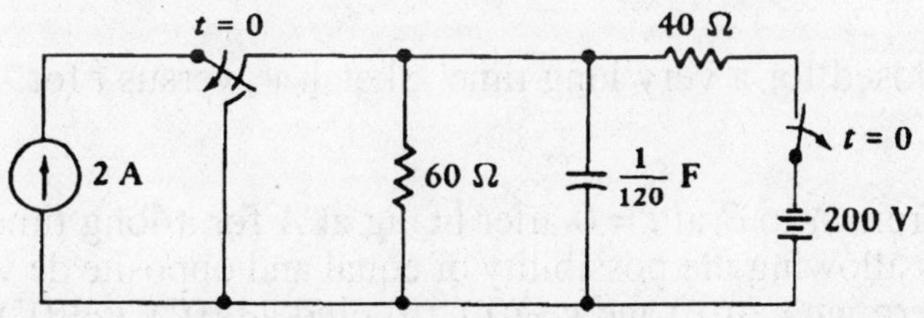


**Fig. 4-16:** See Prob. 19.

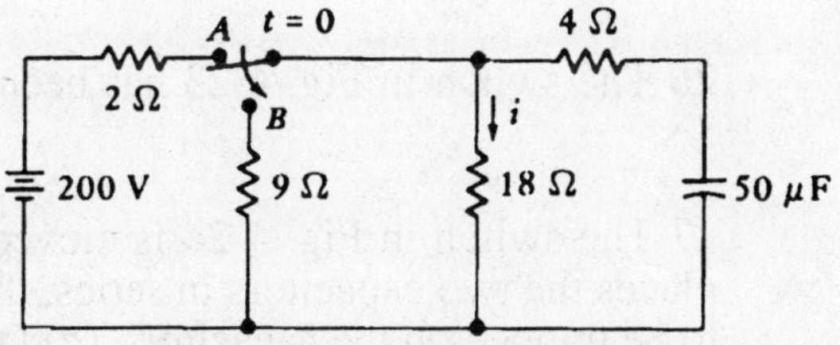


**Fig. 4-17:** See Prob. 20.

**21** At $t$ = 0, the left switch in Fig. 4-18 is closed while the right one moves from *A* to *B*. At $t$ = -2 μs and +2 μs, find: (*a*) *i*; (*b*) the rate at which energy is leaving the capacitor.

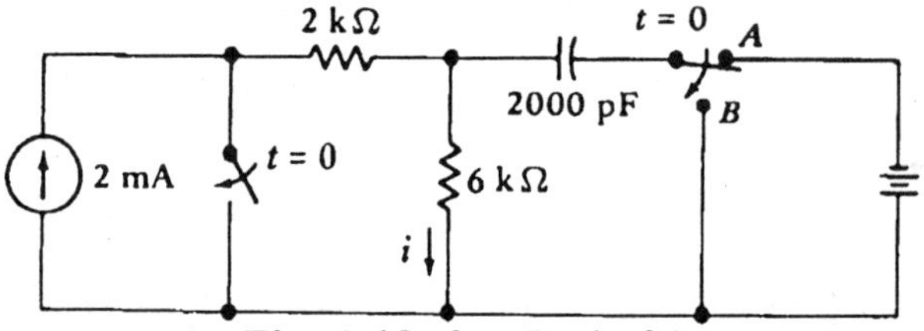


**Fig. 4-18:** See Prob. 21.

**22** In the circuit shown in Fig. 4-19: (*a*) find $v_C(0)$; (*b*) calculate $\tau$; (*c*) obtain an expression for $v_C(\mathrm{t})$.

**23** Select values for $R_1$ and $R_2$ in Fig. 4-20 so that $v_C = 10$ V at $t = 0.5$ ms, and $v_C = 1$ V at $t =$ 2 ms.

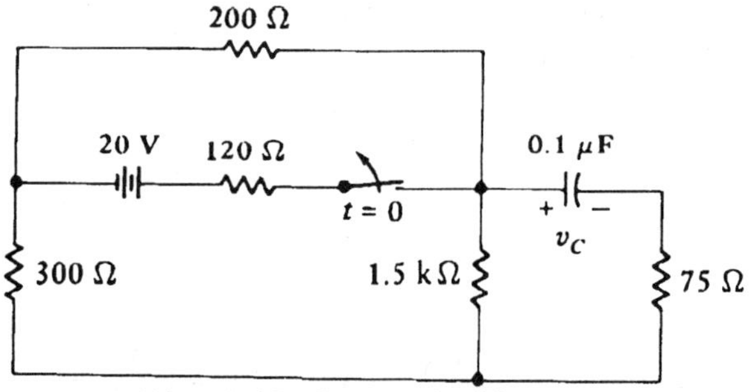


**Fig. 4-19:** See Prob. 22.

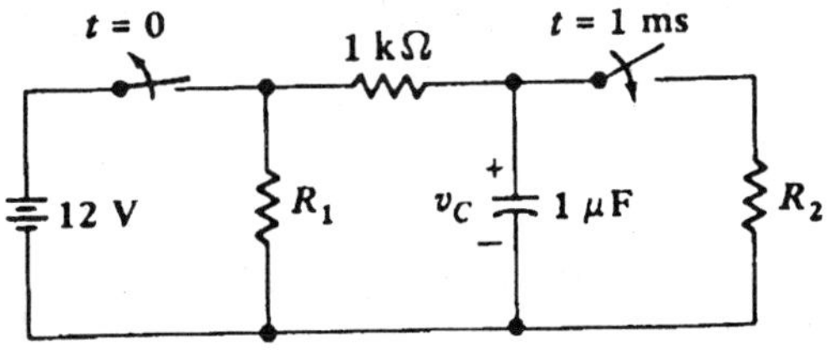


**Fig. 4-20:** See Prob. 23.

**24** Plot $v_R$ versus $t$, $-1 < t < 4$ μs, for the circuit appearing in Fig. 4-21.

**25** Derive an expression for $v_C(\mathrm{t})$ in Fig. 4-22 that is valid for: (*a*) $t < 0$; (*b*) $t > 0$.

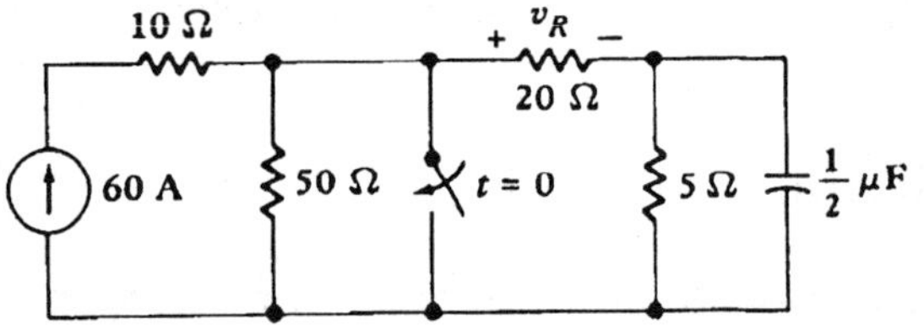


**Fig. 4-21:** See Prob. 24.

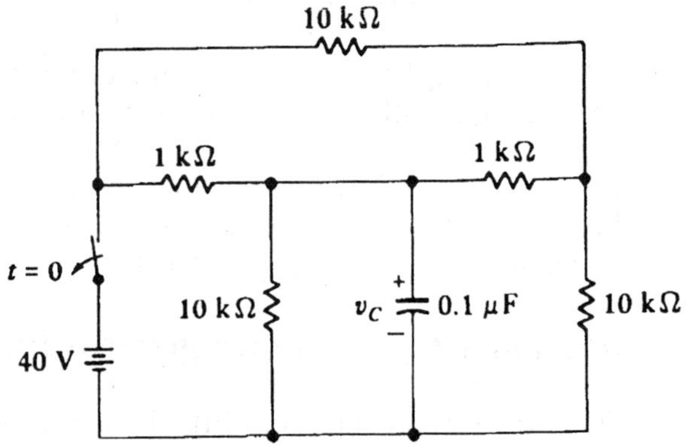


**Fig. 4-22:** See Prob. 25.

**26** The switch in Fig. 4-23 has been closed for a very long time. Sketch $v_x$ versus $t$ for $-1 < t <$ 3 s.

**27** The switch in Fig. 4-24 is moved from $A$ to $B$ at $t = 0$ after being at $A$ for a long time. This places the two capacitors in series, thus allowing the possibility of equal and opposite dc voltages to be trapped on the capacitors. (*a*) Determine $v_{C1}(0^-)$ and $v_{C2}(0^-)$. (*b*) Find $v_{C1}(0^+)$, $v_{C2}(0^+)$, and $v_R(0^+)$. (*c*) Determine the time constant of $v_R(t)$. (*d*) Find $v_R(t)$, $t > 0$. (*e*) Find $i(t)$. (*f*) Find $v_{C1}(t)$ and $v_{C2}(t)$ from $i(t)$ and the initial values. (*g*) Show that the stored energy at $t = \infty$ plus the total energy dissipated in the 2.5-kΩ resistor is equal to the energy stored in the capacitors at $t = 0$.

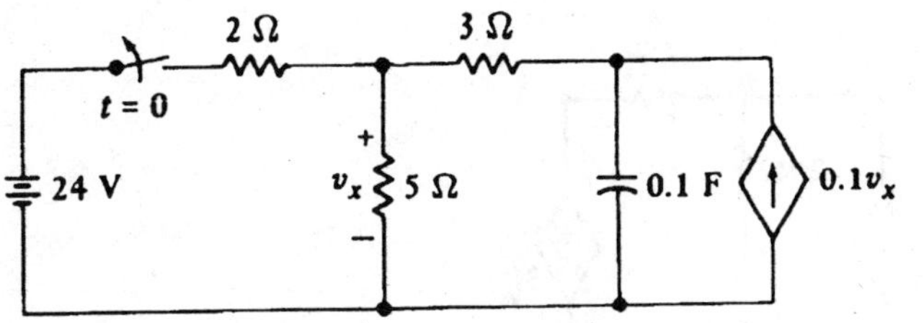


**Fig. 4-23:** See Prob. 26.

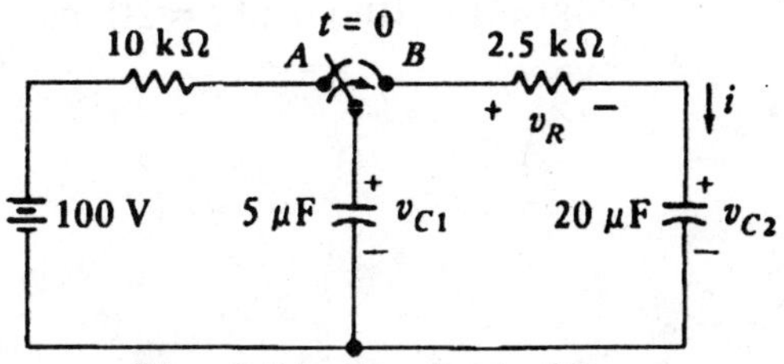


**Fig. 4-24:** See Prob. 27.

**28** The switch in Fig. 4-25 opens at $t = 0$, having been closed for a long time. Find $v_{30}(t)$, $t > 0$.

**29** The circuit shown in Fig. 4-26 has been operating for a long time in the configuration shown. At $t = 0$, a loose piece of copper wire falls on the circuit and short-circuits the upper terminals of four elements, as shown. Find $i_L(t)$, $t > 0$.

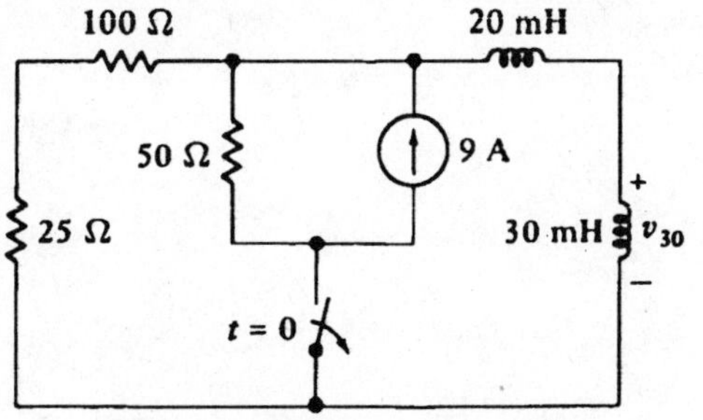


**Fig. 4-25:** See Prob. 28.

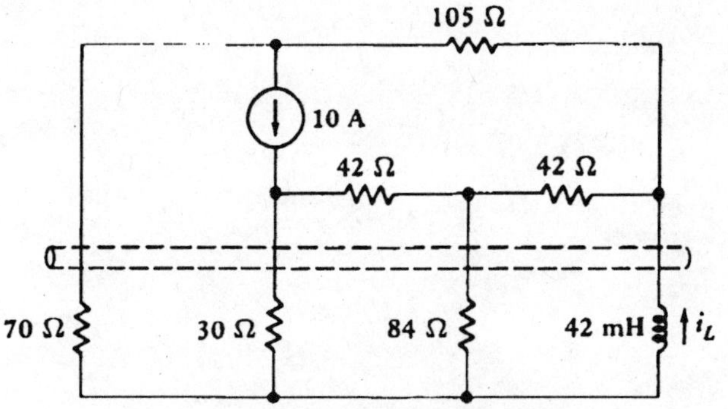


**Fig. 4-26:** See Prob. 29.

**30** (*a*) Find $v_C(0^-)$ as a function of *k* for the circuit shown in Fig. 4-27. (*b*) Find $v_C(t)$ as a function of *k* for $t > 0$. (*c*) Evaluate your expressions for $k = 0$ and $k = 10^{-3}$.

**31** In the circuit of Fig. 4-28, let $i_s = 50$ A for $t < 0$ and zero for $t > 0$. Find $v(t)$ for all *t*.

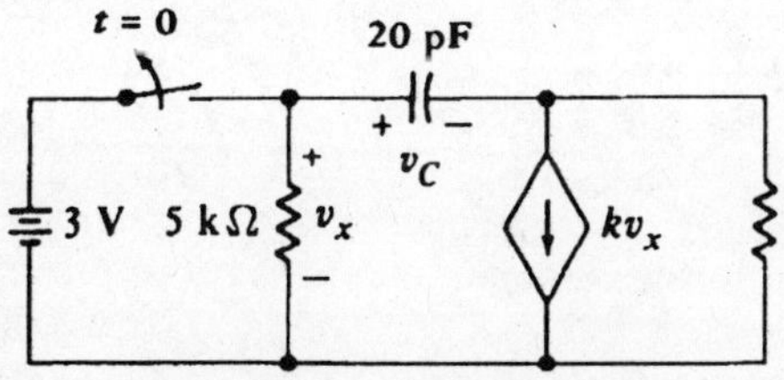

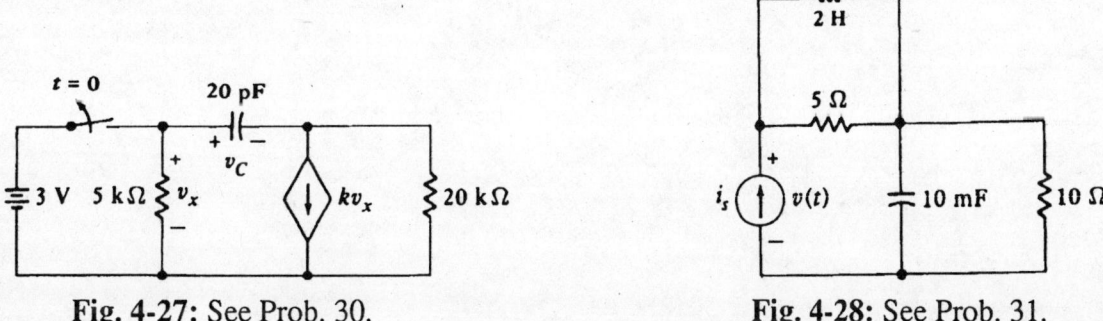

**Fig. 4-27:** See Prob. 30.

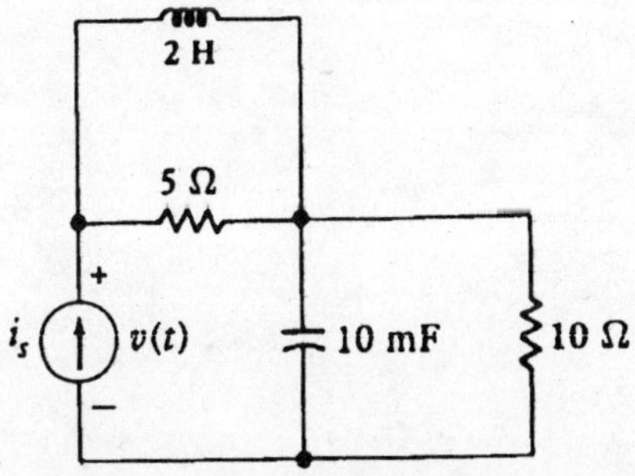


**Fig. 4-28:** See Prob. 31.

**32** (SPICE) Let $i_L(0) = 20$ A in the circuit shown in Fig. 4-29. Use TSTEP = 1 ms in a SPICE program to find $i_L$ at $t = 20$ ms.

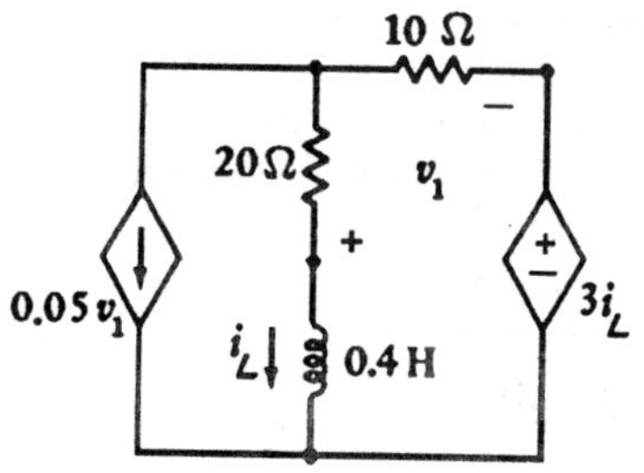


Fig. 4-29: See Prob. 32.

# Chapter 4

**4-1**

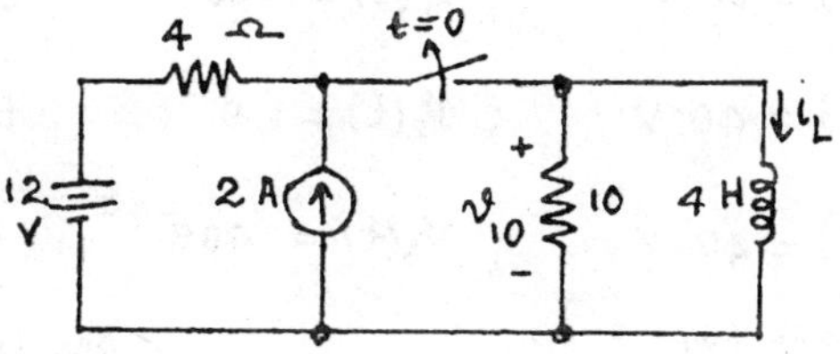


(a) $i_L(0^-) = i_L(0^+) = i_L(0) = \frac{12}{4} + 2 = \underline{5\ A}$

$W_L(0^+) = \frac{1}{2} L i_L^2(0^+) = \frac{1}{2} \times 4 \times 5^2 = \underline{50\ J}$

(b) $i_L(t) = i_L(0) e^{-Rt/L} = 5e^{-10t/4} = \underline{5e^{-2.5t}\ A}, \ t > 0$

(c) $v_{10}(t) = v(0^+) e^{-Rt/L} = \underline{-50e^{-2.5t}\ V}, \ t > 0$

**4-2** (a)

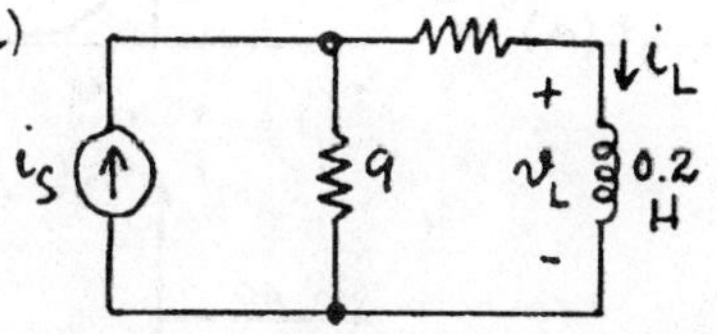


$i_S = 12A,\ t < 0;\ i_S = 0,\ t > 0$

$i_L(0^-) = 9A \therefore i_L(t) = 9e^{-12t/0.2}$

$i_L(t) = \underline{9e^{-60t}\ A}, \ t > 0$

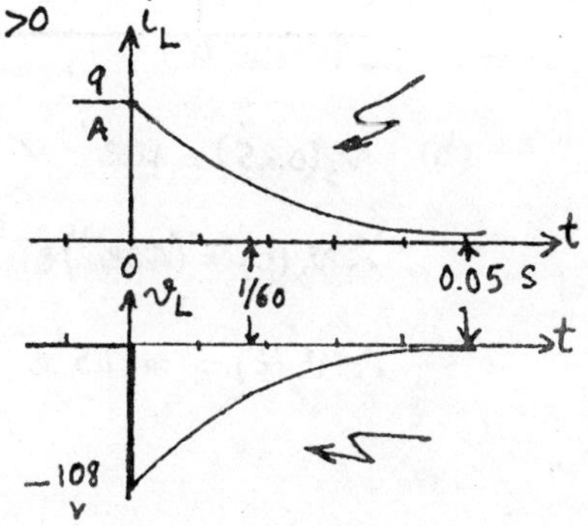


(b) $v_L(t) = -(3+9)\, i_L = \underline{-108e^{-60t}}\ V \quad t > 0$

**4-3** (a)

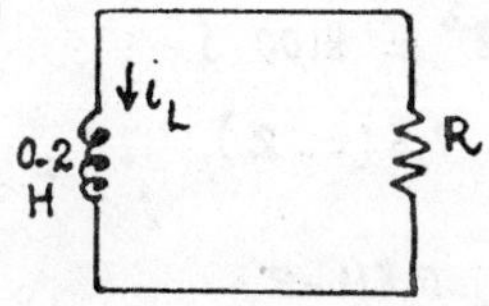


$i_L(0) = 5A\ ;\ R = 25\ \Omega$

$i_L(t) = \underline{5e^{-125t}\ A};\ t > 0$

$1 = 5e^{-125t_1}$

$\therefore t_1 = \underline{12.876\ ms}$

(b) $R \to 25 \| 5 = \frac{25}{6}\ \Omega$

$i_L(t) = 1e^{-(25/6)(t-t_1)/0.2}$

$i_L(t) = \underline{1e^{-20.83(t-t_1)}\ A}, \ t > t_1$

$10^{-3} = e^{-20.83(t_2 - 0.012876)}$

$\therefore t_2 = \underline{344.4\ ms}$

**4-4**

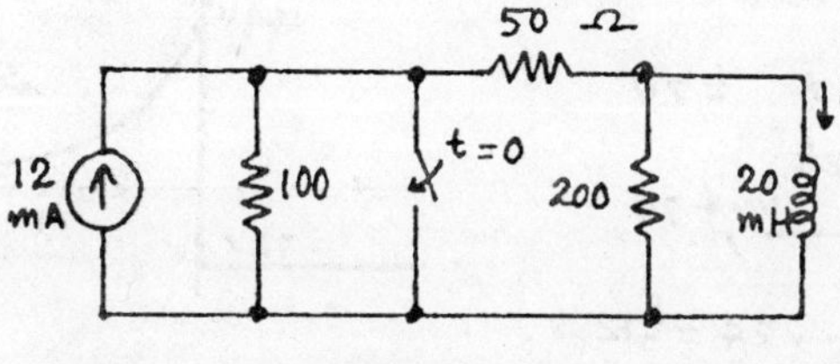


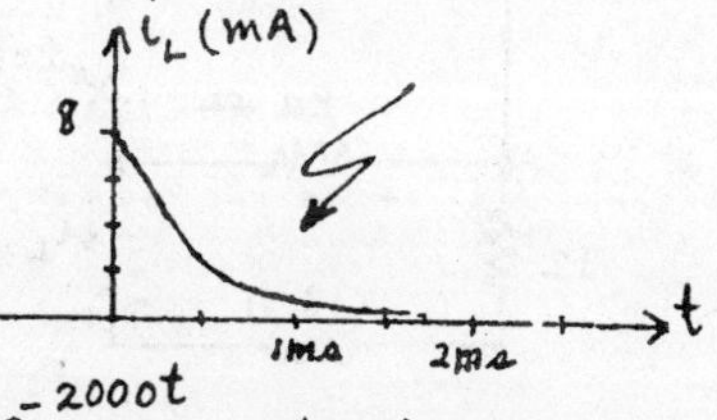


$i_L(0) = 12 \times \frac{100}{150} = 8\ mA \quad \therefore i_L(t) = 8e^{-40t/0.02} = \underline{8e^{-2000t}\ mA}, \ t > 0$

**4-5**

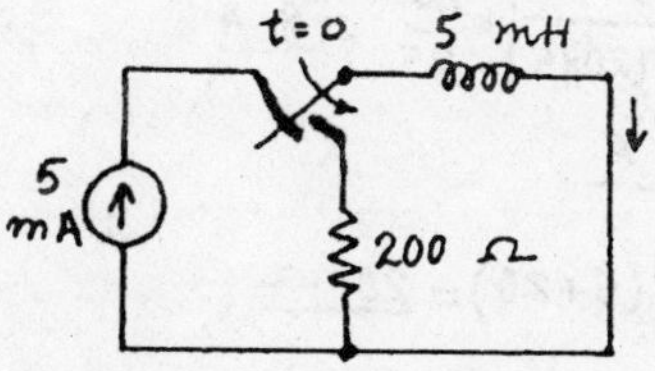


$i_L(t) = 5e^{-200t/0.005} = 5e^{-4\times10^4 t}\ mA,\ t > 0$

$\therefore W_L(t) = \frac{1}{2} L i_L^2 = \frac{1}{2} \times 5\times10^{-3} \times 25\times10^{-6}\, e^{-8\times10^4 t}\ J$

$W_L(2\times10^{-5}) = \underline{12.619\ nJ}$

**4-6** (a)

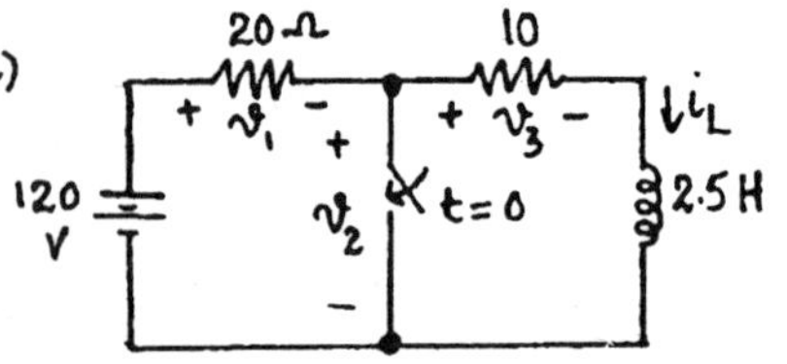


$i_L = \frac{120}{30} e^{-10t/2.5} = \underline{4e^{-4t}\ \text{A}}, \quad t > 0$

$v_1(0^-) = 80$ V ; $v_1(t) = 120$ V $\quad t > 0$.

$v_2(0^-) = 40$ V ; $v_2(t) = 0 \quad t > 0$

$v_3(0^-) = 40$ V ; $v_3(t) = 40e^{-4t}$ V $\quad t > 0$

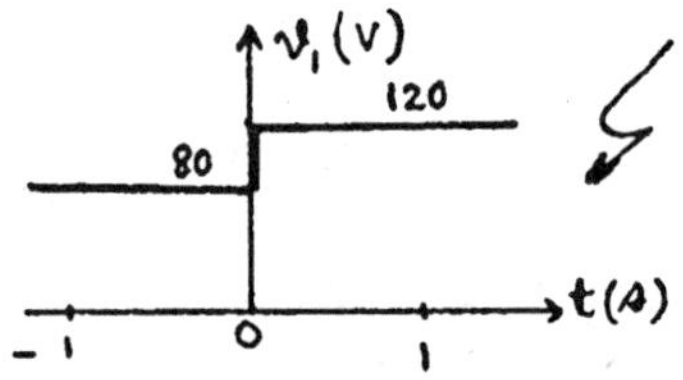


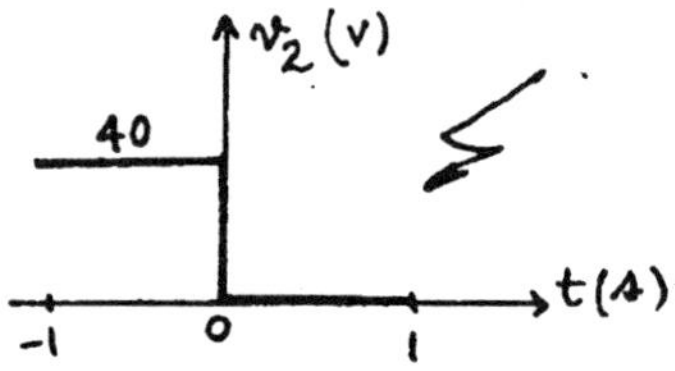


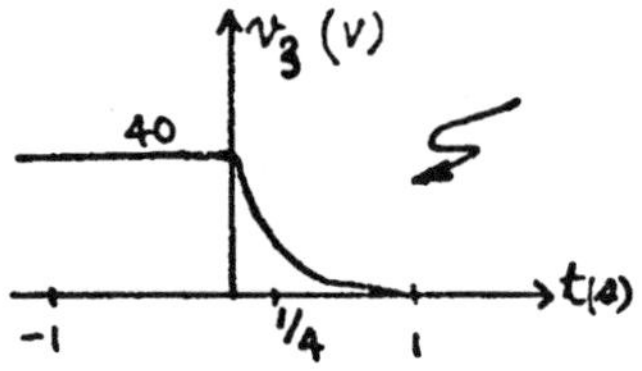


(b) $v_3(0.25) = 40e^{-1}$ V

$\therefore v_3(t) = (40e^{-1})e^{-5(t-0.25)/2.5}$

$\therefore v_3(t) = 14.715\, e^{-2(t-0.25)} \quad t > 0.25$

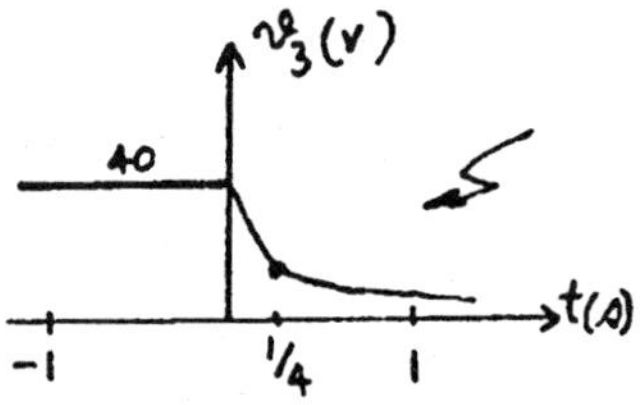


**4-7**

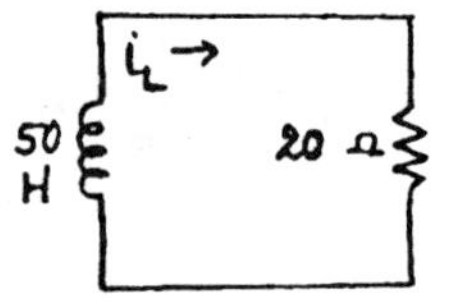


(a) $i_L(2) = 18 \quad \therefore i_L(t) = 18e^{-0.4(t-2)}$ A

$w_L(2) = \frac{1}{2} L i_L^2(2) = \frac{1}{2} \times 50 \times 18^2 = 8100$ J

$w_L(t) = 16{,}200 = \frac{1}{2} \times 50 \times 18^2 e^{-0.8(t-2)} \quad \therefore t = \underline{1.1336\ \text{s}}$

(b) $w_R(t) = R i_L^2(t)$ $\therefore$ $1000 = 20 \times 18^2 e^{-0.8(t-2)} \quad \therefore t = \underline{4.336\ \text{s}}$

**4-8**

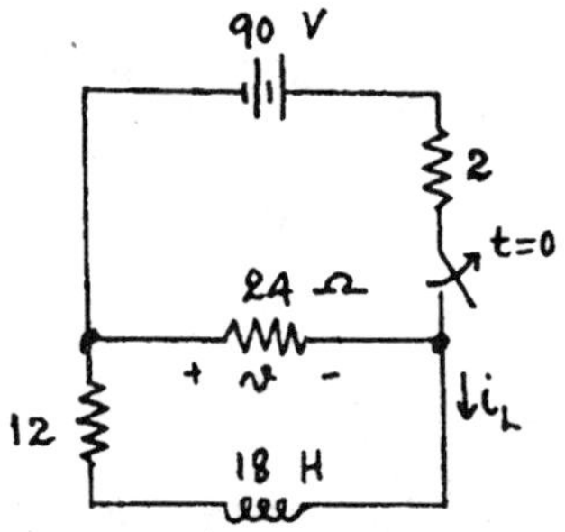


(a) $i_L(0) = \frac{90}{2 + (12 // 24)} \times \frac{2}{3} = 6$ A

$i_L(t) = 6e^{-2t}$ A; $t > 0$

$\underline{v(t) = 144e^{-2t}\ \text{V}}; t > 0$

(b) $v(t) = \frac{-90}{10} \times \frac{1}{3} \times 24 = -72$ V

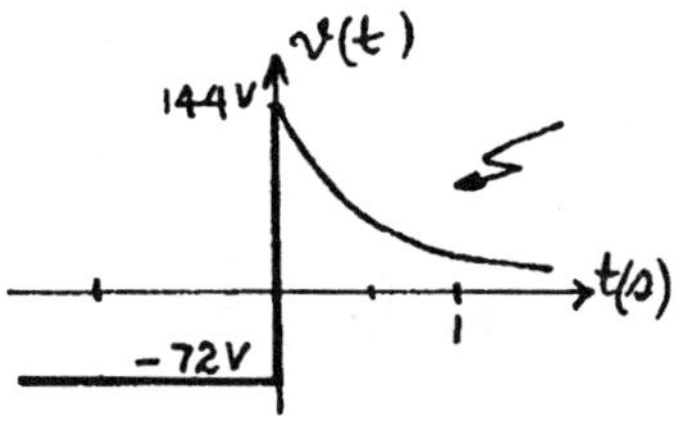


**4-9**

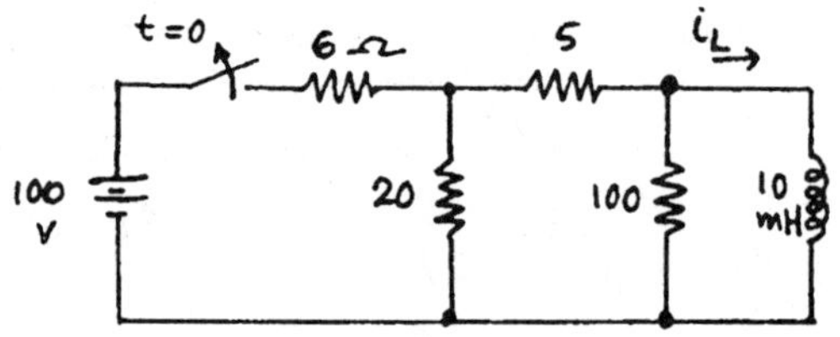


(a) $i_L(0^-) = \frac{100}{6 + (20 \| 5)} \times \frac{20}{25} = \underline{8\ \text{A}}$

(b) $i_L(0^+) = \underline{8\ \text{A}}$

(c) $R_{eq} = 100 \| (5 + 20) = \underline{20\ \Omega}$

(d) $\tau = \frac{L}{R_{eq}} = \frac{0.01}{20} = \underline{0.5\ \text{ms}}$

(e) $i_L(t) = i_L(0)e^{-t/\tau} = \underline{8e^{-2000t}\ \text{A}} \quad t > 0$

**4-10**

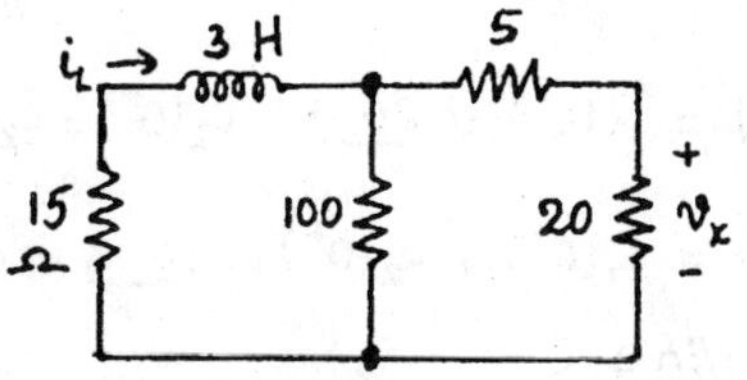


$i_L(0) = 10$ A $\therefore v_x(0) = 160$ V

$\therefore v_x(t) = \underline{160e^{-35t/3}}$ V, $t > 0$

$\therefore \tau = \dfrac{3}{35} = 85.71$ ms

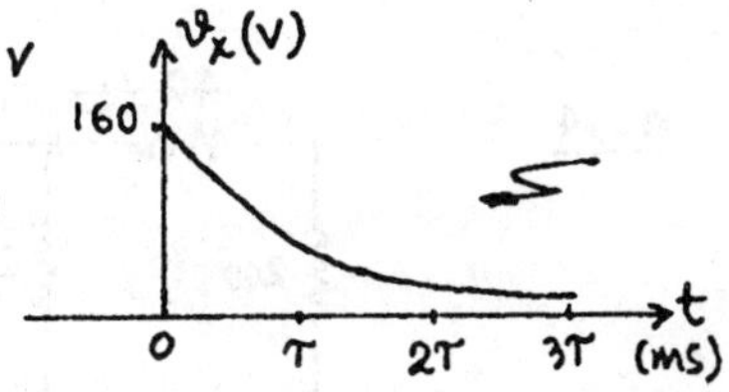


**4-11**

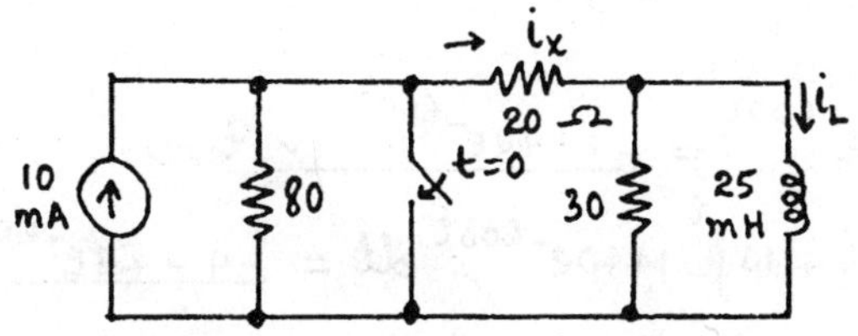


$i_L(0) = 10 \times \dfrac{80}{100} = 8$ mA $\quad \therefore i_x(0^+) = 4.8$ mA

$i_x(t) = 4.8e^{-12t/0.025} = \underline{4.8e^{-480t}\text{ mA}}, t > 0$

$i_x(t) = \underline{8\text{ mA}}, t < 0$

$i_x(-2\text{ms}) = \underline{8\text{ mA}}$; $i_x(0^-) = \underline{8\text{ mA}}$

$i_x(0^+) = \underline{4.8\text{ mA}}$; $i_x(2\text{ ms}) = \underline{1.8379\text{ mA}}$

$i_x(4\text{ ms}) = 4.8e^{-1.92} = \underline{0.7037\text{ mA}}$

**4-12**

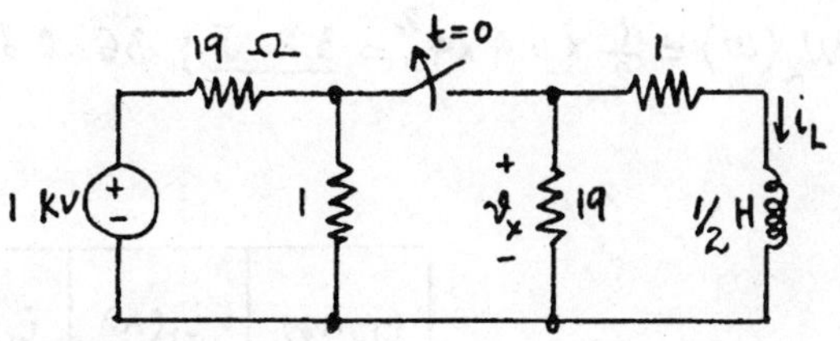


$i_L(0) = \dfrac{1000}{19 + (1\|19\|1)} \times \dfrac{1}{1 + 1 + 1/19} = 25$ A

$v_x(0^+) = -475$ V $\therefore v_x = -475e^{-40t}$ V, $t > 0$

$v_x(-10\text{ms}) = v_x(0^-) = \underline{25\text{ V}}$

$v_x(0^+) = \underline{-475\text{ V}}$ ; $v_x(10\text{ ms}) = -475e^{-0.4} = \underline{-318.4\text{ V}}$ ; $v_x(20\text{ ms}) = \underline{-213.4\text{ V}}$

**4-13**

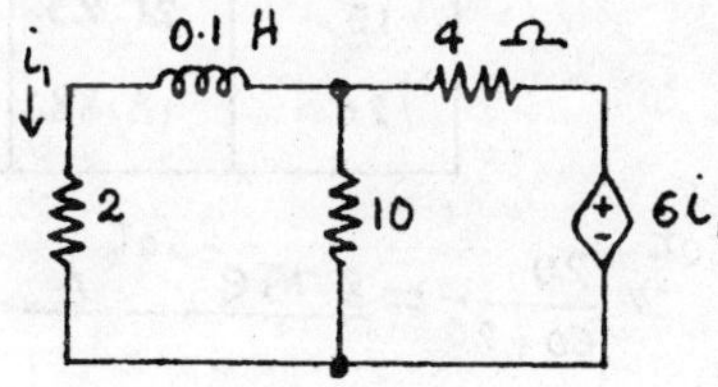


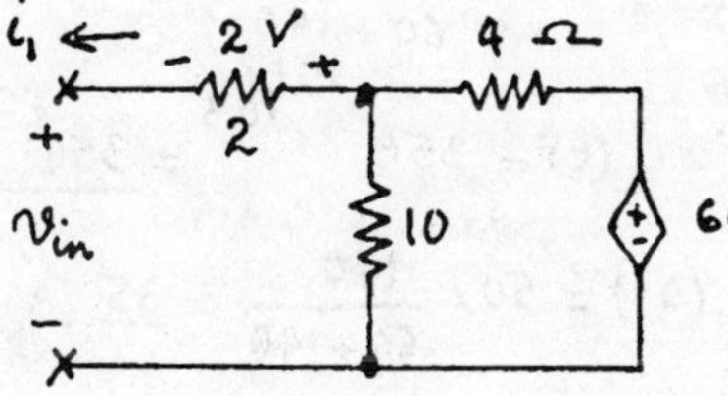


Let $i_1 = 1 \quad \therefore 6i_1 = 6$ V

$1 + \dfrac{2 + v_{in}}{10} + \dfrac{2 + v_{in} - 6}{4} = 0$

$0.2 + 0.35v_{in} = 0$ , $v_{in} = -R_{Th} = -0.5714$

$R_{th} = 0.5714\ \Omega$

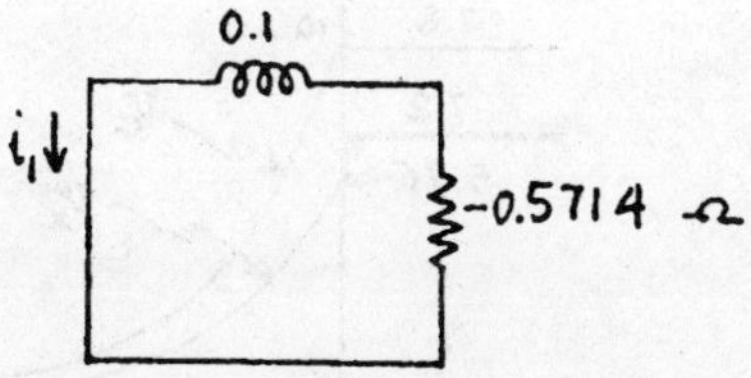


$\therefore i_1(t) = \underline{5e^{-5.71t}\text{ A}} \quad t > 0$

**4-14**

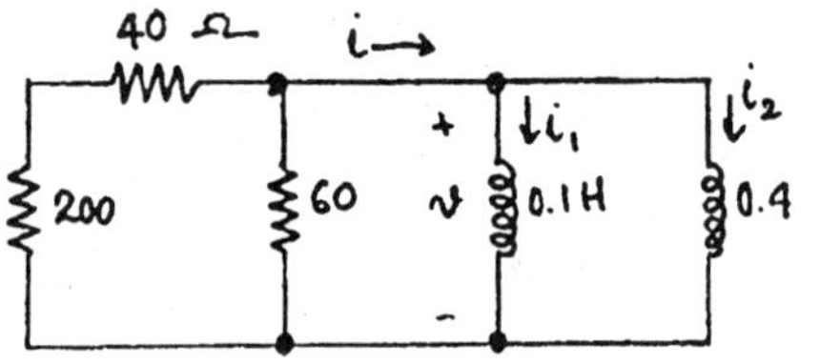


(a) $i_1(0^+) = i_1(0^-) = \underline{20\ A}$; $i_2(0^+) = i_2(0^-) = \underline{10\ A}$

$i(0^+) = i_1(0^+) + i_2(0^+) = \underline{30\ A}$

(b) $\tau = \dfrac{0.1 \| 0.4}{48} = \dfrac{1}{600}\ s$

(c) $i(t) = \underline{30e^{-600t}\ A}$, $t > 0$

(d) $v(t) = L_{eq} i' = 0.08 \times 30(-600)e^{-600t} = \underline{-1440e^{-600t}\ V}, t > 0$

(e) $i_1(t) = i_1(0) + \dfrac{1}{L}\int_0^t v(t)dt = 20 + 10\int_0^t -1440e^{-600t}dt = \underline{-4 + 24e^{-600t}\ A}, t > 0$

$i_2(t) = 10 + 2.5\int_0^t -1440e^{-600t}dt = 10 + \left[6e^{-600t}\right]_0^t = \underline{4 + 6e^{-600t}\ A}$ $t > 0$

(f) $W(0) = \dfrac{1}{2} \times 0.1 \times 20^2 + \dfrac{1}{2} \times 0.4 \times 10^2 = \underline{40\ J}$; $W_R = \int_0^\infty Ri^2(t)dt = 48\int_0^\infty 900e^{-1200t}dt = \underline{36\ J}$

$w_1(\infty) = \dfrac{1}{2} \times 0.1(-4)^2 = \underline{0.8\ J}$; $W_2(\infty) = \dfrac{1}{2} \times 0.4 \times 4^2 = \underline{3.2\ J}$; $36 + 0.8 + 3.2 = 40$ ✓

**4-15**

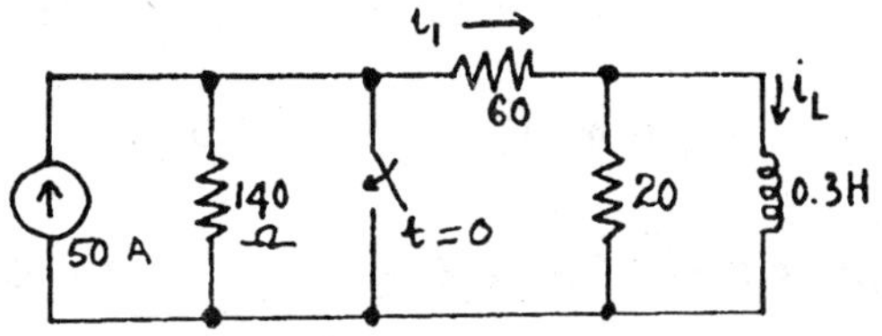


| t(ms) | $i_L$(A) | $i_1$(A) |
|---|---|---|
| -10 | 35 | 35 |
| $0^-$ | 35 | 35 |
| $0^+$ | 35 | 8.75 |
| 10 | 21.23 | 5.307 |
| 20 | 12.88 | 3.219 |

$i_L(0) = 50 \times \dfrac{140}{60 + 140} = 35\ A$; $R_{eq} = 15\ \Omega$

$\therefore i_L(t) = 35e^{-15t/0.3} = \underline{35e^{-50t}\ A}$, $t > 0$

$i_1(0^-) = 50 \times \dfrac{140}{60 + 140} = \underline{35\ A}$; $i_1(t) = 35e^{-50t} \times \dfrac{20}{60 + 20} = \underline{8.75e^{-50t}\ A}$, $t > 0$

**4-16**

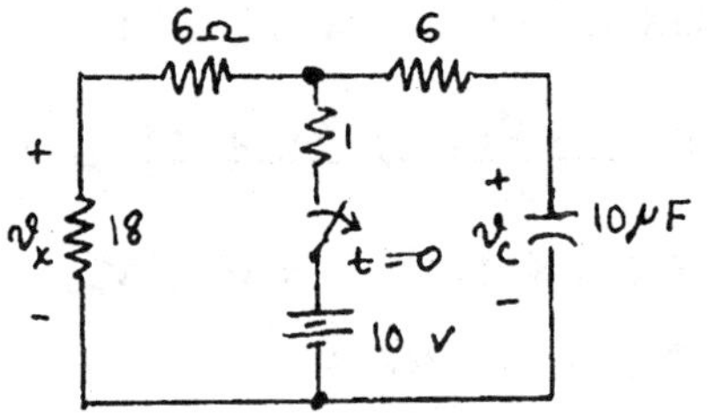


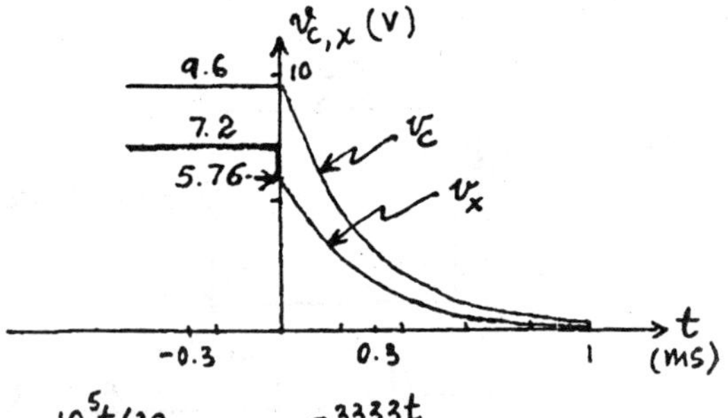


$v_c(0) = \dfrac{10}{6 + 18 + 1} \times 24 = 9.6\ V$, $\therefore v_c(t) = 9.6e^{-10^5 t/30} = 9.6e^{-3333t}\ V$, $t > 0$

$v_x(0^-) = 7.2\ V$ ; $v_x(0^+) = 9.6 \times \dfrac{18}{6 + 6 + 18} = 5.76\ V \therefore v_x(t) = 5.76e^{-3333t}\ V, t > 0$

4-17

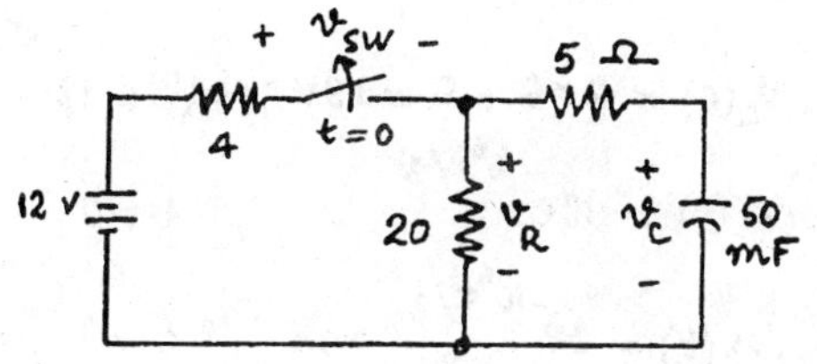

(a) $v_c(0) = 10$ V. $\therefore v_c(t) = 10e^{-0.8t}$ V , $t>0$

$v_c(1) = \underline{4.493 \text{ V}}$

(b) $v_R(t) = v_R(0^+)e^{-0.8t} = 8e^{-0.8t}$ $\therefore v_R(1) = \underline{3.595 \text{ V}}$

(c) $v_{sw}(1) = 12 - v_R(1) = \underline{8.405^+ \text{ V}}$

4-18

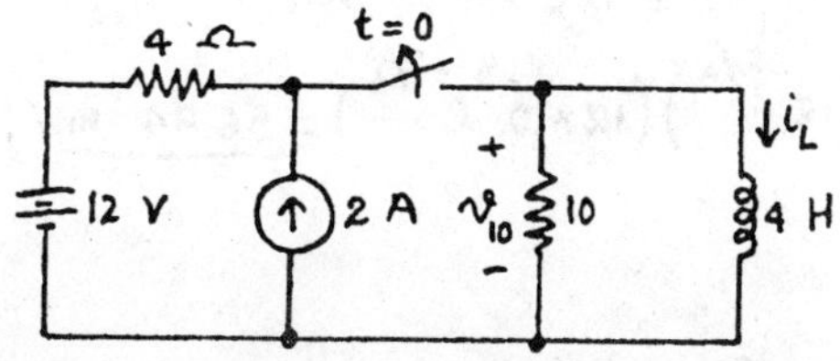

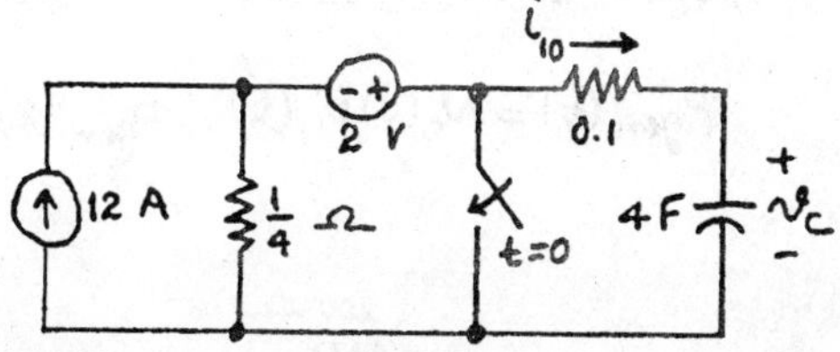

"After being open for a very long time, the switch in Fig 5-13 is closed at $t=0$. (a) Find $v_c(0^+)$ and $w_c(0^+)$. (b) Find $v_c(t)$ (c) Find $i_{10}(t)$."

From solution to Prob. 5-1 : (a) $v_c(0^+) = \underline{5 \text{ V}}$ ; $W_c(0^+) = \underline{50 \text{ J}}$ ; $v_c = \underline{5e^{-2.5t}}$ V, $t>0$

(c) $i_{10}(t) = \underline{-50e^{-2.5t} \text{ A}}$ , $t>0$

4-19

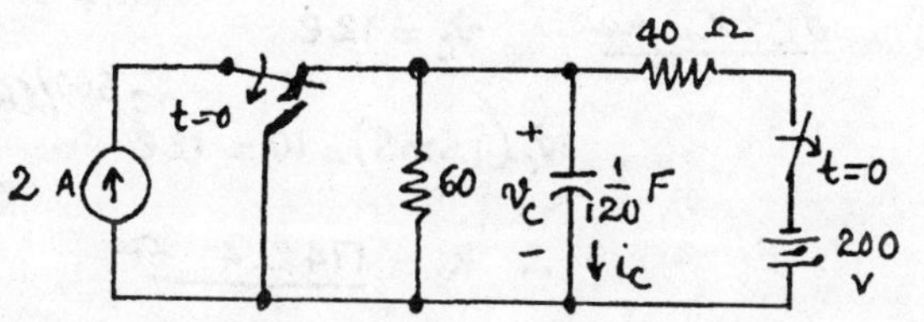

$v_c(0) = 2(40\|60) - 200 \times 0.6 = 48 - 120 = -72$ V

$\therefore v_c(t) = -72e^{-2t}$ V , $t>0$

$i_c(t) = Cv_c' = \frac{1}{120}(-72)(-2)e^{-2t} = 1.2e^{-2t}$ A , $t>0$

$q = \int_{0.3}^{1} i_c dt = 1.2\int_{0.3}^{1} e^{-2t} dt = \underline{0.2481 \text{ C}}$

4-20

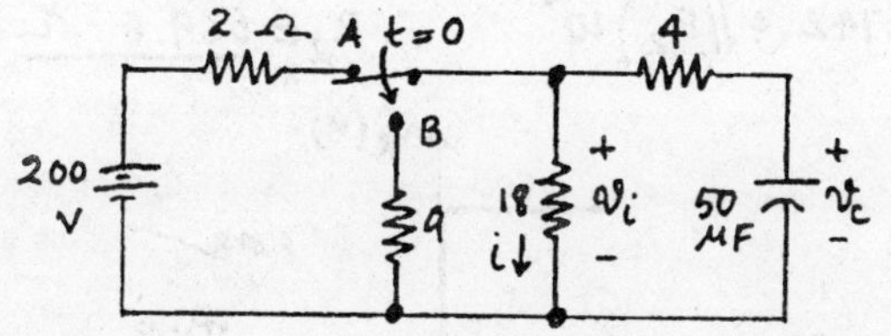

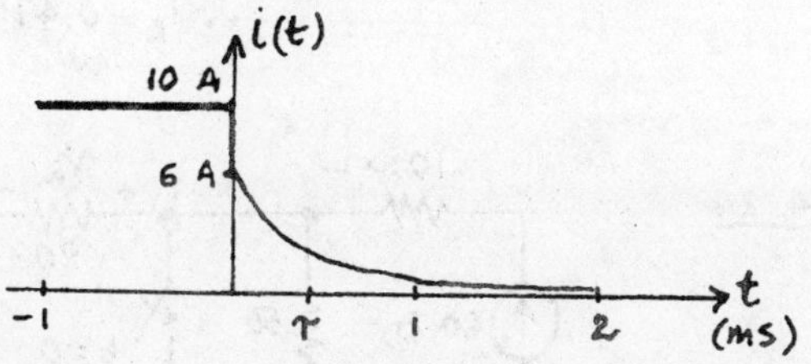

$v_c(0) = 200 \times \frac{18}{20} = 180$ V ; $R_{eq} = 4 + (9\|18) = 10\ \Omega$

$v_c(t) = 180e^{-t/10\times 50\times 10^{-6}} = 180e^{-2000t}$ V, $t>0$

$v_i(t) = \frac{6}{4+6}v_c = 108e^{-2000t}$ V, $t>0$ $\therefore i(t) = \underline{6e^{-2000t} \text{ A}}$, $t>0$ ; $i = \underline{10 \text{ A}}$, $t<0$

**4-21** (a)

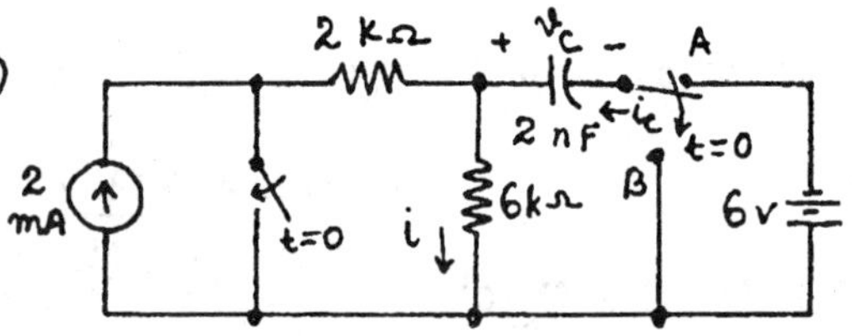


$v_c(0) = 2\times6+6 = 18\text{V}$ ; $v_c(t) = 18e^{-t/1.5\times10^3\times2\times10^{-9}}$

$v_c(t) = 18e^{-10^6t/3}$ V , $t>0$

$\therefore i(t) = 3e^{-10^6t/3}$ mA , $t>0$ ; $i = 2$ mA , $t<0$

$\therefore i(2\mu s) = 3e^{-2/3} = \underline{1.5403 \text{ mA}}$ and $i(-2\mu s) = \underline{2 \text{ mA}}$

(b) $i_c(t) = -Cv_c'(t) = -2\times10^{-9}\left(-\frac{18}{3}\times10^6\right)e^{-10^6t/3} = 12e^{-10^6t/3}$ mA

$P_{gen}(t) = v_c(t)i_c(t) \therefore P_{gen}(2\mu s) = \left(18^{-2/3}\right)\left(12\times10^{-3}e^{-2/3}\right) = \underline{56.94 \text{ mW}}$; $P_{gen}(-2\mu s) = \underline{0}$

**4-22**

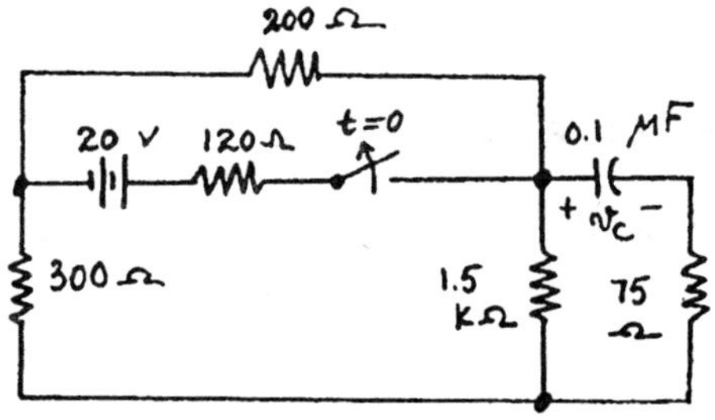


(a) $v_c(0) = \frac{20}{120+(1800\|200)} \times \frac{200}{1800+200} \times 1500$

$v_c(0) = \underline{10 \text{ V}}$

(b) $\tau = \left[75+(500\|1500)\right]10^{-7} = \underline{45\ \mu s}$

(c) $v_c(t) = \underline{10e^{-22,222t} \text{ V}}$ , $t>0$

**4-23**

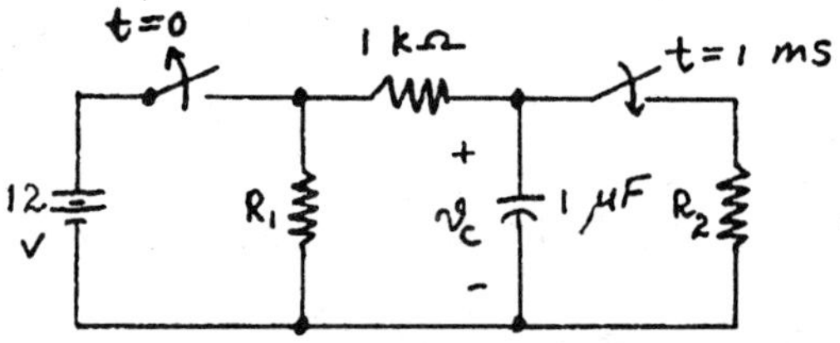


$v_c(0.5\text{mS}) = 10\text{V}$ ; $v_c(2\text{mS}) = 1$ V

$\underline{0<t<1 \text{ ms}}$ $v_c = 12e^{-10^6t/(R_1+1000)}$ V

$v_c(0.5\text{mS}) = 10 = 12e^{-500/(R_1+1000)}$

$\therefore R_1 = \underline{1742.4\ \Omega}$

also, $v_c(1 \text{ ms}) = 12e^{-1000/2742.4} = 8.333$ V

$\underline{t>1 \text{ ms}}$ $v_c = 8.333e^{-(t-0.001)/\tau_2}$ ; $v_c(2 \text{ ms}) = 1 = 8.333e^{-0.001/\tau_2}$

$\therefore \tau_2 = 0.4716 \text{ mS} = (2742.4\|R_2)10^{-6}$ $\therefore R_2 = \underline{569.6\ \Omega}$

**4-24**

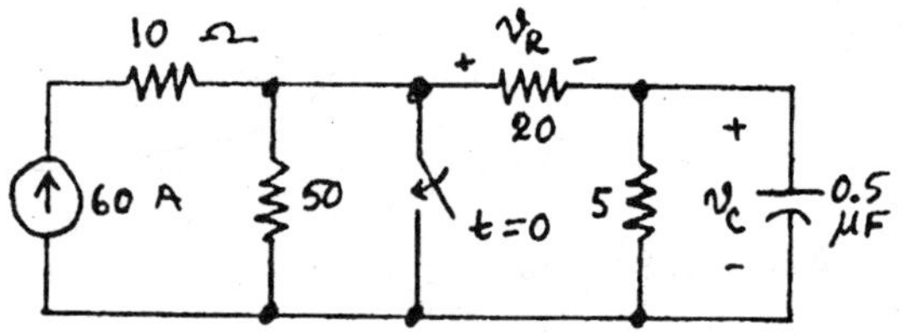


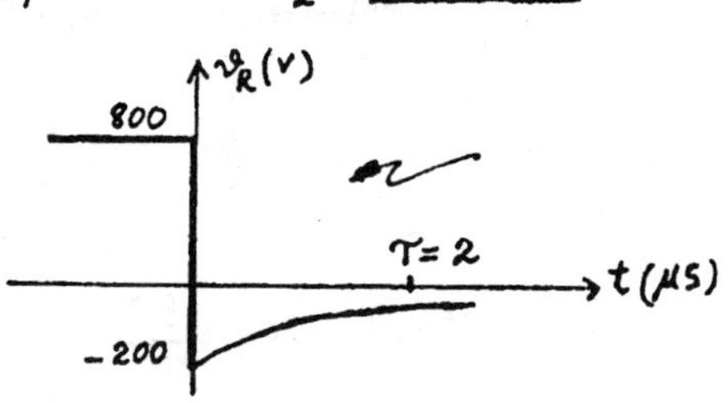


$v_c(0) = 60\times\frac{50}{25+50}\times5 = 200$ V , $v_c(t) = 200e^{-t/4\times0.5\times10^{-6}} = 200e^{-500,000t}$ V , $t>0$

$v_R(t) = -v_c = -200e^{-500,000t}$ V, $t>0$ and $v_R = 60\times\frac{50}{75}\times20 = 800$ V, $t<0$

**4-25** (a)

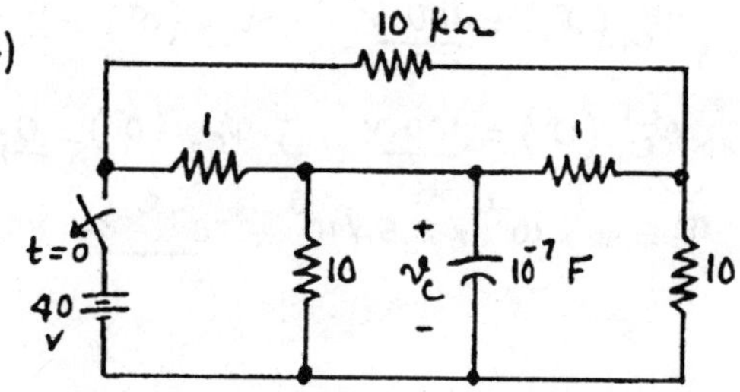


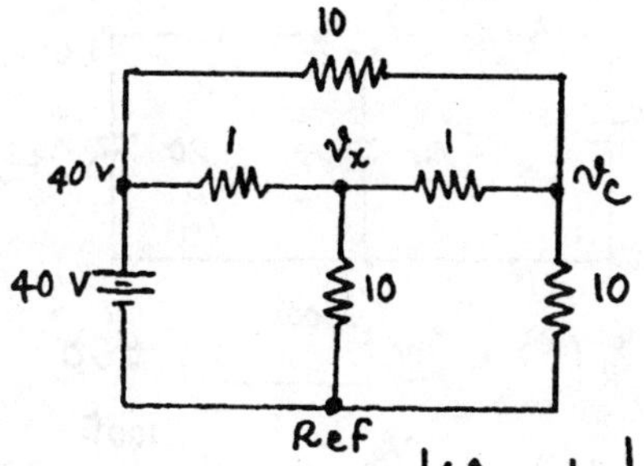


$$v_c - 40 + 0.1v_c + v_c - v_x = 0 \qquad \therefore 2.1\,v_c - v_x = 40$$

$$v_x - v_c + 0.1v_x + 0.1v_x - 0.1\times 40 = 0 \;\therefore -v_c + 1.2v_x = 4$$

$$v_c = \frac{\begin{vmatrix} 40 & -1 \\ 4 & 1.2 \end{vmatrix}}{\begin{vmatrix} 2.1 & -1 \\ -1 & 1.2 \end{vmatrix}} = \underline{34.2105\text{ V}}, \quad t<0$$

(b)

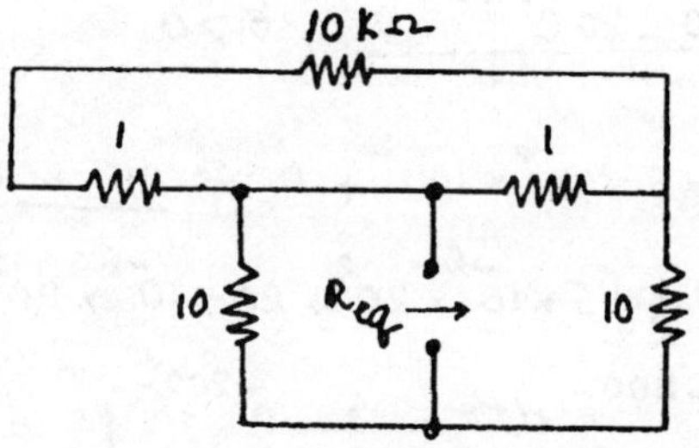


$$R_{eq} = (10 + 1\|1)\|10 = 5.2191\text{ k}\Omega$$

$$\therefore v_c = 34.2105e^{-10^4 t/5.2191} = \underline{34.2105e^{-1916.0t}\text{ V}}, \quad t>0$$

**4-26**

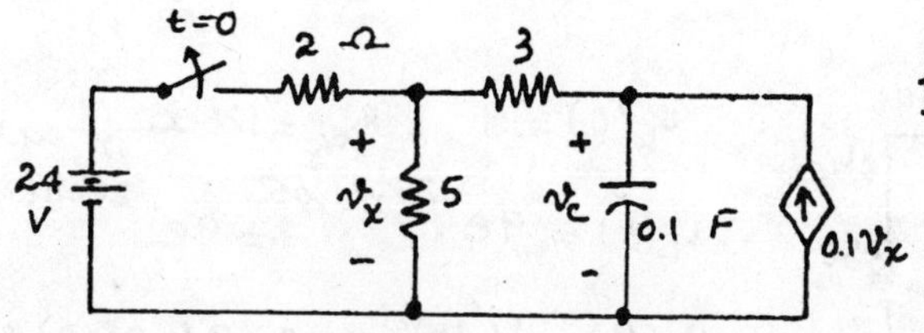


t<0

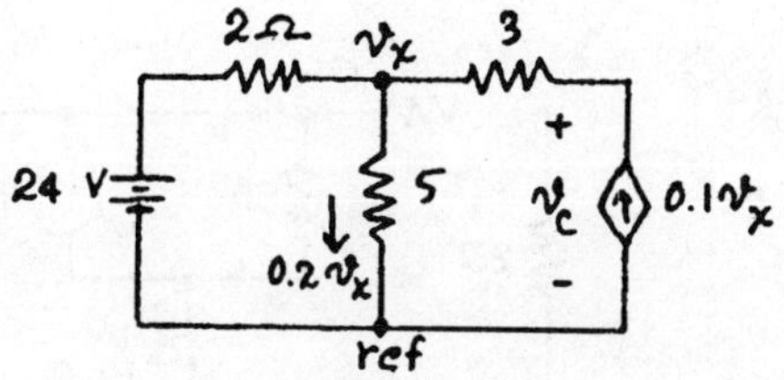


$$24 = 2(0.2v_x - 0.1v_x) + v_x$$

$$\therefore v_x = \underline{20\text{ V}},\ t<0$$

$$v_c = v_x + 3(0.1v_x) = 26\text{ V}$$

t>0

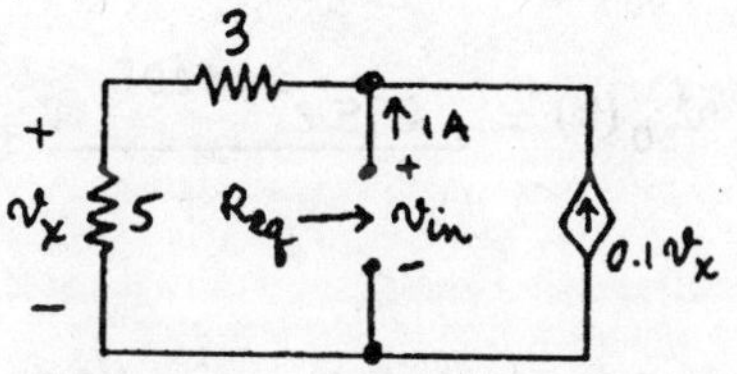


$$v_c(0) = 26\text{ V}$$

$$v_x = (1 + 0.1v_x)5 \quad \therefore v_x = 10\text{ V}$$

$$v_{in} = (1 + 0.1v_x)8 \quad \therefore v_{in} = 16\text{ V}$$

$$\therefore R_{eq} = \frac{16}{1} = 16\ \Omega$$

$$\therefore v_c = 26e^{-10t/16}\text{ V}$$

$$v_x = \frac{5}{8}v_c = \underline{16.25e^{-0.625t}\text{ V}} \quad t>0$$

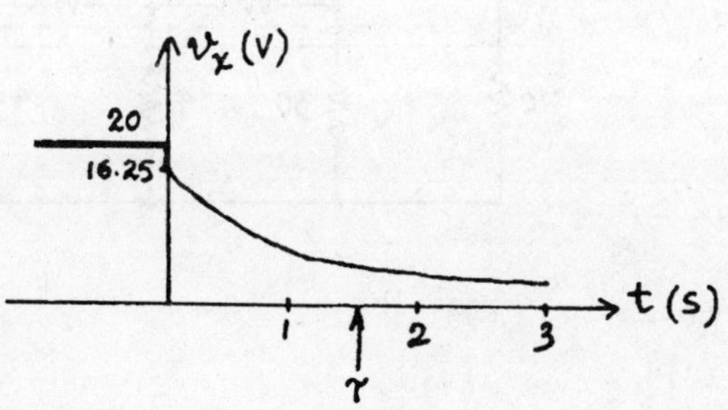

4-27

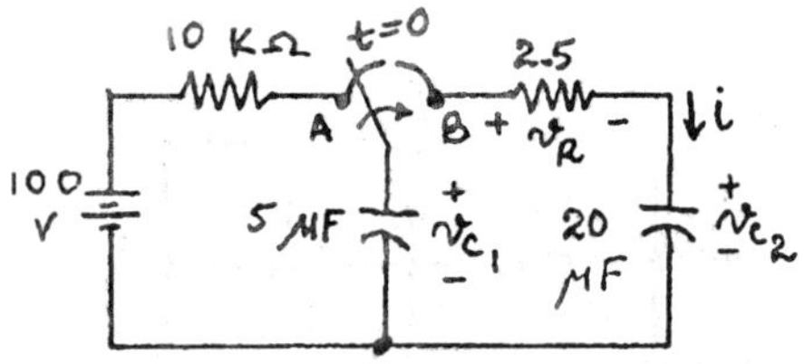


(a) $v_{c_1}(0^-) = \underline{100\ V}$ ; $v_{c2}(0^-) = \underline{0}$

(b) $v_{c_1}(0^+) = \underline{100\ V}$ ; $v_{c_2}(0^-) = \underline{0}$; $v_R(0^+) = \underline{100\ V}$

(c) $\tau = 4\times 10^{-6} \times 2.5\times 10^{3} = \underline{10^{-2}\ s}$

(d) $v_R(t) = \underline{100 e^{-100t}\ v}$, $t>0$

(e) $i(t) = \dfrac{v_R}{2500} = \underline{40 e^{-100t}\ mA}$, $t>0$

(f) $v_{c_1}(t) = v_{c_1}(0) + \dfrac{1}{C}\int_0^t i\,dt = 100 - \dfrac{10^6}{5}\int_0^t 0.04 e^{-100t}\,dt = \underline{20 + 80 e^{-100t}\ V}$ , $t>0$

$v_{c_2}(t) = 0 + \dfrac{10^6}{20}\int_0^t 0.04 e^{-100t}\,dt = \underline{20 - 20 e^{-100t}\ V}$ , $t>0$

(g) $w_c(0^+) = \frac{1}{2} C_1 v_{c_1}^2(0^+) + \frac{1}{2} C_2 v_{c_2}^2(0^+) = \frac{1}{2}\times 5\times 10^{-6}\times 100^2 + 0 = \underline{25\ mJ}$

$W_c(\infty) = \frac{1}{2}\left[C_1 v_{c_1}^2(\infty) + C_2 v_{c2}^2(\infty)\right] = \frac{1}{2}\left[5\times 10^{-6}\times 20^2 + 20\times 10^{-6}\times 20^2\right] = \underline{5\ mJ}$

$W_R = \int_0^\infty R i^2(t)\,dt = 2500\int_0^\infty 40^2\times 10^{-6} e^{-200t}\,dt = \frac{4}{-200} e^{-200t}\Big]_0^\infty = \underline{20\ mJ}$

4-28

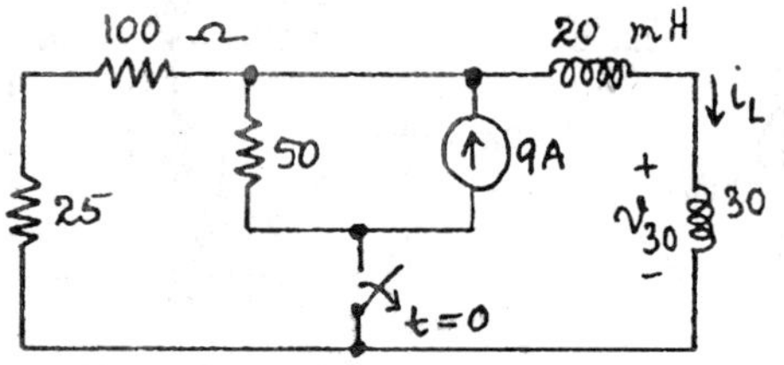


$i_L(0) = 9$ ; $R_{eq} = 125\ \Omega$ ; $L_{eq} = 50\ mH$

$\therefore i_L(t) = 9 e^{-125t/0.05} = 9 e^{-2500t}\ A$, $t>0$

$v_{30}(t) = L i_L'(t) = 0.03\times 9(-2500) e^{-2500t}$

$v_{30}(t) = \underline{-675 e^{-2500t}\ V}$, $t>0$

4-29

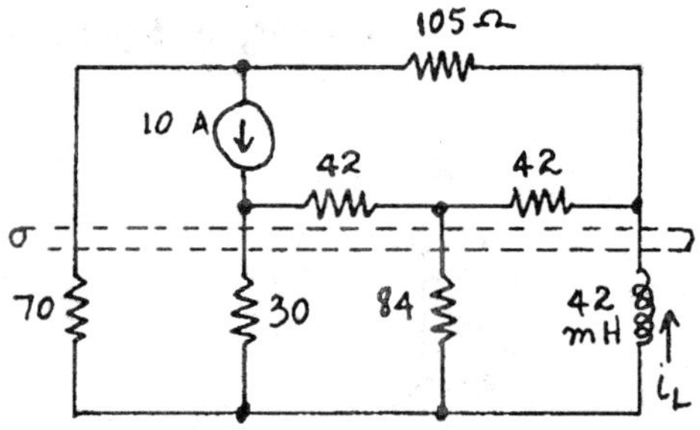


$\underline{t<0}$ $L \to$ S.C. ; $42 \| 84 = 28\ \Omega$

$28 + 42 = 70\ \Omega$, $70\|30 = 21\ \Omega$

$10\times 21 = 210\ V$; $\dfrac{210}{70} = 3\ A$

$3\times\dfrac{84}{84+42} = 2\ A$; $10\times\dfrac{70}{70+105} = 4\ A$

$\therefore i_L(0) = 4 - 2 = 2\ A$

$R_{eq} = 70\|30\|84 = 16.8\ \Omega$

$\therefore i_L(t) = 2 e^{-16.8t/0.042} = \underline{2 e^{-400t}}\ A$, $t>0$

**4-30** (a)

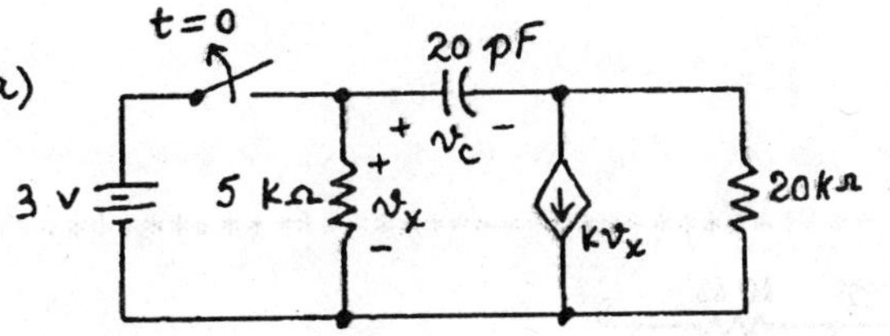


$\underline{t<0}$ $\quad v_x(0^-) = 3\text{ V}$

$\therefore v_c(0^-) = \underline{3 + 60{,}000K}\text{ V}$

(b) $\underline{t>0}$

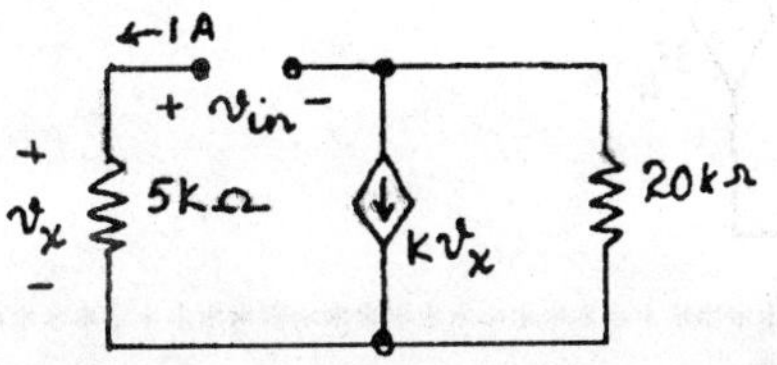


$v_x = 1 \times 5000 = 5000\text{ V}$

$\therefore v_{in} = 5000 + (1 + 5000K)2\times10^4\text{ V}$

$v_{in} = 25{,}000 + 10^8K = R_{eq}$

$v_c(t) = (3 + 60{,}000K)e^{-10^{12}t/20(25000+10^8K)} = \underline{(3 + 60{,}000K)e^{-10^6t/(0.5+2000K)}}\text{ V}$

(c) $\underline{K=0}$ $\quad v_c(t) = \underline{3e^{-2\times10^6t}\text{ V}}$ $(t>0)$

$\underline{K = 10^{-3}}$ $\quad v_c(t) = \underline{63e^{-400{,}000t}\text{ V}}$ $(t>0)$

**4-31**

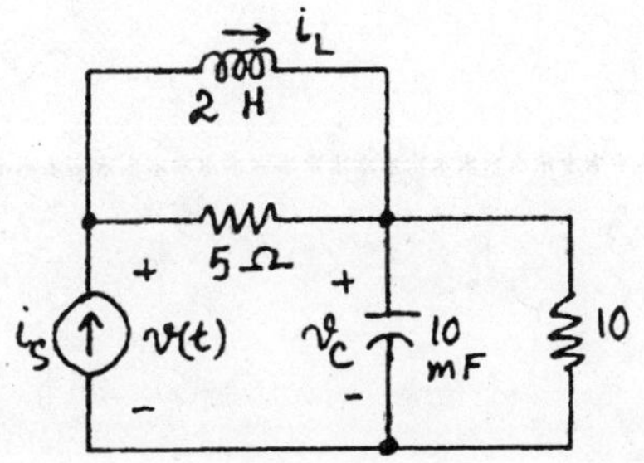


$i_s = 50\text{ A},\ t<0\ ;\ i_s = 0,\ t>0$

$i_L(0) = 50\text{ A}\ ;\ v_c(0) = 500\text{ V}$

$\therefore i_L(t) = 50e^{-\frac{5}{2}t} = 50e^{-2.5t}\text{ A},\ t>0$

$v_c(t) = 500e^{-t/10\times10\times10^{-3}} = 500e^{-10t}\text{ V},\ t>0$

For $\underline{t<0}$ $\quad v(t) = \underline{500\text{ V}}$

For $\underline{t>0}$ $\quad v(t) = v_c - 5i_L = \underline{500e^{-10t} - 250e^{-2.5t}\text{ V}}$

# 4-32

***********************************************************************

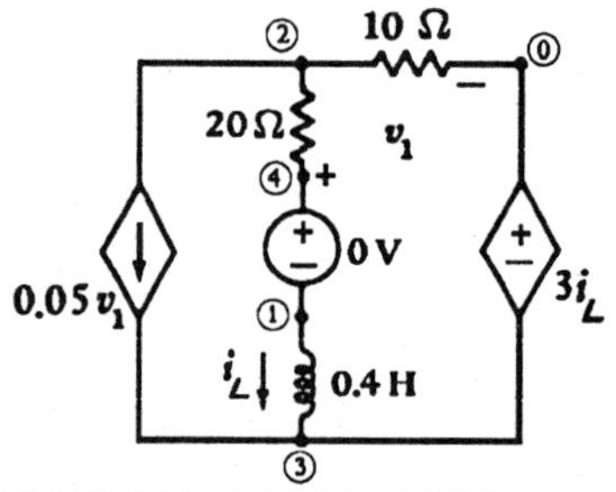


***********************************************************************

**PROGRAM**

```
R24  2  4  20
R20  2  0  10
L13  1  3  0.4  IC = 20
V41  4  1
G23  2  3  1  0  0.05
H03  0  3  V41  3
.TRAN  2M  26M  UIC
.PRINT TRAN I(V41)
.END
```

***********************************************************************

**ANSWERS: Source currents**

| TIME | I(V41) | |
|---|---|---|
| 0.000E+00 | 2.000E+01 | |
| 2.000E-03 | 1.837E+01 | |
| 4.000E-03 | 1.687E+01 | |
| 6.000E-03 | 1.550E+01 | |
| 8.000E-03 | 1.424E+01 | |
| 1.000E-02 | 1.308E+01 | |
| 1.200E-02 | 1.201E+01 | |
| 1.400E-02 | 1.103E+01 | |
| 1.600E-02 | 1.013E+01 | |
| 1.800E-02 | 9.307E+00 | |
| **2.000E-02** | **8.548E+00** | **A** = $i_L(0.02)$ |
| 2.200E-02 | 7.851E+00 | |
| 2.400E-02 | 7.212E+00 | |
| 2.600E-02 | 6.624E+00 | |

# CHAPTER 5: THE APPLICATION OF THE UNIT-STEP FORCING FUNCTION

## Section 5-2 The unit-step forcing function

**1** At $t = 5$ s, determine the value of: *(a)* $u(t - 6)$; *(b)* $u(6 - t)$; *(c)* $2 - u(t^2 - 10t + 24)$; *(d)* $(t - 2)u(t - 2)$; *(e)* $2^{(t-3)}(-1)^{[1+u(t-3)]}$; *(f)* $-tu(-t) + tu(t)$; *(g)* $i_x$ in Fig. 5-1.

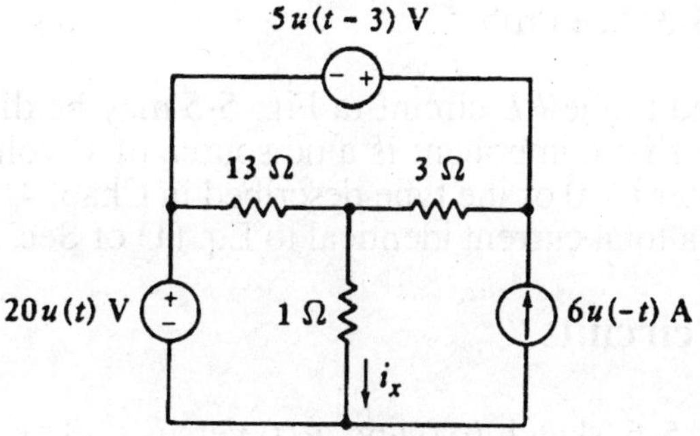


**Fig. 5-1:** See Prob. 1.

**2** Sketch the time function $f(t) = [u(t + 5) - u(t - 5)][u(\cos \frac{1}{2}\pi t) + u(\sin \pi t)]$.

**3** Find the value of each of the following time functions at $t = 1.5$ s: *(a)* $2u(t + 1) - 3u(t + 2) + 4u(t - 3)$; *(b)* $2u(1 - t) + 5[u(t)]^2$; *(c)* $4u(\cos 3t)$; *(d)* $4\cos[3u(t)]$. *(e)* A voltage is 10 V for $t < 0$, $2e^{-0.4t}$ V for $0 < t < 5$ s, and zero for $t > 5$ s. Express it as a single function of time using unit-step forcing functions.

**4** The "general network" in each part of Fig. 5-2 consists of a voltage source, $10u(2 - t)$ V, + reference connected to the upper terminal, in series with a 5-Ω resistor. If $V = 5$ V and $t_0 = -1$ s in the external circuits, sketch the current entering the network at the upper terminal in each case.

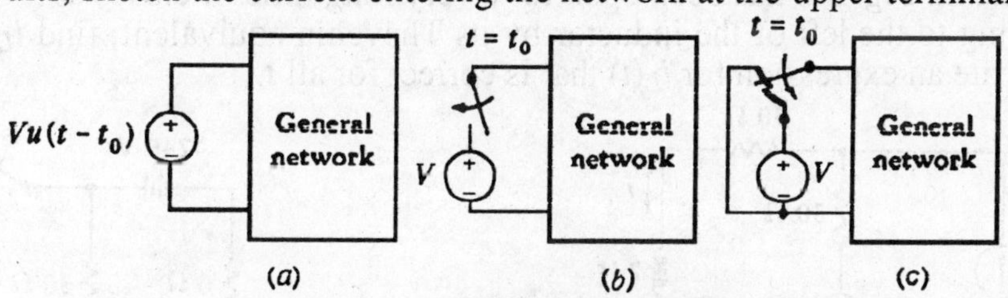


**Fig. 5-2:** See Prob. 4.

## Section 5-3 A first look at the driven *RL* circuit
## Section 5-4 The natural and the forced response

**5** Find $v_x$ in the network of Fig. 5-3 at 1-s intervals from $t = -0.5$ s to $t = 3.5$ s.

**6** The switch in Fig. 5-4 is at position *A* for $t < 0$. At $t = 0$ it moves to *B*, and then to *C* at $t = 4$ s and on to *D* at $t = 6$ s, where it remains. Sketch $v(t)$ as a function of time and express it as a sum of step forcing functions.

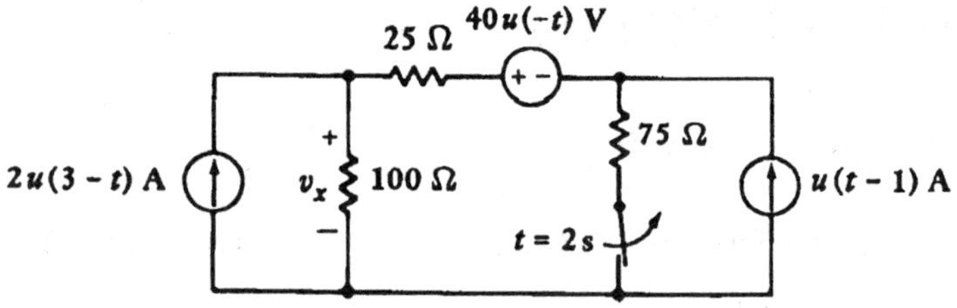


**Fig. 5-3:** See Prob. 5.

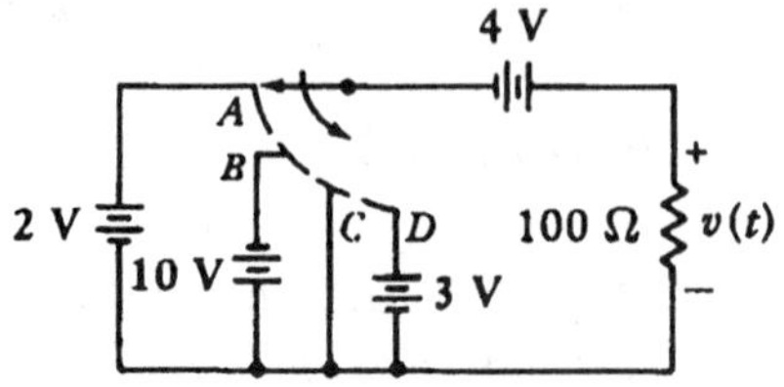


**Fig. 5-4:** See Prob. 6.

**7** The excitation applied to the *RL* circuit in Fig. 5-5 may be divided into two parts as follows: $Vu(t) = V - Vu(-t)$. The first component is a dc source of $V$ volts, while the second produces a source-free response after $t = 0$ of the type described in Chap. 4. Show that superposition of these two responses leads to a total current identical to Eq. (1) of Sec. 5-3.

## Section 5-5 *RL* circuits

**8** For the circuit of Fig. 5-6, sketch $i_L(t)$ and $v_L(t)$ versus $t$, $-3\tau < t < 3\tau$.

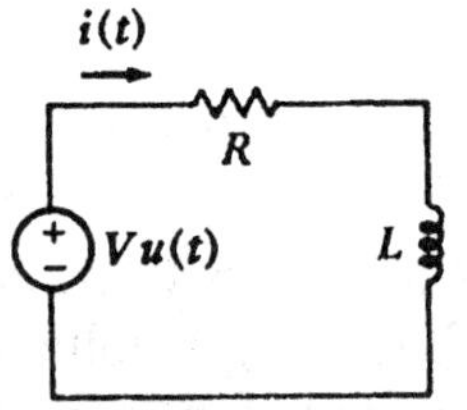

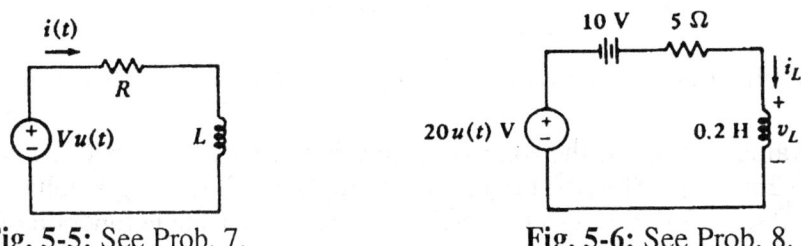

**Fig. 5-5:** See Prob. 7.

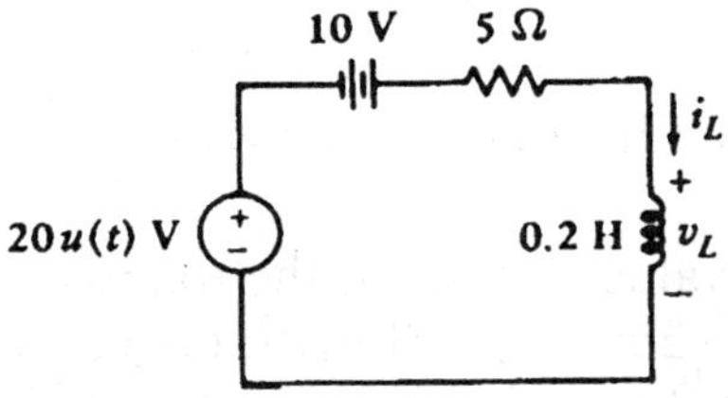


**Fig. 5-6:** See Prob. 8.

**9** Replace the network to the left of the inductor in Fig. 5-7 with its Thévenin equivalent and then determine $i(t)$ for $t > 0$.

**10** The switch in Fig. 5-8 has been open for a very long time. It closes at $t = 0$. (*a*) After replacing everything to the left of the inductor by its Thévenin equivalent, find $i_L(t)$ and sketch it for $t > 0$. (*b*) Write an expression for $i_L(t)$ that is correct for all $t$.

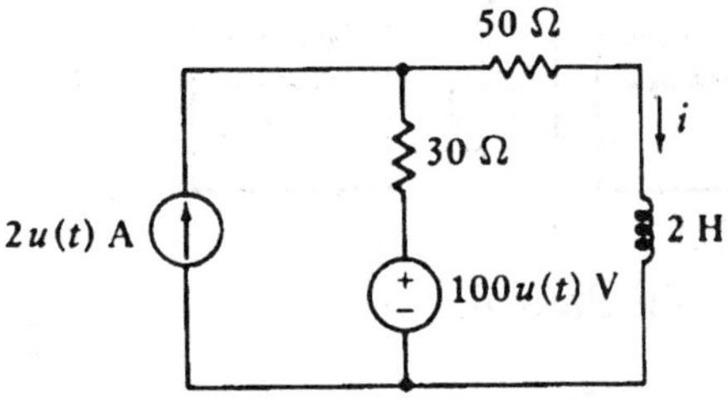

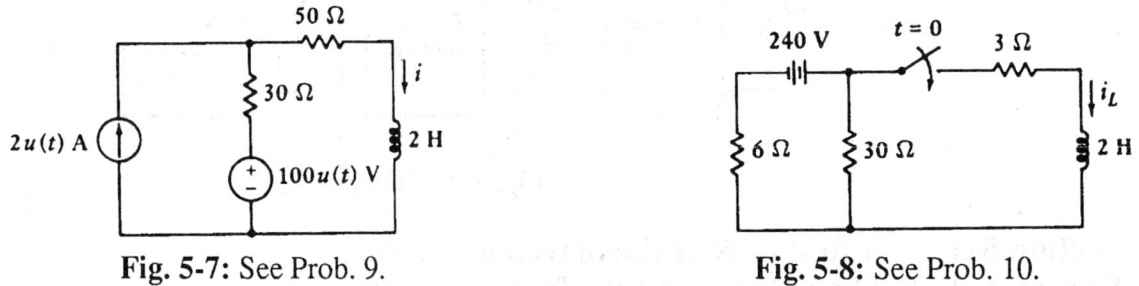

**Fig. 5-7:** See Prob. 9.

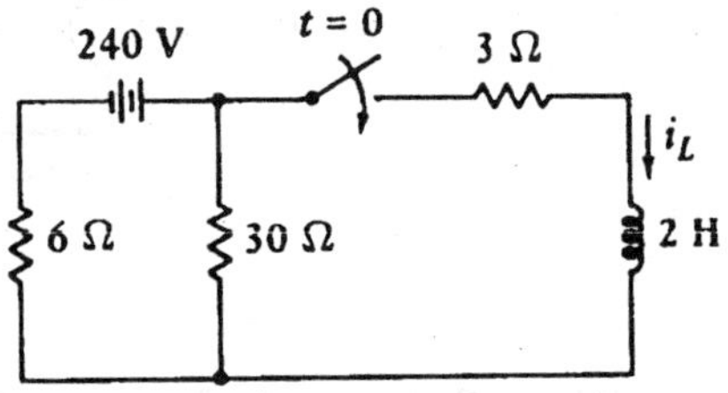


**Fig. 5-8:** See Prob. 10.

**11** Equation (3) of Sec. 5-4 represents the general solution of the driven *RL* series circuit, where $Q$ *is* a function of time and $A$ and $P$ are constants. Let $R = 200\ \Omega$ and $L = 5$ H, and find $i(t)$ if the voltage forcing function $LQ(t)$ is: (*a*) $5u(t)$ V; (*b*) $5 + 5u(t)$ V; (*c*) $5u(t) \sin 40t$ V.

**12** The switch in Fig. 5-9 has been closed for a long time. It is opened at $t = 0$. Find an expression for $i_L(t)$ for $t < 0$ and $t > 0$, and sketch it over the time interval, $-1 < t < 1$ s.

**13** The switch in Fig. 5-10 is thrown down at $t = 0$. Find: (*a*) $i$ at $t = 0.1$ s; (*b*) the maximum voltage magnitude that will appear across the inductor.

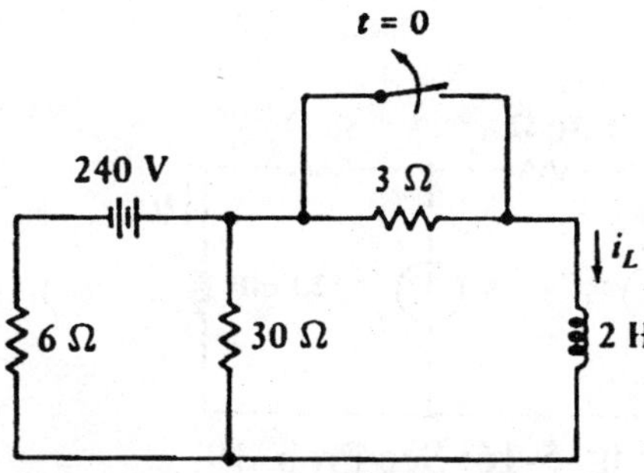


**Fig. 5-9:** See Prob. 12.

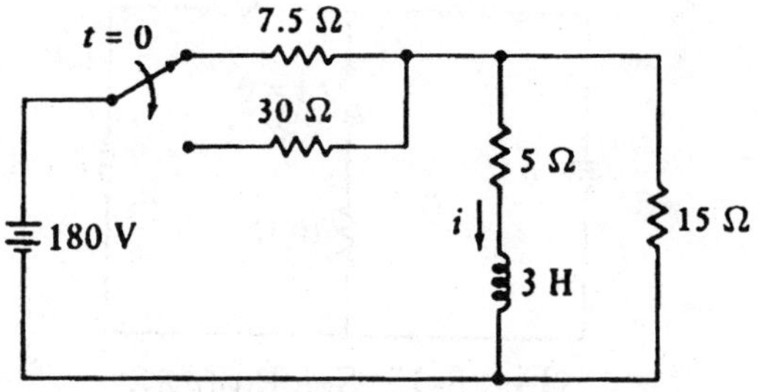


**Fig. 5-10:** See Prob. 13.

**14** Determine an expression for $i_L$ in Fig. 5-11 for all nonnegative $t$ and then make an accurate sketch of $i_L$ versus $t$, showing scale values on both axes.

**15** Find $i_L$(t) for $t > 0$ in the circuit shown in Fig. 5-12.

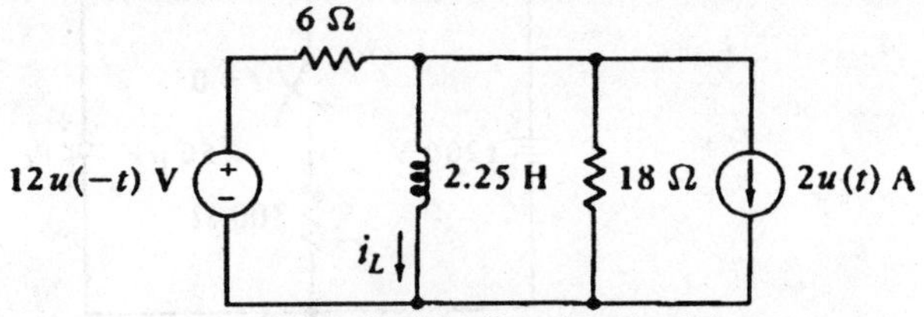


**Fig. 5-11:** See Prob. 14.

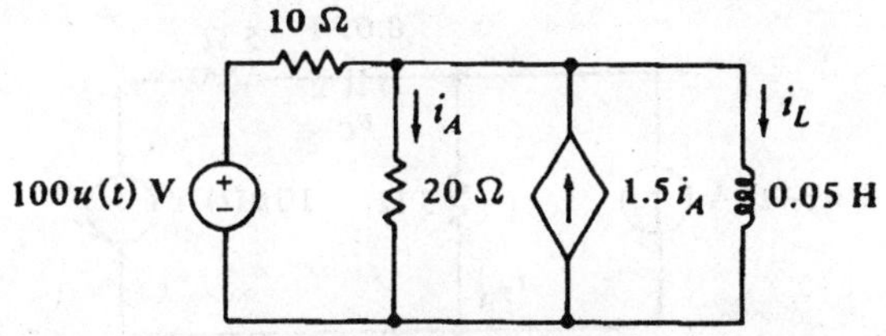


**Fig. 5-12:** See Prob. 15.

**16** If the switch in Fig. 5-13 is thrown down at $t = 0$, then: (*a*) at what value of $t$ will $i$ become zero, and (*b*) what will be the value of $v$ when $i = 0$?

**17** Calculate values for both $i_L$ and $v_R$, 16 s before and 16 s after the switch in Fig. 5-14 closes.

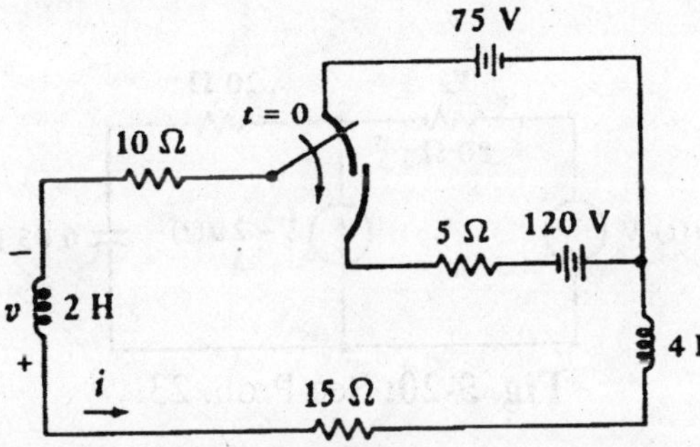

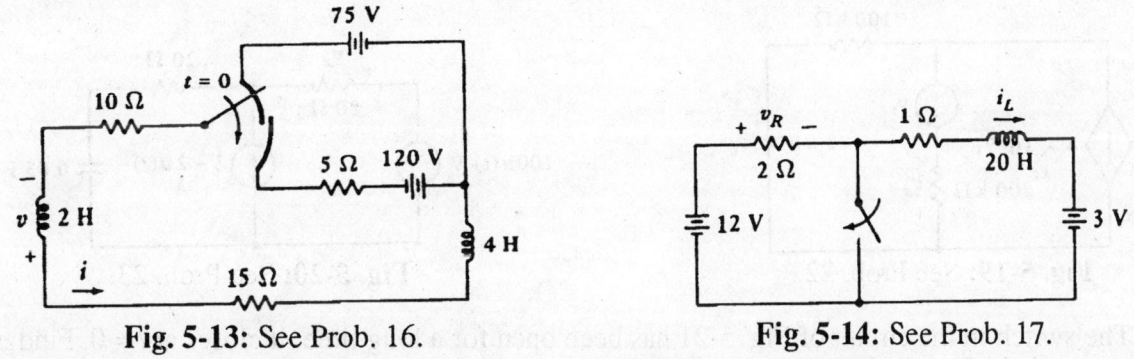

**Fig. 5-13:** See Prob. 16.

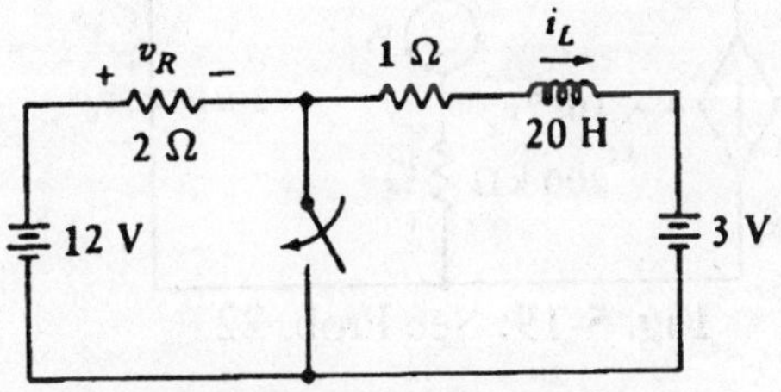


**Fig. 5-14:** See Prob. 17.

**18** Find $i_L$ in the circuit of Fig. 5-15 at $t =$ : (*a*) -5 ms; (*b*) 5 ms; (*c*) 10 ms.

**19** In Fig. 5-16, let $v_s = 2u(t)$ V and $i_s = 80u(t - 0.0002)$ mA. Find $i_L(t)$ for all $t$.

## Section 5-6 *RC* circuits

**20** (*a*) Determine an expression for $v_C(t)$ in the circuit shown in Fig. 5-17. Graph it as a function of $t$ for both negative and positive values of $t$. (*b*) Repeat for $i_5(t)$.

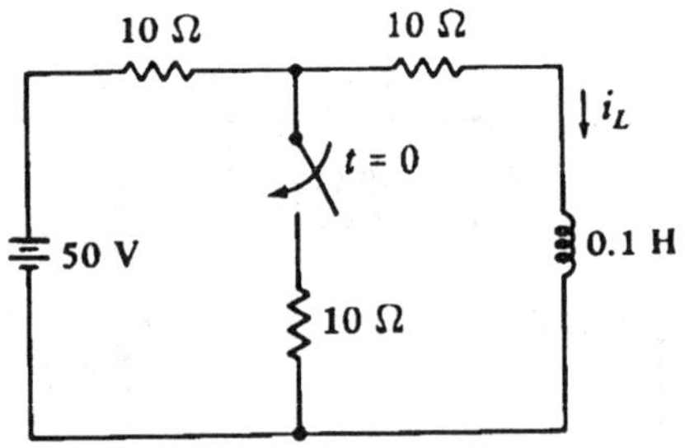


**Fig. 5-15:** See Prob. 18.

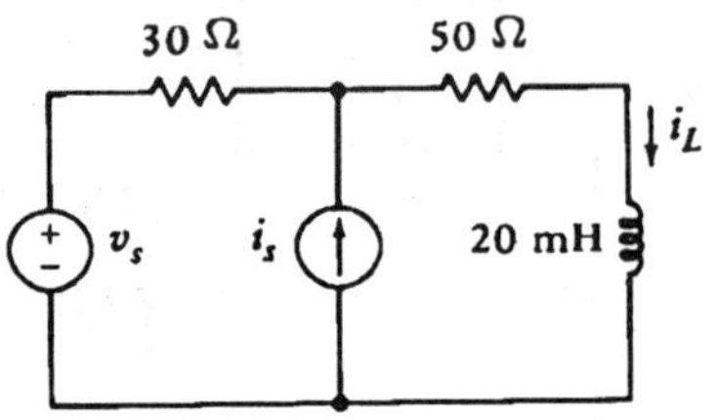


**Fig. 5-16:** See Prob. 19.

**21** For the circuit shown in Fig. 5-18, the switch has been open for a long time. (*a*) Calculate $v_C(-2 \text{ ms})$ and $v_C(2 \text{ ms})$. (*b*) If the switch is now kept closed for a long time, find $v_C$ two milliseconds after it suddenly opens.

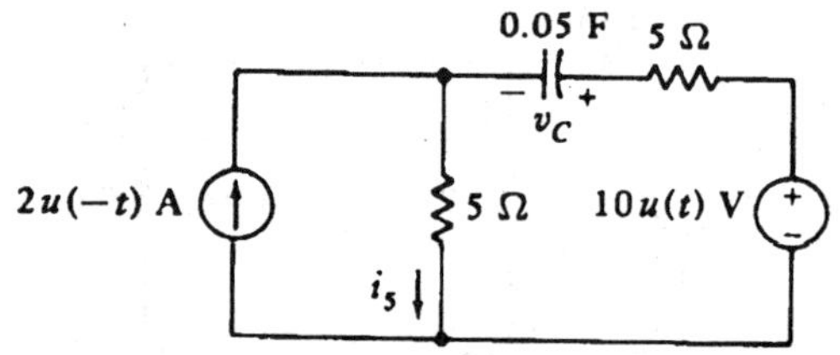


**Fig. 5-17:** See Prob. 20.

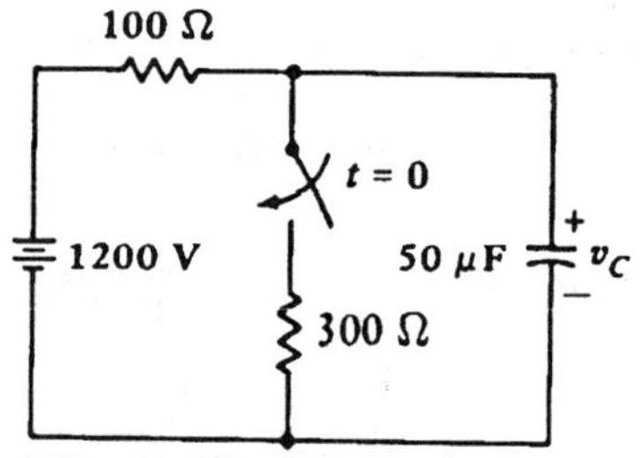


**Fig. 5-18:** See Prob. 21.

**22** Let $v_s = 9 - 12u(t)$ V for the circuit shown in Fig. 5-19. (*a*) Find $v_C(t)$ for all values of time. (*b*) Sketch $v_C$ versus *t*. (*c*) At what value of *t* is $v_C = 0$?

**23** In the circuit of Fig. 5-20, obtain a single expression for $v_x(t)$ that is valid for all *t*.

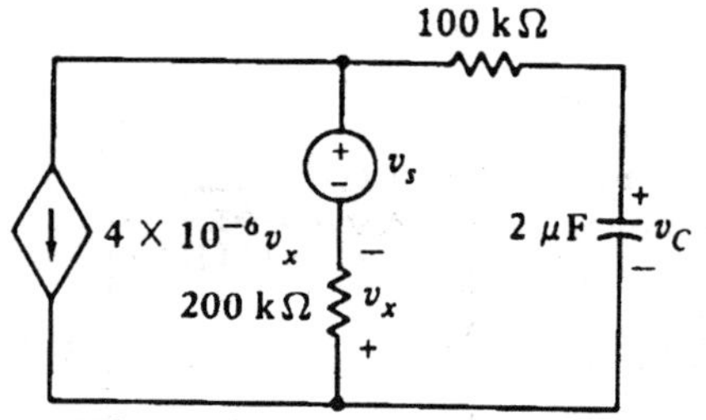


**Fig. 5-19:** See Prob. 22.

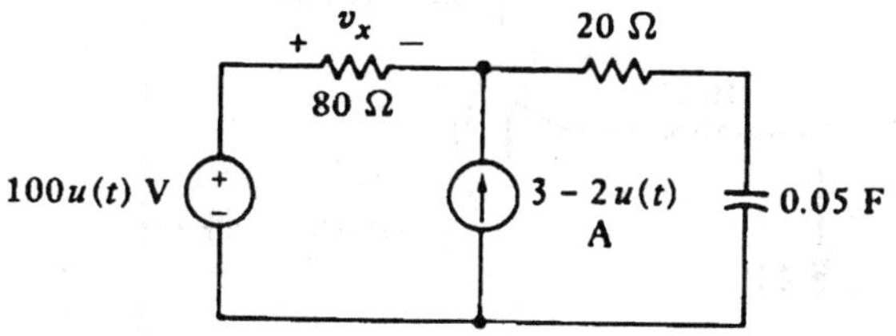


**Fig. 5-20:** See Prob. 23.

**24** The switch in the circuit of Fig. 5-21 has been open for a long time; it closes at $t = 0$. Find expressions valid for all *t* for: (*a*) $v_C(t)$; (*b*) $v_2(t)$.

**25** The switch in Fig. 5-22 has been at *A* for a long time. At $t = 0$ it is moved to *B*. Find $v_C(t)$ for $t < 0$ and $t > 0$.

**26** The switch in Fig. 5-22 has been at *B* for a long time. At $t = 0$ it is moved to *A*. Find $v_C(t)$ for $t < 0$ and $t > 0$.

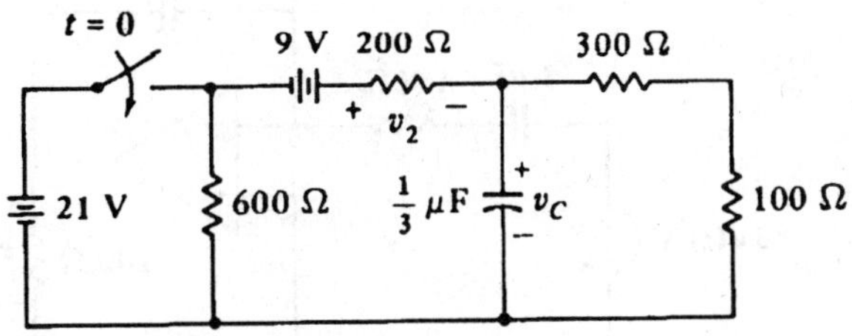


**Fig. 5-21:** See Prob. 24.

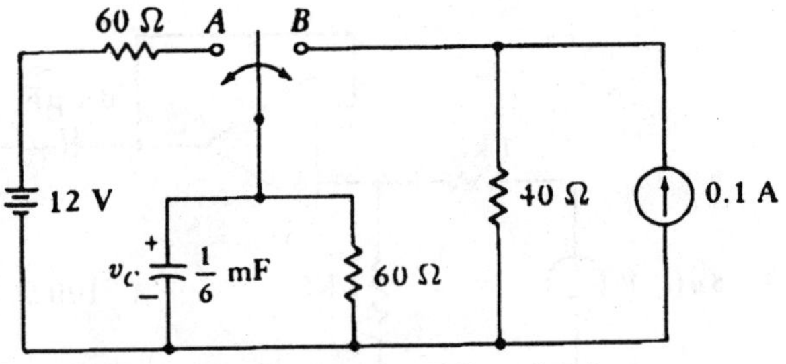


**Fig. 5-22:** See Probs. 25 and 26.

**27** The switch in Fig. 5-23 has been open for a long time. (*a*) Obtain an expression for $v_C(t)$ as a function of $t$, $t > 0$. (*b*) At what time is $v_C(t) = 0$?

**28** The current source, $5u(t)$ A, the parallel combination of 20 Ω and 0.8 H, and the parallel combination of 20 Ω and $C$ are all in series. Sketch the magnitude of the voltage across the current source as a function of time if $C =$: (*a*) 5 mF; (*b*) 2 mF.

**29** Find the instant of time at which the capacitor voltage in the circuit of Fig. 5-24 is equal to zero.

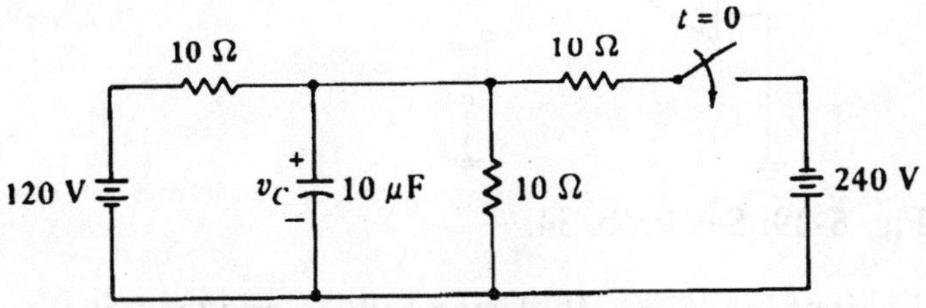


**Fig. 5-23:** See Prob. 27.

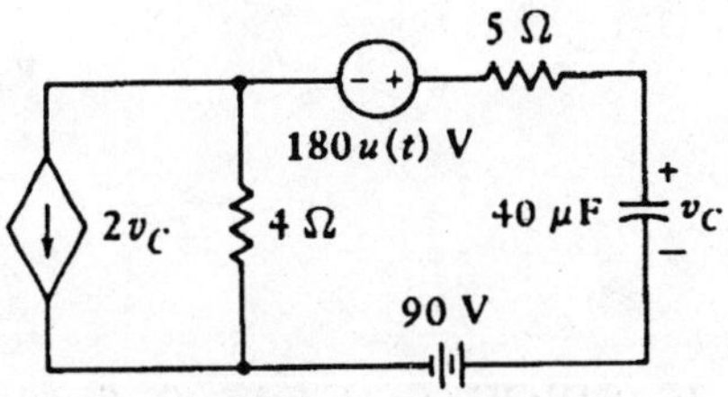


**Fig. 5-24:** See Prob. 29.

**30** The switch of Fig. 5-25 has been at $A$ for a long time. It is moved to $B$ at $t = 0$ and back to $A$ at $t = 1$ s. Find $R_1$ and $R_2$ so that $v_C = 7.5$ V at $t = 1$ and 1V at $t = 1.001$ s.

**31** If the op-amp in Fig. 5-26 is assumed to be ideal, find $v_o(t)$ for all $t$.

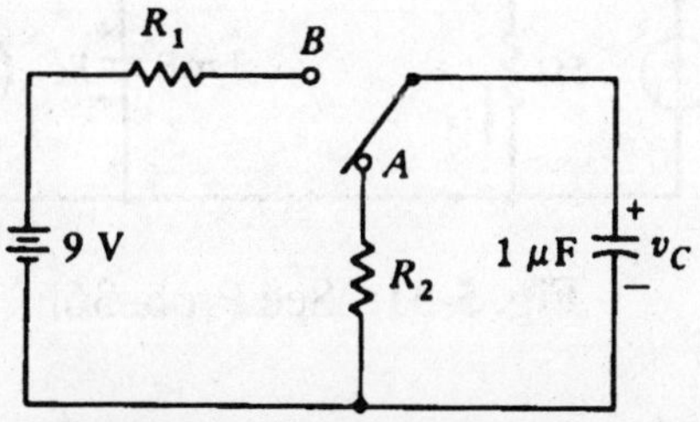

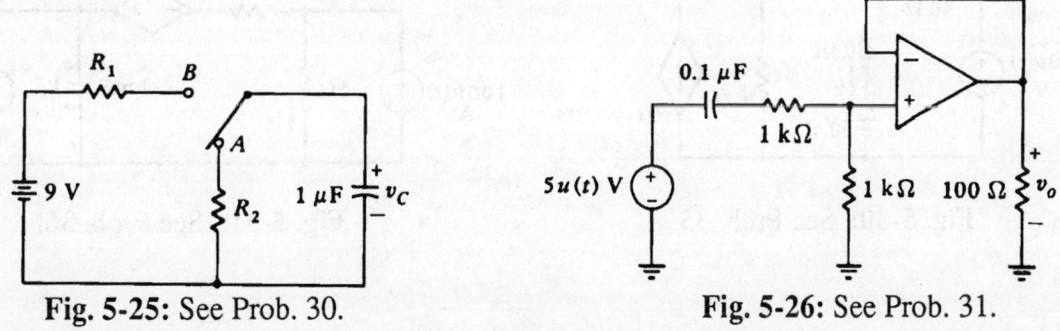

**Fig. 5-25:** See Prob. 30.

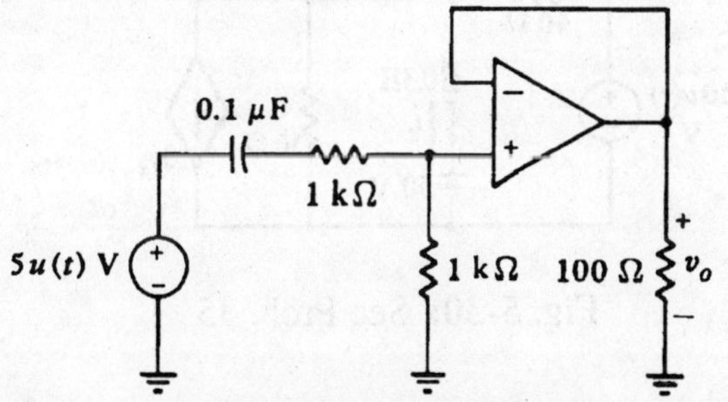


**Fig. 5-26:** See Prob. 31.

**32** The op-amp shown in Fig. 5-27 is assumed to be ideal. Find $v_x(t)$.

**33** If the op-amp shown in Fig. 5-28 is ideal, find an expression for $v_o(t)$ valid for all $t$.

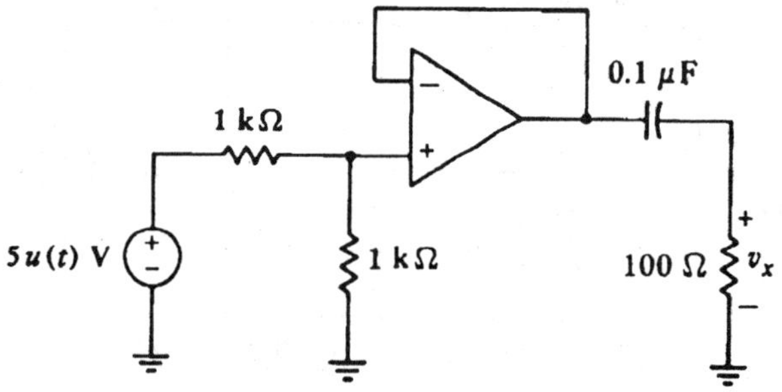


Fig. 5-27: See Prob. 32.

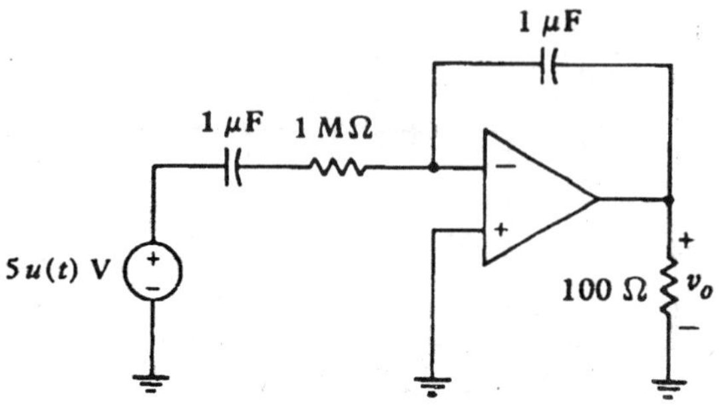


Fig. 5-28: See Prob. 33.

**34** Assume that the op-amp of Fig. 5-29 is ideal and find $v_{out}$ if $v_{in}(t) = u(t)$ V and $v_{out}(0) = 0$.

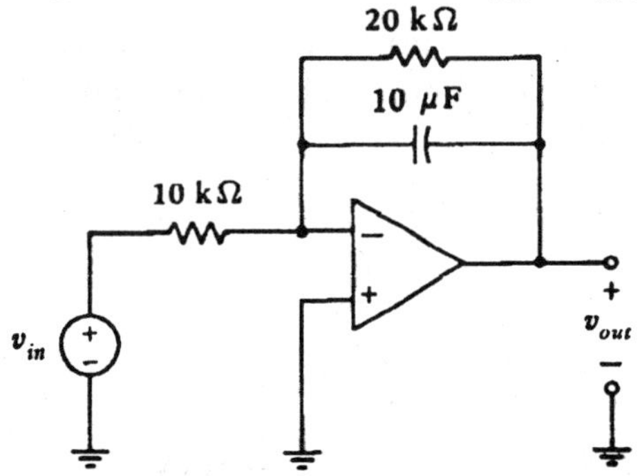


Fig. 5-29: See Prob. 34.

**35** (SPICE) (*a*) Find $i_L(0)$ for the *RL* circuit of Fig. 5-30. (*b*) Using SPICE, and the initial value found in part *a*, determine $i_L$ at $t = 80$ ms. Let TSTEP = 1.25 ms.

**36** (SPICE) (*a*) Find $v_C(0)$ for the *RC* circuit of Fig. 5-31. (*b*) Using SPICE and the initial value found in part *a*, determine $v_C$ at $t = 40$ ms. Let TSTEP = 1.25 ms.

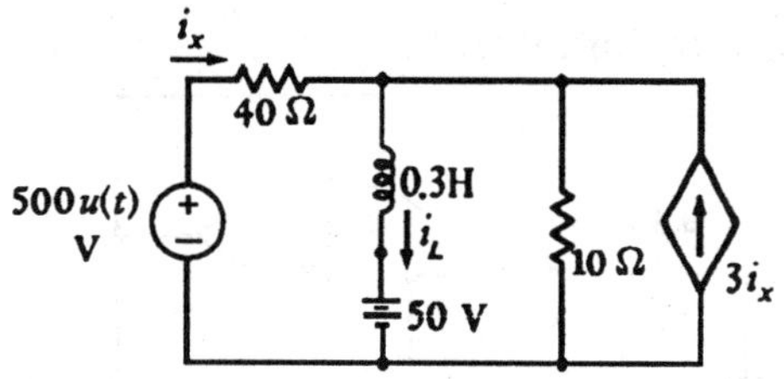


Fig. 5-30: See Prob. 35.

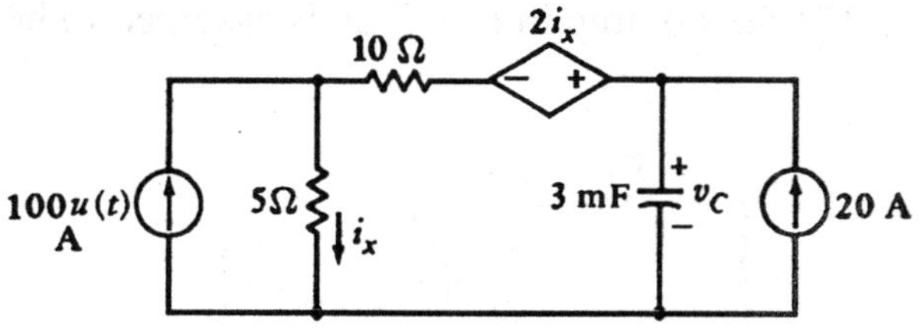


Fig. 5-31: See Prob. 36.

# Chapter 5

**5-1** (a) $u(5-6) = \underline{0}$  (b) $u(6-5) = \underline{\underline{1}}$

(c) $2 - u(25-50+24) = \underline{2}$  (d) $(5-2)\,u(5-2) = \underline{3}$

(e) $2^{5-3}(-1)^{1+u(5-3)} = 4(-1)^2 = \underline{4}$  (f) $-5u(-5) + 5u(5) = \underline{5}$

(g)

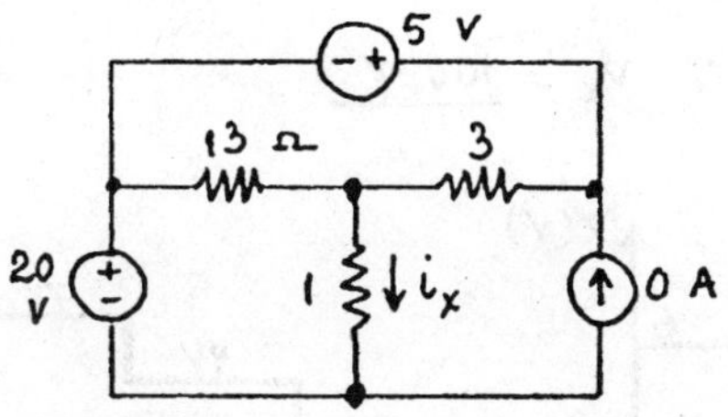


$$\frac{20}{1+39/16} + \frac{5}{3+13/14} \times \frac{13}{14} = \underline{7\ \text{A}}$$

**5-2** $f(t) = \left[u(t+5) - u(t-5)\right]\left[u\left(\cos\frac{\pi t}{2}\right) + u(\sin \pi t)\right]$

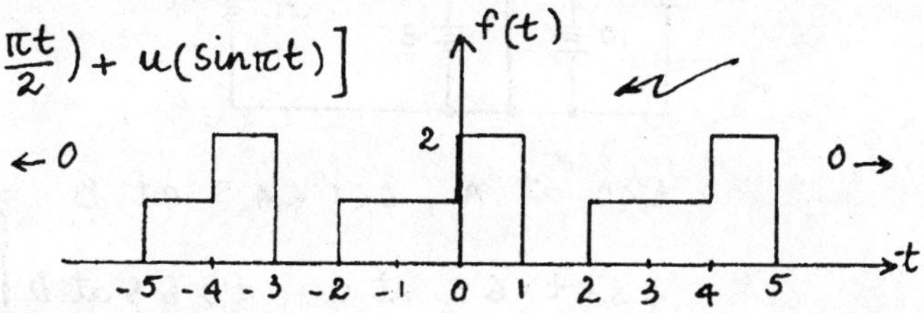


**5-3** (a) $2u(2.5) - 3u(3.5) + 4u(-1.5) = 2 - 3 - 0 = \underline{-1}$

(b) $2u(-0.5) + 5[u(1.5)]^2 = 0 + 5 = \underline{5}$

(c) $4u(\cos 4.5) = 4u(-0.2108) = \underline{0}$  (d) $4\cos[3u(1.5)] = 4\cos 3 = \underline{-3.960}$

(e) $\underline{10u(-t) + 2e^{-0.4t}}\,[u(t) - u(t-5)] = \underline{10 + (2e^{-0.4t} - 10)\,u(t) - 2e^{-0.4t}\,u(t-5)}$

**5-4**

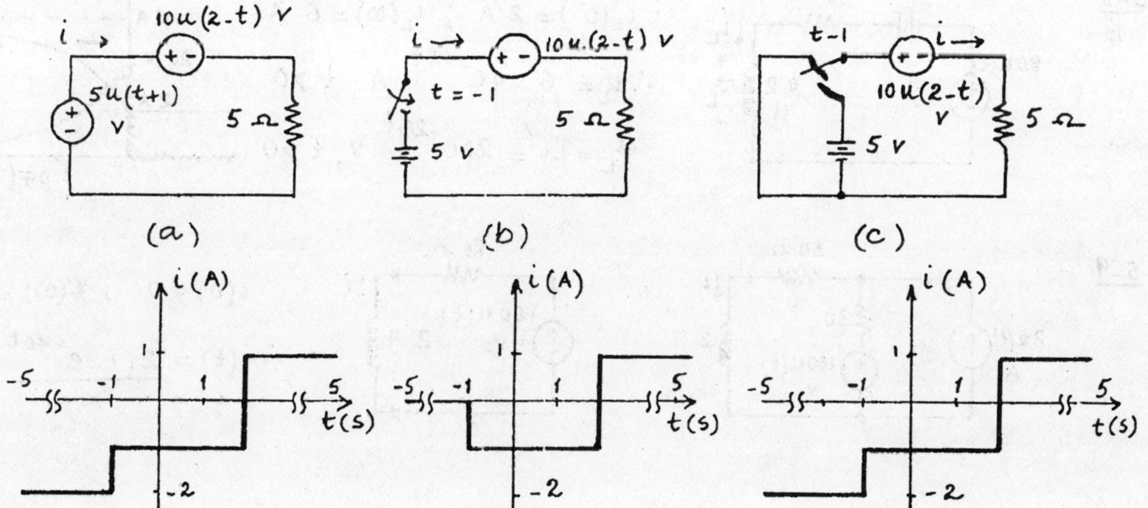

5.5

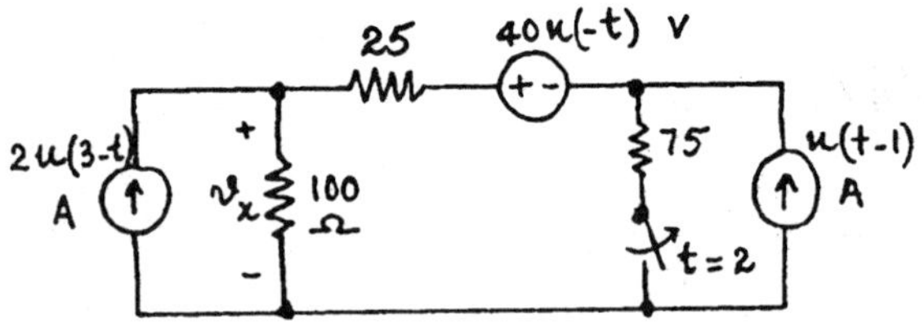


$t = -0.5\ s$ : 2 A , 40 V on

$\therefore v_x = 2 \times \frac{1}{2} \times 100 + 40 \times \frac{1}{2} = \underline{120\ V}$

$t = 0.5\ s$ : 2 A on $\therefore v_x = \underline{100\ V}$

$t = 1.5\ s$ : 2 A , 1 A on $\therefore v_x = 100 + 37.5 = \underline{137.5\ V}$

$t = 2.5\ s$ : 2 A , 1 A on , switch open $\therefore v_x = 200 + 100 = \underline{300\ V}$

$t = 3.5$ : 1 A on , switch open $\therefore v_x = \underline{100\ V}$

5.6

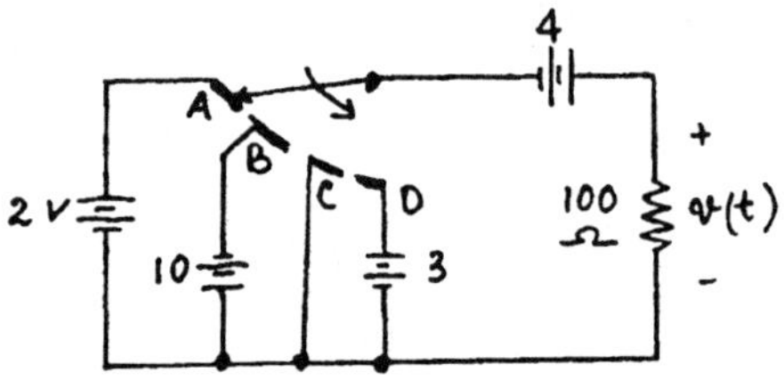


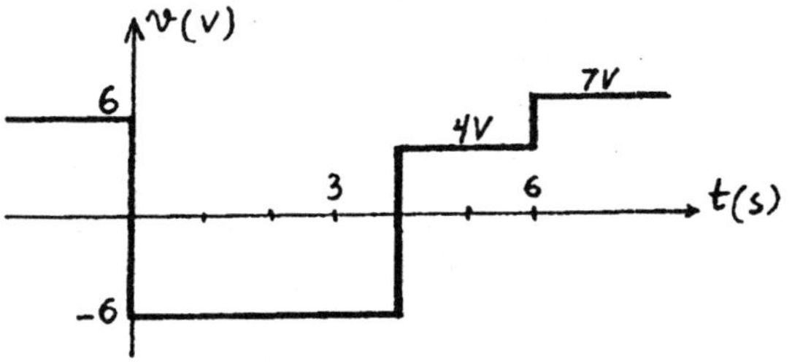


t<0 at A , 0<t<4 s at B

4 s<t<6 s at C , t>6 s at D

$v(t) = \underline{6 - 12u(t) + 10u(t-4) + 3u(t-6)}$

5.7

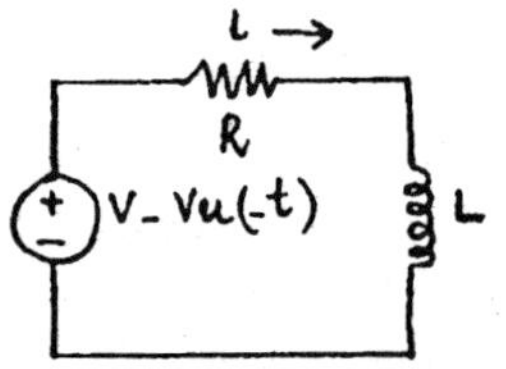


From V : $i_1 = \frac{V}{R}$ ; From $-Vu(-t)$; $i_2 = \frac{-Vu(-t)}{R} - \frac{Vu(t)}{R} e^{-Rt/L}$

$\therefore i = i_1 + i_2 = \frac{V}{R}\left[1 - u(-t) - u(t)e^{-Rt/L}\right]$

or $i = \frac{V}{R}\left(1 - e^{-Rt/L}\right)u(t)$ ✓

5.8

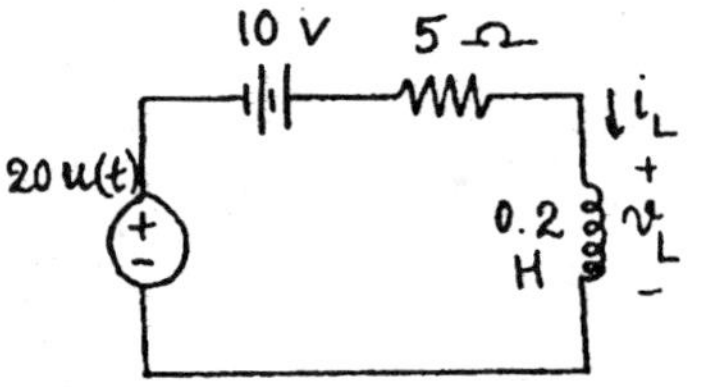


$i_L(0^-) = 2\ A$ ; $i_L(\infty) = 6\ A$

$\therefore i_L = 6 - 4e^{-25t}$ A , t>0

$v_L = Li' = 20e^{-25t}$ V, t>0

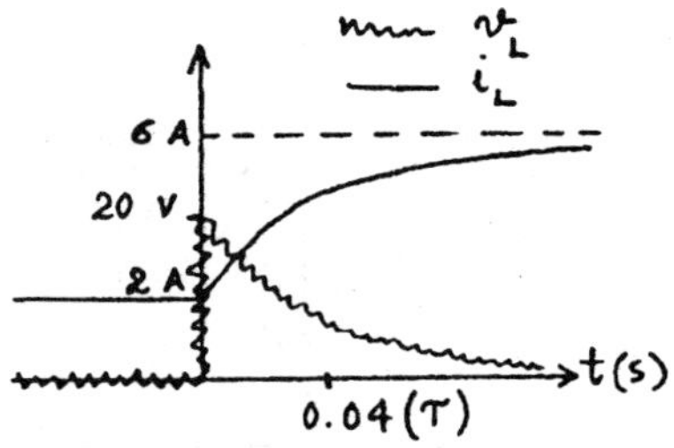


5.9

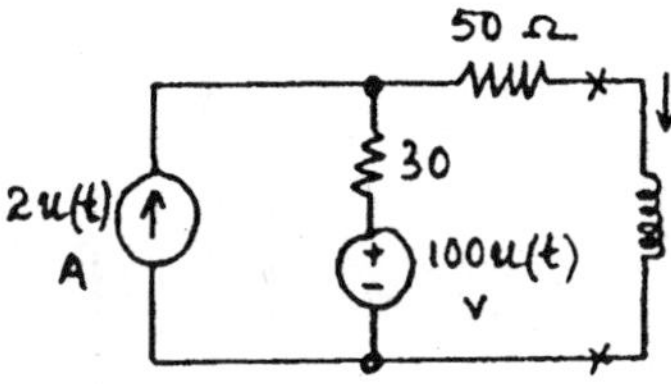


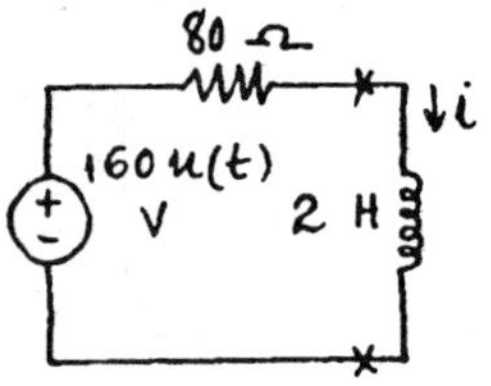


$i(0) = 0$ , $i(\infty) = 2$ A

$\therefore i(t) = \underline{2(1 - e^{-40t})}$ A , t>0

5-10

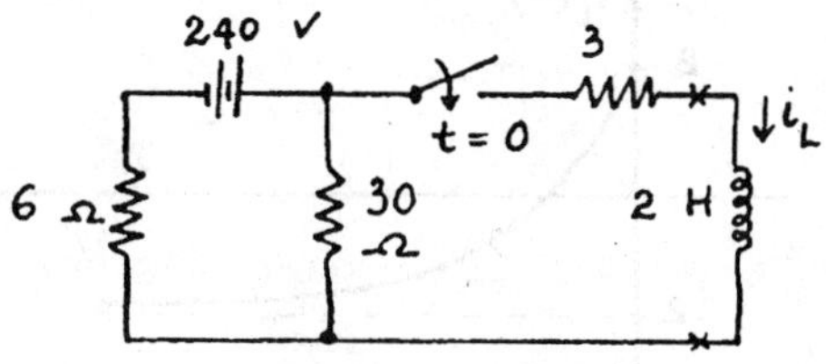


(a)

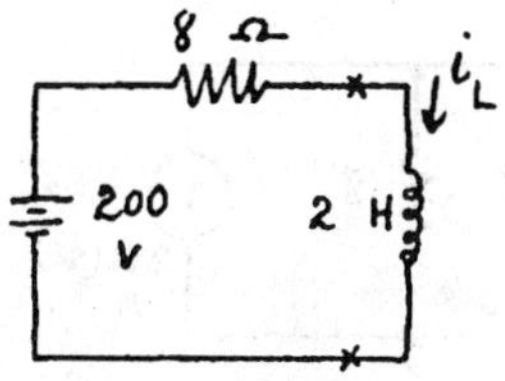


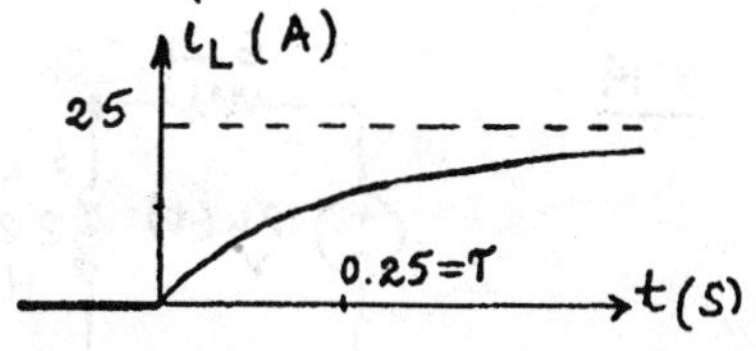


$i_L = \underline{25(1 - e^{-4t})\ \ A}, \ t > 0$

(b) $i_L(t) = \underline{25(1 - e^{-4t})u(t)\ \ A}$

5-11 (a) $i = e^{-Pt}\int Q e^{Pt}\,dt + Ae^{-Pt}$ ; $Lq(t) = 5u(t)$ V $\therefore i = e^{-40t}\int u(t)e^{40t}\,dt + Ae^{-40t}$

$\therefore i(t) = e^{-40t}\left(\frac{1}{40}\right)e^{40t} + Ae^{-40t}, \ t > 0$ ; $i(0) = 0$ $\therefore i(t) = \underline{\frac{1}{40}(1 - e^{-40t})\ A}, \ t > 0$

(b) $Lq(t) = 5 + 5u(t)$ V $\therefore i = e^{-40t}\int [1 + u(t)]e^{40t}\,dt + Ae^{-40t}$

$\underline{t < 0}$ : $i = e^{-40t}\int e^{40t}\,dt + Ae^{-40t} = \frac{1}{40} + Ae^{-40t}$ ; $i(0) = \frac{1}{40}$ $\therefore i = \underline{\frac{1}{40}\ A}$

$\underline{t > 0}$ : $i = e^{-40t}\int 2e^{40t}\,dt + Ae^{-40t} = \frac{2}{40} + Ae^{-40t}$ ; $i(0) = \frac{1}{40}$ $\therefore i = \underline{\frac{2}{40} - \frac{1}{40}e^{-40t}\ A}$

(c) $Lq(t) = 5u(t)\sin 40t$ V $\therefore i = e^{-40t}\int \sin 40t\, e^{40t}\,dt + Ae^{-40t}, \ t > 0$

$i = e^{-40t}\left[\frac{1}{3200}e^{40t}(40\sin 40t - 40\cos 40t)\right] + Ae^{-40t}, \ t > 0$ ; $i(0) = 0$

$\therefore i = \frac{1}{80}(\sin 40t - \cos 40t) + \frac{1}{80}e^{-40t}$ A, $t > 0$ ←

5-12

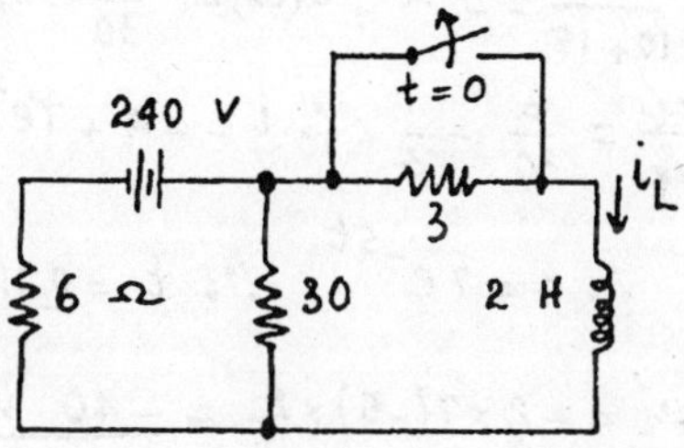


$i_L(0) = \frac{240}{6} = 40$ A

$i_L(\infty) = \frac{240}{6 + 3\|30} \times \frac{30}{33} = 25$ A

$\tau = \frac{L}{R_{eq}} = \frac{2}{3 + 5} = \frac{1}{4}$

$\therefore i_L = \underline{40\ A}, \ t < 0$

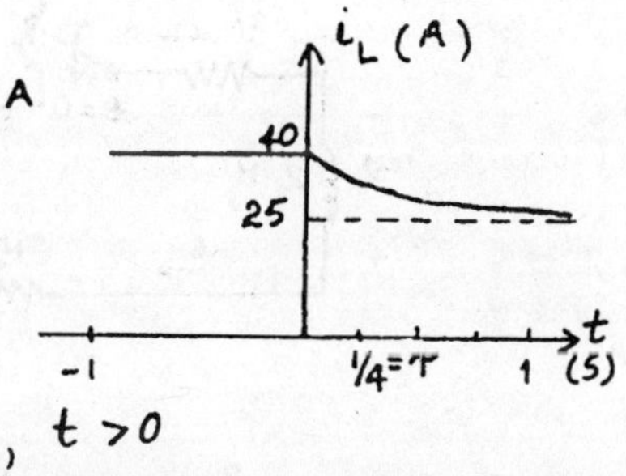


$i_L = \underline{25 + 15e^{-4t}\ \ A}, \ t > 0$

5-13

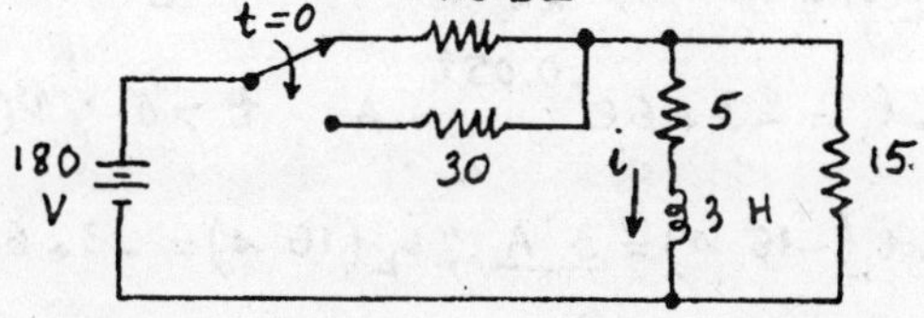


(a) $i(0) = \frac{180}{7.5 + 5\|15} \times \frac{3}{4} = 12$ A

$i(\infty) = \frac{180}{30 + 5\|15} \times \frac{3}{4} = 4$ A

$\tau = \frac{3}{5 + 15\|30} = 0.2\,s$ $\therefore i(t) = 4 + 8e^{-5t}$ A, $t > 0$ $\therefore i(0.1\ s) = \underline{8.852\ A}$

(b) $\overset{+}{\underset{-}{}}v_L = Li' = 3(-5)8e^{-5t} = -120e^{-5t}$ $\therefore |v_L|_{max} = \underline{120\ V}$

5-14

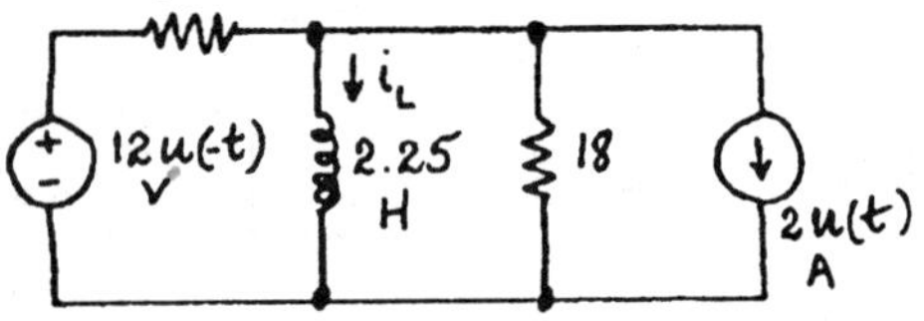


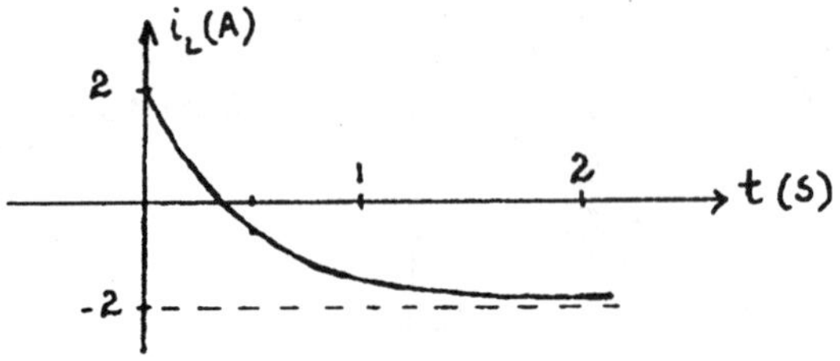


$i_L(0) = 2$ A ; $i_L(\infty) = -2$ A ; $\tau = \dfrac{2.25}{6 \| 18} = 0.5$ $\therefore\ i_L = \underline{-2 + 4e^{-2t}\ \text{A}}$, $t > 0$

5-15

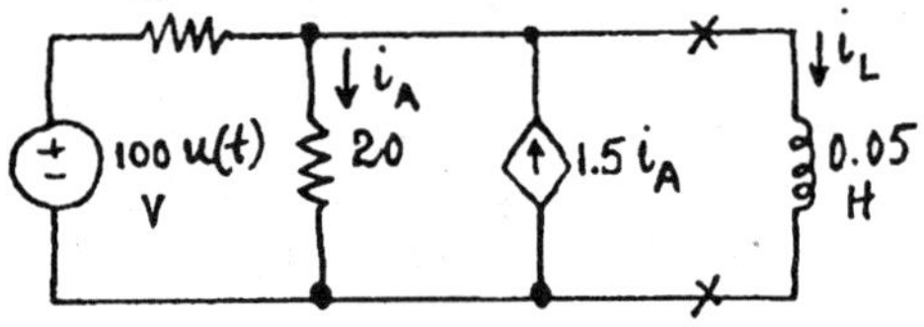


At x - x

SC : $i_A = 0$ $\therefore\ 1.5\,i_A = 0$ $\therefore\ i_{sc} = \dfrac{100u(t)}{10} = 10u(t)$

OC : $\dfrac{v_{oc} - 100\,u(t)}{10} + \dfrac{v_{oc}}{20} - 1.5\,i_A = 0$; $i_A = 0.05\,v_{oc}$

$\therefore\ 0.1\,v_{oc} - 10\,u(t) + 0.05\,v_{oc} - 0.075\,v_{oc} = 0$ $\therefore\ v_{oc} = \dfrac{10\,u(t)}{0.075} = 133.33\,u(t)$

$\therefore\ R_{Th} = \dfrac{v_{oc}}{i_{sc}} = \dfrac{133.33}{10} = 13.333\ \Omega$

$i_L(0) = 0$ and $i_L(\infty) = 10$ A ; $\tau = \dfrac{0.05}{13.333} = \dfrac{1}{266.7}$

$\therefore\ i_L(t) = \underline{10(1 - e^{-266.7t})\,u(t)\ \text{A}}$

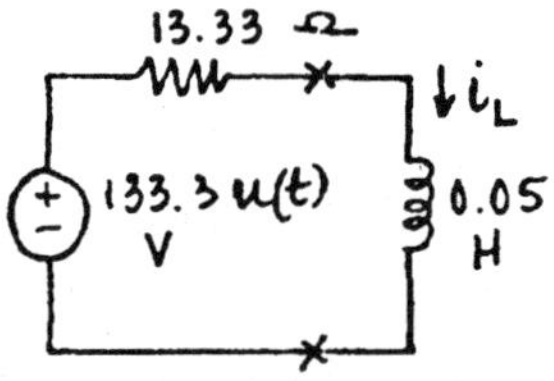


5-16

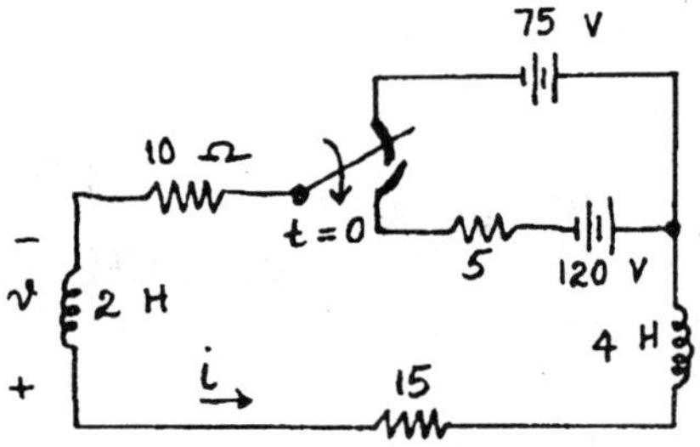


(a) $i(0) = \dfrac{75}{10 + 15} = 3$ A ; $i(\infty) = \dfrac{-120}{30} = -4$ A

$\tau = \dfrac{L_{eq}}{R_{eq}} = \dfrac{6}{30} = \dfrac{1}{5}$ $\therefore\ i = -4 + 7e^{-5t}$ A , $t > 0$

$i = 0$ $\therefore\ 4 = 7e^{-5t}$ $\therefore\ t = \underline{0.11192\ s}$

(b) $v = -Li' = -2 \times 7(-5) \times \dfrac{4}{7} = \underline{-40\ \text{V}}$ at $t = 0.11192$ s

5-17

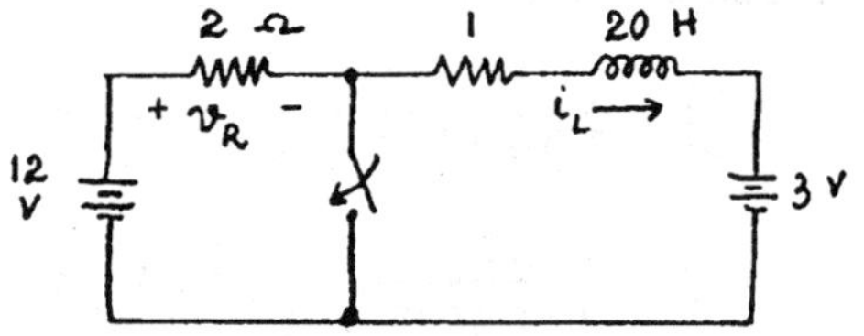


$i_L(0) = 3$ A ; $v_R(0^-) = 6$ V ; $i_L(\infty) = -3$ A

$\therefore\ i_L = -3 + 6e^{-0.05t}$ A , $t > 0$ ; $v(0^+) = 12$ V

$\therefore\ i_L(-16\ s) = \underline{3\ \text{A}}$ ; $i_L(16\ s) = -3 + 6e^{-0.8}$

$i_L(16\ s) = \underline{-0.3040\ \text{A}}$

$v_R(-16\ s) = \underline{6\ \text{V}}$ ; $v_R(16\ s) = \underline{12\ \text{V}}$

5-18

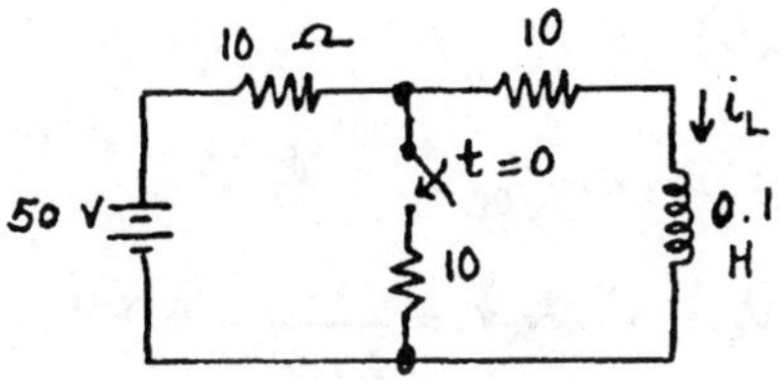


$i_L(0) = 2.5$ A ; $i_L(\infty) = \frac{50}{10+5} \times \frac{1}{2} = \frac{5}{3}$ A

$i_L(t) = \frac{5}{3} + \left(\frac{5}{2} - \frac{5}{3}\right)e^{-15t/0.1} = \frac{5}{3} + \frac{5}{6}e^{-150t}$ A

(a) $i_L(-5 \text{ ms}) = \underline{2.5 \text{ A}}$

(b) $i_L(5 \text{ ms}) = \frac{5}{3} + \frac{5}{6}e^{-0.75} = \underline{2.060 \text{ A}}$

(c) $i_L(10 \text{ ms}) = \frac{5}{3} + \frac{5}{6}e^{-1.5} = \underline{1.8526 \text{ A}}$

5-19

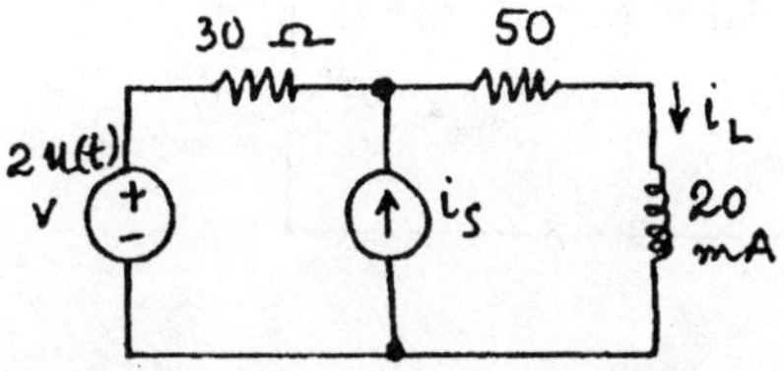


$i_s = 80u(t - 0.0002)$ mA

$\underline{t < 0}$ : $i_L = \underline{0}$

$\underline{0 < t < 0.0002 \text{ s}}$ : $i_L(\infty) = \frac{2}{80}$ A ; $\tau = \frac{20 \times 10^{-3}}{30 + 50} = \frac{1}{4000}$

$\therefore i_L(t) = \underline{25(1 - e^{-4000t}) \text{ mA}}$

$\therefore i_L(0.0002) = 13.767$ mA

$\underline{t > 0.0002 \text{ s}}$ : $i_L(\infty) = 25 + 80 \times \frac{30}{80} = 55$ mA

$\therefore i_L = 55 + (13.767 - 55)e^{-4000(t - 0.0002)} = \underline{55 - 41.23e^{-4000(t - 0.0002)} \text{ mA}}$

5-20

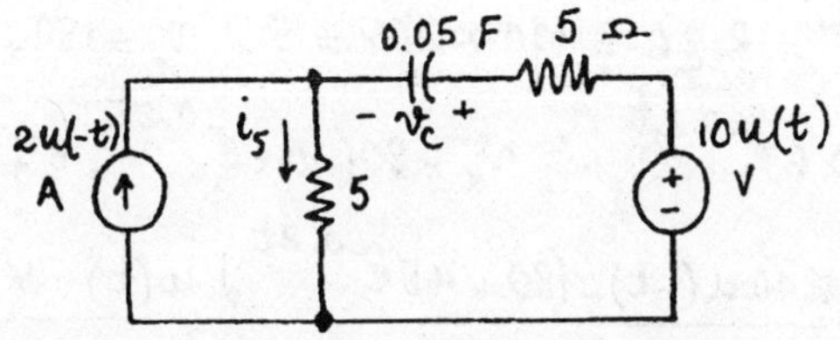


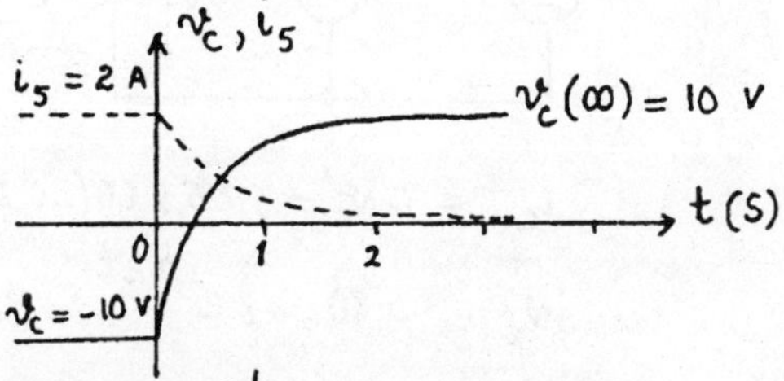


(a) $v_c(0) = -10$ V ; $v_c(\infty) = 10$ V $\therefore v_c(t) = \underline{10 - 20e^{-2t} \text{ V}}$, $t > 0$

(b) $t < 0$, $i_s = 2$ A ; $t > 0$: $i_s = C v_c'(t) = 0.05(-20)(-2)e^{-2t} = \underline{2e^{-2t} \text{ A}}$, $t > 0$

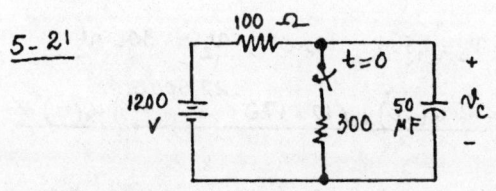

5-21




(a) $v_c(0) = 1200$ V ; $v_c(\infty) = 900$ V ; $R_{eq} = 100 \| 300$ Ω

$\therefore v_c(t) = 900 + 300e^{-266.7t}$ V , $t > 0$

$\therefore v_c(-2 \text{ ms}) = \underline{1200 \text{ V}}$ ; $v_c(2 \text{ ms}) = 900 + 300e^{-0.533}$

$\therefore v_c(2 \text{ ms}) = \underline{1076.0 \text{ V}}$

(b) Let $t' = 0$ at $t = 2$ ms ; $v_c(t' = 0) = 900$ V

$v_c(\infty) = 1200$ V $\therefore v_c(t') = 1200 - 300e^{-200t'}$ V, $t' > 0$

$\therefore v_c(t' = 2 \text{ ms}) = 1200 - 300e^{-0.4} = \underline{998.9 \text{ V}}$

5-22 (a)

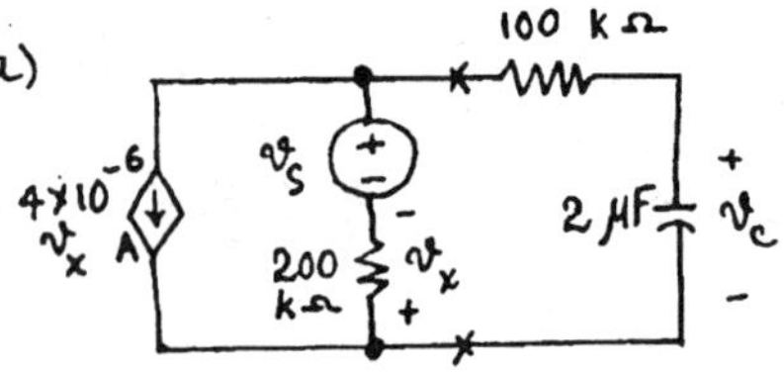


let $v_s = 9 - 12u(t)$ V

OC : $v_x = 0 \quad \therefore v_{oc} = v_s$

SC : $v_x = v_s \quad \therefore i_{sc}\downarrow = \frac{v_s}{2\times10^5} - 4\times10^{-6} v_s = 10^{-6} v_s$

$R_{TR} = \frac{v_{oc}}{i_{sc}} = 10^6\ \Omega$ ; $v_c(0) = 9$ V ; $v_c(\infty) = -3$ V

$\therefore v_c = \underline{-3 + 12e^{-t/2.2}\ \text{V}}$, $t>0$ ; $v_c = 9$ V, $t<0$

(b)

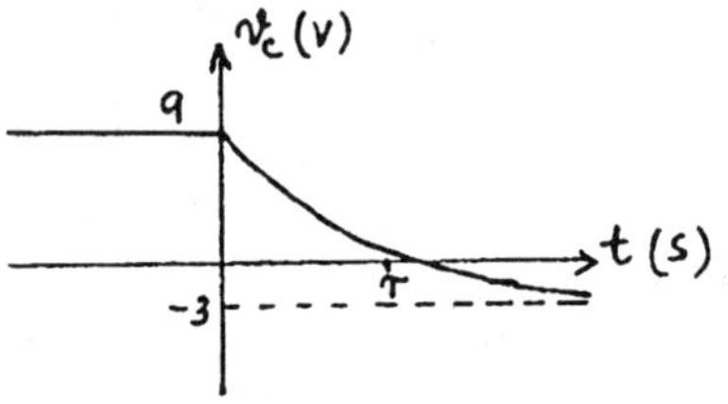


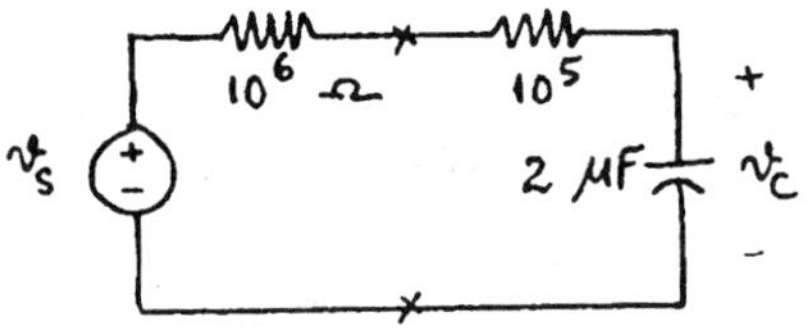


(c) $v_c = 0 \quad \therefore e^{-t/2.2} = \frac{3}{12} = 0.25$

$\therefore t = \underline{3.05\ s}$

5-23

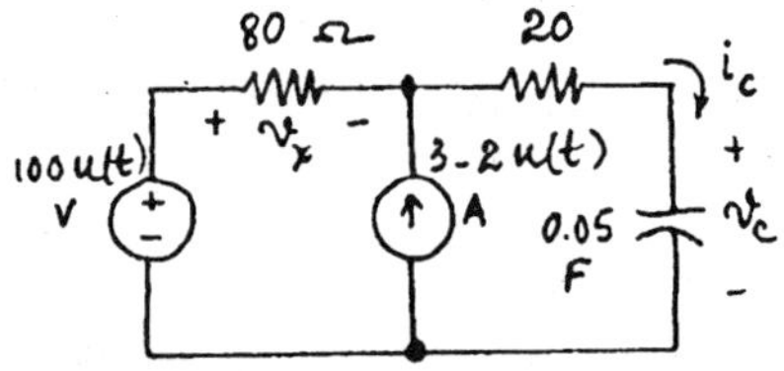


$t<0$ : $v_c = 3\times80 = 240$ V , $v_x = -240$ V

$t>0$ : $v_{cf} = 100 + (3-2)80 = 180$ V ; $v_{xf} = -80$ V

$\tau = R_{eq}C = 100\times0.05 = 5 \quad \therefore v_c = 180 + 60e^{-0.2t}$

$i_c = Cv_c' = 0.05\times60(-0.2)e^{-0.2t} = -0.6e^{-0.2t} \quad \therefore v_x = 80(-0.6e^{-0.2t} - 3 + 2)$

$v_x = -80 - 48e^{-0.2t} \quad \therefore v_x(t) = \underline{-240u(-t) - (80 + 48e^{-0.2t})u(t)\ \text{V}}$

5-24

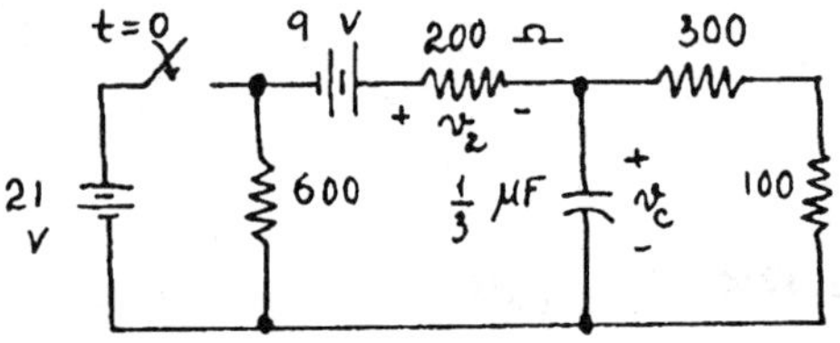


(a) $v_c(0) = 9\times\frac{4}{12} = 3$ V ; $v_c(\infty) = \frac{21\times4}{6} + \frac{9\times4}{6}$

$v_c(\infty) = 20$ V $\therefore v_c(t) = \underline{3u(-t) + (20 - 17e^{-22{,}500t})u(t)}$ V

(b) $t<0$ , $v_2 = 1.5$ V ; $t>0$ , $v_2 = 30 - v_c$

$\therefore v_2(t) = \underline{1.5u(-t) + (10 + 17e^{-22{,}500t})u(t)\ \text{V}}$

5-25

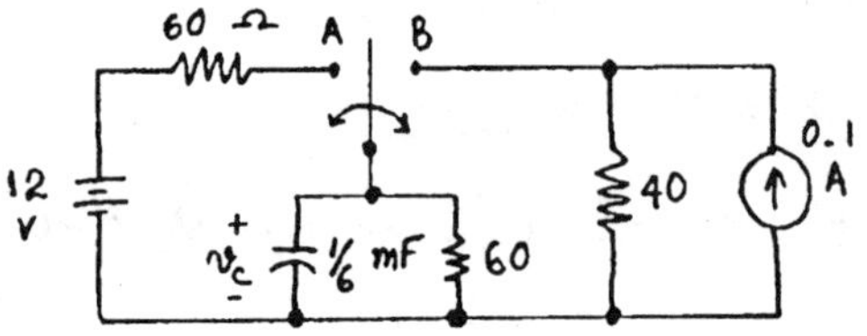


$v_c(t) = \underline{6\ \text{V}}$ for $t<0 \quad \therefore v_c(0) = 6$ V

$v_c(\infty) = 0.1\times\frac{4}{10}\times60 = 2.4$ V ; $R_{eq} = 60\|40\ \Omega$

$\therefore v_c(t) = 2.4 + 3.6e^{-6000t/24}$

$v_c(t) = \underline{2.4 + 3.6e^{-250t}\ \text{V}}$ , $t>0$

**5-26**

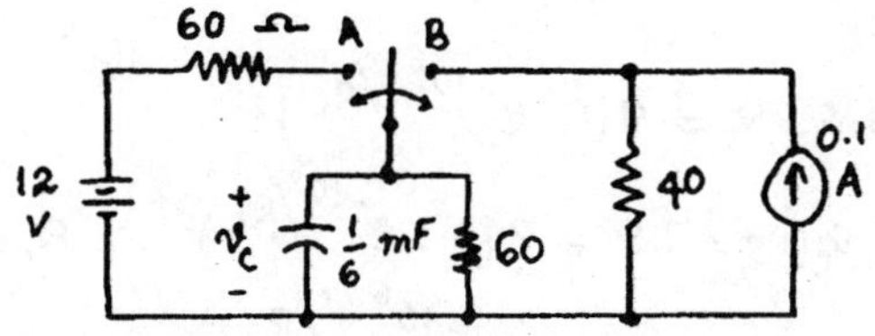


$v_c(t) = \underline{2.4\ V}$ for $t<0$ $\therefore v_c(0) = 2.4$ V

$v_c(\infty) = 6$ V ; $R_{eq} = 60 \| 60$ Ω

$\therefore v_c(t) = 6 - 3.6e^{-6000t/30} = \underline{6 - 3.6e^{-200t}\ V}$

$t > 0$

**5-27**

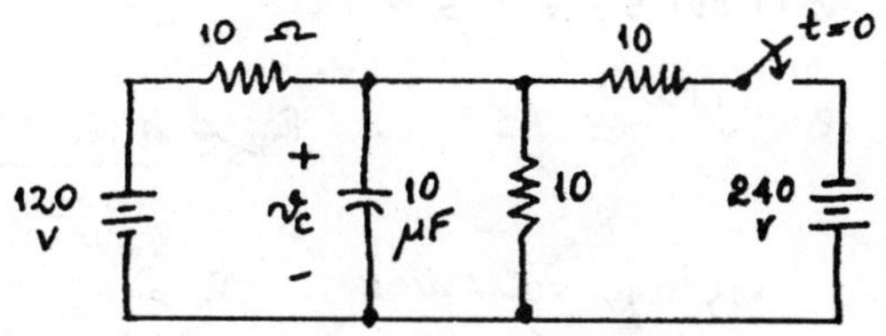


(a) $v_c(0) = 120 \times \frac{1}{2} = 60$ V

$v_c(\infty) = \frac{120}{15} \times \frac{1}{2} \times 10 - \frac{240}{15} \times \frac{1}{2} \times 10 = -40$ V

$R_{eq} = 10 \| 10 \| 10$ $\therefore v_c = -40 + 100e^{-10^5 t/(10/3)}$

$v_c(t) = \underline{-40 + 100e^{-30,000t}\ V}$ , $t > 0$

(b) $v_c = 0$ $\therefore e^{-30,000t} = 2.5$ $\therefore t = \underline{30.54\ \mu s}$

**5-28**

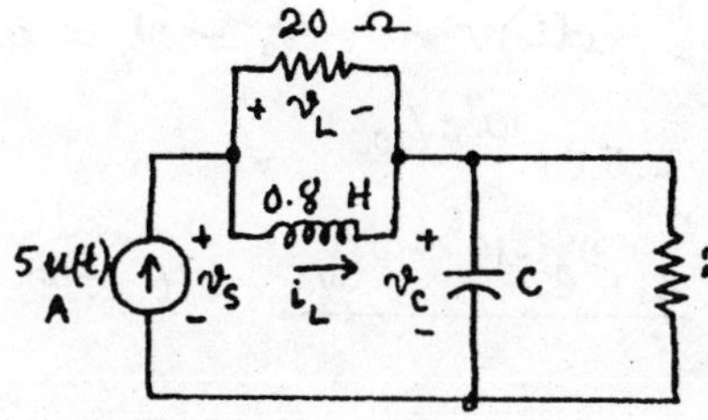


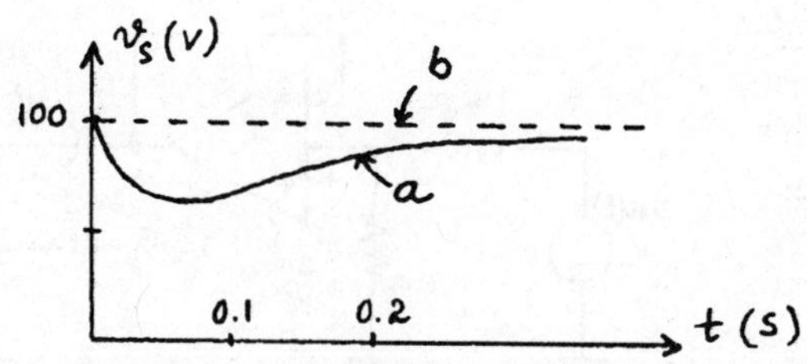


(a)

$\tau_c = R_{eq} C = \frac{1}{10}$ ; $v_c(0) = 0$ $\therefore v_c = 100(1 - e^{-10t})$ ; $\tau_L = \frac{0.8}{20}$ ; $i_L(0) = 0$

$\therefore i_L = 5(1 - e^{-25t})$ A ; $v_L = L i_L' = 100e^{-25t}$ V $\therefore v_s = \underline{100 - 100e^{-10t} + 100e^{-25t}\ V}$

(b) $\tau_c = \frac{1}{25}$ $\therefore v_c = 100(1 - e^{-25t})$ ; $v_L = 100e^{-25t}$ $\therefore v_s = \underline{100\ V}$

**5-29**

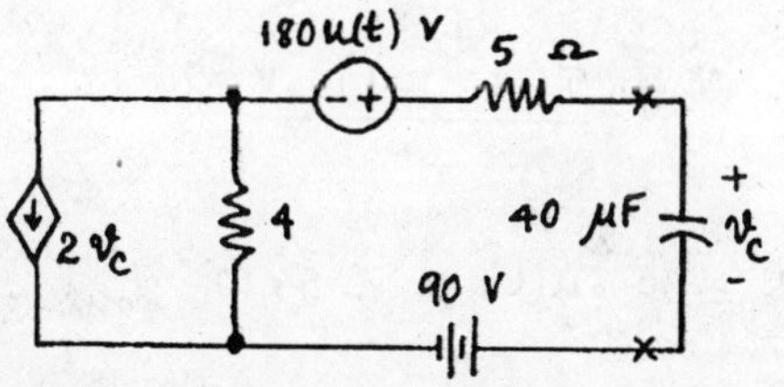


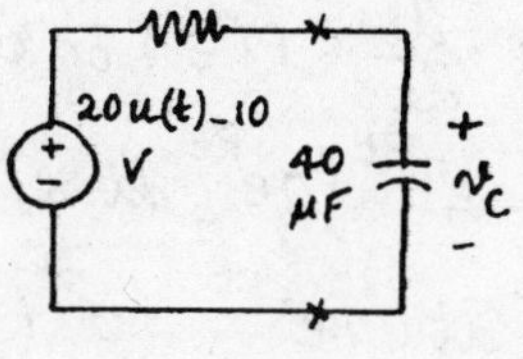


$\underline{OC}$ : $v_{c,oc} = 180u(t) - 8v_{c,oc} - 90$

$v_{c,oc} = 20u(t) - 10$ V

$\underline{SC}$ : $v_{c,sc} = 0$ $\therefore i_{sc}\downarrow = \frac{180u(t) - 90}{4 + 5}$

$i_{sc} = 20u(t) - 10$ V ; $R_{Th} = \frac{v_{c,oc}}{i_{sc}} = 1$ Ω

$v_c(0) = -10$ V ; $v_c(\infty) = 10$ V

$\therefore v_c = 10 - 20e^{-10^6 t/40} = \underline{10 - 20e^{-25,000t}\ V}$

$t > 0$

$v_c = 0$ $\therefore t = \frac{\ln 2}{25,000} = \underline{27.73\ \mu s}$

5-30

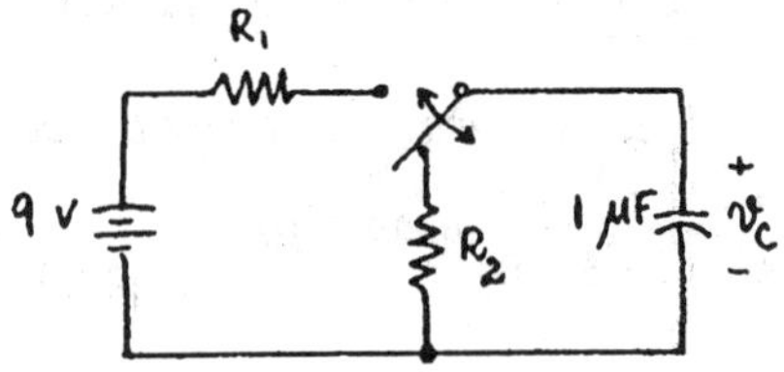


$0<t<1\,s: \ v_c = 9(1-e^{-10^6 t/R_1})$

$v_c(1\,s) = 7.5 = 9(1-e^{-10^6/R_1})$ ∴ $R_1 = \underline{558.1\ k\Omega}$

$t>1\,s: \ v_c = 7.5e^{-10^6 t'/R_2}$ ; $t' = t-1$

$v_c(1.001\,s) = 1 = 7.5e^{-10^6\times 10^{-3}/R_2}$

$e^{1000/R_2} = 7.5$ ∴ $R_2 = \underline{496.3\ \Omega}$

5-31

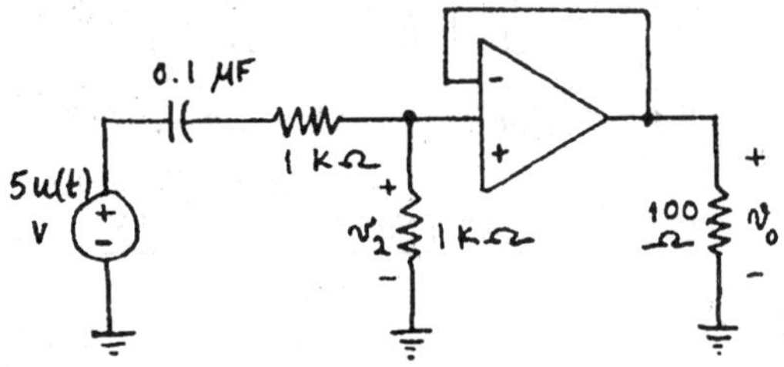


Voltage follower : $v_o = v_2$

$R_{in} = \infty$ ∴ $v_2 = 2.5e^{-10^7 t/2000}$ V

∴ $v_o = \underline{2.5e^{-5000t}\ \text{V}}$ , $t>0$

5-32

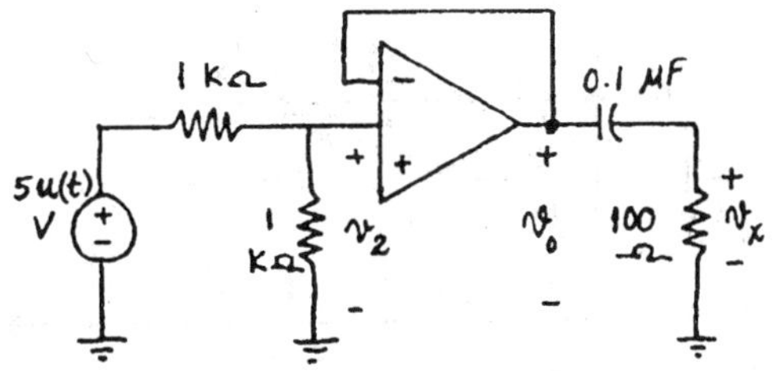


voltage follower ; $v_2 = v_o = 2.5u(t)$

∴ $v_x = 2.5e^{-10^7 t/100}$ V

$v_x = \underline{2.5e^{-10^5 t}\ \text{V}}$ , $t>0$

5-33

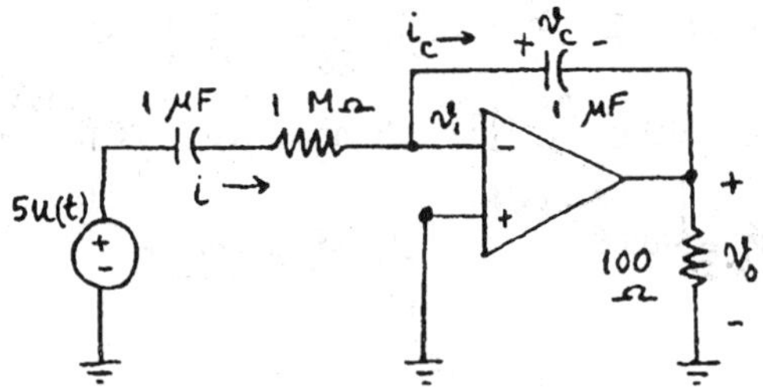


$v_1 = 0$ ∴ $i = \frac{5}{10^6}e^{-t} = 5\times 10^{-6}e^{-t}$ A, $t>0$

$i_c = i = 10^{-6}v_c' = 5\times10^{-6}e^{-t}$ ; $v_o = -v_c$

∴ $10^{-6}v_o' = -5\times10^{-6}e^{-t}$ ∴ $v_o = \int_0^t -5e^{-t}\,dt$

$v_o = \underline{5(e^{-t}-1)\ \text{V}}$

5-34

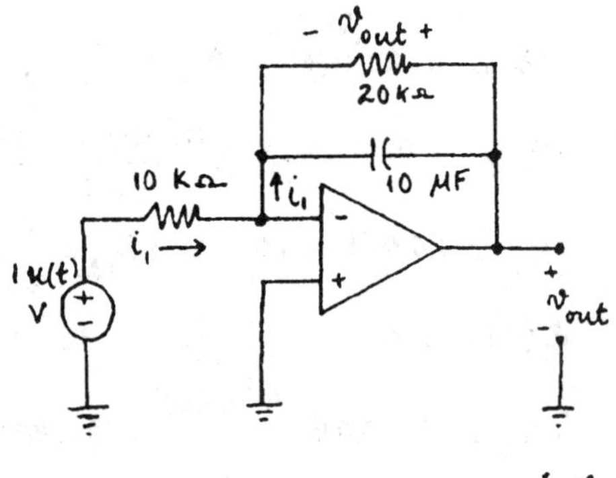


$i_1 = 10^{-4}u(t) = -5\times10^{-5}v_{out} - 10^{-5}v_{out}'$

∴ $v_{out}' + 5v_{out} + 10u(t) = 0$

From Sect 6.4 : $\frac{di}{dt} + Pi = Q$ or $v' + Pv = Q$

Sol'n is : $v = e^{-Pt}\int Qe^{Pt}\,dt + Ae^{-Pt}$

∴ $v_{out} = e^{-5t}\int -10u(t)e^{5t}\,dt + Ae^{-5t} = e^{-5t}(-2e^{5t}) + Ae^{-5t}$

$v_{out}(0) = 0$ ∴ $v_{out} = \underline{-2+2e^{-5t}\ \text{V}}$ , $t>0$

## 5-35

(*a*)

$i_x = -50/40 = -1.25$ , $3i_x = -3.75$,
$i_{10} = 50/10 = 5$,
$i_L = i_x - i_{10} + 3i_x = \underline{\underline{-10}}$ A

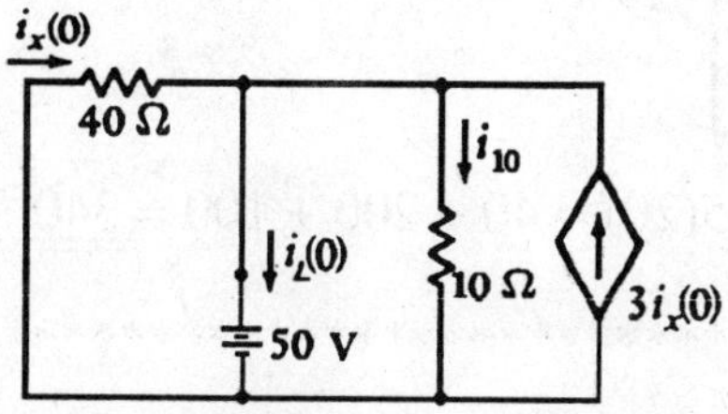


**********************************************************************

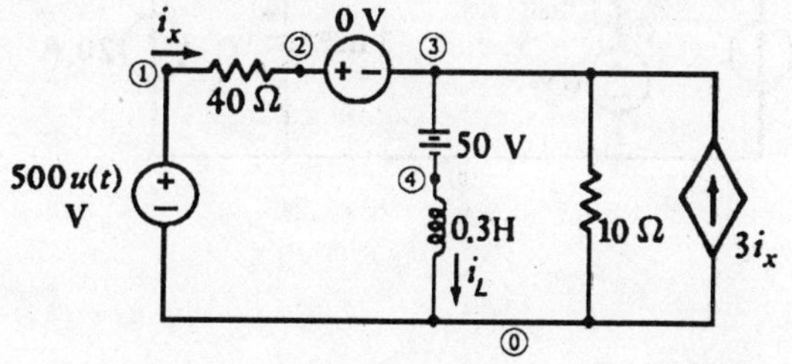


(*b*)

**********************************************************************

**PROGRAM**

```
R12  1  2  40
V23  2  3
V34  3  4  50
L40  4  0  0.3  IC=-10
R30  3  0  10
V10  1  0  PWL(0  0  1N  500  0.1  500)
F03  0  3  V23  3
.TRAN 1.25M  0.1  UIC
.PRINT TRAN I(V34)
.END
```

**********************************************************************

**ANSWERS: Source currents**

| **TIME** | **I(V34)** |
|---|---|
| 0.000E+00 | -1.000E+01 |
| 1.250E-03 | -8.970E+00 |
| 2.500E-03 | -7.963E+00 |
| ....... | |
| 7.500E-02 | 2.568E+01 |
| ....... | |
| 7.875E-02 | 2.654E+01 |
| 8.000E-02 | 2.682E+01 A = $i_L(0.08)$ |
| 8.125E-02 | 2.709E+01 |
| ..... | |

*5-36*

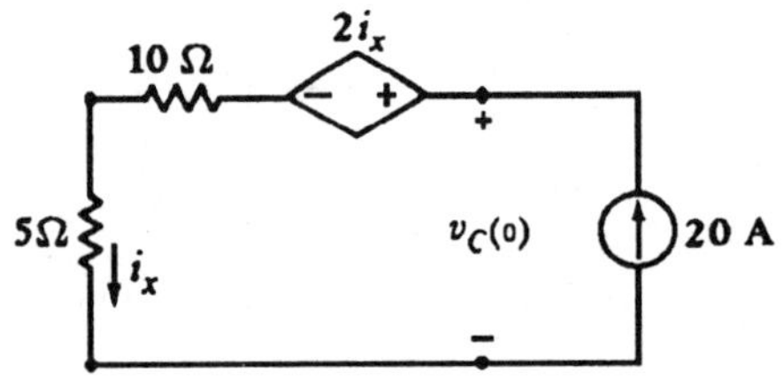


(*a*)

$i_x(0) = 20$ A, and so $v_C(0) = 2(20) + 10(20) + 5(20) = 40 + 200 + 100 = \underline{\underline{340\ \text{V}}}$

********************************************************************

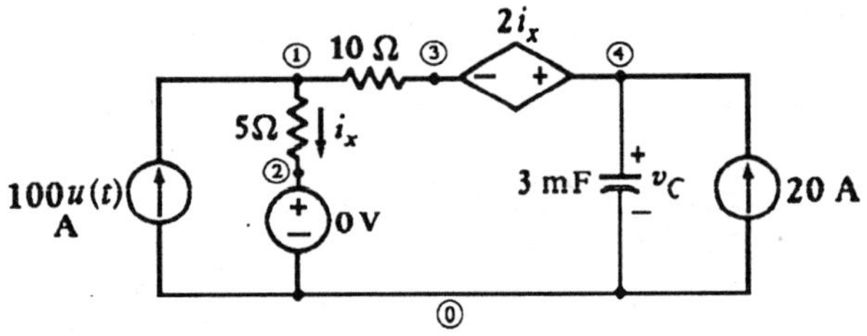


(*b*)

**PROGRAM**

```
I01  0  1  PWL(0  0  1N  100  1  100)
R12  1  2  5
V20  2  0
R13  1  3  10
H43  4  3  V20  2
C40  4  0  3M  IC=340
I04  0  4  20
.TRAN  1.25M  40M
.PRINT TRAN V(4)
.END
```

********************************************************************

**ANSWERS: Node voltage**
**TIME V(4)**

```
0.000E+00  3.400E+02
1.250E-03  3.569E+02
.....
3.750E-02  7.044E+02
3.875E-02  7.126E+02
4.000E-02  7.205E+02 V = vC(0.04)
```

# CHAPTER 6: THE *RLC* CIRCUIT

## Section 6-2 The source-free parallel circuit
## Section 6-3 The overdamped parallel *RLC* circuit

**1** Select element values for a parallel *RLC* circuit so that $s_1 = -200\ \mathrm{s}^{-1}$, $s_2 = -500\ \mathrm{s}^{-1}$, and the initial current through the resistor (in mA) is numerically equal to the initial voltage across the capacitor (in V).

**2** The inductor current in the circuit shown in Fig. 6-1 is $i = 2e^{-5t} - 5e^{-10t}$ A. If $L = 0.2$ H, find: (*a*) $v(t)$; (*b*) $i_R(t)$; (*c*) $i_C(t)$.

**3** Element values in Fig. 6-1 are $R = 10\ \Omega$, $L = \frac{1}{32}$ H, and $C = 50\ \mu$F. If $i(0) = -2$ A and $v(0) = 40$ V, find $i(t)$ for $t > 0$.

**4** Find $v_L(t)$ for the circuit shown in Fig. 6-2.

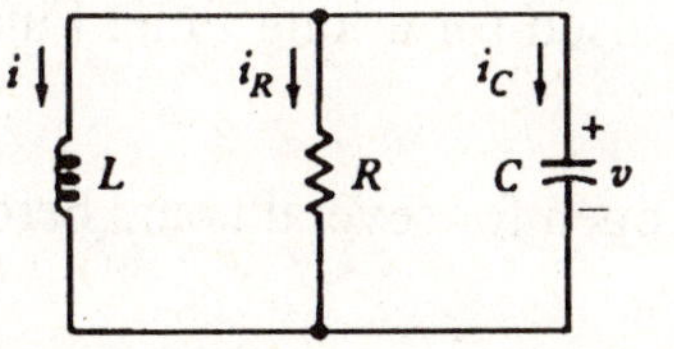


**Fig. 6-1:** See Probs. 2, 3, 8, 10, and 11.

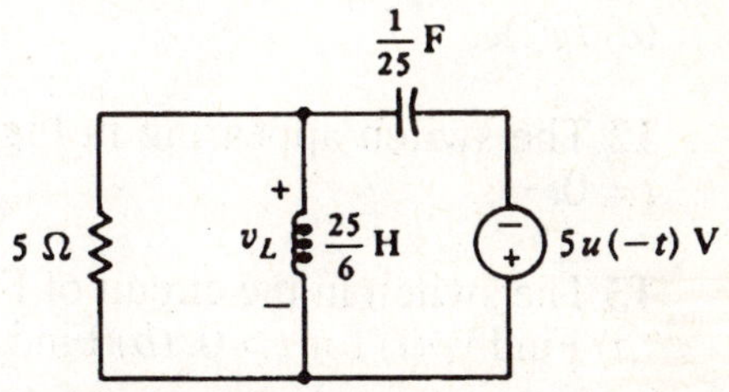


**Fig. 6-2:** See Prob. 4.

**5** For the *RLC* circuit of Fig. 6-3: (*a*) find $i_R(t)$; (*b*) find the time at which $i_R = 250$ mA.

**6** The switch in Fig. 6-4 has been closed for a long time. (*a*) Find $i_L(t)$ for all *t*. (*b*) Sketch $i_L(t)$ for the time interval, $-1 < t < 1$ ms.

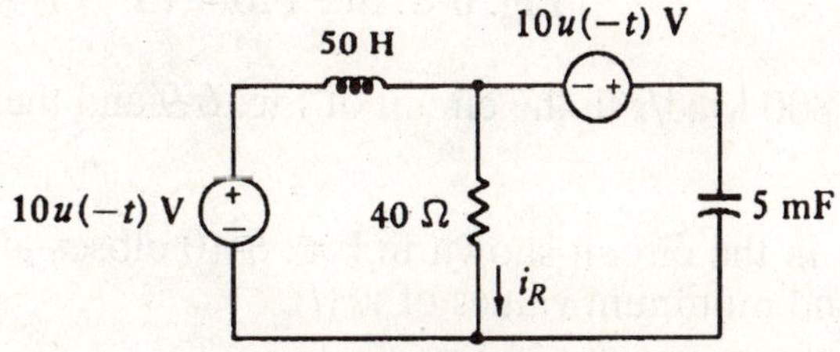


**Fig. 6-3:** See Prob. 5.

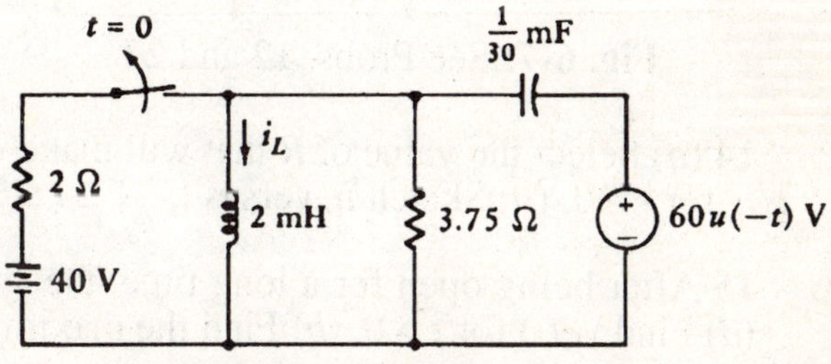


**Fig. 6-4:** See Prob. 6.

**7** After being open for an hour, the switch in Fig. 6-5 closes at $t = 0$. (*a*) Find $v_C(t)$ for $t > 0$. (*b*) Find $i_L(t)$ for $t > 0$. (*c*) Find the settling time $t_s$ for $v_C$. (*d*) Find $t_s$ for $i_L$.

## Section 6-4 Critical damping

**8** Suppose the circuit of Fig. 6-1 is critically damped, with $v(0) = 100$ V and $i(0) = 4$ A. Then if $R = 10\ \Omega$ and $\alpha = 100$ Np/s: (*a*) find *L* and *C*; (*b*) find $v(t)$ for $t > 0$ and sketch the response as a function of time. (*c*) At what time is $v = 0$? (*d*) Find the settling time $t_s$.

**9** (*a*) In the circuit shown in Fig. 6-6, what value of *R* will cause critical damping? (*b*) Using this value of *R*, find $i_C(t)$ for $t > 0$. (*c*) What is $t_s$ for $i_C$?

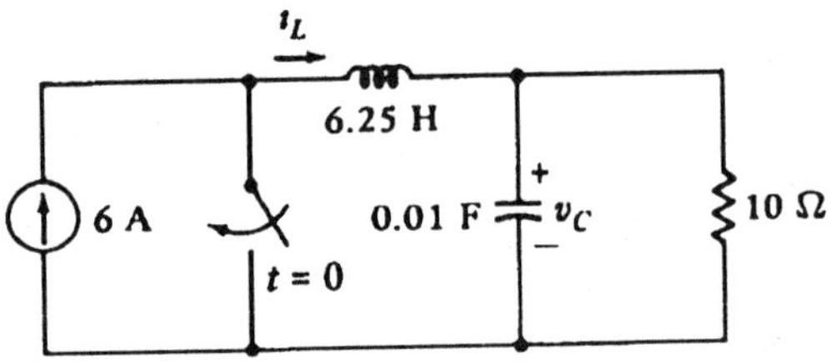


**Fig. 6-5:** See Prob. 7.

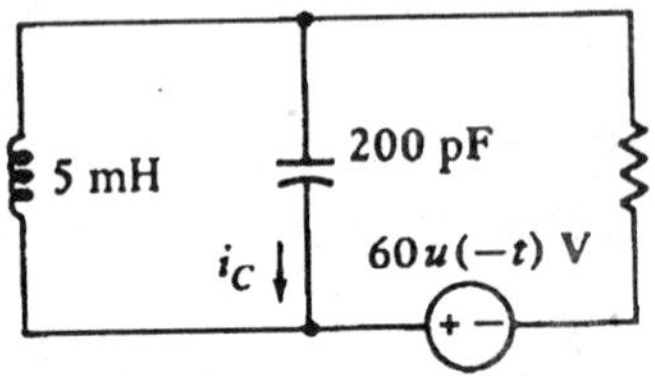


**Fig. 6-6:** See Prob. 9.

**10** Let $v(0) = 0$ and $i(0) = 5$ A in the circuit shown in Fig. 6-1. If $L = 0.1$ H, $\omega_0 = 100$ rad/s, and the circuit is critically damped: (*a*) find $R$ and $C$; (*b*) find $v(t)$ for $t > 0$. (*c*) Find $|v|_{max}$ and the time $t_m$ at which it occurs. (*d*) Find the settling time $t_s$.

## Section 6-5 The underdamped parallel *RLC* circuit

**11** Let $i = 10e^{-4t} \sin 28t$ A in the circuit shown in Fig. 6-1. If $R = 200$ Ω, find: (*a*) $v(t)$; (*b*) $i_C(t)$; (*c*) $i_R(t)$.

**12** The switch appearing in Fig. 6-7 has been closed for a long time. Find $i(t)$ after it opens at $t = 0$.

**13** The switch in the circuit of Fig. 6-8 has been open for several hours before it is closed at $t = 0$. (*a*) Find $v_C(t)$ for $t > 0$. (*b*) Find $i_{SW}$ for $t > 0$.

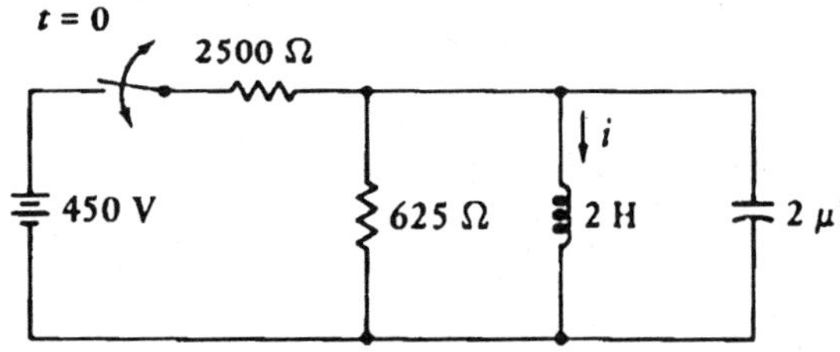


**Fig. 6-7:** See Probs. 12 and 27.

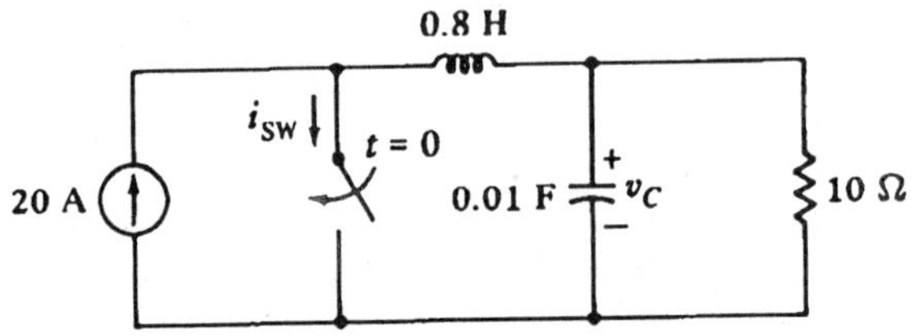


**Fig. 6-8:** See Prob. 13.

**14** (*a*) Select the value of $R$ that will make $\omega_d = 800$ krad/s in the circuit of Fig. 6-9 and then find $i_L$ for $t > 0$. (*b*) Sketch $i_L$ versus $t$, $-1 \leq t \leq 7$ μs.

**15** After being open for a long time, the switch in the circuit shown in Fig. 6-10 closes at $t = 0$. (*a*) Find $v_C(t)$ for $t > 0$. (*b*) Find the maximum and minimum values of $v_C(t)$.

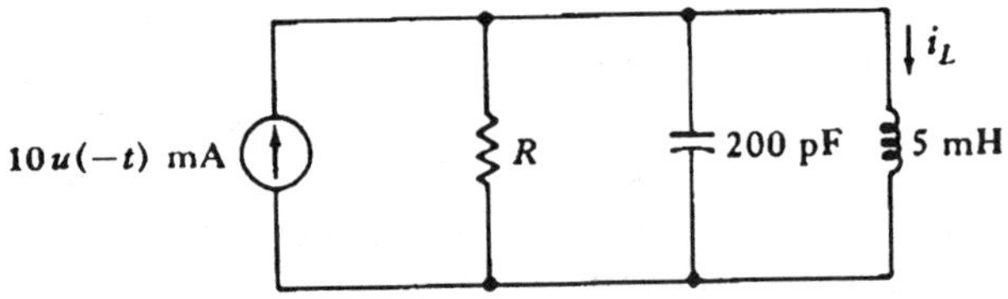

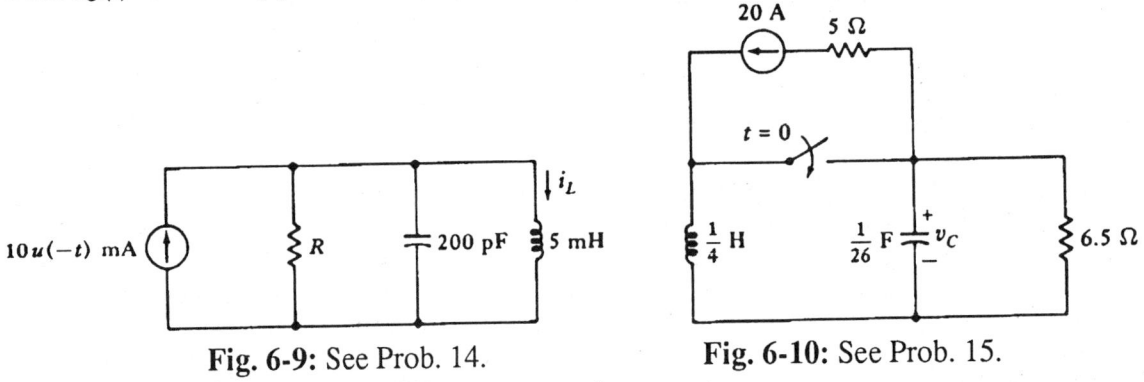

**Fig. 6-9:** See Prob. 14.

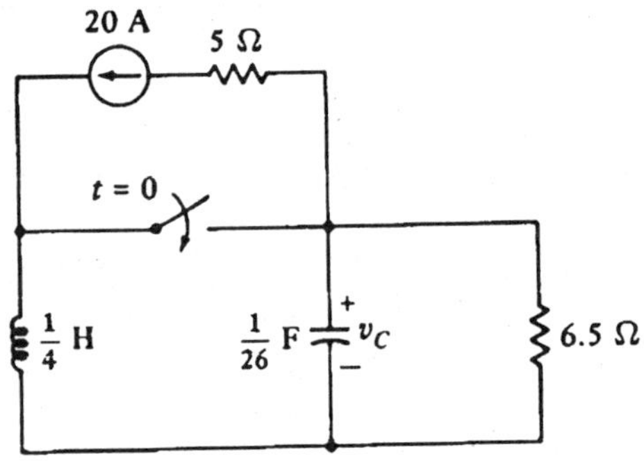


**Fig. 6-10:** See Prob. 15.

**16** The switch in the circuit of Fig. 6-11 has been closed for a very long time. It opens at $t = 0$ just as the source turns off. (*a*) Find $v_R(t)$ for $t > 0$. (*b*) Find $v_1(t)$ for $t > 0$.

## Section 6-6 The source-free series *RLC* circuit

**17** If the switch in the circuit of Fig. 6-12 is closed at $t = -10^6$ s and opened at $t = 0$, find $v(t)$ for $t > 0$.

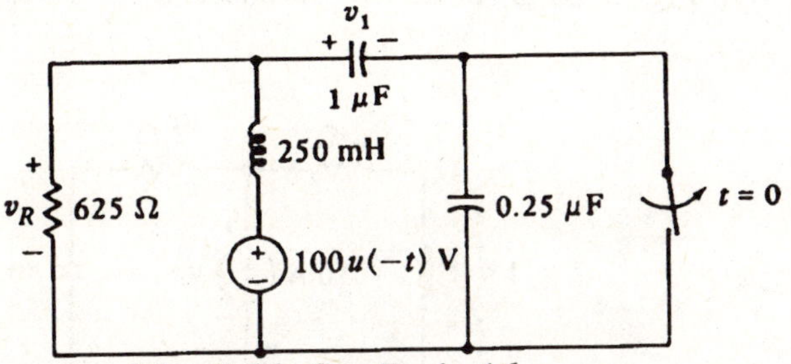


**Fig. 6-11:** See Prob. 16.

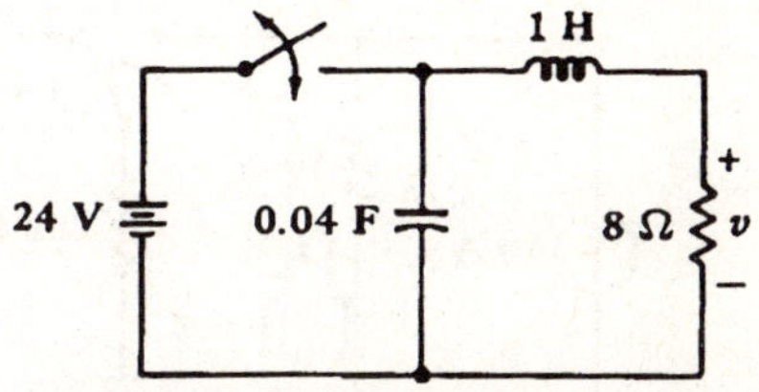


**Fig. 6-12:** See Prob. 17.

**18** (*a*) Find $v(t)$ for $t > 0$ in the circuit of Fig. 6-13. (*b*) What is the settling time $t_s$ for $v(t)$?

**19** (*a*) Find $i_L(t)$ in the circuit of Fig. 6-14. (*b*) What is the value of $d^2i_L/dt^2$ at $t = 0^+$ ?

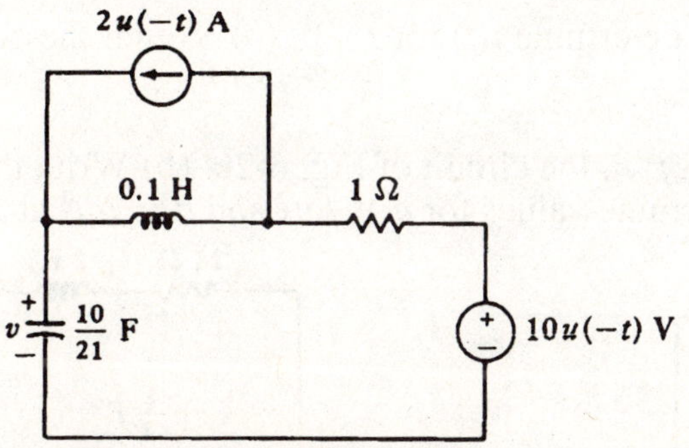


**Fig. 6-13:** See Prob. 18.

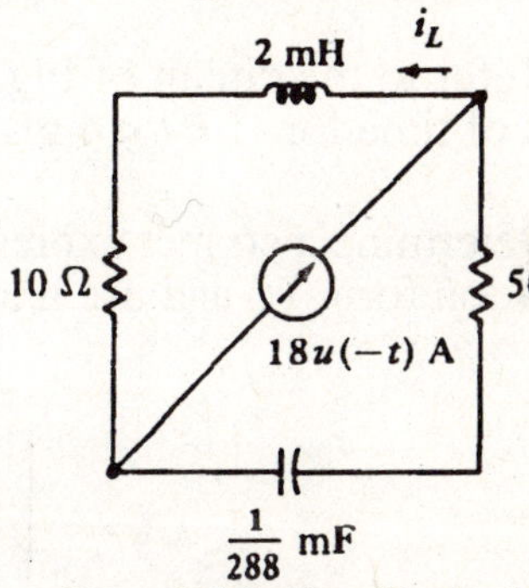


**Fig. 6-14:** See Prob. 19.

**20** Find $v_C(t)$ for the circuit shown in Fig. 6-15 and sketch the response.

**21** For the circuit of Fig. 6-16: (*a*) find $i_L(t)$ for all *t*; (*b*) find $v_C(t)$ for $t > 0$.

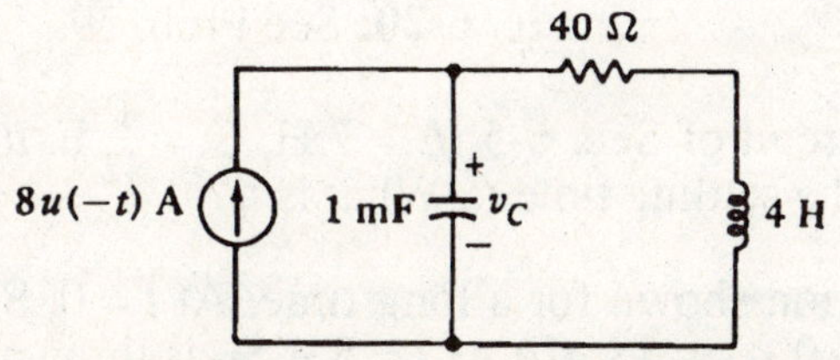


**Fig. 6-15:** See Probs. 20 and 25.

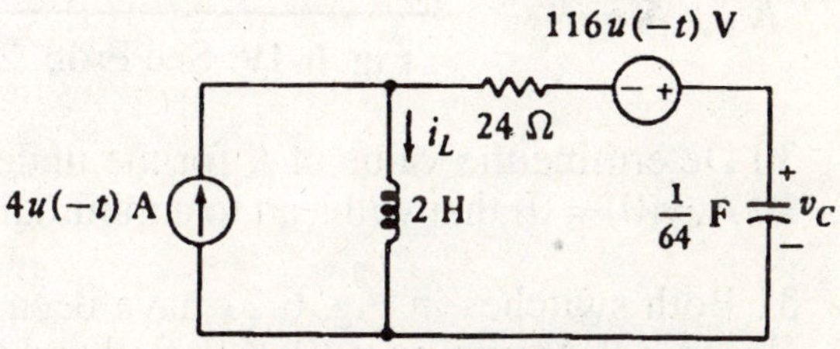


**Fig. 6-16:** See Probs. 21 and 24.

**22** Sketch $v_C$ versus $t$, $0 < t < 3$ s, for a series *RLC* circuit in which $\alpha = 1$ Np/s, $\omega_d = 25$ rad/s, $v_C(0^+) = 0$, and $v_C'(0^+) = 100$ V/s.

## Section 6-7 The complete response of the *RLC* circuit

**23** (*a*) Find $i_L(t)$ for $t > 0$ in the circuit shown in Fig. 6-17, and (*b*) determine its minimum and maximum values.

**24** Change $4u(-t)$ A in Fig. 6-16 to $4u(t)$ A and rework Prob. 21.

**25** The source in the circuit shown in Fig. 6-15 is changed from $8u(-t)$ to $4u(t) + 4$ A. (*a*) Find $v_C(t)$ for $t > 0$. (*b*) Find $v_{C,max}$.

**26** At $t = 0$, the strength of the current source in Fig. 6-18 is suddenly increased from 10 to 20 A. Find $v_s$ for $t > 0$.

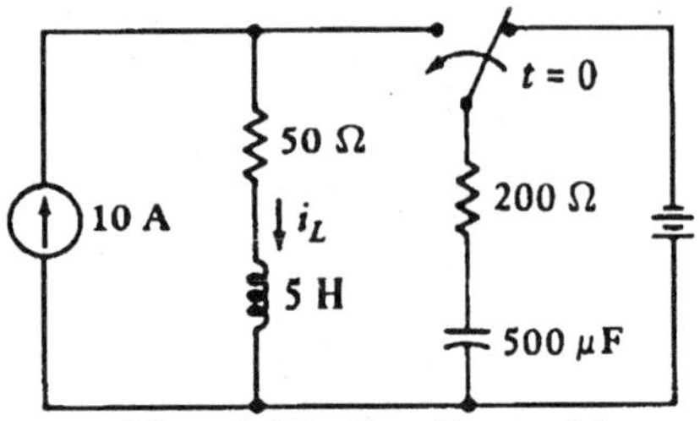


**Fig. 6-17:** See Prob. 23.

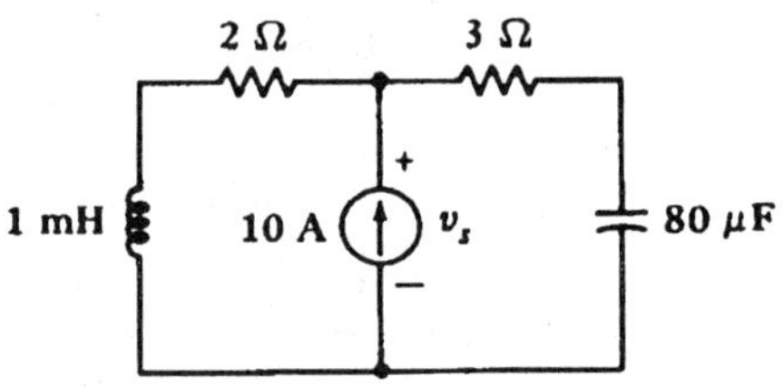


**Fig. 6-18:** See Prob. 26.

**27** Assume that the switch in Fig. 6-7 has been open for a long time and is closed at $t = 0$. Find $i(t)$ for $t > 0$.

**28** (*a*) Refer to the circuit of Fig. 6-19, and determine $i(t)$ for $t > 0$. (*b*) Sketch the current as a function of time for $-1 < t < 4$ ms.

**29** (*a*) Determine a correct expression for $v_C(t)$ in the circuit of Fig. 6-20. (*b*) Write the differential equation for $v_C(t)$ and use it to help determine values for $d^2v_C/dt^2$ and $d^3v_C/dt^3$ at $t = 0^+$.

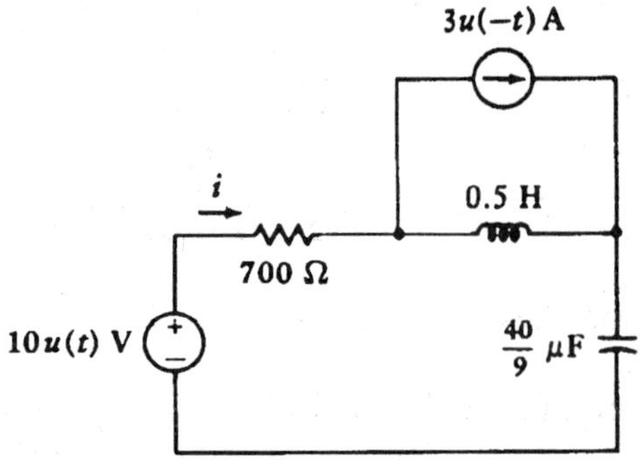


**Fig. 6-19:** See Prob. 28.

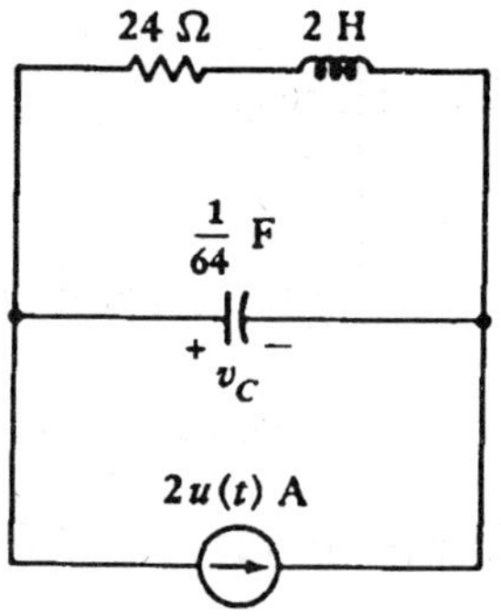


**Fig. 6-20:** See Prob. 29.

**30** Determine the value of $R$ for the underdamped circuit of Sec. 6-5 ($L = 7$ H, $C = \frac{1}{42}$ F, $i(0) = 10$ A, $v(0) = 0$) that will lead to a minimum value of the settling time $t_s$. What is $t_s$?

**31** Both switches in Fig. 6-21 have been in the position shown for a long time. At $t = 0$, $\mathbf{S_A}$ is closed. (*a*) Assuming $v_C(0) = 0$, find and sketch $v_C(t)$, $0 < t < 5$ s. (*b*) At $t = 5$ s, $\mathbf{S_B}$ is thrown to the right. Find and sketch $v_C(t)$, $5 < t < 8$ s.

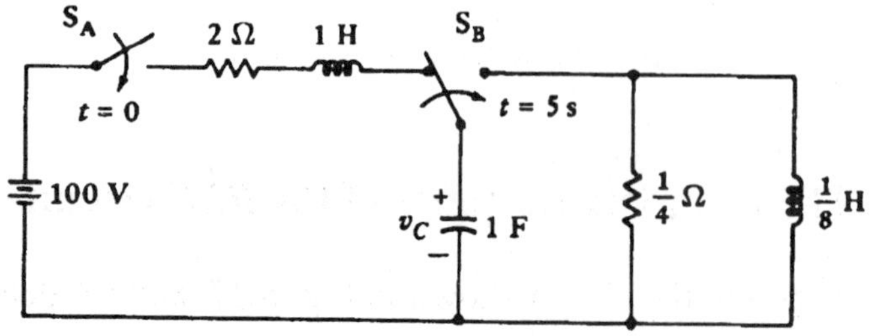


**Fig. 6-21:** See Prob. 31.

**32** Refer to the simple parallel *RC* circuit of Fig. 4-9 in your text, with $R = 10\ \text{k}\Omega$, $C = 5\ \mu\text{F}$, and $v(0) = 10$ V. Write the differential equation for $v$, $t > 0$, and design a circuit using an op-amp integrator to provide $v(t)$ as the output. First use an integrator having $R_1 = 1\ \text{M}\Omega$, $C_f = 1\ \mu\text{F}$, and then repeat for $R_1 = 10\ \text{k}\Omega$, $C_f = 5\ \mu\text{F}$.

## Section 6-8 The lossless *LC* circuit

**33** After being in the position shown for a long time, the switch in the circuit of Fig. 6-22 is moved down at $t = 0$. Construct an op-amp circuit whose output is $v(t)$ for $t > 0$.

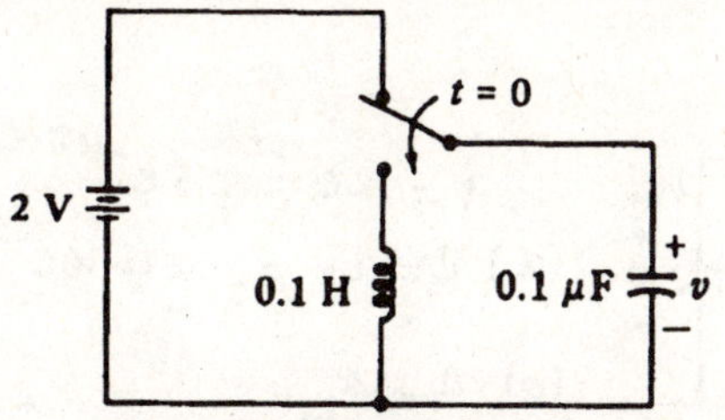

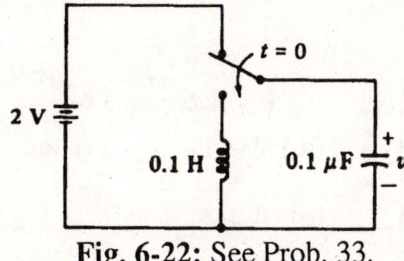

**Fig. 6-22:** See Prob. 33.

**34** (SPICE) Given the initial conditions $i(0) = 10$ A and $v(0) = 0$, the solution for $v(t)$ in the circuit of Fig. 6-8 in your text (repeated here as Fig. 6-23) was found to be expressed by $v(t) = 210\sqrt{2}\,e^{-2t}\sin\sqrt{2}t$ V, for $t > 0$. Use SPICE to find $v(t)$ at $t = 0.6$ s and 2.4 s if TSTEP is selected as 0.02 s.

**35** (SPICE) Modify Prob. 34 by installing a 0.1-Ω resistor between the bottom of the inductor and the reference node. Let $v(0) = 0$ and $i(0) = 10$ A as before. (*a*) Use SPICE and TSTEP = 0.02 s to find values for $v(t)$ at $t = 0.6$ s and 2.4 s. (*b*) Compare these results with those obtained by the exact solution when the 0.1-Ω resistor is not in the circuit.

**36** (SPICE) Use a SPICE analysis to find reasonably accurate values for $v_C$ in the circuit of Fig. 6-24 at $t = 10$, 40 and 100 ms.

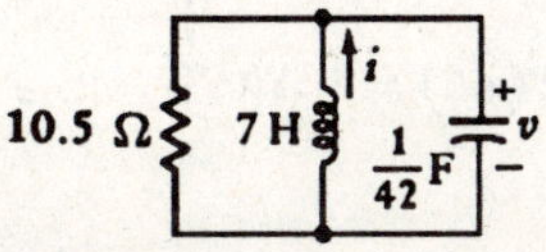


**Fig. 6-23:** See Probs. 34 and 35.

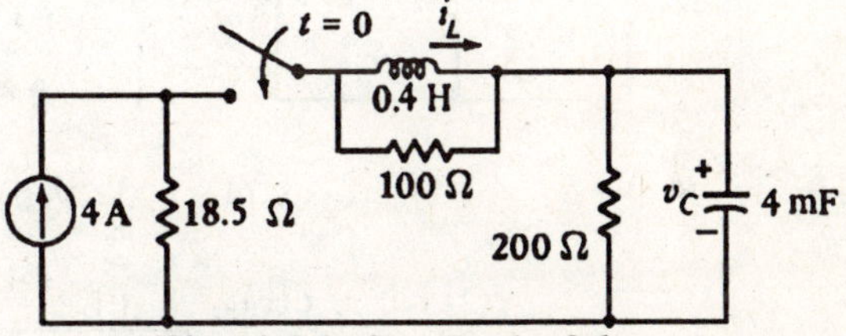


**Fig. 6-24:** See Prob. 36.

# Chapter 6

**6-1**

$R = \frac{v_c}{i_R} = \frac{1}{1\times10^{-3}} = \underline{1000\ \Omega}$ ; $s_1 = -200 = -\alpha + \sqrt{\alpha^2 - \omega_0^2}$ , $s_2 = -500 = -\alpha - \sqrt{\alpha^2 - \omega_0^2}$

adding $s_1, s_2$ $\therefore -2\alpha = -700$ $\therefore \alpha = 350$; $2\sqrt{\alpha^2 - \omega_0^2} = 300$ $\therefore \omega_0 = \sqrt{10^5}$

$\alpha = 350 = \frac{1}{2RC}$ $\therefore C = \frac{1}{2(1000)(350)} = \underline{1.4286\ \mu F}$ ; $L = \frac{1}{\omega_0^2 C} = \underline{7\ H}$

**6-2**

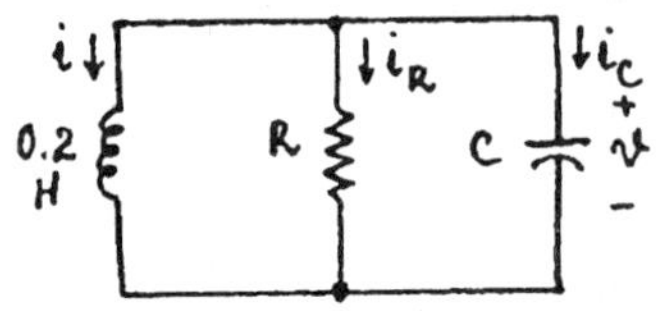


$i = 2e^{-5t} - 5e^{-10t}$

(a) $v = Li' = 0.2(-10e^{-5t} + 50e^{-10t}) = \underline{-2e^{-5t} + 10e^{-10t}\ V}$

(b) $s_1 + s_2 = -15 = -2\alpha$ $\therefore \alpha = 7.5 = \frac{1}{2RC}$ $\therefore R = \frac{1}{15C}$

$s_1 - s_2 = 5 = 2\sqrt{\alpha^2 - \omega_0^2}$ $\therefore \omega_0^2 = 50$ $\therefore \frac{1}{LC} = 50$ $\therefore C = \frac{1}{50\times0.2} = 0.1\ F$

$R = \frac{1}{(15)(0.1)} = \frac{2}{3}\ \Omega$

$i_R = \frac{v}{R} = \underline{-3e^{-5t} + 15e^{-10t}\ V}$ ; (c) $i_c = Cv' = 0.1(10e^{-5t} - 100e^{-10t})$

$i_c = \underline{e^{-5t} - 10e^{-10t}\ A}$

**6-3**

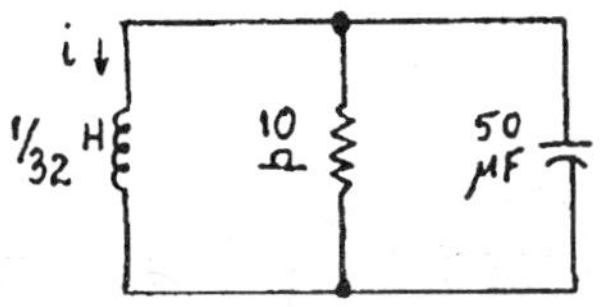


$i(0) = -2\ A$ , $v(0) = 40\ V$ ; $\alpha = \frac{1}{2RC} = 10^3$

$\omega_0^2 = \frac{1}{LC} = 64\times10^4$ ; $s_{1,2} = -\alpha \pm \sqrt{\alpha^2 - \omega_0^2} = -400, -1600\ s^{-1}$

$i = Ae^{-400t} + Be^{-1600t}$ ; $i(0) = -2 = A + B$

$i'(0) = \frac{1}{L}v(0) = -400A - 1600B = 32(40) = 1280$ $\therefore A = -1.6$, $B = -0.4$

$\therefore i(t) = \underline{-1.6e^{-400t} - 0.4e^{-1600t}\ A}$

**6-4**

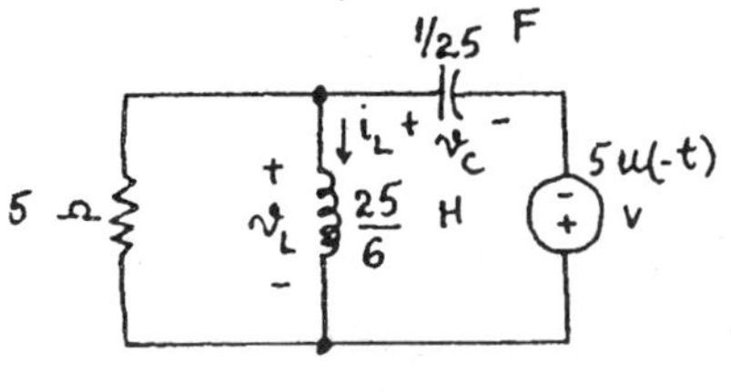


$\alpha = \frac{1}{2RC} = 2.5$ ; $\omega_0^2 = \frac{1}{LC} = 6$ ; $s_{1,2} = -2.5 \pm 0.5$

$s_{1,2} = -2$ and $-3\ s^{-1}$ $\therefore v_L = v_c = Ae^{-2t} + Be^{-3t}$

$v_c(0) = 5$ ; $i_L(0) = 0$ $\therefore A + B = 5$

$\rightarrow i_c(0) = Cv_c'(0) = \frac{1}{25}(-2A - 3B) = -\frac{v_c(0)}{R} - i_L(0) = -1$ $\therefore -A - 1.5B = -12.5$ $\therefore A = -10$, $B = 15$

$\therefore v_L(t) = \underline{-10e^{-2t} + 15e^{-3t}\ V}$ $t > 0$

**6-5** (a)

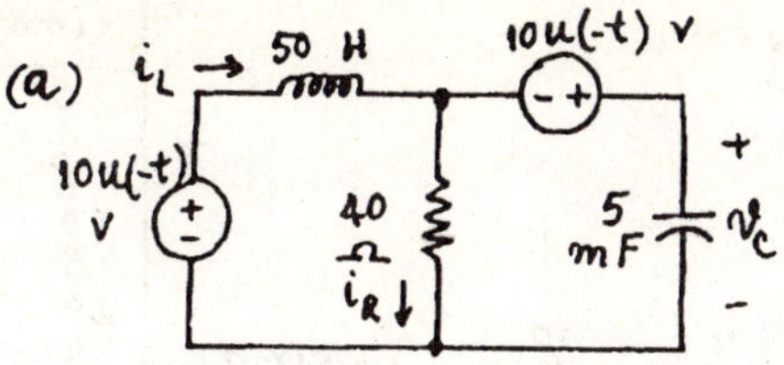


$i_L(0) = \frac{10}{40} = 0.25$ A ; $v_c(0) = 20$ V ; $\alpha = \frac{1}{2RC} = 2.5$

$\omega_0^2 = \frac{1}{LC} = \frac{1000}{50\times 5} = 4$ ; $s_{1,2} = -2.5 \pm 1.5 = -1, -4\ s^{-1}$

$v_c = Ae^{-t} + Be^{-4t}$ ; $v_c(0) = 20 = A + B$

$i_c = Cv_c'(0^+) = i_L(0^+) - i_R(0^+) = 5\times10^{-3}(-A - 4B) = 0.25 - \frac{20}{40}$ $\therefore A + 4B = 50$ $\Big\}$ $\therefore A = 10$, $B = 10$

$\therefore v_c = 10e^{-t} + 10e^{-4t}$ ; $i_R = \frac{v_c}{R} = \underline{0.25(e^{-t} + e^{-4t})}$ , $t > 0$

(b) $i_R = .250 = 0.25(e^{-t} + e^{-4t})$ or $e^{-t} + e^{-4t} = 1$ by trial error or solve $\underline{t = 0.3223\ s}$

**6-6**

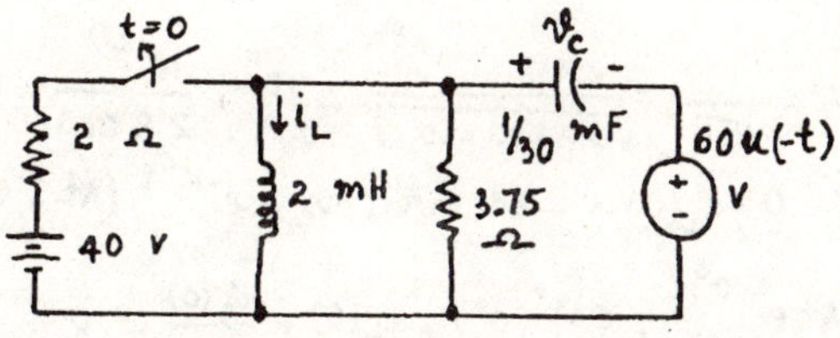


(a) $i_L(0) = 20$ A ; $v_c(0) = -60$ V

$\alpha = \frac{30\times10^3}{2\times3.75} = 4000$ ; $\omega_0^2 = \frac{30\times10^3}{2\times10^{-3}} = 15\times10^6$

$s_{1,2} = -4000 \pm 1000 = -3000, -5000\ s^{-1}$

$\therefore i_L = Ae^{-3000t} + Be^{-5000t}$ ; $i_L(0) = 20 = A + B$

$v_L(0) = -60 = Li_L'(0) = 2\times10^{-3}(-3000A - 5000B)$

or $-A - 5/3\,B = -10$ $\Big\}$ $\therefore A = 35$, $B = -15$

(b)

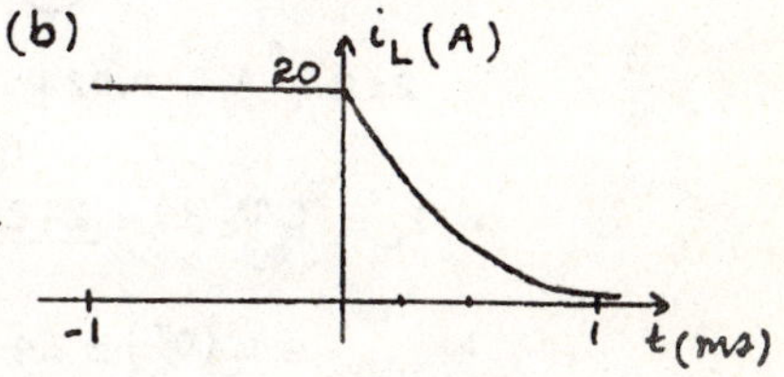


$\therefore i_L = \underline{20\ A}$ , $t < 0$ ; $i_L = \underline{35e^{-3000t} - 15e^{-5000t}\ A}$, $t > 0$

**6-7**

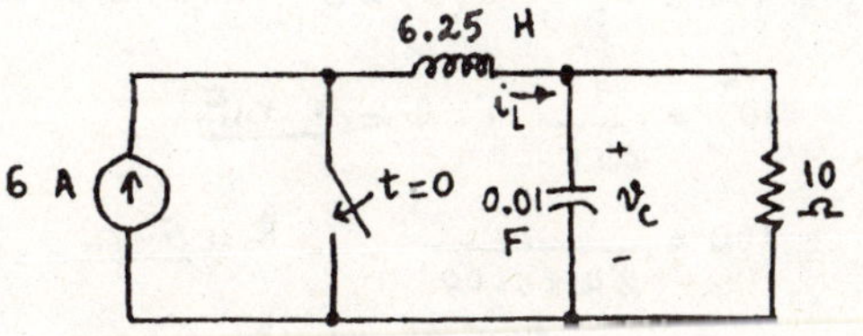


(a) $i_L(0) = 6$ A ; $v_c(0) = 60$ V ; $\alpha = \frac{1}{2RC} = 5$

$\omega_0^2 = \frac{100}{6.25} = 16$ ; $s_{1,2} = -5 \pm 3 = -2, -8\ s^{-1}$

$\therefore v_c = Ae^{-2t} + Be^{-8t}$ $\therefore A + B = 60$

$i_c = Cv_c' = 0.01(-2A - 8B) = 6 - \frac{60}{10}$ $\therefore -A - 4B = 0$ $\Big\}$ $\therefore A = 80$, $B = -20$

$\therefore v_c = \underline{80e^{-2t} - 20e^{-8t}\ V}$ , $t > 0$

(b) $i_L = 0.01v_c' + 0.1v_c = -1.6e^{-2t} + 1.6e^{-8t} + 8e^{-2t} - 2e^{-8t} = \underline{6.4e^{-2t} - 0.4e^{-8t}\ A}$, $t > 0$

(c) $v_{c,max} = v_c(0) = 60$ V $\therefore 0.6 = 80e^{-2t_s} - 20e^{-8t_s}$ $\therefore t_s = \underline{2.4464\ s}$

(d) $i_{L,max} = i_L(0) = 6$ $\therefore 0.06 = 6.4e^{-2t_s} - 0.4e^{-8t_s}$ $\therefore t_s = \underline{2.33485\ s}$

**6-8**

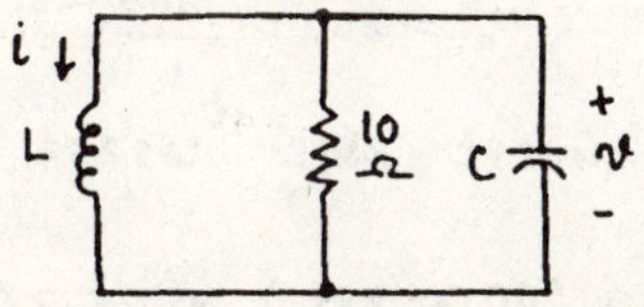


$v(0) = 100$ V ; $i(0) = 4$ A ; $\alpha = 100 = \omega_0 = \frac{1}{2RC} = \frac{1}{20C}$

(a) $C = \frac{1}{2000} = \underline{0.5\ mF}$ ; $100^2 = \frac{10^4}{L\times5}$ $\therefore$ $L = \underline{0.2\ H}$

(b) $v(t) = e^{-100t}(At + B)$ $\therefore B = 100$

(cont. on next page)

6-8 (cont.) $Cv'(0^+) = -i_R(0^+) - i(0) = -\frac{100}{10} - 4$

$0.5\times10^{-3}(A - 100\times100) = -14$

$\therefore A = -18000$

$\therefore v(t) = \underline{e^{-100t}(-18000t + 100)}\,,\ t>0$

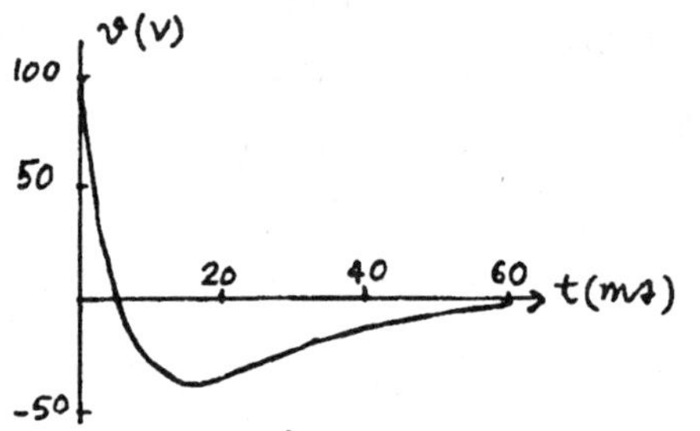


| t(ms) | v(V) |
|---|---|
| 0 | 100 |
| 2.5 | 42.8 |
| 5 | 6.1 |
| 10 | -29.4 |
| 15 | -37.9 |
| 20 | -35.2 |
| 30 | -21.9 |
| 60 | -2.4 |

(c) $e^{-100t}(-18000t + 100) = 0 \quad \therefore t = \frac{100}{18000} = 5.55^{+}\times10^{-3} = \underline{5.56\ \text{ms}}$

(d) $-1 = e^{-100t}(-18000t + 100) \quad \therefore \underline{t = 70.66\,\text{ms}}$

6-9

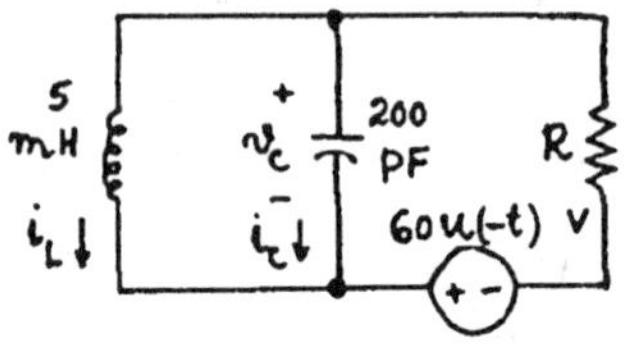


(a) $\alpha = \omega_0 = \frac{1}{\sqrt{LC}} = \frac{1}{\sqrt{5\times200\times10^{-15}}} = 10^6\,;\ \frac{10^{12}}{2R\,200} = 10^6 \ \therefore R = \underline{2500\ \Omega}$

(b) $v_c(0) = 0\,;\ i_L(0) = -24\ \text{mA}\,;\ v_c = e^{-10^6t}(At + B) \quad \therefore B = 0$

$v_c = Ate^{-10^6t}\,;\ Cv_c'(0^+) + i_L(0) + \frac{v_c(0)}{R} = 0$

$\therefore 2\times10^{-10}A - 0.024 + 0 = 0\,,\ A = 1.2\times10^8\,,\ v_c = 1.2\times10^8\,te^{-10^6t}\ \text{V}\,,\ t>0$

$\therefore i_c = Cv_c'(t) = \underline{24e^{-10^6t}(1 - 10^6t)\ \text{mA}}\,,\ t>0$

(c) $i_{c,max} = i_c(0^+) = 24\ \text{mA} \quad \therefore -0.24 = 24e^{-10^6t_s}(1 - 10^6t_s) \quad \therefore t_s = \underline{6.26654\ \mu s}$

6-10

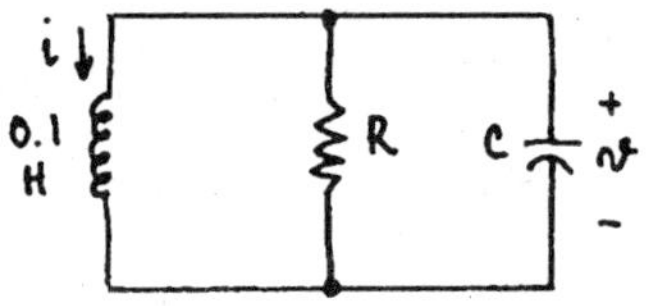


Crit. d. $\omega_0 = 100 = \alpha\,;\ v(0) = 0\,;\ i(0) = 5\ \text{A}$

(a) $\omega_0^2 = \frac{1}{LC} = 10^4 = \frac{1}{0.1C} \quad \therefore C = \underline{1\ \text{mF}}$

$\omega_0 = \frac{1}{2RC} = 100 = \frac{1}{2R\times0.001} \quad \therefore R = \underline{5\ \Omega}$

(b) $v = Ate^{-100t}$ since $v(0) = 0\,;\ Cv'(0) = -i(0) - \frac{v(0)}{R} \quad \therefore 10^{-3}\times A\times1 = -5 - 0$

$\therefore A = -5000\,;\ v = \underline{-5000te^{-100t}\ \text{V}}\,,\ t>0$

(c) $\frac{d}{dt}(te^{-100t}) = 0 \quad \therefore -100t_m + 1 = 0\,;\ t_m = \underline{0.01\ s}\,;\ v_m = \underline{18.3940\ \text{V}}$

(d) $0.183940 = 5000t_se^{-100t_s} \quad \therefore t_s = \underline{0.07638\ s}$

6-11

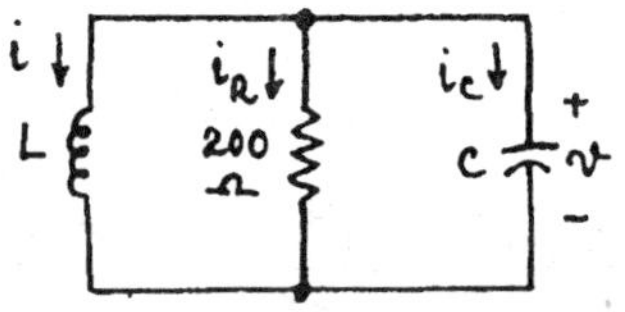


(a) $i = 10e^{-4t}\sin 28t\,;\ \alpha = 4 = \frac{1}{2\times200C} \quad \therefore C = \frac{1}{1600}\ \text{F}$

$\omega_d^2 = 28^2 = \omega_0^2 - 16 \quad \therefore \omega_0^2 = 800 = \frac{1600}{L} \quad \therefore L = 2\ \text{H}$

$v = Li' = 2(-40e^{-4t}\sin 28t + 280e^{-4t}\cos 28t)$

$v = \underline{e^{-4t}(-80\sin 28t + 560\cos 28t)\ \text{V}}\,,\ t>0$

(cont. on next page)

6-11 (cont.)

(b) $i_c = -i - \frac{v}{R} = -10e^{-4t}\sin 28t + 0.4e^{-4t}\sin 28t - 2.8e^{-4t}\cos 28t$

$$i_c = e^{-4t}(-9.6\sin 28t - 2.8\cos 28t)\ \text{A},\quad t>0$$

(c) $i_R = \frac{v}{R} = e^{-4t}(-0.4\sin 28t + 2.8\cos 28t)\ \text{A},\quad t>0$

6-12

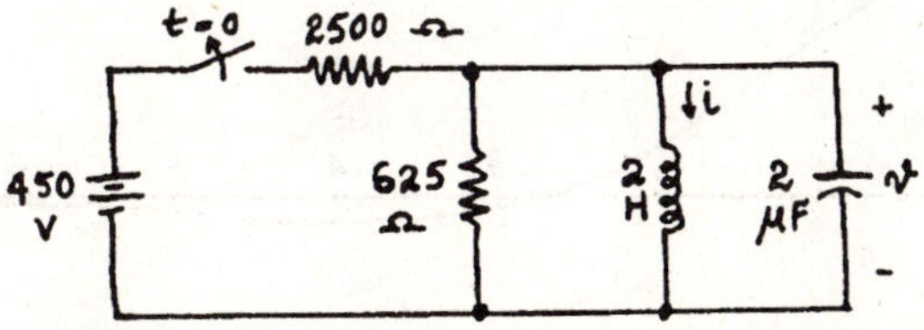


$v(0) = 0\ ;\ i(0) = \frac{450}{2500} = 0.18\ \text{A}$

$\alpha = \frac{1}{2RC} = \frac{10^6}{1250\times 2} = 400$

$\omega_0^2 = \frac{1}{LC} = \frac{10^6}{2\times 2} = 250{,}000\ ;\ \omega_d = \sqrt{\alpha^2 - \omega_0^2}$

$\omega_d = \sqrt{25-16}\times 10^2 = 300$

$i(t) = e^{-400t}(A\cos 300t + B\sin 300t)\quad \therefore A = 0.18$

$v_L(0) = v(0) = 2i'(0^+) = 0 = 300B - 400\times 0.18\quad \therefore B = 0.2$

$$i(t) = e^{-400t}(0.18\cos 300t + 0.24\sin 300t)\ \text{A},\quad t>0$$

6-13 (a)

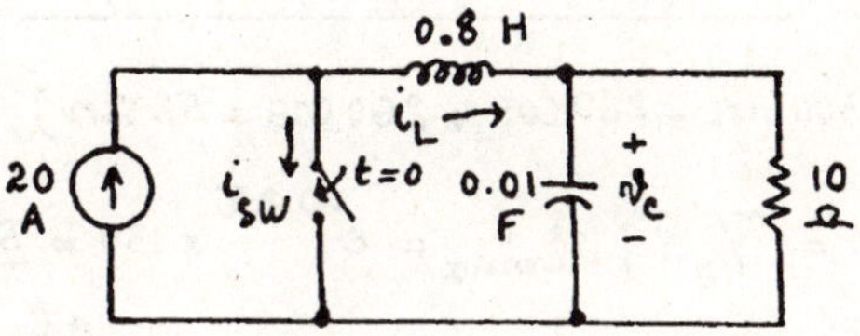


$i_L(0) = 20\ ;\ v_c(0) = 200\ ;\ \alpha = \frac{1}{2RC} = \frac{100}{20} = 5$

$\omega_0^2 = \frac{1}{LC} = \frac{100}{0.8} = 125\ ;\ \omega_d = \sqrt{\omega_0^2 - \alpha^2} = 10$

$v_c = e^{-5t}(A\cos 10t + B\sin 10t)\quad \therefore A = 200$

$0.01v_c'(0^+) = 0.01(10B - 1000) = 20 - \frac{200}{10} = 0\quad \therefore B = 100$

$$\therefore v_c = e^{-5t}(200\cos 10t + 100\sin 10t)\ \text{V},\quad t>0$$

(b) $i_{sw} = 20 - i_L\ ;\ i_L = \frac{v_c}{10} + 0.01v_c'\quad \therefore i_L = e^{-5t}(20\cos + 10\sin - 20\sin + 10\cos - 10\cos - 5\sin)$

$i_L = e^{-5t}(20\cos 10t - 15\sin 10t)\quad \therefore i_{sw} = 20 + e^{-5t}(15\sin 10t - 20\cos 10t)\ \text{A},\ t>0$

6-14 (a)

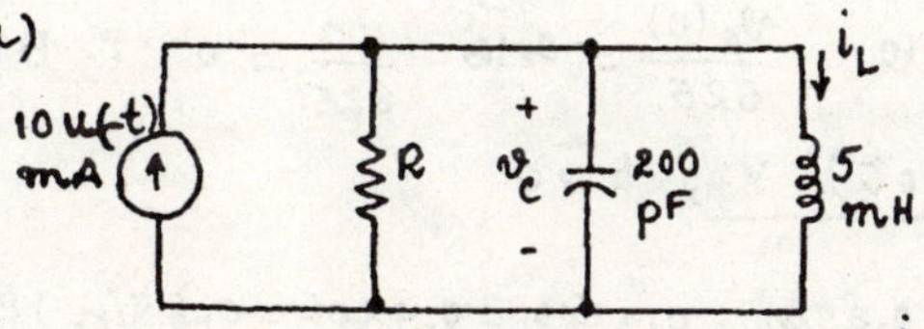


$\omega_d = 8\times 10^5 = \sqrt{10^{12} - \alpha^2}\quad \therefore \alpha = 6\times 10^5 = \frac{1}{2RC}$

$\therefore R = \frac{1}{2\alpha C} = \frac{10^{-5+12}}{12\times 200} = 4167\ \Omega\ ;\ v_c(0) = 0$

$i_L(0) = 10\ \text{mA}\quad \therefore i_L = e^{-6\times 10^5 t}(A\cos 8\times 10^5 t + B\sin 8\times 10^5 t)$

$\therefore A = 10^{-2}\ ;\ Li_L'(0^+) = v_c(0) = 0\quad \therefore 8\times 10^5 B = 6\times 10^5 A$

$\therefore B = 7.5\times 10^{-3}\quad \therefore i_L = e^{-6\times 10^5 t}(10\cos 8\times 10^5 t + 7.5\sin 8\times 10^5 t)\ \text{mA},\quad t>0$

(cont. on next page)

6-14 (cont.)

| $t\,(\mu s)$ | $i_L\,(mA)$ |
|---|---|
| 0 | 10 |
| 0.5 | 8.99 |
| 1 | 6.78 |
| 1.5 | 4.32 |
| 2 | 2.17 |
| 2.5 | 0.59 |
| 3 | -0.38 |
| 3.5 | -0.80 |
| 4 | -0.95 |
| 5 | -0.61 |
| 6 | -0.18 |
| 7 | +0.05 |

(b)

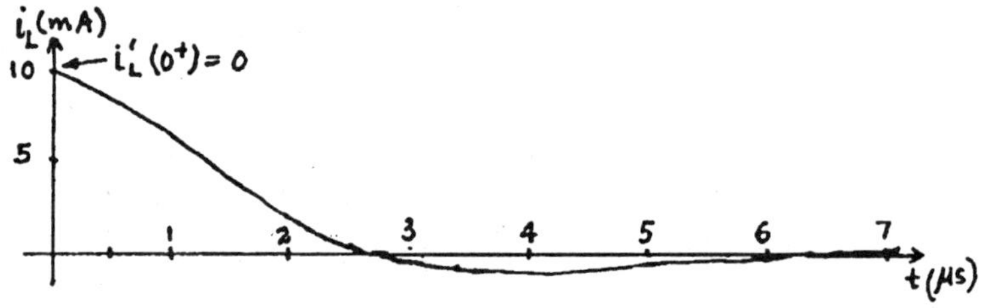


6-15 (a)

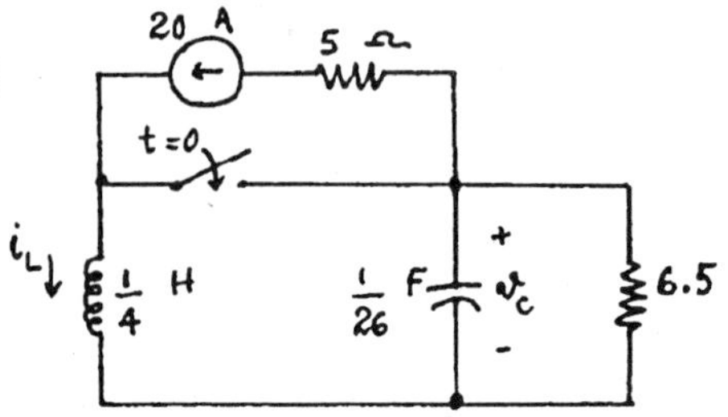


$\alpha = \dfrac{1}{2RC} = \dfrac{26}{2\times 6.5} = 2 \;;\; \omega_0^2 = \dfrac{1}{LC} = 26\times 4 = 104$

$\omega_d = \sqrt{\omega_0^2 - \alpha^2} = 10 \;;\; v_c(0) = -130\text{ V} \;;\; i_L(0) = 20\text{ A}$

$\therefore v_c = e^{-2t}(-130\cos 10t + B\sin 10t)$

$\frac{1}{26} v_c'(0^+) = -20 + 20 = 0 = 10B + 260 \quad \therefore B = -26$

$\therefore v_c = e^{-2t}(-130\cos 10t - 26\sin 10t)\text{ V}, \quad t > 0$

(b) $v_{min} = -130\text{ V}$, at $t = 0$ ; $v_c' = e^{-2t}(1300\sin - 260\cos + 260\cos + 52\sin)$

$\therefore \sin 10t = 0 \quad \therefore t = 0\ (v_{c\,min})$ and $t = \pi/10$ ; $v_{c\,max} = e^{-0.2\pi}\times 130 = 69.353\text{ V}$ at $t = \pi/10$

6-16 (a)

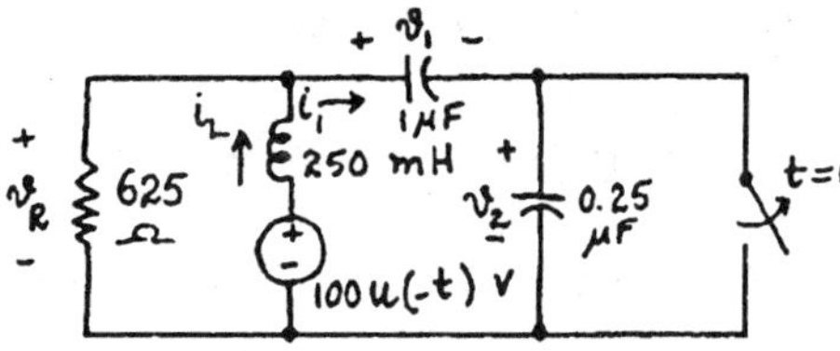


$\alpha = \dfrac{1}{2RC} = \dfrac{10^6}{2\times 625(0.25 /\!/ 1)} = 4000 \;;\; \omega_0^2 = \dfrac{1}{LC}$

$\omega_0^2 = \dfrac{10^6}{0.2\times 0.25} = 20\times 10^6 \;;\; \omega_d = \sqrt{\omega_0^2 - \alpha^2} = 2000$

$v_R(0^+) = v_1(0) = 100\text{ V}, \quad i_L(0) = \dfrac{100}{625} = 0.16\text{ A}$

$v_R = e^{-4000t}(100\cos 2000t + B\sin 2000t)$

$0.2\times 10^{-6}\, v_R'(0^+) = 2000B - 4000\times 100 = i_L(0) - \dfrac{v_R(0)}{625} = 0.16 - \dfrac{100}{625} = 0 \quad \therefore B = 200$

$\therefore v_R = e^{-4000t}(100\cos 2000t + 200\sin 2000t)\text{ V}, \quad t > 0$

(b) $v_2(0) = 0 \;;\; i_1 = 0.2\times 10^{-6} v_R' = 0.2\times 10^{-6}(-0.2\sin + 0.4\cos - 0.4\cos - 0.8\sin)e^{-4000t}$

$\therefore i_1 = -0.2e^{-4000t}\sin 2000t \quad \therefore v_1 = 100 + \dfrac{1}{10^{-6}}\int_0^t -0.2e^{-4000t}\sin 2000t\, dt$

$v_1 = 100 + 10^6\times 0.2\left[e^{-4000t}\dfrac{1}{(16+4)10^6}(4000\sin 2000t + 2000\cos 2000t)\right]_0^t$

$v_1 = 80 + e^{-4000t}(40\sin 2000t + 20\cos 2000t)\text{ V}, \quad t > 0$

6-17

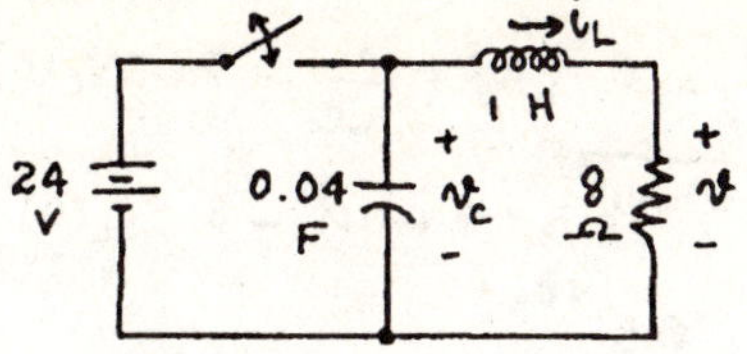


$\alpha = \frac{R}{2L} = \frac{8}{2} = 4$ ; $\omega_0^2 = \frac{1}{LC} = 25$ ; $\omega_d = \sqrt{\omega_0^2 - \alpha^2} = 3$

$v_C(0) = 24$ V ; $i_L(0) = 3$ A

$i_L = e^{-4t}(3\cos 3t + B\sin 3t)$

$1 \times i_L'(0^+) = 3B - 12 = 0$ $\therefore B = 4$ $\therefore i_L = e^{-4t}(3\cos 3t + 4\sin 3t)$ A , $t > 0$

$v = 8i_L = \underline{e^{-4t}(24\cos 3t + 32\sin 3t) \text{ V}}$ , $t > 0$

6-18 (a)

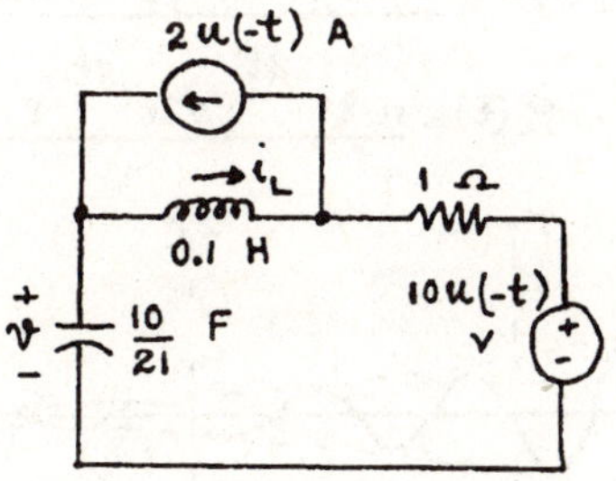


$\alpha = \frac{R}{2L} = \frac{1}{0.2} = 5$ ; $\omega_0^2 = \frac{1}{LC} = \frac{21}{1} = 21$

$s_{1,2} = -\alpha \pm \sqrt{\alpha^2 - \omega_0^2} = -5 \pm 2 = -3$ and $-7$

$v(0) = 10$ V ; $i_L(0) = 2$ A ; $v = Ae^{-3t} + Be^{-7t}$

$$\left.\begin{array}{l} v(0) = A + B = 10 \\ -i_L(0^+) = \frac{10}{21}(-3A - 7B) = -2 \end{array}\right\} \therefore \begin{array}{l} A = 16.45 \\ B = -6.45 \end{array}$$

$\therefore v = \underline{16.45e^{-3t} - 6.45e^{-7t} \text{ V}}$ , $t > 0$

(b) $16.45e^{-3t_s} - 6.45e^{-7t_s} = 0.1$ $\therefore t_s = \underline{1.70083 \text{ s}}$

6-19 (a)

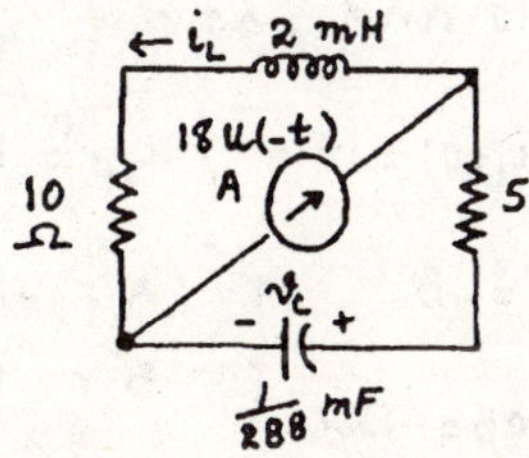


$\alpha = \frac{R}{2L} = \frac{60 \times 10^3}{2 \times 2} = 15{,}000$ ; $\omega_0^2 = \frac{1}{LC} = \frac{288}{2} \times 10^6 = 144 \times 10^6$

$s_{1,2} = -\alpha \pm \sqrt{\alpha^2 - \omega_0^2} = -15{,}000 \pm 9000 = -6000$ and $-24{,}000$ $s^{-1}$

$i_L(0) = 18$ A ; $v_C(0) = 180$ V ; $i_L = Ae^{-6000t} + Be^{-24{,}000t}$

$$\left.\begin{array}{l} i_L(0) = A + B = 18 \\ Li_L'(0) = 2\times 10^{-3}(-6A - 24B)\times 10^3 = -60 \times 18 + 180 \end{array}\right\} \therefore \begin{array}{l} A = -1 \\ B = 19 \end{array} \therefore i_L = \underline{-e^{-6000t} + 19e^{-24{,}000t} \text{ A}}$$

$t > 0$

(b) $60i_L + 0.002i_L' + 288 \times 10^3 \int_0^t i_L\,dt - 180 = 0$ $\therefore 60i_L' + 0.002i_L'' + 288 \times 10^3 i_L = 0$

$\therefore 60(6000 - 19 \times 24000) + 0.002i_L''(0^+) + 288 \times 10^3 \times 18 = 0$ $\therefore i_L''(0^+) = \underline{1.0908 \times 10^{10}} \text{ A/s}^2$

6-20

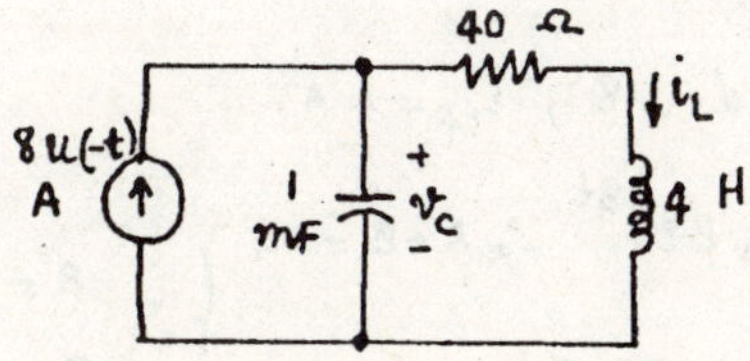


$\alpha = \frac{R}{2L} = \frac{40}{8} = 5$

$\omega_0^2 = \frac{1}{LC} = \frac{1000}{4} = 250$

$\omega_d = \sqrt{\omega_0^2 - \alpha^2} = 15$

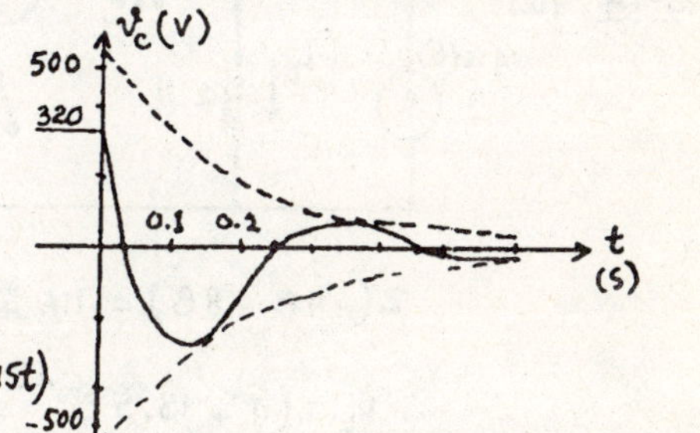


$v_C(0) = 320$ V ; $i_L(0) = 8$ A $\therefore v_C = e^{-5t}(320\cos 15t + B\sin 15t)$

$10^{-3}v_C'(0) = -8 = .015B - 5(320)$ $\therefore B = -426.7$

$\therefore v_C = \underline{e^{-5t}(320\cos 15t - 426.7\sin 15t) \text{ V}}$ , $t > 0$ or $v_C = \underline{533.3e^{-5t}\cos(15t + 53.13^\circ) \text{ V}}$

$t > 0$

**6-21** (a)

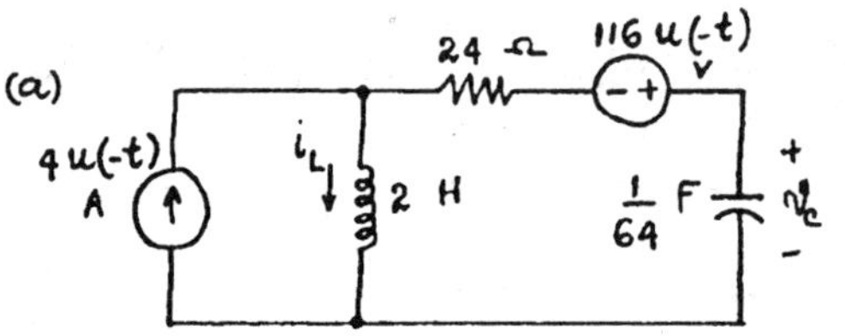


$\alpha = \frac{R}{2L} = \frac{24}{2\times 2} = 6$ ; $\omega_0^2 = \frac{1}{LC} = 32$ ; $v_c(0) = 116$ V

$s_{1,2} = -\alpha \pm \sqrt{\alpha^2 - \omega_0^2} = -4$ and $-8$ ; $i_L(0) = 4$ A

$\therefore i_L = Ae^{-4t} + Be^{-8t}$

$i_L(0) = A + B = 4$

$2(-4A - 8B) = v_c(0) - 4\times 24 \;\therefore\; -8A - 16B = 20$

$\therefore A = 10.5,\; B = -6.5$ ; $i_L = 10.5e^{-4t} - 6.5e^{-8t}$

$i_L(t) = \underline{4u(-t) + (10.5e^{-4t} - 6.5e^{-8t})u(t)\text{ A}}$

(b) $v_c = 2i_L' + 24i_L = -84e^{-4t} + 104e^{-8t} + 252e^{-4t} - 156e^{-8t} \;\therefore\; v_c(t) = \underline{168e^{-4t} - 52e^{-8t}\text{ V}}$, $t > 0$

**6-22**

$v_c(t) = e^{-t}(A\cos 25t + B\sin 25t)$

$v_c(0) = 1(A + 0) = 0 \;\therefore\; A = 0$

$v_c = Be^{-t}\sin 25t$ ; $v_c'(0^+) = 25B = 100$

$\therefore B = 4$ ; $v_c(t) = 4e^{-t}\sin 25t$, $T = \frac{2\pi}{25} = 0.2513$ s

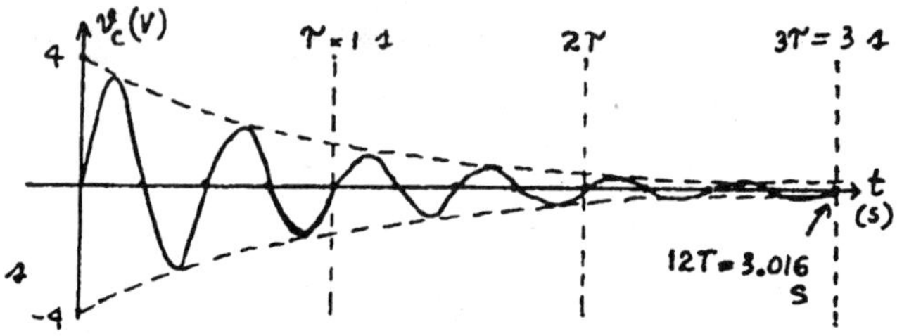


**6-23** (a)

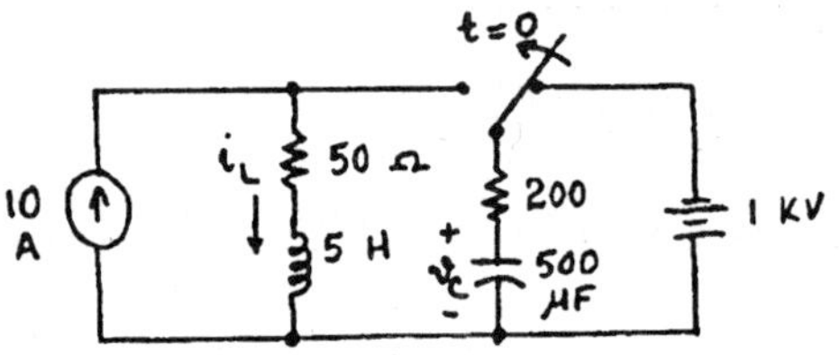


$\alpha = \frac{R_{eq}}{2L} = \frac{250}{10} = 25$ ; $\omega_0^2 = \frac{1}{LC} = \frac{10^6}{5\times 500} = 400$

$s_{1,2} = -25 \pm 15 = -10$ and $-40$

$v_c(0) = -1000$ V , $i_L(0) = 10$ A , $i_{Lf} = 10$ A

$\therefore i_L = 10 + Ae^{-10t} + Be^{-40t}$ ; $i_L(0) = 10 + A + B = 10 \;\therefore\; A = -B$

$5(-10A - 40B) = -50i_L(0) + [10 - i_L(0)]200 + v_c(0) \;\therefore\; -50A - 200B = -1500$

$\therefore A = -10,\; B = 10$

$\therefore i_L = \underline{10 - 10e^{-10t} + 10e^{-40t}\text{ A}}$, $t > 0$

(b) $i_{L,max} = \underline{10\text{ A}}$ , $i_L' = 100e^{-10t} - 400e^{-40t} = 0$ , $1 = 4e^{-30t}$ , $t = 0.04621$ s $\therefore i_{L,min} = \underline{5.2753\text{ A}}$

**6-24** (a)

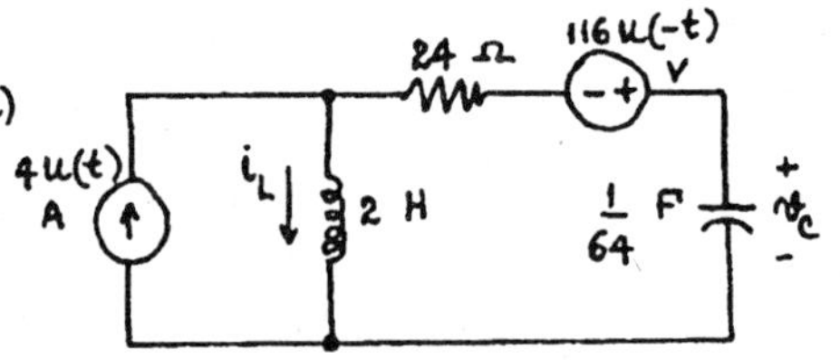


$\alpha = 6$ ; $\omega_0^2 = 32$ ; $i_L(0) = 0$ ; $v_c(0) = 116$ V

$s_{1,2} = -4$ and $-8$ ; $i_{Lf} = 4$ A

$i_L = 4 + Ae^{-4t} + Be^{-8t} \;\therefore\; A + B = -4$

$2(-4A - 8B) = 116 + 4\times 24 = 212 \;\therefore\; -A - 2B = 26.5$

$\therefore A = 18.5,\; B = -22.5$

$\therefore i_L = \underline{(4 + 18.5e^{-4t} - 22.5e^{-8t})u(t)\text{ A}}$

(b) $v_c = 24(i_L - 4) + 2i_L' \;\therefore\; v_c = 444e^{-4t} - 540e^{-8t} - 148e^{-4t} + 360e^{-8t} = \underline{296e^{-4t} - 180e^{-8t}\text{ V}}$, t>

**6-25** (a)

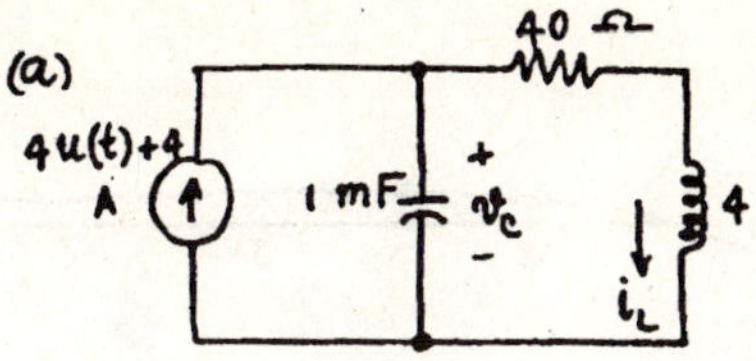


$\alpha = 5$ ; $\omega_d = 15$ ; $i_L(0) = 4$ A ; $v_c(0) = 160$ V ; $v_{cf} = 320$ V

$v_c = 320 + e^{-5t}(A\cos 15t + B\sin 15t)$ ; $160 = 320 + A$ $\therefore A = -160$

$v_c'(0^+) = 10^3 \times 4 = 15B + 800$ $\therefore B = \frac{3200}{15} = 213.3$

$\therefore v_c = \underline{320 + e^{-5t}(-160\cos 15t + 213.3\sin 15t)\ \text{V}}$ , $t > 0$

(b) $v_c' = e^{-15t}(2400\sin + 3200\cos + 800\cos - 1066.7\sin) = 0$ $\therefore 1333.3\sin 15t = -4000\cos 15t$

$\tan 15t_m = -3$ $\therefore t_m = 0.12617\,s$ $\therefore v_{c,max} = \underline{454.6\ \text{V}}$

**6-26**

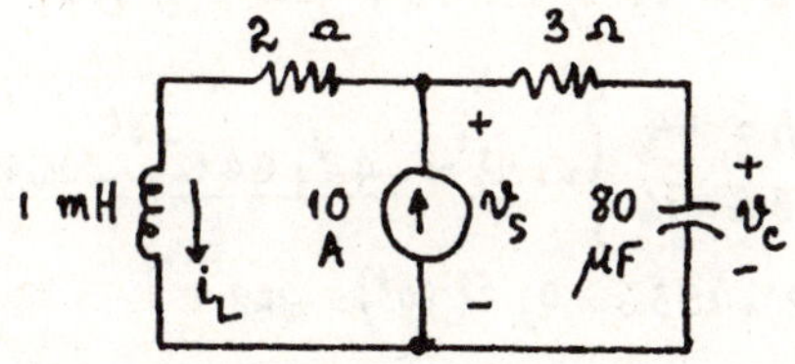


$\alpha = \frac{R_{eq}}{2L} = \frac{5}{2} \times 10^3 = 2500$ ; $\omega_0^2 = \frac{1}{LC} = \frac{10^9}{80} = 12.5 \times 10^6$

$\omega_d = 2500$ ; $i_L(0) = 10$ A ; $v_c(0) = 20$ V ; $i_{Lf} = 20$ A

$\therefore i_L = 20 + e^{-2500t}(A\cos 2500t + B\sin 2500t)$ ; $A = -10$

$10^{-3} i_L'(0^+) = 10^{-3}(2.5B - 2.5A)10^3 = 20 + 30 - 20 = 30$ $\therefore B = 2$

$\therefore i_L = 20 + e^{-2500t}(-10\cos 2500t + 2\sin 2500t)$ ; $v_s = 2i_L + 10^{-3} i_L'$

$v_s = 40 + e^{-2500t}(10\cos 2500t + 20\sin 2500t + 4\sin 2500t)$

$\therefore v_s = \underline{40 + e^{-2500t}(10\cos 2500t + 24\sin 2500t)\ \text{V}}$ , $t > 0$

**6-27**

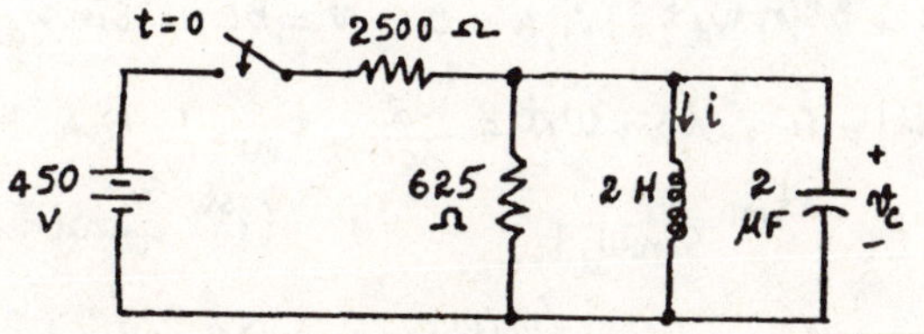


$\alpha = \frac{1}{2R_{eq}C} = \frac{10^6}{2(625 \| 2500)2} = 500$ ; $\omega_0^2 = \frac{1}{LC}$

$\omega_0^2 = \frac{10^6}{2 \times 2} = 250{,}000$ ; $\alpha = \omega_0$ $\therefore$ Crit. damp

$v_c(0) = 0$ ; $i(0) = 0$ ; $i_f = \frac{450}{2500} = 0.18$ A

$\therefore i = 0.18 + e^{-500t}(At + B)$ $\therefore B = -0.18$ ; $i'(0^+) = \frac{v_c(0)}{2} = 0 = A - 500B$

$\therefore A = 500(-0.18) = -90$ $\therefore i = \underline{0.18 + e^{-500t}(-90t - 0.18)\ \text{A}}$ , $t > 0$

**6-28** (a)

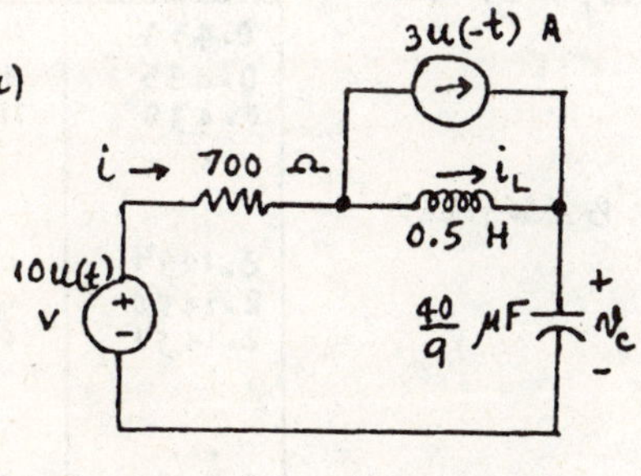


$\alpha = \frac{R}{2L} = \frac{700}{1} = 700$ ; $\omega_0^2 = \frac{1}{LC} = \frac{10^6}{20/9} = 450{,}000$

$s_{1,2} = -700 \pm 200 = -500$ and $-900$ ; $i_L(0) = -3$ A

$\therefore i(0^+) = -3$ A ; $v_c(0) = 0$ ; $i(\infty) = 0$ ; $i = Ae^{-500t} + Be^{-700t}$

$-3 = A + B$

$\frac{1}{2}(-500A - 900B) = 2100 + 10 + 0$ $\therefore -A - 1.8B = 8.44$ $\Big\}$ $\therefore$ $A = 3.8$, $B = -6.8$ $\Big\}$ $\therefore i = \underline{3.8e^{-500t} - 6.8e^{-900t}\ \text{A}}$

$t > 0$

(cont. on next page)

6-28 (cont.) (b)

| t (ms) | i (A) |
|---|---|
| 0 | −3.0 |
| 0.2 | −2.24 |
| 0.5 | −1.376 |
| 1 | −0.460 |
| 1.5 | 0.032 |
| 2 | 0.274 |
| 2.5 | 0.372 |
| 3 | 0.391 |
| 4 | 0.328 |

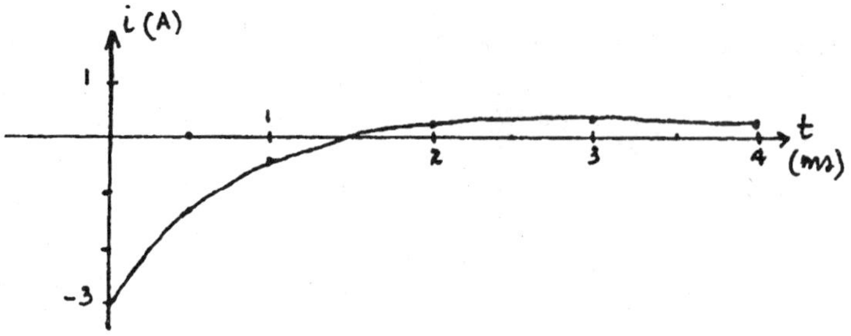


6-29 (a)

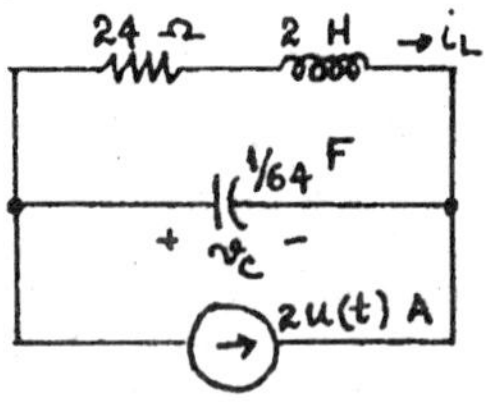


$\alpha = \frac{24}{4} = 6$ ; $\omega_o^2 = 32$ ; $s_{1,2} = -4$ and $-8$ ; $i_L(0) = 0$

$v_c(0) = 0$ ; $v_{c,f} = -48$ V $\therefore v_c = -48 + Ae^{-4t} + Be^{-8t}$ V

$\left.\begin{matrix} A + B = 48 \\ -4A - 8B = 64(-2) \end{matrix}\right\} \therefore \left.\begin{matrix} A = 64 \\ B = -16 \end{matrix}\right\} \therefore v_c = \underline{-48 + 64e^{-4t} - 16e^{-8t}}$ V, $t > 0$

(b) $v_c + 2 \times \frac{1}{64} v_c'' + 24\left(\frac{1}{64} v_c' + 2\right) = 0 \therefore v_c'' + 12 v_c' + 32 v_c + 1536 = 0$; $v_c'(0^+) = -128$

$\therefore v_c''(0^+) = -12(-128) - 1536 = \underline{0}$ ; $v_c''' + 12 v_c'' + 32 v_c' = 0$; $v_c'''(0^+) + 0 + 32(-128) = 0 \therefore v_c'''(0^+) = \underline{4096 \text{ V/s}^3}$

6-30

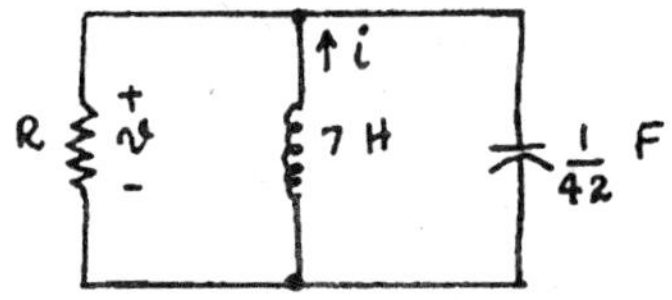


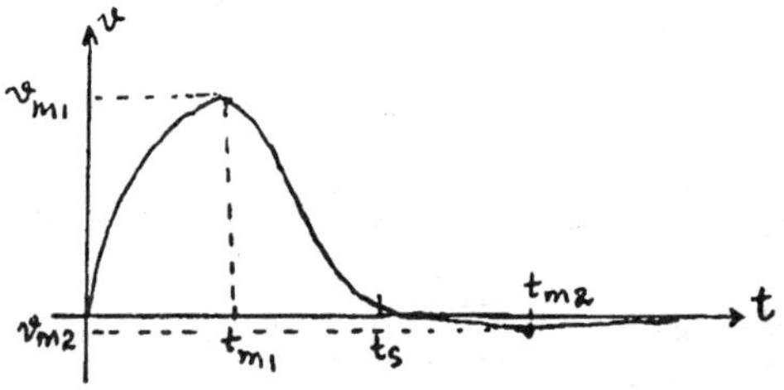


$v(0) = 0$; $i(0) = 10$ A; $v = e^{-\alpha t}(A\cos\omega_d t + B\sin\omega_d t)$ $\therefore A = 0$, $v = Be^{-\alpha t}\sin\omega_d t$

$v' = e^{-\alpha t}\left[-\alpha B\sin\omega_d t + \omega_d B\cos\omega_d t\right] = 0 \therefore \tan\omega_d t = \frac{\omega_d}{\alpha}$, $t_{m1} = \frac{1}{\omega_d}\tan^{-1}\frac{\omega_d}{\alpha}$ (1st quadrant)

$t_{m2} = t_{m1} + \frac{1}{2}T_d = t_{m1} + \frac{\pi}{\omega_d}$ ; $v_{m1} = Be^{-\alpha t_{m1}}\sin\omega_d t_{m1}$

$v_{m2} = -Be^{-\alpha t_{m1} - \alpha\pi/\omega_d}\sin\omega_d t_{m1} \therefore \frac{v_{m2}}{v_{m1}} = -e^{-\alpha\pi/\omega_d}$ ; let $\left|\frac{v_{m2}}{v_{m1}}\right| = \frac{1}{100}$

$\therefore e^{\alpha\pi/\omega_d} = 100$, $\alpha = \frac{\omega_d}{\pi}\ln 100$ ; $\alpha = \frac{1}{2RC} = \frac{21}{R}$, $\omega_o^2 = \frac{1}{LC} = 6 \therefore \omega_d = \sqrt{6 - 441/R^2}$

$\therefore \frac{21}{R} = \frac{\ln 100}{\pi R}\sqrt{6R^2 - 441}$ $\therefore R = \sqrt{1/6\left[441 + \left(\frac{21\pi}{\ln 100}\right)^2\right]} = 10.3781\ \Omega$

To keep $\left|\frac{v_{m2}}{v_{m1}}\right| < 0.01$, choose $R = \underline{10.3780\ \Omega}$

$v'(0^+) = \omega_d B = B\sqrt{6 - \left(\frac{21}{10.378}\right)^2} = 42\left(10 + \frac{0}{10.3780}\right) \therefore B = 304.268$

$\alpha = \frac{21}{10.378} = 2.02351$ ; $\omega_d = \sqrt{6 - \left(\frac{21}{10.378}\right)^2} = 1.380363$

$\therefore v = 304.268e^{-2.02351t}\sin 1.380363t$ V

$t_{m1} = 0.434$ s , $v_{m1} = 71.2926$ V

Computed values show $t_s = \underline{2.145 \text{ sec}}$ ; $v_{m2} = 0.7126 < 0.01 v_{m1}$

| t (sec) | v (V) |
|---|---|
| 0.434 | 71.2926 |
| 0.435 | 71.2922 |
| 0.436 | 71.2915 |
| ⋮ | ⋮ |
| 2.1449 | 0.7133 |
| 2.1450 | 0.7126 |
| 2.1451 | 0.7119 |
| ⋮ | ⋮ |
| 2.708 | 0.712796 |
| 2.709 | 0.712801 |
| 2.710 | 0.712801 |
| 2.711 | 0.712797 |
| ⋮ | ⋮ |

**6-31** (a)

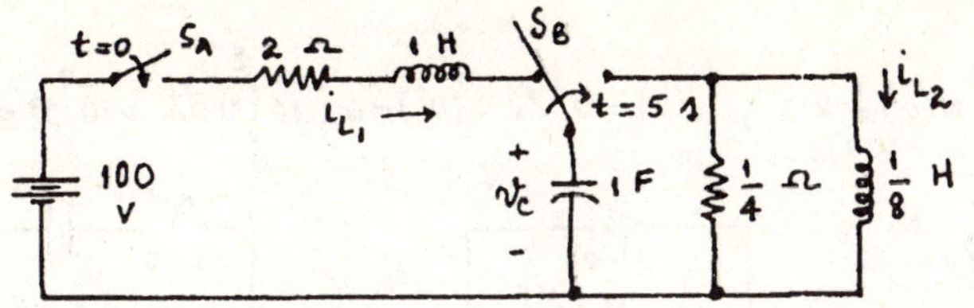


$\alpha = \frac{R}{2L} = \frac{2}{2\times 1} = 1$ ; $\omega_0^2 = 1$ ; $i_{L_1}(0) = 0$

$v_c(0) = 0$ (given) ; $v_{c,f} = 100$ V

Crit. damp $\therefore v_c = 100 + e^{-t}(At + B)$

$\therefore B = -100$ ; $v_c'(0^+) = 1\times 0 = 0 = A + 100 \;\therefore A = -100$ ; $v_c = \underline{100 - 100e^{-t}(t+1)\text{ V}}$ , $0 < t < 5$ s

(b) $\alpha = \frac{1}{2RC} = 2$ ; $\omega_0^2 = 8$ ; $\omega_d = 2$ ; $v_c(5) = 95.957$ V ; $i_{L_2}(5) = 0$

$v_c(\infty) = 0 \;\therefore v_c = e^{-2(t-5)}\left[A\cos 2(t-5) + B\sin 2(t-5)\right]$

$\therefore A = 95.957$ ; $v_c'(5) = -\frac{95.957}{0.25} = 2B - 2\times 95.957 \;\therefore B = -95.957$

$\therefore v_c = \underline{e^{-2(t-5)}(95.957)\left[\cos 2(t-5) - \sin 2(t-5)\right]\text{ V}}$ , $t > 5$

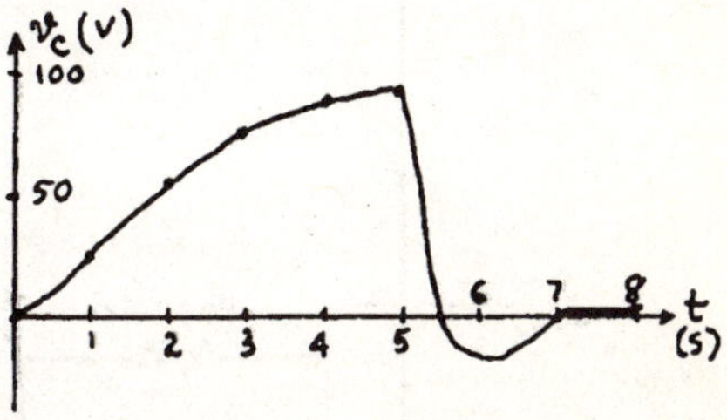


**6-32**

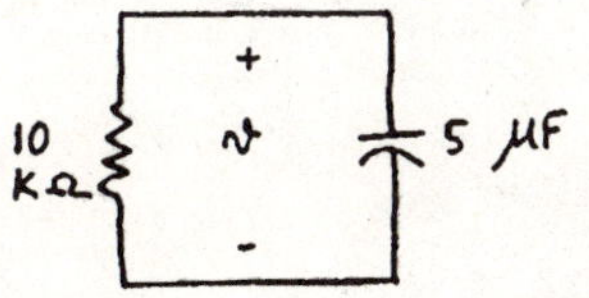


$Cv' + \frac{v}{R} = 0 \;\therefore 5\times 10^{-6+4}\, v' + v = 0$ , $\underline{v' = -20v}$

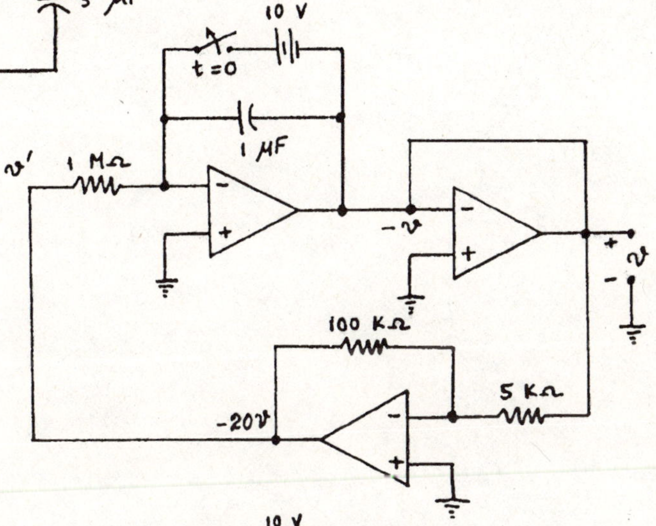


and

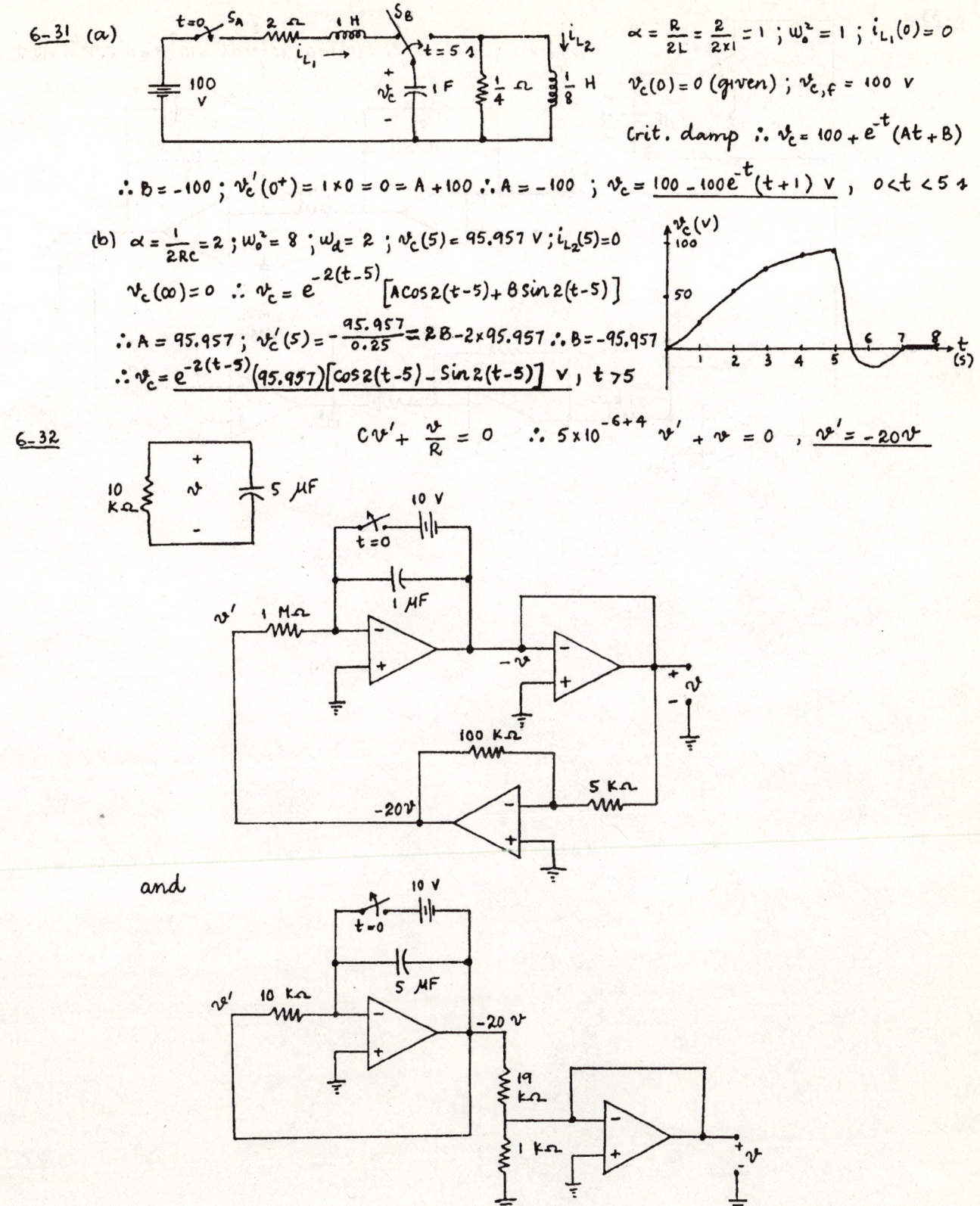

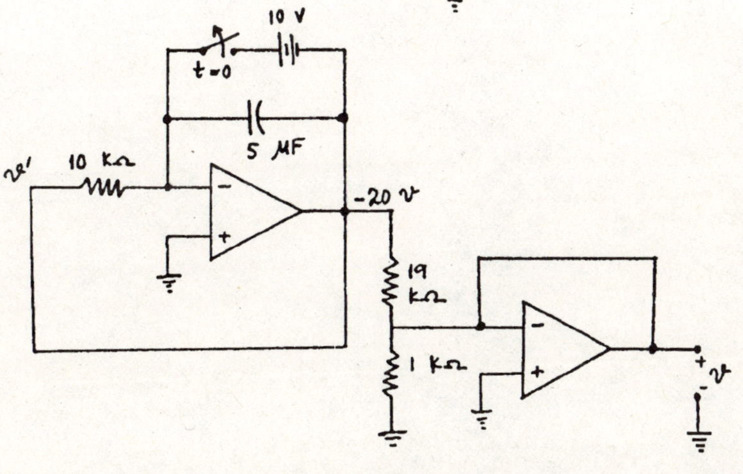

6-33

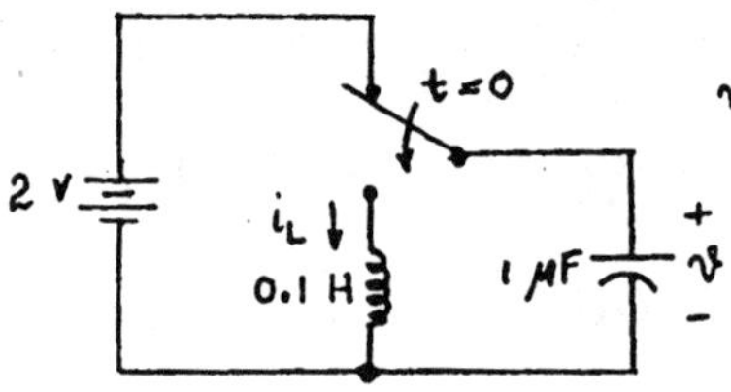


$$v(0) = 2\text{ V};\ i_L(0) = 0\ \therefore v'(0^+) = 0;\ 10\int_0^t v\,dt + 10^{-7}v' = 0\ \therefore v'' = -10^8 v$$

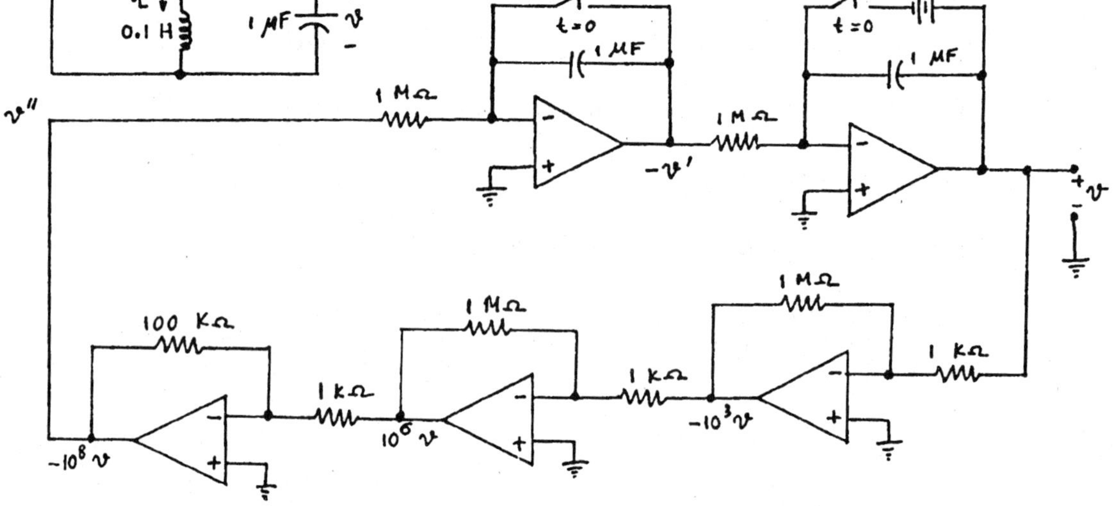

# 6-34

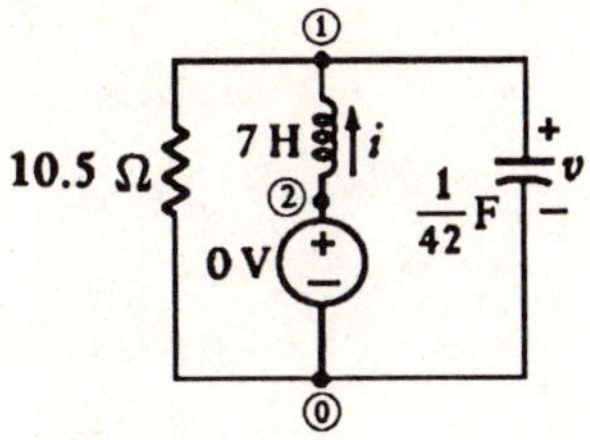


If $v(t) = 210\sqrt{2}\, e^{-2t} \sin \sqrt{2}t$,
then $v(0.6) =$ 67.1152 V and $v(2.4) =$ -0.61065 V.

***********************************************************************

**PROGRAM**

```
R10  1  0  10.5
L21 2 1 7 IC=10
V02 0 2
C10 1 0 23.80952M  IC=0
.TRAN 20M  2.4  UIC
.PRINT TRAN V(1)
.END
```

***********************************************************************

**ANSWER: Node voltage**

| TIME | V(1) |
|---|---|
| 0.000E+00 | 3.861E-04 |
| 2.000E-02 | 8.030E+00 |
| 4.000E-02 | 1.536E+01 |
| 6.000E-02 | 2.230E+01 |
| ..... | |
| 5.600E-01 | 6.906E+01 |
| 5.800E-01 | 6.820E+01 |
| 6.000E-01 | 6.724E+01  V = $v(0.6)$ |
| 6.200E-01 | 6.617E+01 |
| 6.400E-01 | 6.508E+01 |
| 6.600E-01 | 6.386E+01 |
| ..... | |
| 2.360E+00 | -5.371E-01 |
| 2.380E+00 | -5.863E-01 |
| 2.400E+00 | -6.324E-01  V = $v(2.4)$ |

*6-35*

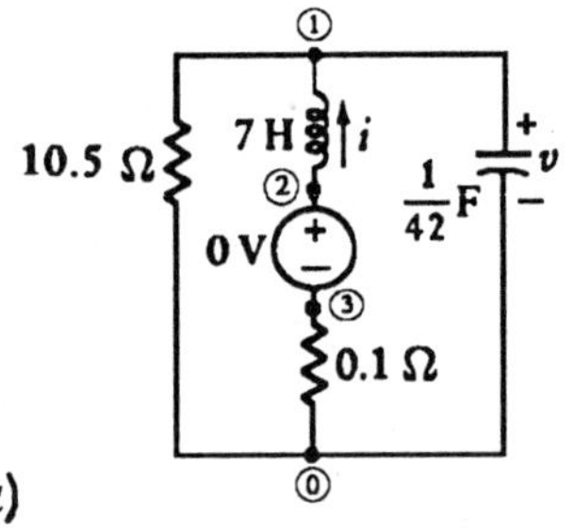


(*a*)

************************************************************************

**PROGRAM**

```
R10 1 0 10.5
L21 2 1 7 IC=10
V32 3 2
R30 3 0 0.1
C10 1 0 23.80952M IC=0
.TRAN 20M 2.4 UIC
.PRINT TRAN V(1)
.END
```

************************************************************************

**ANSWERS: Node voltage**

| TIME | V(1) | |
|---|---|---|
| 0.000E+00 | 3.875E-04 | |
| 2.000E-02 | 8.029E+00 | |
| ..... | | |
| 5.200E-01 | 7.015E+01 | |
| 5.400E-01 | 6.954E+01 | |
| ..... | | |
| 5.600E-01 | 6.868E+01 | |
| 5.800E-01 | 6.781E+01 | |
| 6.000E-01 | 6.683E+01 | V = $v(0.6)$ |
| 6.200E-01 | 6.575E+01 | |
| 6.400E-01 | 6.466E+01 | |
| ..... | | |
| 6.600E-01 | 6.342E+01 | |
| 6.800E-01 | 6.219E+01 | |
| ..... | | |
| 2.380E+00 | -6.304E-01 | |
| 2.400E+00 | -6.731E-01 | V = $v(2.4)$ |

************************************************************************

(*b*) From Prob. 6-34 solution, exact answers were $v(0.6)$ = 67.1152 V and $v(2.4)$ = -0.61065 V. Exact and approximate values are pretty close when $t = 0.6$ s, but accuracy is comparatively poor at $t = 2.4$ s.

# 6-36

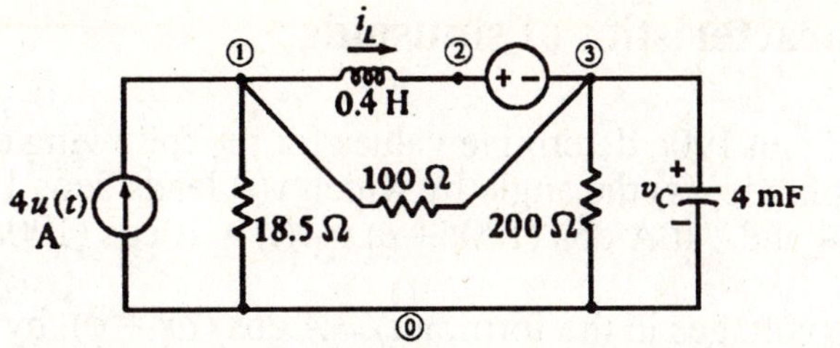


****************************************************************************

**PROGRAM:**

```
I01  0  1  PWL(0 0 1U 4 1 4)
R10  1  0  18.5
R13  1  3  100
V23  2  3
R30  3  0  200
L12  1  2  0.4  IC=0
C30  3  0  4M  IC=0
.TRAN  0.5M  0.1  UIC
.PRINT TRAN  V(3)
.END
```

****************************************************************************

| TIME | V(3) | |
|---|---|---|
| 1.000E-02 | 2.958E+00 | $v_C(10\ \text{ms}) = 2.958$ V |
| 4.000E-02 | 2.042E+01 | $v_C(40\ \text{ms}) = 20.42$ V |
| 1.000E-01 | 5.193E+01 | $v_C(100\ \text{ms}) = 51.93$ V |

# CHAPTER 7: THE SINUSOIDAL FORCING FUNCTION

## Section 7-2 Characteristics of sinusoids

**1** If $y(t) = 26\cos 190t - 18\sin 190t$, determine values for the following characteristics of $y(t)$: (*a*) $f$; (*b*) $T$; (*c*) $\omega$; (*d*) the amplitude; (*e*) the angle by which $y(t)$ leads $5\cos 190t$; (*f*) the angle by which $y(t)$ lags $8\sin 190t$; (*g*) $A$ and $\theta$ if $A\cos(190t + \theta) = y(t) + 20\cos(190t + 160°)$.

**2** Express each of these voltages in the form, $v(t) = A\cos(\omega t + \phi)$, by giving values for $A$ and $\phi$: (*a*) $6\sin 20t$; (*b*) $4\sin(20t - 20°)$; (*c*) $8\cos 5t + 3\sin 5t$. (*d*) What is the smallest positive value of $t$ for which $5\cos(100t + 40°) = 0$? (*e*) What is the smallest positive value of $t$ for which $-3\cos 20t + 2\sin 20t$ reaches its maximum value, and what is the maximum value?

**3** (*a*) A 100-Hz sinusoid has a maximum positive value of 50 at $t = 0.1234$ s. Express it in the form, $A\cos\omega t + B\sin\omega t$. (*b*) The sinusoid shown in Fig. 7-1 passes through zero every 2 ms and has a value of -12 at $t = 2.4$ ms. Express it in the form $C\cos(\omega t + \phi)$.

**4** Carry out the exercise threatened in the text by substituting the assumed current response of Eq. (3), $i(t) = A\cos(\omega t - \theta)$, directly into the differential equation, $L(di/dt) + Ri = V_m\cos\omega t$, to show that values for $A$ and $\theta$ are obtained which agree with Eq. (4).

## Section 7-3 Forced response to sinusoidal forcing functions

**5** In Fig. 7-2, let $v_s = 20\cos(500t + 100°)$ V and $v_R = K\cos(500t + 45°)$ V (+ reference on left). If $L = 80$ mH, find $R$, $K$, $i(t)$, and $v_L(t)$ (+ reference on top).

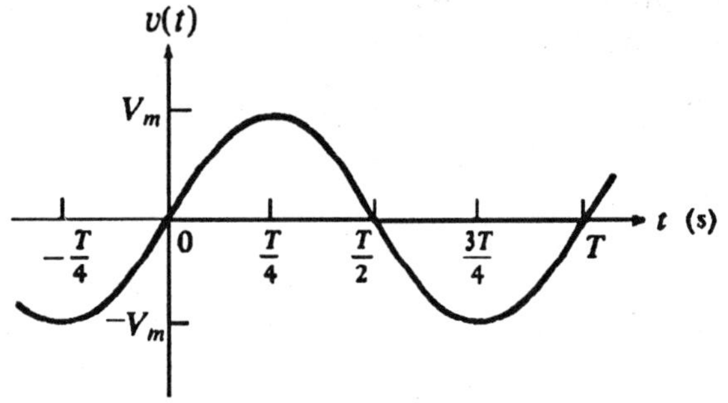


**Fig. 7-1:** See Prob. 3.

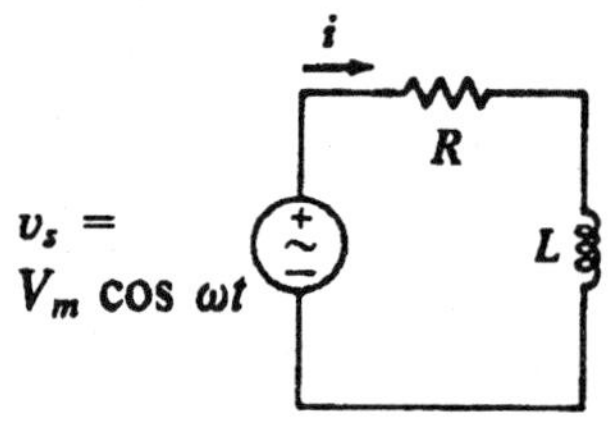


**Fig. 7-2:** See Prob 5.

**6** A 0.2-H inductor is in series with a 10-Ω resistor, a 30-Ω resistor, and a voltage source, $v_s = 100\cos 300t$ V. Determine the amplitude of the voltage across the series combination of the inductor and the 10-Ω resistor.

**7** A voltage source, $v_s = 10\cos(800t - 30°)$ V, a 20-Ω resistor, and a 50-mH inductor are in series. Determine the amplitude of the maximum instantaneous power: (*a*) generated by the source; (*b*) dissipated by the resistor; (*c*) absorbed by the inductor.

**8** Use Thévenin's theorem on the circuit of Fig. 7-3 to help find: (*a*) $i_L$; (*b*) $v_{60}$.

**9** Use repeated source transformations on the circuit shown in Fig. 7-4 and find $i_L(t)$.

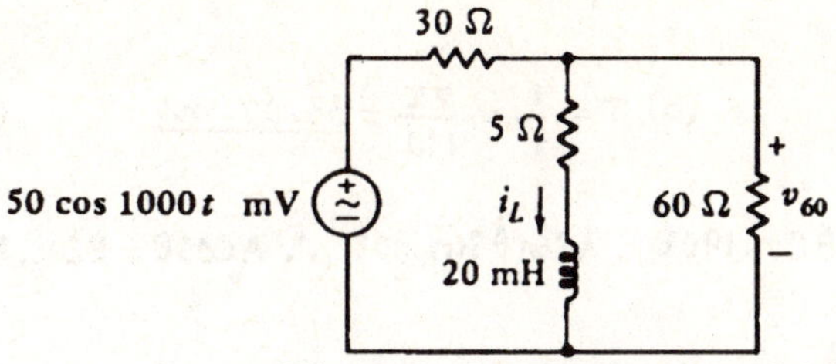


Fig. 7-3: See Probs. 8 and 10.

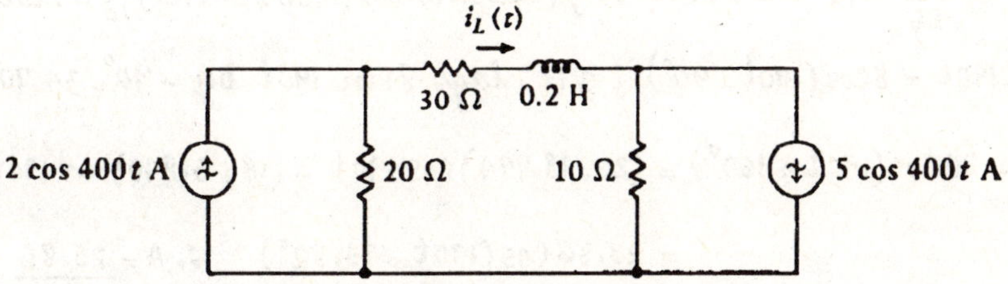


Fig. 7-4: See Probs. 9 and 11.

**10** Replace the 5-Ω resistor in Fig. 7-3 by a dependent voltage source, + reference at the top, and labeled $0.1v_{60}$. Now rework Prob. 8*a*.

**11** Change the right-hand current source in Fig. 7-4 to 5cos 200*t* A, and rework Prob. 9.

**12** Assume that the op-amp in Fig. 7-5 is ideal ($R_i = \infty$, $R_o = 0$, and $A = \infty$). Note also that the integrator input has two signals applied to it, $-V_m \cos \omega t$ and $v_{out}$. If the product $R_1C_1$ is set equal to the ratio $L/R$ in the series $RL$ circuit of Fig. 7-2, show that $v_{out}$ equals the voltage across $R$ (+ reference on the left) in that circuit.

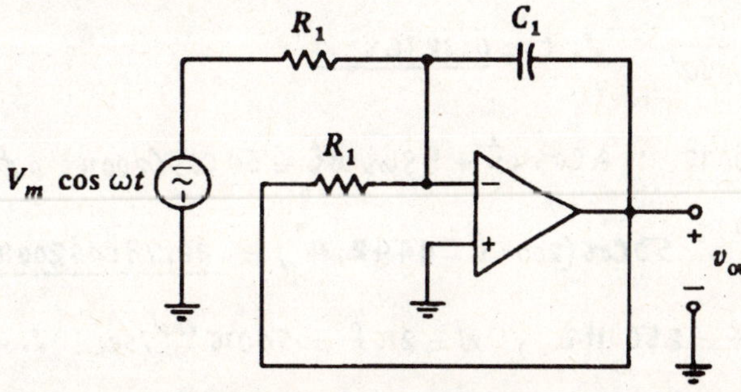


Fig. 7-5: See Prob. 12.

**13** A voltage source $V_m \cos \omega t$, a resistor $R$, and a capacitor $C$ are all in series. (*a*) Write an integrodifferential equation in terms of the loop current $i$ and then differentiate it to obtain the differential equation for this circuit. (*b*) Assume a suitable general form for the forced response $i(t)$, substitute it into the differential equation, and determine the exact form of the forced response.

# Chapter 7

7-1 (a) $f = \frac{190}{2\pi} = \underline{30.24\ Hz}$ (b) $T = \frac{1}{f} = \frac{2\pi}{190} = \underline{33.07\ ms}$ (c) $\omega = \underline{190\ rad/s}$

(d) $A\cos(190t + \theta) = A\cos\theta\cos 190t - A\sin\theta\sin 190t \quad \therefore A\cos\theta = 26,\ A\sin\theta = 18$

$A = \sqrt{26^2 + 18^2} = \underline{31.62}$

(e) $\tan\theta = \frac{+18}{+26} \quad \therefore \theta = 34.70^\circ \therefore y(t) = 31.62\cos(190t + 34.70^\circ)$; $y(t)$ leads $5\cos 190t$ by $\underline{34.70^\circ}$

(f) $8\sin 190t = 8\cos(190t - 90^\circ)$ ; $y(t)$ lags $8\sin 190t$ by $-90^\circ - 34.70^\circ = \underline{-124.70^\circ}$ or $\underline{235.30^\circ}$

(g) $y(t) + 20\cos(190t + 160^\circ) = (26 - 18.794)\cos 190t - (18 + 6.840)\sin 190t$

$= 25.86\cos(190t + 73.82^\circ) \quad \therefore A = \underline{25.86}\ ;\ \theta = \underline{73.82^\circ}$

7-2 (a) $6\sin 20t = 6\cos(20t - 90^\circ) \quad \therefore A = \underline{6},\ \phi = \underline{-90^\circ}$

(b) $4\sin(20t - 20^\circ) = 4\cos(20t - 110^\circ) \quad \therefore A = \underline{4},\ \phi = \underline{-110^\circ}$

(c) $8\cos 5t + 3\sin 5t = 8.544\cos(5t - 20.56^\circ) \quad \therefore A = 8.544,\ \phi = -20.56^\circ$

(d) $5\cos(100t + 40^\circ) = 0 \quad \therefore 100t + \frac{40}{180}\pi = \frac{\pi}{2} \quad \therefore t = \underline{8.727\ ms}$

(e) $-3\cos 20t + 2\sin 20t = -3.606\cos(20t + 33.69^\circ) = 3.606\cos(20t - 146.31^\circ) \quad \therefore A = \underline{3.606}$

$20t = 146.31^\circ \times \frac{\pi}{180^\circ} \quad \therefore t = \underline{0.12768\ s}$

7-3 (a) $\omega = 2\pi f = 200\pi$ ; $A\cos\omega t + B\sin\omega t = 50\cos(200\pi t + \phi) \therefore 200\pi \times 0.1234 = -\phi$

$\phi = -4442.4^\circ \therefore 50\cos(200\pi t - 4442.4^\circ) = \underline{-26.79\cos 200\pi t + 42.22\sin 200\pi t}$

(b) $T = 4\ ms,\ f = 250\ Hz,\ \omega = 2\pi f = 500\pi\ rad/sec \quad \therefore v(t) = C\sin 500\pi t$

$-12 = C\sin(500\pi \times 2.4 \times 10^{-3}) \quad \therefore C = 20.42,\ v(t) = \underline{20.42\cos(500\pi t - 90^\circ)}$

7-4 $L[-\omega A\sin(\omega t - \theta)] + RA\cos(\omega t - \theta) = V_m\cos\omega t$

$\therefore -\omega LA\cos\theta\sin\omega t + \omega LA\sin\theta\cos\omega t + RA\cos\theta\cos\omega t + RA\sin\theta\sin\omega t = V_m\cos\omega t$

$\therefore \omega LA\cos\theta = RA\sin\theta,\ \tan\theta = \frac{\omega L}{R}$ ✓; $\omega LA\sin\theta + RA\cos\theta = V_m$

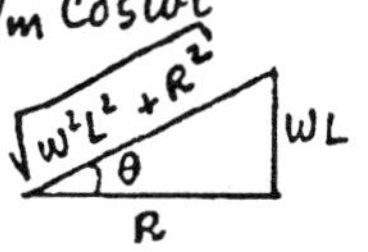


$\therefore \omega LA\left(\frac{\omega L}{\sqrt{R^2 + \omega^2L^2}}\right) + RA\left(\frac{R}{\sqrt{R^2 + \omega^2L^2}}\right) = A\sqrt{R^2 + \omega^2L^2} = V_m$ ✓

**7-5**

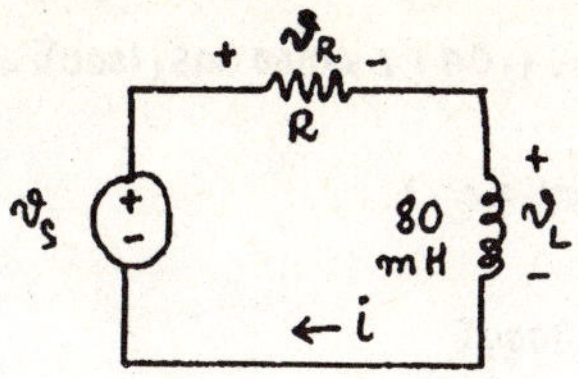


$v_s = 20\cos(500t + 100°)$ , $v_R = K\cos(500t + 45°)$ V

$\tan^{-1}\dfrac{wL}{R} = 100° - 45° = 55°$ ∴ $R = \dfrac{500 \times 80 \times 10^{-3}}{\tan 55°} = \underline{28.01\ \Omega}$

$v_R = \dfrac{20 \times 28.01}{\sqrt{28.01^2 + (500 \times 0.08)^2}}\cos(500t + 100° - 55°)$ ∴ $K = \underline{11.472}$ V

$i = \dfrac{v_R}{R} = \dfrac{11.472}{28.01}\cos(500t + 45°) = \underline{0.410\cos(500t + 45°)\ \text{A}}$

$v_L = Li' = 0.08 \times 0.41 \times 500[-\sin(500t + 45°)] = \underline{16.38\cos(500t + 135°)\ \text{V}}$

**7-6**

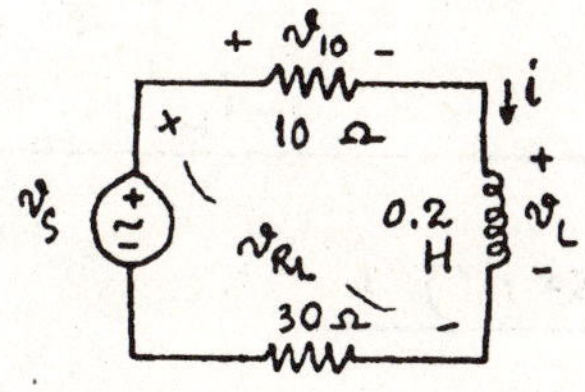


$v_s = 100\cos 300t$ ; $\tan^{-1}\dfrac{wL}{R} = 56.31°$ ∴ $i = \dfrac{100}{\sqrt{40^2 + 60^2}}\cos(300t - 56.31°)$

$i = 1.3868\cos(300t - 56.31°)$ A ; $v_{10} = 13.868\cos(300t - 56.31°)$ V

$v_L = Li' = 0.2 \times 1.3868(-300)\cos(300t - 146.31°)$

$v_L = 83.205\cos(300t + 33.69°)$ V

$v_{RL} = v_{10} + v_L = 7.693\cos 300t + 11.539\sin 300t + 69.23\cos 300t - 46.15\sin 300t$

$v_{RL} = 76.92\cos 300t - 34.62\sin 300t = \underline{84.35}\cos(300t + 24.23°)$ V

**7-7** (a)

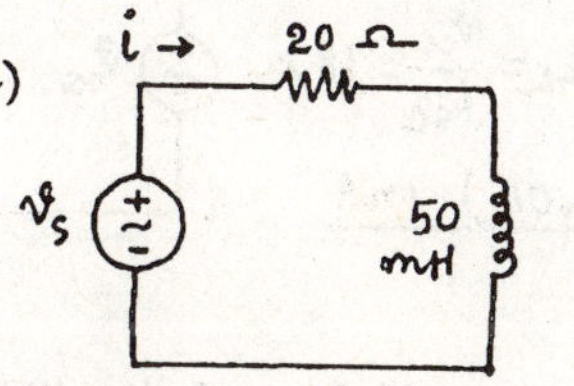


$v_s = 10\cos(800t - 30°)$ V

∴ $i = \dfrac{10}{\sqrt{20^2 + 40^2}}\cos(800t - 30° - \tan^{-1}\dfrac{40}{20}) = 0.2236\cos(800t - 93.43°)$ A

$p_s = v_s i = 10 \times 0.2236\cos(800t - 30°)\cos(800t - 93.43°)$

$\cos A\cos B = \frac{1}{2}\cos(A + B) + \frac{1}{2}\cos(A - B)$ ∴ $p_s = \dfrac{2.236}{2}[\cos(1600t - 123.43°) + \cos 63.43°]$

∴ $p_{s,max} = 1.118(1 + 0.4472) = \underline{1.6181\ \text{W}}$

(b) $p_R = Ri^2 = 20 \times 0.2236^2\cos^2(800t - 93.43°)$ ∴ $p_{R,max} = 20 \times 0.2236^2 = \underline{1\ \text{W}}$

(c) $v_L = 0.05i' = 0.05 \times 0.2236(-800)\cos(800t + 176.57°) = 8.944\cos(800t - 3.43°)$

∴ $p_L = v_L i = 8.944 \times 0.2236\cos(800t - 3.43°)\cos(800t - 93.43°)$

$= \frac{1}{2} \times 2[\cos(1600t - 96.87°) + \cos 90°]$ ∴ $p_{L,max} = \underline{1\ \text{W}}$

**7-8** (a)

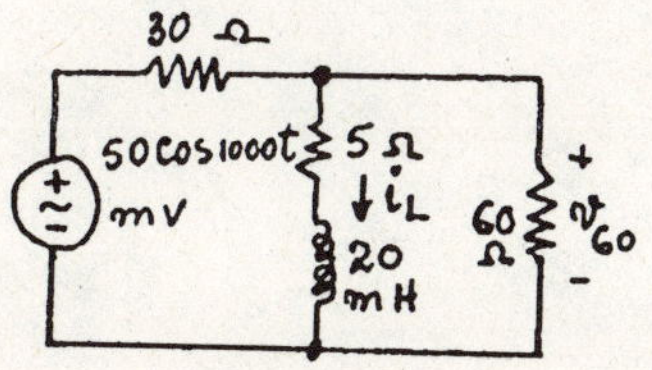


$(30 \| 60) + 5 = 25\ \Omega$ ; $v_s = 50\cos 1000t\ \dfrac{60}{30 + 60}$

$v_s = \dfrac{100}{3}\cos 1000t$ mV

$i_L = \dfrac{100/3}{\sqrt{25^2 + 20^2}}\cos(1000t - \tan^{-1}\dfrac{20}{25})$

$i_L = \underline{1.0412\cos(1000t - 38.66°)\ \text{mA}}$

⇒

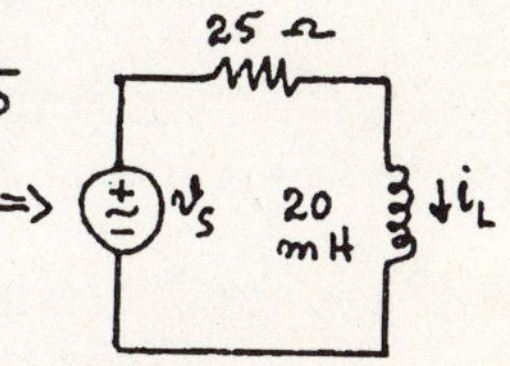

**7-8 (Cont) (b)** $v_{60} = 5i_L + 0.02 i_L' = 5.206\cos(1000t - 38.66°) + 0.02 \times 1.0412 \times 1000 \cos(1000t - 38.66° + 90°)$

$$v_{60} = 5.206\cos(1000t - 38.66°) + 20.82\cos(1000t + 51.34°)$$

$$= (4.065 + 13.006)\cos 1000t + (3.252 - 16.258)\sin 1000t$$

$$v_{60} = 17.071\cos 1000t - 13.006\sin 1000t = \underline{21.46\cos(1000t + 37.30°)\ \text{V}}$$

**7-9**

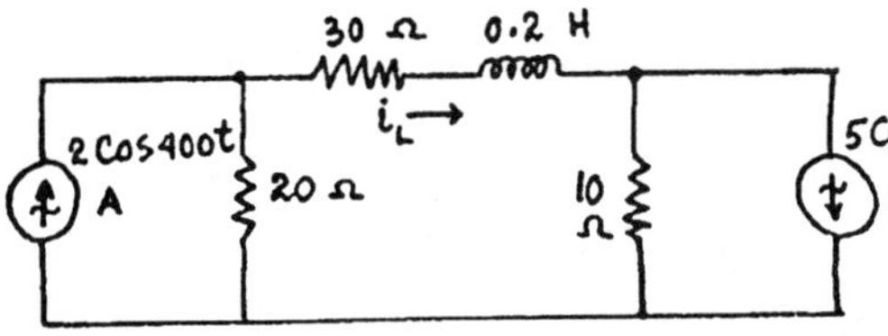


⇒

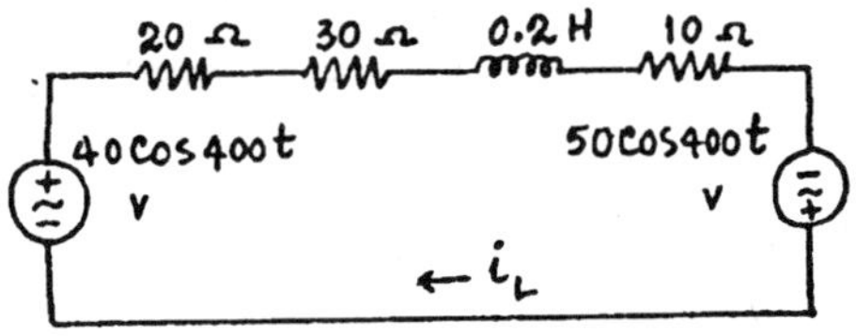


$$i_L(t) = \frac{90}{\sqrt{60^2 + 80^2}}\cos\left(400t - \tan^{-1}\frac{80}{60}\right) = \underline{0.9\cos(400t - 53.13°)\ \text{A}}$$

**7-10**

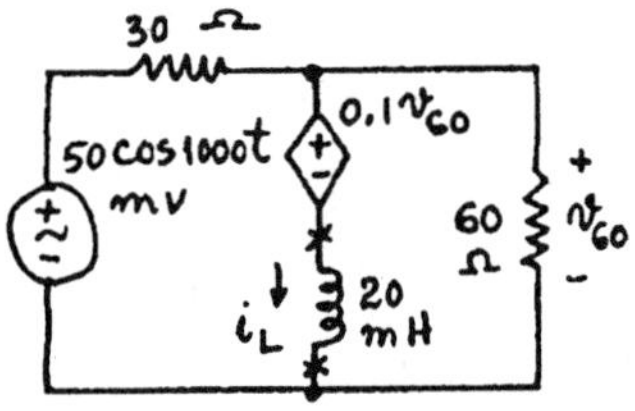


O.C. $v_{60} = \frac{2}{3} \times 50\cos 1000t = \frac{100}{3}\cos 1000t$ mV

$v_{oc} + 0.1 v_{60} = v_{60}$ ∴ $v_{oc} = 30\cos 1000t$ mV

S.C. $0.1 v_{60} = v_{60}$ ∴ $v_{60} = 0$

∴ $i_{sc}\downarrow = \frac{.050}{30}\cos 1000t$ ; $R_{Th} = \frac{v_{oc}}{i_{sc}} = 18\ \Omega$

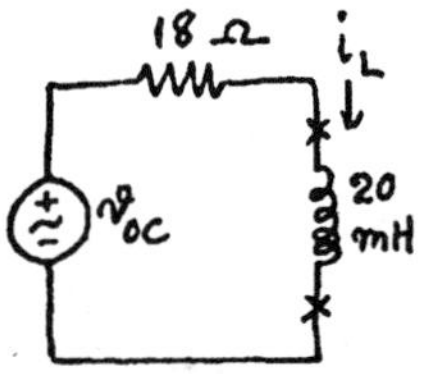


$$∴ i_L = \frac{30}{\sqrt{18^2 + 20^2}}\cos\left(1000t - \tan^{-1}\frac{20}{18}\right) = \underline{1.1149\cos(1000t - 48.01°)\ \text{mA}}$$

**7-11**

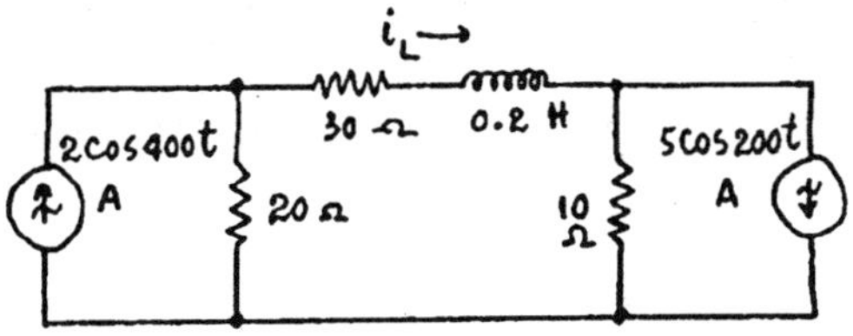


⇒

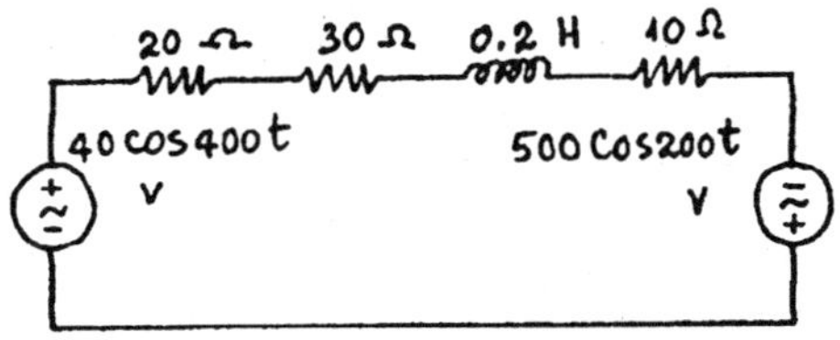


$$∴ i_L = \frac{40}{\sqrt{60^2 + 80^2}}\cos\left(400t - \tan^{-1}\frac{80}{60}\right) + \frac{50}{\sqrt{60^2 + 40^2}}\cos\left(200t - \tan^{-1}\frac{40}{60}\right)\ \text{A}$$

$$i_L = \underline{0.4\cos(400t - 53.13°) + 0.6934\cos(200t - 33.69°)\ \text{A}}$$

**7-12**

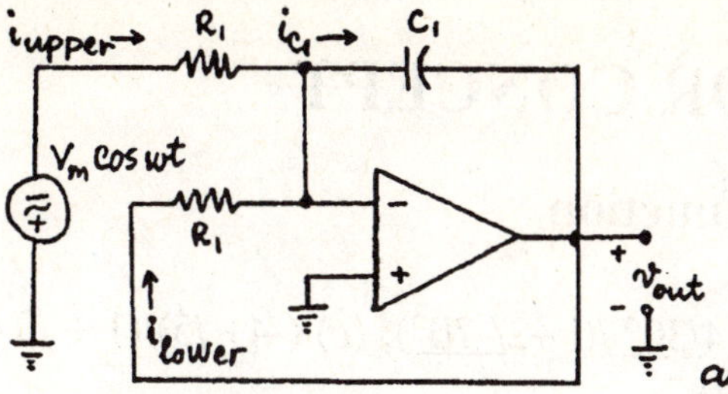


$R_i = \infty$ , $R_o = 0$ , $A = \infty$ , $R_1 C_1 = L/R$

$i_{upper} = -\dfrac{V_m \cos wt}{R_1}$ ; $i_{Lower} = \dfrac{v_{out}}{R_1}$

$\therefore i_{C_1} = i_{upper} + i_{Lower} = \dfrac{1}{R_1}(v_{out} - V_m \cos wt)$

and $i_{C_1} = -C_1 v'_{out}$

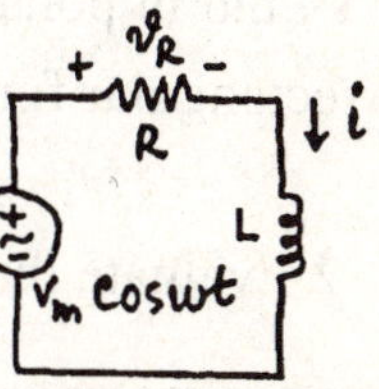


$\therefore V_m \cos wt = v_{out} + R_1 C_1 v'_{out} = v_{out} + \dfrac{L}{R} v'_{out}$

For RL circuit, $V_m \cos wt = v_R + L\dfrac{d}{dt}\left(\dfrac{v_R}{R}\right) = v_R + \dfrac{L}{R} v'_R$

By comparasion , $\underline{v_R = v_{out}}$ ✓

**7-13**

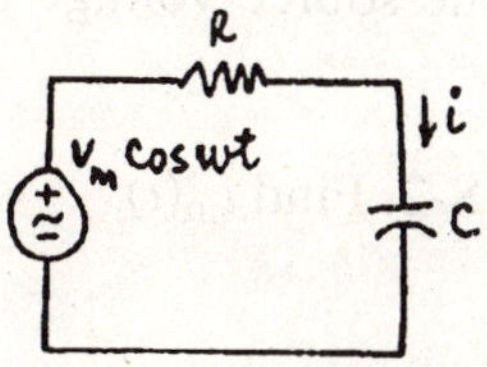


(a) $V_m \cos wt = Ri + \dfrac{1}{C}\int i\,dt$ (ignore initial cond.)

$\therefore \underline{-wV_m \sin wt = Ri' + 1/C\, i}$

(b) $i = A\cos(wt + \phi)$

$\therefore -wV_m \sin wt = -RwA\sin(wt+\phi) + \dfrac{A}{C}\cos(wt+\phi)$

$= -RwA\cos\phi \sin wt - RwA\sin\phi\cos wt + \dfrac{A}{C}\cos\phi\cos wt - \dfrac{A}{C}\sin\phi\sin wt$

$\therefore RwA\sin\phi = \dfrac{A}{C}\cos\phi$ $\quad \therefore \tan\phi = \dfrac{1}{wCR}$

and $-wV_m = -RwA\dfrac{wCR}{\sqrt{1+w^2C^2R^2}} - \dfrac{A}{C}\dfrac{1}{\sqrt{1+w^2C^2R^2}} = -\dfrac{A}{C}\left[\dfrac{R^2w^2C^2+1}{\sqrt{1^2+w^2C^2R^2}}\right]$

$wV_m = \dfrac{A}{C}\sqrt{1+w^2C^2R^2}$ $\quad \therefore A = \dfrac{wCV_m}{\sqrt{1+w^2C^2R^2}}$ ,

$\therefore i = \dfrac{wCV_m}{\sqrt{1+w^2C^2R^2}}\cos\left(wt + \tan^{-1}\dfrac{1}{wCR}\right)$ ←

# CHAPTER 8: THE PHASOR CONCEPT

## Section 8-2 The complex forcing function

**1** Find in polar form: (*a*) $5\angle 27^\circ - 3\angle 112^\circ$; (*b*) $4\angle 30^\circ/(1 + 2\angle -70^\circ)$; (*c*) $(-4 + j5)/(1 + j0.4)$. Find in rectangular form: (*d*) $(2\angle 30^\circ + 1\angle 60^\circ)^2$; (*e*) $8e^{j0.5}$.

**2** Evaluate the following in polar form: (*a*) $(36\angle 21^\circ)\left(\frac{1}{j}\right) + 19\angle -108^\circ$; (*b*) $11j + \sqrt{80\angle 83^\circ}$. Evaluate in rectangular form: (*c*) $\dfrac{1}{\dfrac{1}{7 - j3.1} + \dfrac{1}{4 + j9.2}}$; (*d*) $\mathrm{Re}[6.11\angle 26^\circ] + \mathrm{Im}[4.18\angle -99^\circ] + [5.50\angle 412^\circ]^*$.

**3** In Fig. 8-1, let $i(t)$ be the complex current, $0.8e^{j20t}$ A, while the source voltage $v_s(t)$ is $(30 - j40)e^{j20t}$ V, and find the complex input voltage $v_{in}(t)$.

**4** Let $i_s(t) = j2e^{j50t}$ A and $v(t) = 240e^{j50t}$ V in the network of Fig. 8-2. Find $i_{in}(t)$.

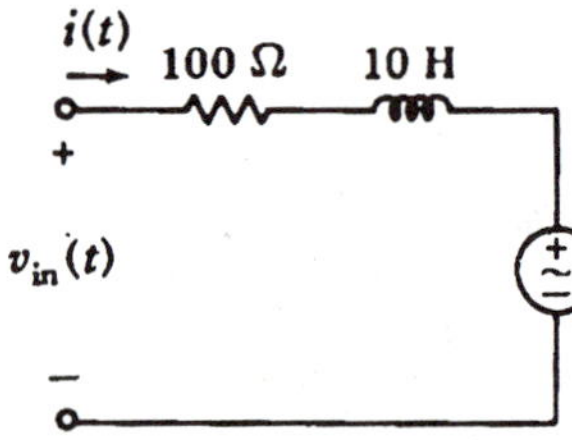

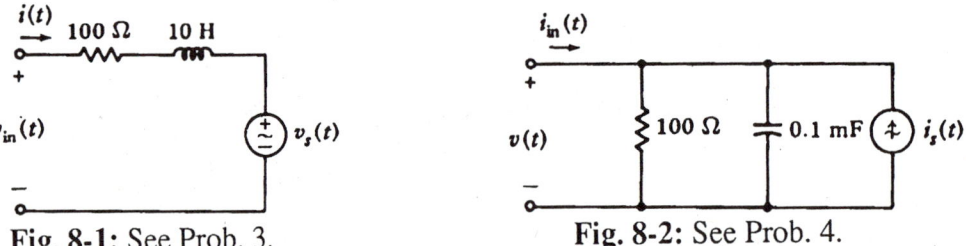

**Fig. 8-1:** See Prob. 3.

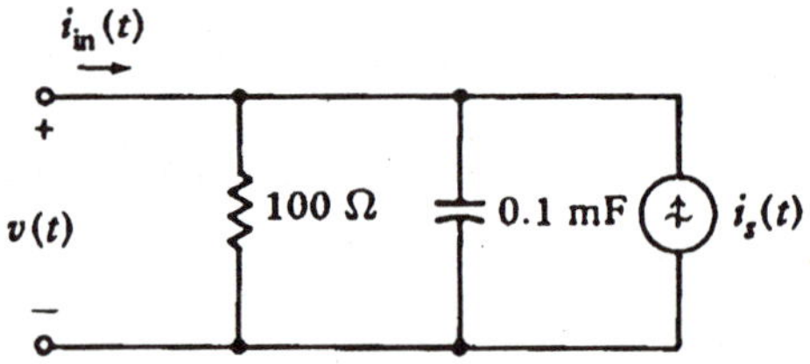


**Fig. 8-2:** See Prob. 4.

**5** In a linear network, similar to that shown in Fig. 8-3, a sinusoidal voltage source, $v_s = 40\cos 1000t$ V, produces a current response, $i_o = 2.5\cos(1000t - 24^\circ)$ A. Find $i_o(t)$ if $v_s(t)$ =: (*a*) $20\sin 1000t$ V; (*b*) $20\cos(1000t - 40^\circ)$ V; (*c*) $20e^{j27^\circ}e^{j1000t}$ V; (*d*) $(10 - j6)e^{j1000t}$ V.

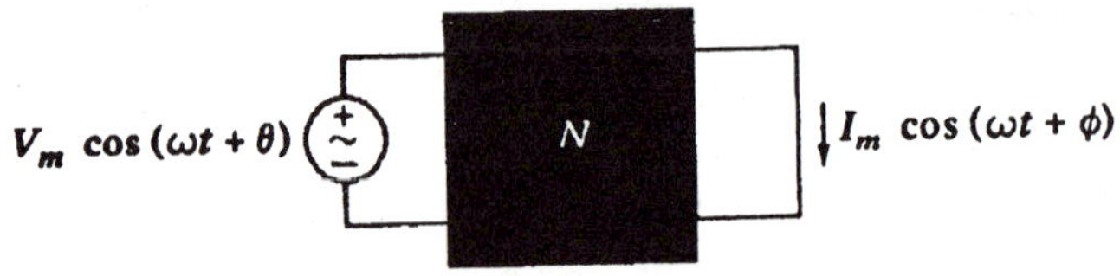


**Fig. 8-3:** See Prob. 5.

## Section 8-3 The phasor

**6** Express as a phasor: (*a*) $v(t) = 165\cos(120\pi t + 30^\circ)$ V; (*b*) $i(t) = 2\cos 10^6 t + 3\sin 10^6 t$ mA. Find the instantaneous value at $t = 1$ ms of: (*c*) $\mathbf{V} = 40 - j70$ V, $\omega = 400$ rad/s; (*d*) $\mathbf{I} = -8.1 - j2.4$ A, $f = 60$ Hz.

**7** Let $\mathbf{I} = 10\angle -130^\circ$ A and find $i$ at $t = 1$ ms if $\omega$ =: (*a*) 1200 rad/s; (*b*) 600 rad/s. (*c*) Let $i_1 = 5\cos(100t + 50^\circ)$, $i_2 = 4\cos 100t$, and $i_3 = 3\sin(100t + 70^\circ)$ A, and $i_x = i_1 + i_2 + i_3$. By transforming $i_1$, $i_2$, and $i_3$ into phasors, find $\mathbf{I}_x$ and $i_x(t)$.

**8** If $\omega = 200$ rad/s, find the instantaneous value at $t = 1$ ms of: (*a*) $\mathbf{I} = 2\angle 70^\circ$ A; (*b*) $\mathbf{I} = 4 - j1$ A. Find the phasor voltage corresponding to: (*c*) $v(t) = 6\sin(500t - 50^\circ)$ V; (*d*) $v(t) = -3\cos 50t + 6\sin 50t$ V; (*e*) $8\cos(100t - 100^\circ) - 6\cos(100t - 40^\circ)$ V.

**9** In Fig. 8-4, let $\mathbf{I}_1 = 6 + j1$ A, $\mathbf{I}_2 = 2 - j5$ A, and $\omega = 1$ krad/s. At $t = 2.5$ ms, find the instantaneous value of: (*a*) $v_A$; (*b*) $i_A$; (*c*) the power being absorbed by element *A*.

**10** A purple- and yellow-striped box contains a voltage source $\mathbf{V}_{s1}$ and a current source $\mathbf{I}_{s2}$. The voltage between an available pair of terminals is labeled $\mathbf{V}_{AB}$. If $\mathbf{V}_{s1} = 10\angle 20^\circ$ V and $\mathbf{I}_{s2} = 0.2\angle -30^\circ$ A, then $\mathbf{V}_{AB} = -50 + j80$ V. However, if $\mathbf{V}_{s1} = 20\angle -80^\circ$ V and $\mathbf{I}_{s2} = 0.5\angle 30^\circ$ A, then $\mathbf{V}_{AB} = 40 + j60$ V. Find $\mathbf{V}_{AB}$ if $\mathbf{V}_{s1} = 10 - j30$ V and $\mathbf{I}_{s2} = 0.3 - j0.1$ A.

## Section 8-4 Phasor relationships for *R*, *L*, and C
## Section 8-5 Impedance

**11** A resistor *R*, an inductor *L*, and an ideal source $v_s = 100\cos \omega t$ V are in series. If $\omega = 200$ rad/s, the phasor current is $5 - j2$ A. What will the phasor current be if $\omega = 400$ rad/s?

**12** Let $\omega = 500$ rad/s for the network of Fig. 8-5 and find the phasor voltage **V**.

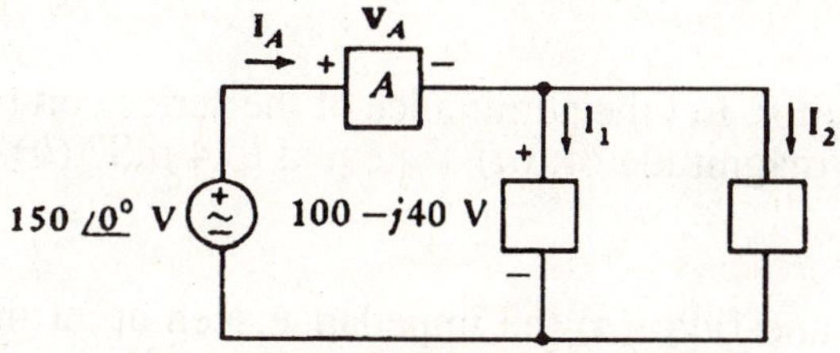

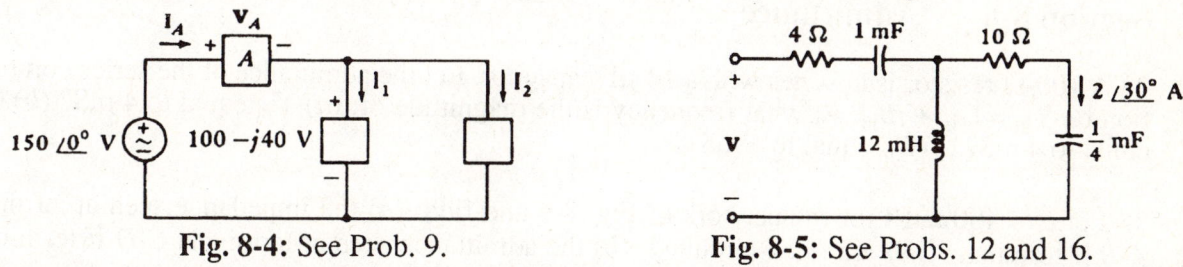

**Fig. 8-4:** See Prob. 9.

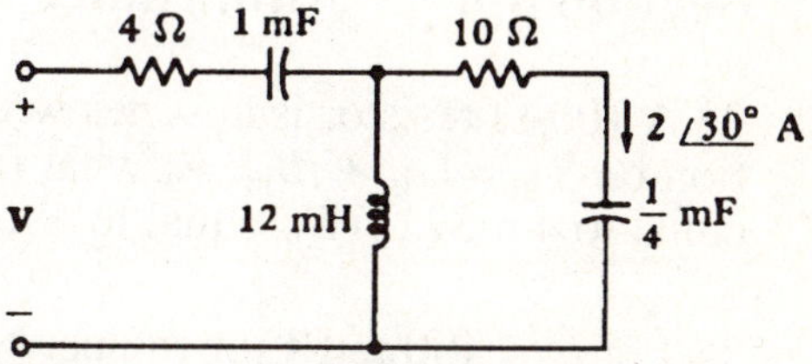


**Fig. 8-5:** See Probs. 12 and 16.

**13** Find $v_1(t)$, $v_2(t)$, and $v_3(t)$ in the circuit shown in Fig. 8-6 if $\omega = 5$ krad/s.

**14** Find **V** in Fig. 8-7 if the box contains: (*a*) 8 Ω in series with 5 mH; (*b*) 8 Ω in series with 50 μF; (*c*) 8 Ω, 5 mH, and 50 μF in series; (*d*) 8 Ω, 5 mH, and 50 μF in series, but $\omega = 4$ krad/s.

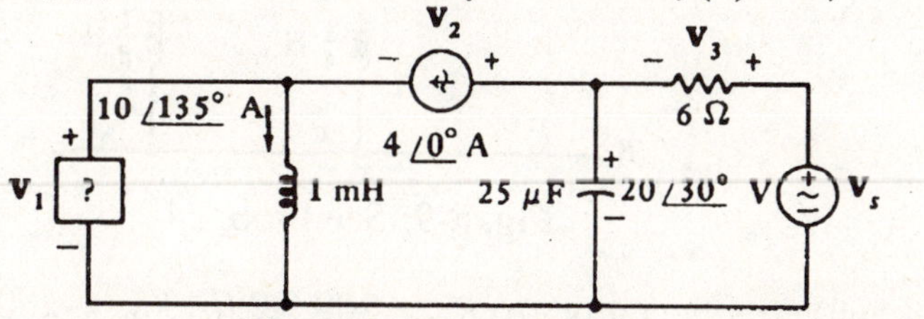


**Fig. 8-6:** See Prob. 13.

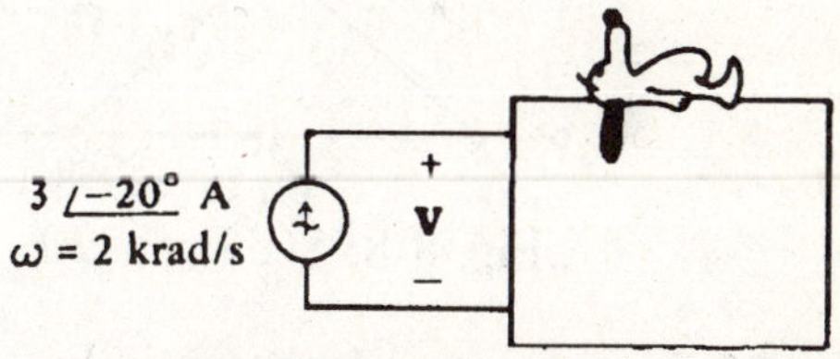


**Fig. 8-7:** See Prob. 14.

**15** A 200-Ω resistor, a 25-μF capacitor, and an inductor *L* are in parallel. The phasor voltage across the combination is $100\angle 0^\circ$ V. (*a*) Find *L* if the current entering the plus-marked terminal is $\mathbf{I} = 0.5 - j2$ A and $\omega = 1$ krad/s. (*b*) Find *L* if $\omega = 100$ rad/s and $|\mathbf{I}| = 1$ A.

**16** Find the input impedance of the network shown in Fig. 8-5 if $\omega = 500$ rad/s.

**17** A 200-Ω resistor, a 25-μF capacitor, and an inductor *L* are in parallel. (*a*) Find the impedance of the parallel combination at $\omega = 100$ rad/s if $L = 10$ H. (*b*) If the magnitude of the impedance is 80 Ω at $\omega = 100$ rad/s, find *L*. (*c*) At what values of $\omega$ is the impedance magnitude equal to 150 Ω if $L = 10$ H?

**18** A 2-kΩ resistor is in parallel with a 4-μF capacitor. If the impedance of the parallel combination is $\mathbf{Z}_{in}$, at what frequency is: (*a*) $|\mathbf{Z}_{in}| = 1500$ Ω? (*b*) the angle of $\mathbf{Z}_{in} = -30^\circ$? (*c*) $\text{Re}[\mathbf{Z}_{in}] = 200$ Ω? (*d*) $|X_{in}| = 400$ Ω?

**19** A certain two-terminal network has an input impedance of $5 - j8\ \Omega$. At a frequency of 4 krad/s, what inductance should be placed in parallel with the network to cause the input impedance to: (*a*) have zero reactance? (*b*) have a magnitude of $4\ \Omega$?

**20** Find $\mathbf{Z}_{in}$ at $\omega = 5$ rad/s for the network of Fig. 8-8 if terminals *a-b* are: (*a*) open-circuited; (*b*) short-circuited.

**21** A 10-Ω resistor, a 2-mH inductor, and a 0.2-μF capacitor are in series. The impedance of the series combination is $\mathbf{Z}_{in}$. (*a*) At what value of $\omega$ is $|\mathbf{Z}_{in}|$ a minimum? (*b*) At what values of $\omega$ is $|\mathbf{Z}_{in}| = 2|\mathbf{Z}_{in}|_{min}$?

**22** A 2-Ω resistor and a 0.5-F capacitor are in series, and a 3-H inductor is in parallel with the series combination. If the impedance of the parallel network is designated $\mathbf{Z}_{in}(\omega) = R_{in}(\omega) + jX_{in}(\omega)$, find $R_{in}(\omega)$ and $X_{in}(\omega)$.

## Section 8-6 Admittance

**23** A 100-Ω resistor is in series with a 10-μF capacitor. Let the admittance of the series combination be $\mathbf{Y}_{in} = G_{in} + jB_{in}$. At what frequency is the magnitude of: (*a*) $\mathbf{Y}_{in}$ equal to 4 mS? (*b*) $G_{in}$ equal to 4 mS? (*c*) $B_{in}$ equal to 4 mS?

**24** Let $\omega = 400$ rad/s for the network of Fig. 8-9 and find: (*a*) the impedance seen at terminals *A-B* if terminals *C-D* are open-circuited; (*b*) the admittance seen at terminals *C-D* if terminals *A-B* are short-circuited.

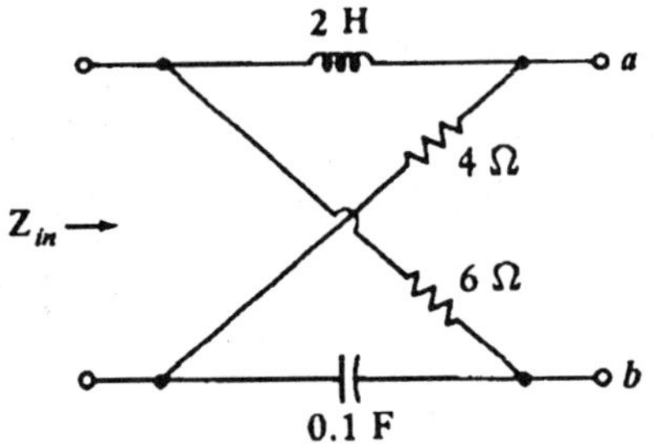


**Fig. 8-8:** See Prob. 20.

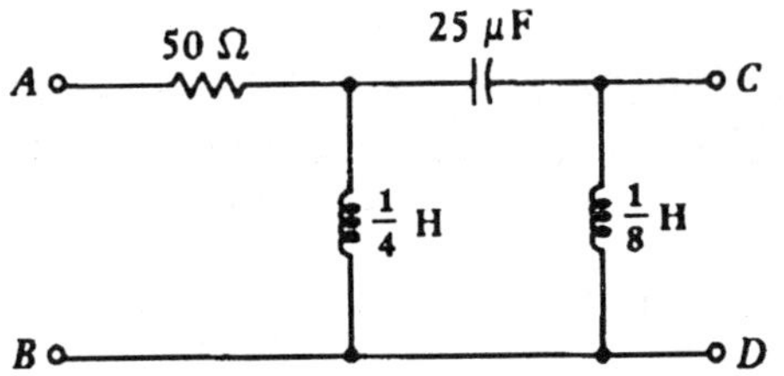


**Fig. 8-9:** See Prob. 24.

**25** For the network shown in Fig. 8-10, at what frequency is: (*a*) $R_{in} = 200\ \Omega$? (*b*) $X_{in} = -200\ \Omega$? (*c*) $G_{in} = 1/200$ S? (*d*) $B_{in} = 1/240$ S?

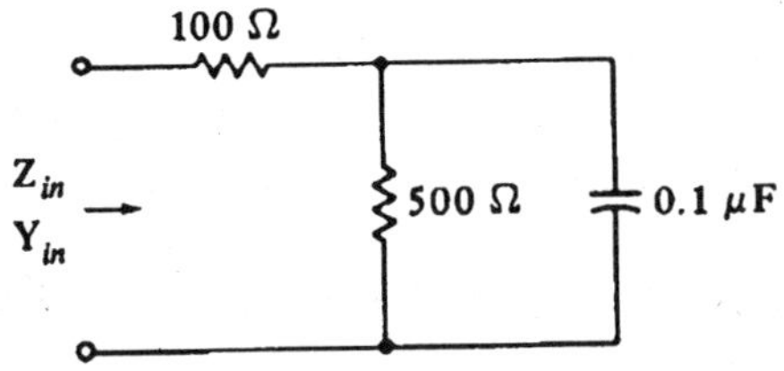

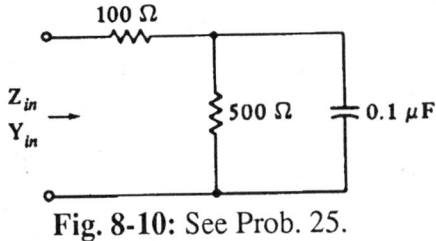

**Fig. 8-10:** See Prob. 25.

**26** Three admittances, $\mathbf{Y}_1 = 2 - j4\ \mu S$, $\mathbf{Y}_2 = 4 - j1\ \mu S$, and $\mathbf{Y}_3 = 1 + j3\ \mu S$ are in series, and the current $\mathbf{I} = 20\underline{/40^\circ}\ \mu A$ flows through the series combination. Find the amplitude of the voltage across each admittance and also across the series combination.

**27** The admittance of the parallel combination of 100 Ω and 0.2 H at $\omega = 1500$ rad/s is the same as the admittance of $R_1$ and $L_1$ in series at that frequency. (*a*) Find $R_1$ and $L_1$. (*b*) Repeat for $\omega = 500$ rad/s.

**28** A network is composed of 5 kΩ, 0.1 H, and 0.1 μF in parallel. (*a*) At what frequency does the input admittance of the network have its minimum magnitude? (*b*) At what frequency is the phase angle of the input admittance 45°? (*c*) -45°?

**29** (SPICE) Let $f = 60$ Hz in the circuit shown in Fig. 8-11. Use a SPICE analysis to find the phasor voltage $\mathbf{V}_x$ if element $X$ is a: (*a*) 200-Ω resistor; (*b*) 0.3-H inductor; (*c*) 90-μF capacitor.

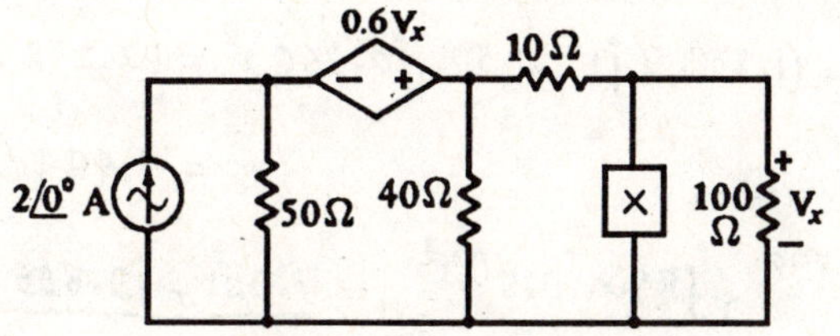


**Fig. 8-11:** See Prob. 29.

# Chapter 8

**8-1** (a) $5\angle 27^\circ - 3\angle 112^\circ = 4.455 + j2.270 + 1.124 - j2.782 = 5.579 - j0.512 = \underline{5.602\angle -5.240^\circ}$

(b) $\dfrac{4\angle 30^\circ}{1+2\angle -70^\circ} = \dfrac{4\angle 30^\circ}{1.684 - j1.879} = \dfrac{4\angle 30^\circ}{2.524\angle -48.14^\circ} = \underline{1.5851\angle 78.14^\circ}$

(c) $\dfrac{-4+j5}{1+j0.4} = \underline{5.945^+\angle 106.86^\circ}$

(d) $(2\angle 30^\circ + 1\angle 60^\circ)^2 = (1.732 + j1 + 0.5 + j0.866)^2 = (2.232 + j1.866)^2 = (2.909\angle 39.90^\circ)^2$

$= 8.464\angle 79.79^\circ = \underline{1.5 + j8.330}$

(e) $8e^{j0.5} = 8\cos 0.5^{rad} + j8\sin 0.5^{rad} = \underline{7.021 + j3.835^+}$

**8-2** (a) $(36\angle 21^\circ)\dfrac{1}{j} + 19\angle -108^\circ = 36\angle -69^\circ - 5.871 - j18.070 = \underline{52.15^+\angle -82.25^{+\circ}}$

(b) $j11 + \sqrt{80}\angle 41.5^\circ = j11 + 6.699 + j5.927 = \underline{18.204\angle 68.41^\circ}$

(c) $\dfrac{1}{\dfrac{1}{7-j3.1} + \dfrac{1}{4+j9.2}} = \dfrac{1}{0.11943 + j0.05289 + 0.03975 - j0.09141} = \dfrac{1}{0.16377\angle -13.605^\circ}$

$= \underline{5.935^- + j1.4362}$

(d) $\text{Re}[6.11\angle 26^\circ] + \text{Im}[4.18\angle -99^\circ] + 5.50\angle -412^\circ = 5.492 - 4.129 + 3.386 - j4.334$

$= \underline{4.749 - j4.334}$

**8-3**

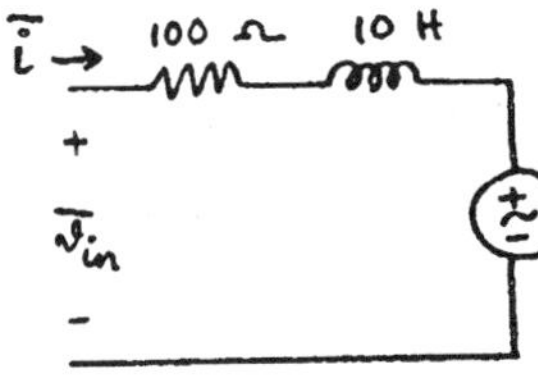


$\bar{i} = 0.8e^{j20t}$ A ; $\bar{v}_s = (30 - j40)e^{j20t}$ V

$\bar{v}_{in}(t) = e^{j20t}(80 + j20 \times 0.8 \times 10 + 30 - j40)$

$\bar{v}_{in}(t) = (110 + j120)e^{j20t} = \underline{162.79e^{j(20t + 47.49^\circ)}\ \text{V}}$

**8-4**

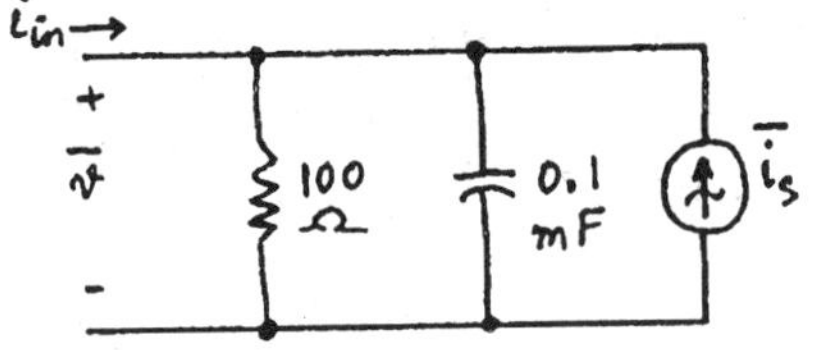


$\bar{v} = 240e^{j50t}$, $\bar{i}_s = j2e^{j50t}$

$\bar{i}_{in} = e^{j50t}(2.4 + j50 \times 10^{-4} \times 240 - j2)$

$= (2.4 - j0.8)e^{j50t} = \underline{2.530e^{j(50t - 18.435^\circ)}\ \text{A}}$

**8-5** (a)

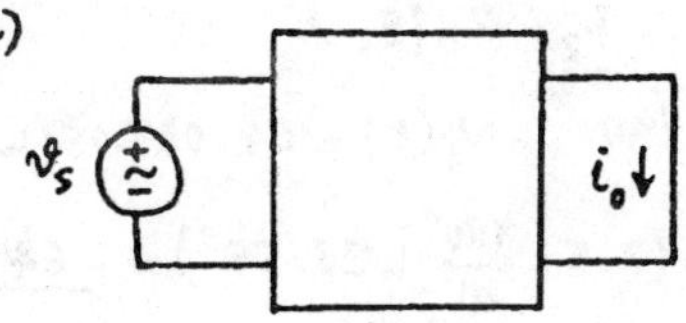


$v_s = 40\cos 1000t$ V , $i_o = 2.5\cos(1000t - 24°)$ A

Let $v_s = 20\sin 1000t = 20\cos(1000t - 90°)$

$\therefore i_o = 2.5 \times \frac{20}{40}\cos(1000t - 24° - 90°) = \underline{1.25\cos(1000t - 114°)\text{ A}}$

(b) $v_s = 20\cos(1000t - 40°)$ $\therefore i_o = \underline{1.25\cos(1000t - 64°)\text{ A}}$

(c) $\bar{v}_s = 20e^{j27°}e^{j1000t}$ $\therefore \bar{i}_o = 1.25e^{j(-24° + 27°)}e^{j1000t} = \underline{1.25e^{j3°}e^{j1000t}\text{ A}}$

(d) $\bar{v}_s = (10 - j6)e^{j1000t} = 11.662e^{-j30.96°}e^{j1000t}$ $\therefore \bar{i}_o = 11.662 \times \frac{2.5}{40}e^{-j54.96°}e^{j1000t}$

$\therefore \bar{i}_o = \underline{0.7289e^{-j54.96°}e^{j1000t}\text{ A}}$

**8-6** (a) $v(t) = 165\cos(120\pi t + 30°) \Rightarrow \bar{V} = \underline{165\angle 30°\text{ V}}$

(b) $i(t) = 2\cos 10^6 t + 3\sin 10^6 t$ mA $\Rightarrow \bar{I} = 2 - j3 = \underline{3.606\angle -56.31°\text{ mA}}$

(c) $\bar{V} = 40 - j70 = 80.62\angle -60.25°$ at $\omega = 400 \Rightarrow v(t) = 80.62\cos(400t - 60.25°)$

$\therefore v(1\text{ ms}) = \underline{64.10\text{ V}}$

(d) $\bar{I} = -8.1 - j2.4$ at $f = 60$ Hz $\therefore i(t) = 8.448\cos(120\pi t - 163.50°)$ A

$\therefore i(1\text{ ms}) = 8.448\cos(21.6° - 163.50°) = \underline{-6.648\text{ A}}$

**8-7** (a) $\bar{I} = 10\angle -130°$ A , $\omega = 1200$ $\therefore i(1\text{ ms}) = 10\cos(1200 \times 10^{-3}\text{ rad} - 130°) = \underline{4.811\text{ A}}$

(b) $\omega = 600$ $\therefore i(1\text{ ms}) = 10\cos\left(0.6 \times \frac{180°}{\pi} - 130°\right) = \underline{-0.9797\text{ A}}$

(c) $i_1 = 5\cos(100t + 50°)$ $\therefore \bar{I}_1 = \underline{5\angle 50°}$ ; $i_2 = 4\cos 100t$ $\therefore \bar{I}_2 = 4\angle 0°$

$i_3 = 3\sin(100t + 70°)$ $\therefore \bar{I}_3 = 3\angle -20°$ $\therefore \bar{I}_x = 5\angle 50° + 4 + 3\angle -20° = \underline{10.418\angle 15.615°\text{ A}}$

$\therefore i_x(t) = \underline{10.418\cos(100t + 15.615°)\text{ A}}$

**8-8**

(a) $\omega = 200$ ; $\bar{I} = 2\angle 70°$ $\therefore i(1\text{ ms}) = 2\cos\left(0.2 \times \frac{180°}{\pi} + 70°\right) = \underline{0.2970\text{ A}}$

(b) $\bar{I} = 4 - j1$ $\therefore i = 4\cos 200t + \sin 200t$ , $i(1\text{ ms}) = 4\cos\frac{36°}{\pi} + \sin\frac{36°}{\pi} = \underline{4.119\text{ A}}$

(c) $v = 6\sin(500t - 50°)$ $\therefore \bar{V} = \underline{6\angle -140°\text{ V}}$

(d) $v = -3\cos 50t + 6\sin 50t$ $\therefore \bar{V} = -3 - j6 = \underline{6.708\angle -116.57°\text{ V}}$

(e) $v = 8\cos(100t - 100°) - 6\cos(100t - 40°) \Rightarrow \bar{V} = 8\angle -100° - 6\angle -40° = \underline{7.211\angle -146.10°\text{ V}}$

**8-9** (a)

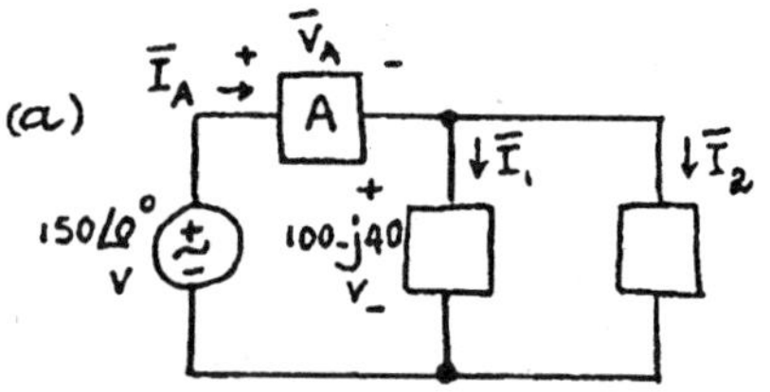


$\omega = 1000$; $\bar{I}_1 = 6 + j1$, $\bar{I}_2 = 2 - j5$ A

$\bar{V}_A = 150 - 100 + j40 = 50 + j40 \therefore v_A(t) = 64.03\cos(\omega t + 38.66°)$

$v_A(2.5\text{ ms}) = 64.03\cos\left(2.5 \times \frac{180°}{\pi} + 38.66°\right) = \underline{-64.00\text{ V}}$

(b) $\bar{I}_A = \bar{I}_1 + \bar{I}_2 = 8 - j4 = 8.944\angle -26.57° \therefore i_A(2.5\text{ ms}) = 8.944\cos\left(\frac{450°}{\pi} - 26.57°\right) = \underline{-4.015\text{ A}}$

(c) $p_A(2.5\text{ ms}) = [v_A(2.5\text{ ms})][i_A(2.5\text{ ms})] = (-64.00)(-4.015) = \underline{257.0\text{ W}}$

**8-10** $\bar{V}_{AB} = \bar{C}_1\bar{V}_{S1} + \bar{C}_2\bar{I}_{S2}$

$$\left.\begin{aligned}\therefore -50 + j80 &= \bar{C}_1(10\angle 20°) + \bar{C}_2(0.2\angle -30°)\\ \text{and } 40 + j60 &= \bar{C}_1(20\angle -80°) + \bar{C}_2(0.5\angle 30°)\end{aligned}\right\} \therefore \bar{C}_1 = \frac{\begin{vmatrix} -50+j80 & 0.2\angle -30° \\ 40+j60 & 0.5\angle 30° \end{vmatrix}}{\begin{vmatrix} 10\angle 20° & 0.2\angle -30° \\ 20\angle -80° & 0.5\angle 30° \end{vmatrix}} = \frac{56.81\angle 163.91°}{8.865\angle 58.88°}$$

$$\therefore \bar{C}_1 = 6.408\angle 105.03°\ ;\ \bar{C}_2 = \frac{\begin{vmatrix} 10\angle 20° & -50+j80 \\ 20\angle -80° & 40+j60 \end{vmatrix}}{8.865\angle 58.88°} = \frac{1353.6\angle -155.47°}{8.865\angle 58.88°} = 152.69\angle 145.66°$$

$\therefore \bar{V}_{AB} = (6.408\angle 105.03°)(10 - j30) + (152.69\angle 145.66°)(0.3 - j0.1) = \underline{205.2\angle 47.04°\text{ V}}$

**8-11**

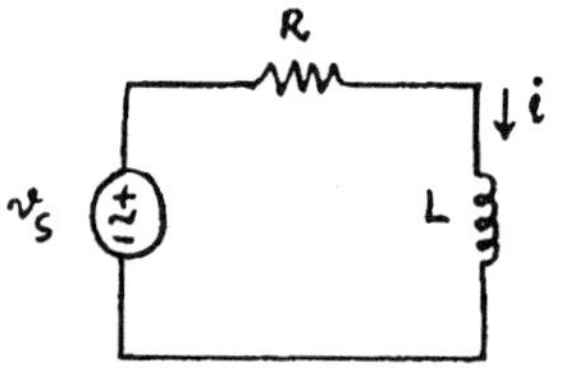


$v_s = 100\cos\omega t$ V, $\bar{I} = 5 - j2$ A

$\underline{\omega = 200}$ : $\bar{I} = 5 - j2 = \frac{100}{R + j200L}$ $\therefore 0.01R + j2L = \frac{5 + j2}{29}$

$\therefore R = \frac{500}{29}$ ; $L = \frac{1}{29}$

$\underline{\omega = 400}$ : $I = \frac{100}{\frac{500}{29} + j\frac{400}{29}} = \underline{4.529\angle -38.66°} = \underline{3.537 - j2.829\text{ A}}$

**8-12**

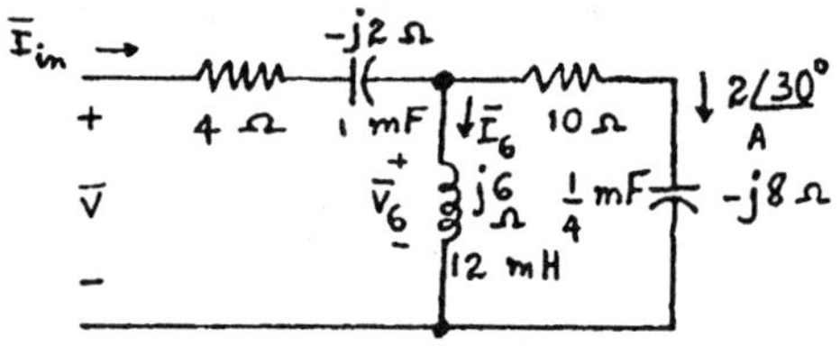


$\omega = 500$ rad/s

$\bar{V}_6 = (2\angle 30°)(10 - j8) \therefore I_6\downarrow = \left(\frac{1}{3}\angle -60°\right)(10 - j8)$

$\bar{I}_{in} = 2\angle 30° + \left(\frac{1}{3}\angle -60°\right)(10 - j8) = 3.399\angle -71.31°$ A

$\therefore \bar{V} = (4 - j2)(3.399\angle -71.31°) + (2\angle 30°)(10 - j8)$

$\bar{V} = \underline{29.96\angle -39.14°\text{ V}}$

**8-13**

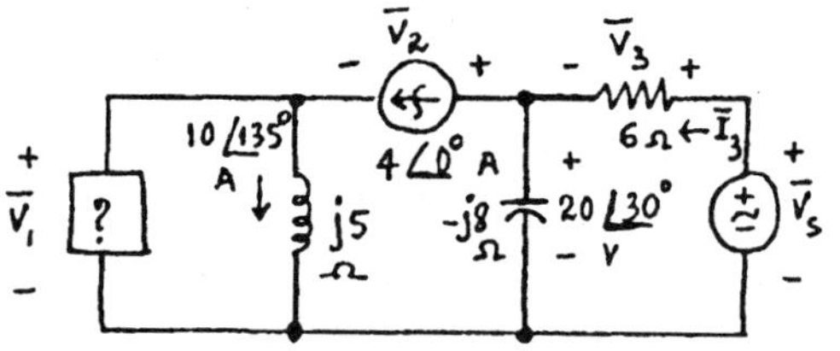


$\omega = 5000$, $\bar{V}_1 = (10\angle 135°)j5 = 50\angle 225°$ V

$\therefore v_1(t) = \underline{50\cos(5000t - 135°)\text{ V}}$

$\bar{V}_2 = 20\angle 30° + 50\angle 45° = 69.51\angle 40.73°$ V

$\therefore v_2(t) = \underline{69.51\cos(5000t + 40.73°)\text{ V}}$

$\bar{I}_3 = 4\angle 0° + \frac{20\angle 30°}{-j8} = 3.500\angle 38.21°$ ; $\bar{V}_3 = 6\bar{I}_3 = 21\angle 38.21° \therefore v_3(t) = \underline{21\cos(5000t + 38.21°)\text{ V}}$

**8-14**

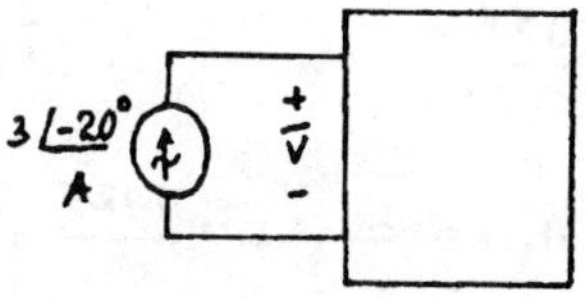


$\omega = 2000$ rad/s

(a) 8 Ω 5 mH $\quad \bar{V} = (3\angle{-20°})(8+j10) = \underline{38.42\angle{31.34°}\text{ V}}$

(b) 8 Ω 50 μF $\quad \bar{V} = (3\angle{-20°})(8-j10) = \underline{38.42\angle{-71.34°}\text{ V}}$

(c) 8 Ω 5 mF 50 μF $\quad \bar{V} = (3\angle{-20°})(8) = \underline{24.00\angle{-20.00°}\text{ V}}$

(d) $\bar{V} = (3\angle{-20°})(8+j20-j5) = \underline{51.00\angle{41.93°}\text{ V}}$

**8-15**

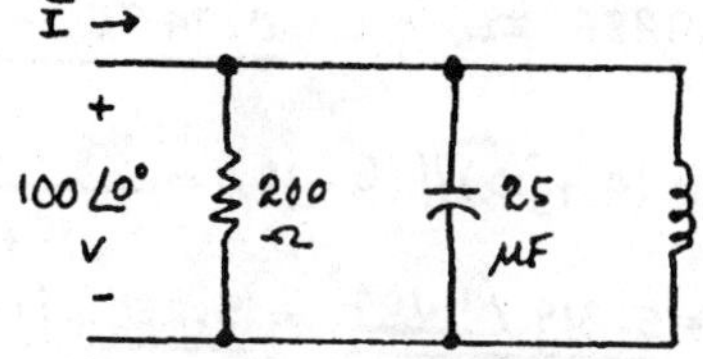


(a) $\bar{I} = 0.5 - j2$

(b) $|\bar{I}| = 1$

(a) $\omega = 1000;\ \bar{I} = \frac{100}{200} + j25\times10^{-6}\times1000\times100 + \frac{100}{j1000L} = 0.5 + j\left(2.5 - \frac{0.1}{L}\right)$

$\therefore 2.5 - \frac{0.1}{L} = -2, \quad L = \underline{1/45 \text{ H}}$

(b) $\omega = 100;\ 1 = \left|0.5 + j25\times10^{-6}\times100\times100 - j\frac{1}{L}\right| = \left|0.5 + j\left(0.25 - \frac{1}{L}\right)\right|$

$\therefore 1 = 0.25 + \left(0.25 - \frac{1}{L}\right)^2;\ 0.25 - \frac{1}{L} = \pm\sqrt{0.75} \quad \therefore L = \underline{0.8960 \text{ H}}$

**8-16**

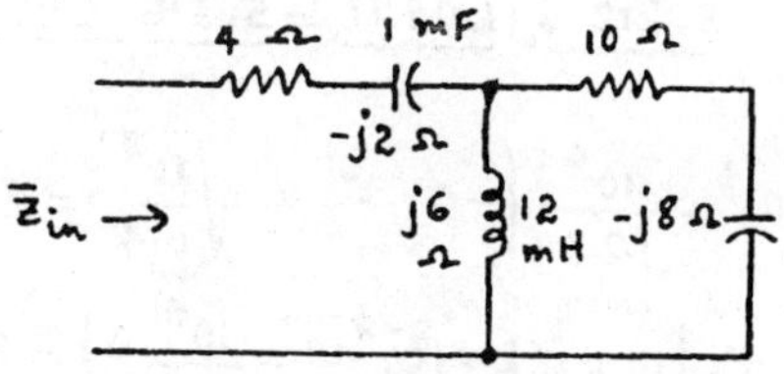


$\omega = 500, \quad \bar{Z}_{in} = 4 - j2 + \frac{j6(10-j8)}{10-j2} = 4 - j2 + \frac{24+j30}{5-j1}$

$\bar{Z}_{in} = \underline{7.462} + j\underline{4.692} = \underline{8.814\angle{32.16°}\ \Omega}$

**8-17** (a)

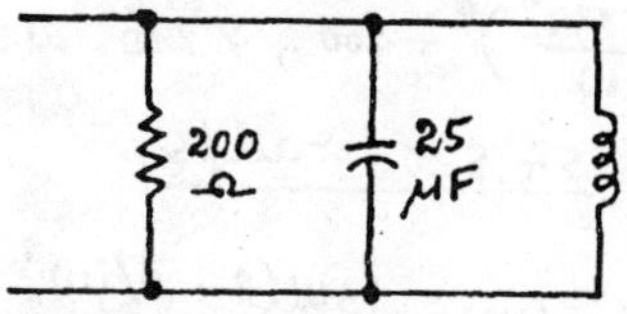


$\omega = 100, \ L = 10 \text{ H} \quad \therefore \bar{Z} = \frac{1}{1/200 + j0.0025 - j0.001}$

$\bar{Z} = \frac{1}{0.005 + j0.0015} = \underline{191.57\angle{-16.699°}}$

$\bar{Z} = \underline{183.49 - j55.05^{-}\ \Omega}$

(b) $\omega = 100 \quad \therefore \frac{1}{80} = \left|0.005 + j\left(0.0025 - \frac{0.01}{L}\right)\right|, \ \frac{1}{6400} = 0.25\times10^{-4} + \left(0.0025 - \frac{0.01}{L}\right)^2$

$\therefore 0.0025 - \frac{0.01}{L} = \pm 1.1456\times10^{-2} \quad \therefore L = \underline{0.7165^{+} \text{ H}}$

(c) $L = 10$ H $\quad \therefore \frac{1}{150} = \left|0.005 + j\omega 25\times10^{-6} - j\frac{0.1}{\omega}\right|, \ \frac{1}{22{,}500} = \frac{1}{40{,}000} + \left(\frac{\omega}{40{,}000} - \frac{1}{10\omega}\right)^2$

$\therefore \omega = \underline{20.33}$ and $\underline{196.72 \text{ rad/s}}$

**8-18**

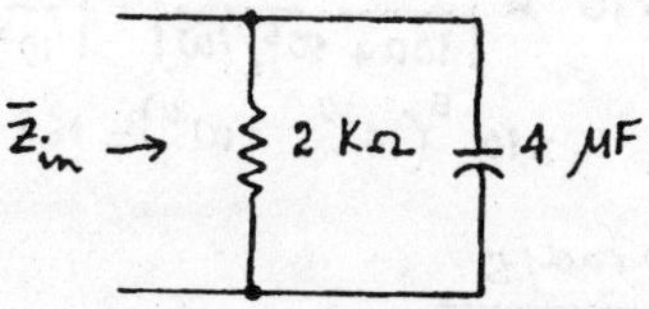


(a) $|\bar{Z}_{in}| = 1500 \ \therefore \frac{1}{1500} = \left|\frac{1}{2000} + j4\times10^{-6}\omega\right|$

$\frac{1}{225} = \frac{1}{400} + 16\times10^{-8}\omega^2 \quad \therefore \omega = \underline{110.24 \text{ rad/s}}$

(b) $\angle\bar{Z}_{in} = -30° \ \therefore \tan 30° = 4\times10^{-6}\omega\times2000 \ \therefore \omega = \underline{72.17 \text{ rad/s}}$

(c) $\text{Re}[\bar{Z}_{in}] = 200;\ \bar{Z}_{in} = \frac{2\times10^{3+6}/j4\omega}{2\times10^3 + 10^6/j4\omega} = \frac{2\times10^9}{10^6 + j8\times10^3\omega} = \frac{2000}{1 + j\omega/125} \times \frac{1 - j\omega/125}{1 - j\omega/125}$

$\therefore 200 = \frac{2000}{1 + \omega^2/125^2} \quad \therefore \omega = \underline{375 \text{ rad/s}}$

(d) $|X_{in}| = 400 = \frac{2000\,\omega}{1 + \omega^2/125^2} \quad \therefore 1 + \omega^2/125^2 - 5\times\frac{\omega}{125} = 0 \quad \therefore \omega = \underline{26.09}$ and $\underline{598.9 \text{ rad/s}}$

**8-19**

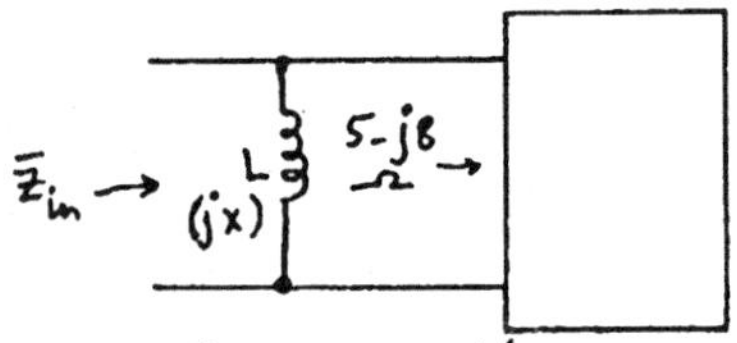


$\omega = 4000$ rad/s

(a) $\bar{Z}_{in} = \dfrac{jx(5-j8)}{5+j(x-8)} = \dfrac{8x+j5x}{5+j(x-8)}$, $X_{in}=0 \therefore \angle\bar{Z}_{in}=0$

$\therefore \dfrac{8x}{5} = \dfrac{5x}{x-8}$, $x = 8+\dfrac{25}{8} = 11.125\ \Omega \therefore L = \dfrac{11.125}{4} = \underline{2.781\ \text{mH}}$

(b) $|\bar{Z}_{in}| = 4 \therefore x^2\dfrac{8^2+5^2}{25+x^2-16x+64} = 4^2 = \dfrac{89x^2}{89-16x+x^2}$

$\therefore x = 2.9986\ \Omega$ ; $L = \underline{0.7496}$ mH

**8-20**

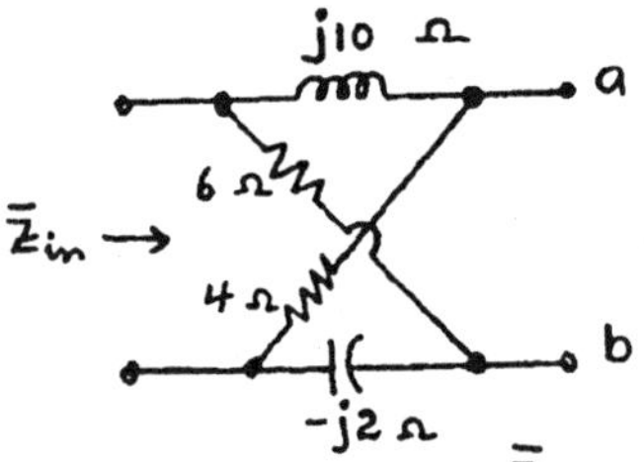


(a) a-b O.C. : $\bar{Z}_{in} = (4+j10)//(6-j2) = \dfrac{24+20+j52}{10+j8}$

$\therefore \bar{Z}_{in} = \underline{5.319\ \angle 11.104^\circ} = \underline{5.220+j1.0244\ \Omega}$

(b) a-b S.C. : $\bar{Z}_{in} = (6\|j10)+(4\|-j2) = \dfrac{j60}{6+j10}+\dfrac{-j8}{4-j2}$

$\bar{Z}_{in} = \dfrac{600+j360}{136}+\dfrac{16-j32}{20} = \underline{5.212+j1.0471} = \underline{5.316\ \angle 11.360^\circ}\ \Omega$

**8-21**

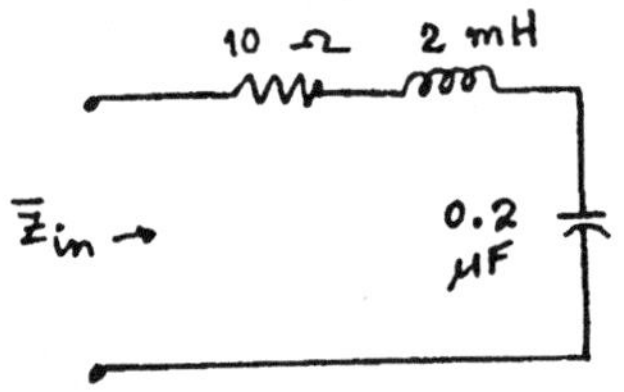


(a) $|X_{in}| = \left|j\omega 2\times10^{-3}+\dfrac{10^6}{j0.2\omega}\right| = 0 \therefore \omega = \sqrt{\dfrac{10^9}{0.4}} = \underline{5\times10^4\ \text{rad/s}}$

(b) $|\bar{Z}_{in}|_{min} = 10 \therefore \left|10+j\left(2\times10^{-3}\omega-\dfrac{10^6}{0.2\omega}\right)\right| = 20$

$\therefore 100+\left(2\times10^{-3}\omega-\dfrac{5\times10^6}{\omega}\right)^2 = 400$, $2\times10^{-3}\omega-\dfrac{5\times10^6}{\omega} = \pm17.321$

$\therefore \omega = \underline{45{,}860}$ or $\underline{54{,}520}$ rad/s

**8-22**

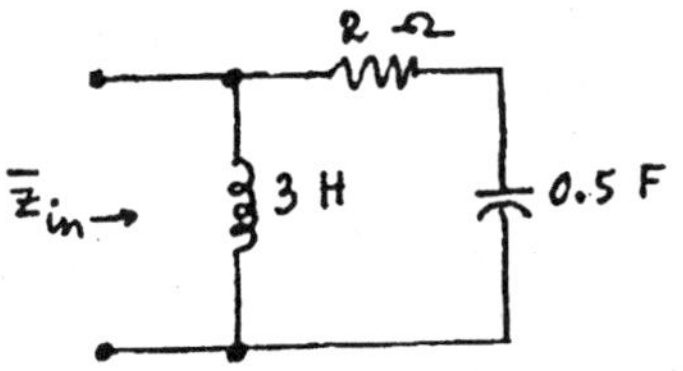


$\bar{Z}_{in}(\omega) = R_{in}(\omega)+jX_{in}(\omega) = \dfrac{j3\omega(2+2/j\omega)}{2+j(3\omega-2/\omega)}$

$= \dfrac{(6+j6\omega)\omega}{2\omega+j(3\omega^2-2)} \times \dfrac{2\omega-j(3\omega^2-2)}{2\omega-j(3\omega^2-2)}$

$= \dfrac{12\omega^2+6\omega^2(3\omega^2-2)+j12\omega^3-j6\omega(3\omega^2-2)}{4\omega^2+9\omega^4-12\omega^2+4}$

$\therefore R_{in}(\omega) = \dfrac{18\omega^4}{9\omega^4-8\omega^2+4}$ $\qquad X_{in}(\omega) = \dfrac{12\omega-6\omega^3}{9\omega^4-8\omega^2+4}$

**8-23** (a)

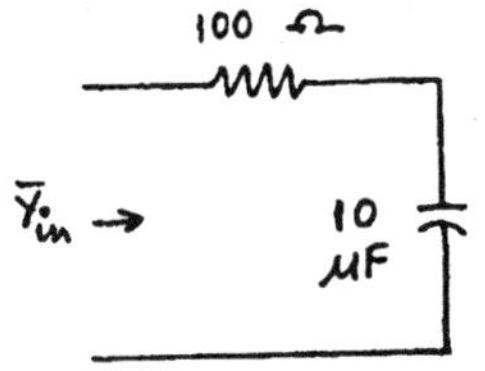


$\bar{Y}_{in} = G_{in}+jB_{in}$ ; $|\bar{Y}_{in}| = 4\times10^{-3} = \left|\dfrac{1}{100+10^5/j\omega}\right| = \left|\dfrac{j\omega}{10^5+j100\omega}\right|$

$\therefore \dfrac{\omega}{\sqrt{10^{10}+10^4\omega^2}} = 4\times10^{-3}$, $\omega^2 = 16\times10^{-6}(10^{10}+4\omega^2) = 16\times10^4+0.16\omega^2$

$\therefore \omega = \underline{436.4\ \text{rad/s}}$

(b) $\bar{Y}_{in} = \dfrac{j\omega(10^5-j100\omega)}{10^{10}+10^4\omega^2}$ $\therefore G_{in} = \dfrac{100\omega^2}{10^{10}+10^4\omega^2} = 4\times10^{-3}$ $\quad\therefore \omega = \underline{816.5\ \text{rad/s}}$

(c) $B_{in} = \dfrac{10^5\omega}{10^{10}+10^4\omega^2} = \dfrac{10\omega}{\omega^2+10^6} = 4\times10^{-3} \therefore 10\omega = 4000+\dfrac{\omega^2}{250}$, $\omega^2-2500\omega+10^6=0$

$\omega = \underline{500}$ and $\underline{2000}$ rad/s

**8-24**

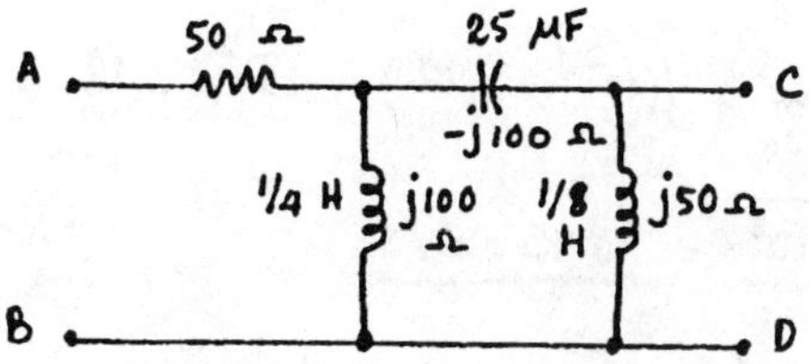


(a) C-D O.C.: $\bar{Z}_{in\,A-B} = 50 + \dfrac{j100(-j50)}{j50}$

$\bar{Z}_{in\,A-B} = \underline{50 - j100} = \underline{111.80\angle -63.43^\circ}\ \Omega$

(b) A-B S.C.: $50 // j100 = \dfrac{50(j100)}{50 + j100} = \dfrac{j100}{1+j2} \times \dfrac{1-j2}{1-j2}$

$\therefore 50//j100 = 40 + j20\ \Omega\ ; 40 + j20 - j100 = 40 - j80\ \Omega\ ; (40-j80)//j50 = \dfrac{4000 + j2000}{40 - j30}$

$\dfrac{400 + j200}{4 - j3} \times \dfrac{4+j3}{4+j3} = (16 + j8)(4 + j3) = 40 + j80\ ,\ \bar{Y}_{in\,C-D} = \dfrac{1}{40 + j80} = \underline{11.18\angle -63.43^\circ} = \underline{5 - j10}\ \text{mS}$

**8-25**

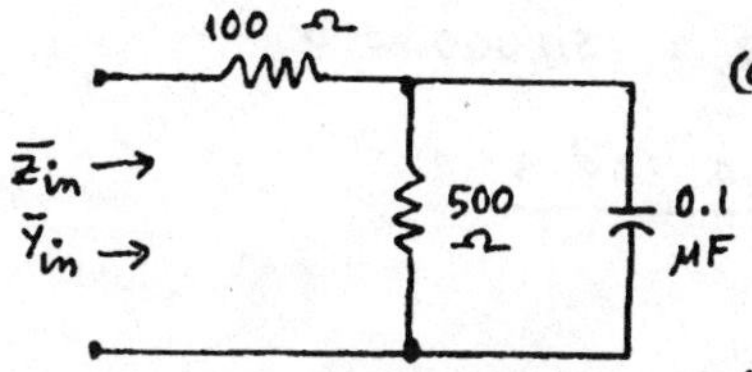


(a) $\bar{Z}_{in} = 100 + \dfrac{500(10^7/j\omega)}{500 + 10^7/j\omega} = 100 + \dfrac{500}{1 + j5\times10^{-5}\omega}$

$= 100 + \dfrac{500 - j\omega/40}{1 + 25\times10^{-10}\omega^2}\ \therefore \dfrac{500}{1 + 25\times10^{-10}\omega^2} = 100\ \therefore \omega = \underline{40\ \text{Krad/s}}$

(b) $X_{in} = -200 = \dfrac{-\omega/40}{1 + 25\times10^{-10}\omega^2}\ ;\ 200 + 5\times10^{-7}\omega^2 - 0.025\omega = 0$

$\therefore \omega = \underline{10}$ and $\underline{40\ \text{Krad/s}}$

(c) $\bar{Y}_{in} = \dfrac{1 + j5\times10^{-5}\omega}{600 + j5\times10^{-3}\omega} \times \dfrac{600 - j5\times10^{-3}\omega}{600 - j5\times10^{-3}\omega} = \dfrac{600 + 25\times10^{-8}\omega^2 + j(0.03\omega - 0.005\omega)}{36\times10^4 + 25\times10^{-6}\omega^2}$

$\therefore G_{in} = \dfrac{1}{200} = \dfrac{600 + 25\times10^{-8}\omega^2}{36\times10^4 + 25\times10^{-6}\omega^2}\ \therefore \omega = \underline{97.98\ \text{Krad/s}}$

(d) $B_{in} = 1/240 = \dfrac{0.025\omega}{36\times10^4 + 25\times10^{-6}\omega^2}\ ,\ 36\times10^4 + 25\times10^{-6}\omega^2 - 6\omega = 0\ \therefore \omega = \underline{120\ \text{Krad/s}}$

**8-26**

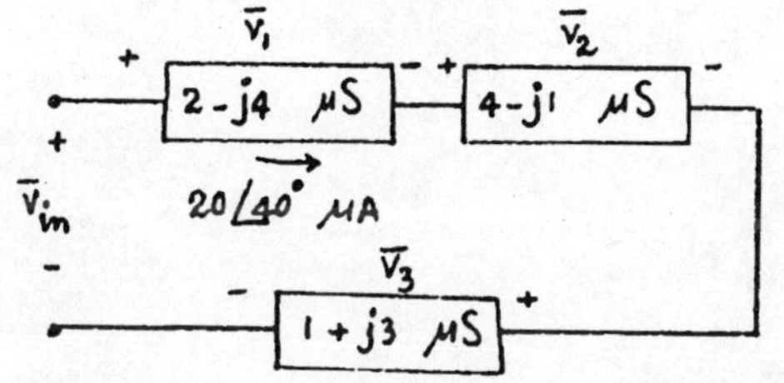


$\bar{V}_1 = \dfrac{20}{2 - j4} = \dfrac{10(1+j2)}{5} = 2 + j4\quad |\bar{V}_1| = \underline{4.472\ \text{V}}$

$\bar{V}_2 = \dfrac{20}{4 - j1} \times \dfrac{4 + j1}{4 + j1} = \dfrac{80 + j20}{17}\quad |\bar{V}_2| = \underline{4.851\ \text{V}}$

$\bar{V}_3 = \dfrac{20}{1 + j3} = 2 - j6\quad \therefore |\bar{V}_3| = \underline{6.325\ \text{V}}$

$\bar{V}_{in} = 2 + j4 + \dfrac{80 + j20}{17} + 2 - j6 = \dfrac{148 - j14}{17}\ ,\ |\bar{V}_{in}| = \underline{8.745\ \text{V}}$

**8-27** (a)

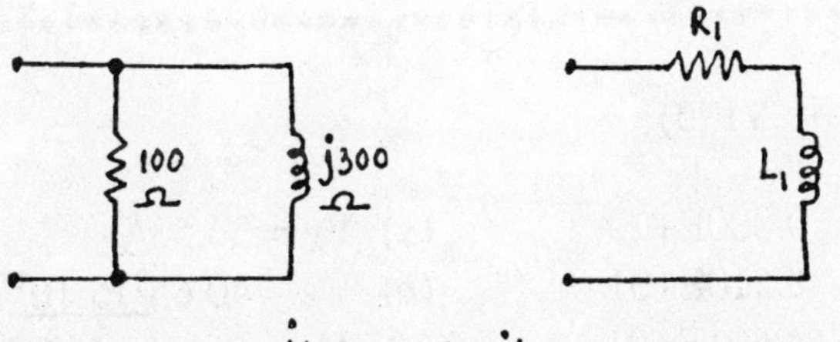


$\omega = 1500$ rad/s

$R_1 + j1500L_1 = \dfrac{j300}{1+j3} \times \dfrac{1-j3}{1-j3} = 90 + j30$

$\therefore R_1 = \underline{90\ \Omega}\ ,\ L_1 = \dfrac{30}{1500} = \underline{0.02\ \text{H}}$

(b) $R_1 + j500L_1 = \dfrac{j100}{1+j1} \times \dfrac{1-j1}{1-j1} = 50 + j50\quad \therefore R_1 = \underline{50\ \Omega}\ ,\ L_1 = \underline{0.1\ \text{H}}$

8-28

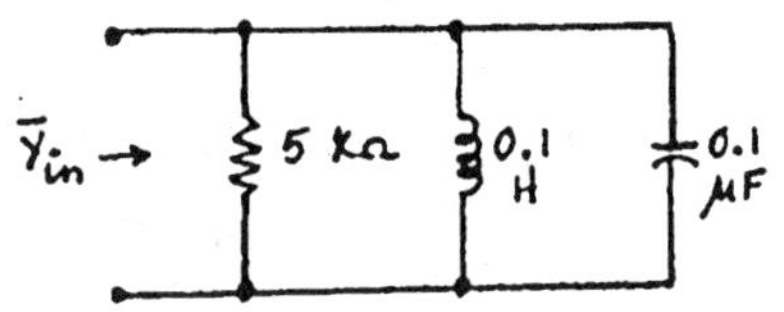


(a) $\bar{Y}_{in} = \frac{1}{5000} + j(10^{-7}\omega - \frac{10}{\omega})$ ; $10^{-7}\omega - \frac{10}{\omega} = 0$

$\therefore \omega = \sqrt{10^8} = \underline{10 \text{ krad/s}}$

(b) $\tan^{-1} 1 = 45° \therefore 10^{-7}\omega - \frac{10}{\omega} = \frac{1}{5000}$

$\therefore 5 \times 10^{-4}\omega^2 - \omega - 50{,}000 = 0$

$$\omega = \frac{1 \pm \sqrt{1 + 20\times 10^{-4} \times 5 \times 10^{4}}}{10^{-3}} = \underline{11{,}050 \text{ rad/s}}$$

(c) $5 \times 10^{-4}\omega^2 + \omega - 50{,}000 = 0$

$\therefore \omega = \underline{9{,}050 \text{ rad/s}}$

8-29

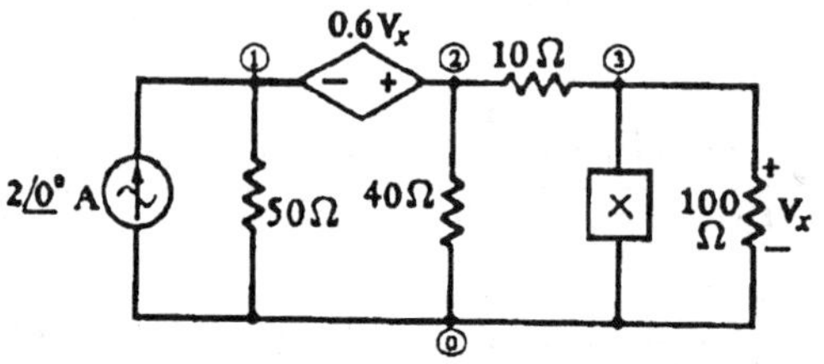


*****************************************************************************

PROGRAM

```
        I01 0 1 AC 2
        R10 1 0 50
        E21 2 1 3 0 0.6
        R20 2 0 40
        R23 2 3 10
        R30 3 0 100
(a)     R30X  3  0  200
(b)     L30 3 0 0.3
(c)     C30  3  0  90U
        .AC LIN 1 60 60
        .PRINT AC VM(3) VP(3)
        .END
```

*****************************************************************************

ANSWERS:

| FREQ | VM(3) | VP(3) |
|---|---|---|
| 6.000E+01 | 3.653E+01 | 0.000E+00 |
| 6.000E+01 | 4.065E+01 | 1.510E+01 |
| 6.000E+01 | 2.925E+01 | -4.601E+01 |

(a) $V_X = 36.53\underline{/0°}$ V

(b) $V_X = 40.65\underline{/15.10°}$ V

(c) $V_X = 29.25\underline{/-46.01°}$ V

# CHAPTER 9: THE SINUSOIDAL STEADY-STATE RESPONSE

## Section 9-2 Nodal, mesh, and loop analysis

**1** Use nodal analysis on the circuit shown in Fig. 9-1 to find the phasor voltage $\mathbf{V}_{RL}$.

**2** (*a*) Find $\mathbf{V}_1$ in Fig. 9-2. (*b*) To what identical values should the capacitive impedances be changed so that $\mathbf{V}_1$ is 180° out of phase with the source voltage?

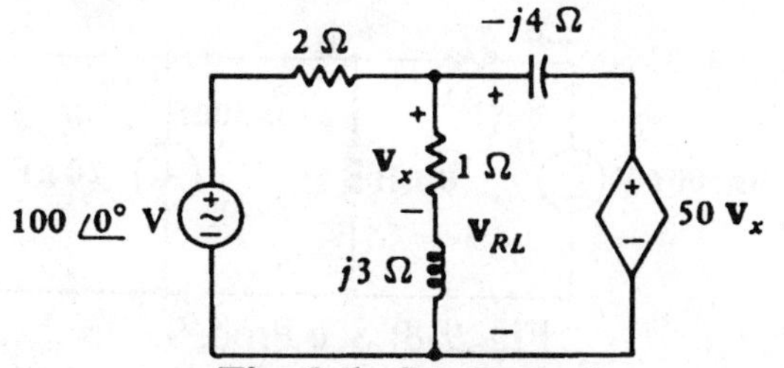


**Fig. 9-1:** See Prob. 1.

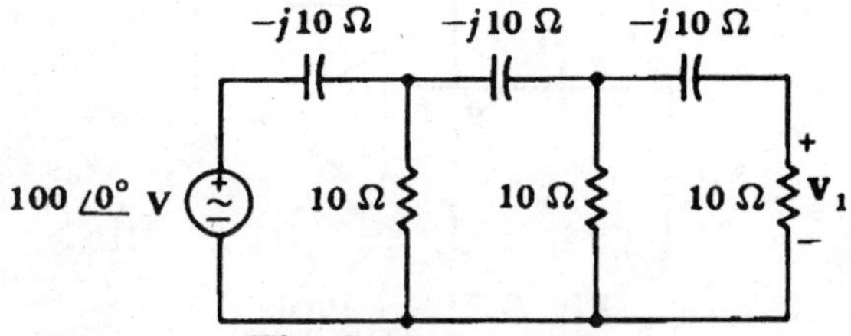


**Fig. 9-2 :** See Prob. 2.

**3** If $i_s(t) = 10^{-3}\cos 10^4 t$ A in the circuit shown in Fig. 9-3, find $v_1(t)$.

**4** Find $v_3(t)$ in the circuit of Fig. 9-4 by using nodal analysis.

**5** Write three mesh equations and solve them to determine $i_3(t)$ for the circuit shown in Fig. 9-4.

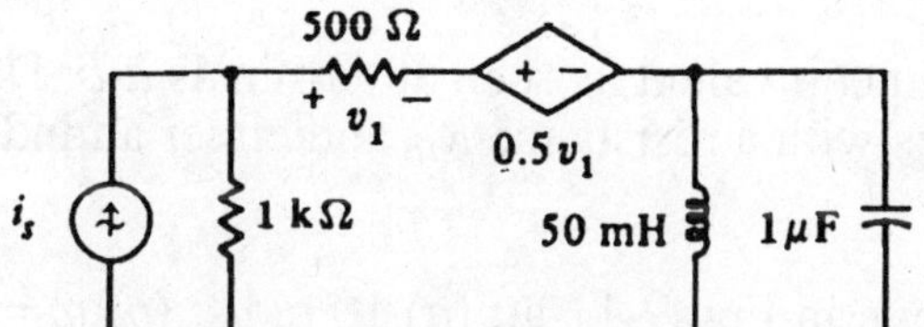

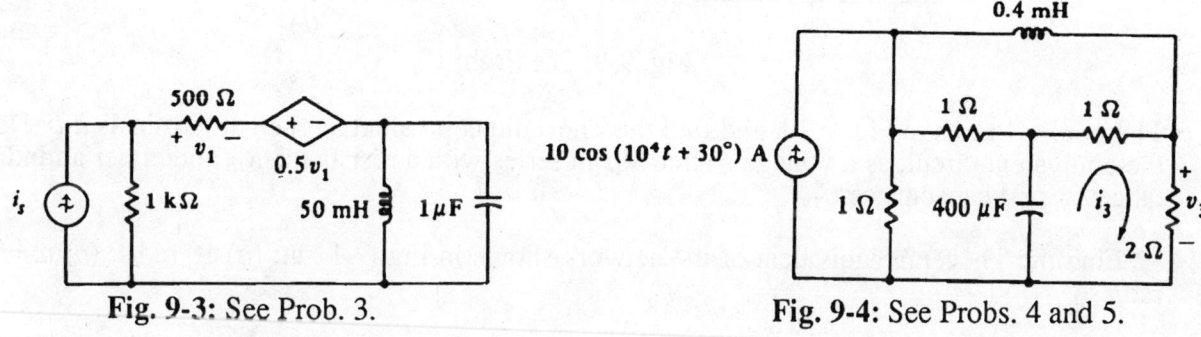

**Fig. 9-3:** See Prob. 3.

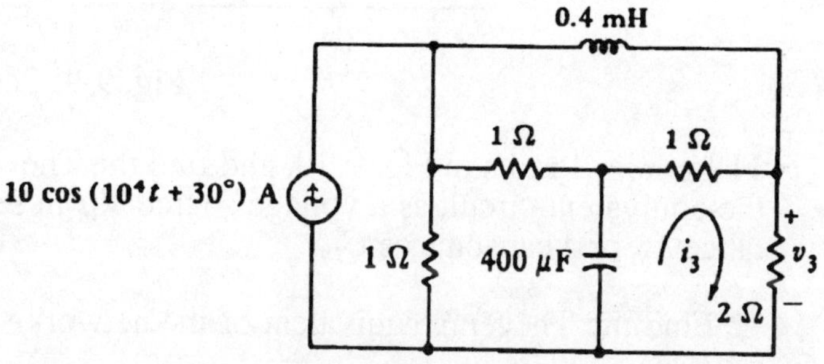


**Fig. 9-4:** See Probs. 4 and 5.

**6** If $\omega = 100$ rad/s for the source in Fig. 9-5, find $\mathbf{I}_1$.

**7** (*a*) Draw a tree for the circuit shown in Fig. 9-6 so that $i_x$ is a link current, assign a complete set of link currents, and find $i_x(t)$. (*b*) Construct another tree in which $v_x$ is a tree-branch voltage, assign a complete set of tree-branch voltages, and find $v_x(t)$.

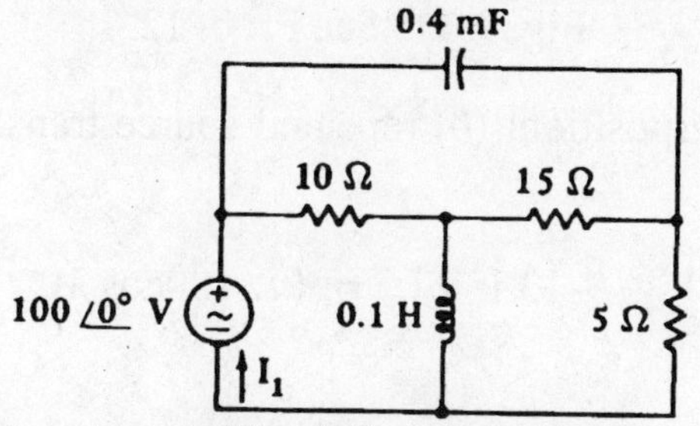


**Fig. 9-5:** See Prob. 6.

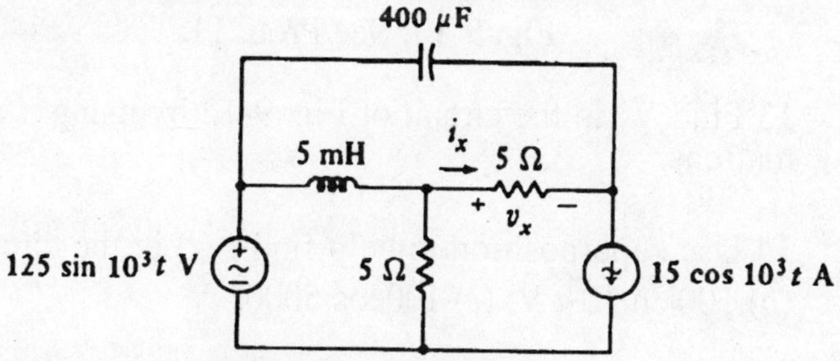


**Fig. 9-6:** See Prob. 7.

**8** The op-amp shown in Fig. 9-7 has an infinite input impedance, zero output impedance, and a large but finite (positive, real) gain, $\mathbf{A} = -\mathbf{V}_o/\mathbf{V}_i$. (*a*) Construct a basic differentiator by letting $\mathbf{Z}_f = R_f$, find $\mathbf{V}_o/\mathbf{V}_s$, and then show that $\mathbf{V}_o/\mathbf{V}_s \to -j\omega C_1 R_f$ as $\mathbf{A} \to \infty$. (*b*) Let $\mathbf{Z}_f$ represent $C_f$ and $R_f$ in parallel, find $\mathbf{V}_o/\mathbf{V}_s$, and then show that $\mathbf{V}_o/\mathbf{V}_s \to -j\omega C_1 R_f/(1 + j\omega C_f R_f)$

## Section 9-3 Superposition, source transformations, and Thévenin's theorem

**9** Use superposition to find $\mathbf{I}_x$ and $i_x(t)$ in the circuit of Fig. 9-8.

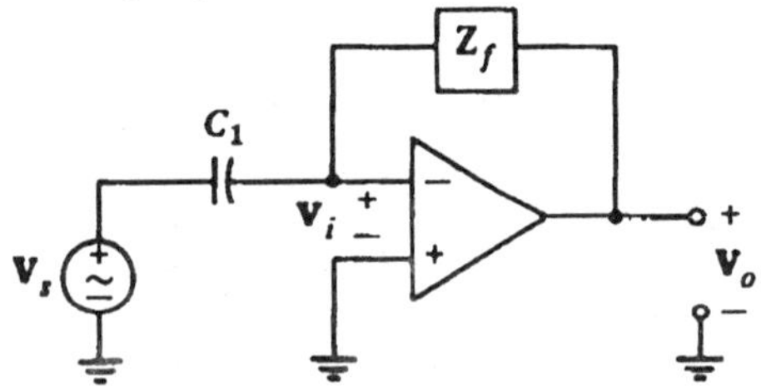


**Fig. 9-7:** See Prob. 8.

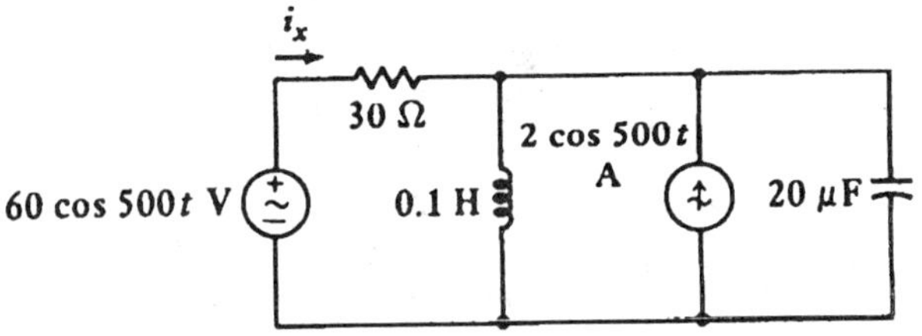


**Fig. 9-8:** See Prob. 9.

**10** Find the Thévenin equivalent of the network shown in: (*a*) Fig. 9-9*a*; (*b*) Fig. 9-9*b*.

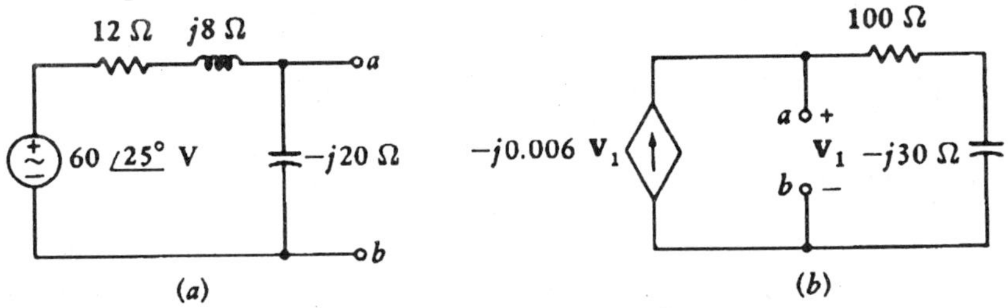


**Fig. 9-9:** See Prob. 10.

**11** Let $\omega = 1$ rad/s in Fig. 9-10, and find the Thévenin equivalent as seen at terminals *a*-*b*. Draw the equivalent circuit as a voltage source $\mathbf{V}_{th}$ in series with a resistance $R_{th}$ and either an inductance $L_{th}$ or a capacitance $C_{th}$.

**12** Find the Thévenin equivalent of the network shown in Fig. 9-11 at: (*a*) 10 rad/s; (*b*) $\omega = 20$ rad/s.

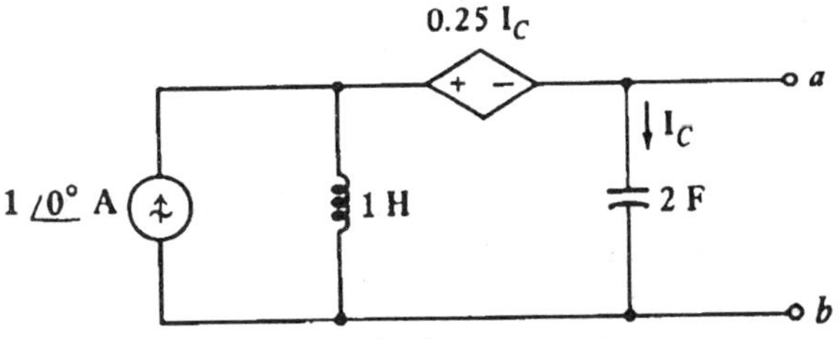


**Fig. 9-10:** See Prob. 11.

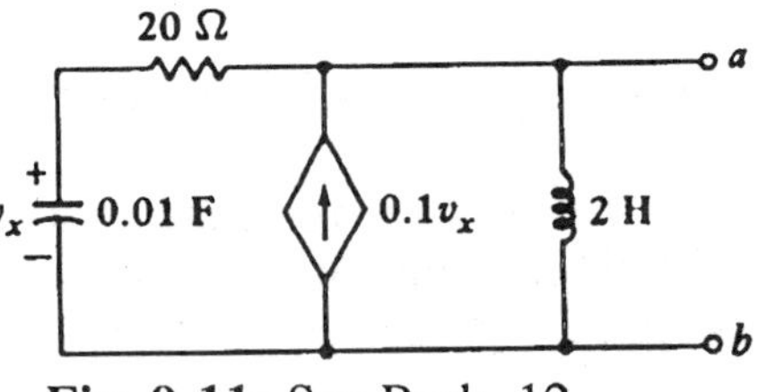


**Fig. 9-11:** See Prob. 12.

**13** Find $\mathbf{V}_x$ in the circuit of Fig. 9-12 by using: (*a*) superposition; (*b*) repeated source transformations.

**14** Use superposition to help find $i_x(t)$ in the circuit of Fig. 9-13 if $v_s(t)$ =: (*a*) $100\cos 10^4 t$ V; (*b*) $100\sin 10^4 t$ V; (*c*) $100\cos 5000t$ V.

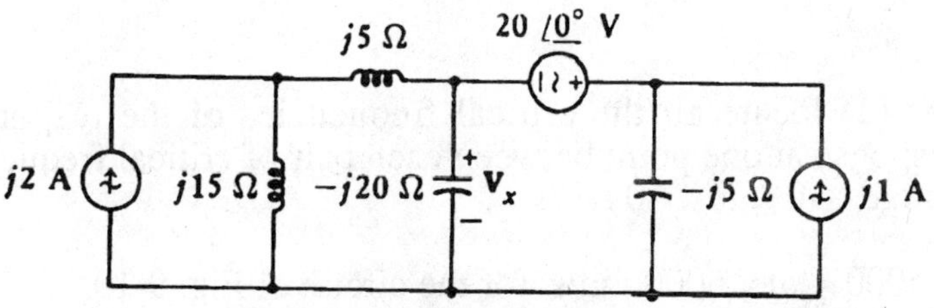


Fig. 9-12: See Prob. 13.

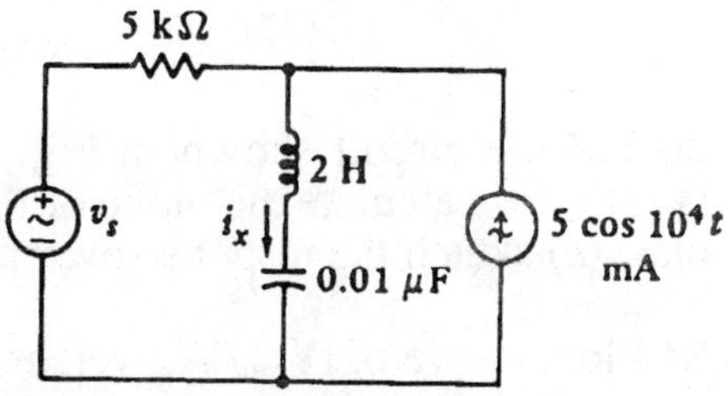


Fig. 9-13: See Prob. 14.

## Section 9-4 Phasor diagrams

**15** Let $\omega = 50$ rad/s for the source of Fig. 9-14. (*a*) Construct a phasor diagram showing the three currents. Use a scale of 1 A/in. (*b*) Construct another phasor diagram showing the three voltages, using a scale of 1 V/in.

**16** In Fig. 9-15, let $\mathbf{I}_s$ be a current source with an amplitude of 3 A at 1000 rad/s. Using this current as the reference phasor, find the four indicated voltages and show them on an accurate phasor diagram. Identify the equality, $\mathbf{V}_1 + \mathbf{V}_2 = \mathbf{V}_3 + \mathbf{V}_4$.

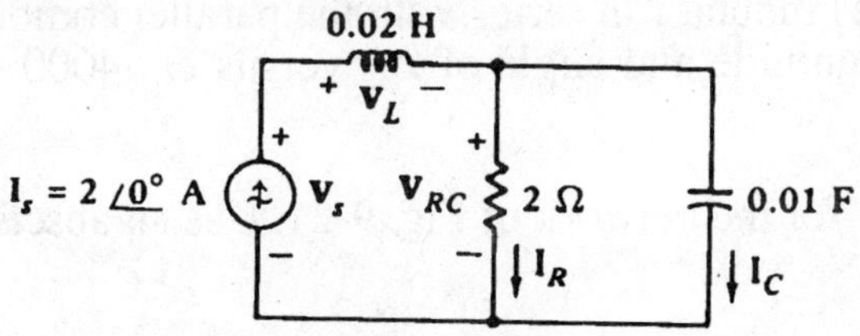


Fig. 9-14: See Prob. 15.

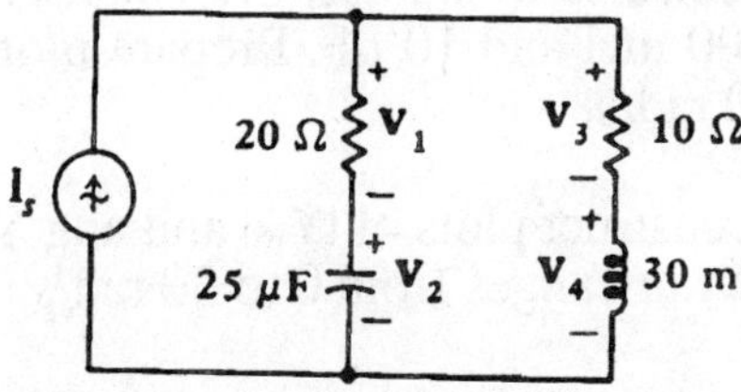


Fig. 9-15: See Prob. 16.

**17** For the circuit shown in Fig. 9-16, let $\mathbf{V}_s = 100\underline{/0°}$ V, $|\mathbf{V}_1| = 80$ V, and $|\mathbf{V}_2| = 50$ V. (*a*) Use a phasor diagram to determine $\mathbf{V}_1$ and $\mathbf{V}_2$. There are two possible answers; give both. (*b*) If $\mathbf{Z}_1$ is an inductive impedance and $\mathbf{Z}_2$ is a capacitive impedance, find $\mathbf{V}_1$ and $\mathbf{V}_2$.

**18** In the circuit of Fig. 9-17, it is known that $|\mathbf{V}_1| = 50$ V, $|\mathbf{V}_2| = 125$ V, and $\mathbf{V}_3 = 200\underline{/-45°}$ V. Use graphical methods (straightedge, ruler, protractor, compass, and all that fun stuff) to find the angle of $\mathbf{V}_1$. Although there may seem to be two answers to this problem, one is right, and the other is wrong.

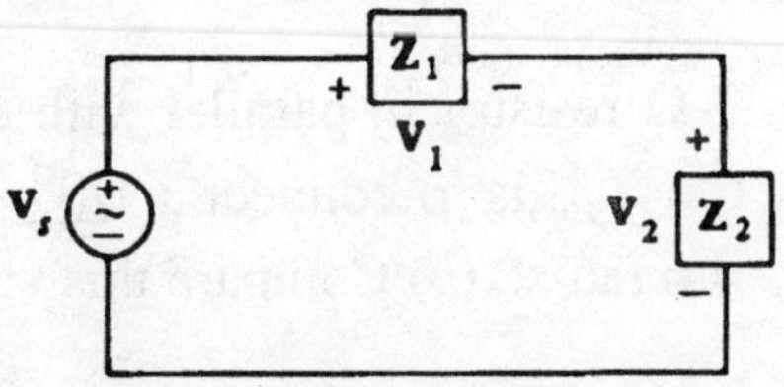

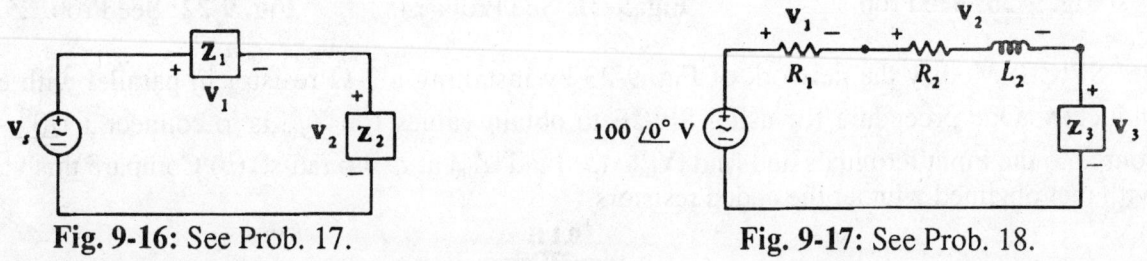

Fig. 9-16: See Prob. 17.

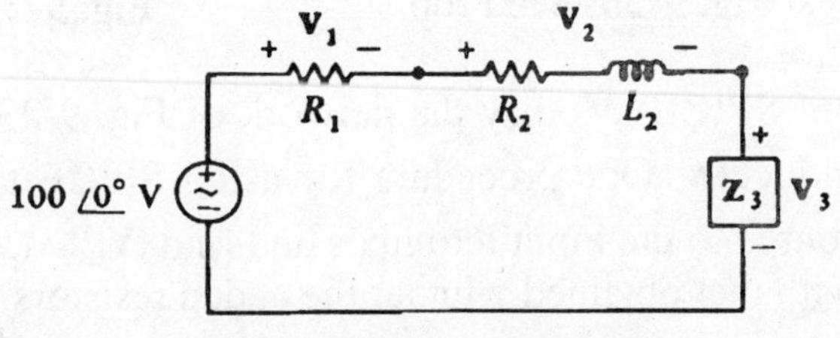


Fig. 9-17: See Prob. 18.

## Section 9-5 Response as a function of $\omega$

**19** A 4-Ω resistor is in parallel with a 125-μF capacitor. The parallel combination is in series with a 2-mH inductor and a sinusoidal voltage source $\mathbf{V}_s$. Calculate a few points, and then plot a curve of the magnitude of the ratio of the capacitor voltage to the source voltage. Cover the frequency interval, $0 \le \omega \le 5$ krad/s.

**20** For the circuit shown in Fig. 9-18: (*a*) locate all the critical frequencies of the response $|\mathbf{V}_C/\mathbf{I}_s|$; (*b*) calculate the value of the response at one point between each pair of critical frequencies; (*c*) sketch the response over the range, $-10 < \omega < 10$ rad/s.

**21** Plot a curve of $|\mathbf{V}_{out}/\mathbf{V}_{in}|$ versus $\omega$, $-5000 < \omega < 5000$ rad/s, for the circuit of Fig. 9-19.

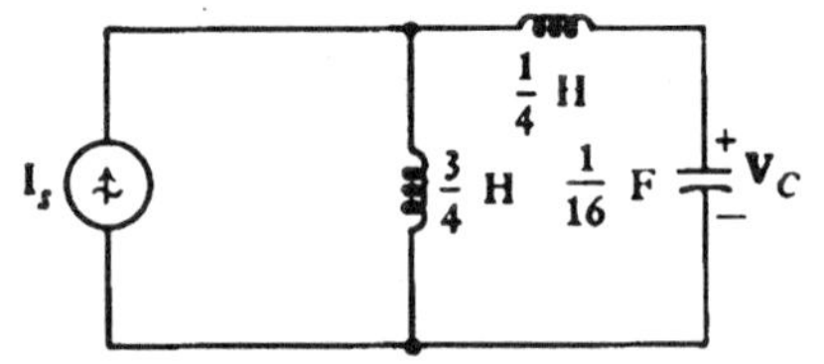


**Fig. 9-18:** See Prob. 20.

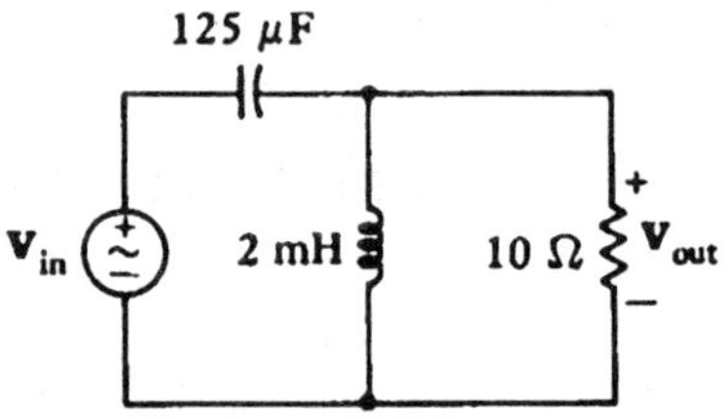


**Fig. 9-19:** See Prob. 21.

**22** Refer to Fig. 9-20 and determine all the critical frequencies; then prepare magnitude and phase plots of the response: (*a*) $\mathbf{V}_{in}/\mathbf{I}_s$; (*b*) $\mathbf{I}_C/\mathbf{I}_s$.

**23** A two-terminal network consists of an 80-mH inductor in series with the parallel combination of 100-mH and 10 $\mu$F. Prepare plots of the magnitude and angle of $\mathbf{Z}_{in}$ versus $\omega$, $-4000 < \omega < 4000$ rad/s.

**24** Construct plots of $|\mathbf{Y}_{in}|$ and ang $\mathbf{Y}_{in}$ versus $\omega$ for the network of Fig. 9-21. Use an abscissa on which $\omega$ ranges from 0 to 30 rad/s.

**25** Prepare sketches of $|\mathbf{Z}_{in}|$ and ang $\mathbf{Z}_{in}$ versus $\omega$ for the two-terminal network shown in Fig. 9-22.

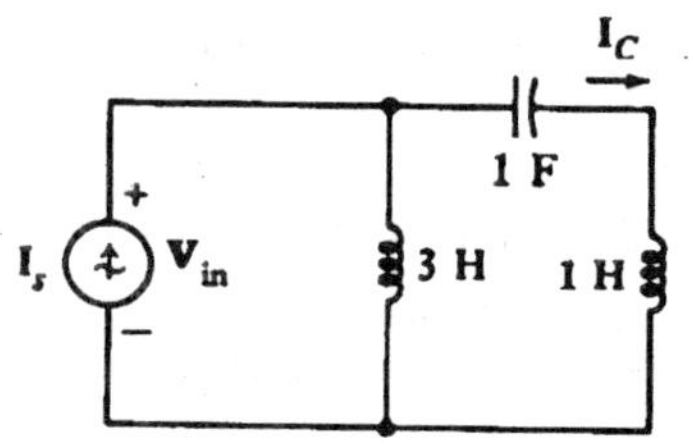

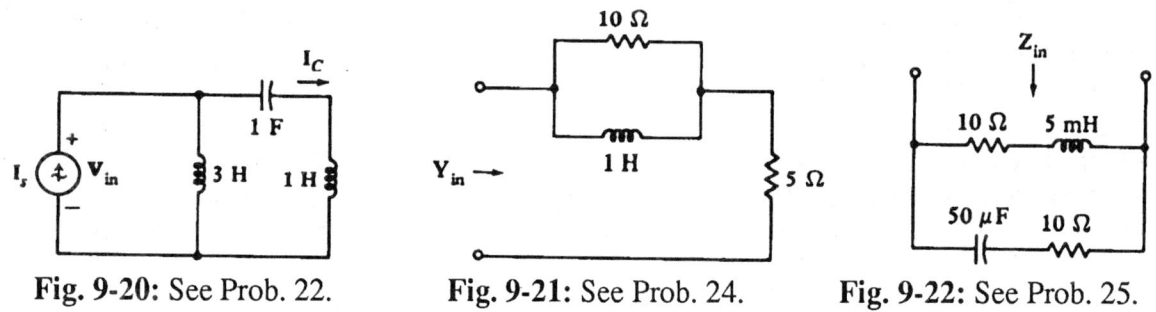

**Fig. 9-20:** See Prob. 22.

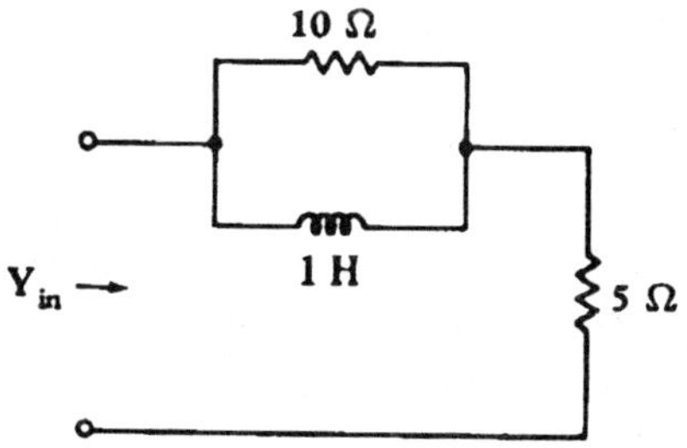


**Fig. 9-21:** See Prob. 24.

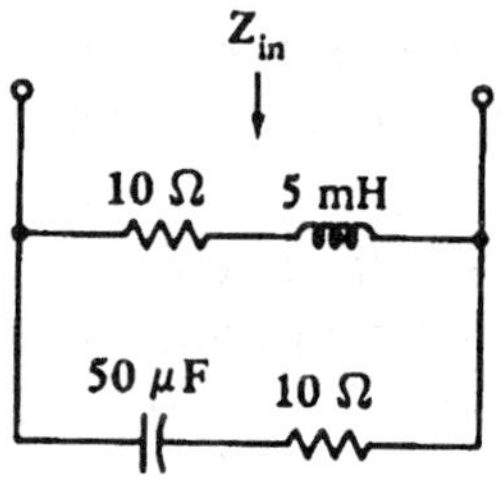


**Fig. 9-22:** See Prob. 25.

**26** (SPICE) Modify the network of Fig. 9-23 by installing a 2-Ω resistor in parallel with each inductor. One procedure for using SPICE to obtain values for $|\mathbf{Z}_{in}|$ is to connect a $1\underline{/0^\circ}$ - A source to the input terminals and find $|\mathbf{V}_{in}|$. (*a*) Find $|\mathbf{Z}_{in}|$ at $\omega = 6$ rad/s. (*b*) Compare this value with that obtained without the added resistors.

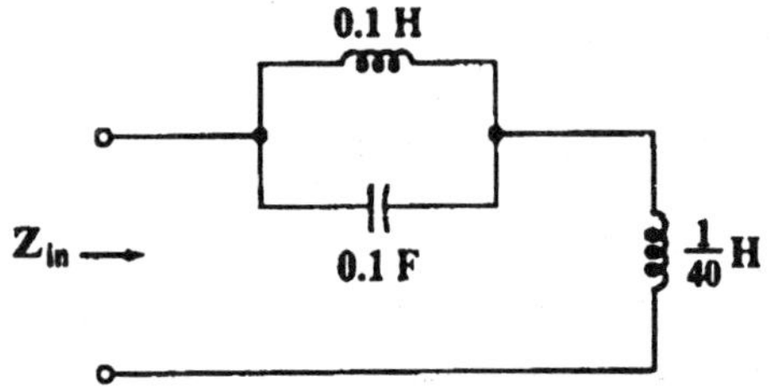


**Fig. 9-23:** See Prob. 26.

# Chapter 9

**9-1**

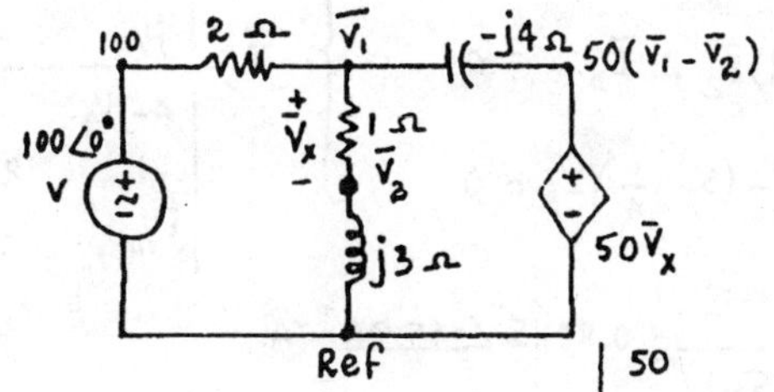


$$\frac{\bar V_1-100}{2}+\frac{\bar V_1-\bar V_2}{1}+\frac{\bar V_1-50(\bar V_1-\bar V_2)}{-j4}=0$$

$$\therefore (1.5-j12.25)\bar V_1+(-1+j12.5)\bar V_2=50$$

$$\frac{\bar V_2-\bar V_1}{1}+\frac{\bar V_2}{j3}=0 \quad \therefore -\bar V_1+\left(1-j\tfrac{1}{3}\right)\bar V_2=0$$

$$\bar V_1=\bar V_{RL}=\frac{\begin{vmatrix}50 & -1+j12.5\\ 0 & 1-j1/3\end{vmatrix}}{\begin{vmatrix}1.5-j12.25 & -1+j12.5\\ -1 & 1-j\frac{1}{3}\end{vmatrix}}=\frac{50-j50/3}{-\frac{10.75}{3}-j0.25}=\underline{14.673\angle 157.57^\circ\ \text{V}}$$

**9-2** (a)

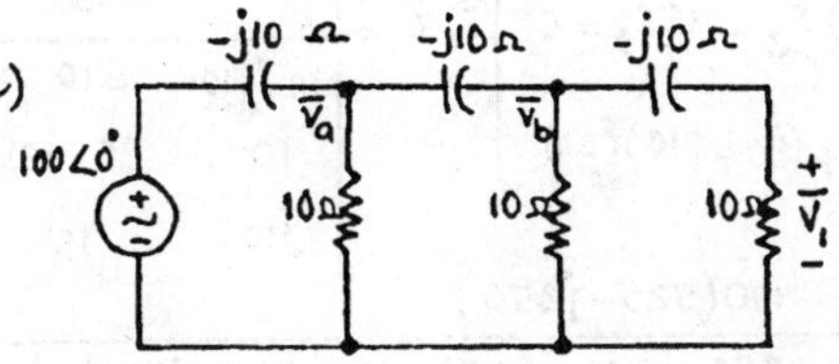


Let $\bar V_1=10$ V $\therefore \bar V_b=10-j10$, $\frac{\bar V_b}{10}=1-j1$

$1-j1+\frac{\bar V_1}{10}=2-j1 \;\therefore \bar V_a=(2-j1)(-j10)+10-j10=-j30$

$\frac{-j30}{10}=-j3$, $-j3+2-j1=2-j4 \;\therefore \bar V_s=(2-j4)(-j10)-j30$

$\bar V_s=-40-j50 \;\therefore \bar V_1=10\times\frac{100}{-40-j50}=\underline{15.617\angle 128.66^\circ\ \text{V}}$

(b) let each $\bar Z_c=-jX$, assume $\bar V_1=10$ V $\therefore \bar V_b=10-jX$, $\frac{10-jX}{10}=1-j\frac{X}{10}$, $1-\frac{jX}{10}+1=2-j\frac{X}{10}$

$\therefore \bar V_a=-jX\left(2-j\frac{X}{10}\right)+10-jX=10-\frac{X^2}{10}-j3X$, $\frac{\bar V_a}{10}=1-\frac{X^2}{100}-j\frac{3X}{10}$, $\bar I_s=3-\frac{X^2}{100}-j\frac{4X}{10}$

$\therefore \bar V_s=-\frac{4X^2}{10}-j3X+j\frac{X^3}{100}+10-\frac{X^2}{10}-j3X=10-0.5X^2-j6X+j\frac{X^3}{100} \quad \therefore -6X+\frac{X^3}{100}=0$

$X=\sqrt{600}=24.49 \quad \therefore \bar Z_c=\underline{-j24.49\ \Omega}$

**9-3**

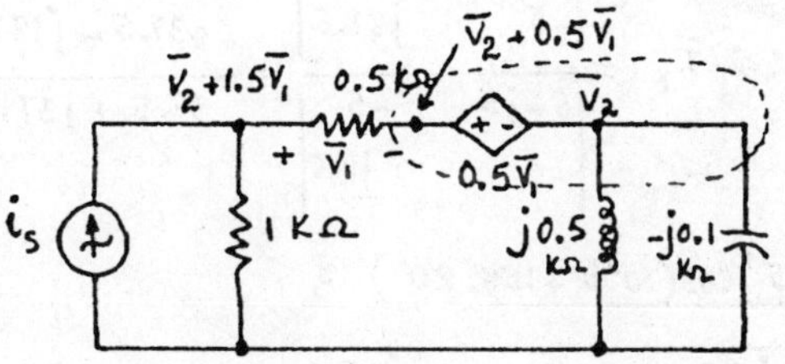


$$\frac{\bar V_2+1.5\bar V_1}{1}+\frac{\bar V_1}{0.5}=1 \quad \therefore \bar V_2=1-3.5\bar V_1$$

$$-\frac{\bar V_1}{0.5}+\frac{\bar V_2}{j0.5}+\frac{\bar V_2}{-j0.1}=0 \quad \therefore \bar V_2-5\bar V_2-j1\bar V_1=0$$

$$\therefore \bar V_2=\frac{j1}{-4}\bar V_1=-j0.25\bar V_1=1-3.5\bar V_1$$

$i_s=10^{-3}\cos 10^4 t$ A, $\bar V_1=\frac{1}{3.5-j0.25}=0.2850\angle 4.09^\circ \;\therefore v_1(t)=\underline{0.2850\cos(10^4 t+4.09^\circ)\ \text{V}}$

**9-4**

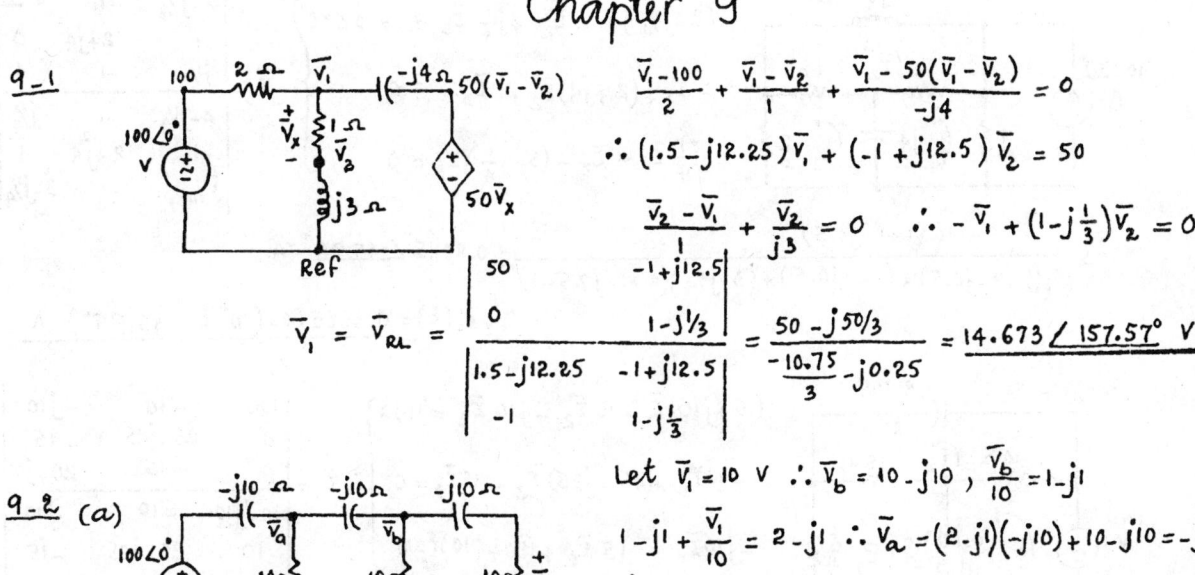

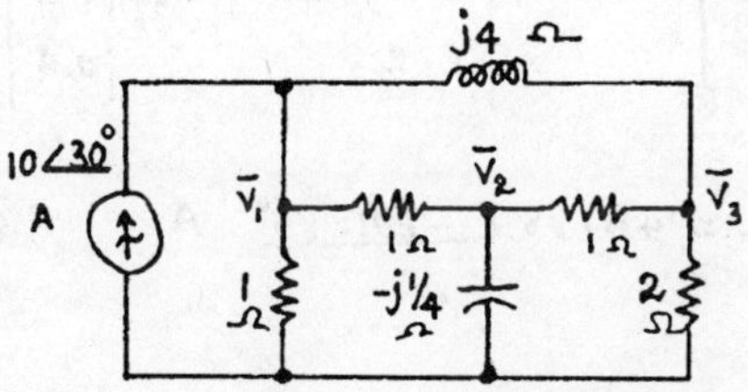


$$\left.\begin{aligned}&\frac{\bar V_1}{1}+\frac{\bar V_1-\bar V_2}{1}+\frac{\bar V_1-\bar V_3}{j4}=10\angle 30^\circ\\ &\frac{\bar V_2-\bar V_1}{1}+\frac{\bar V_2-\bar V_3}{1}+j4\bar V_2=0\\ &\frac{\bar V_3-\bar V_1}{j4}+\frac{\bar V_3-\bar V_2}{1}+0.5\bar V_3=0\end{aligned}\right\}\therefore \bar V_3=\frac{\begin{vmatrix}2-j1/4 & -1 & 10\angle 30^\circ\\ -1 & 2+j4 & 0\\ j1/4 & -1 & 0\end{vmatrix}}{\begin{vmatrix}2-j1/4 & -1 & j1/4\\ -1 & 2+j4 & -1\\ j1/4 & -1 & 1.5-j1/4\end{vmatrix}}$$

$$\therefore \bar V_3=\frac{10(1+1-j0.5)}{j1/4(2-j0.5)+(-2+j1/4+j1/4)+(1.5-j1/4)(5+j7.5-1)}=1.6169\angle -45.96^\circ\ \text{V}$$

$$\therefore v_3(t)=\underline{1.6169\cos(10^4 t-45.96^\circ)\ \text{V}}$$

**9-5**

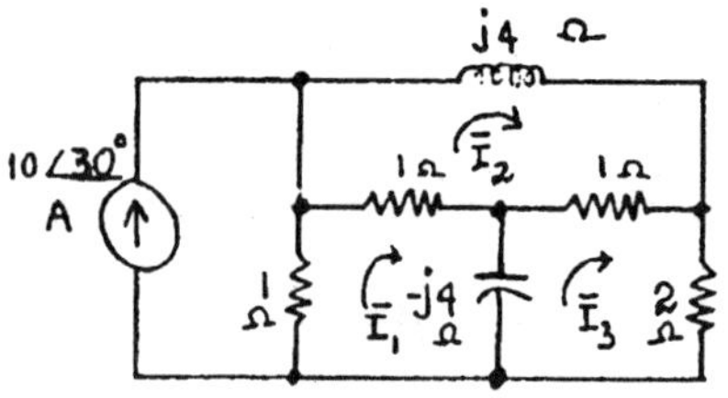


$$\left.\begin{aligned}(2-j\tfrac{1}{4})\bar{I}_1 - \bar{I}_2 + j\tfrac{1}{4}\bar{I}_3 &= 1\times 10\angle 30^\circ\\ -\bar{I}_1 + (2+j4)\bar{I}_2 - \bar{I}_3 &= 0\\ j\tfrac{1}{4}\bar{I}_1 - \bar{I}_2 + (3-j\tfrac{1}{4})\bar{I}_3 &= 0\end{aligned}\right\} \therefore \bar{I}_3 = \frac{\begin{vmatrix}2-j\frac{1}{4} & -1 & 10\angle 30^\circ\\ -1 & 2+j4 & 0\\ j\frac{1}{4} & -1 & 0\end{vmatrix}}{\begin{vmatrix}2-j\frac{1}{4} & -1 & j\frac{1}{4}\\ -1 & 2+j4 & -1\\ j\frac{1}{4} & -1 & 3-j\frac{1}{4}\end{vmatrix}}$$

$$\therefore \bar{I}_3 = \frac{10(1+1-j0.5)}{j\frac{1}{4}(2-j0.5)+(-2+j0.5)+(3-j\frac{1}{4})(4+1+j7.5-1)} = 0.8085^{-}\angle -45.96^\circ \text{ A}$$

$$\therefore i_3(t) = \underline{0.8085\cos(10^4 t - 45.96^\circ)\text{ A}}$$

**9-6**

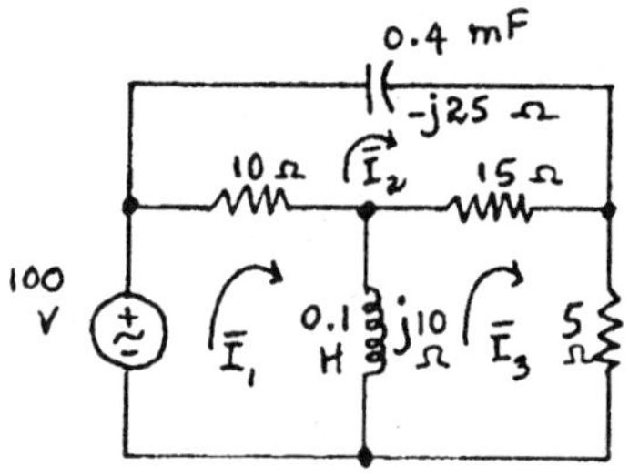


$\omega = 100$ rad/s

$$\left.\begin{aligned}(10+j10)\bar{I}_1 - 10\bar{I}_2 - j10\bar{I}_3 &= 100\\ -10\bar{I}_1 + (25-j25)\bar{I}_2 - 15\bar{I}_3 &= 0\\ -j10\bar{I}_1 - 15\bar{I}_2 + (20+j10)\bar{I}_3 &= 0\end{aligned}\right\} \therefore \bar{I}_1 = \frac{\begin{vmatrix}100 & -10 & -j10\\ 0 & 25-j25 & -15\\ 0 & -15 & 20+j10\end{vmatrix}}{\begin{vmatrix}10+j10 & -10 & -j10\\ -10 & 25-j25 & -15\\ -j10 & -15 & 20+j10\end{vmatrix}}$$

$$\therefore \bar{I}_1 = \frac{100(525-j250)}{(10+j10)(500+250-j250-225)+10(-200-j250)-j10(400+j250)}$$

$$\bar{I}_1 = \frac{100(525-j250)}{8250-j3750} = \underline{6.417\angle -1.02^\circ \text{ A}}$$

**9-7** (a)

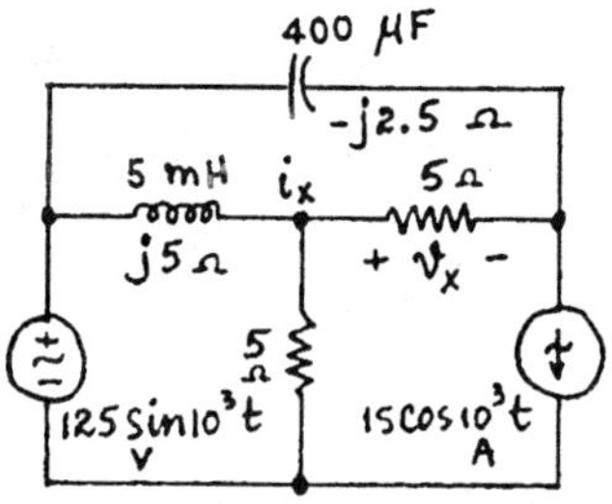


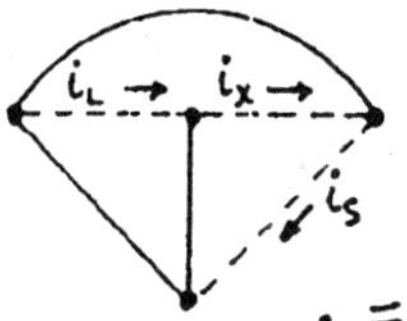


$$j5\bar{I}_L + 5(\bar{I}_L - \bar{I}_x) = -j125$$

$$5\bar{I}_x - j2.5(\bar{I}_x - 15) - j125 + 5(\bar{I}_x - \bar{I}_L) = 0$$

$$\therefore \bar{I}_x = \frac{\begin{vmatrix}5+j5 & -j125\\ -5 & j87.5\end{vmatrix}}{\begin{vmatrix}5+j5 & -5\\ -5 & 10-j2.5\end{vmatrix}} = \frac{-437.5-j187.5}{37.5+j37.5}$$

$\bar{I}_x = 8.975^{+}\angle -201.80^\circ$ A $\quad \therefore i_x(t) = \underline{8.975^{+}\cos(10^3 t + 158.20^\circ)\text{ A}}$

(b)

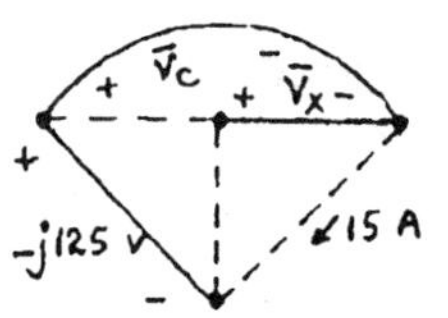


$$\left.\begin{aligned}\frac{\bar{V}_x}{5} + \frac{\bar{V}_x - \bar{V}_c}{j5} + \frac{\bar{V}_x - \bar{V}_c - j125}{5} &= 0\\ 15 - \frac{\bar{V}_x}{5} - \frac{\bar{V}_c}{-j2.5} &= 0\end{aligned}\right\} \therefore \bar{V}_x = \frac{\begin{vmatrix}j25 & -0.2+j0.2\\ 15 & j0.4\end{vmatrix}}{\begin{vmatrix}0.4-j0.2 & -0.2+j0.2\\ 0.2 & j0.4\end{vmatrix}}$$

$$\therefore \bar{V}_x = \frac{-10+3-j3}{0.08+j0.16+0.04-j0.04} = \frac{-7-j3}{0.12+j0.12} = 44.88\angle -201.80^\circ \text{ A}$$

$$\therefore v_x(t) = \underline{44.88\cos(10^3 t + 158.20^\circ)\text{ V}}$$

9-8 (a)

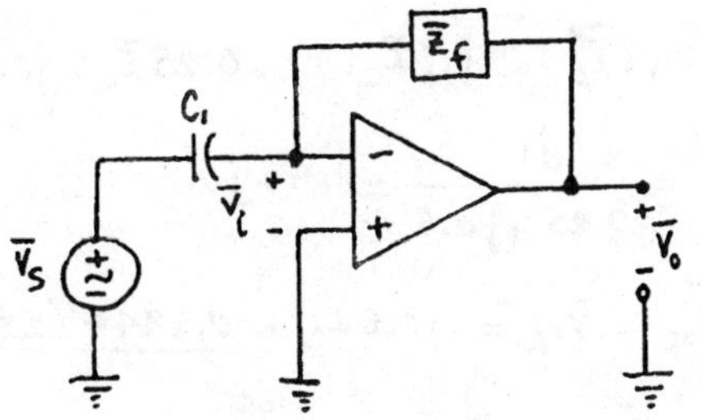


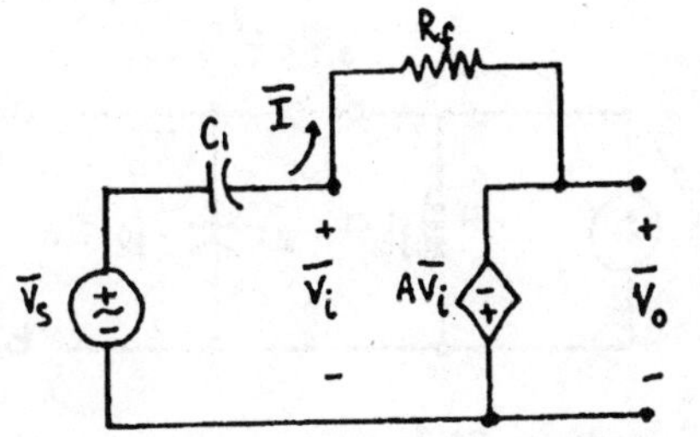

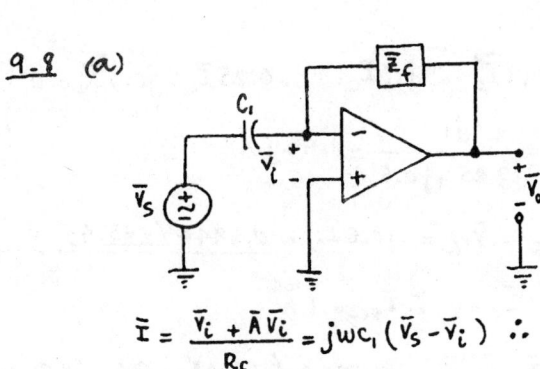

$$\bar{I} = \frac{\bar{V}_i + A\bar{V}_i}{R_f} = j\omega C_1(\bar{V}_s - \bar{V}_i) \quad \therefore \bar{V}_i\left(\frac{A+1}{R_f} + j\omega C_1\right) = j\omega C_1\bar{V}_s \quad \therefore \bar{V}_i = \frac{j\omega C_1\bar{V}_s}{\frac{A+1}{R_f} + j\omega C_1}$$

$$\bar{V}_o = -A\bar{V}_i \quad \therefore \frac{\bar{V}_o}{\bar{V}_s} = \frac{-j\omega C_1 A}{\frac{A+1}{R_f} + j\omega C_1} = \frac{-j\omega C_1 R_f A}{A+1+j\omega C_1 R_f} \leftarrow \text{as } A \to \infty, \ \frac{\bar{V}_o}{\bar{V}_s} \to -j\omega C_1 R_f \ ✓$$

(b)

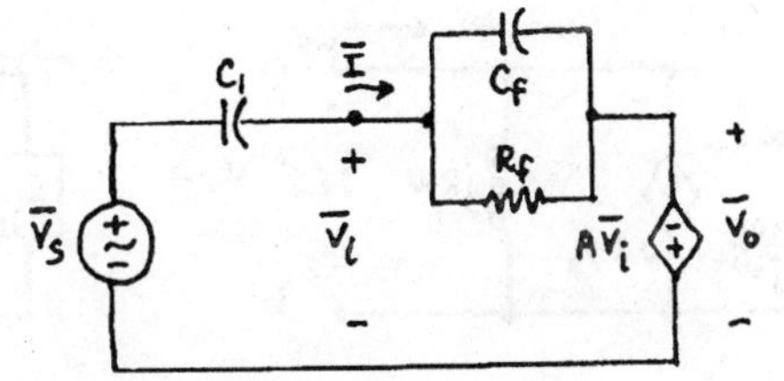


$$R_f \| C_f = \frac{R_f/j\omega C_f}{R_f + 1/j\omega C_f} = \frac{R_f}{1 + j\omega C_f R_f}$$

$$\bar{I} = \frac{\bar{V}_i + A\bar{V}_i}{R_f} \times (1 + j\omega C_f R_f) = j\omega C_1(\bar{V}_s - \bar{V}_i)$$

$$\therefore \bar{V}_i\left[\frac{(A+1)(1+j\omega C_f R_f)}{R_f} + j\omega C_1\right] = j\omega C_1 \bar{V}_s \ ; \quad \bar{V}_o = -A\bar{V}_i$$

$$\therefore \frac{\bar{V}_o}{\bar{V}_s} = \frac{-j\omega C_1 A}{\frac{(A+1)(1+j\omega C_f R_f)}{R_f} + j\omega C_1} = \frac{-j\omega C_1 R_f A}{(A+1)(1+j\omega C_f R_f) + j\omega C_1 R_f} \leftarrow$$

$$\text{as } A \to \infty, \ \frac{V_o}{V_s} \to \frac{-j\omega C_1 R_f}{1 + j\omega C_f R_f} \ ✓$$

9-9

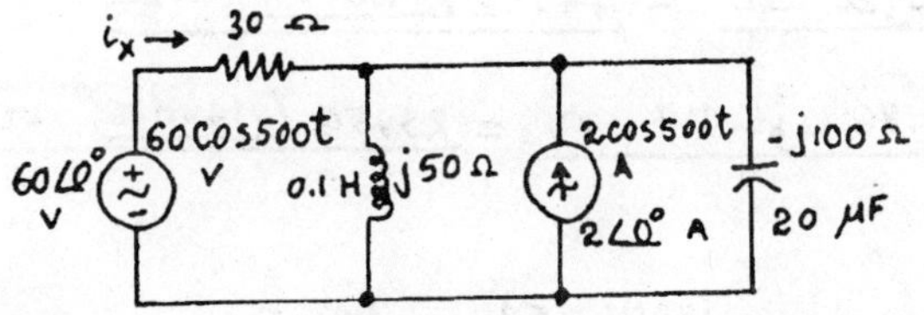


$$\bar{I}_x = \frac{60}{30 + j50 \| (-j100)} - \frac{2 \times 1/30}{1/30 + 1/j50 + 1/-j100}$$

$$= \frac{60}{30 + j100} - \frac{2}{1 - j0.6 + j0.3} = 2\angle{-146.6^\circ} \text{ A}$$

$$\therefore i_x(t) = \underline{2\cos(500t - 146.60^\circ) \text{ A}}$$

9-10 (a)

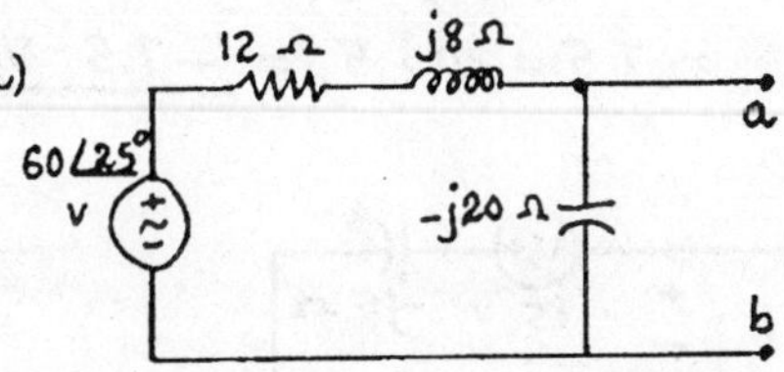


$$\bar{Z}_{Th} = \frac{(12+j8)(-j20)}{12-j12} = \underline{16.667 - j3.333} \ \Omega$$

$$\bar{V}_{oc} = 60\angle 25^\circ \frac{-j20}{12-j20} = \underline{70.71\angle{-20^\circ}} \text{ V}$$

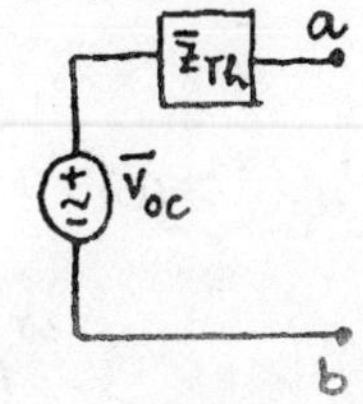


(b)

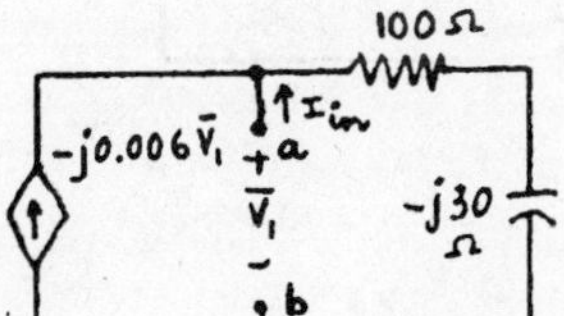


Apply $\bar{V}_1 = 1\angle 0^\circ$ V

$$\therefore \bar{I}_{in} = j0.006 + \frac{1}{100 - j30} = \frac{1.18 + j0.6}{100 - j30}$$

$$\therefore \bar{Z}_{Th} = \frac{1}{\bar{I}_{in}} = \frac{100 - j30}{1.18 + j0.6} = \underline{57.06 - j54.44 \ \Omega}$$

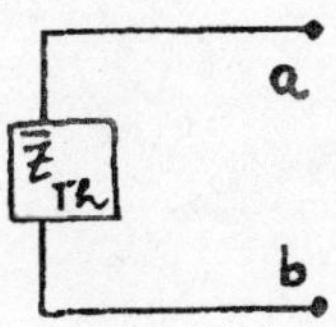

**9-11**

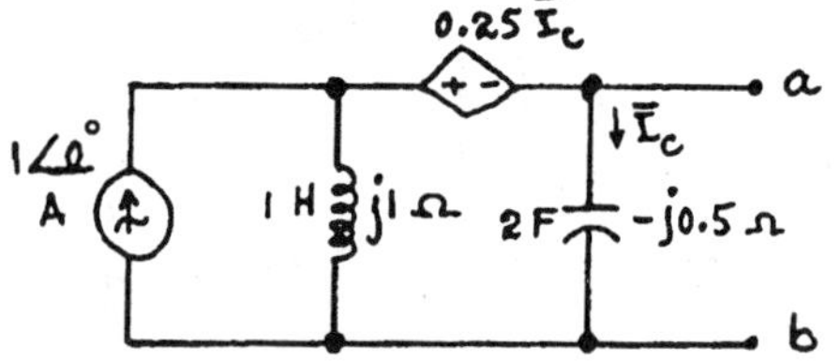


$\omega = 1$ rad/sec

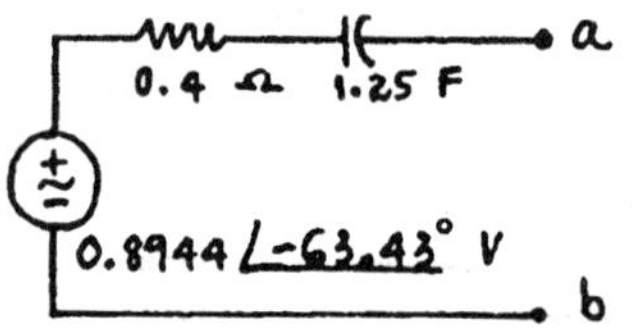


OC: $\circlearrowright 1$, $\circlearrowright \bar{I}_c$ $\therefore j1(\bar{I}_c - 1) + 0.25\bar{I}_c - j0.5\bar{I}_c = 0$

$$\bar{I}_c = \frac{j1}{0.25 + j0.5} = 1.6 + j0.8$$

$$\therefore \bar{V}_{oc} = \bar{V}_{Th} = -j0.5\bar{I}_c = \underline{0.8944\angle -63.43^\circ \text{ V}}$$

SC: $\bar{I}_c = 0$ $\therefore \bar{I}_{sc} = 1$ A

$$\therefore \bar{Z}_{Th} = \bar{V}_{oc} = 0.8944\angle -63.43^\circ = \underline{0.4 - j0.8\ \Omega}$$

$$\therefore C = \frac{1}{(0.8)(1)} = 1.25 \text{ F}$$

**9-12**

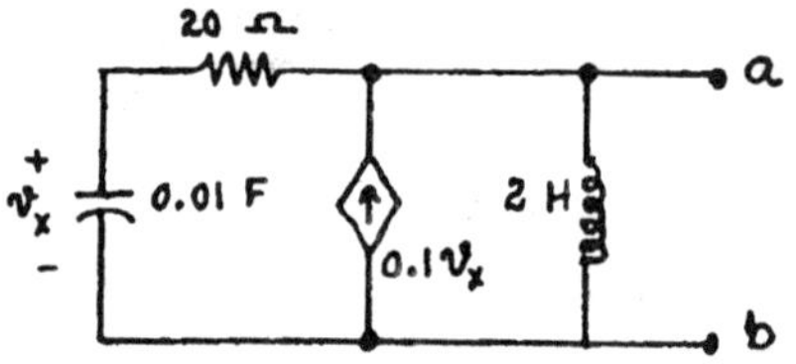


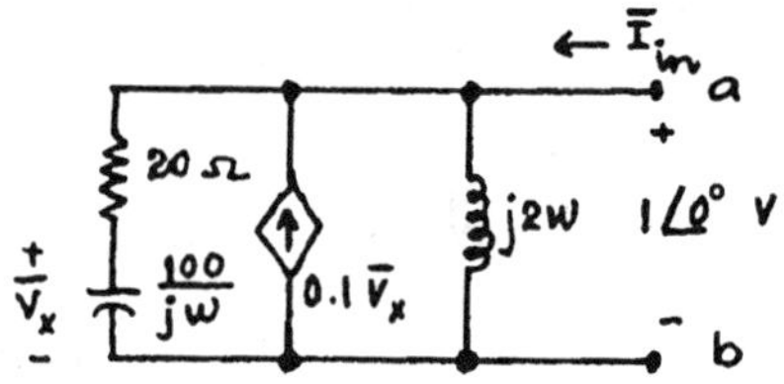


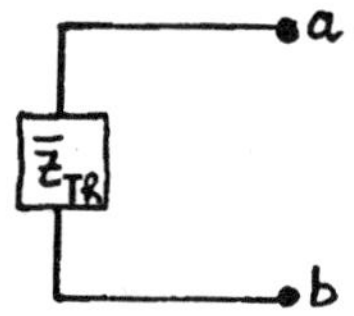


$$\bar{V}_x = 1 \times \frac{100/j\omega}{20 + 100/j\omega} = \frac{5}{5 + j\omega} \quad \therefore \bar{I}_{in} = \frac{1}{j2\omega} - 0.1 \times \frac{5}{5 + j\omega} + \frac{1}{20 + 100/j\omega}$$

$$\therefore \bar{I}_{in} = \frac{1}{j2\omega} + \frac{-0.5 + j0.05\omega}{5 + j\omega} = \frac{5 - 0.1\omega^2}{j2\omega(5 + j\omega)} \quad \therefore \bar{Z}_{Th} = \frac{j2\omega(5 + j\omega)}{5 - 0.1\omega^2}$$

(a) $\omega = 10$, $\bar{Z}_{Th} = \dfrac{j20(5 + j10)}{-5} = \underline{40 - j20\ \Omega} = \underline{44.72\angle -26.57^\circ\ \Omega}$

(b) $\omega = 20$, $\bar{Z}_{Th} = \dfrac{j40(5 + j20)}{-35} = \underline{22.86 - j5.714\ \Omega} = \underline{23.56\angle -14.036^\circ\ \Omega}$

**9-13** (a)

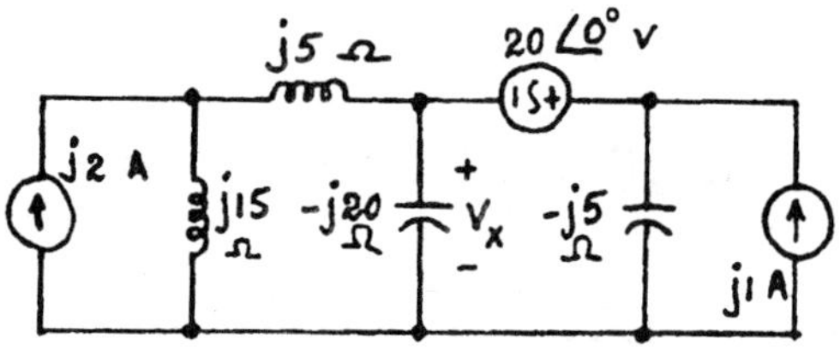


$$-j20 \| (-j5) = -j4\ \Omega$$

$$\therefore \bar{V}_x = j2 \times \frac{j15}{j15 + j5 - j4}(-j4) - 20 + j1(-j5)$$

$$= 7.5 - 20 + 5 = \underline{-7.5 \text{ V}}$$

(b)

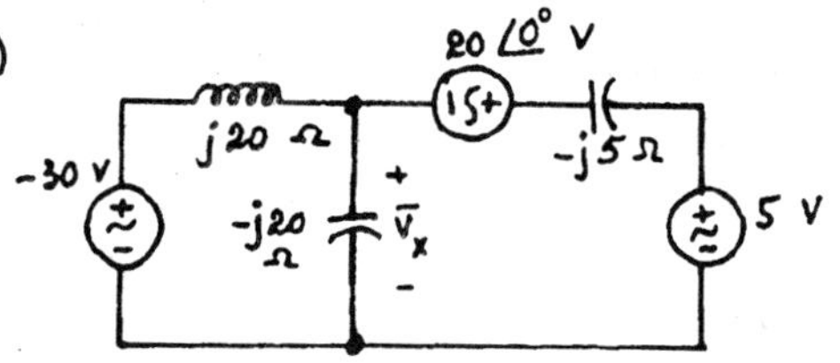


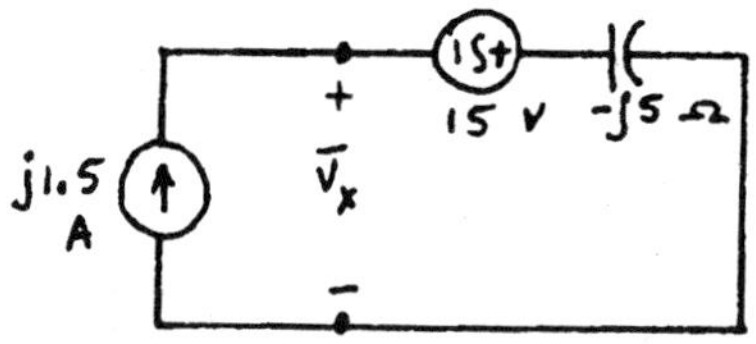


$$\bar{V}_x = -15 - j5(j1.5) = \underline{-7.5 \text{ V}}$$

**9-14** (a)

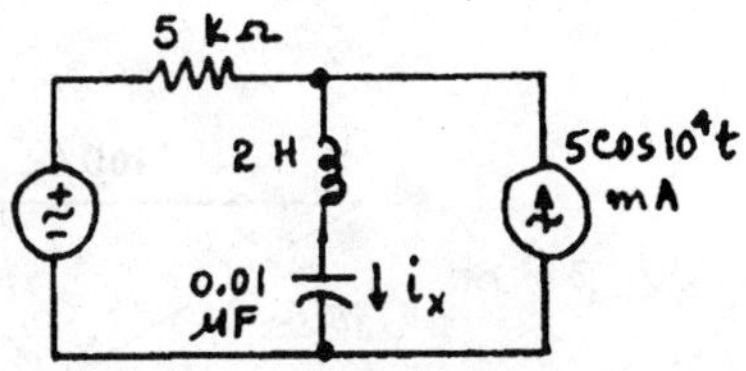


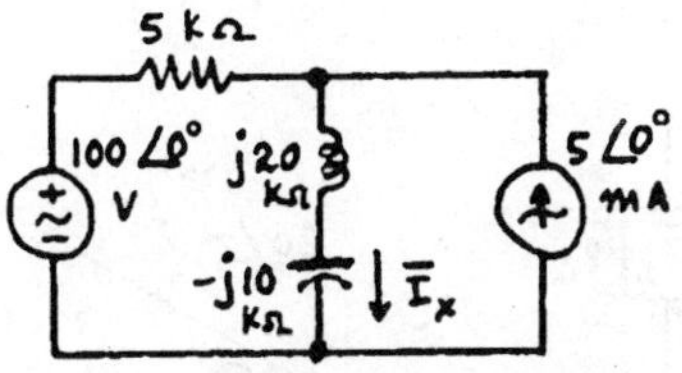


$v_s = 100\cos 10^4 t$, $\bar{V}_s = 100\angle 0^\circ$

$\bar{I}_x = \dfrac{100}{5+j10} + 5 \times \dfrac{5}{5+j10}$

$\bar{I}_x = 5 - j10$ mA

$I_x = 11.180\angle -63.43^\circ$ mA $\quad \therefore\ i_x(t) = \underline{11.180\cos(10^4 t - 63.43^\circ)\ \text{mA}}$

(b) $v_s = 100\sin 10^4 t$, $\bar{V}_s = -j100$ $\quad \therefore\ \bar{I}_x = \dfrac{-j100 + 25}{5+j10} = 9.220\angle -139.40^\circ$ mA

$\therefore\ i_x(t) = \underline{9.220\cos(10^4 t - 139.40^\circ)\ \text{mA}}$

(c) $v_s = 100\cos 5000t$ ; $\omega_1 = 10^4$, $\omega_2 = 5000$ $\quad \therefore\ \bar{I}_{x1} = 5 \times \dfrac{5}{5+j10} = \dfrac{25}{5+j10} = 2.236\angle -63.43^\circ$ mA

$\bar{I}_{x2} = \dfrac{100}{5+j10-j20} = \dfrac{100}{5-j10} = 8.944\angle 63.43^\circ$ mA

$\therefore\ i_x(t) = \underline{2.236\cos(10^4 t - 63.43^\circ) + 8.944\cos(5000t + 63.43^\circ)\ \text{mA}}$

**9-15**

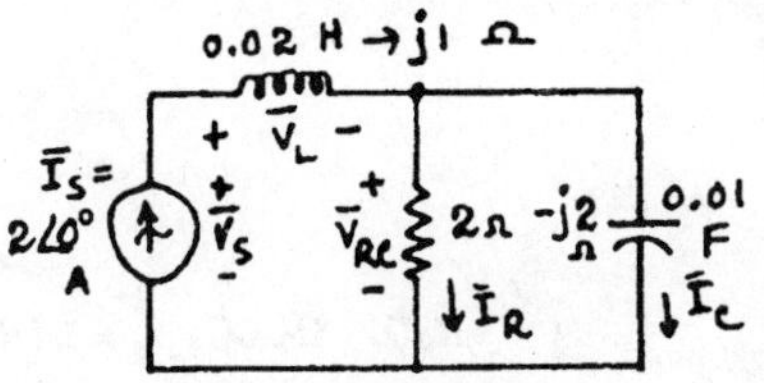


$\omega = 50$ rad/sec

$\bar{I}_s = 2$, $\bar{I}_R = 1 - j1$, $\bar{I}_C = 1 + j1$ A $\quad 2 \| -j2 = 1 - j1\ \Omega$; $\bar{V}_s = 2$ V

$\bar{V}_L = j2$, $\bar{V}_{RC} = 2 - j2$ V

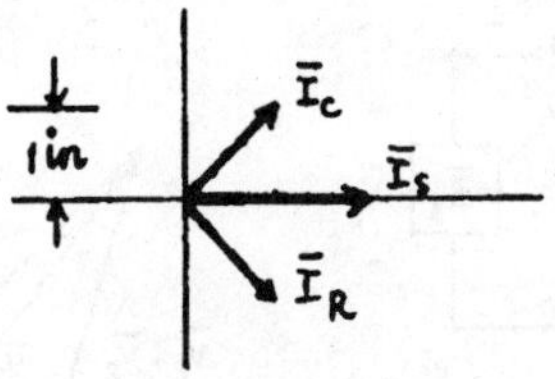


(a)

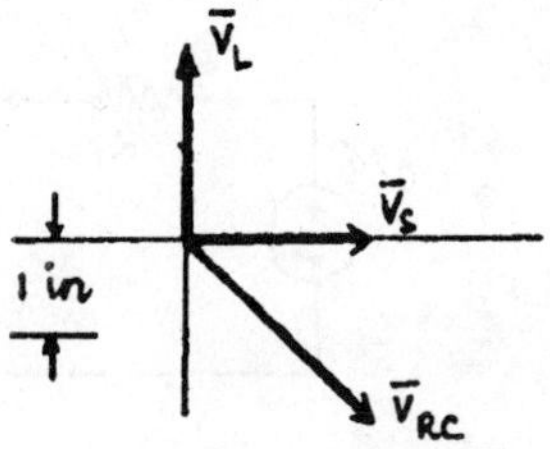


(b)

**9-16**

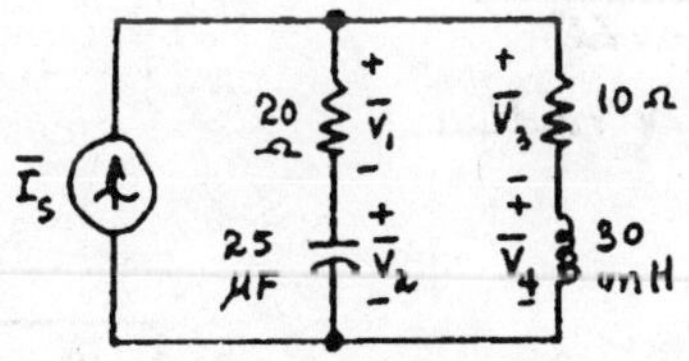


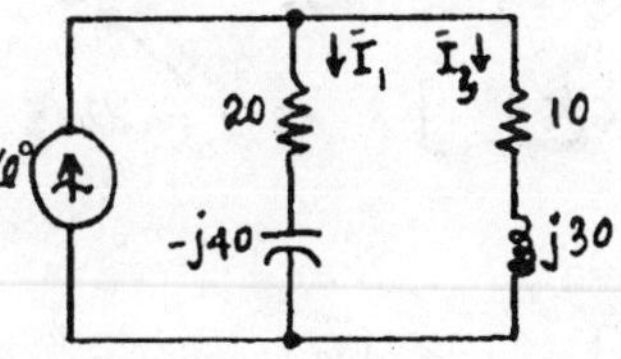


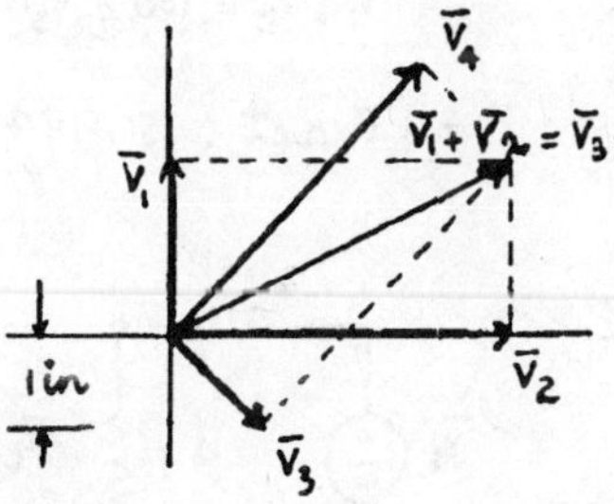


$\bar{I}_1 = 3 \times \dfrac{10 + j30}{30 - j10} = j3$, $\bar{I}_3 = 3 - j3$

$\bar{V}_1 = \underline{j60\ \text{V}}$, $\bar{V}_2 = \underline{120\ \text{V}}$, $\bar{V}_3 = \underline{30 - j30\ \text{V}}$, $\bar{V}_4 = \underline{90 + j90\ \text{V}}$

**9-17** (a)

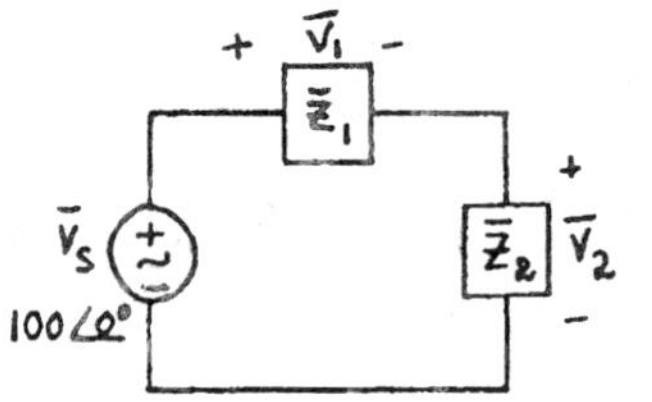

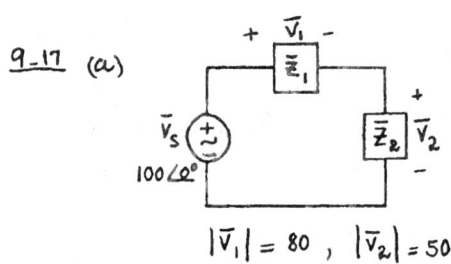

$|\bar{V}_1| = 80$ , $|\bar{V}_2| = 50$

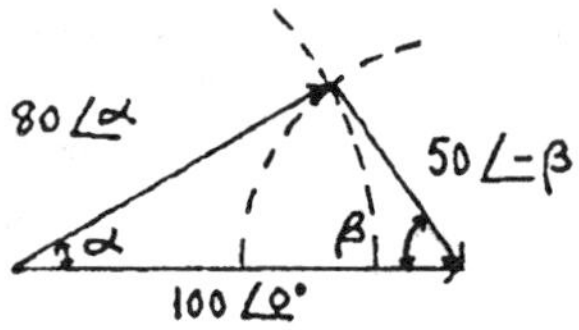


or

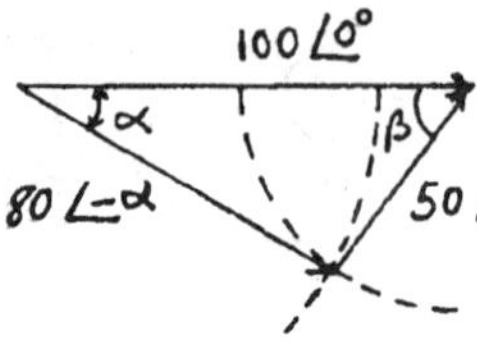


$\alpha \doteq 30°$ , $\beta \doteq 52°$ , $\bar{V}_1 \doteq \underline{80\angle 30°}$ V , $\bar{V}_2 \doteq \underline{50\angle -52°}$

or $\bar{V}_1 \doteq \underline{80\angle -30°}$ V , $\bar{V}_2 \doteq \underline{50\angle 52°}$

[Exact soln : $80\cos\alpha + 50\cos\beta = 100$ , $80\sin\alpha = 50\sin\beta$

$\cos\beta = \sqrt{1-\sin^2\beta} = \sqrt{1-2.56\sin^2\alpha} = \sqrt{2.56\cos^2\alpha - 1.56}$

$\therefore\ 80\cos\alpha + 50\sqrt{2.56\cos^2\alpha - 1.56} = 100$ $\therefore\ \cos\alpha = \frac{139}{160}$ $\therefore\ \alpha = 29.69°$ , $\beta = 52.41°$

$\therefore\ \bar{V}_1 = 80\,\underline{\angle \pm 29.69°}$ , $\bar{V}_2 = 50\,\underline{\angle \mp 52.41°}$ ]

(b) $\bar{Z}_1$ and $\bar{Z}_2$ have same $\bar{I}$ $\therefore \angle\bar{Z}_1$ or $\angle\bar{V}_1 > \angle\bar{Z}_2$ or $\angle\bar{V}_2$

$\therefore\ \bar{V}_1 \doteq \underline{80\angle 30°\ \text{V}}$ $\bar{V}_2 \doteq \underline{50\angle -52°\ \text{V}}$

**9-18**

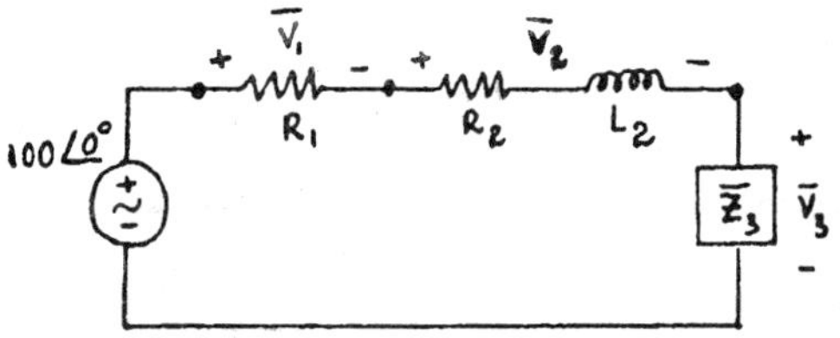


$|\bar{V}_1| = 50$ , $|\bar{V}_2| = 125$ , $\bar{V}_3 = 200\angle -45°$

$\bar{V}_1 + \bar{V}_2 = 100 - \bar{V}_3$ , Get $\bar{V}_1 \doteq 50\angle 52°$

[Exact : $51.984\ 336\angle 64°$ ]

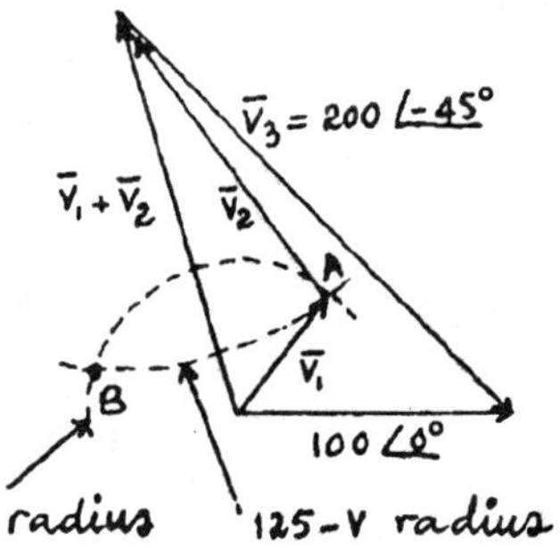


Note that $\bar{V}_2 = \bar{I}(R_2 + j\omega L_2)$

and $\bar{V}_1 = \bar{I}R_1$

$\therefore\ \bar{V}_2$ leads $\bar{V}_1$

$\therefore$ A is OK, B is not

**9-19**

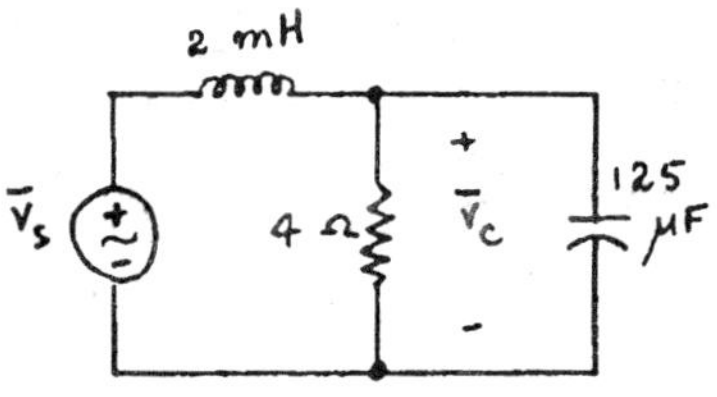


$4 \| \frac{8000}{j\omega} = \frac{8000}{2000 + j\omega}$

$\frac{\bar{V}_c}{\bar{V}_s} = \frac{8000/(2000+j\omega)}{j\omega/500 + 8000/(2000+j\omega)}$

$\frac{\bar{V}_c}{\bar{V}_s} = \frac{8000}{8000 - \omega^2/500 + j4\omega}$

$\left|\frac{\bar{V}_c}{\bar{V}_s}\right| = \frac{8000}{\sqrt{(8000 - \omega^2/500)^2 + 16\omega^2}}$

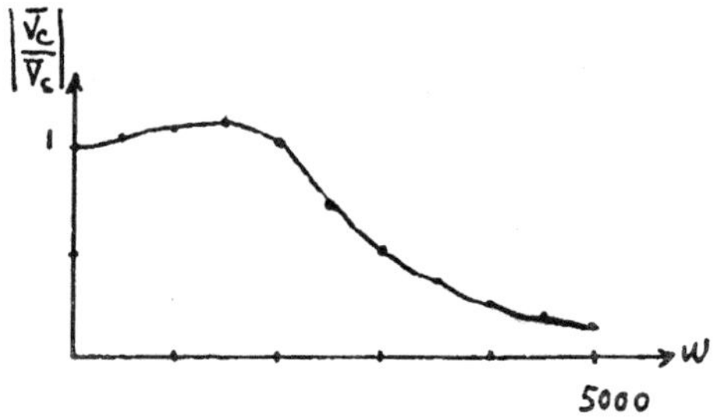


| ω | $\lvert \bar{V}_c/\bar{V}_s \rvert$ |
|---|---|
| 0 | 1 |
| 500 | 1.031 |
| 1000 | 1.109 |
| 1500 | 1.152 |
| 2000 | 1.000 |
| 2500 | 0.730 |
| 3000 | 0.512 |
| 3500 | 0.370 |
| 4000 | 0.277 |
| 4500 | 0.215 |
| 5000 | 0.172 |

**9-20**

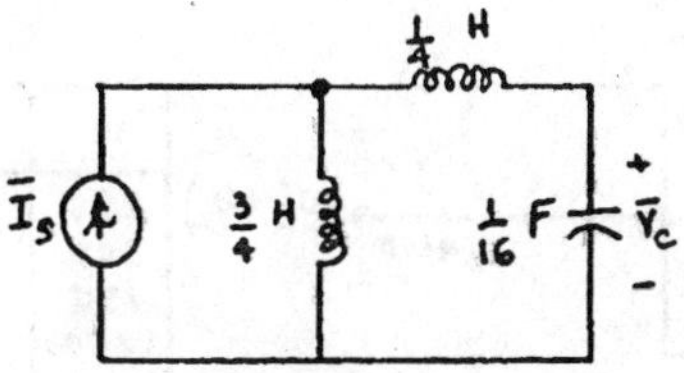


(a) $\dfrac{\bar{V}_c}{\bar{I}_s} = \dfrac{j0.75\omega}{j\omega - j16/\omega}\left(-j\dfrac{16}{\omega}\right)$

$= \dfrac{-j12\omega}{\omega^2 - 16} \quad \therefore \left|\dfrac{\bar{V}_c}{\bar{I}_s}\right| = \dfrac{12|\omega|}{|\omega^2 - 16|}$

O's: $\omega = 0, \infty$ ; x's: $\omega = \pm 4$

(c)

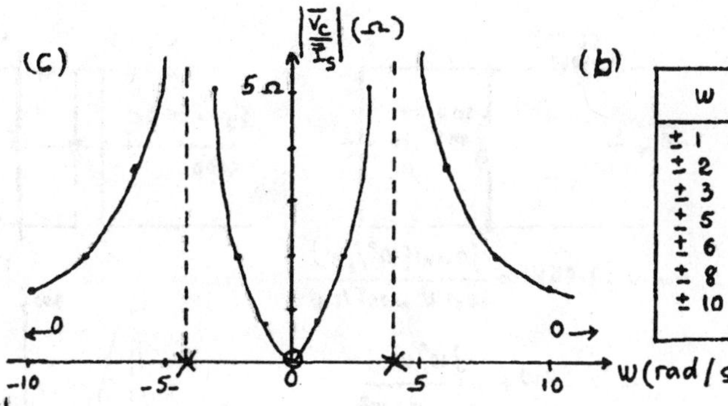


(b)

| $\omega$ | $\lvert\bar{V}_c/\bar{I}_s\rvert$ |
|---|---|
| ± 1 | 0.800 |
| ± 2 | 2.00 |
| ± 3 | 5.14 |
| ± 5 | 6.67 |
| ± 6 | 3.60 |
| ± 8 | 2.00 |
| ± 10 | 1.43 |

**9-21**

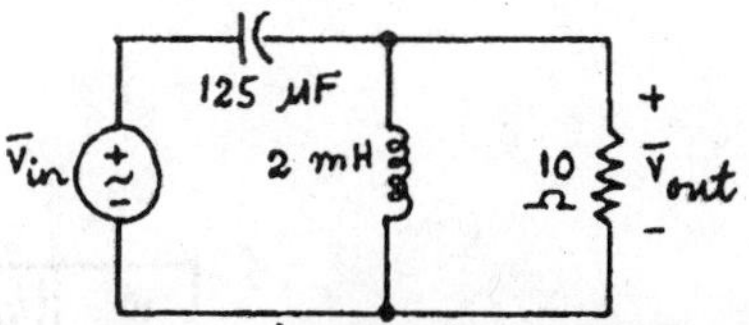


$\dfrac{\bar{V}_{out}}{\bar{V}_{in}} = \dfrac{j0.02\omega/(10 + j0.002\omega)}{8000/j\omega + j0.02\omega/(10 + j0.002\omega)}$

$= \dfrac{-0.02\omega^2}{80000 - 0.02\omega^2 + j16\omega}$

$|\bar{V}_{out}/\bar{V}_{in}| = 0.02\omega^2 / \sqrt{(80000 - 0.02\omega^2)^2 + 256\omega^2}$

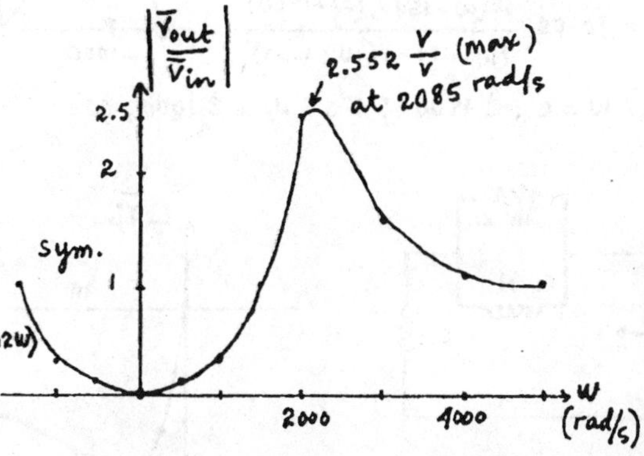


| $\omega$ | $\lvert\bar{V}_{out}/\bar{V}_{in}\rvert$ |
|---|---|
| 0 | 0 |
| 500 | 0.066 |
| 1000 | 0.322 |
| 1500 | 1.060 |
| 2000 | 2.50 |
| 2500 | 2.08 |
| 3000 | 1.623 |
| 3500 | 1.406 |
| 4000 | 1.286 |
| 5000 | 1.169 |

**9-22**

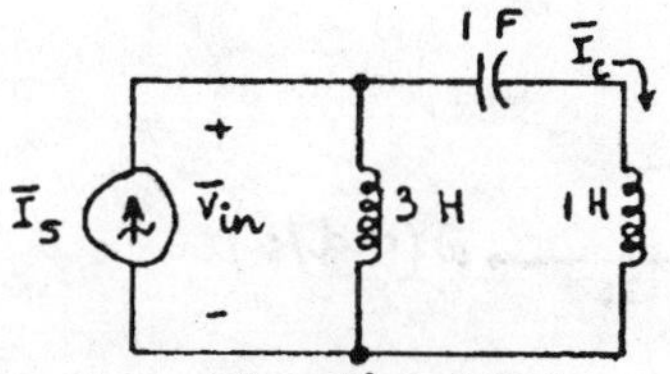


(a) $\dfrac{\bar{V}_{in}}{\bar{I}_s} = \dfrac{j3\omega(j\omega + 1/j\omega)}{j4\omega + 1/j\omega}$

$= \dfrac{j3\omega(1 - \omega^2)}{1 - 4\omega^2}$

O's: $\omega = 0, \pm 1$ ; x's: $\omega = \pm 0.5, \infty$

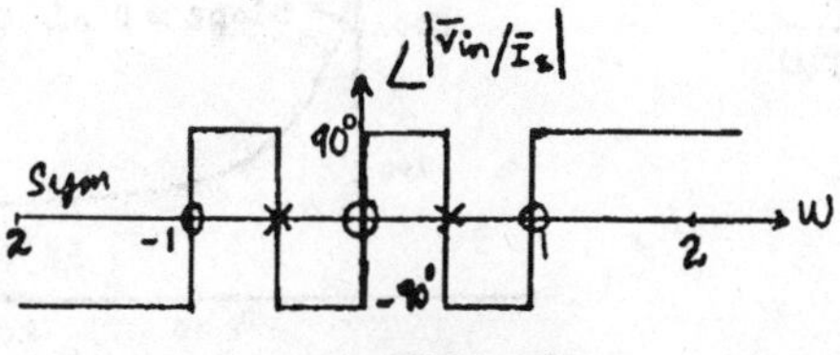


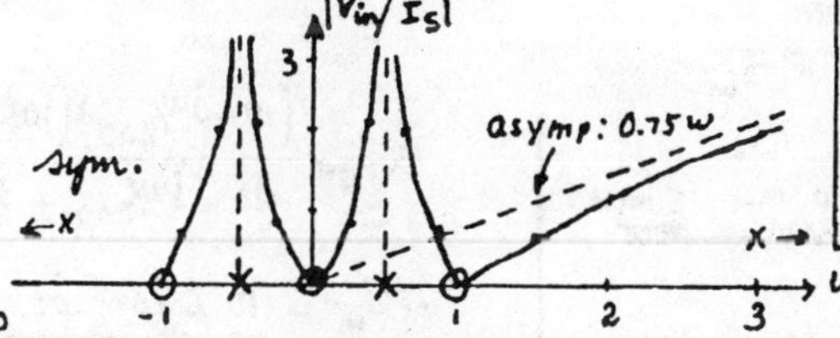


| $\omega$ | $\lvert\bar{V}_{in}/\bar{I}_s\rvert$ | $\angle$ |
|---|---|---|
| 1/8 | 0.394 | 90° |
| 1/4 | 0.938 | 90° |
| 3/8 | 2.21 | 90° |
| 5/8 | 2.03 | -90° |
| 3/4 | 0.788 | -90° |
| 7/8 | 0.298 | -90° |
| 1.5 | 0.703 | 90° |
| 2 | 1.200 | 90° |
| 3 | 2.06 | 90° |

(b) $\dfrac{\bar{I}_c}{\bar{I}_s} = \dfrac{j3\omega}{j4\omega + 1/j\omega}$

$= \dfrac{-3\omega^2}{1 - 4\omega^2}$

O's: $\omega = 0$ ; x's: $\omega = \pm 0.5$

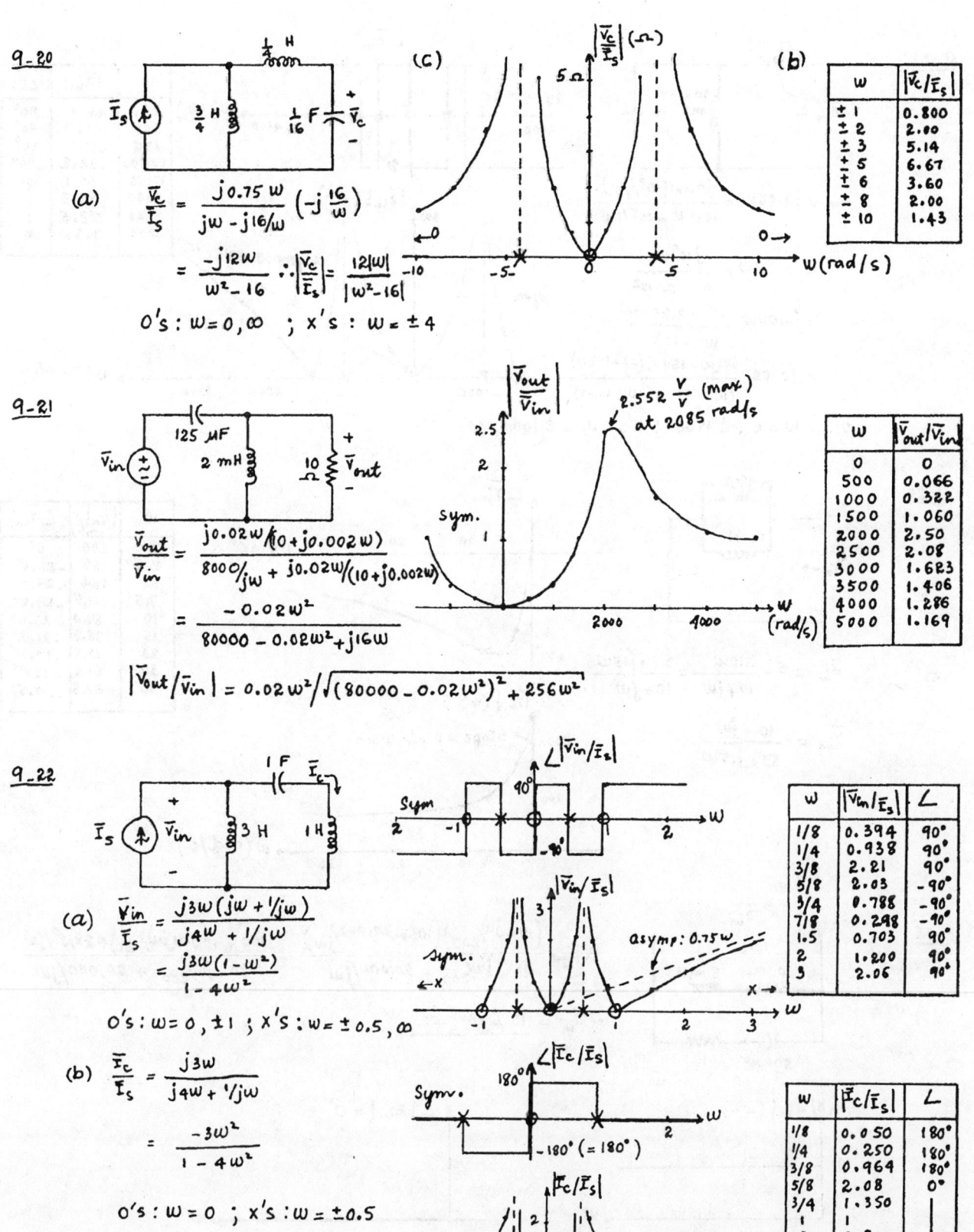

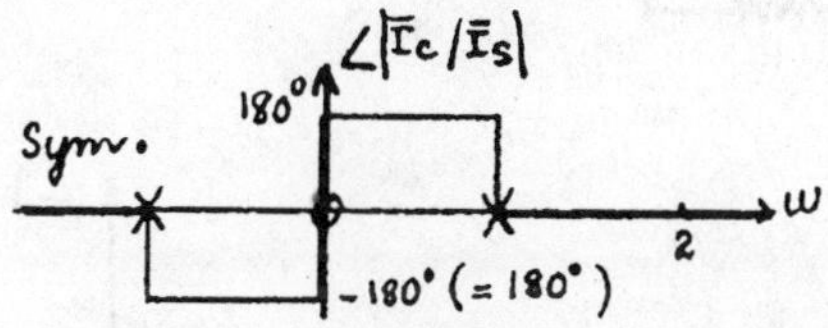


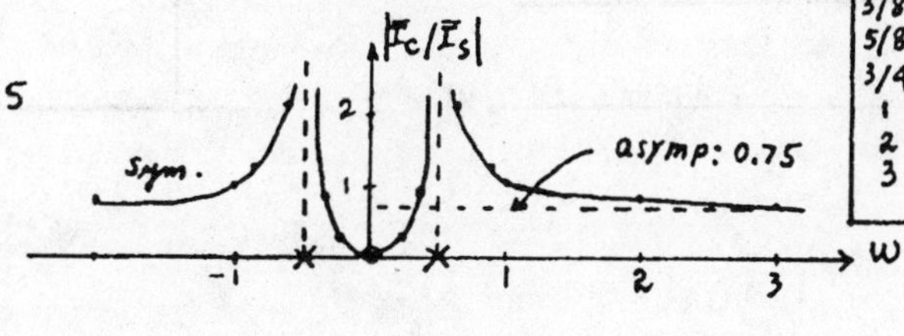


| $\omega$ | $\lvert\bar{I}_c/\bar{I}_s\rvert$ | $\angle$ |
|---|---|---|
| 1/8 | 0.050 | 180° |
| 1/4 | 0.250 | 180° |
| 3/8 | 0.964 | 180° |
| 5/8 | 2.08 | 0° |
| 3/4 | 1.350 | ↓ |
| 1 | 1 | |
| 2 | 0.8 | |
| 3 | 0.771 | |

**9-23**

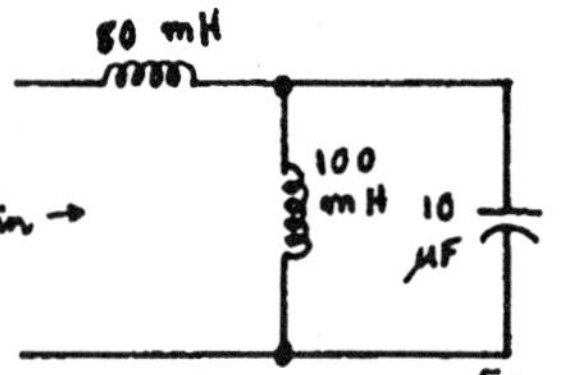


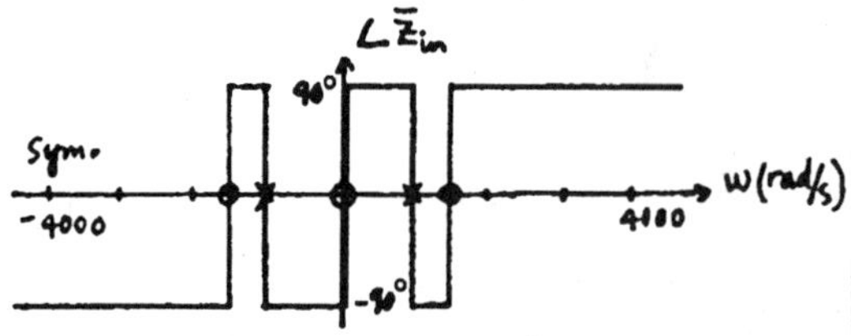


| w | $\lvert\bar{Z}_{in}\rvert$ | ∠$\bar{Z}_{in}$ |
|---|---|---|
| 250 | 46.7 | 90° |
| 500 | 106.7 | 90° |
| 750 | 231 | 90° |
| 1250 | 122.2 | -90° |
| 1750 | 55.2 | 90° |
| 2000 | 93.3 | ↓ |
| 3000 | 202.5 | |
| 4000 | 293 | |

$$\bar{Z}_{in} = j0.08w + \frac{j0.1w(10^5/jw)}{j0.1w + 10^5/jw}$$

$$= j0.08w + \frac{j10^4 w}{10^5 - 0.1w^2}$$

$$= j0.08w\,\frac{w^2 - 2.25\times10^6}{w^2 - 10^6}$$

$$= j0.08\,\frac{w(w-1500)(w+1500)}{(w-1000)(w+1000)}$$

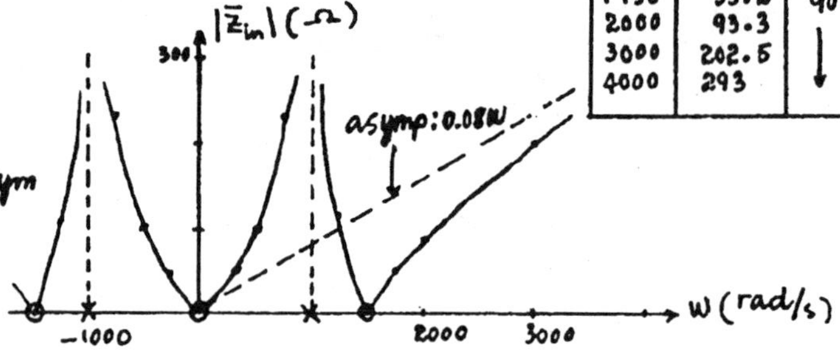


0's : $w = 0, \pm 1500$ ; x's : $w = \pm 1000, \infty$

**9-24**

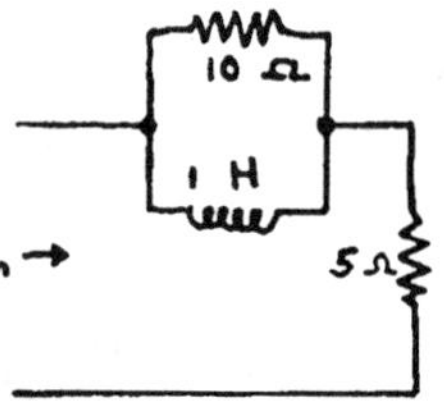


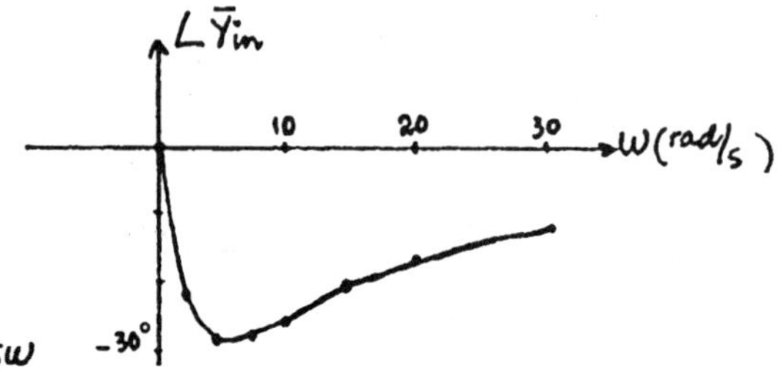


| w | $\lvert\bar{Y}_{in}\rvert$ (mS) | ∠$\bar{Y}_{in}$ |
|---|---|---|
| 0 | 200 | 0° |
| 2.5 | 165 | -22.8° |
| 5 | 124 | -29.8° |
| 7.5 | 101.5 | -29.2° |
| 10 | 89.4 | -26.6° |
| 15 | 78.2 | -21.2° |
| 20 | 73.5 | -17.1° |
| 30 | 69.8 | -12.1° |
| 50 | 67.8 | -7.5° |

$$\bar{Z}_{in} = 5 + \frac{j10w}{10+jw} = \frac{50+j15w}{10+jw}$$

$$\bar{Y}_{in} = \frac{10+jw}{50+j15w}$$

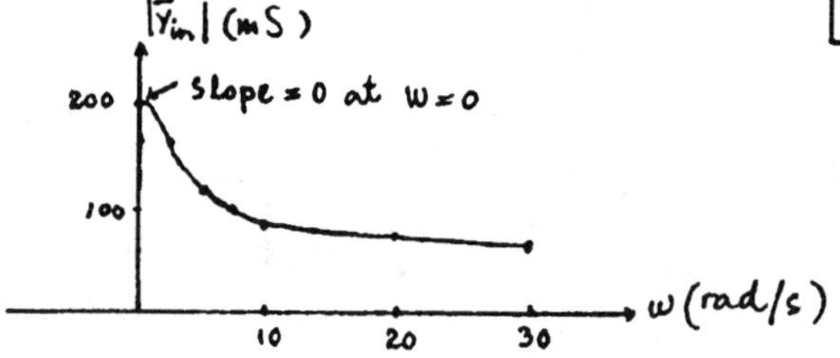


**9-25**

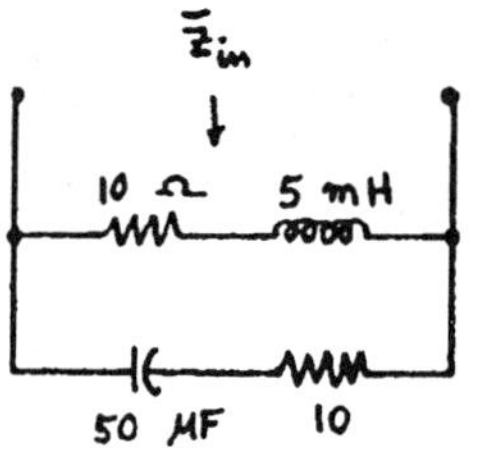


$$\bar{Z}_{in} = \frac{(10 + jw/200)(10 + 20{,}000/jw)}{20 + jw/200 + 20{,}000/jw} = \frac{100 + 100 + (jw/20) + 2\times10^5/jw}{20 + jw/200 + 20{,}000/jw}$$

$$\therefore \bar{Z}_{in} = \underline{10 + j0\ \ \Omega}$$

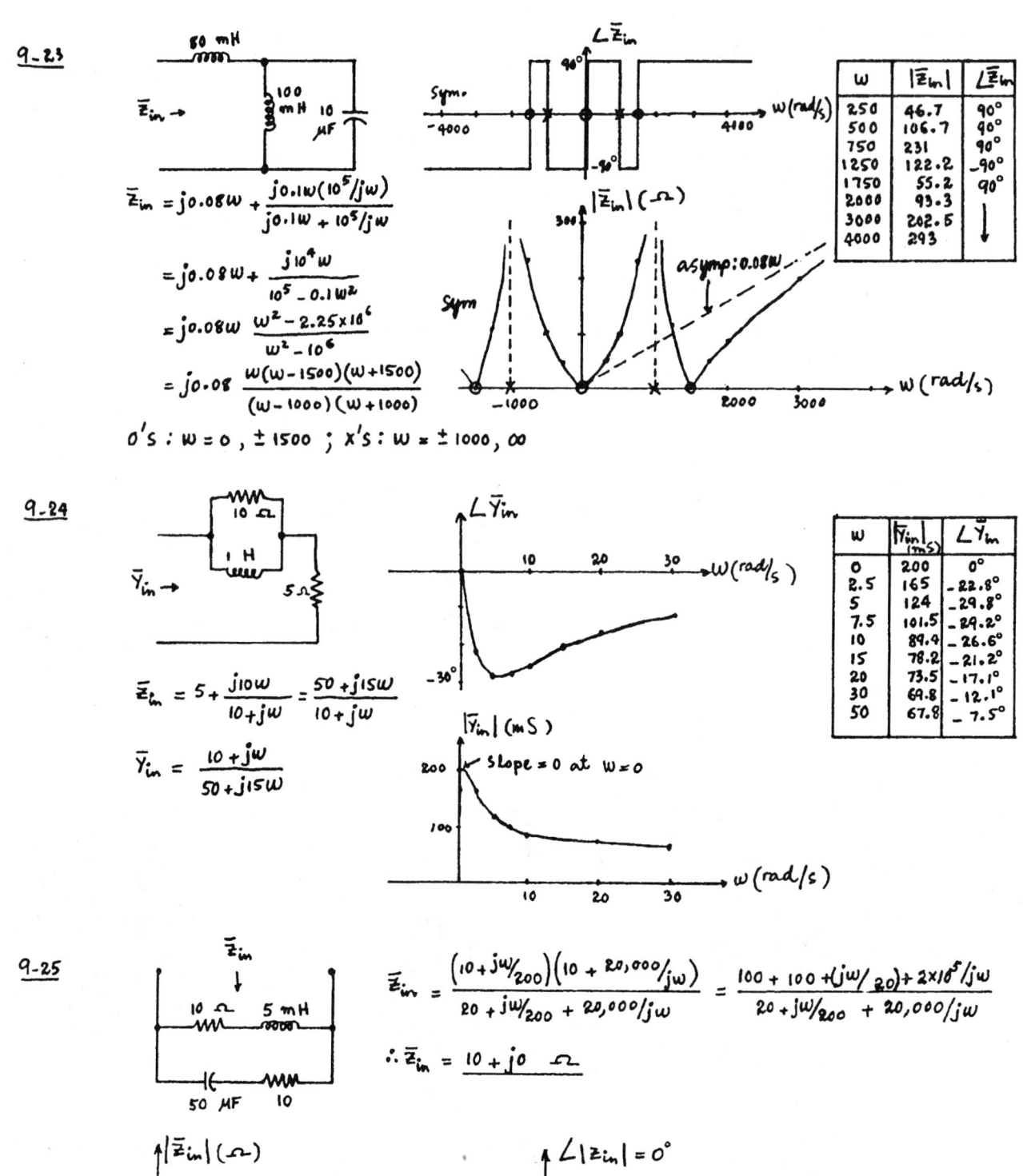

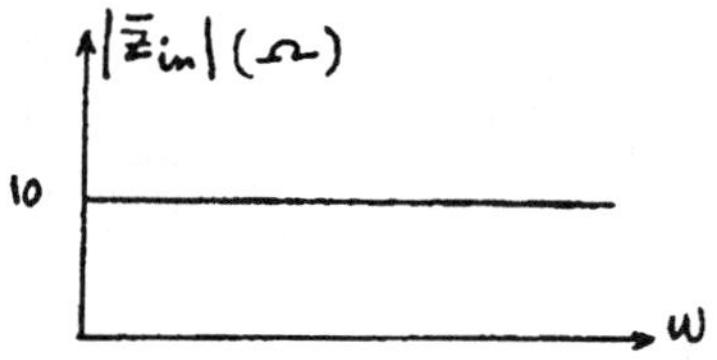


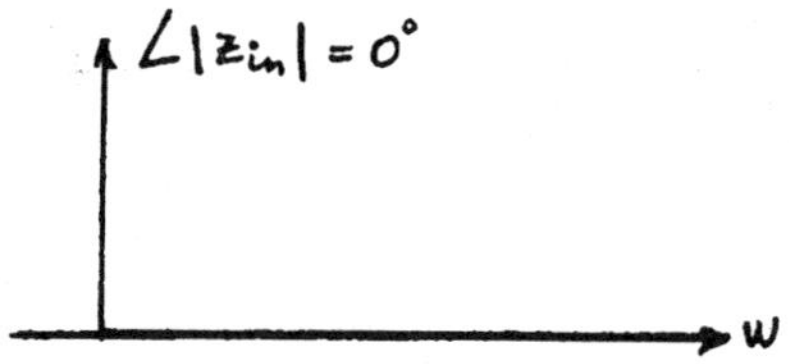

## 9-26

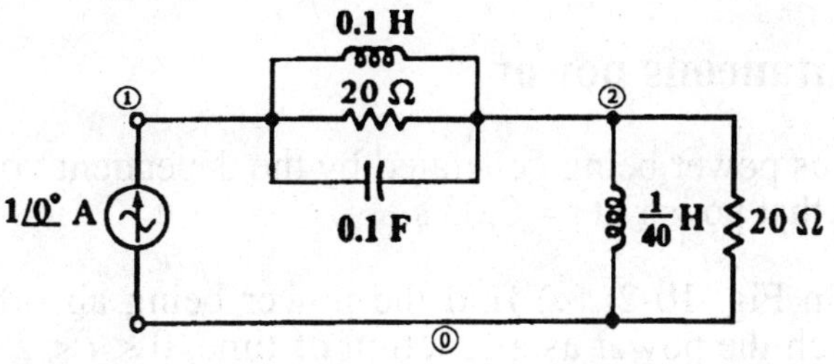


****************************************************************************

**PROGRAM:**

```
I01 0 1 AC 1
R12 1 2 20
L12 1 2 0.1
C12 1 2 0.1
L20 2 0 0.025
R20 2 0 20
.AC LIN 1 0.95493 0.95493
.PRINT AC VM(1)
.END
```

****************************************************************************

| FREQ | VM(1) |
|---|---|
| 9.549E-01 | 1.086E+00 = $\lvert Z_{in}\rvert$ = <u>1.086 Ω</u> (*a*) |

(*b*) $|Z_{in}| = \left| \dfrac{1}{\dfrac{1}{j0.6} + j0.6} + j0.15 \right| = \ldots = \underline{1.0875\ \Omega}$

# CHAPTER 10: AVERAGE POWER AND RMS VALUES

## Section 10-2 Instantaneous power

**1** (*a*) Find the instantaneous power being generated by the dependent voltage source in the circuit of Fig. 10-1. (*b*) Evaluate that power at $t = 0.03$ s.

**2** For the circuit shown in Fig. 10-2: (*a*) find the power being absorbed by the inductor as a function of time; (*b*) sketch the power as a function of time, $0 < t < 2$ ms; (*c*) and calculate the value of the power at $t = 1$ ms.

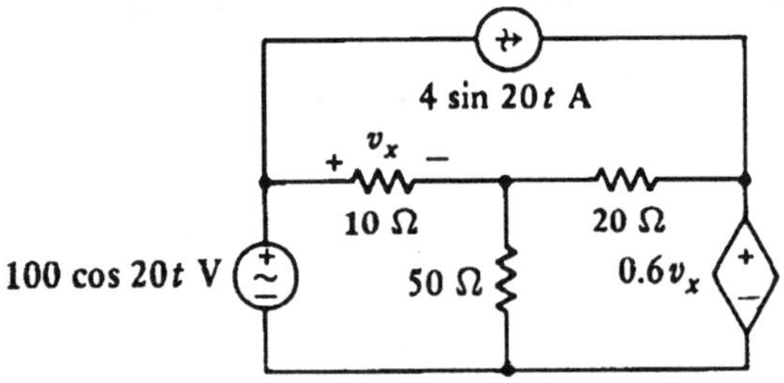


**Fig. 10-1:** See Prob. 1.

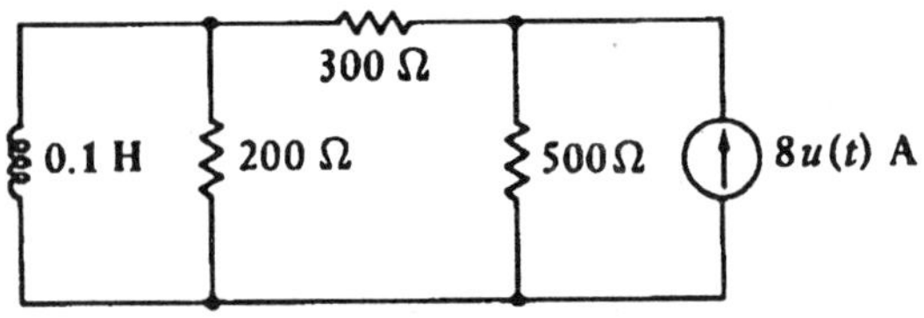


**Fig. 10-2:** See Prob. 2.

**3** A voltage source, $10\cos 1000t$ V, a 0.4-Ω resistor, a l-mH inductor, and a 1.25-mF capacitor are all in series. Find the power being absorbed by each of the four elements at $t = 1$ ms.

**4** A 100-V battery, an open switch, a 10-Ω resistor, and an uncharged l-mF capacitor are in series. Let the switch close at $t = 0$. (*a*) Obtain an expression for the instantaneous power being absorbed by the capacitor as a function of time. (*b*) Sketch it. (*c*) Find the time at which this power is a maximum and also the value of this maximum power.

## Section 10-3 Average power

**5** Using the time interval, $10 < t < 12$ ms, find the average value of: (*a*) $40\cos 2000\pi t$; (*b*) $40\cos^2 2000\pi t$; (*c*) the periodic waveform illustrated in Fig. 10-3; (*d*) the square of that waveform.

**6** (*a*) In Fig. 10-4, how much average power is delivered to the 8-Ω load? (*b*) How much of this power comes from the $20\underline{/0^\circ}$-V source?

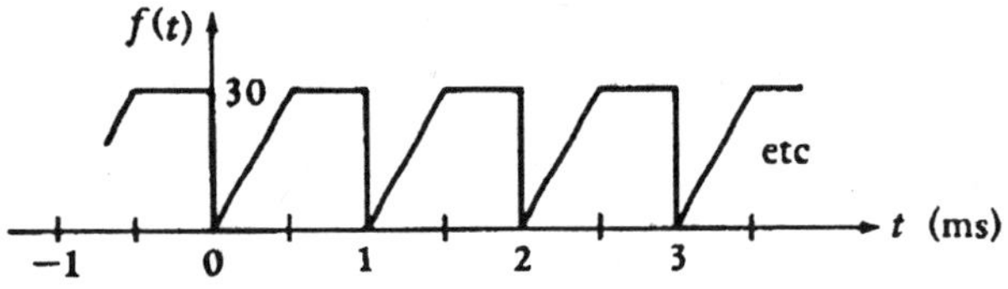


**Fig. 10-3:** See Prob. 5.

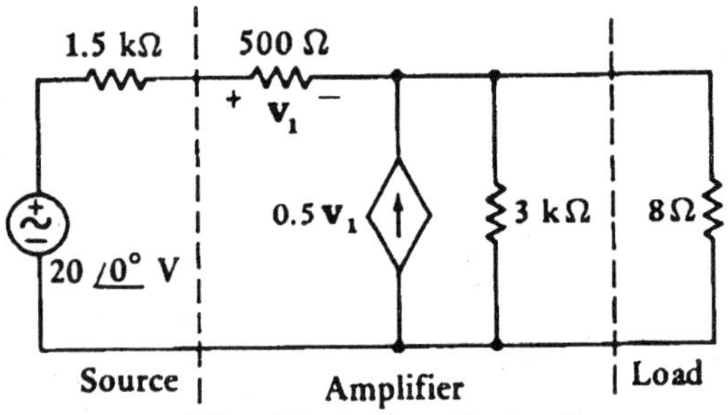


**Fig. 10-4:** See Prob. 6.

**7** Find the average power absorbed by each of the five elements in the circuit of Fig. 10-5.

**8** A square wave of voltage is 100 V for $0 < t < 1$ ms, 0 for $1 < t < 2$ ms, and continues in this pattern. As some help to a busy student, this leads to a current $i$ in the $RC$ circuit of Fig. 10-6 that is $73.11e^{-1000t}$ mA for $0 < t < 1$ ms, $-73.11e^{-1000(t\,-\,0.001)}$ mA for $1 < t < 2$ ms, and repeats. Find the average value of the: (*a*) voltage square wave; (*b*) current; (*c*) power delivered to the resistor; (*d*) power supplied by the source.

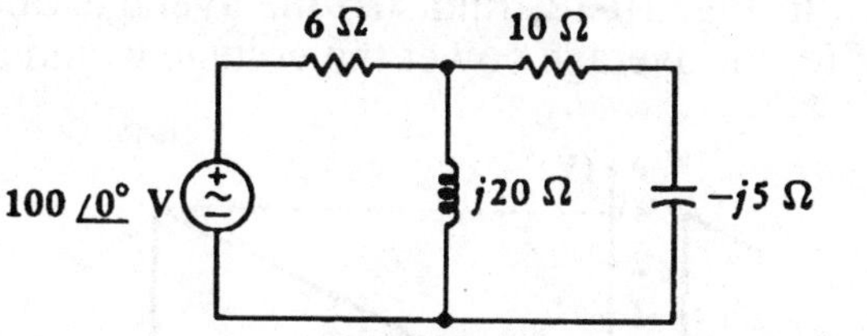


**Fig. 10-5:** See Prob. 7.

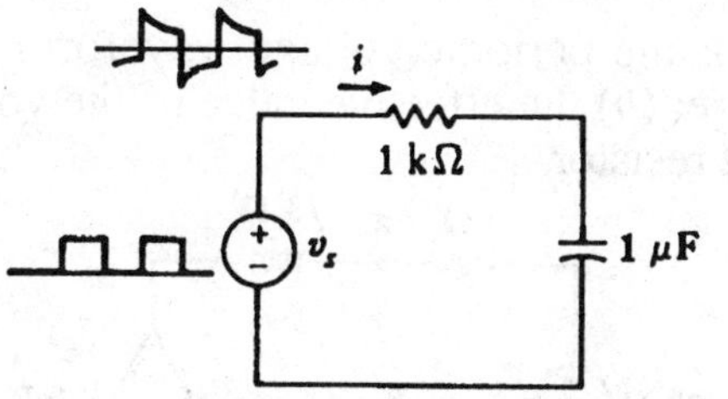


**Fig. 10-6:** See Prob. 8.

**9** A sinusoidal voltage source, $100\cos\omega t$ V, a 1-k$\Omega$ resistor, and a 1-$\mu$F capacitor are in series. Find the average power delivered to the resistor if $\omega =$ : (*a*) 400 rad/s; (*b*) 3000 rad/s.

**10** A frequency-domain Thévenin equivalent circuit consists of a sinusoidal source $\mathbf{V}_{th}$ in series with an impedance $\mathbf{Z}_{th} = R_{th} + jX_{th}$. Specify the conditions on a load, $\mathbf{Z}_L = R_L + jX_L$, if it is to receive a maximum average power subject to the constraint that: (*a*) $X_{th} = 0$; (*b*) $R_L$ and $X_L$ may be selected independently; (*c*) $R_L$ is fixed (not equal to $R_{th}$); (*d*) $X_L$ is fixed (independent of $X_{th}$); (*e*) $X_L = 0$.

**11** Find the average power dissipated in the 5-$\Omega$ resistor of Fig. 10-7 for the following source voltages: (*a*) $A = C = 0, B = D = 10$ V; (*b*) $A = C = 10$ V, $B = D = 0$; (*c*) $A = B = C = 10$ V, $D = 0$; (*d*) $A = B = C = D = 10$ V.

**12** In Fig. 10-8, find the value of $\mathbf{Z}_L$ to which a maximum average power can be delivered, and the value of $P_{L,\,av,\,max}$.

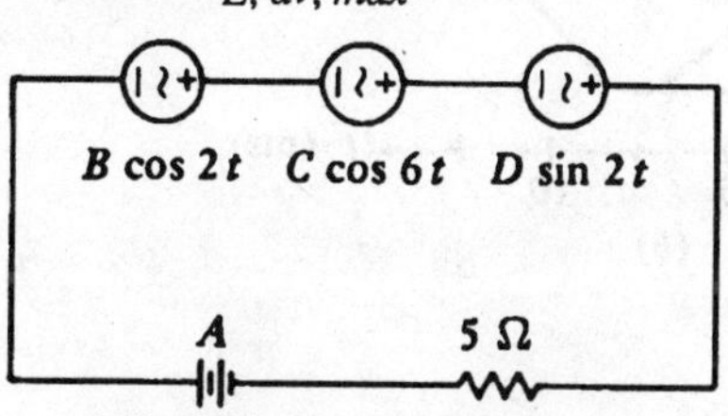


**Fig. 10-7:** See Prob. 11.

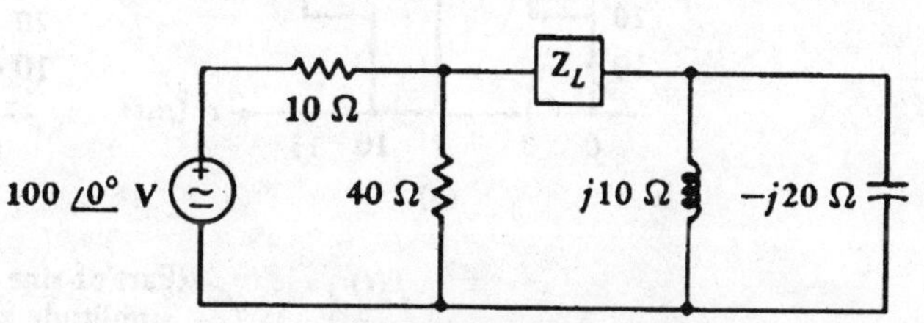


**Fig. 10-8:** See Prob. 12.

**13** Select values of $R$ and $C$ in Fig. 10-9 so that a maximum average power is delivered to $R$, and then calculate the value of that power.

**14** For the circuit of Fig. 10-10, make a sketch of the average power delivered to $R_L$ versus the value of $R_L$, $0 \le R_L \le 40\ \Omega$.

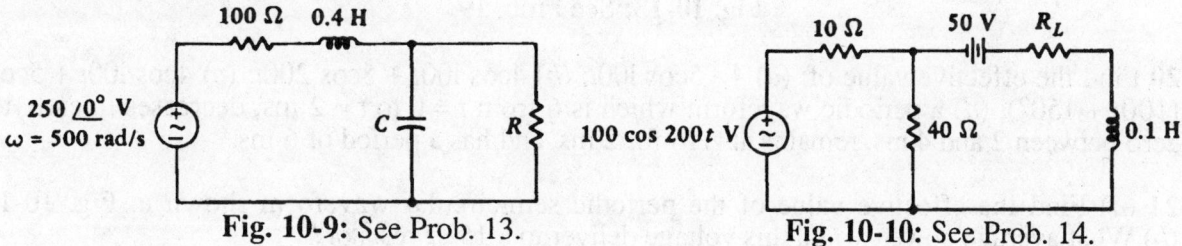

**Fig. 10-9:** See Prob. 13.

**Fig. 10-10:** See Prob. 14.

**15** With reference to the circuit of Fig. 10-11: (*a*) what load impedance should be connected between $x$ and $y$ to absorb a maximum average power? (*b*) What is $P_{L,\, av,\, max}$?

## Section 10-4 Effective values of current and voltage

**16** For the periodic voltage waveform shown in Fig. 10-12, find: (*a*) the average value of the voltage; (*b*) the effective value of the voltage; (*c*) the average power the voltage would deliver to a 6-Ω resistor.

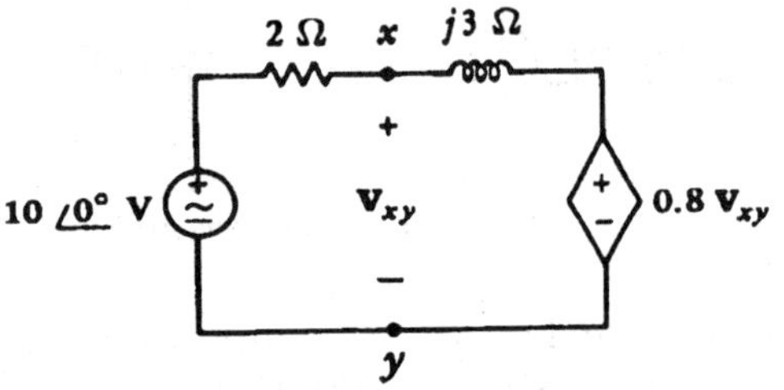


**Fig. 10-11:** See Prob. 15.

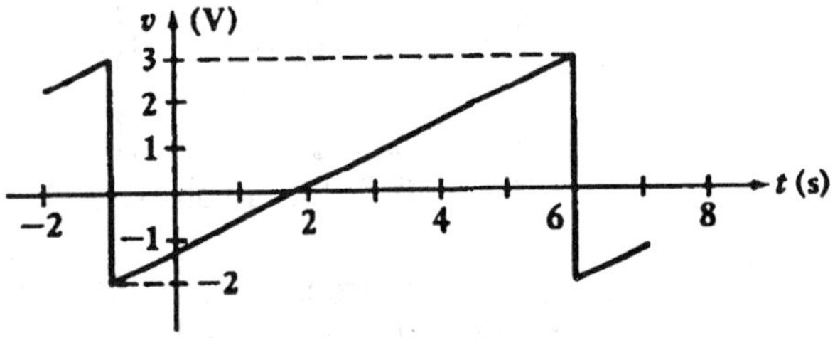


**Fig. 10-12:** See Prob. 16.

**17** A periodic current waveform is $3e^{-t/4}$ A for $0 < t < 3$ s, -3 A for $3 < t < 5$ s, and zero for $5 < t < 7$ s. The period is 7 s. Find: (*a*) the average value of the current; (*b*) the effective value of the current.

**18** A periodic current $i_1(t)$ is 1 A for $0 < t < 3$s, 5 A for $3 < t < 5$ s, -2 A for $5 < t < 7$ s, and zero for $7 < t < 8$ s. If the period is 8 s, find: (*a*) $i_{1,\, av}$; (*b*) $|i_1|_{av}$; (*c*) $i_{1,\, eff}$.

**19** The three waveforms shown in Fig. 10-13 all have $T$ = 10 ms. Find the effective value of each.

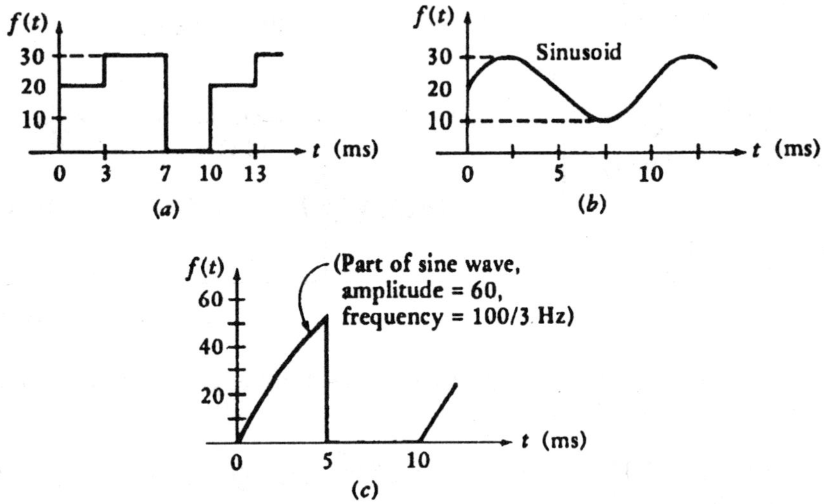


**Fig. 10-13:** See Prob. 19.

**20** Find the effective value of: (*a*) 4 - 5cos 100$t$; (*b*) 4cos 100$t$ + 5cos 200$t$; (*c*) 4cos 100$t$ + 5cos (100$t$ + 150°); (*d*) a periodic waveform which is 6 from $t = 0$ to $t = 2$ ms, decreases linearly to zero between 2 and 4 ms, remains at zero for 2 ms, and has a period of 6 ms.

**21** (*a*) Find the effective value of the periodic semicircular waveform shown in Fig. 10-14. (*b*) What average power would this voltage deliver to a 15-Ω resistor?

## Section 10-5 Apparent power and power factor

**22** A voltage source, $\mathbf{V}_s = 88 + j66$ V rms, in series with 3 Ω of resistance, is supplying two loads in parallel: $\mathbf{Z}_A = 80 + j60$ Ω and $\mathbf{Z}_B = 50 - j100$ Ω. Find: (*a*) the average power delivered to $\mathbf{Z}_A$; (*b*) the average power delivered to $\mathbf{Z}_B$; (*c*) the average power supplied by the ideal source; (*d*) the power factor at which the source is operating.

**23** (*a*) Find the apparent power and the real power being supplied by the source in the circuit of Fig. 10-15. (*b*) Find the power factor of the entire load. (*c*) To what value should the inductance be changed to cause the power factor to be 0.92 lagging?

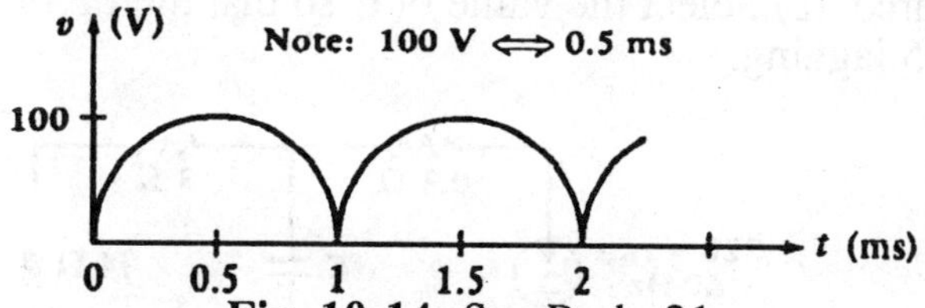

**Fig. 10-14:** See Prob. 21.

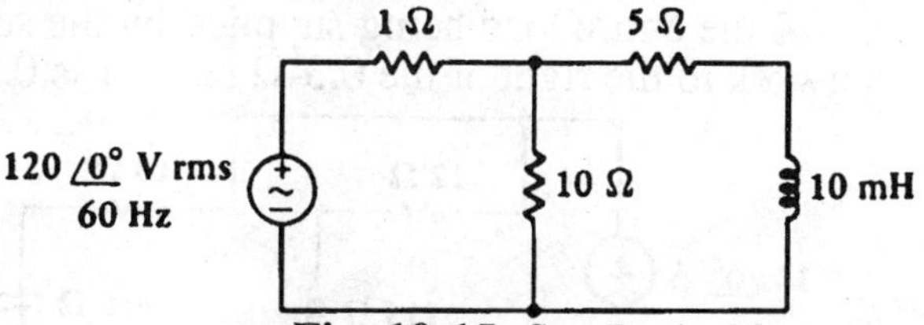

**Fig. 10-15:** See Prob. 23.

**24** (*a*) Find the power factor of the load to the right of *a*-*b* in the circuit shown in Fig. 10-16. (*b*) What value of capacitance should be connected across *a*-*b* to achieve a lagging PF of 0.95?

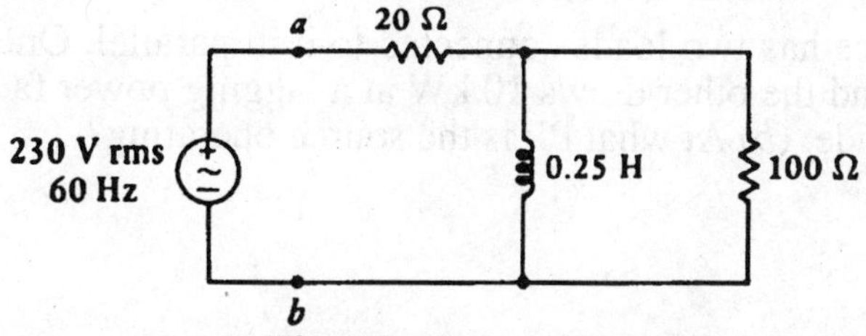

**Fig. 10-16:** See Prob. 24.

**25** A source, $v_s(t) = 200\cos 100t$ V, is in series with 8 Ω, 100 mH, and 2.5 mF. Find: (*a*) the PF at which the source is operating; (*b*) the value of capacitance that would have to be connected directly across the source so that it operates with a power factor of 0.9 leading.

**26** A certain two-terminal network has across it a steady-state sinusoidal voltage whose frequency is 60 Hz and whose effective voltage is 120 V. The impedance of the network is $4 + j5$ Ω. (*a*) At what power factor is it operating? (*b*) What average power is it absorbing? (*c*) What value of capacitance should be placed in parallel with it so that the power factor becomes 0.9 leading?

**27** The following three electrical devices are all connected in parallel: 3 kVA at 0.707 PF leading, 7 kVA at 0.8 PF lagging, and a series *RL* network with $R = 2$ Ω and $L = 10.61$ mH. This composite load is supplied with power from an ideal 100-V rms, 60-Hz, sinusoidal voltage source. Find: (*a*) the power factor at which the source is operating; (*b*) the effective value of the source current; (*c*) the capacitance that must be placed in parallel with the composite load so that the power factor will be 0.95 lagging.

**28** A two-terminal network has a voltage, $v_1(t) = 4\cos(10t - 30°)$ V, across its terminals, and a current, $i_1(t) = 2\cos(10t + 20°)$ A, entering the terminal at which the positive voltage reference is located. Find: (*a*) $V_{1,\,eff}$; (*b*) $\mathbf{I}_{1,\,rms}$; (*c*) the average power absorbed by the network; (*d*) the power factor at which the network is operating; (*e*) the apparent power drawn by the network; (*f*) the complex power drawn by the network.

## Section 10-6 Complex power

**29** (*a*) Find the complex power absorbed by each of the five circuit elements in Fig. 10-17. (*b*) At what PF is the source operating?

**30** A source of $500\angle 0°$ V is in series with an impedance of $-j10\ \Omega$ and a load $\mathbf{Z}_L$. If the source is generating a complex power of $4 + j1.5$ kVA: (*a*) what complex power is delivered to $\mathbf{Z}_L$, and (*b*) what is the load PF?

**31** Let $C = 0$ in the circuit shown in Fig. 10-18, and find the PF of: (*a*) the network to the right of $C$; (*b*) the entire load being supplied by the source. (*c*) Select the value of $C$ so that the PF of the network to the right of the 0.3-Ω resistor is 0.95 lagging.

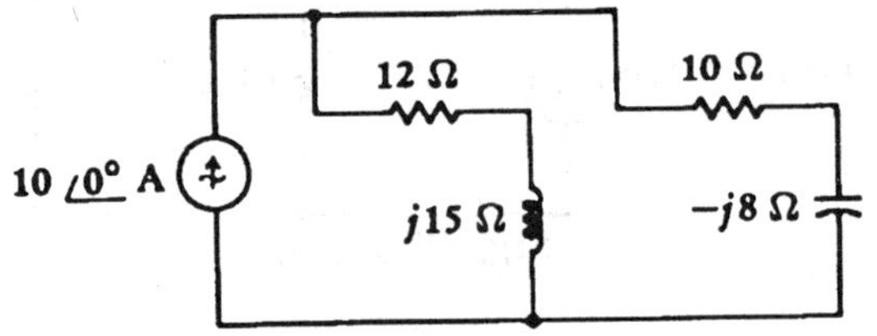

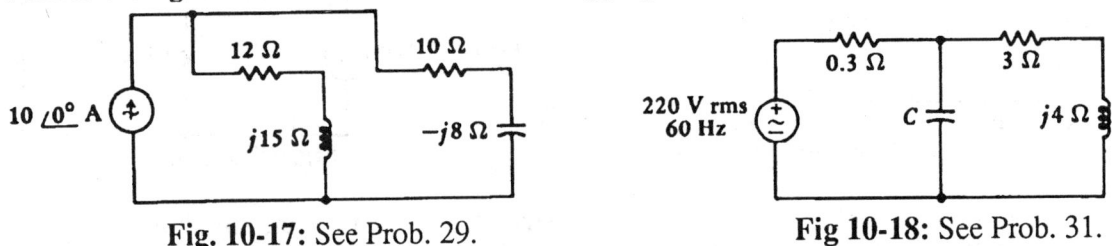

**Fig. 10-17:** See Prob. 29.

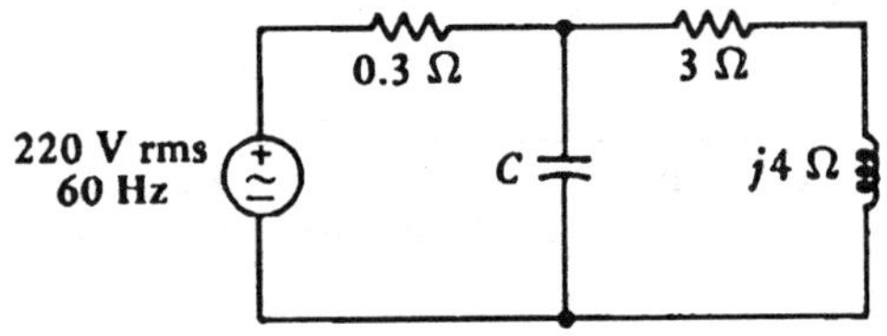


**Fig 10-18:** See Prob. 31.

**32** A 230-V rms source has two loads connected to it in parallel. One draws 8 kVA at a lagging power factor of 0.9, and the other draws 10 kW at a lagging power factor of 0.8. (*a*) Find the rms source current amplitude. (*b*) At what PF is the source operating?

# Chapter 10

**10-1** (a)

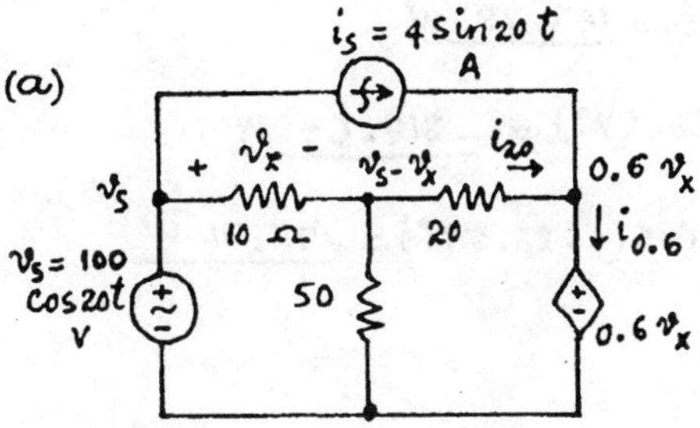


Nodal: $-0.1 v_x + 0.02(v_s - v_x) + 0.05(v_s - 1.6 v_x) = 0$

$\therefore 0.07 v_s = 0.2 v_x \quad \therefore v_x = 0.35 v_s = 35 \cos 20t$ V

$i_{20} = \dfrac{v_s - v_x - 0.6 v_x}{20} = 2.2 \cos 20t$ A

$i_{0.6} = i_s + i_{20} = 4 \sin 20t + 2.2 \cos 20t$

$P_{\Diamond, gen} = -0.6 \times 35 \cos 20t (4\sin 20t + 2.2 \cos 20t) = -\cos 20t (84 \sin 20t + 46.2 \cos 20t)$

$= \underline{-(42 \sin 40t + 23.1 + 23.1 \cos 40t)} = -[23.1 + 47.93 \cos(40t - 61.19°)]$ W

(b) $P_{\Diamond, gen}(0.03) = -\cos 0.6 (84 \sin 0.6 + 46.2 \cos 0.6) = \underline{-70.62 \text{ W}}$

**10-2** (a)

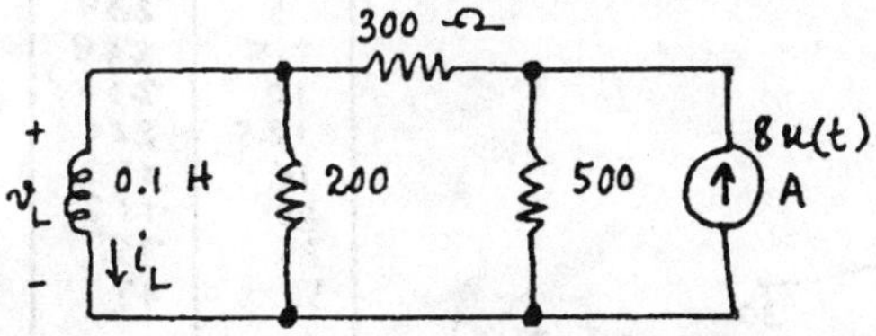


$i_L(0) = 0$, $i_L(\infty) = 5$ A, $R_{eq} = 800 \| 200 = 160$ Ω

$\therefore i_L = 5(1 - e^{-1600t})$ A, $t > 0$

$v_L = 0.1 \times 5 \times 1600 e^{-1600t} = 800 e^{-1600t}$ V, $t > 0$

$\therefore P_L(t) = \underline{4000 e^{-1600t} (1 - e^{-1600t})}$ W

$P_L(t) = 4000 e^{-1600t} (1 - e^{-1600t})$ W

(b)

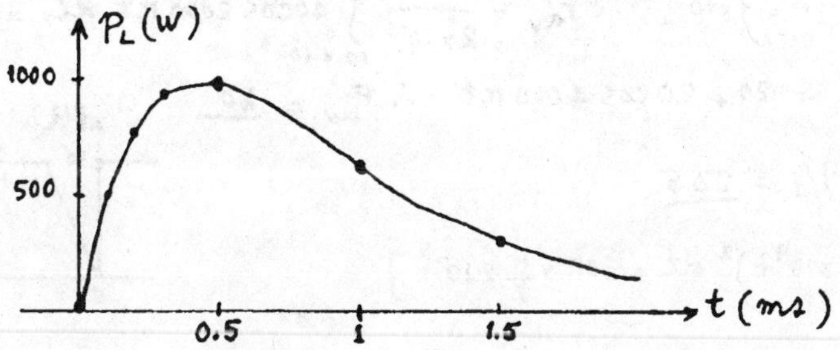


| t (ms) | $P_L$(W) |
|---|---|
| 0 | 0 |
| 0.1 | 504 |
| 0.2 | 795 |
| 0.3 | 944 |
| 0.5 | 990 |
| 0.75 | 842 |
| 1 | 645 |
| 1.5 | 330 |
| 2 | 156 |

(c) $P_L(1 \text{ ms}) = \underline{645 \text{ W}}$

**10-3**

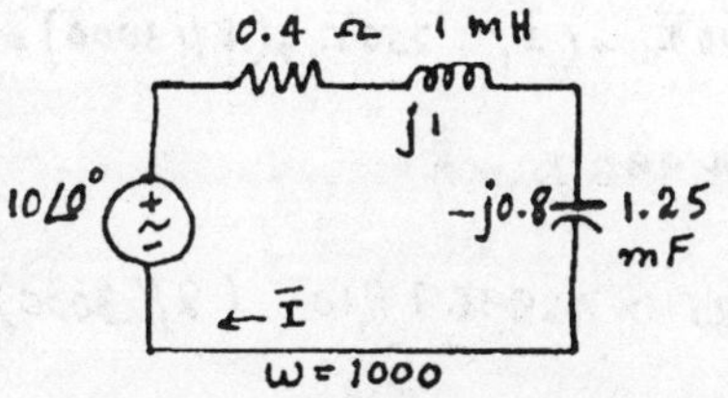


$\bar{I} = \dfrac{10}{0.4 + j0.2} = 22.36 \angle -26.57° \quad \therefore i = 22.36 \cos(1000t - 26.57°)$ A

$v_R = 0.4 i = 8.944 \cos(1000t - 26.57°)$ V

$\bar{V}_L = 22.36(1) \angle 63.43° \quad \therefore v_L = 22.36 \cos(1000t + 63.43°)$ V

$\bar{V}_C = 22.36(0.8) \angle -26.57 - 90° = 17.889 \angle -116.57° \quad \therefore v_c = 17.889 \cos(1000t - 116.57°)$

(cont. on next page)

**10-3 (cont.)** At $t = 1\ ms$ : $p_s = -22.36(10)(\cos 1^{rad})\cos(1^{rad} - 26.57°) = \underline{-103.85^{+}\ W}$

$p_R = 0.4\, i^2 = 0.4 \times 22.36^2 \cos^2(1 - 26.57°) = \underline{147.78\ W}$

$p_L = v_L i = 22.36^2 \cos(1 + 63.43°)\cos(1 - 26.57°) = \underline{-219.64\ W}$

$p_C = v_C i = 17.889 \times 22.36 \cos(1 - 116.57°)\cos(1 - 26.57°) = \underline{175.71\ W}$

Note : $\Sigma p = 0$ ✓

**10-4** (a)

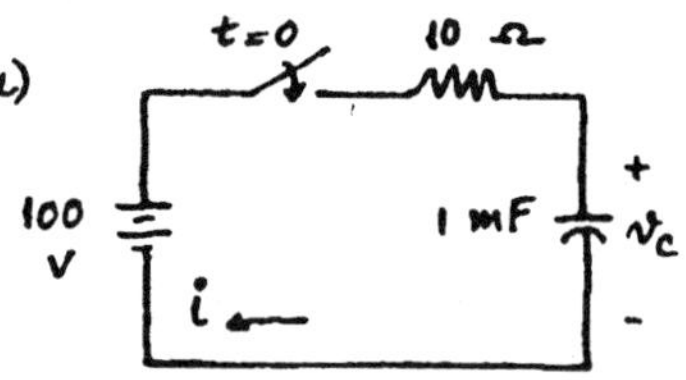


$v_c = 100(1 - e^{-100t})\ V,\ t > 0,\ i = C v_c' = 10 e^{-100t}\ A$

$p_c = v_c i = \underline{1000\, e^{-100t}(1 - e^{-100t})\, u(t)\ W}$

(b)

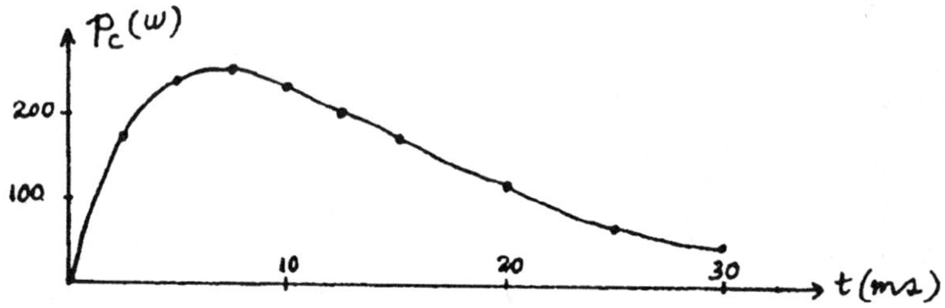


| t(ms) | $p_c$(W) |
|---|---|
| 0 | 0 |
| 2.5 | 173 |
| 5 | 239 |
| 7.5 | 249 |
| 10 | 233 |
| 12.5 | 204 |
| 15 | 173 |
| 20 | 117 |
| 25 | 75 |
| 30 | 47 |

(c) $p_c' = 1000[e^{-100t}\, 100 e^{-100t} + (1 - e^{-100t})(-100 e^{-100t})] = 0 \therefore 200 e^{-200t} = 100 e^{-100t}$

$\therefore \underline{t = 6.93\ ms}$ , $p_{c,max} = 1000(1 - 1/2)\,1/2 = \underline{250\ W}$

**10-5** (a) $f(t) = 40\cos 2000\pi t$ , $T = 1 \times 10^{-3}$ $\therefore F_{av} = \frac{1}{2 \times 10^{-3}} \int_{10 \times 10^{-3}}^{12 \times 10^{-3}} 40\cos 2000\pi t\, dt = \underline{0}$

(b) $f(t) = 40\cos^2 2000\pi t = 20 + 20\cos 4000\pi t$ $\therefore F_{av} = \underline{20}$

(c) $F_{av} = (15 \times \frac{1}{2} + 30 \times \frac{1}{2})/1 = \underline{22.5}$

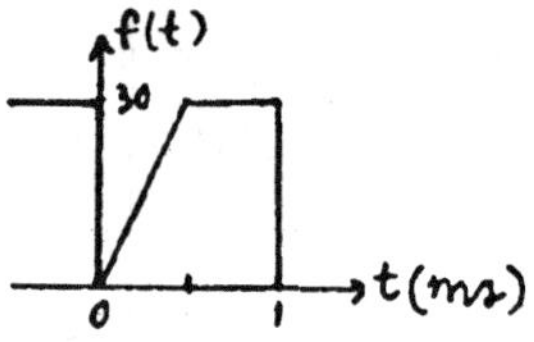


(d) $F_{av} = \frac{1}{10^{-3}}\left[\int_0^{0.5 \times 10^{-3}} (6 \times 10^4 t)^2\, dt + 30^2 \times \frac{1}{2} \times 10^{-3}\right]$

$= 10^3\left[36 \times 10^8 \times \frac{1}{3} \times \frac{1}{8} \times 10^{-9} + 0.45\right] = 10^3(0.15 + 0.45) = \underline{600}$

**10-6** (a)

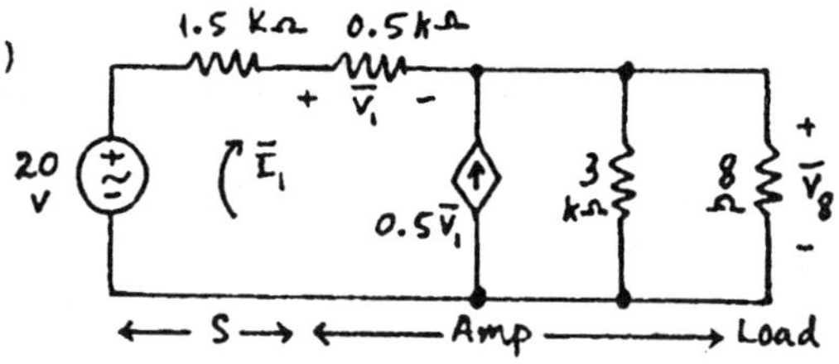


$20 = 2000\bar{I}_1 + (\bar{I}_1 + 250\bar{I}_1)(8 // 3000) = 4002.7\bar{I}_1$

$\therefore \bar{I}_1 = 4.9967\ mA$

$\therefore \bar{V}_8 = 251 \times 4.9967 \times 10^{-3}(8 // 3000) = 10.007\ V$

$\therefore P_{av, 8\Omega} = \frac{1}{2}\frac{(10.007)^2}{8} = \underline{6.258\ W}$

(b) $P_{S,av} = \frac{1}{2} \times 20 \times 4.9967 \times 10^{-3} = \underline{0.049967\ W}$

**10-7**

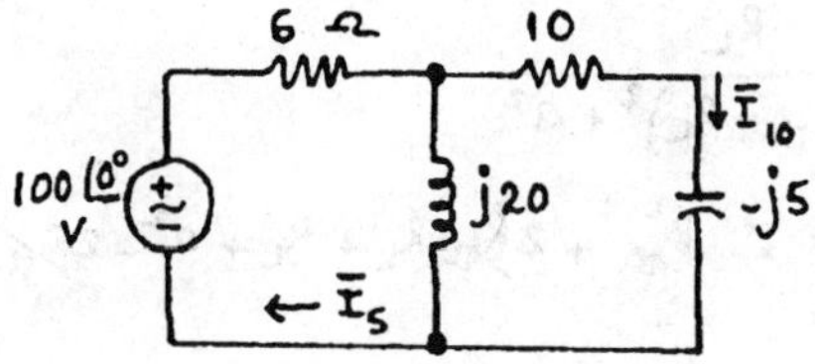


$$j20\,//\,(10-j5)=\frac{100+j200}{10+j15}\qquad \bar{Z}_{in}=6+\frac{100+j200}{10+j15}=\frac{16+j29}{1+j1.5}$$

$$\bar{I}_S=\frac{100+j150}{16+j29}=5.443\angle -4.803^\circ\ \text{A}$$

$$\therefore P_{av,s}=-\frac{1}{2}\times 100\times 5.443\cos(4.803^\circ)=\underline{-271.2\ \text{W}}$$

$$P_{av,6}=\frac{1}{2}\times 6\times 5.443^2=\underline{88.9\ \text{W}}\ ;\quad \bar{I}_{10}=\bar{I}_S\,\frac{j20}{10+j15}=6.038\angle 28.9^\circ$$

$$\therefore P_{av,10}=\frac{1}{2}\times 10\times 6.038^2=\underline{182.3\ \text{W}}\ ;\quad P_{av,L}=P_{av,C}=\underline{0}\ ;\quad \Sigma P_{av}=0\ ✓$$

**10-8**

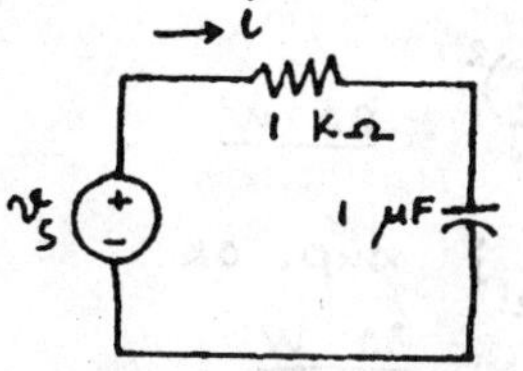


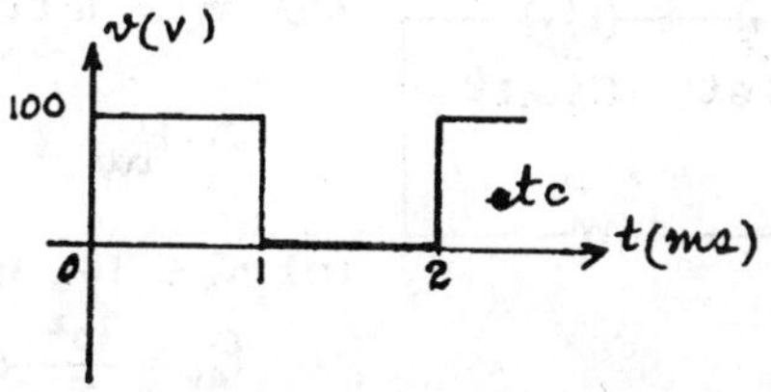


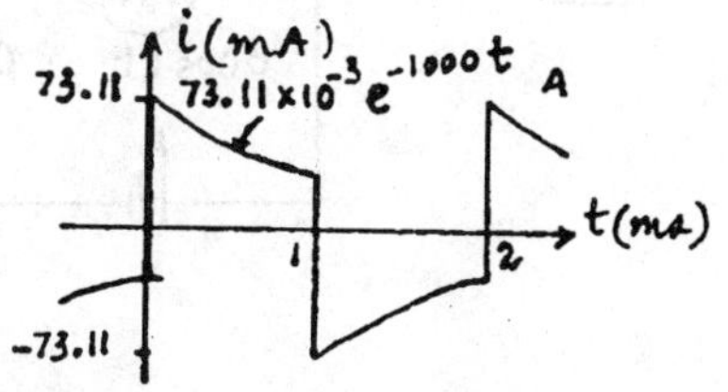


(a) $V_{s,av}=\frac{1}{2}(100)(1)=\underline{50\ \text{V}}$ (b) $\bar{I}_{av}=\underline{0}$

(c) $p_R(t)=10^3\times 73.11^2\times 10^{-6}e^{-2000t}=5.344e^{-2000t}$, $0<t<1$ ms

$$\therefore P_{av,R}=\frac{1}{10^{-3}}\int_0^{0.001}5.344e^{-2000t}\,dt=\underline{2.311\ \text{W}}\ ;\quad P_{av,C}=0$$

(d) $p_s(t)_{gen}=\begin{cases}7.311e^{-1000t}\ \text{W}, & 0<t<1\text{ ms}\\ 0 & 1<t<2\text{ ms}\end{cases}$ $\quad\therefore P_{av,s}=\frac{1}{2\times10^{-3}}\int_0^{0.001}7.311e^{-1000t}\,dt=\underline{2.311\ \text{W}}$

$[3.655(1-e^{-1})]$

**10-9**

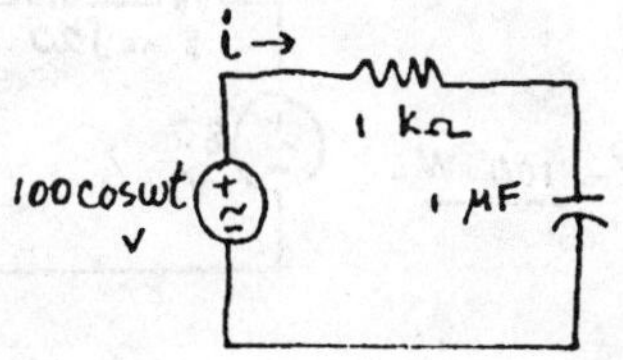


(a) $\omega=400:\ \bar{I}=\frac{100}{1000-j2500}=\frac{1}{10-j25}\quad\therefore |\bar{I}|^2=\frac{1}{725}$

$$P_{av,R}=\frac{1}{2}\times\frac{1}{725}\times 1000=\underline{0.6897\ \text{W}}$$

(b) $\omega=3000:\ \bar{I}=\frac{100}{1000-j\frac{1000}{3}}=0.09+j0.03$

$$P_{av,R}=\frac{1}{2}(0.09^2+0.03^2)10^3=\underline{4.500\ \text{W}}$$

**10-10**

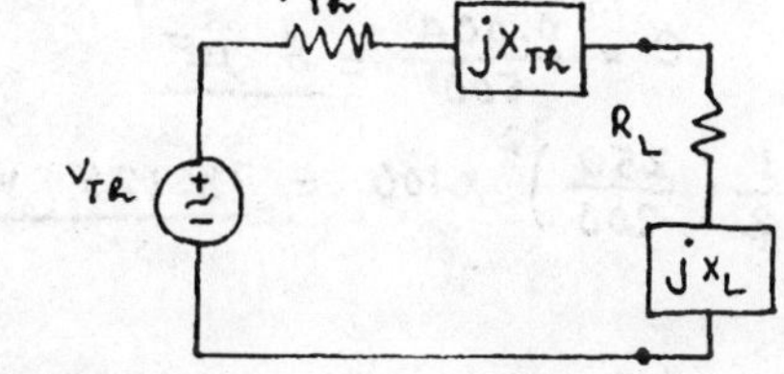


(a) $X_{Th}=0\quad\therefore \bar{Z}_L=\underline{R_{Th}+j0}$

(b) $R_L, X_L$ indep. $\therefore \bar{Z}_L=\bar{Z}^*_{Th}=\underline{R_{Th}-jX_{Th}}$

(c) $R_L$ fixed $\quad P_L=\frac{1}{2}\,\frac{|V_{Th}|^2}{(R_{Th}+R_L)^2+(X_{Th}+X_L)^2}\times R_L$

$$\therefore \bar{Z}_L=\underline{R_L-jX_{Th}}$$

(cont. on next page)

10-10 (cont.) (d) $X_L$ fixed. let $X_{Th} + X_L = a \quad \therefore \quad \frac{2P_L}{|V_{Th}|^2} = \frac{R_L}{(R_{Th}+R_L)^2 + a^2}$

$$\frac{d\left(\frac{2P_L}{|V_{Th}|^2}\right)}{dR_L} = \frac{(R_{Th}+R_L)^2 + a^2 - 2R_L(R_{Th}+R_L)}{[(R_{Th}+R_L)^2 + a^2]^2} = 0 \quad \therefore \quad R_{Th}^2 + 2R_{Th}R_L + R_L^2 + a^2 - 2R_{Th}R_L - 2R_L^2 = 0$$

$$\therefore R_L = \sqrt{R_{Th}^2 + a^2} = \underline{\sqrt{R_{Th}^2 + (X_{Th}+X_L)^2}} \quad (\text{Think of } X_{Th}+X_L \text{ as } X'_{Th})$$

(e) $X_L = 0 \quad \therefore \quad R_L = \underline{\sqrt{R_{Th}^2 + X_{Th}^2} = |\bar{Z}_{Th}|}$

10-11

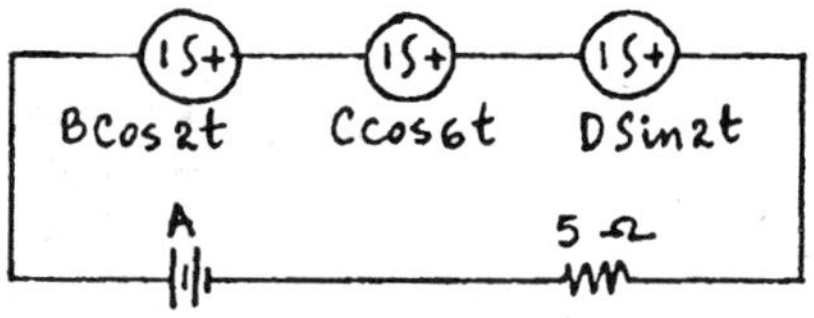


(a) $v_s = 10\cos 2t + 10\sin 2t = 10\sqrt{2}\cos(2t - 45°)$

$$\therefore P_{av} = \frac{1}{2}\frac{(10\sqrt{2})^2}{5} = \underline{20\ W}$$

(b) $v_s = 10 + 10\cos 6t$ ; sup. OK.

$$P_{av} = \frac{10^2}{5} + \frac{1}{2}\frac{10^2}{5} = \underline{30\ W}$$

(c) $v_s = 10 + 10\cos 2t + 10\cos 6t$, sup. OK $\therefore P_{av} = \frac{10^2}{5} + \frac{1}{2}\frac{10^2}{5} + \frac{1}{2}\frac{10^2}{5} = \underline{40\ W}$

(d) $v_s = 10 + 10\cos 2t + 10\sin 2t + 10\cos 6t = 10 + 10\sqrt{2}\cos(2t - 45°) + 10\cos 6t$

$$\therefore P_{av} = \frac{10^2}{5} + \frac{1}{2}\frac{(10\sqrt{2})^2}{5} + \frac{1}{2}\frac{10^2}{5} = \underline{50\ W}$$

10-12

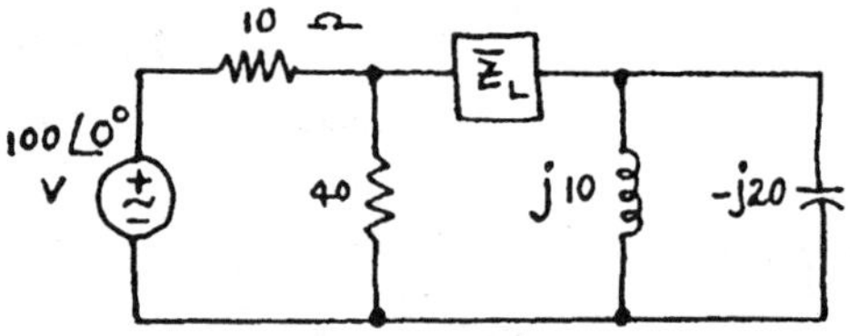


$\bar{Z}_{Th} = (40 \| 10) + j10 \| -j20 = 8 + j20\ \Omega$ ; $\bar{V}_{oc} = 100 \times \frac{40}{50} = 80\ V$

$\therefore \bar{Z}_L = \underline{8 - j20\ \Omega}$

$\therefore P_{av,L(max)} = \frac{1}{2}\left(\frac{80}{16}\right)^2 8 = \underline{100\ W}$

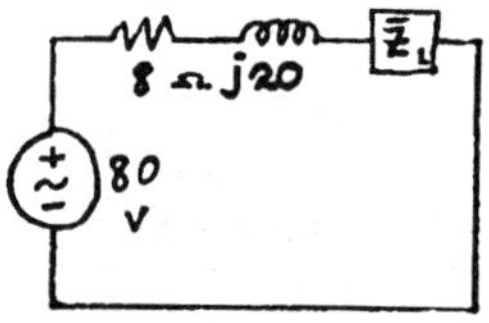


10-13

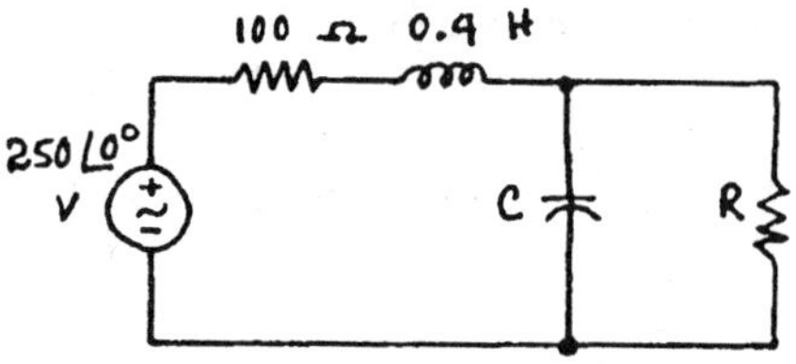


W = 500

$\bar{Z}_{Th} = 100 + j200 \quad \therefore \bar{Z}_L = 100 - j200 = \frac{1}{j500C + 1/R}$

$$\therefore \frac{1}{R} + j500C = \frac{1}{100 - j200} = 0.002 + j0.004$$

$\therefore R = \underline{500\ \Omega}$ , $C = \frac{0.004}{500} = \underline{8\ \mu F}$

$$P_{av(max)} = \frac{1}{2}\left(\frac{250}{200}\right)^2 \times 100 = \underline{78.125\ W}$$

**10-14**

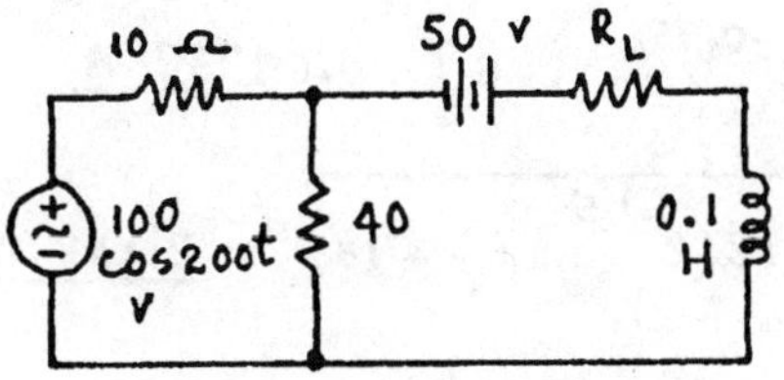


Sup. OK. $P_{dc} = \left(\frac{50}{8+R_L}\right)^2 R_L = \frac{2500 R_L}{(R_L+8)^2}$

$P_{ac} = \frac{1}{2}\left|\frac{80}{8+R_L+j20}\right|^2 R_L = \frac{3200 R_L}{(R_L+8)^2+400}$

$\therefore P_{av} = \frac{2500 R_L}{(R_L+8)^2} + \frac{3200 R_L}{(R_L+8)^2+400}$

$P_{dc}$ max at $R_L = 8\ \Omega$

$P_{ac}$ max at $R_L = |8+j20| = 21.54\ \Omega$

$P_{av}$ max at $R_L = 12.954\ \Omega$

$P_{av,max} = 123.16$ W

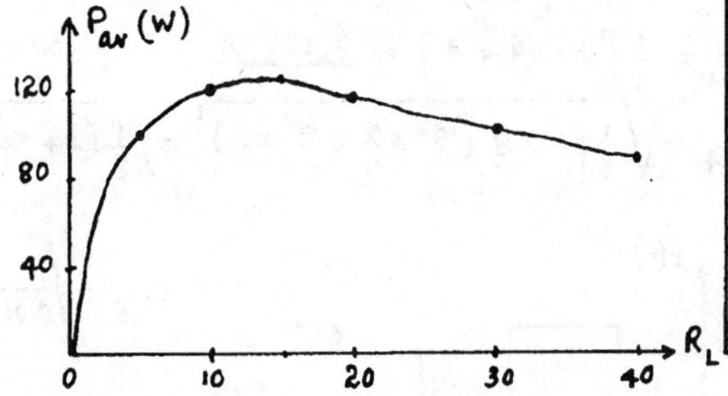


| $R_L$ | $P_{av}$ |
|---|---|
| 0 | 0 |
| 1 | 37.5 |
| 3 | 80.4 |
| 5 | 102.1 |
| 10 | 121.4 |
| 15 | 122.6 |
| 20 | 117.8 |
| 30 | 104.0 |
| 40 | 90.7 |

**10-15**

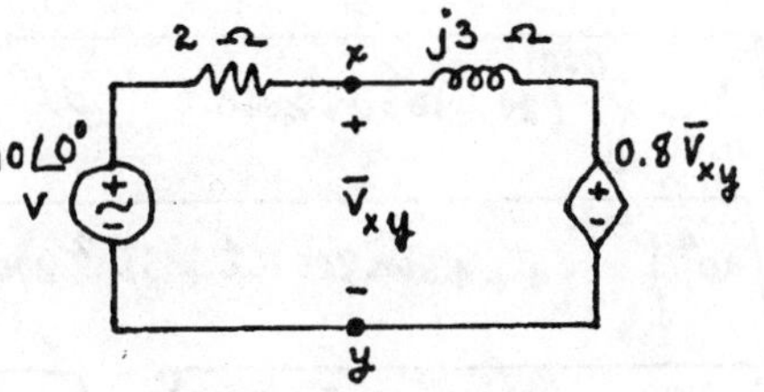


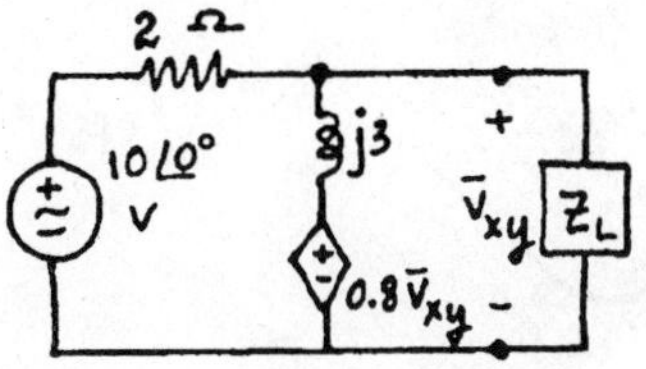


OC: $\circlearrowright \bar{I}\ \therefore 10 = 2\bar{I} + \bar{V}_{xy}$

$\bar{V}_{xy} = j3\bar{I} + 0.8\bar{V}_{xy}$

$\therefore \bar{I} = \frac{\bar{V}_{xy}}{j15},\ 10 = \left(\frac{2}{j15}+1\right)\bar{V}_{xy}$

$\therefore \bar{V}_{xy} = 9.912\angle 7.595° = \bar{V}_{oc}$

SC: $\bar{V}_{xy} = 0\ \therefore \downarrow \bar{I}_{sc} = \frac{10}{2} = 5$ A $\therefore \bar{Z}_{Th} = \frac{\bar{V}_{oc}}{\bar{I}_{sc}} = \frac{9.912\angle 7.595°}{5} = 1.9825\angle 7.595°\ \Omega$

(a) $\bar{Z}_L = \bar{Z}_{Th}^* = 1.9825\angle -7.595° = \underline{1.9651 - j0.2620\ \Omega}$

(b) $\bar{I}_L = \frac{\bar{V}_{oc}}{2R_{Th}} = \frac{9.912\angle 7.595°}{2 \times 1.9651} = 2.522\angle 7.595°\ \therefore P_{av,L(max)} = \frac{1}{2} \times 2.522^2 \times 1.9651 = \underline{6.250\ \text{W}}$

**10-16**

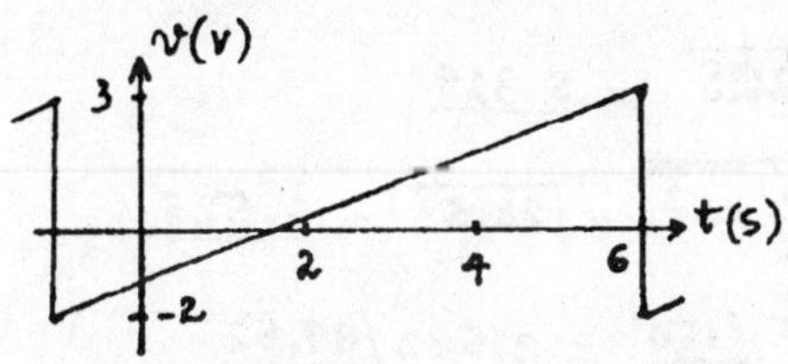


(a) $v = mt + v_0 = \frac{5}{7}t + \frac{5}{7} - 2 = \frac{5t}{7} - \frac{9}{7}$

$V_{av} = \frac{1}{7}\int_{-1}^{6} \frac{5t-9}{7}\,dt = \frac{1}{49}(2.5t^2 - 9t)_{-1}^{6} = \underline{0.5\ \text{V}}$

(b) $V_{eff} = \sqrt{\frac{1}{7}\int_{-1}^{6}\frac{(5t-9)^2}{49}\,dt} = \frac{1}{7\sqrt{7}}\sqrt{\left(\frac{25}{3}t^3 - 45t^2 + 81t\right)_{-1}^{6}}$

$= \underline{1.5275\ \text{V}}$

(c) $V_{eff} = 1.5275$ V, $R = 6\ \Omega\ \therefore P_{av} = \frac{1.5275^2}{6} = \underline{0.3889\ \text{W}}$

**10-17** (a) $I_{av} = \frac{1}{7}\left[\int_0^3 3e^{-t/4}\,dt - \int_3^5 3\,dt\right] = \frac{1}{7}\left[-12(e^{-0.75}-1)-6\right] = \underline{0.04737\ A}$

(b) $I_{eff} = \sqrt{\frac{1}{7}\left[\int_0^3 9e^{-t/2}\,dt + \int_3^5 9\,dt\right]} = \sqrt{\frac{1}{7}\left[-18(e^{-1.5}-1)+18\right]} = \underline{2.138\ A}$

**10-18** (a) $i_{1,av} = \frac{1}{8}\left[1\times3 + 5\times2 - 2\times2\right] = \underline{1.125\ A}$

(b) $|i_1|_{av} = \frac{1}{8}\left[3+10+4\right] = \underline{2.125\ A}$

(c) $i_{1,eff} = \sqrt{\frac{1}{8}(1^2\times3 + 5^2\times2 + 2^2\times2)} = \sqrt{\frac{1}{8}(3+50+8)} = \sqrt{\frac{61}{8}} = \underline{2.761\ A}$

**10-19** (a)

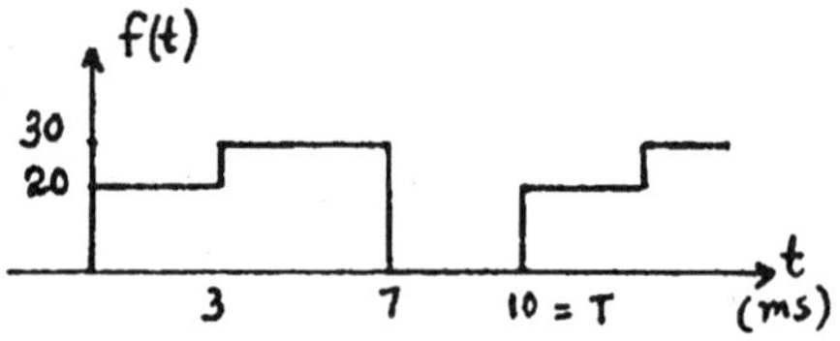


$F_{eff} = \sqrt{\frac{1}{0.01}(20^2\times3\times10^{-3} + 30^2\times4\times10^{-3})} = \sqrt{480} = \underline{21.91}$

(b)

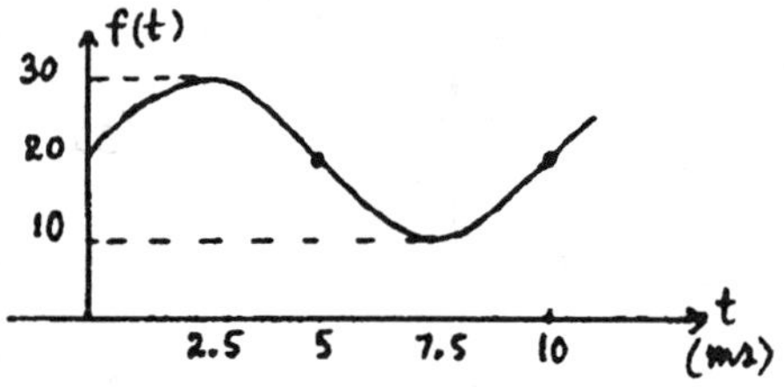


$F_{eff} = \sqrt{\frac{1}{0.01}\int_0^{0.01}(20+10\sin 200\pi t)^2\,dt}$

$= \sqrt{10^4\int_0^{0.01}(4+4\sin 200\pi t + \sin^2 200\pi t)\,dt}$

$= \sqrt{10^4(4\times0.01 + 0 + 0.5\times0.01)} = \sqrt{450} = \underline{21.21}$

(c)

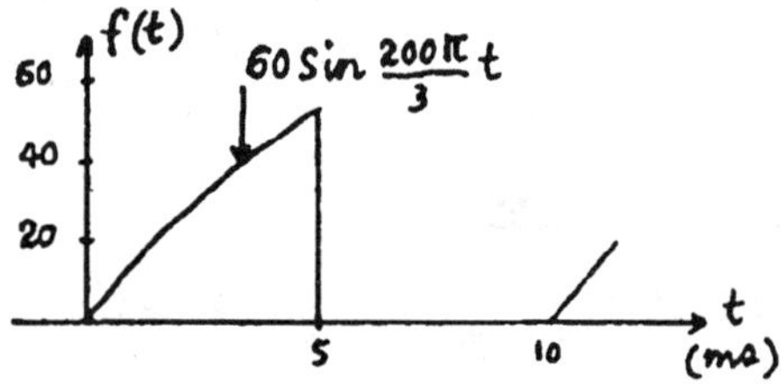


$F_{eff} = \sqrt{\frac{1}{0.01}\int_0^{0.005}\left(60\sin\frac{200\pi t}{3}\right)^2 dt}$

$= \sqrt{600^2\int_0^{0.005}\left(0.5 - 0.5\cos\frac{400\pi t}{3}\right)dt}$

$= 600\sqrt{0.5\times0.005 - 0.5\times\frac{3}{400\pi}\sin\frac{400\pi\times0.005}{3}} = \underline{22.98}$

**10-20** (a) $f = 4 - 5\cos 100t$, $F_{eff} = \sqrt{4^2 + \frac{1}{2}\times5^2} = \sqrt{28.5} = \underline{5.339}$

(b) $f = 4\cos 100t + 5\cos 200t$, $F_{eff} = \sqrt{\frac{1}{2}\times4^2 + \frac{1}{2}\times5^2} = \sqrt{20.5} = \underline{4.528}$

(c) $f = 4\cos 100t + 5\cos(100t + 150^\circ)$, $\overline{F} = 4 + 5\angle 150^\circ = 2.522\angle 97.52^\circ$

$F_{eff} = \frac{2.522}{\sqrt{2}} = \underline{1.7831}$

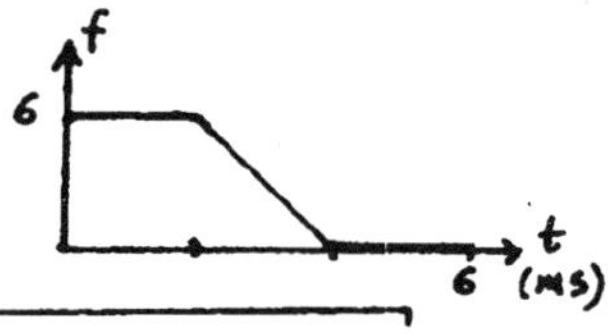


(d) $F_{eff} = \sqrt{\frac{1000}{6}\left[\int_0^{0.002} 36\,dt + \int_{0.002}^{0.004}\left[3\times10^3(4\times10^{-3}-t)\right]^2 dt\right]}$

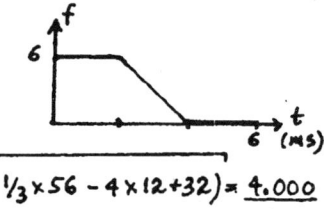

$= \sqrt{12 + 1.5\times10^9\left(\frac{1}{3}t^3 - 4\times10^{-3}t^2 + 16\times10^{-6}t\right)\Big|_{0.002}^{0.004}} = \sqrt{12 + 1.5(1/3\times56 - 4\times12 + 32)} = \underline{4.000}$

**10-21** (a)

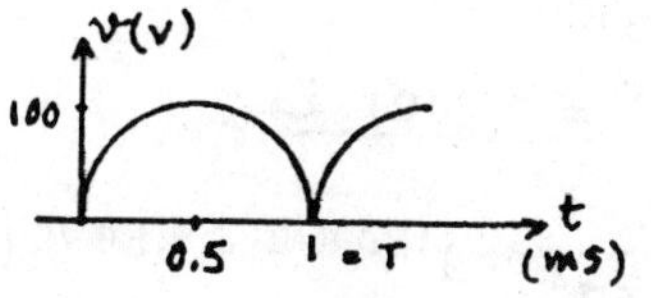


Eqn: $\left(\frac{v}{100}\right)^2 + \left(\frac{t - 5\times10^{-4}}{5\times10^{-4}}\right)^2 = 1$; $v^2 = 10^4\left[1 - \left(\frac{t-0.0005}{0.0005}\right)^2\right]$

$\therefore V_{eff} = \sqrt{\frac{1}{10^{-3}}\int_0^{0.001} 10^4\left[1-(2000t-1)^2\right]dt}$

$V_{eff} = \sqrt{10^3\int_0^{0.001}(\cancel{10^4} - 4\times10^{10}t + 4\times10^7 t - \cancel{10^4})\,dt} = \sqrt{10^3\left(-\frac{4}{3}\times10+20\right)} = \sqrt{\frac{20{,}000}{3}} = \underline{81.65\text{ V}}$

(b) $P = \frac{81.65^2}{15} = \underline{444.4\text{ W}}$

**10-22**

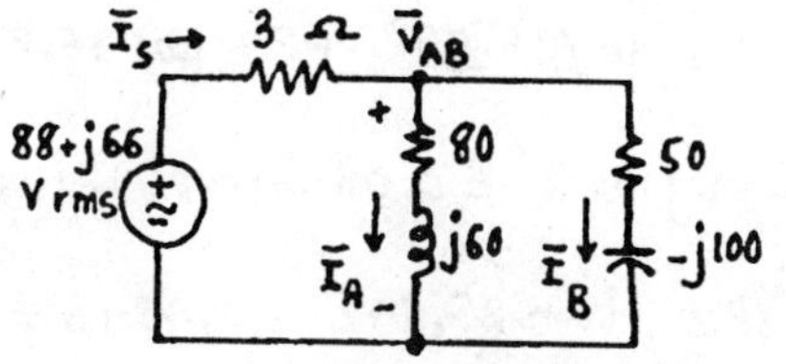


$(80+j60)\,//\,(50-j100) = 82.199\angle{-9.462^\circ} = 81.081 - j13.514\ \Omega$

$\bar{I}_s = \frac{88+j66}{85.160\angle{-9.131^\circ}} = 1.2917\angle{46.000^\circ}$

$\bar{V}_{AB} = 1.2917\angle{46^\circ}\,(82.199\angle{-9.462^\circ}) = 106.18\angle{36.538^\circ}$

$\bar{I}_A = \frac{106.18\angle{36.538^\circ}}{80+j60} = 1.0618\angle{-0.332^\circ}$, $\bar{I}_B = \frac{106.18\angle{36.538^\circ}}{50-j100} = 0.9497\angle{99.97^\circ}$

(a) $P_A = |\bar{I}_A|^2\,80 = 1.0618^2 \times 80 = \underline{90.19\text{ W}}$

(b) $P_B = |\bar{I}_B|^2\,50 = 0.9497^2 \times 50 = \underline{45.10\text{ W}}$

(c) $\bar{V}_s = 88 + j66 = 110\angle{36.87^\circ}$ $\therefore P_s = 110 \times 1.2917\cos(36.87^\circ - 46^\circ) = \underline{140.29\text{ W}}$

[Check: $P_{3\Omega} = 1.2917^2 \times 3 = 5.01\text{ W}$, $P_T = 5.01 + 90.19 + 45.10 = 140.29\text{ W}$ ✓]

(d) $PF = \cos(36.87^\circ - 46^\circ) = \underline{0.9873\ lead}$

**10-23** (a)

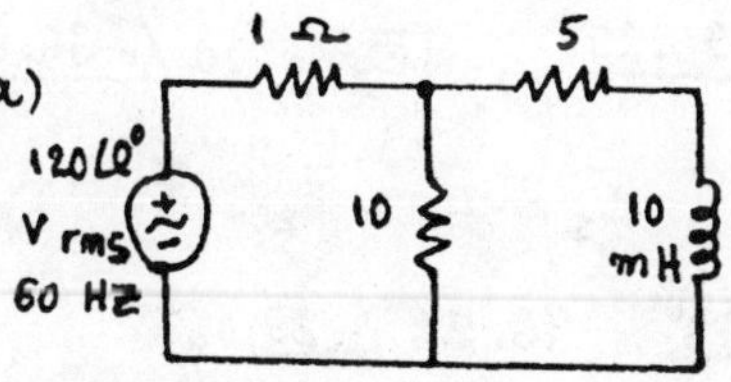


$\bar{Z}_{in} = 1 + \frac{10(5+j1.2\pi)}{15+j1.2\pi} = 4.985\angle{18.43^\circ}$

$\bar{I}_s = \frac{120}{4.985\angle{18.43^\circ}} = 24.07\angle{-18.43^\circ}\text{ A}$

$\therefore AP_s = 120 \times 24.07 = \underline{2.888\text{ kVA}}$, $P_s = 2.888\cos18.43^\circ = \underline{2.74\text{ kW}}$

(b) $PF = \cos 18.43^\circ = \underline{0.9487\ lag}$

(c) $\bar{Z}_{in} = 1 + \frac{10(5+jx)}{15+jx} = \frac{65+j11x}{15+jx} \times \frac{15-jx}{15-jx} = \frac{975+11x^2+j100x}{x^2+225}$

$\tan(\cos^{-1}0.92) = \frac{100x}{975+11x^2} = 0.4260$ $\therefore x = 5.649$ or $15.692$

$L = \frac{x}{120\pi} = \underline{14.984}$ or $\underline{41.62\text{ mH}}$

**10-24**

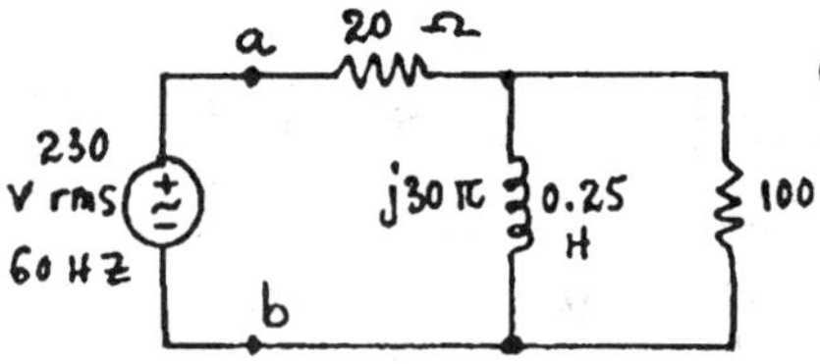


(a) $\bar{Z}_{in} = 20 + \dfrac{j300\pi}{10 + j3\pi} = 83.58\angle 36.67°$

$PF = \cos 36.67° = \underline{0.8021\ lag}$

(b) $\bar{Y}_{in} = \dfrac{1}{83.58\angle 36.67°} + j120\pi C = 9.597 - j7.145 + j120 \times 10^3 \pi C$ (mS)

$PF = 0.95\ lag \therefore \theta = \cos^{-1} 0.95 = 18.195°$, $\tan\theta = 0.3287 \therefore \dfrac{12\times10^4\pi C - 7.145}{9.597} = -0.3287$

$\therefore 12\times10^4 \pi C - 7.145 = -$ (lag) $0.3287 \times 9.597 \quad \therefore C = \underline{10.58\ \mu F}$

**10-25**

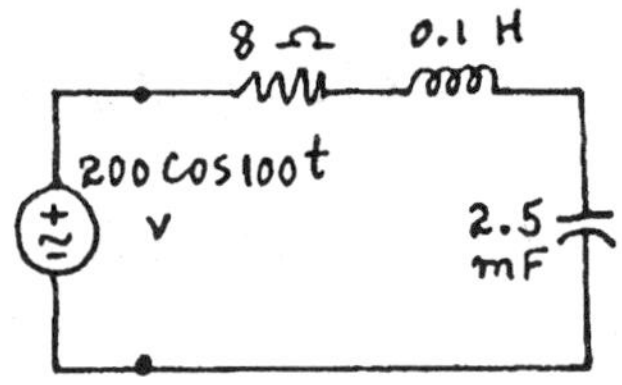


(a) $\bar{Z}_{in} = 8 + j10 - j4 = 10\angle 36.87°$, $PF = \cos 36.87° = \underline{0.8\ lag}$

(b) $\bar{Y}_{in} = \dfrac{1}{10\angle 36.87°} + j100C = 0.08 - j0.06 + j100C$

$\cos^{-1} 0.9 = +25.842°$, $\tan 25.842° = 0.4843 = \dfrac{100C - 0.06}{0.08}$

$\therefore C = \underline{987.5^{-}\ \mu F}$

**10-26**

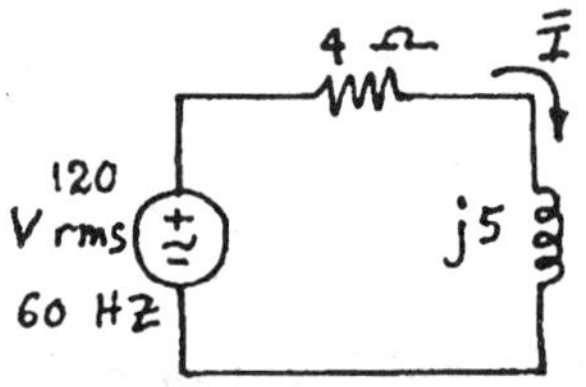


(a) $PF = \cos(\tan^{-1}\frac{5}{4}) = \underline{0.6247\ lag}$

(b) $\bar{I} = \dfrac{120}{4 + j5} \quad \therefore P = \dfrac{120^2}{4^2 + 5^2} \times 4 = \underline{1404.9\ W}$

(c) $\bar{Y}_{in} = \dfrac{1}{4 + j5} + j120\pi C = \dfrac{4}{41} - j\dfrac{5}{41} + j120\pi C$

$\cos^{-1} 0.9 = +25.84°$, $\tan 25.84° = 0.4843 = \dfrac{120\pi C - 5/41}{4/41} \quad \therefore C = \underline{448.8\ \mu F}$

**10-27** (a) $3000\angle -\cos^{-1} 0.707 = 100\bar{I}_1^* \quad \therefore \bar{I}_1^* = 30\angle -45°$, $\bar{I}_1 = 30\angle 45°$

$7000\angle +\cos^{-1} 0.8 = 100\bar{I}_2^* \quad \therefore \bar{I}_2^* = 70\angle 36.87°$, $\bar{I}_2 = 70\angle -36.87°$

$\bar{Z} = 2 + j4 \quad \therefore \bar{I}_3 = \dfrac{100}{2 + j4} = 22.36\angle -63.43°$, $\bar{I}_s = \bar{I}_1 + \bar{I}_2 + \bar{I}_3$

$\therefore \bar{I}_s = 30\angle 45° + 70\angle -36.87° + 22.36\angle -63.43° = 96.28\angle -25.06°$

$PF = \cos 25.06° = \underline{0.9058\ lag}$

(b) $I_{s,\,eff} = \underline{96.28\ A\ rms}$

(c) $\bar{I}_s = 96.28\angle -25.06° = 87.21 - j40.79$, $\cos^{-1} 0.95 = -18.195°$

$\tan(-18.195°) = -0.3287 = \dfrac{-40.79 + 100 \times 120\pi C}{87.21} \quad \therefore C = \underline{322\ \mu F}$

10-28

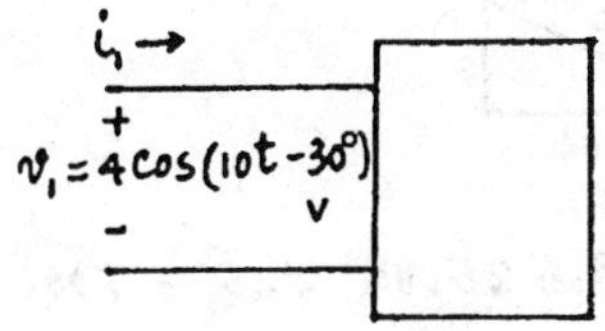


(a) $i_1 = 2\cos(10t + 20°)$ A , $V_{1,eff} = \frac{4}{\sqrt{2}} = \underline{2.828 \text{ V rms}}$

(b) $\bar{I}_{1,rms} = \frac{2}{\sqrt{2}}\angle 20° = \underline{1.4142 \angle 20° \text{ A rms}}$

(c) $P = \frac{1}{2} \times 4 \times 2 \cos 50° = \underline{2.571 \text{ W}}$

(d) $PF = \cos 50° = \underline{0.6428 \text{ lead}}$

(e) $AP = \frac{1}{2} \times 4 \times 2 = \underline{4 \text{ VA}}$

(f) $\bar{S} = 4\angle{-50°} = \underline{2.571 - j3.064 \text{ VA}}$

10-29 (a)

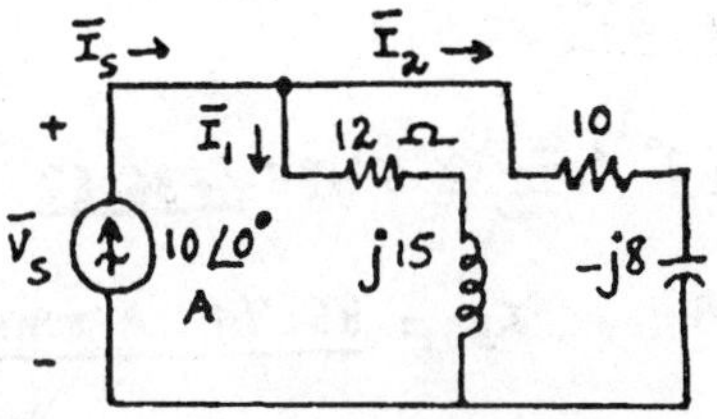


$\bar{I}_1 = 10 \times \frac{10-j8}{22+j7}$ , $\bar{I}_2 = 10 \times \frac{12+j15}{22+j7}$ , $\bar{S}_{12\Omega} = \frac{1}{2}|\bar{I}_1|^2 12$

$\therefore \bar{S}_{12\Omega} = 6 \times 100 \times \frac{164}{484+49} = \underline{184.62 + j0 \text{ VA}}$

$\bar{S}_{j15} = \frac{1}{2}|\bar{I}_1|^2 j15 = j7.5 \times \frac{164}{533} \times 100 = \underline{0 + j230.77 \text{ VA}}$

$\bar{S}_{10} = \frac{1}{2}|\bar{I}_2|^2 10 = 5 \times \frac{144+225}{533} \times 100 = \underline{346.15 + j0 \text{ VA}}$ ; $\bar{S}_{-j8} = \frac{1}{2}|\bar{I}_2|^2(-j8)$

$\therefore \bar{S}_{-j8} = -j400 \times \frac{369}{533} = \underline{0 - j276.92 \text{ VA}}$ ; $\bar{S}_s = -\frac{1}{2}\bar{V}_s \bar{I}_s^* = -\frac{1}{2}\left[(12+j15) \times 10 \times \frac{10-j8}{22+j7}\right]10$

$\therefore \bar{S}_s = -50 \times \frac{240+j54}{22+j7} = \underline{-530.77 + j46.15^+ \text{ VA}}$   [Note: $\Sigma \bar{S} = 0$ ]

(b) $\cos\left(\tan^{-1}\frac{46.15}{530.77}\right) = \underline{0.9962 \text{ lead}}$

10-30

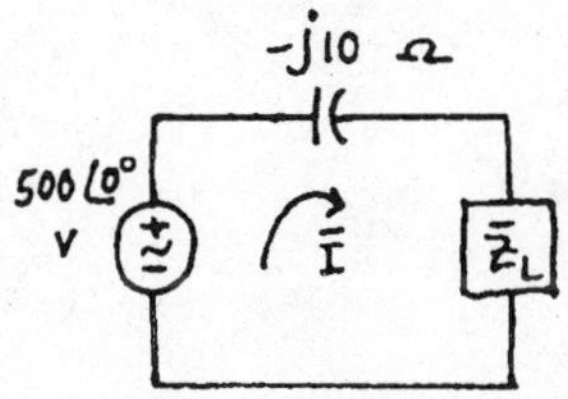


(a) $\bar{S}_{s,gen} = 4000 + j1500 = \frac{1}{2} \times 500 \bar{I}^*$   $\therefore \bar{I}^* = 16 + j6$ A

$\therefore \bar{I} = 16 - j6$ A , $\bar{S}_{-j10} = \frac{1}{2} \times (16^2 + 6^2)(-j10) = -j1460$ VA

$\bar{S}_L = \bar{S}_{s,gen} - \bar{S}_{-j10} = 4000 + j1500 + j1460 = \underline{4000 + j2960 \text{ VA}}$

(b) $PF_L = \cos\left(\tan^{-1}\frac{2960}{4000}\right) = \underline{0.8038 \text{ lag}}$

10-31

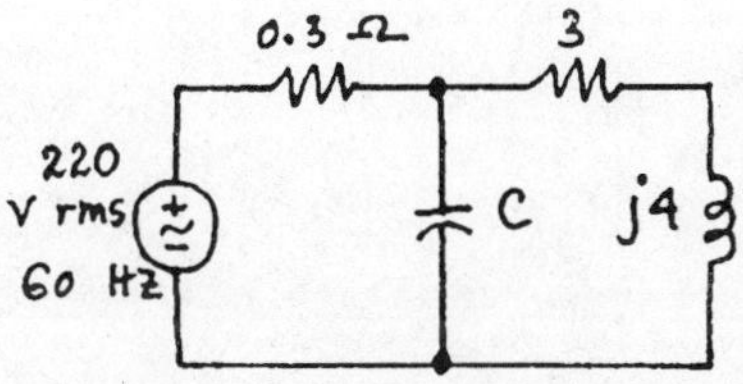


(a) $C = 0$ , $\bar{Z}_R = 3 + j4 = 5\angle 53.13°$ $\therefore PF = \cos 53.13° = \underline{0.6 \text{ lag}}$

(b) $C = 0$ , $\bar{Z}_{in} = 3.3 + j4$ $\therefore PF = \cos\left(\tan^{-1}\frac{4}{3.3}\right) = \underline{0.6364 \text{ lag}}$

(c) $\bar{Y}_{in} = j\omega C + \frac{1}{3+j4} = j\omega C + 0.12 - j0.16$

$\tan(\cos^{-1} 0.95) = -0.3287 = \frac{\omega C - 0.16}{0.12}$

$\therefore C = \underline{319.8 \ \mu F}$

10-32 (a)

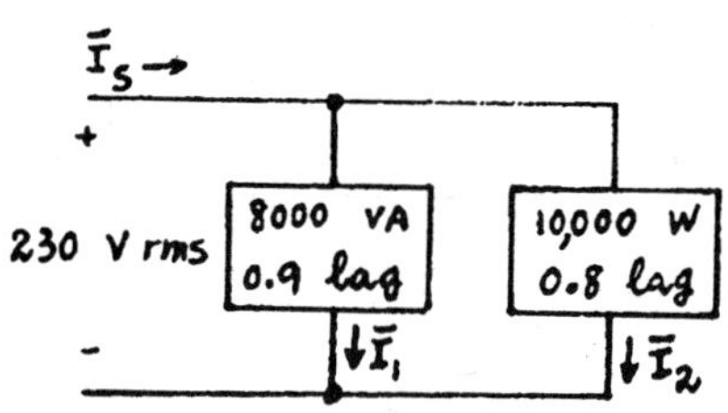


Load #1 :

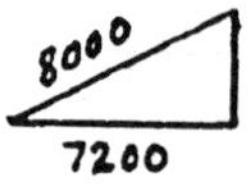


$\cos^{-1} 0.9 = 25.84° \therefore \bar{S}_1 = 8000 \angle 25.84°$ VA

$$\bar{I}_1^* = \frac{\bar{S}_1}{\bar{V}} = \frac{8000 \angle 25.84°}{230 \angle 0°} = 34.78 \angle 25.84° \text{ A}$$

$\therefore \bar{I}_1 = 34.78 \angle -25.84°$ A

Load #2 :

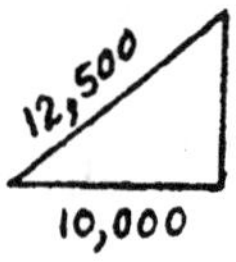


$\cos^{-1} 0.8 = 36.87° \therefore \bar{S}_2 = 12{,}500 \angle 36.87°$

$$\bar{I}_2^* = \frac{12{,}500 \angle 36.87°}{230 \angle 0°} = 54.35 \angle 36.87° \therefore \bar{I}_2 = 54.35 \angle -36.87°$$

$\bar{I}_s = \bar{I}_1 + \bar{I}_2 = 88.74 \angle -32.57°$ A , $I_s = 88.74$ A rms

(b) PF = cos 32.57° = 0.8427 lag

# CHAPTER 11: POLYPHASE CIRCUITS

## Section 11-2 Single-phase three-wire systems

**1** Some current values for the circuit of Fig. 11-1 are $\mathbf{I}_{AD} = 11$ A, $\mathbf{I}_{BE} = -12$ A, $\mathbf{I}_{EF} = 7$ A, $\mathbf{I}_{CF} = 13$ A, $\mathbf{I}_{DG} = 16$ A, $\mathbf{I}_{EH} = -49$ A, $\mathbf{I}_{AE} = -4$ A, and $\mathbf{I}_{HI} = -27$ A. Find $\mathbf{I}_{BC}$, $\mathbf{I}_{EG}$, and $\mathbf{I}_{IE}$.

**2** Given the voltages, $\mathbf{V}_{ax} = 161.8\angle 144^\circ$ V, $\mathbf{V}_{ay} = 161.8\angle 180^\circ$ V, $\mathbf{V}_{wz} = 161.8\angle -108^\circ$ V, and $\mathbf{V}_{yw} = 161.8\angle 36^\circ$ V, draw a phasor diagram showing $\mathbf{V}_{wx}$, $\mathbf{V}_{xy}$, $\mathbf{V}_{yz}$, $\mathbf{V}_{za}$, and $\mathbf{V}_{aw}$.

**3** The 110/220 V rms, 60 Hz, single-phase, three-wire system shown in Fig. 11-2 supplies power to three loads: load *AN* draws 1100 W at unity PF, load *NB* requires 2200 VA at PF = 0.707 lagging, and load *AB* uses a complex power of $5500\angle -30^\circ$ VA. If $\mathbf{V}_{an} = 110\angle 0^\circ$ V, find: (*a*) $\mathbf{I}_{aA}$; (*b*) $\mathbf{I}_{nN}$.

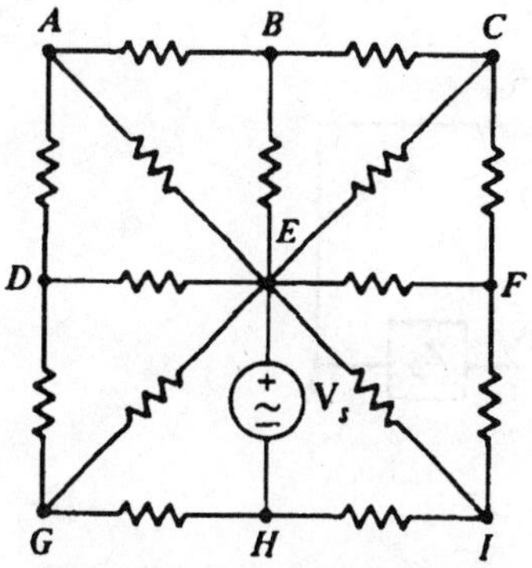


**Fig. 11-1:** See Prob. 1.

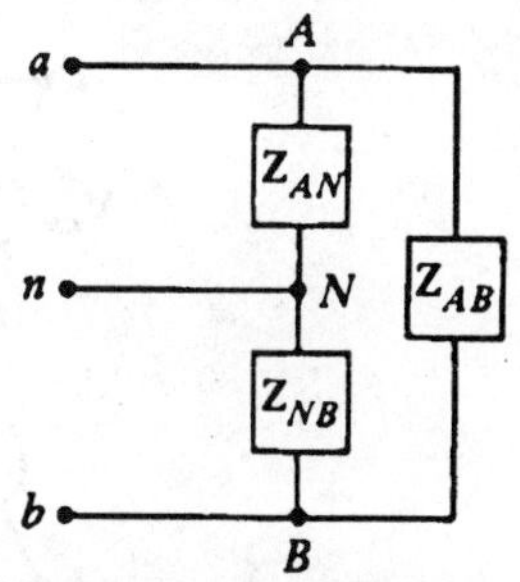


**Fig. 11-2:** See Probs. 3 and 5.

**4** A single-phase, three-wire source provides 220 and 440 V rms to the following loads: $10 + j0\ \Omega$ in one 220-V phase, $3 + j4\ \Omega$ in the other 220-V phase, and $10\angle 36.87^\circ\ \Omega$ at 440 V. Find: (*a*) the effective value of each line current; (*b*) the effective value of the neutral current; (*c*) the total power absorbed by the composite load.

**5** A single-phase, three-wire power source is supplying a load through lines that have line resistances of 1 Ω each, and a neutral wire resistance of 2 Ω. Voltages *at the load* are 120 and 240 V rms. The loads (as in Fig. 11-2) are $\mathbf{Z}_{AN} = 3 + j4\ \Omega$, $\mathbf{Z}_{NB} = 12 - j5\ \Omega$, and $\mathbf{Z}_{AB} = 8 + j0\ \Omega$. Find the rms amplitudes of: (*a*) $\mathbf{I}_{nN}$; (*b*) $\mathbf{I}_{aA}$; (*c*) the source voltage $\mathbf{V}_{an}$.

**6** (*a*) If $\mathbf{Z}_{AN} = 20 + j10\ \Omega$, find the power supplied by each source in the circuit of Fig. 11-3. (*b*) Let $\mathbf{Z}_{AN} = 10 + j20\ \Omega$ and find the power supplied by the upper source.

**7** In the three-wire single-phase system of Fig. 11-4, let $V_1 = 120$ V rms at 60 Hz. (*a*) What size should *C* be to provide a unity power-factor load? (*b*) How many kVA does *C* handle?

## Section 11-3 Three-phase Y-Y connection

**8** The balanced three-phase, three-wire circuit of Fig. 11-5 operates with $\mathbf{V}_{BC} = 100\angle 20^\circ$ V rms with positive phase sequence. The total load draws 3 kW at PF = 0.8 lagging. If $R_w = 0.6\ \Omega$, find: (*a*) the total power lost in line resistances; (*b*) $\mathbf{V}_{ab}$.

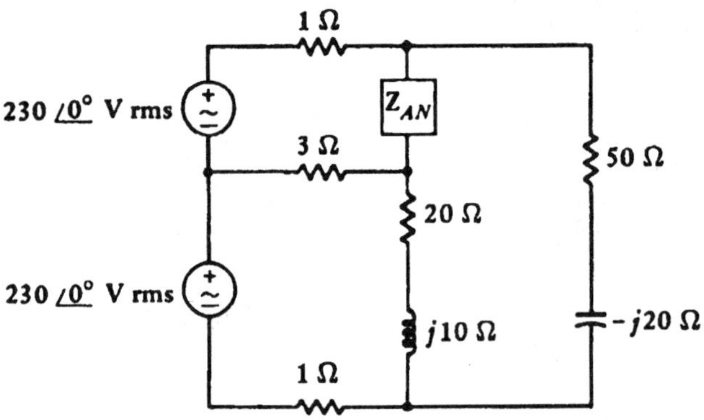


Fig. 11-3: See Prob. 6.

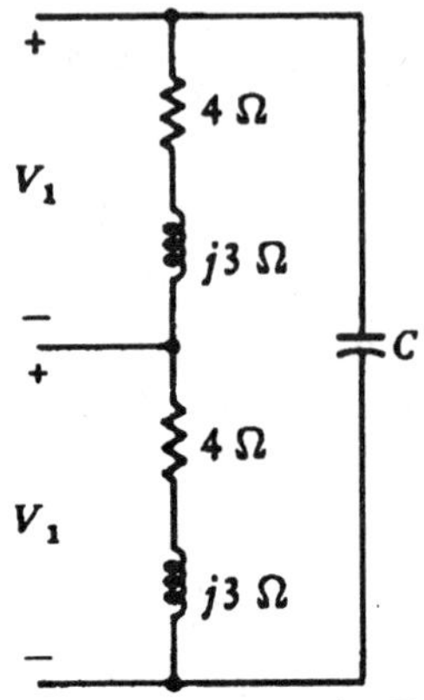


Fig. 11-4: See Prob. 7.

**9** Refer to the balanced Y-Y circuit of Fig. 11-5 and let $\mathbf{V}_{an} = 4600\underline{/0^\circ}$ V rms with (+) phase sequence. Let the source supply $\mathbf{S} = 240 + j60$ kVA with $R_w = 3.2\ \Omega$. Find: (*a*) $\mathbf{V}_{AN}$; (*b*) $\mathbf{I}_{aA}$; (*c*) $\mathbf{Z}_p$; (*d*) the transmission efficiency.

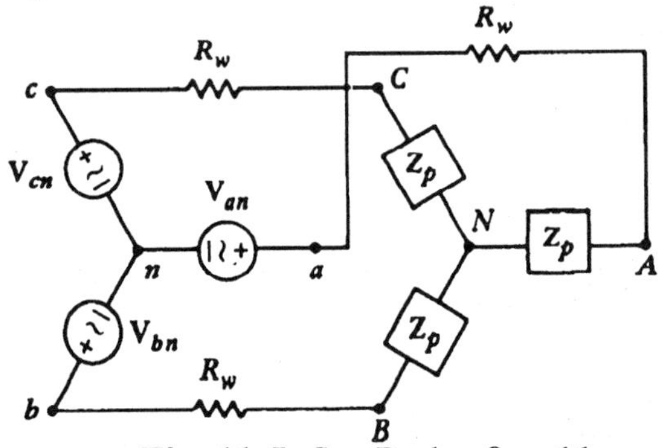


Fig. 11-5: See Probs. 8 to 11.

**10** In the system shown in Fig. 11-5, $\mathbf{Z}_p = 20 + j8\ \Omega$, and (+) phase sequence is assumed. If $\mathbf{I}_{aA} = 20\underline{/-46^\circ}$ A rms, and the source is operating with PF = 0.94 lagging, find: (*a*) $R_w$; (*b*) the complex power supplied by the source; (*c*) $\mathbf{V}_{AB}$; (*d*) $\mathbf{V}_{ab}$.

**11** In Fig. 11-5, let $\mathbf{Z}_p$ represent a capacitive impedance of $80\underline{/-24^\circ}\ \Omega$ in parallel with 0.25 H. If $\mathbf{V}_{an} = 230\underline{/0^\circ}$ V rms at 60 Hz, and $R_w = 2.5\ \Omega$, determine: (*a*) $\mathbf{I}_{aA}$; (*b*) the total power delivered to the load; (*c*) the total power lost in the wire resistance; (*d*) the PF at which the source operates.

**12** An unbalanced three-phase, four-wire system is shown in Fig. 11-6. Assume positive phase sequence with $\mathbf{V}_{an} = 100\underline{/0^\circ}$ V rms, and let the loads be $\mathbf{Z}_{AN} = 10 - j5\ \Omega$, $\mathbf{Z}_{BN} = 8 + j5\ \Omega$, and $\mathbf{Z}_{CN} = 12 + j0\ \Omega$. Find $\mathbf{I}_{nN}$.

**13** A Y-connected load has an impedance of $25 + j12\ \Omega$ in each phase. It is desired to raise the load power factor to 0.95 lagging. (*a*) What size capacitor should be placed in parallel with each phase impedance if $f = 60$ Hz? (*b*) Let the line voltage at the load have an amplitude of 500 V rms. How many kVAR (capacitive) are required at the load to obtain PF = 0.95 lagging?

## Section 11-4 The delta (Δ) connection

**14** The balanced three-wire, three-phase circuit of Fig. 11-7 has $R_w = 0$ and $\mathbf{V}_{an} = 288.7\angle{-30^\circ}$ V rms. Each phase of the load absorbs 2.4 kW at 0.8 PF leading. Assume *abc* sequence. Find: (*a*) $\mathbf{V}_{ab}$; (*b*) $\mathbf{Z}_p$; (*c*) $\mathbf{I}_{cC}$; (*d*) the total load power.

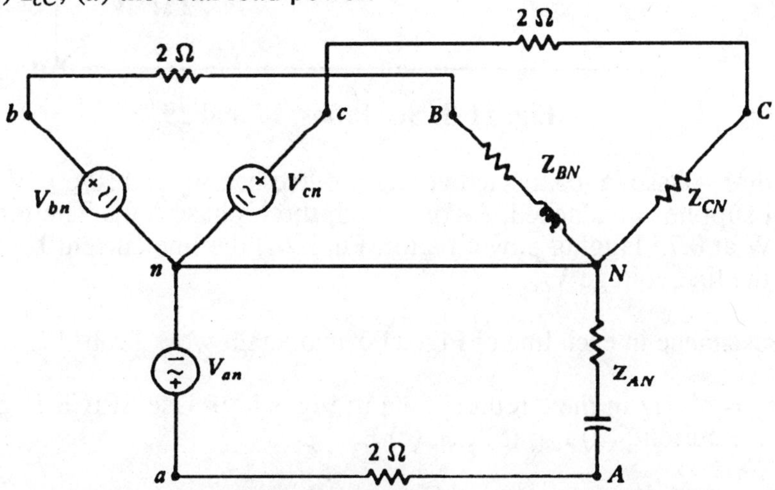


**Fig. 11-6:** See Prob. 12.

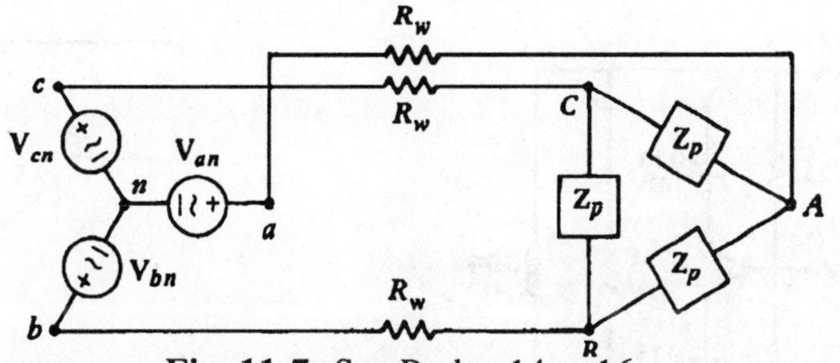


**Fig. 11-7:** See Probs. 14 to 16.

**15** The balanced three-phase load shown in Fig. 11-7 requires a total power of 6 kW at PF = 0.83 lagging. If $R_w = 0.8\ \Omega$ and $\mathbf{V}_{CA} = 100\angle{20^\circ}$ V rms with (+) phase sequence, find: (*a*) $\mathbf{V}_{ab}$; (*b*) the total complex power being supplied by the source.

**16** Let $R_w = 4\ \Omega$ in the balanced three-phase circuit of Fig. 11-7. The load draws a total complex power, $\mathbf{S} = 2400 + j2100$ VA, and an additional 300 W is dissipated in the lines. Determine the rms amplitude of: (*a*) $\mathbf{I}_{aA}$; (*b*) $\mathbf{I}_{AB}$; (*c*) $\mathbf{V}_{an}$.

**17** The source in Fig. 11-8 is balanced and has positive phase sequence with $f = 60$ Hz. Find $\mathbf{I}_{aA}$, $\mathbf{I}_{bB}$, $\mathbf{I}_{cC}$, and the power furnished by each phase of the source.

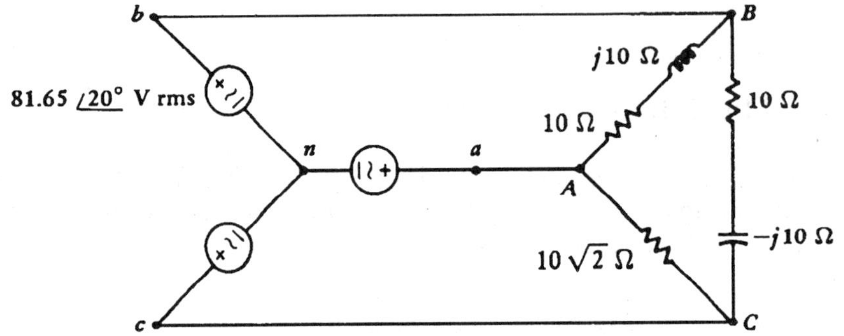


**Fig. 11-8:** See Probs. 17 and 19.

**18** A balanced, three-phase, Y-connected voltage source has $\mathbf{V}_{an} = 110\angle 60^\circ$ V rms with (+) phase sequence, and it supplies a balanced, Δ-connected, three-phase load. The total power drawn by the load is 12 kW at 0.75 lagging power factor. Find: (*a*) the line current $\mathbf{I}_{bB}$; (*b*) the load phase current $\mathbf{I}_{CA}$; (*c*) the line voltage $\mathbf{V}_{bc}$.

**19** Add 1 Ω of resistance in each line of Fig. 11-8 and again work Prob. 17.

**20** (SPICE) Let $f = 60$ Hz in the circuit shown in Fig. 11-9. Use SPICE to determine the (rms) phasor value of the current: (*a*) $\mathbf{I}_{aA}$; (*b*) $\mathbf{I}_{bB}$; (*c*) $\mathbf{I}_{nN}$.

**21**(SPICE) In the three-phase system of Fig. 11-10, assume a balanced source with (+) phase sequence. Let $\mathbf{V}_{AB} = 220\angle 0^\circ$ V rms at $f = 60$ Hz. Find $\mathbf{V}_{BN}$.

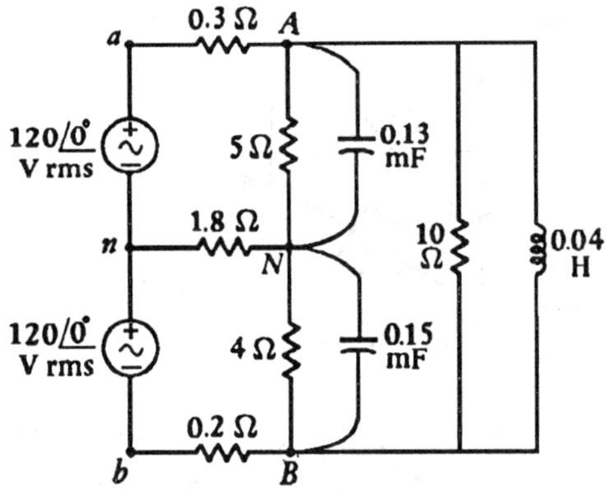


**Fig. 11-9:** See Prob. 20.

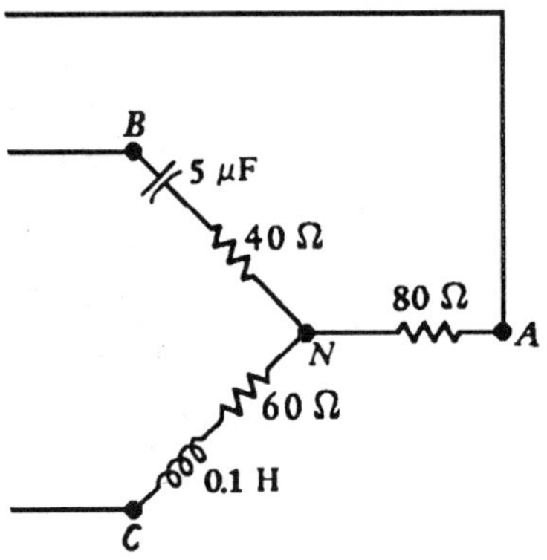


**Fig. 11-10:** See Prob. 21.

# Chapter 11

**11-1**

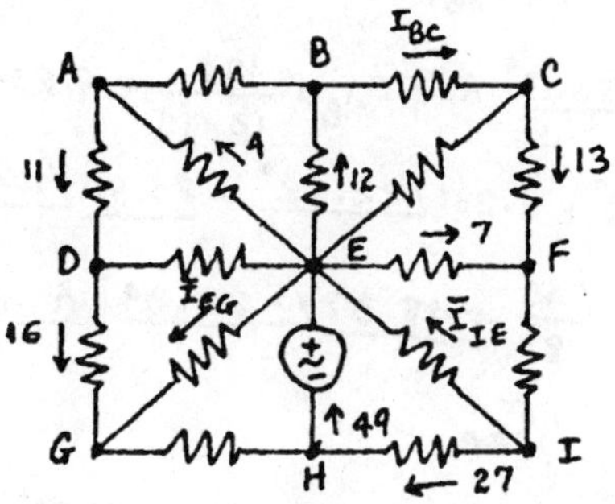


$\bar{I}_{AB} = 4 - 11 = -7 \quad \therefore \bar{I}_{BC} = 12 - 7 = \underline{5\ A}$

$\bar{I}_{HG} = 27 - 49 = -22 \quad \therefore \bar{I}_{EG} = -16 + 22 = \underline{6\ A}$

$\bar{I}_{FI} = 7 + 13 = 20 \quad \therefore \bar{I}_{IE} = 20 - 27 = \underline{-7\ A}$

**11-2**

$\bar{V}_{ax} = 161.8\angle 144^\circ \; ; \; \bar{V}_{ay} = 161.8\angle 180^\circ \; ; \; \bar{V}_{wz} = 161.8\angle -108^\circ \; ; \; \bar{V}_{yw} = 161.8\angle 36^\circ$

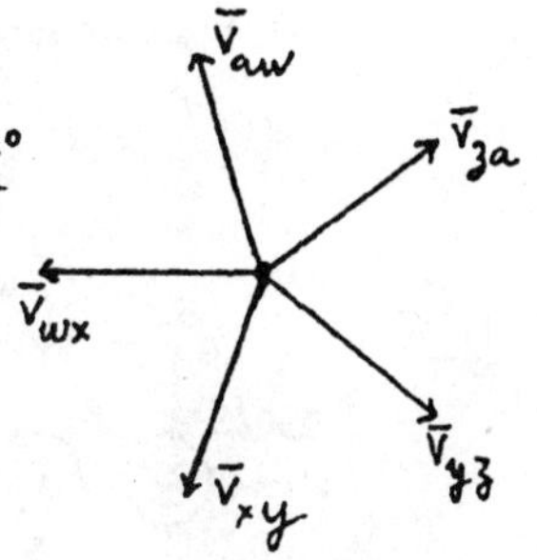


$\bar{V}_{wx} = \bar{V}_{wy} + \bar{V}_{ya} + \bar{V}_{ax} = 161.8(-1\angle 36^\circ + 1 + 1\angle 144^\circ) = \underline{100\angle 180^\circ}$

$\bar{V}_{xy} = \bar{V}_{xw} + \bar{V}_{wy} = 100 - 161.8\angle 36^\circ = \underline{100\angle -108^\circ}$

$\bar{V}_{yz} = \bar{V}_{yw} + \bar{V}_{wz} = 161.8(1\angle 36^\circ + 1\angle -108^\circ) = \underline{100\angle -36^\circ}$

$\bar{V}_{za} = \bar{V}_{zy} + \bar{V}_{ya} = 100\angle 144^\circ + 161.8 = \underline{100\angle 36^\circ}$

$\bar{V}_{aw} = \bar{V}_{ax} + \bar{V}_{xw} = 161.8\angle 144^\circ + 100 = \underline{100\angle 108^\circ}$

**11-3** (a)

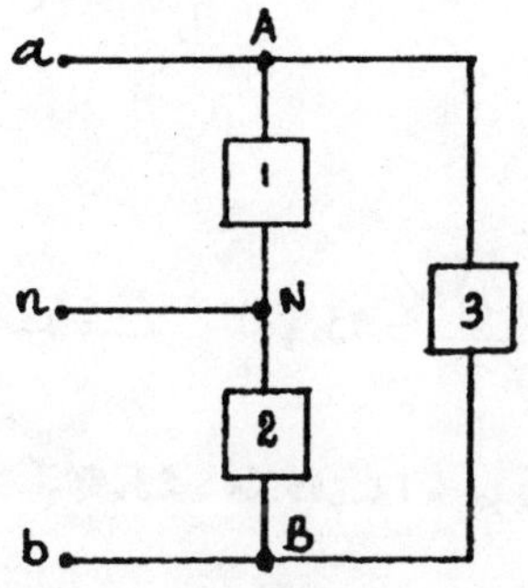


$\bar{V}_{an} = \bar{V}_{nb} = 110\angle 0^\circ$ V rms

$\bar{S}_1 = 1100 + j0 = 110\,\bar{I}^*_{AN} \quad \therefore \bar{I}_{AN} = 10\angle 0^\circ$ A

$\bar{S}_2 = 2200\angle 45^\circ = 110\,\bar{I}^*_{NB} \quad \therefore \bar{I}_{NB} = 20\angle -45^\circ$ A

$\bar{S}_3 = 5500\angle -30^\circ = 220\,\bar{I}^*_{AB} \quad \therefore \bar{I}_{AB} = 25\angle 30^\circ$ A

$\bar{I}_{aA} = 10 + 25\angle 30^\circ = \underline{34.03\angle 21.55^\circ \text{ A rms}}$

(b) $\bar{I}_{nN} = \bar{I}_{NB} - \bar{I}_{AN} = 20\angle -45^\circ - 10 = \underline{14.736\angle -73.68^\circ \text{ A rms}}$

**11-4**

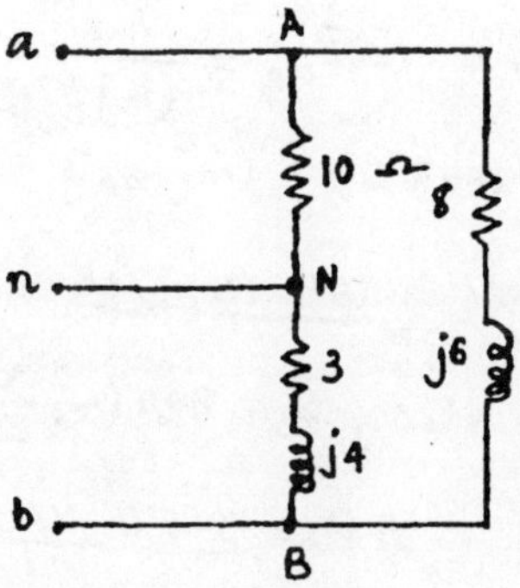


$\bar{V}_{an} = \bar{V}_{nb} = 220\angle 0^\circ \quad \therefore \bar{I}_{AN} = 22\angle 0^\circ \; , \; \bar{I}_{NB} = 44\angle -53.13^\circ$

$\bar{V}_{AB} = 440\angle 0^\circ \quad \therefore \bar{I}_{AB} = 44\angle -36.87^\circ$

(a) $\bar{I}_{aA} = \bar{I}_{AN} + \bar{I}_{AB} = 22 + 44\angle -36.87^\circ = \underline{63.00\angle -24.78^\circ \text{ A rms}}$

$\bar{I}_{bB} = -44\angle -53.1^\circ - 44\angle -36.9^\circ = \underline{87.12\angle 135^\circ \text{ A rms}}$

(b) $\bar{I}_{nN} = \bar{I}_{NB} - \bar{I}_{AN} = 44\angle -53.13^\circ - 22 = \underline{35.47\angle -82.87^\circ \text{ A rms}}$

(c) $P_{tot} = 22^2 \times 10 + 44^2 \times 3 + 44^2 \times 8 = \underline{26.136 \text{ kW}}$

11-5

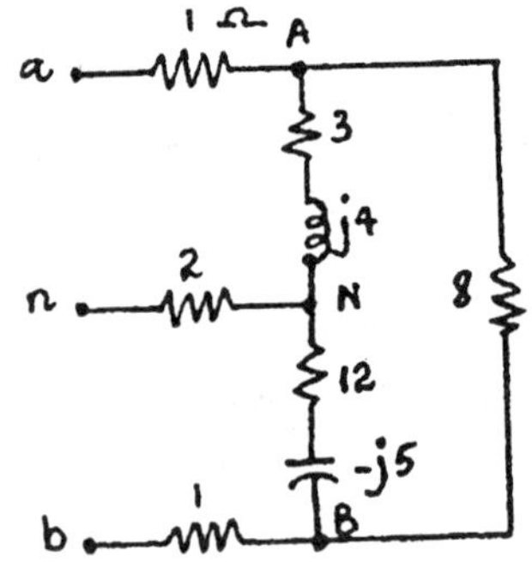


$|\bar{V}_{AN}| = 120 \text{ V} = |\bar{V}_{NB}| \;;\; |\bar{V}_{AB}| = 240 \text{ V}$

(a) $\bar{I}_{AN} = \dfrac{120}{3+j4} = 24\angle{-53.13^\circ} \text{ A}$ , $\bar{I}_{NB} = \dfrac{120}{12-j5} = 9.231\angle{22.62^\circ} \text{ A}$

$\therefore \bar{I}_{nN} = -24\angle{-53.13^\circ} + 9.231\angle{22.62^\circ} = \underline{23.50\angle{104.49^\circ} \text{ A rms}}$

(b) $\bar{I}_{aA} = 24\angle{-53.13^\circ} + \dfrac{240}{8} = \underline{48.37\angle{-23.39^\circ} \text{ A rms}}$

(c) $\bar{V}_{an} = 2\bar{I}_{Nn} + 120 + 1\,\bar{I}_{aA}$

$\therefore \bar{V}_{an} = -2 \times 23.50\angle{104.49^\circ} + 120 + 48.37\angle{-23.39^\circ} = \underline{187.66\angle{-20.17^\circ} \text{ V rms}}$

11-6

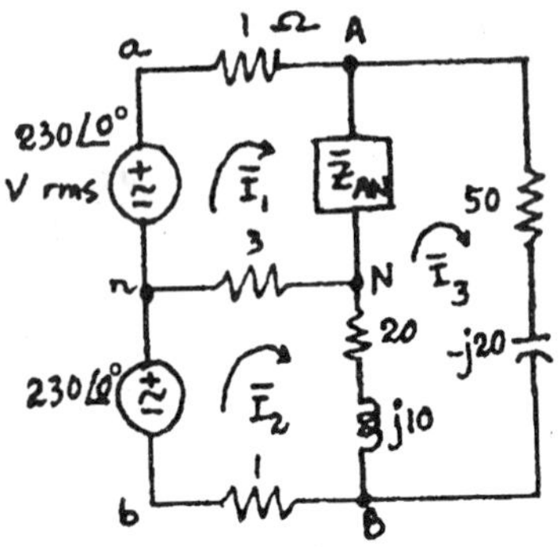


(a) $\bar{Z}_{AN} = 20 + j10 \;\; \therefore \text{sym} \;\; \therefore \bar{I}_{nN} = 0$

Load on $460\angle{0^\circ}$ V is $2 + \dfrac{(40+j20)(50-j20)}{90} = \dfrac{258+j20}{9}$

$\therefore \bar{I}_{aA} = \dfrac{460}{(258+j20)/9}$

$\therefore P_{230} = \dfrac{460^2 \times 9^2}{258^2 + 20^2} \times \dfrac{258}{9} \times \dfrac{1}{2} = \underline{3668.7 \text{ W}}$

(b) $\bar{Z}_{AN} = 10 + j20$ use mesh currents

$$\bar{I}_1 = \frac{230\begin{vmatrix} 1 & -3 & -10-j20 \\ 1 & 24+j10 & -20-j10 \\ 0 & -20-j10 & 80+j10 \end{vmatrix}}{\begin{vmatrix} 14+j20 & -3 & -10-j20 \\ -3 & 24+j10 & -20-j10 \\ -10-j20 & -20-j10 & 80+j10 \end{vmatrix}} = \frac{230(176+j117)}{1896+j2967} = 13.805^+ \angle{-23.805^\circ}{}^+$$

$\therefore P_{upper} = 230 \times 13.805 \cos 23.805^\circ = \underline{2905^+ \text{ W}}$

11-7

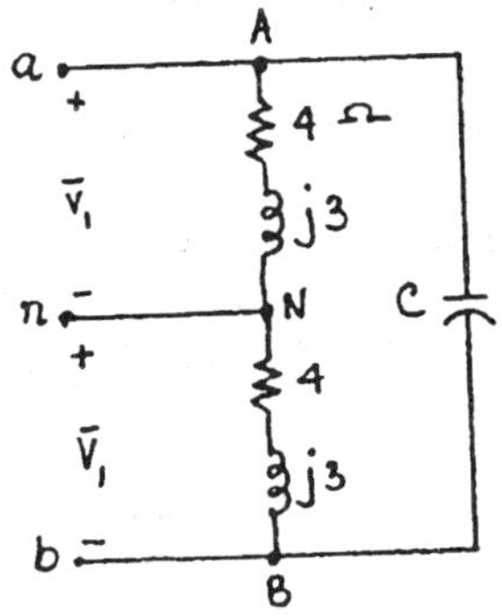


$\bar{V}_1 = 120 \text{ V rms}$ , 60 Hz

(a) Balanced $\therefore \bar{I}_{nN} = 0 \;\; \therefore \bar{I}_{aA} = \dfrac{\bar{V}_1}{4+j3} + j2\bar{V}_1\omega C$

$\therefore \bar{I}_{aA} = \bar{V}_1\left(\dfrac{4-j3}{25} + j240\pi C\right) ; \; \text{Im}[\bar{I}_{aA}] = 0$

$\therefore 240\pi C = \dfrac{3}{25}$ , $C = \underline{159.16 \; \mu\text{F}}$

(b) $\bar{I}_c = j240\pi C\,\bar{V}_1 = j\dfrac{3}{25} \times 120 \;;\; \bar{S}_c = 240\left(-j\dfrac{3}{25} \times 120\right)$

$\therefore \bar{S}_c = -j3456 \text{ VA} \qquad \therefore AP = \underline{3456 \text{ VA}}$

**11-8**

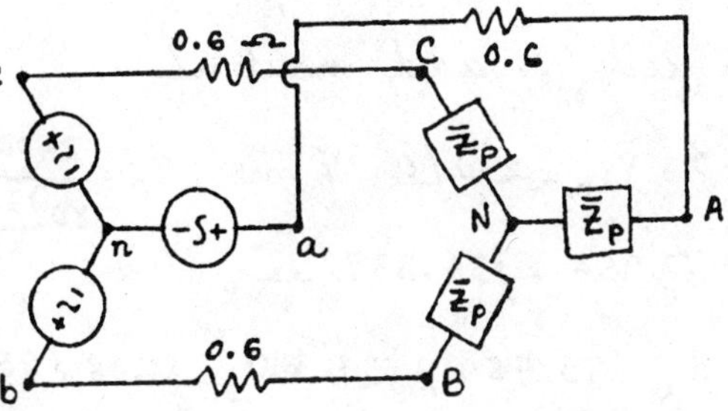


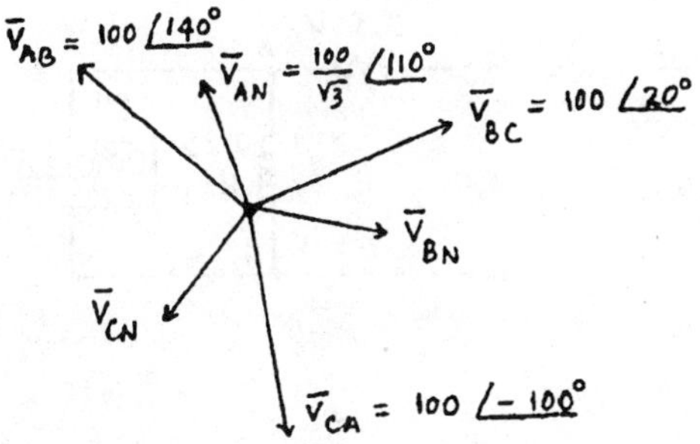


Total : 3 KW , 0.8 lag ; $\bar{V}_{BC} = 100\angle 20^\circ$ V rms, (+) seq

(a) $\bar{V}_{AN} = \frac{100}{\sqrt{3}}\angle 110^\circ$ ; $\left(\frac{100}{\sqrt{3}}\angle 110^\circ\right)\bar{I}_{AN}^{*} = \frac{1}{3} \times \frac{3000}{0.8}\angle 36.87^\circ$ $\therefore$ $\bar{I}_{AN} = 21.65^{+}\angle 73.13^\circ$

$\therefore P_{wire} = 3 \times 21.65^2 \times 0.6 = \underline{843.75 \text{ W}}$

(b) $\bar{V}_{ab} = 0.6\,\bar{I}_{AN} + \bar{V}_{AB} - 0.6\,\bar{I}_{BN}$ ; $\bar{I}_{BN} = \bar{I}_{AN}\angle -120^\circ = 21.65\angle -46.87^\circ$

$\therefore \bar{V}_{ab} = 0.6 \times 21.65\angle 73.13^\circ + 100\angle 140^\circ - 0.6 \times 21.65\angle -46.87^\circ = \underline{118.77\angle 133.47^\circ \text{ V rms}}$

**11-9** (a)

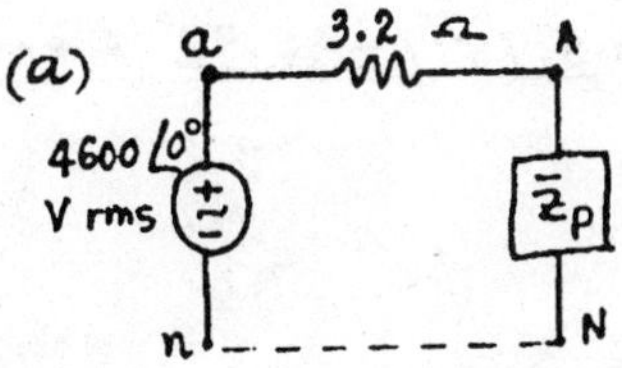


Balanced $\therefore$ may add neutral

$\frac{1}{3}\bar{S} = 80{,}000 + j20{,}000 = 4600\,\bar{I}_{aA}^{*}$ $\therefore$ $\bar{I}_{aA} = 17.927\angle -14.036^\circ$ A rms

$\bar{V}_{AN} = 4600 - 3.2 \times 17.927\angle -14.036^\circ = \underline{4544\angle 0.175^\circ \text{ V rms}}$

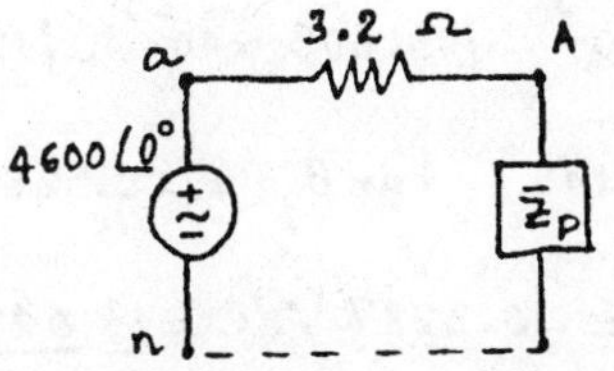


(b) $\bar{I}_{aA} = \underline{17.927\angle -14.036^\circ \text{ A rms}}$

(c) $\bar{Z}_p = \frac{\bar{V}_{AN}}{\bar{I}_{aA}} = \frac{4544\angle 0.175^\circ}{17.927\angle -14.036^\circ} = \underline{245.7 + j62.23\ \Omega}$

(d) eff $= \frac{245.7}{245.7 + 3.2} \times 100\% = \underline{98.71\%}$

**11-10** (a)

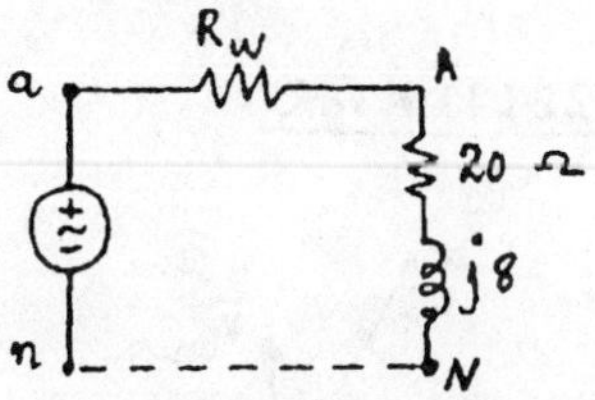


Balanced $\therefore$ may add neutral

$\bar{I}_{aA} = 20\angle -46^\circ$ A rms ; (+) seq , $PF_S = 0.940$ lag

$\therefore \theta = 19.948^\circ$ , $\tan\theta = 0.3630 = \frac{8}{R_W + 20}$ $\therefore R_W = \underline{2.042\ \Omega}$

(b) $\bar{S} = 3 \times 20^2(22.042 + j8) = \underline{26.45^{-} + j9.6 \text{ KVA}}$

(c) $\bar{V}_{AN} = (20\angle -46^\circ)(20 + j8) = 430.8\angle -24.20^\circ$ ; $\bar{V}_{an} = (20\angle -46^\circ)(22.042 + j8)$

$\therefore \bar{V}_{an} = 469.0\angle -26.05^\circ$ ; $\bar{V}_{AB} = \sqrt{3}\,\bar{V}_{AN}\angle 30^\circ = \underline{746.2\angle 5.80^\circ \text{ V rms}}$

(d) $\bar{V}_{ab} = \sqrt{3}\,\bar{V}_{an}\angle 30^\circ = \underline{812.3\angle 3.95^\circ \text{ V rms}}$

11-11

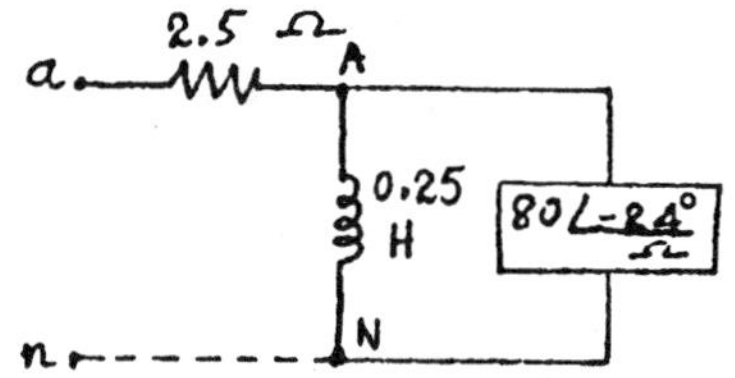


Balanced ∴ add neutral

$f = 60$ ; $\bar{V}_{an} = 230\angle 0°$ V rms ; $\bar{Z}_p = \dfrac{2400\pi\angle 90° - 24°}{80\angle -24° + j30\pi}$

$\bar{Z}_p = 70.954 + j34.337\ \Omega$

$\bar{Z}_p + R_w = 73.454 + j34.337 = \underline{81.08\angle 25.05°}\ \Omega$

(a) $\bar{I}_{aA} = \dfrac{\bar{V}_{an}}{\bar{Z}_p + R_w} = \dfrac{230}{81.08\angle 25.05°} = \underline{2.837\angle -25.05°\ \text{A rms}}$

(b) $P_{tot} = 3 \times 2.837^2 \times 70.954 = \underline{1712.8\ \text{W}}$

(c) $P_{tot,w} = 3 \times 2.837^2 \times 2.5 = \underline{60.35\ \text{W}}$

(d) $PF_s = \cos 25.05° = \underline{0.9059\ \text{lag}}$

11-12

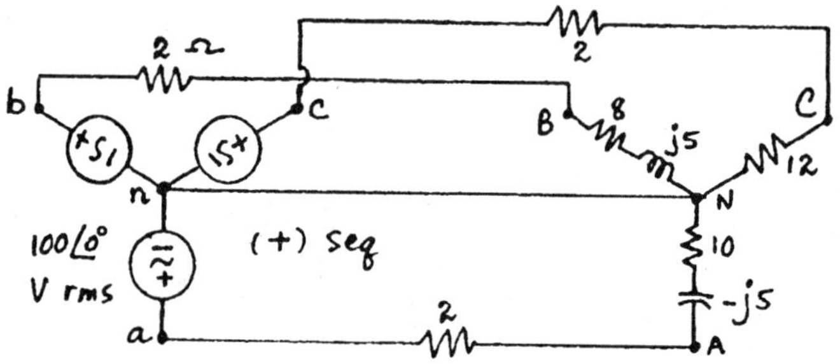


$\bar{I}_{nN} = -\left(\dfrac{100}{12-j5} + \dfrac{100\angle -120°}{10+j5} + \dfrac{100\angle 120°}{14}\right)$

$\bar{I}_{nN} = \underline{5.767\angle -46.98°\ \text{A rms}}$

11-13 (a)

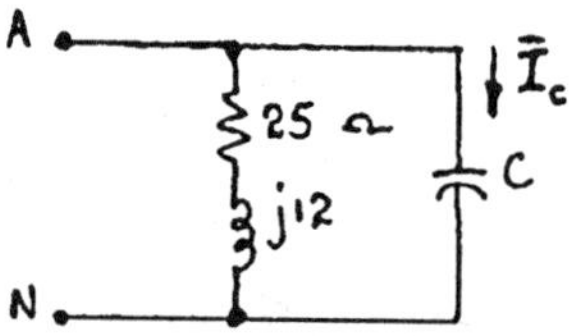


$\bar{Y}_p = \dfrac{1}{25+j12} + j\omega C = 32.51\times 10^{-3} - j15.605\times 10^{-3} + j120\pi C$

PF = 0.95 lag ∴ $\theta_y = -18.195°$, $\tan\theta_y = -0.3287$

$\tan\theta_y = \dfrac{120\pi C \times 10^3 - 15.605}{32.51} = -0.3287$ ∴ $C = \underline{13.049\ \mu F}$

(b) $X_c = \dfrac{10^6}{120\pi \times 13.049} = 203.3\ \Omega$ ; $I_c = \dfrac{500}{\sqrt{3} \times 203.3} = 1.4201$ A

$Q_c = 3 \times 1.4201^2 \times 203.3 = 1229.7$ VAR $= \underline{1.2297\ \text{KVAR}}$

11-14

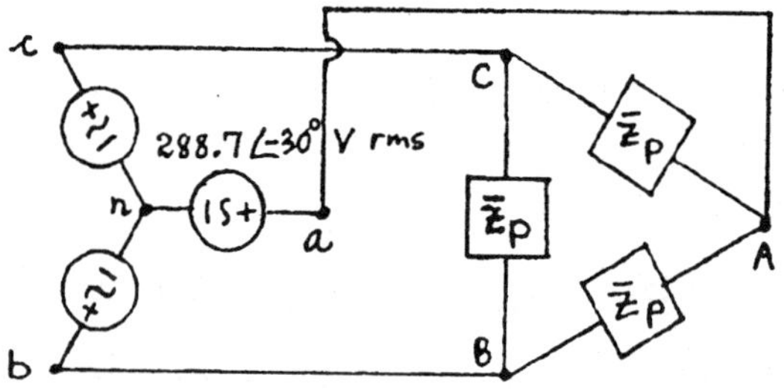


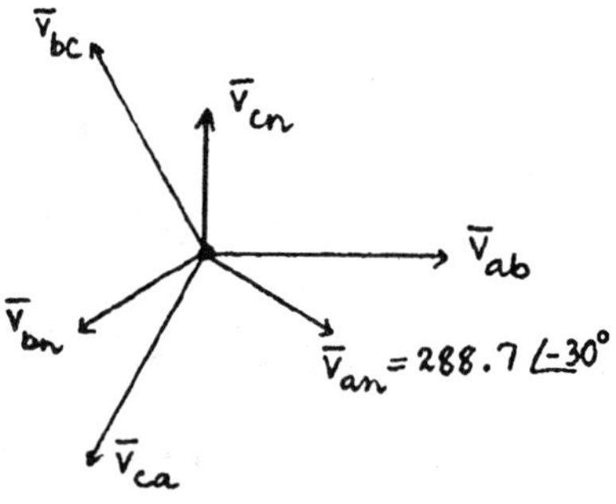


Per phase: 2400 W, 0.8 lead

(cont. on next page)

**11-14 (cont.)**

(a) $\bar{V}_{ab} = 288.7\sqrt{3}\angle 0^\circ = \underline{500\angle 0^\circ \text{ V rms}}$

(b) $\bar{S}_p = \frac{2400}{0.8}\angle -36.87^\circ = 500\angle 0^\circ\, \bar{I}^*_{AB}$ $\therefore \bar{I}_{AB} = 6\angle 36.87^\circ$ A rms ; $\bar{Z}_p = \frac{\bar{V}_{ab}}{\bar{I}_{AB}}$

$\bar{Z}_p = \frac{500\angle 0^\circ}{6\angle 36.87^\circ} = 83.33\angle -36.87^\circ = \underline{66.67 - j50\ \Omega}$

(c) $\bar{I}_{cC} = \bar{I}_{CA} - \bar{I}_{BC} = 6\angle 36.87^\circ + 120^\circ - 6\angle 36.87^\circ - 120^\circ = \underline{10.392\angle 126.87^\circ \text{ A rms}}$

(d) $P_{tot} = 3 \times 2400 = \underline{7.20 \text{ KW}}$

**11-15**

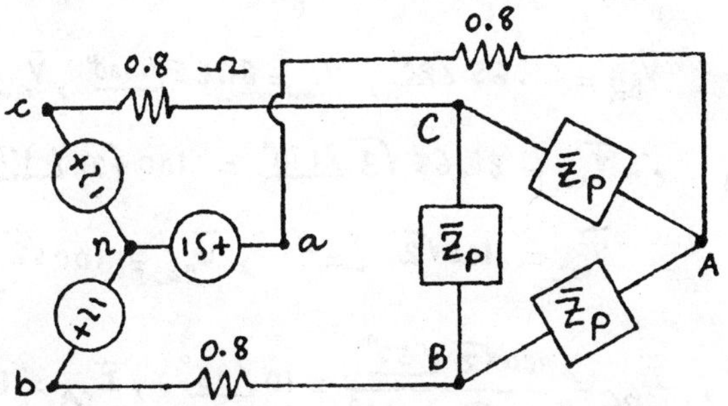


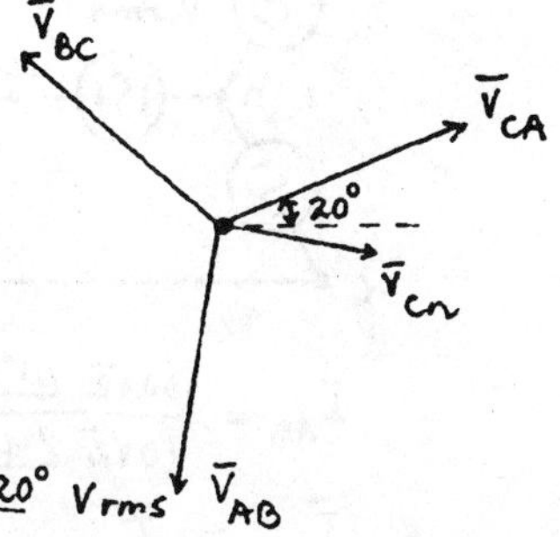


Per phase 2000 W, 0.83 lag ; $\bar{V}_{CA} = 100\angle 20^\circ$ Vrms

(a) $\bar{S}_p = 2000 + j2000\tan(\cos^{-1}0.83) = 2000 + j1344.0 = 100\angle 20^\circ\, \bar{I}^*_{CA}$

$\therefore \bar{I}_{CA} = 24.10\angle -13.901^\circ$, $\bar{I}_{BC} = 24.10\angle 106.10^\circ$ $\therefore \bar{I}_{cC} = \bar{I}_{CA} - \bar{I}_{BC} = 41.74\angle -43.90^\circ$

$\bar{I}_{bB} = 41.74\angle 76.10^\circ$, $\bar{I}_{aA} = 41.74\angle -163.90^\circ$

$\bar{V}_{ab} = 0.8(\bar{I}_{aA} - \bar{I}_{bB}) + \bar{V}_{AB} = 0.8(41.74)(1\angle -163.90^\circ - 1\angle 76.10^\circ) + 100\angle -100^\circ$

$\therefore \bar{V}_{ab} = \underline{151.47\angle -112.29^\circ \text{ V rms}}$

(b) $P_w = 3 \times 41.74^2 \times 0.8 = 4180.5^+$ W $\therefore \bar{S}_{source} = 4181 + 6000 + j3 \times 1344.0$

$= \underline{10,181 + j4032 \text{ VA}}$

**11-16**

(a)

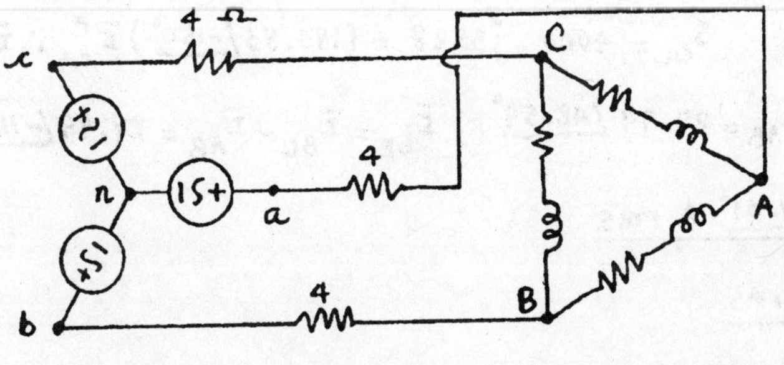


$\bar{S}_{Load} = 2400 + j2100$ VA

$\bar{S}_w = 300 + j0$ VA

$\frac{300}{3} = 4|\bar{I}_L|^2$ $\therefore I_L = 5$ A rms

$I_L = |\bar{I}_{aA}| = \underline{5 \text{ A rms}}$

(b) $I_p = |\bar{I}_{AB}| = \frac{I_L}{\sqrt{3}} = \frac{5}{\sqrt{3}} = \underline{2.887 \text{ A rms}}$ (cont. on next page)

11-16 (cont.) (c) Assume $\bar{I}_{AB} = 2.887\angle 0°$ and (+) seq $\therefore \frac{2400}{3} + j\frac{2100}{3} = \bar{V}_{AB} \times 2.887$

$\therefore \bar{V}_{AB} = 368.2\angle 41.19°$, $\bar{I}_{CA} = \bar{I}_{AB}\angle 120° = 2.887\angle 120°$; $\bar{I}_{aA} = \bar{I}_{AB} - \bar{I}_{CA}$

$\therefore \bar{I}_{aA} = 2.887\angle 0° + 2.887\angle -60° = 5\angle -30°$ A ; $\bar{V}_{ab} = 4(\bar{I}_{aA} - \bar{I}_{bB}) + \bar{V}_{AB}$

$\therefore \bar{V}_{ab} = 4(5\angle -30° - 5\angle -150°) + 368.2\angle 41.19° = 395.0\angle 37.88°$ $\therefore |\bar{V}_{an}| = \frac{395}{\sqrt{3}} = \underline{228 \text{ V rms}}$

11-17

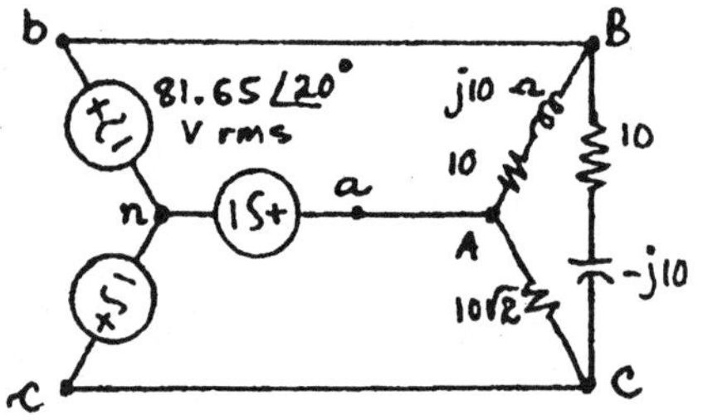


$f = 60$, (+) seq

$\bar{V}_{bn} = 81.65\angle 20°$, $\bar{V}_{cn} = 81.65\angle -100°$, $\bar{V}_{an} = 81.65\angle 140°$

$\therefore \bar{V}_{ab} = 81.65\sqrt{3}\angle 170° = 100\sqrt{2}\angle 170°$

$\bar{V}_{bc} = 100\sqrt{2}\angle 50°$, $\bar{V}_{ca} = 100\sqrt{2}\angle -70°$

$\bar{I}_{AB} = \frac{100\sqrt{2}\angle 170°}{10\sqrt{2}\angle 45°} = 10\angle 125°$, $\bar{I}_{BC} = \frac{100\sqrt{2}\angle 50°}{10\sqrt{2}\angle -45°} = 10\angle 95°$, $\bar{I}_{CA} = 10\angle -70°$

$\bar{I}_{aA} = \bar{I}_{AB} - \bar{I}_{CA} = 10\angle 125° - 10\angle -70° = \underline{19.829\angle 117.5° \text{ A rms}}$

$\bar{I}_{bB} = 10\angle 95° - 10\angle 125° = \underline{5.176\angle 20°}$ A rms, $\bar{I}_{cC} = 10\angle -70° - 10\angle 95° = \underline{19.829\angle -77.5° \text{ A rms}}$

$P_{an} = 81.65 \times 19.829\cos(140° - 117.5°) = \underline{1495.8 \text{ W}}$, $P_{bn} = 81.65 \times 5.176\cos(20° - 20°) = \underline{422.7 \text{ W}}$

$P_{cn} = 81.65 \times 19.829\cos(-100° + 77.5°) = \underline{1495.8 \text{ W}}$ (Check: $\Sigma P = 3414$ W; $10^2 \times 10 + 10^2 \times 10 + 10^2 \times 10\sqrt{2} = 3414$ W)

11-18

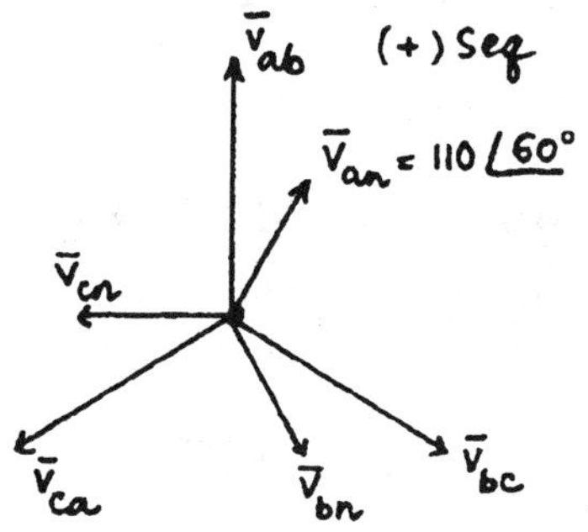


$\bar{S} = \frac{12{,}000}{0.75}\angle \cos^{-1}0.75 = 16{,}000\angle 41.41° = 12000 + j10{,}583$

$\bar{S}_p = \frac{1}{3}\bar{S} = 4000 + j3528$ VA

$\bar{V}_{bc} = \bar{V}_{BC} = 110\sqrt{3}\angle -30° = 190.53\angle -30°$ V rms

$\bar{V}_{ca} = \bar{V}_{CA} = 190.53\angle -150°$

$\bar{S}_{CA} = 4000 + j3528 = (190.53\angle -150°)\bar{I}^*_{CA}$ $\therefore \bar{I}_{CA} = 27.99\angle +168.59°$

$\therefore \bar{I}_{BC} = 27.99\angle -71.41°$, $\bar{I}_{AB} = 27.99\angle 48.59°$; $\bar{I}_{bB} = \bar{I}_{BC} - \bar{I}_{AB} = 27.99\angle -71.41° - 27.99\angle 48.59°$

(a) $\bar{I}_{bB} = \underline{48.48\angle -101.41° \text{ A rms}}$

(b) $\bar{I}_{CA} = \underline{27.99\angle 168.59° \text{ A rms}}$

(c) $\bar{V}_{bc} = \underline{190.53\angle -30° \text{ V rms}}$

11-19

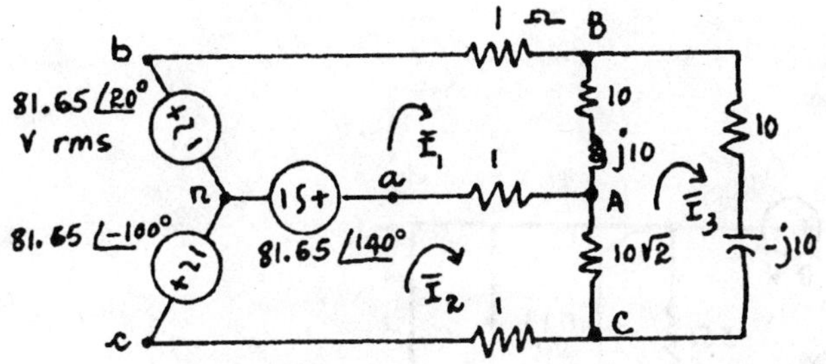


$$\bar{I}_1 = \frac{\begin{vmatrix} 141.42\angle -10^\circ & -1 & -14.142\angle 45^\circ \\ 141.42\angle 110^\circ & 16.142 & -14.142 \\ 0 & -14.142 & 34.142 \end{vmatrix}}{\begin{vmatrix} 15.620\angle 39.81^\circ & -1 & -14.142\angle 45^\circ \\ -1 & 16.142 & -14.142 \\ -14.142\angle 45^\circ & -14.142 & 34.142 \end{vmatrix}} = \frac{23{,}004\angle 20^\circ}{3896.35\angle 0^\circ} = \underline{5.904\angle 20^\circ \text{ A}} = \bar{I}_{bB}$$

$$\bar{I}_2 = \frac{\begin{vmatrix} 15.620\angle 39.81^\circ & 141.42\angle -10^\circ & -14.142\angle 45^\circ \\ -1 & 141.42\angle 110^\circ & -14.142 \\ -14.142\angle 45^\circ & 0 & 34.142 \end{vmatrix}}{3896.35} = \frac{63{,}889\angle 99.63^\circ}{3896.35} = 16.397\angle 99.63^\circ \text{ A}$$

$$\bar{I}_{cC} = -\bar{I}_2 = \underline{16.397\angle -80.37^\circ \text{ A}}$$

$$\bar{I}_{aA} = -\bar{I}_{bB} - \bar{I}_{cC} = -5.904\angle 20^\circ - 16.397\angle -80.37^\circ = \underline{16.397\angle 120.37^\circ \text{ A}}$$

$$P_{an} = 81.65 \times 16.397 \cos(140^\circ - 120.37^\circ) = \underline{1261.0 \text{ W}}$$

$$P_{bn} = 81.65 \times 5.904 \cos(20^\circ - 20^\circ) = \underline{482.1 \text{ W}}$$

$$P_{cn} = 81.65 \times 16.397 \cos(-100^\circ + 80.37^\circ) = \underline{1261.0 \text{ W}}$$

## *11-20*

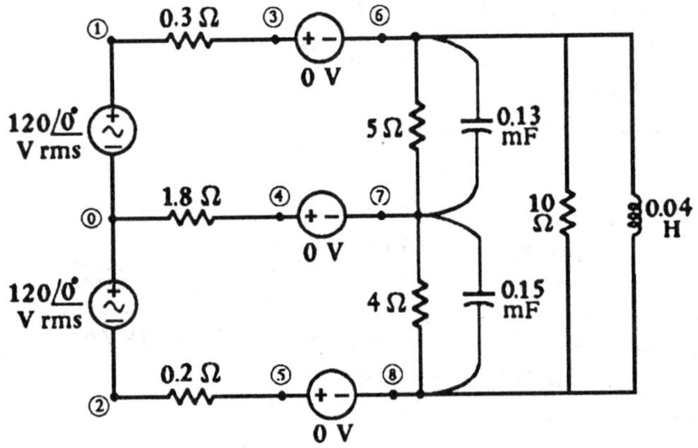


******************************************************************************

**PROGRAM:**

```
V10 1 0 AC 120
V02 0 2 AC 120
R13 1 3 0.3
R04 0 4 1.8
R25 2 5 0.2
V36 3 6
V47 4 7
V58 5 8
R67 6 7 5
C67 6 7 0.13M
R78 7 8 4
C78 7 8 0.15M
R68 6 8 10
L68 6 8 0.04
.AC LIN 1 60 60
.PRINT AC IM(V36) IP(V36) IM(V47) IP(V47) IM(V58) IP(V58)
.END
```

******************************************************************************

| FREQ | IM(V36) | IP(V36) | IM(V47) | IP(V47) | IM(V58) | IP(V58) |
|---|---|---|---|---|---|---|
| 6.000E+01 | 4.514E+01 | -1.013E+01 | 3.498E+00 | 2.401E+00 | 4.856E+01 | 1.708E+02 |

(*a*) $\mathbf{I}_{aA} = \mathbf{I}_{36} = 45.14\underline{/-10.13^\circ}$ A

(*b*) $\mathbf{I}_{bB} = \mathbf{I}_{58} = 48.56\underline{/170.8^\circ}$ A

(*c*) $\mathbf{I}_{nN} = \mathbf{I}_{47} = 3.498\underline{/2.401^\circ}$ A

# 11-21

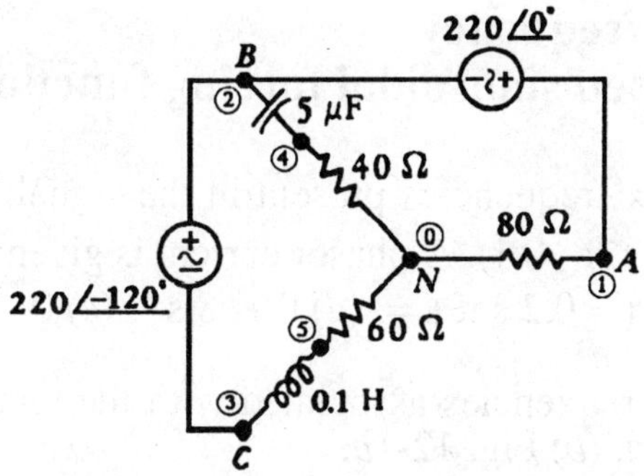


**************************************************************************

**PROGRAM:**

```
V12  1  2  AC  220  0
V23  2  3  AC  220  -120
R10  1  0  80
R40  4  0  40
R50  5  0  60
C42  4  2  5U
L35  3  5  0.1
.AC  LIN  1  60  60
.PRINT  AC  VM(2)  VP(2)
.END
```

**************************************************************************

| FREQ | VM(2) | VP(2) |
|---|---|---|
| 6.000E+01 | 2.253E+02 | -1.523E+02 |

Therefore, $\mathbf{V}_{BN} = 225.3\underline{/-152.3^\circ}$ V rms

# CHAPTER 12: COMPLEX FREQUENCY

## Section 12-2 Complex frequency
## Section 12-3 The damped sinusoidal forcing function

**1** Identify each of the complex frequencies present in the signal: (*a*) $2(e^{-3t} + 12\sin 8t)\cos 10t$; (*b*) $2e^{-3t}(5 + 3e^{-8t})$; (*c*) $4 + 3\sin(7t - 0.1)$. A phasor current is given as $2.1\underline{/50°}$ A. Find the instantaneous value of the current at $t = 0.2$ s if $\mathbf{s}$ =: (*d*) $0 + j8\ \text{s}^{-1}$; (*e*) $0 - j8\ \text{s}^{-1}$; (*f*) $-8\ \text{s}^{-1}$.

**2** Determine all the complex frequencies associated with the total response $v(t)$ for $t > 0$ in the circuit shown in: (*a*) Fig. 12-1*a*; (*b*) Fig. 12-1*b*.

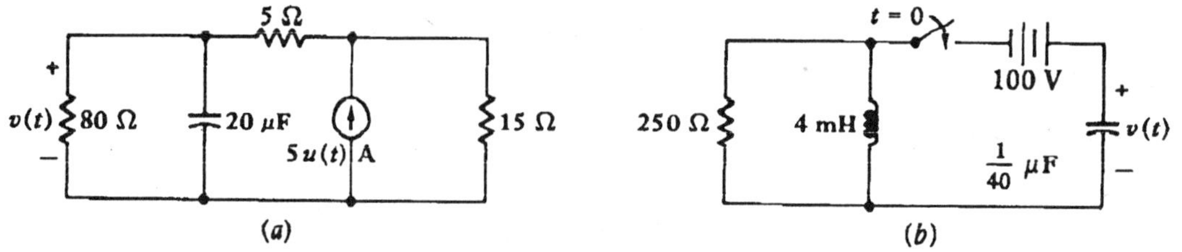


**Fig. 12-1:** See Prob. 2.

**3** The complex phasor current $\mathbf{I}$ is specified by the equation, $(\mathbf{s} + 250)\mathbf{I} + (6000\mathbf{I}/\mathbf{s}) = 20$. (*a*) Express $\mathbf{I}$ as the ratio of two polynomials in $\mathbf{s}$. (*b*) Calculate $\mathbf{I}$ in polar form for $\mathbf{s} = -200 + j100\ \text{s}^{-1}$. (*c*) Compute $i(t)$ at $t = 2$ ms if $\mathbf{s} = -200 + j100\ \text{s}^{-1}$.

**4** A current, $i = 8e^{-3t}\cos 4t$ A, flows through the series combination of 5 Ω and 2 H. Using the passive sign convention, the voltage across the series network is designated $v(t)$. (*a*) Find $v(t)$. (*b*) Calculate $v(t)$ at $t = 0.3$ s. (*c*) What value of inductance would cause the answer to part *b* to be zero?

## Section 12-4 Z(s) and Y(s)

**5** (*a*) Find $\mathbf{Z}(\mathbf{s})$ for the network of Fig. 12-2. (*b*) Calculate $\mathbf{Z}$ at $\mathbf{s} = -1 + j1\ \text{s}^{-1}$. (*c*) If $\mathbf{s} = \sigma + j0$, at what value of $\sigma$ is $|\mathbf{Z}|$ a minimum?

**6** (*a*) Determine $\mathbf{Z}(\mathbf{s})$ as a ratio of two polynomials in $\mathbf{s}$ for the network shown in Fig. 12-3. (*b*) Evaluate $\mathbf{Z}(-5 + j2)$ in rectangular form. (*c*) At what value or values of $\mathbf{s}$ is $\mathbf{Z}(\mathbf{s}) = 0$?

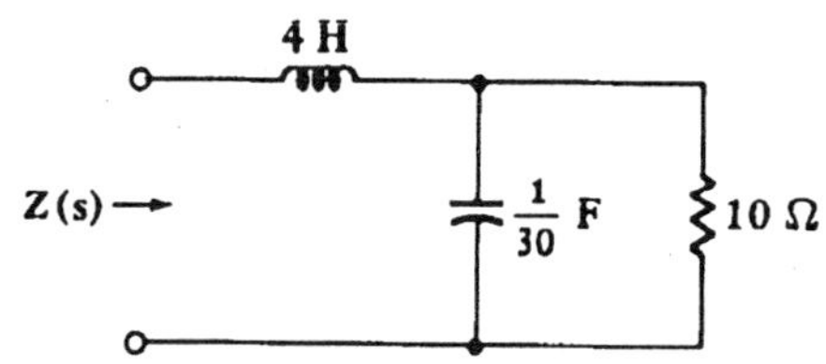

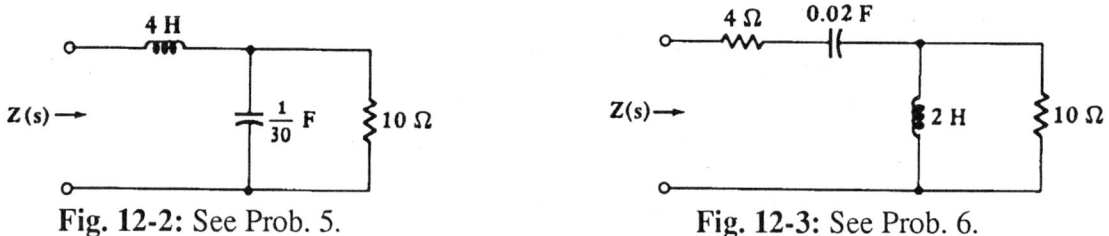

**Fig. 12-2:** See Prob. 5.

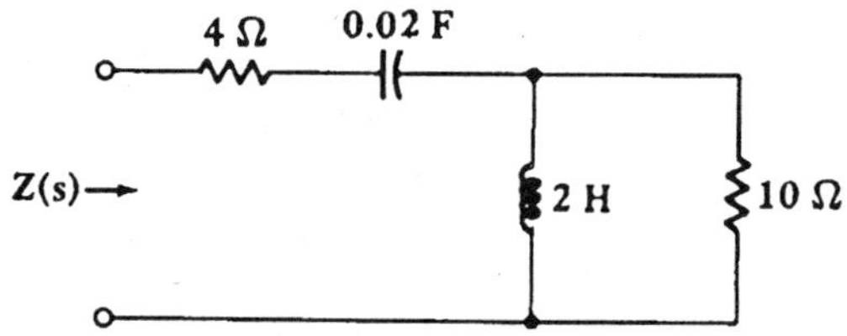


**Fig. 12-3:** See Prob. 6.

**7** (*a*) The series combination of a 2-Ω resistor and an 8-H inductor is in parallel with the series combination of a 5-Ω resistor and a 50-mF capacitor. Find the impedance of the parallel combination, $\mathbf{Z}_p(\mathbf{s})$, as the ratio of two polynomials in $\mathbf{s}$. (*b*) Find $\mathbf{Z}_p(-0.5)$. (*c*) Find $\mathbf{Z}_p(j2)$. (*d*) Find the admittance of the *RL* branch, $\mathbf{Y}_{RL}(\mathbf{s})$, as the ratio of two polynomials in $\mathbf{s}$. (*e*) Find $\mathbf{Y}_{RC}(\mathbf{s})$ of the *RC* branch. (*f*) Show that $\mathbf{Y}_{RL} + \mathbf{Y}_{RC} = 1/\mathbf{Z}_p$.

**8** If $i_{s1}(t) = 2e^{-t} \cos 2t$ A and $i_{s2}(t) = e^{-t} \sin 2t$ A in Fig. 12-4, find $v_L(t)$.

**9** For the network shown in Fig. 12-5: (*a*) find $\mathbf{Z}(\mathrm{s})$ as a ratio of polynomials in s; (*b*) let $\mathrm{s} = j\omega$ and $\mathbf{Z}(j\omega) = R(j\omega) + jX(j\omega)$, and find $X(j\omega)$ as a ratio of polynomials in $\omega$; (*c*) find $\mathbf{Y}(\mathrm{s})$; and (*d*) find $G(j\omega)$, where $\mathbf{Y}(j\omega) = G(j\omega) + jB(j\omega)$ and $\mathrm{s} = j\omega$.

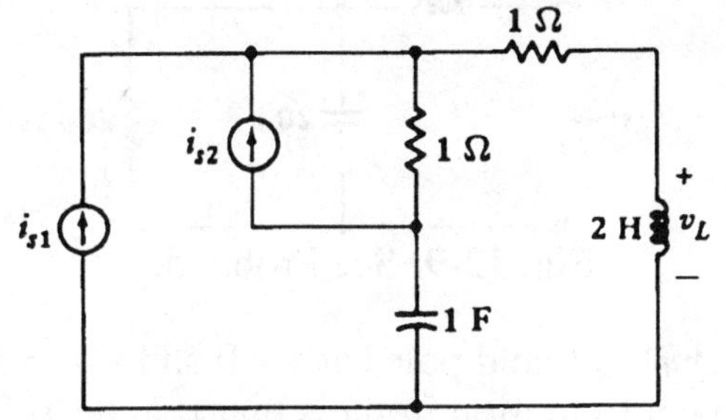


**Fig. 12-4:** See Prob. 8.

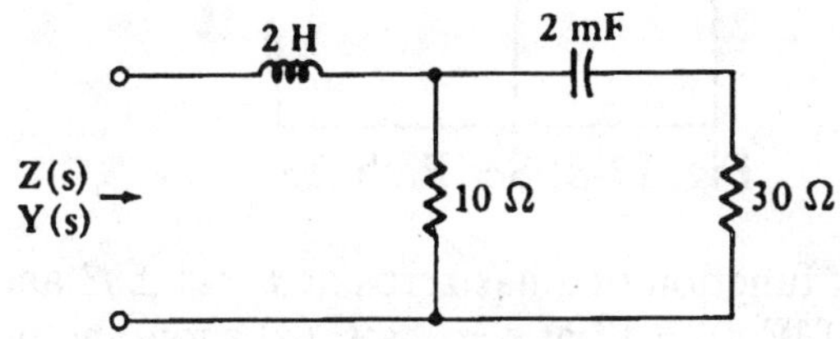


**Fig. 12-5:** See Prob. 9.

**10** (*a*) Let $v_s = 100e^{-2t} \cos 5t$ V in Fig. 12-6 and find $i_R(t)$. (*b*) Repeat if $v_s = 100e^{-2t} \cos \omega t$ V.

## Section 12-5 Frequency response as a function of $\sigma$

**11** An input impedance has zeros at $\mathrm{s} = -2\ \mathrm{s}^{-1}$ and $\mathrm{s} = \infty$, poles at $\mathrm{s} = 0$ and $\mathrm{s} = -3\ \mathrm{s}^{-1}$, and a value of 16 Ω at $\mathrm{s} = -2.5\ \mathrm{s}^{-1}$. (*a*) Write $\mathbf{Z}(\mathrm{s})$ as a ratio of polynomials in s. (*b*) Plot $|\mathbf{Z}(\sigma)|$ versus $\sigma$ if $\mathrm{s} = \sigma + j0$.

**12** (*a*) For the circuit of Fig. 12-7, determine $\mathbf{I}(\mathrm{s})$ as a function of s. (*b*) State all the critical frequencies of $\mathbf{I}(\mathrm{s})$. (*c*) Plot $|\mathbf{I}(\sigma)|$ versus $\sigma$. (*d*) What size capacitor should be added in series with the 2-Ω resistor to cause $\mathbf{I}(\mathrm{s})$ to have a pole at $\mathrm{s} = -5\ \mathrm{s}^{-1}$? (*e*) Find all critical frequencies of $\mathbf{I}_{new}(\mathrm{s})$.

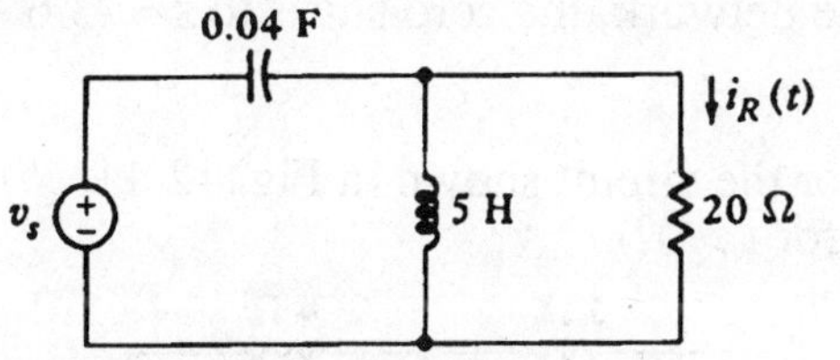

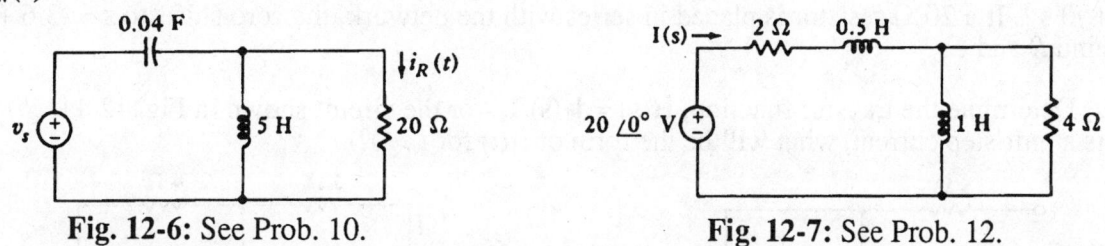

**Fig. 12-6:** See Prob. 10.

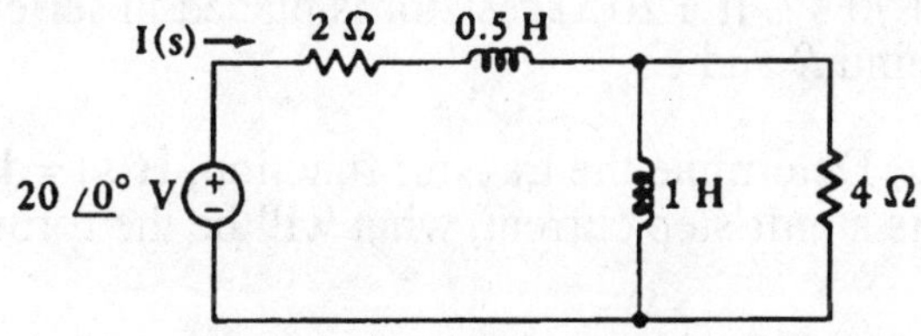


**Fig. 12-7:** See Prob. 12.

**13** (*a*) Find $\mathbf{V}_{out}(\mathrm{s})/\mathbf{I}_{in}(\mathrm{s})$ as the ratio of two polynomials in s for the circuit of Fig. 12-8. (*b*) Identify all the critical frequencies of $\mathbf{V}_{out}(\sigma)/\mathbf{I}_{in}(\sigma)$. (*c*) Plot $|\mathbf{V}_{out}(\sigma)/\mathbf{I}_{in}(\sigma)|$, $-5 \le \sigma \le 2$ Np/s.

## Section 12-6 The complex-frequency plane

**14** The parallel combination of 2 Ω and 0.5 F is in series with a 0.4-H inductor and an ideal voltage source, $v_s(t) = 25e^{-2t}$ V. Find the average power absorbed by the capacitor during the time interval, $0.1 \le t \le 0.4$ s.

**15** (*a*) Find $\mathbf{Z}(\mathrm{s})$ for the network shown in Fig. 12-9. (*b*) Draw the pole-zero constellation for $\mathbf{Z}(\mathrm{s})$. (*c*) Prepare plots of $|\mathbf{Z}(\sigma)|$ versus $\sigma$ with $\omega = 0$, and of $|\mathbf{Z}(j\omega)|$ versus $\omega$ with $\sigma = 0$.

**16** The voltage $\mathbf{V}(s) = 6s/[(s + 2)(s + 5)]$ has its magnitude represented as a plaster model. If the height of the model is 5 cm at $s = 0 + j3$ s$^{-1}$: (*a*) what is the height at $s = -4 + j5$ s$^{-1}$, and (*b*) at what point does the maximum height on the positive $j\omega$ axis occur?

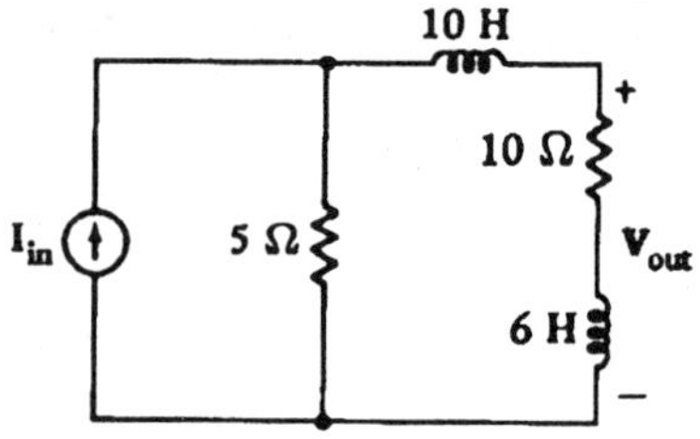


**Fig. 12-8:** See Prob. 13.

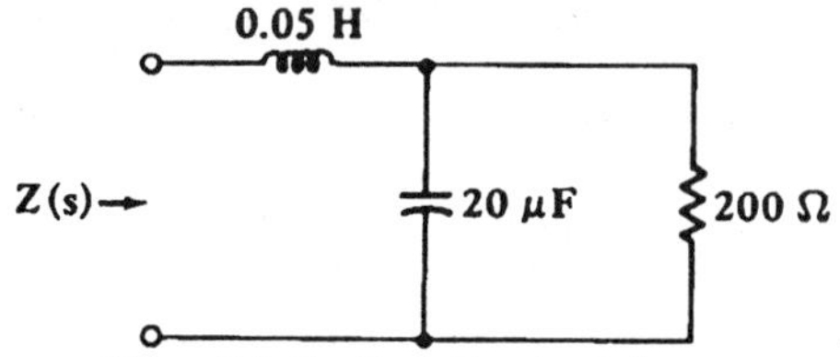


**Fig. 12-9:** See Prob. 15.

**17** A function of s has zeros at $s = -1 \pm j3$ and $s = -2 \pm j1$ s$^{-1}$, and poles at $s = 0$ and $s = -3 \pm j1$ s$^{-1}$. If $|\mathbf{F}(s)| = 12$ at $s = j3$ s$^{-1}$: (*a*) show the pole-zero configuration in the s plane, and (*b*) find $\mathbf{F}(s)$ as the ratio of two polynomials in s.

**18** An input impedance $\mathbf{Z}(s)$ has a zero at the origin and poles at $-1 \pm j10$ s$^{-1}$. As the sinusoidal operating frequency increases from 10 to 11 rad/s, by what factor does the length of the arrow increase that is drawn to the operating frequency from: (*a*) $-1 - j10$? (*b*) 0? (*c*) $-1 + j10$? (*d*) By what factor does $|\mathbf{Z}(s)|$ increase as the operating frequency increases from 10 to 11 rad/s?

**19** For a certain network, $\mathbf{I}_{out}(s)/\mathbf{V}_{in}(s) = \mathbf{H}(s) = (2s^2 + 18s + 36)/(5s^2 + 20s + 100)$ A/V. (*a*) Show the critical frequencies of $\mathbf{H}(s)$ on the complex plane. (*b*) Find $\mathbf{H}(0)$ and $\mathbf{H}(\infty)$. (*c*) Sketch $|\mathbf{H}(\sigma)|$ versus $\sigma$ for $\omega = 0$ and $|\mathbf{H}(j\omega)|$ versus $\omega$ for $\sigma = 0$. (*d*) If a plaster model of $|\mathbf{H}(s)|$ has a height of 2 in at the origin, what is its maximum height along the $j\omega$ axis, approximately? (Don't differentiate, it can ruin your day .)

**20** The three-element network of Fig. 12-10 has an input impedance that includes a zero at $s = -10 + j0$ s$^{-1}$. If a 20-Ω resistor is placed in series with the network, the zero shifts to $s = -3.6 + j0$ s$^{-1}$. Find $R$ and $C$.

**21** (*a*) Determine the transfer function, $\mathbf{H}(s) = \mathbf{I}_1(s)/\mathbf{I}_s$, for the circuit shown in Fig. 12-11. (*b*) If $i_s(t)$ is a unit-step current, what will be the form of $i_1(t)$ for $t > 0$?

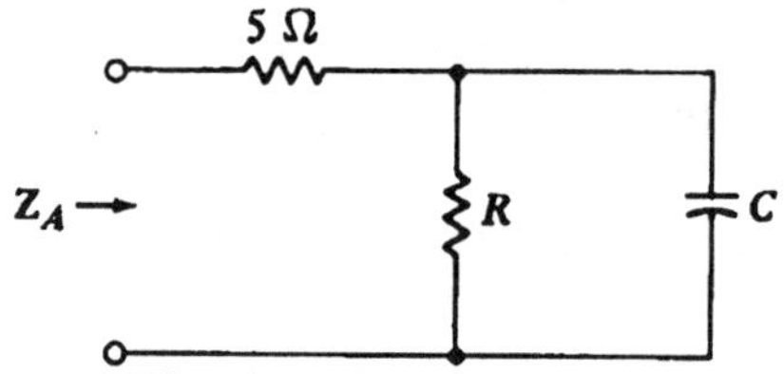

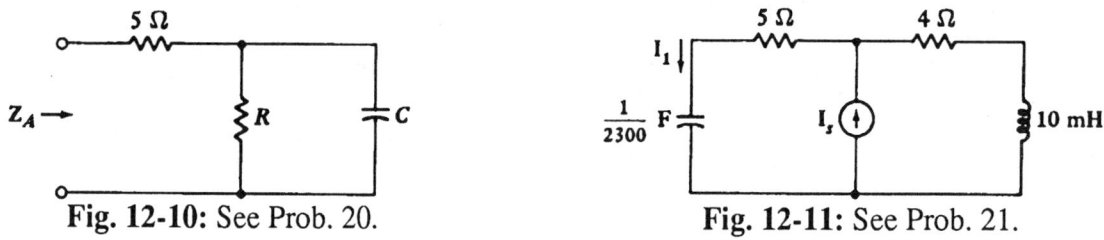

**Fig. 12-10:** See Prob. 20.

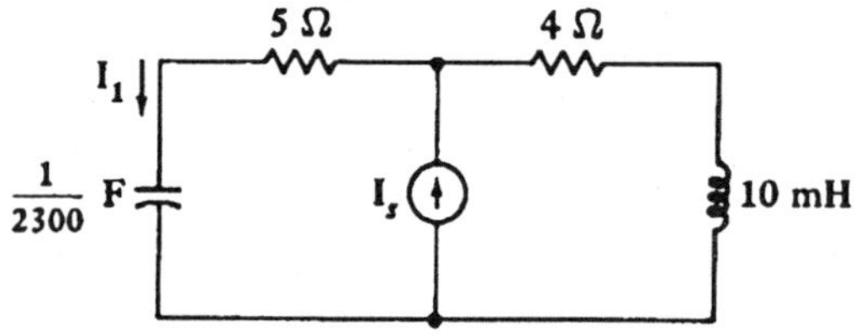


**Fig. 12-11:** See Prob. 21.

## Section 12-7 Natural response and the s-plane

**22** Let $\mathbf{V}_2/\mathbf{V}_1 = 7(s + 8)/(s + 3)$ for the network shown in Fig. 12-12. (*a*) If $v_1(t) = 4e^{-0.2t}$ V, find the forced response $v_{2f}(t)$. (*b*) If $v_1(t) = 6u(t)$ V and $v_2(0^-) = v_2(0^+) = 0$ find the complete response $v_2(t)$. (*c*) If $v_1(t) = 4e^{-0.2t}\,u(t)$ V and $v_2(0^-) = v_2(0^+) = 0$, find the complete response $v_2(t)$.

**23** In the circuit of Fig. 12-13, the transfer function $\mathbf{H}(s) = \mathbf{I}_2/\mathbf{I}_1 = 5(s + 20)(s + 30)/[(s + 10)(s + 15)]$. Let $i_1(t) = u(t)$ A. (*a*) Find $i_{2f}(t)$. (*b*) If $di_2/dt = 60$ A/s at $t = 0^+$, find $i_2(t)$.

**24** (*a*) Find $\mathbf{Z}_{in}(s)$ for the network given in Fig. 12-14. (*b*) If a voltage source were suddenly connected to the input terminals, what would be the form of the input current's natural response?

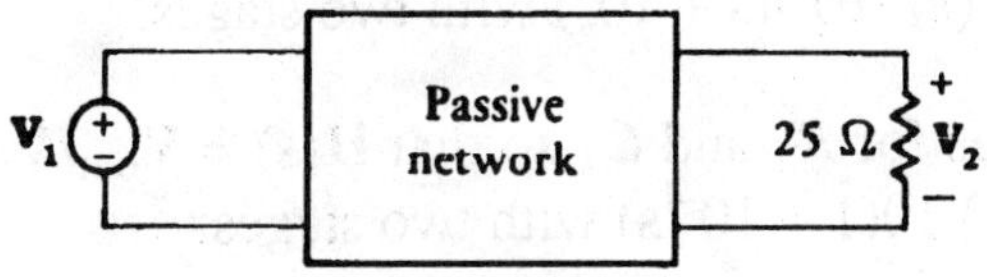


**Fig. 12-12:** See Prob. 22.

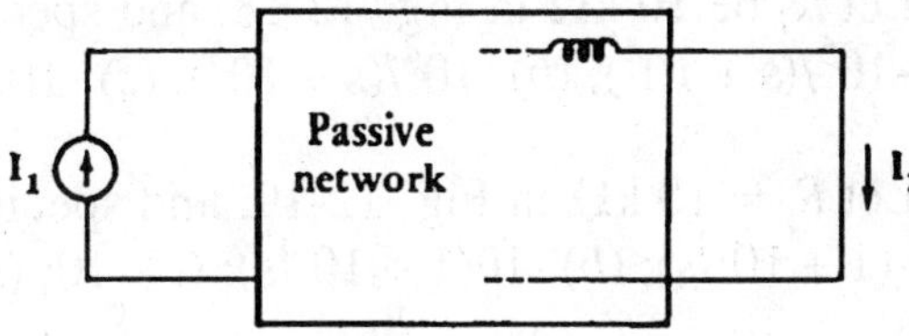


**Fig. 12-13:** See Prob. 23.

**25** (*a*) Find $\mathbf{H}(s) = V_C(s)/\mathbf{V}_s$ for the circuit shown in Fig. 12-15, and calculate its pole frequencies. (*b*) Find the exponential damping coefficient $\alpha$, the resonant frequency $\omega_0$, and the natural resonant frequency $\omega_d$ for the *RLC* circuit. Let $v_s(t) = u(t)$ V, and: (*c*) determine the forced response $v_{Cf}(t)$; (*d*) specify the form of the natural response $v_{Cn}(t)$; (*e*) find $v_C(0^+)$ and $v_C'(0^+)$; (*f*) write the complete response $v_C(t)$.

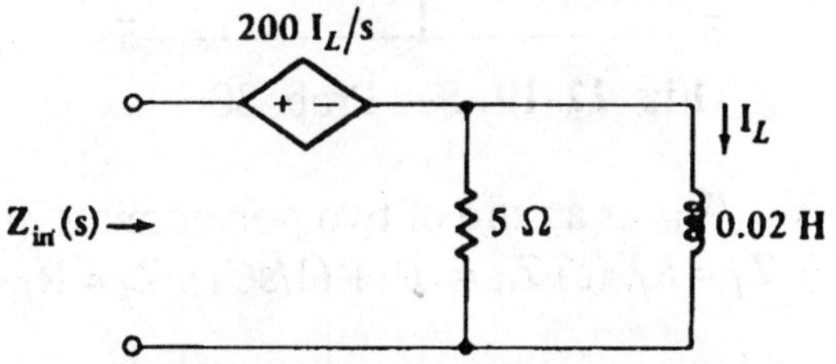

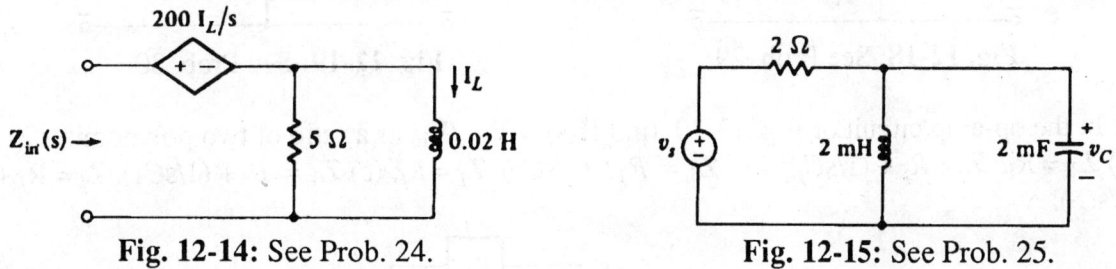

**Fig. 12-14:** See Prob. 24.

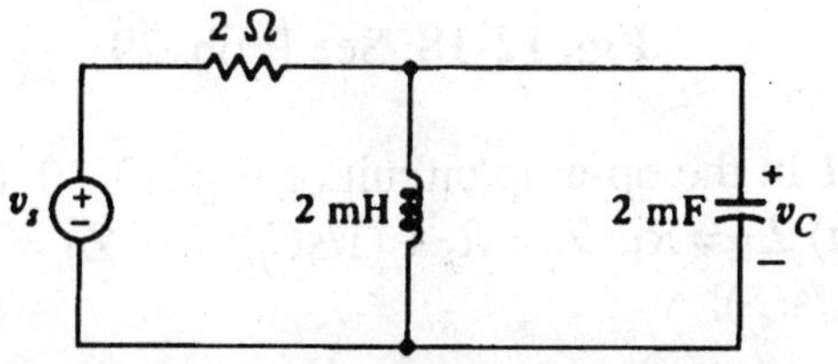


**Fig. 12-15:** See Prob. 25.

**26** The passive network indicated in Fig. 12-16 has the pole-zero configuration shown and also $\mathbf{Z}(s) = 30\ \Omega$ at $s = -1$. Let $i_s(t) = 0.2u(t)$ A. (*a*) What value will $v_{in}(t)$ eventually assume? (*b*) Find $v_{in}(t)$ for $t > 0$ if $v_{in}(0^+) = 0$ and $v_{in}'(0^+) = 9$ V/s.

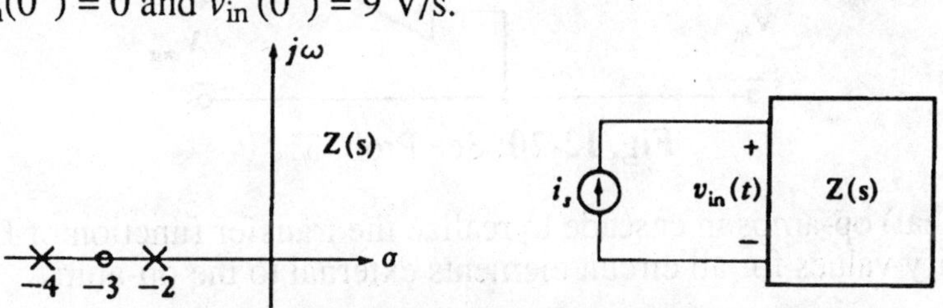


**Fig. 12-16:** See Prob. 26.

**27** (*a*) Refer to Fig. 12-17 and find the transfer functions, $\mathbf{H}_1 = \mathbf{I}_o/\mathbf{V}_{s1}$ and $\mathbf{H}_2 = \mathbf{I}_o/\mathbf{V}_{s2}$. (*b*) If $v_{s1} = 10u(t)$ V and $v_{s2} = 5 + 25u(t)$ V, find $i_o(t)$.

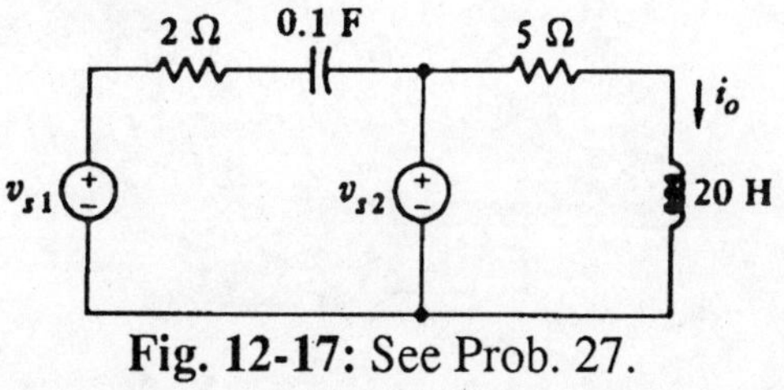


**Fig. 12-17:** See Prob. 27.

**28** An ideal voltage source $v_s(t)$, an open switch, and a 5-Ω resistor are in series with a passive two-terminal network having an input impedance, $\mathbf{Z}_A = (15s + 25)/(s + 5)\ \Omega$. The switch is closed at $t = 0$. If the current leaving the positive terminal of the source is $i_{in} = 2$ A at $t = 0^+$, determine $i_{in}(t)$ for $t > 0$ if $v_s(t)$ =: (*a*) 40 V; (*b*) $40e^{-0.5t}$ V.

## Section 12-8 A technique for synthesizing the voltage ratio, H(s) = $V_{out}/V_{in}$

**29** Let $R_f$ be 10 kΩ in Fig. 12-18, and specify values for $C_f$ and $R_1$ so that **H**(s) = $\mathbf{V}_{out}/\mathbf{V}_{in}$ =: (*a*) $-10^5/(s + 10^4)$; (*b*) $-10^4/(s + 10^5)$; (*c*) -10; (*d*) -1; (*e*) $10^5/(s + 10^4)$ with two stages.

**30** Let $R_f$ = 10 kΩ in Fig. 12-19, and specify values for $R_1$ and $C_1$ so that **H**(s) = $\mathbf{V}_{out}/\mathbf{V}_{in}$ =: (*a*) $-(1 + 10^{-4}s)$; (*b*) $-10(1 + 10^{-4}s)$; (*c*) -10; (*d*) -1; (*e*) $10(1 + 10^{-4}s)$ with two stages.

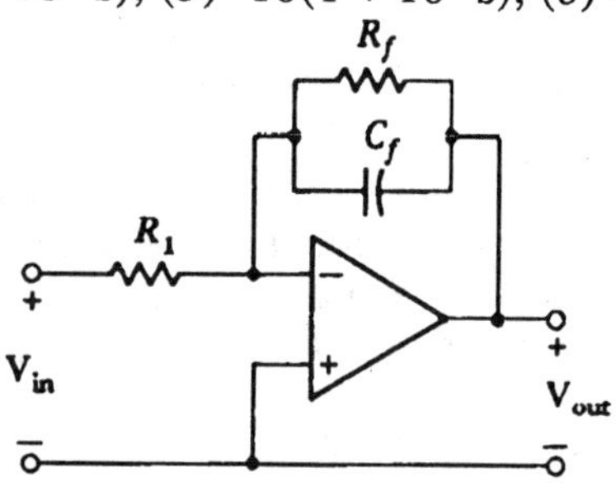


**Fig. 12-18:** See Prob. 29.

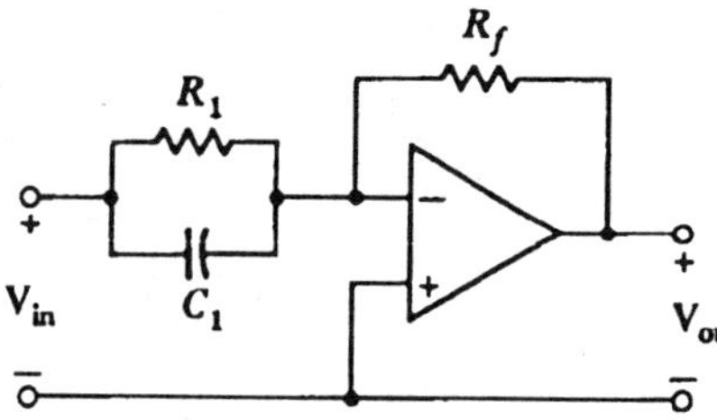


**Fig. 12-19:** See Prob. 30.

**31** In the op-amp circuit of Fig. 12-20, find **H**(s) = $\mathbf{V}_{out}/\mathbf{V}_{in}$ as a ratio of two polynomials in s if: (*a*) $\mathbf{Z}_1 = R_1$, $\mathbf{Z}_f = R_f + (1/sC_f)$; (*b*) $\mathbf{Z}_1 = R_1 + (1/sC_1)$, $\mathbf{Z}_f = R_f$; (*c*) $\mathbf{Z}_1 = R_1 + (1/sC_1)$, $\mathbf{Z}_f = R_f + (1/sC_f)$.

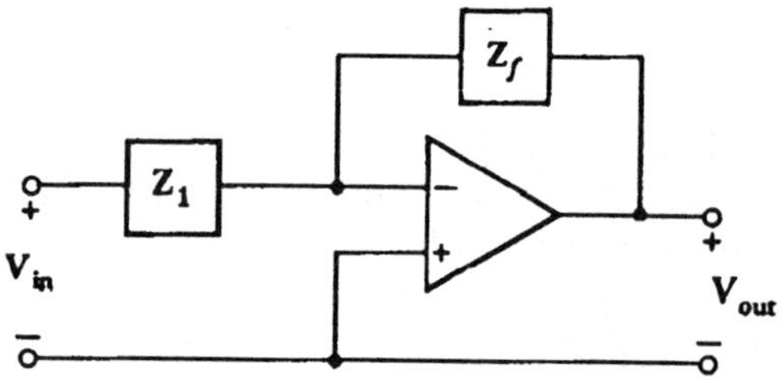


**Fig. 12-20:** See Prob. 31.

**32** Use several (ideal) op-amps in cascade to realize the transfer function of Eq. (29) in Sec. 12-8 of your text. Specify values for all circuit elements external to the op-amps.

# Chapter 12

12-1 (a) $2(e^{-3t} + 12\sin 8t)\cos 10t$ $\therefore \mathbf{s} = \underline{-3 \pm j10}$; $\underline{\pm j2}$, $\underline{\pm j18 \ s^{-1}}$

(b) $2e^{-3t}(5 + 3e^{-8t})$ $\therefore \mathbf{s} = \underline{-3 + j0}$; $\underline{-11 + j0 \ s^{-1}}$

(c) $4 + 3\sin(7t - 0.1)$ $\therefore \mathbf{s} = \underline{0}$; $\underline{0 \pm j7 \ s^{-1}}$

(d) $2.1\angle 50°$, $\mathbf{s} = 0 + j8$, $t = 0.2$ $\therefore i(0.2) = 2.1\cos(8 \times 0.2^{\text{rad}} + 50°) = \underline{-1.6474 \text{ A}}$

(e) $\mathbf{s} = 0 - j8$ $\therefore i(0.2) = 2.1\cos(-8 \times 0.2^{\text{rad}} + 50°) = \underline{1.5686 \text{ A}}$

(f) $\mathbf{s} = -8$ $\therefore i(0.2) = 2.1e^{-1.6}\cos 50° = \underline{0.2725 \text{ A}}$

12-2 (a)

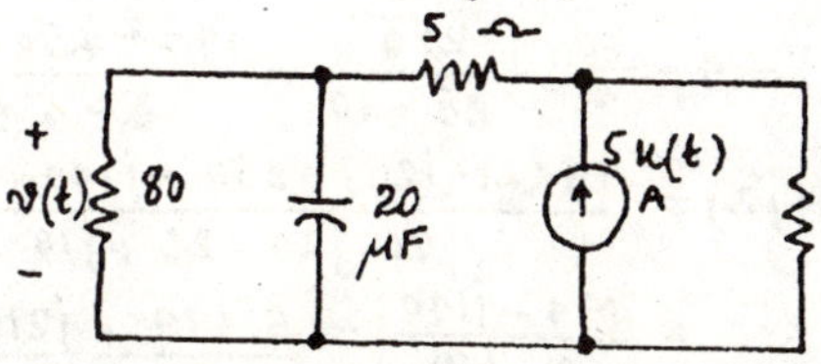


$v_f = 5 \times \frac{15}{15+85} \times 80 = 60$, $\tau = 20 \times 10^{-6} \times 16$

$\therefore v(t) = 60(1 - e^{-3125t})u(t)$

$\therefore \mathbf{s} = \underline{0}$, $\underline{-3125 \ s^{-1}}$

(b)

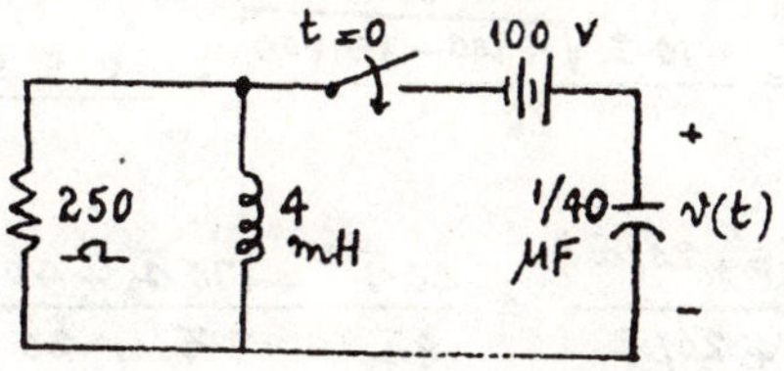


Parallel RLC, $\alpha = \frac{1}{2RC} = \frac{40 \times 10^6}{500} = 80{,}000$

$\omega_0^2 = \frac{40 \times 10^6}{0.004} = 10^{10}$, $\omega_d = \sqrt{10^{10} - 0.64 \times 10^{10}} = 60{,}000$

$\therefore v(t) = 100 + e^{-80{,}000t}(A\sin 60{,}000t + B\cos 60{,}000t)$

$\therefore \mathbf{s} = \underline{0}$, $\underline{-80{,}000 \pm j60{,}000 \ s^{-1}}$ $v(t) = 100 + e^{-80{,}000t}(\frac{400}{3}\sin 60{,}000t - 100\cos 60{,}000t)$

12-3 (a) $(\mathbf{s} + 250)\mathbf{I} + (6000\mathbf{I}/\mathbf{s}) = 20$ $\therefore (\mathbf{s}^2 + 250\mathbf{s} + 6000)\mathbf{I} = 20\mathbf{s}$, $\mathbf{I} = \dfrac{20\mathbf{s}}{\mathbf{s}^2 + 250\mathbf{s} + 6000}$ ←

(b) $\mathbf{s} = -200 + j100$, $\mathbf{s}^2 = 30{,}000 - j40{,}000$ $\therefore \mathbf{I} = \dfrac{20(-200 + j100)}{30{,}000 - j40{,}000 - 50{,}000 + j25{,}000 + 6000}$

$\therefore \mathbf{I} = \dfrac{-4 + j2}{-14 - j15} = \underline{0.2180\angle -73.54° \text{ A}}$

(c) $i(0.002) = 0.2180e^{-200 \times 0.002}\cos(100 \times 0.002^{\text{rad}} - 73.54°) = \underline{68.42 \text{ mA}}$

12-4 (a)

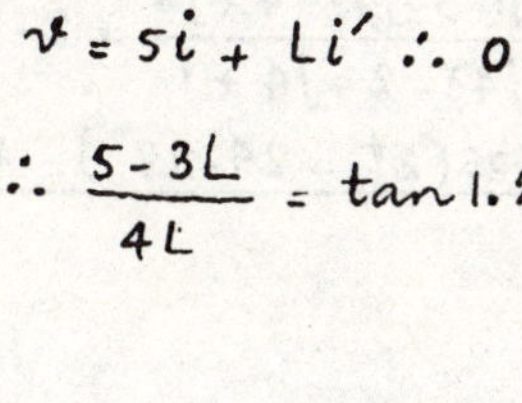


$i = 8e^{-3t}\cos 4t$ A

$v = 5i + 2i' = 8e^{-3t}(5\cos 4t - 8\sin 4t - 6\cos 4t)$

$\therefore v(t) = \underline{-8e^{-3t}\cos 4t - 64e^{-3t}\sin 4t \text{ V}}$

(b) $v(0.3) = -8e^{-0.9}(\cos 1.2 + 8\sin 1.2) = \underline{-25.43 \text{ V}}$

(c) $v = 5i + Li'$ $\therefore 0 = 8e^{-3t}(5\cos 4t - 4L\sin 4t - 3L\cos 4t)$ $\therefore (5 - 3L)\cos 1.2 = 4L\sin 1.2$

$\therefore \frac{5 - 3L}{4L} = \tan 1.2 = 2.572$ $\therefore L = \underline{0.3763 \text{ H}}$

12-5

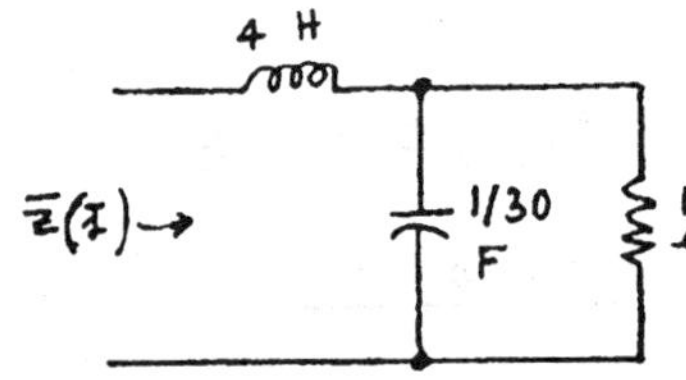


(a) $\bar{Z}(s) = 4s + \dfrac{300/s}{10+30/s} = \dfrac{4s^2+12s+30}{s+3}$ ←

(b) $\bar{Z}(-1+j1) = \dfrac{-j8-12+j12+30}{2+j1} = 8-j2 = \underline{8.246\angle -14.036^\circ\ \Omega}$

(c) $\bar{Z}(\sigma) = 4\sigma + \dfrac{30}{\sigma+3}$, $\dfrac{dZ}{d\sigma} = 4 - \dfrac{30}{(\sigma+3)^2} = 0$

$\therefore (\sigma+3)^2 = 7.5 \quad \therefore \sigma = -0.2614$ and $-5.739$

$\bar{Z}(-0.2614) = 9.909\ \Omega$, $\bar{Z}(-5.739) = -33.91\ \Omega$

$\therefore |\bar{Z}|_{min}$ occurs at $\sigma = \underline{-0.2614\ s^{-1}}$

12-6

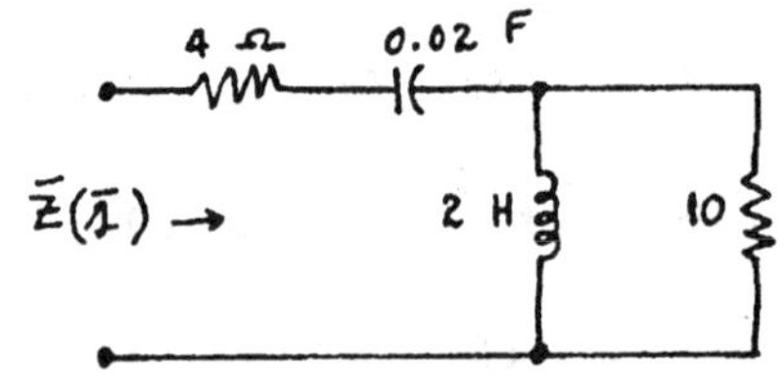


(a) $\bar{Z}(s) = 4 + \dfrac{50}{s} + \dfrac{20s}{2s+10} = \dfrac{14s^2+70s+250}{s^2+5s}$ ←

(b) $\bar{Z}(-5+j2) = \dfrac{14(21-j20)-350+j140+250}{21-j20-25+j10}$

$= \dfrac{194-j140}{-4-j10} = \underline{5.379+j21.55^+\ \Omega}$

(c) $\bar{Z}(s) = 0 \quad \therefore 14s^2+70s+250 = 0 \quad \therefore s = \dfrac{-70 \pm \sqrt{4900-14{,}000}}{28} = \underline{-2.5 \pm j3.407\ s^{-1}}$

12-7

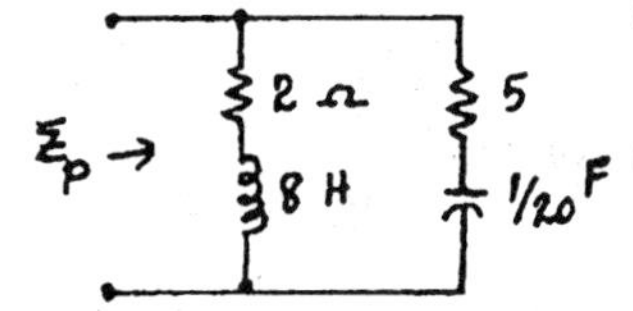


(a) $\bar{Z}_p = \dfrac{(2+8s)(5+20/s)}{7+8s+20/s} = \dfrac{40s^2+170s+40}{8s^2+7s+20}$ ←

(b) $\bar{Z}_p(-0.5) = \dfrac{10-85+40}{2-3.5+20} = \underline{-1.8919\ \Omega}$

(c) $\bar{Z}_p(j2) = \dfrac{-160+40+j340}{-32+20+j14} = \underline{19.554\angle -21.16^\circ\ \Omega}$

(d) $\bar{Y}_{RL}(s) = \dfrac{1}{2+8s}$ ←

(e) $\bar{Y}_{RC}(s) = \dfrac{1}{5+20/s} = \dfrac{s}{5s+20}$ ←

(f) $\bar{Y}_{RL} + \bar{Y}_{RC} = \dfrac{1}{2+8s} + \dfrac{s}{5s+20} = \dfrac{8s^2+7s+20}{(2+8s)(5s+20)} = \dfrac{8s^2+7s+20}{40s^2+170s+40} = \dfrac{1}{\bar{Z}_p}$ ✓

12-8

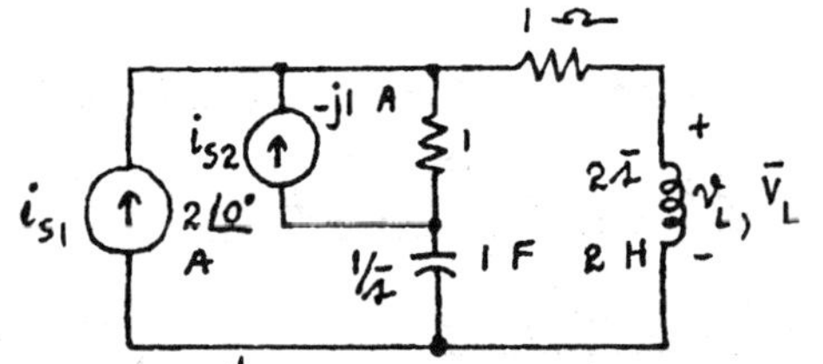


$i_{s1} = 2e^{-t}\cos 2t$, $i_{s2} = e^{-t}\sin 2t$

$\bar{V}_L = 2 \times \dfrac{1+1/s}{1+1/s+1+2s} \times 2s + (-j1)\dfrac{1}{1+1/s+1+2s} \times 2s$

$= \dfrac{4s(s+1)}{2s^2+2s+1} - j2s\,\dfrac{s}{2s^2+2s+1}$

$= \dfrac{(4-j2)s^2+4s}{2s^2+2s+1}$, $s = -1+j2$

$\therefore \bar{V}_L(-1+j2) = \dfrac{(4-j2)(-3-j4)-4+j8}{2(-3-j4)-2+j4+1} = 2.987\angle -24.98^\circ$

$\therefore v_L(t) = \underline{2.987e^{-t}\cos(2t-24.98^\circ)\ \ V}$

**12-9**

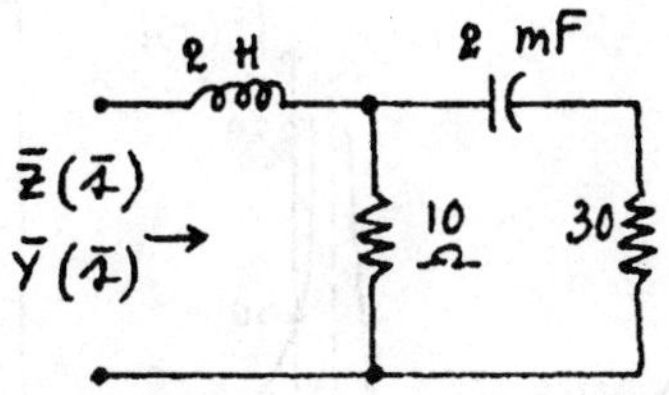


(a) $\mathbf{Z}(s) = 2s + \dfrac{10(30 + 500/s)}{40 + 500/s} = \dfrac{4s^2 + 65s + 250}{2s + 25}$ ←

(b) $\mathbf{Z}(j\omega) = \dfrac{250 - 4\omega^2 + j65\omega}{25 + j2\omega} \times \dfrac{25 - j2\omega}{25 - j2\omega}$

$= \dfrac{6250 + 30\omega^2 + j(1125\omega + 8\omega^3)}{4\omega^2 + 625}$

$\therefore X(j\omega) = \dfrac{1125\omega + 8\omega^3}{4\omega^2 + 625}$ ←

(c) $\mathbf{Y}(s) = 1/\mathbf{Z}(s) = \dfrac{2s + 25}{4s^2 + 65s + 250}$ ←

(d) $\mathbf{Y}(j\omega) = \dfrac{25 + j2\omega}{250 - 4\omega^2 + j65\omega} \times \dfrac{250 - 4\omega^2 - j65\omega}{250 - 4\omega^2 - j65\omega}$ $\therefore G(j\omega) = \dfrac{6250 + 30\omega^2}{62500 + 2225\omega^2 + 16\omega^4}$ ←

**12-10** (a)

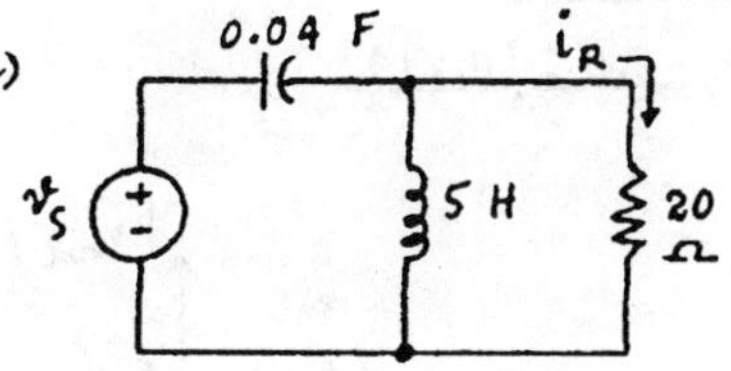


$v_s = 100e^{-2t}\cos 5t \quad \therefore V_s = 100, \ s = -2 + j5$

$\therefore \mathbf{I}_R = \dfrac{100}{25/s + \dfrac{100s}{5s + 20}} \times \dfrac{5s}{5s + 20} = \dfrac{500s}{125 + \dfrac{500}{s} + 100s}$

$= \dfrac{20s^2}{4s^2 + 5s + 20} = \dfrac{20(-21 - j20)}{-84 - j80 - 10 + j25 + 20} = 6.291\angle 6.981^\circ$

$\therefore i_R(t) = \underline{6.291e^{-2t}\cos(5t + 6.981^\circ)\ \text{A}}$

(b) $v_s = 100e^{-2t}\cos\omega t, \quad V_s = 100, \quad s = -2 + j\omega \quad \therefore \mathbf{I}_R = \dfrac{20(4 - \omega^2 - j4\omega)}{16 - 4\omega^2 - j16\omega - 10 + j5\omega + 20}$

$\mathbf{I}_R = \dfrac{20(4 - \omega^2 - j4\omega)}{26 - 4\omega^2 - j11\omega} \quad \therefore \underline{i_R(t) = 20\sqrt{\dfrac{16 + 8\omega^2 + \omega^4}{676 - 87\omega^2 + 16\omega^4}}\, e^{-2t}\cos\left[\omega t + \tan^{-1}\dfrac{-4\omega}{4 - \omega^2} - \tan^{-1}\dfrac{-11\omega}{26 - 4\omega^2}\right]}$

**12-11** (a) $\mathbf{Z}(s) = \dfrac{K(s + 2)}{s(s + 3)}; \ Z(-2.5) = 16 = \dfrac{K(-0.5)}{-2.5(0.5)}$

$\therefore K = 40, \ \mathbf{Z}(s) = \dfrac{40s + 80}{s^2 + 3s}$ ←

(b) $\mathbf{Z}(\sigma) = \dfrac{40\sigma + 80}{\sigma^2 + 3\sigma}$

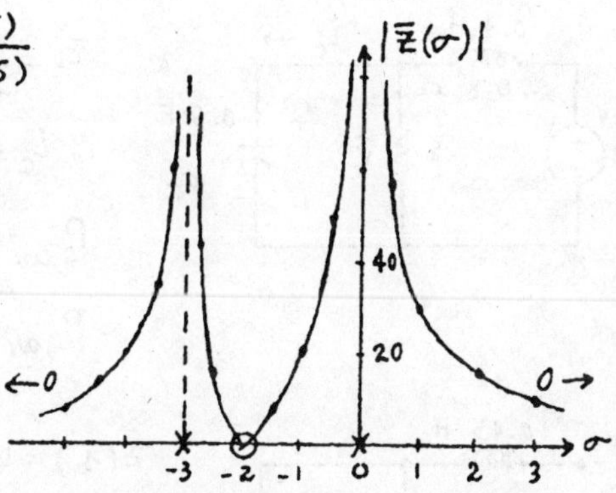


| σ | \|Z(σ)\| |
|---|---|
| 0.5 | 57.1 |
| 1 | 30 |
| 2 | 16 |
| -0.5 | 48 |
| -1.5 | 8.89 |
| -2.5 | 16 |
| -2.75 | 43.6 |
| -3.5 | 34.3 |
| -4 | 20 |
| -1 | 20 |

**12-12**

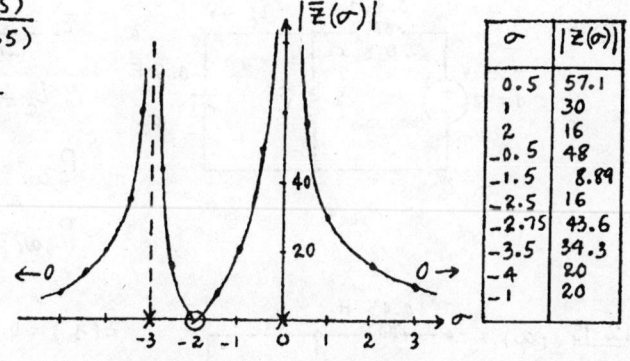

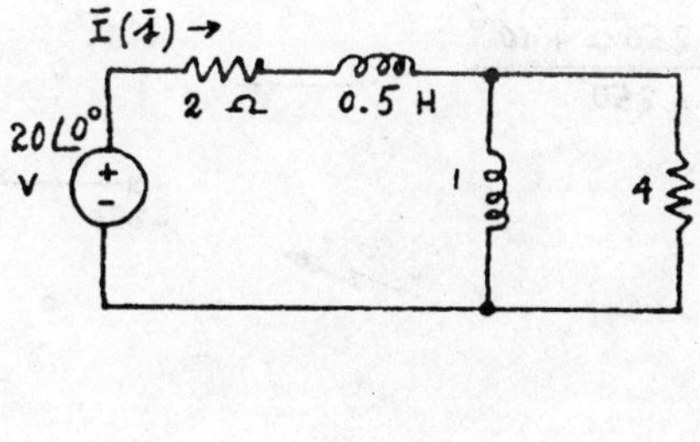


(a) $I(s) = \dfrac{20}{2 + 0.5s + \dfrac{4s}{s + 4}} = \dfrac{20(s + 4)}{0.5s^2 + 8s + 8}$ ←

(b) $s = \dfrac{-8 \pm \sqrt{64 - 16}}{1} = -1.0718, \ -14.928$

Zeros: $\underline{-4}, \underline{\infty}$ Poles: $\underline{-1.0718}, \underline{-14.928}$

(cont. on next page)

**12-12 (cont.)**

(c) $\bar{I}(\sigma) = \dfrac{20(\sigma+4)}{0.5\sigma^2 + 8\sigma + 8}$

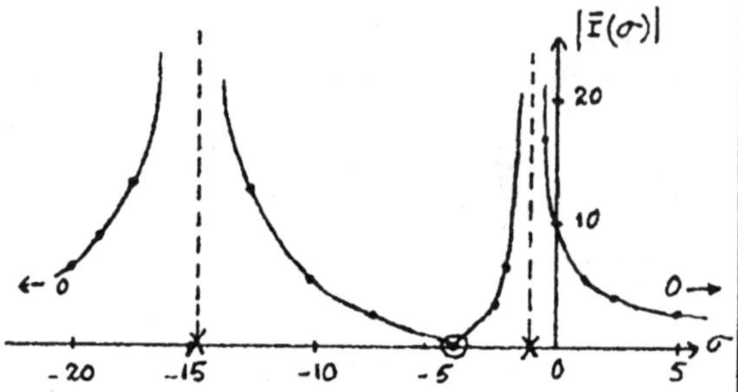

(d) $\bar{I}(\bar{s}) = \dfrac{20}{\frac{1}{\bar{s}C} + 2 + 0.5\bar{s} + \frac{4\bar{s}}{\bar{s}+4}}$

$\bar{I}(-5) = \dfrac{20}{-1/5C + 2 - 2.5 + 20}$

$\therefore \dfrac{-0.2}{C} = -19.5 \;;\; \underline{C = 10.256 \text{ mF}}$



| $\sigma$ | $|\bar{I}(\sigma)|$ |
|---|---|
| 5 | 2.98 |
| 0 | 10 |
| -0.5 | 16.97 |
| -2 | 6.67 |
| -3 | 1.74 |
| -7.5 | 2.93 |
| -10 | 5.45 |
| -12.5 | 12.25 |
| -15 | 44.0 |
| -17.5 | 12.78 |

(e) $\bar{I}_{new} = \dfrac{20}{97.5/\bar{s} + 2 + 0.5\bar{s} + 4\bar{s}/(\bar{s}+4)} = \dfrac{20\bar{s}(\bar{s}+4)}{97.5(\bar{s}+4) + 2\bar{s}(\bar{s}+4) + 0.5\bar{s}^2(\bar{s}+4) + 4\bar{s}^2}$

$= \dfrac{20\bar{s}(\bar{s}+4)}{0.5\bar{s}^3 + 8\bar{s}^2 + 105.5\bar{s} + 390} = \dfrac{40\bar{s}(\bar{s}+4)}{\bar{s}^3 + 16\bar{s}^2 + 211\bar{s} + 780} = \dfrac{40\bar{s}(\bar{s}+4)}{(\bar{s}+5)(\bar{s}^2 + 11\bar{s} + 156)}$

$\bar{s} = \dfrac{-11 \pm \sqrt{121 - 624}}{2} = -5.5 \pm j11.214$

Zeros: $\underline{0, -4, \infty}$
Poles: $\underline{-5, -5.5 \pm j11.214}$

**12-13**



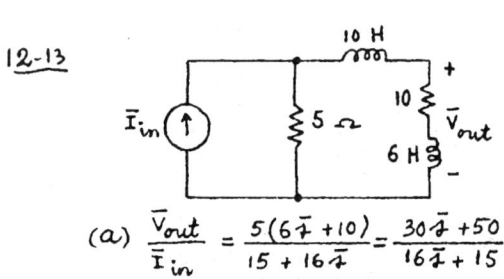

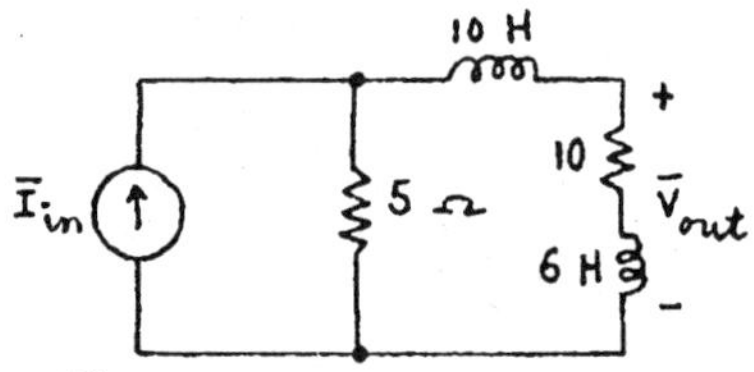

(a) $\dfrac{\bar{V}_{out}}{\bar{I}_{in}} = \dfrac{5(6\bar{s}+10)}{15+16\bar{s}} = \dfrac{30\bar{s}+50}{16\bar{s}+15}$

(b) 0's: $\underline{-5/3}$ ; X's: $\underline{-15/16}$

(c) $\dfrac{\bar{V}_{out}}{\bar{I}_{in}} = \dfrac{15}{8} \times \dfrac{\sigma + 5/3}{\sigma + 15/16}$

asymp = 1.875

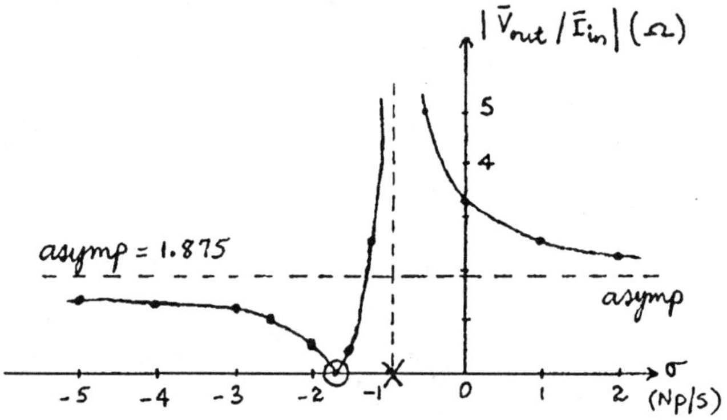

**12-14**



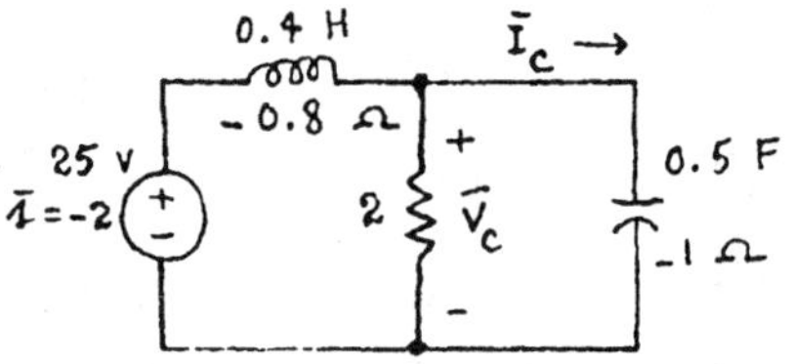

$\bar{I}_c = \dfrac{25}{-0.8 + (2||-1)} \times \dfrac{2}{2-1} = -\dfrac{125}{7} \;\therefore\; \bar{V}_c = \dfrac{125}{7},\; v_c(t) = \dfrac{125}{7}e^{-2t}$

$\therefore i_c = -\dfrac{125}{7}e^{-2t}$ A ; $P_c(t) = -1\left(\dfrac{125}{7}\right)^2 e^{-4t}$ W

$P_{c,av} = \dfrac{1}{0.3}\displaystyle\int_{0.1}^{0.4} -\left(\dfrac{125}{7}\right)^2 e^{-4t} dt = -\dfrac{10}{3}\left(\dfrac{125}{7}\right)^2\left(-\dfrac{1}{4}\right)\left(e^{-1.6} - e^{-0.4}\right)$

$\underline{P_{c,av} = -124.47 \text{ W}}$

**12-15**

(a)

$\bar{Z}(\bar{s}) = 0.5\bar{s} + \dfrac{200(50{,}000/\bar{s})}{200 + 50{,}000/\bar{s}} = 0.05\bar{s} + \dfrac{50{,}000}{\bar{s}+250}$

$= \dfrac{1}{20} \times \dfrac{\bar{s}^2 + 250\bar{s} + 10^6}{\bar{s}+250}$

(b) $\bar{s} = \dfrac{-250 \pm \sqrt{250^2 - 4\times 10^6}}{2} = -125 \pm j992.2$

(cont. on next page)

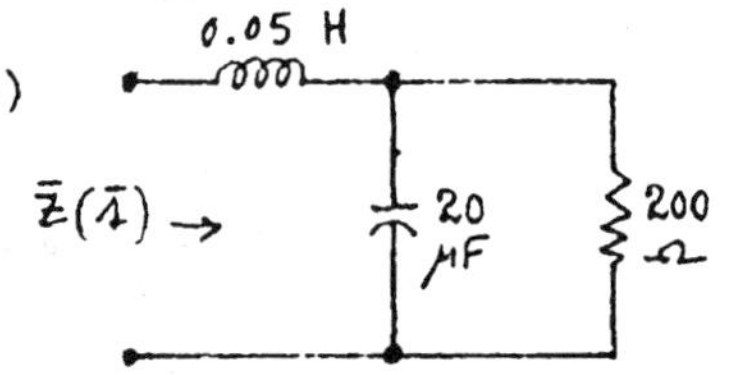

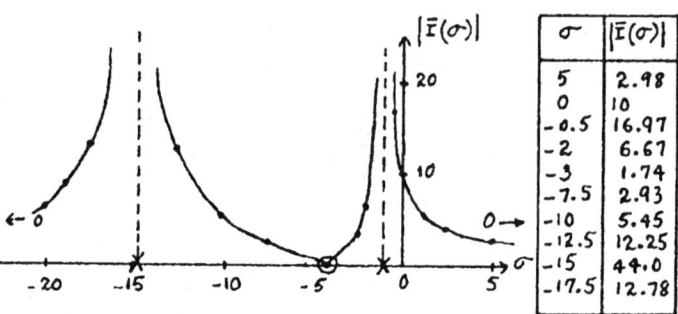

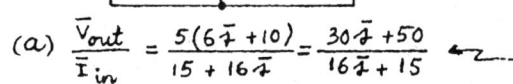

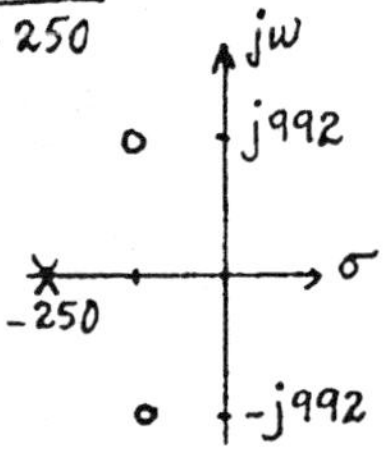

**12-15 (c) (cont.)** $\bar{Z}(\sigma) = \frac{1}{20}\,\frac{\sigma^2 + 250\sigma + 10^6}{\sigma + 250}$

$\bar{Z}(j\omega) = \frac{1}{20}\,\frac{10^6 - \omega^2 + j250\,\omega}{250 + j\omega}$

| $\sigma$ ($\omega=0$) | $\lvert Z(\sigma)\rvert$ ($\Omega$) | $\omega$ ($\sigma=0$) | $\lvert Z(j\omega)\rvert$ ($\Omega$) |
|---|---|---|---|
| 0 | 200 | 0 | 200 |
| 500 | 92 | 250 | 133 |
| 1000 | 90 | 500 | 68 |
| 2000 | 122 | 750 | 30 |
| 5000 | 260 | 1000 | 12.127 |
| −100 | 328 | 1250 | 25 |
| −200 | 990 | 1500 | 43 |
| −300 | 1015 | 1750 | 60 |
| −400 | 353 | 2000 | 75 |
| −500 | 225 | 2500 | 105 |
| −1000 | 117 | 3000 | 133.5 |
| −2000 | 129 | 4000 | 186 |
| −5000 | 260 | 5000 | 240 |

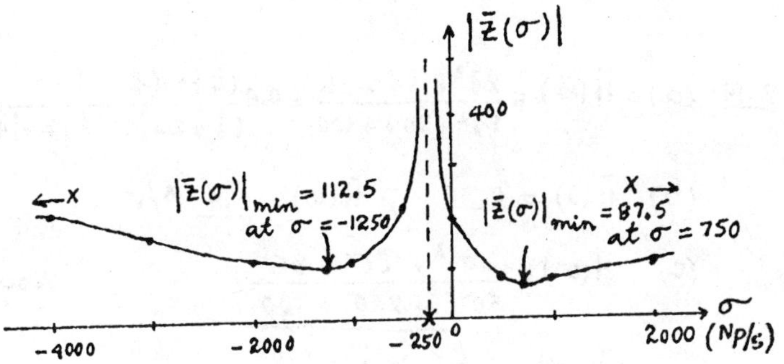


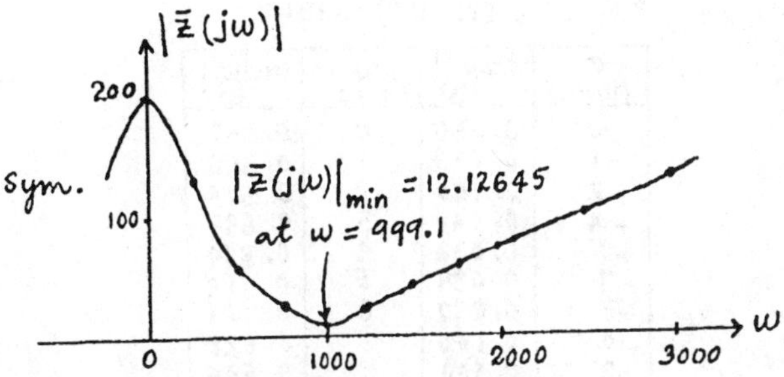


**12-16** (a) $\bar{V}(s) = \frac{6s}{(s+2)(s+5)}$, $\bar{h}(s) = \frac{K6s}{(s+2)(s+5)}$ $\therefore 5 = \left|\frac{j18K}{(2+j3)(5+j3)}\right|$ $\therefore K = 5.840$

$s = -4 + j5$, $\bar{h}(s) = \frac{5.840 \times 6(-4+j5)}{(-2+j5)(1+j5)}$ $\therefore |\bar{h}| = \underline{8.171\ \text{cm}}$

(b) $\bar{h}(j\omega) = \frac{5.840 \times 6\,\omega}{(j\omega+2)(j\omega+5)}$ $\therefore |\bar{h}| = \frac{5.840 \times 6\,\omega}{\sqrt{(\omega^2+4)(\omega^2+25)}}$, $|\bar{h}|^2 = \frac{1228\,\omega^2}{(\omega^2+4)(\omega^2+25)}$

$\therefore \frac{1228}{|\bar{h}^2|} = \omega^2 + 29 + \frac{100}{\omega^2}$ ; $\frac{d(1228/|\bar{h}|^2)}{d\omega} = 2\omega - \frac{200}{\omega^3} = 0$ $\therefore \omega = \underline{3.162}$ rad/s (it is max, not min)

**12-17** (a)

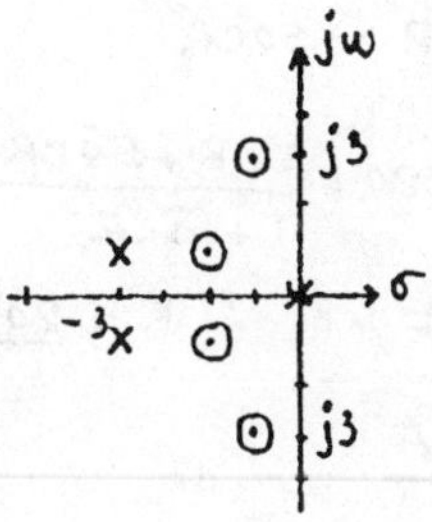


(b) $\bar{F}(s) = \frac{K(s+1+j3)(s+1-j3)(s+2+j1)(s+2-j1)}{s(s+3+j1)(s+3-j1)}$

$\bar{F}(s) = \frac{K(s^2+2s+10)(s^2+4s+5)}{s(s^2+6s+10)}$ $\therefore 12 = \left|\frac{K(1+j6)(-4+j12)}{j3(1+j18)}\right|$

$\therefore K = 8.435$ ; $\bar{F}(s) = \frac{8.435(s^4+6s^3+23s^2+50s+50)}{s^3+6s^2+10s}$

$\bar{F}(s) = \frac{8.435s^4 + 50.61s^3 + 194.00s^2 + 421.7s + 421.7}{s^3 + 6s^2 + 10s}$

**12-18**

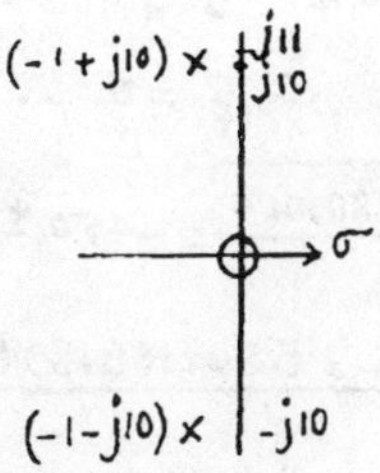


(a) $\left|\frac{j11 - (-1-j10)}{j10 - (-1-j10)}\right| = \left|\frac{1+j21}{1+j20}\right| = \sqrt{\frac{442}{401}} = \underline{1.0499}$

(b) $\left|\frac{j11 - 0}{j10 - 0}\right| = \underline{1.1}$

(c) $\left|\frac{j11+1-j10}{j10+1-j10}\right| = |1+j1| = \sqrt{2} = \underline{1.4142}$

(d) $\sqrt{\frac{401}{442}} \times 1.1 \times \frac{1}{\sqrt{2}} = \underline{0.7409}$

**12-19** (a) $\bar{H}(s) = \dfrac{2s^2 + 18s + 36}{5s^2 + 20s + 100} = 0.4\dfrac{(s+3)(s+6)}{(s+2+j4)(s+2-j4)}$

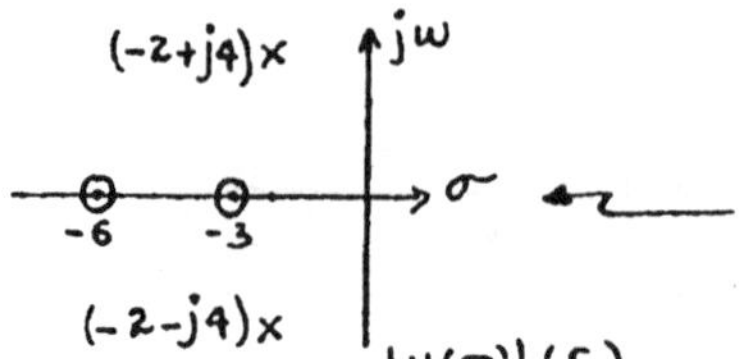


(b) $\bar{H}(0) = \underline{0.36}$ , $\bar{H}(\infty) = \underline{0.4}$ A/V

(c) $H(\sigma) = \dfrac{2\sigma^2 + 18\sigma + 36}{5\sigma^2 + 20\sigma + 100}$

$|\bar{H}(j\omega)| = 0.4\sqrt{\dfrac{(18-\omega^2)^2 + 81\omega^2}{(20-\omega^2)^2 + 16\omega^2}}$

| σ (ω=0) | \|H(σ)\| (S) | ω (σ=0) | \|H̄(jω)\| (S) |
|---|---|---|---|
| 0 | 0.360 | 0 | 0.360 |
| -1 | 0.235 | 1 | 0.396 |
| -2 | 0.100 | 2 | 0.510 |
| -4 | 0.040 | 3 | 0.699 |
| -5 | 0.032 | 4 | 0.874 |
| -7 | 0.039 | 5 | 0.884 |
| -8 | 0.077 | 6 | 0.789 |
| -10 | 0.140 | 8 | 0.628 |
| 2 | 0.500 | 10 | 0.544 |
| 4 | 0.538 | 4.5 | 0.901 |
| 6 | 0.540 | | |
| 8 | 0.531 | | |

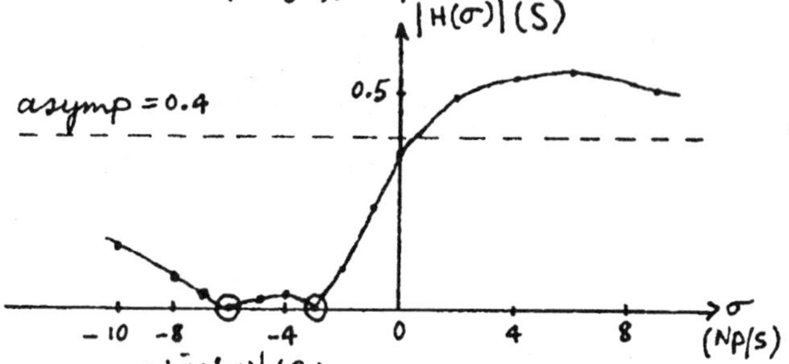


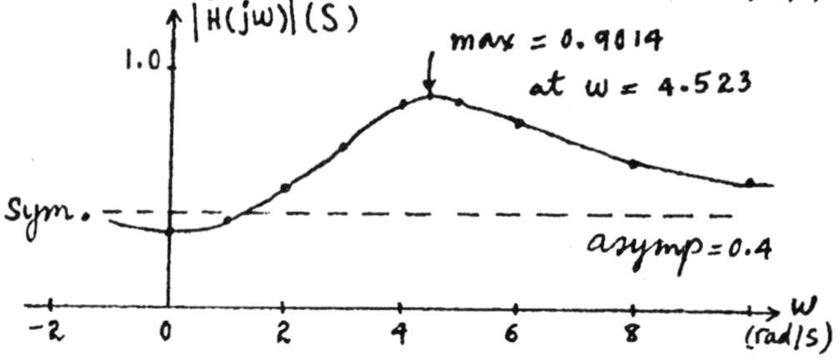


(d) By trial and error, $|\bar{H}(j\omega)|_{max} = \underline{0.9013504}$ at $\omega = \underline{4.523344 \text{ rad/s}}$

$\therefore$ max height $= 0.9013504 \times \dfrac{2}{0.36} = \underline{5.007502 \text{ in}}$

**12-20**

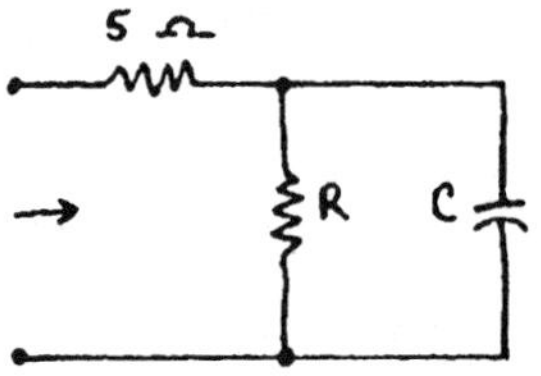


$Z_A = 5 + \dfrac{R/sC}{R + 1/sC} = 5 + \dfrac{R}{1 + sCR} = \dfrac{5 + R + 5sCR}{1 + sCR}$

At $s = -10$, $Z_A = 0$ $\therefore 5 + R = 50CR$

Add 20 Ω in series : $Z_B = 20 + \dfrac{5 + R + 5sCR}{1 + sCR} = \dfrac{25 + R + 25sCR}{1 + sCR}$

At $s = -3.6$, $Z_B = 0$ $\therefore 25 + R = 90CR$ ; $\dfrac{25 + R}{5 + R} = \dfrac{90CR}{50CR} = 1.8$ $\therefore R = \underline{20\ \Omega}$

Also, $25 + 20 = 90(20)C$ $\therefore C = \underline{25 \text{ mF}}$

**12-21**

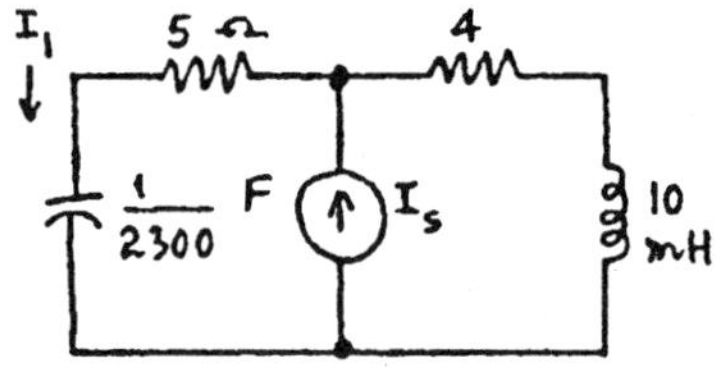


(a) $H(s) = \dfrac{I_1}{I_s} = \dfrac{4 + s/100}{9 + s/100 + 2300/s} = \dfrac{s(s+400)}{s^2 + 900s + 230{,}000}$

(b) $i_s = u(t) \therefore s = 0$ , $I_s = 1$ , $I_{1f} = 0$ $\therefore i_{1f} = 0$

Poles: $s = \dfrac{-900 \pm \sqrt{900^2 - 920{,}000}}{2} = -450 \pm j165.83\ s^{-1}$

$\therefore i_1(t) = \underline{e^{-450t}(A\cos 165.83t + B\sin 165.83t)}$, $t > 0$

**12-22**

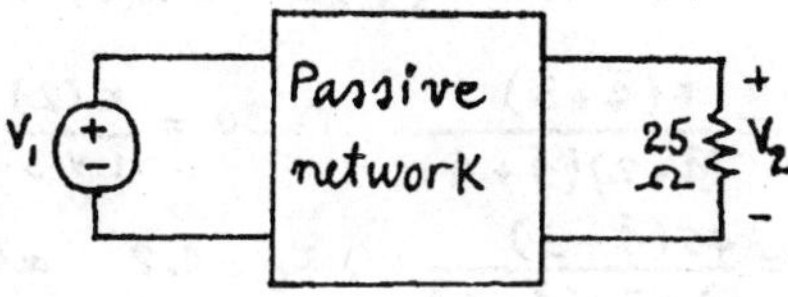

(a) $\dfrac{V_2}{V_1} = \dfrac{7(\bar{s}+8)}{\bar{s}+3}$ ; $v_1(t) = 4e^{-0.2t}$ V $\therefore V_1 = 4, \bar{s} = -0.2$

$\therefore V_2 = \dfrac{7(-0.2+8)}{-0.2+3} \times 4 = 78$ $\therefore v_{2f} = \underline{78e^{-0.2t}}$ V

(b) $v_1 = 6u(t)$ $\therefore V_1 = 6, \bar{s} = 0$ $\therefore V_2 = \dfrac{7(8)(6)}{3} = 112$ V

$v_{2f} = 112$ $\therefore v_2 = 112 + Ae^{-3t}$, $v_2(0) = 0$

$\therefore v_2(t) = \underline{112(1-e^{-3t})u(t)}$ V

(c) $v_1 = 4e^{-0.2t}u(t)$ $\therefore v_2(t) = \underline{78(e^{-0.2t} - e^{-3t})u(t)}$ V

**12-23**

(a) $H(\bar{s}) = \dfrac{I_2}{I_1} = \dfrac{5(\bar{s}+20)(\bar{s}+30)}{(\bar{s}+10)(\bar{s}+15)}$ ; $i_1(t) = u(t)$

$\therefore I_1 = 1, \bar{s} = 0$, $I_2 = 1 \times \dfrac{5(20)(30)}{10(15)} = 20$ $\therefore i_{2f} = \underline{20}$ A

(b) $i_2'(0^+) = 60$ A/s, $i_2(t) = 20 + Ae^{-10t} + Be^{-15t}$

$i_2(0) = 0$ $\therefore A+B = -20$
$i_2'(0^+) = 60 = -10A - 15B$  $\Big\}$ $\therefore A = -48, B = 28$

$\therefore i_2(t) = \underline{(20 - 48e^{-10t} + 28e^{-15t})u(t)}$ A

**12-24** (a) Squirt 1A in $\therefore 1 = I_L + \dfrac{0.02\bar{s}I_L}{5} = I_L(1 + \dfrac{\bar{s}}{250})$

$\therefore I_L = \dfrac{250}{\bar{s}+250}$ $\therefore Z_{in} = V_{in} = \dfrac{200 I_L}{\bar{s}} + \dfrac{\bar{s}I_L}{50} = I_L\left(\dfrac{\bar{s}^2 + 10^4}{50\bar{s}}\right)$

$\therefore Z_{in} = \dfrac{250}{\bar{s}+250} \cdot \dfrac{\bar{s}^2+10^4}{50\bar{s}} = 5\dfrac{\bar{s}^2+10^4}{\bar{s}(\bar{s}+250)}$

(b) $I_{in} = \dfrac{V_{in}}{Z_{in}}$ $\therefore i_{in,nat} = A\cos 100t + B\sin 100t = \underline{C\cos(100t+\phi)}$

**12-25** (a) $Z_{LC} = \dfrac{2\times 10^{-3}\bar{s} / 2\times 10^{-3}\bar{s}}{2\times 10^{-3}\bar{s} + 1/2\times 10^{-3}\bar{s}} = \dfrac{2\times 10^{-3}\bar{s}}{4\times 10^{-6}\bar{s}^2 + 1} = \dfrac{500\bar{s}}{\bar{s}^2 + 250{,}000}$

$\dfrac{V_c}{V_s} = H(\bar{s}) = \dfrac{Z_{LC}}{2+Z_{LC}} = \dfrac{250\bar{s}}{\bar{s}^2 + 250\bar{s} + 250{,}000}$

Poles at $\bar{s} = \dfrac{-250 \pm \sqrt{250^2 - 10^6}}{2} = -125 \pm j484.1$ $\bar{s}^{-1}$

(b) $\alpha = \dfrac{1}{2RC} = \underline{125}$ Np/s ; $\omega_0 = \dfrac{1}{\sqrt{LC}} = \underline{500}$ rad/s , $\omega_d = \sqrt{\omega_0^2 - \alpha^2} = \underline{484.1}$ rad/s

(c) $v_s(t) = u(t)$ $\therefore V_s = 1, \bar{s} = 0$ $\therefore v_{cf} = \underline{0}$

(d) $v_{cn} = e^{-125t}(A\cos 484.1t + B\sin 484.1t)$

(e) $v_c(0^+) = v_c(0^-) = \underline{0}$ ; $i_L(0^+) = 0$ $\therefore i_c(0^+) = \dfrac{1}{2} = 2\times 10^{-3} v_c'(0^+)$ $\therefore v_c'(0^+) = \underline{250}$ V/s

(f) $v_c = 0 + e^{-125t}(A\cos 484.1t + B\sin 484.1t)$ $\therefore A = 0$, $v_c'(0^+) = 250 = 1(484.1B) + 0$

$\therefore B = 0.5164$, $v_c(t) = \underline{0.5164 e^{-125t} \sin 484.1t}$ V, $t > 0$

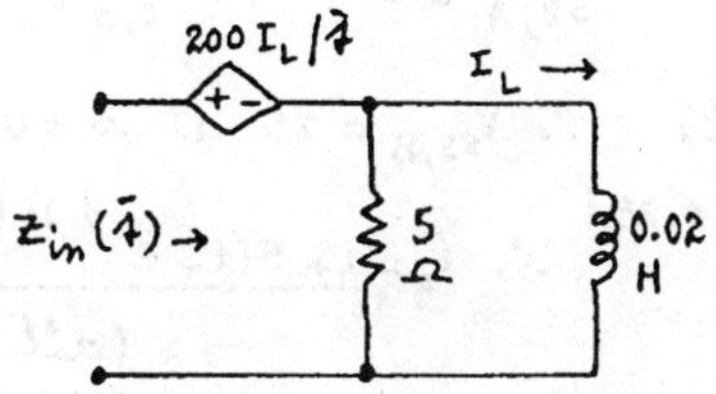

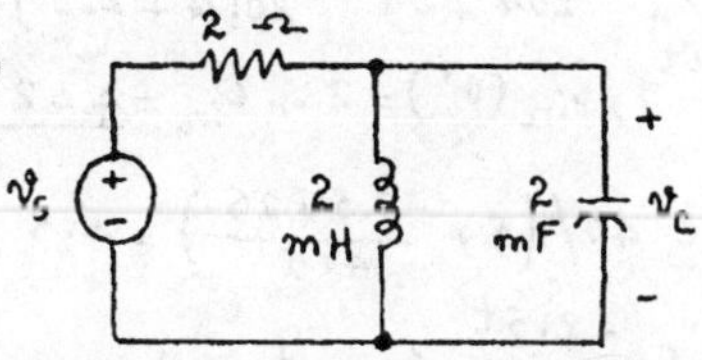

**12-26** (a)

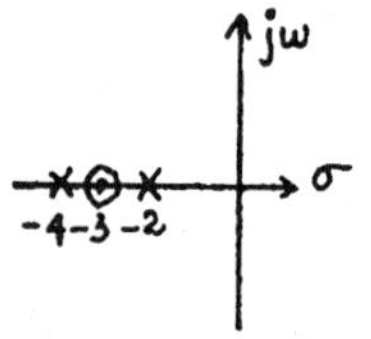

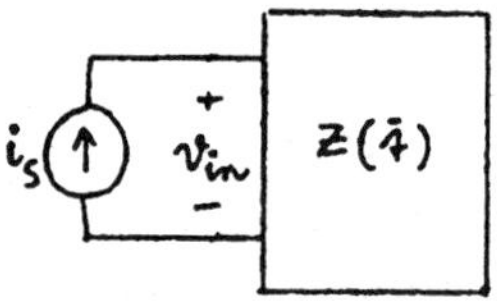

$Z(-1) = 30\ \Omega$, $i_s(t) = 0.2\,u(t)$ A

$Z(\bar{s}) = \dfrac{K(\bar{s}+3)}{(\bar{s}+2)(\bar{s}+4)}$ ∴ $30 = \dfrac{K(2)}{1\times 3}$ ∴ $K = 45$

$\dfrac{V_{in}}{I_s} = \dfrac{45(\bar{s}+3)}{(\bar{s}+2)(\bar{s}+4)}$, $I_s = 0.2$ at $\bar{s}=0$

$V_{in,f} = \dfrac{45\times 3 \times 0.2}{2\times 4} = 3.375$ V

(b) Poles at $\bar{s} = -2$ and $-4$ ∴ $v_{in}(t) = 3.375 + Ae^{-2t} + Be^{-4t}$; $v_{in}(0^+) = 0$, $v'_{in}(0^+) = 9$

∴ $A + B = -3.375$
  $-2A - 4B = 9$ 
∴ $A = -2.25$, $B = -1.125$

∴ $\underline{v_{in} = 3.375 - 2.25e^{-2t} - 1.125e^{-4t}}$ V, $t>0$

{ Extra: possible network is: }

**12-27**

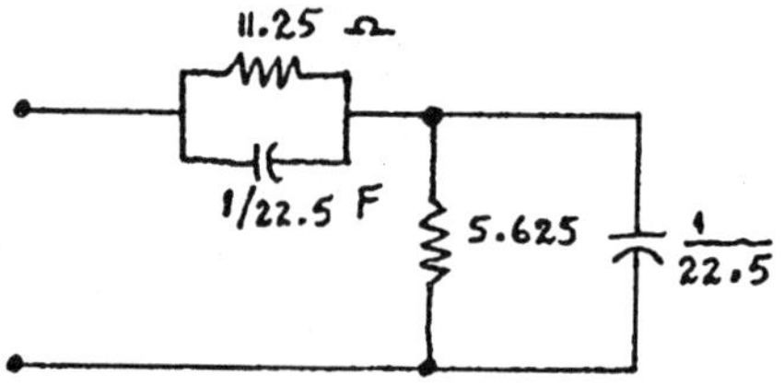

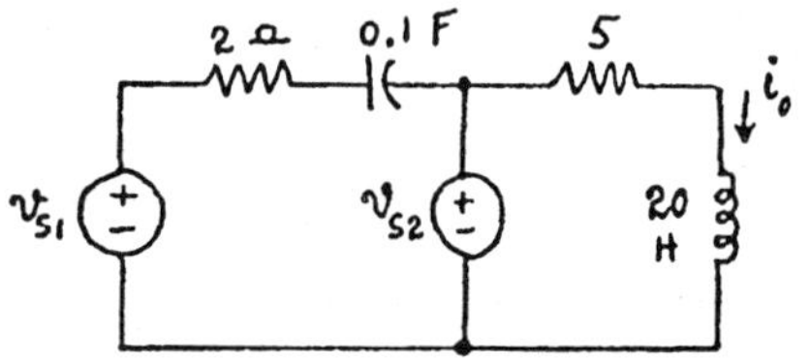

(a) $H_1 = \dfrac{I_0}{V_{s1}} = 0$, $H_2 = \dfrac{I_0}{V_{s2}} = \dfrac{1}{5+20\bar{s}} = \dfrac{0.05}{\bar{s}+0.25}$ ←

(b) Use superp. $V_{s2,A} = 5$ V ∴ $i_{0,A} = 1$ A

$V_{s2,B} = 25\,u(t)$ V ∴ $V_{s2,B} = 25$ at $\bar{s}=0$ ∴ $I_{0,B} = 5$ A

∴ $i_{0,B} = 5 - 5e^{-0.25t}$ ∴ $\underline{i_0 = 1 + 5(1 - e^{-0.25t})u(t)}$ A (all $t$)

**12-28** (a)

$Z_A = \dfrac{15\bar{s}+25}{\bar{s}+5}$; $V_s = 40$ at $\bar{s}=0$ ∴ $I_{in,f} = \dfrac{40}{5+5} = 4$ A

$H(\bar{s}) = \dfrac{I_{in}}{V_s} = \dfrac{1}{5+Z_A} = \dfrac{\bar{s}+5}{20\bar{s}+50} = \dfrac{\bar{s}+5}{20(\bar{s}+2.5)}$

∴ $i_{in} = 4 + Ae^{-2.5t}$, $i_{in}(0^+) = 2$ ∴ $\underline{i_{in} = 4 - 2e^{-2.5t}}$ A, $t>0$

(b) $v_s = 40e^{-0.5t}$ ∴ $V_s = 40$ at $\bar{s} = -0.5$ ∴ $I_{in,f} = 40/\left(5 + \dfrac{-7.5+25}{4.5}\right) = 4.5$ A

∴ $i_{in} = 4.5e^{-0.5t} + Be^{-2.5t}$ ∴ $\underline{i_{in} = 4.5e^{-0.5t} - 2.5e^{-2.5t}}$ A, $t>0$

{ Extra: possible network is }

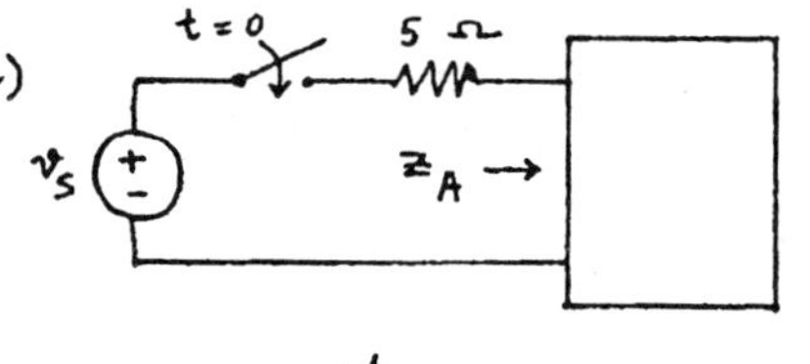

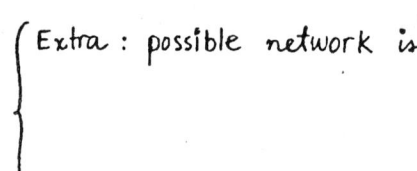

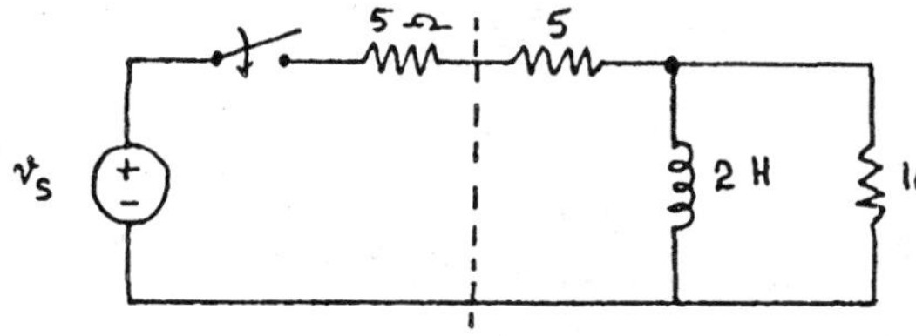

**12-29** (a)

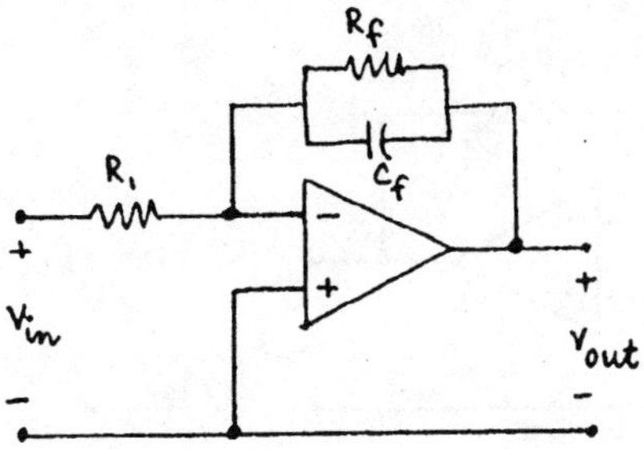


$R_f = 10^4\ \Omega$ ; $\dfrac{V_{out}}{V_{in}} = H(s) = \dfrac{-1/R_1C_f}{s + 1/R_fC_f}$

$H(s) = \dfrac{-10^5}{s+10^4} = \dfrac{-1/R_1C_f}{s + 10^{-4}/C_f}$ $\therefore C_f = \underline{10^{-8}\ F}$

$R_1C_f = 10^{-5} = 10^{-8} R_1$ $\therefore R_1 = \underline{1\ k\Omega}$

(b) $\dfrac{-1/R_1C_f}{s + 10^{-4}/C_f} = \dfrac{-10^{-4}}{s+10^5}$ $\therefore C_f = \underline{10^{-9}\ F}$ , $R_1C_f = 10^{-4}$ $\therefore R_1 = \underline{100\ k\Omega}$

(c) $-10 = \dfrac{-1/R_1}{C_f s + 10^{-4}}$ $\therefore C_f = \underline{0}$ , $R_1 = \underline{1\ k\Omega}$

(d) $-1 = \dfrac{-1/R_1}{C_f s + 10^{-4}}$ $\therefore C_f = \underline{0}$ , $R_1 = \underline{10\ k\Omega}$

(e) $C_{fA} = \underline{10^{-8}\ F}$ , $R_{1A} = \underline{1\ k\Omega}$ ; $C_{fB} = \underline{0}$ , $R_{1B} = \underline{10\ k\Omega}$

**12-30**

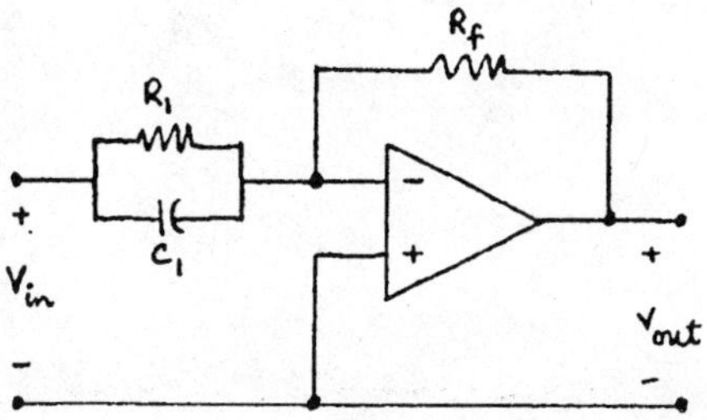


$R_f = 10^4\ \Omega$ , $\dfrac{V_{out}}{V_{in}} = H(s) = -R_fC_1(s + 1/R_1C_1)$

(a) $-(1 + 10^{-4}s) = -10^{-4}(s + 10^4) = -10^4C_1(s + 1/R_1C_1)$

$\therefore C_1 = \underline{10^{-8}\ F}$ , $R_1 = \underline{10\ k\Omega}$

(b) $-10^4C_1(s + 1/R_1C_1) = -10(1 + 10^{-4}s) = -10^{-3}(s + 10^4)$

$\therefore C_1 = \underline{10^{-7}\ F}$ , $R_1 = \underline{1\ k\Omega}$

(c) Let $C_1 = \underline{0}$ $\therefore H(s) = -10 = -\dfrac{R_f}{R_1}$ $\therefore R_1 = \underline{1\ k\Omega}$

(d) $H(s) = -1$ $\therefore C_1 = \underline{0}$ , $R_1 = \underline{10\ k\Omega}$

(e) $C_{1A} = \underline{10^{-7}\ F}$ , $R_{1A} = \underline{1\ k\Omega}$ ; $C_{1B} = \underline{0}$ , $R_{1B} = \underline{10\ k\Omega}$

**12-31** (a)

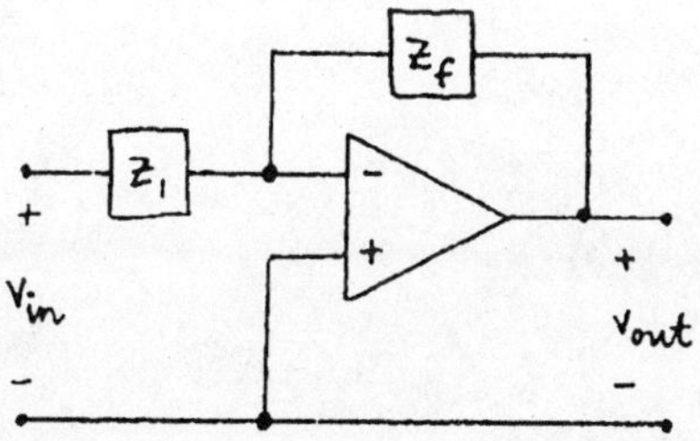


$\dfrac{V_{out}}{V_{in}} = -\dfrac{Z_f}{Z_1}$

$Z_1 = R_1$ , $Z_f = R_f + 1/sC_f$

$\therefore H(s) = \dfrac{V_{out}}{V_{in}} = -\dfrac{R_f + 1/sC_f}{R_1} = -\dfrac{sC_fR_f + 1}{sC_fR_1}$ ←

or $H(s) = -\dfrac{R_f}{R_1}\ \dfrac{s + 1/R_fC_f}{s}$ ←

(b) $Z_1 = R_1 + \dfrac{1}{sC_1}$ , $Z_f = R_f$ $\therefore H(s) = -\dfrac{R_f}{R_1 + 1/sC_1} = -\dfrac{sC_1R_f}{sC_1R_1 + 1}$ ←

(c) $Z_1 = R_1 + \dfrac{1}{sC_1}$ , $Z_f = R_f + \dfrac{1}{sC_f}$ $\therefore H(s) = -\dfrac{R_f + 1/sC_f}{R_1 + 1/sC_1} = -\dfrac{R_f}{R_1}\ \dfrac{s + 1/C_fR_f}{s + 1/C_1R_1}$ ←

12-32

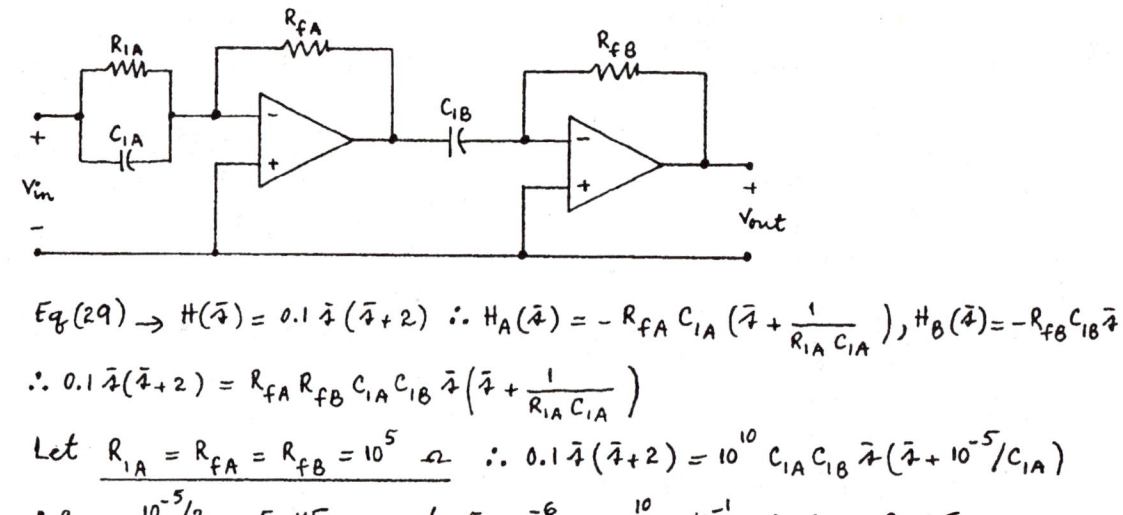

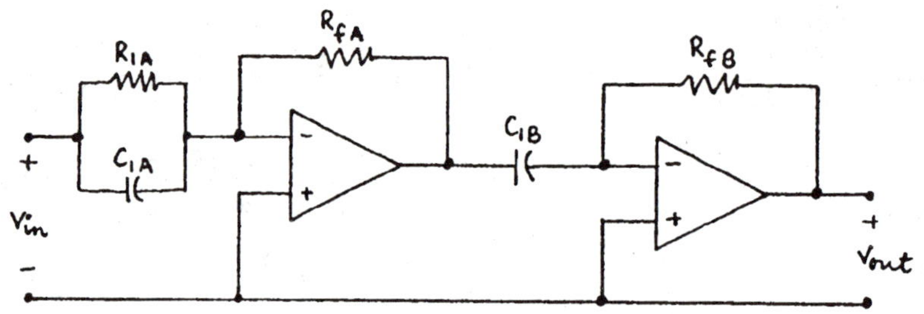


Eq (29) → $H(\bar{s}) = 0.1\,\bar{s}(\bar{s}+2)$ ∴ $H_A(\bar{s}) = -R_{fA}C_{1A}\left(\bar{s} + \dfrac{1}{R_{1A}C_{1A}}\right)$, $H_B(\bar{s}) = -R_{fB}C_{1B}\bar{s}$

∴ $0.1\,\bar{s}(\bar{s}+2) = R_{fA}R_{fB}C_{1A}C_{1B}\,\bar{s}\left(\bar{s} + \dfrac{1}{R_{1A}C_{1A}}\right)$

Let $\underline{R_{1A} = R_{fA} = R_{fB} = 10^5\ \Omega}$ ∴ $0.1\,\bar{s}(\bar{s}+2) = 10^{10}\,C_{1A}C_{1B}\,\bar{s}(\bar{s} + 10^{-5}/C_{1A})$

∴ $C_{1A} = 10^{-5}/2 = \underline{5\ \mu F}$ and $5\times10^{-6}\,C_{1B}\,10^{10} = 10^{-1}$ ∴ $C_{1B} = \underline{2\ \mu F}$

# CHAPTER 13: FREQUENCY RESPONSE

## Section 13-2 Parallel resonance
## Section 13-3 More about parallel resonance

**1** For the resonant circuit shown in Fig. 13-1, find: (*a*) $\omega_0$; (*b*) $f_0$; (*c*) $Q_0$; (*d*) the critical frequencies of $\mathbf{Z}_{in}(\mathbf{s})$; (*e*) $|\mathbf{I}_L|$ if $i = 5\sin \omega_0 t$ mA is connected to the input; (*f*) $|\mathbf{I}_R|$ if $i = 5\sin \omega_0 t$ mA is connected to the input.

**2** Design a parallel resonant circuit that will have admittance zeros at $\mathbf{s} = -20 \pm j160\ \text{s}^{-1}$ and an admittance magnitude of 2 mS at $\mathbf{s} = j160\ \text{s}^{-1}$.

**3** Find values for $R$, $L$, and $C$ in the circuit of Fig. 13-2 if: (*a*) $\mathbf{I} = 2\underline{/0°}\ \mu\text{A}$, $|\mathbf{V}|_{max} = 125$ mV, $Q_0 = 40$, and $f_0 = 1$ MHz; (*b*) $\mathbf{Z}_{in}$ seen by the current source is $8 + j0$ kΩ at $\omega = 2000$ rad/s and $4\underline{/\theta}$ kΩ at $\omega = 1800$ rad/s.

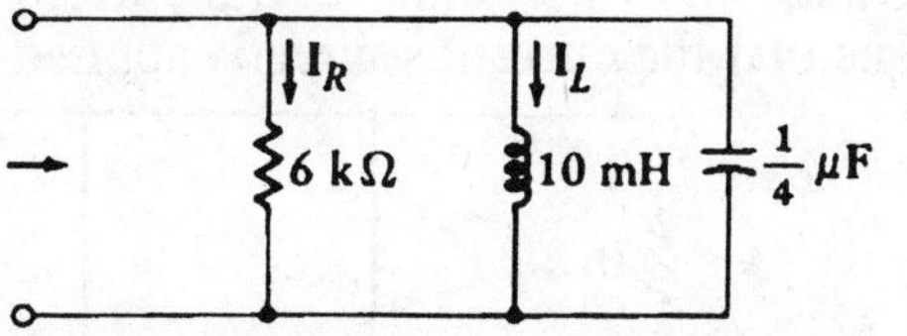

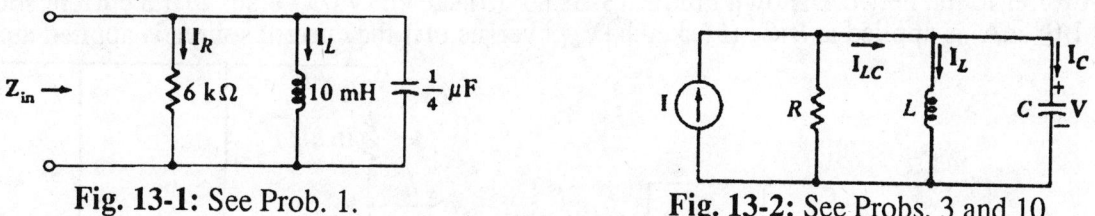

**Fig. 13-1:** See Prob. 1.

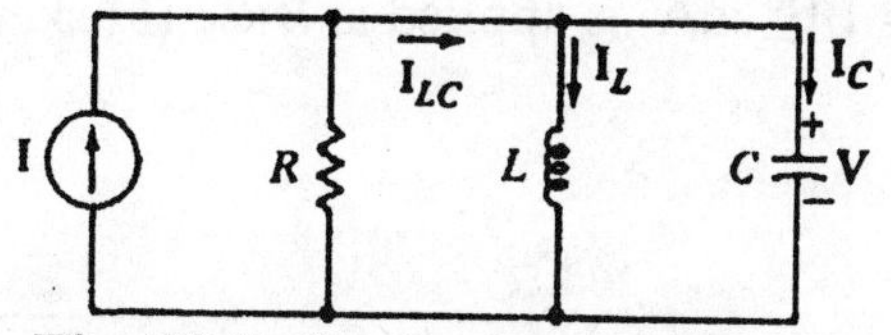


**Fig. 13-2:** See Probs. 3 and 10.

**4** The pole-zero plot for the input admittance of a parallel resonant circuit shows zeros at $-100 \pm j1600\ \text{s}^{-1}$, and a pole at the origin. If the value of $C$ is 20 μF, find: (*a*) $Q_0$; (*b*) $\omega_0$; (*c*) $\alpha$; (*d*) $\zeta$; (*e*) $R$; (*f*) $L$. (*g*) At what frequencies is $|\mathbf{Y}_{in}| = 2/R$?

**5** (*a*) Find the exact resonant frequency of the network shown in Fig. 13-3 and (*b*) the value of $\mathbf{Z}_{in}$ at that frequency.

**6** Design a parallel resonant circuit so that a variable capacitor can adjust the resonant frequency over the AM broadcast band, 535 to 1605 kHz, with $Q_0 \le 45$ at any frequency in the band. Let $R = 25$ kΩ and give values for $L$, $C_{min}$, and $C_{max}$.

**7** First derive an expression for the input admittance of the network appearing in Fig. 13-4, and then use it to determine $\omega_0$ and $Q_0$.

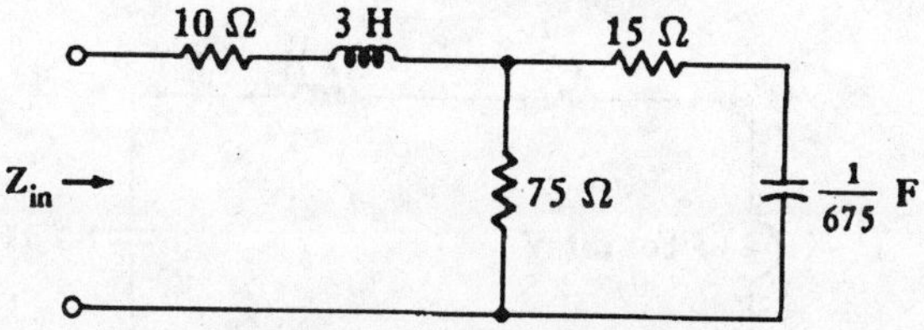

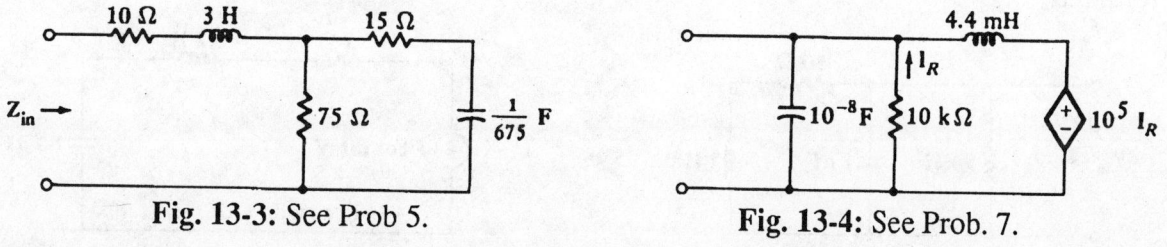

**Fig. 13-3:** See Prob 5.

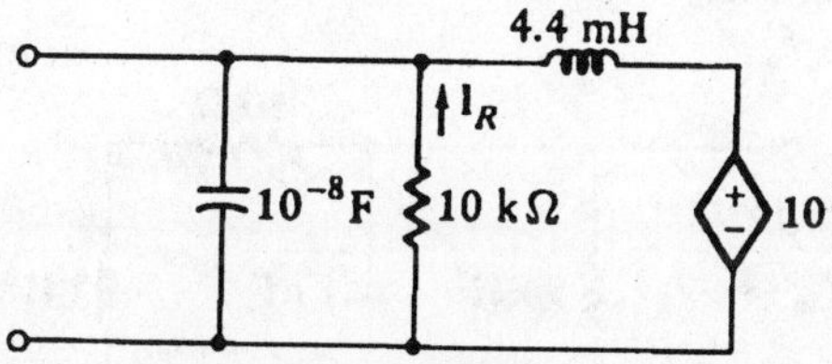


**Fig. 13-4:** See Prob. 7.

**8** A parallel resonant circuit has $\omega_0 = 10$ krad/s, $Q_0 = 8$, and an impedance at resonance of 500 Ω. (*a*) Find $\mathcal{B}$. (*b*) Determine $N$ at 9.2 krad/s. (*c*) Use approximate methods to determine $\mathbf{Z}_{in}$ at 9.2 krad/s. (*d*) Determine the true value of $\mathbf{Z}_{in}$ at that frequency. (*e*) Calculate the percentage error for $|\mathbf{Z}_{in}|$ and ang $\mathbf{Z}_{in}$.

**9** Determine the element values, $Q_0$ and $\mathcal{B}$, for a parallel $RLC$ circuit for which $C = 50\ \mu$F and: (*a*) $\omega_0 = 100$ and $\omega_2 = 120$ rad/s; (*b*) $\omega_0 = 100$ and $\omega_1 = 80$ rad/s; (*c*) $\omega_1 = 80$ and $\omega_2 = 120$ rad/s.

**10** (*a*) Use approximate methods to prepare a sketch of $|\mathbf{V}|$ versus $\omega$, $0.9\omega_0 \leq \omega \leq 1.1\omega_0$, for the parallel resonant circuit of Fig. 13-2 if $R = 8$ kΩ, $L = 0.25$ H, $C = 0.25$ μF, and $\mathbf{I} = 2\underline{/0^\circ}$ mA. Find $\mathbf{V}$ at 3250 rad/s by: (*b*) approximate methods; (*c*) exact methods.

**11** For the circuit shown in Fig. 13-5: (*a*) use approximate methods to plot a curve of $|\mathbf{V}_L|$ versus $\omega$, $920 \leq \omega \leq 1080$ rad/s; (*b*) find the exact value of $\mathbf{V}_L$ at $\omega = 1050$ rad/s.

**12** A parallel resonant circuit has $Q_0 = 20$ and is resonant at $\omega_0 = 10$ krad/s. If $\mathbf{Z}_{in} = 5$ kΩ at $\omega = \omega_0$, what is the width of the frequency band about resonance for which $|\mathbf{Z}_{in}| \geq 3$ kΩ?

**13** A parallel resonant circuit is driven by a current source such that the voltage across it has an amplitude of 100 V at resonance. If $f_0 = 500$ Hz and $Q_0 = 25$, use approximate relationships to find: (*a*) the voltage amplitude at 495 Hz; (*b*) the frequency at which $|\mathbf{V}_{out}| = 50$ V; (*c*) the frequency at which the phase angle of the voltage is 50° greater than it is at resonance.

**14** Refer to the network shown in Fig. 13-6 and: (*a*) sketch $|\mathbf{V}_{AG}|$ versus $\omega$ if a current source, $\mathbf{I}_s = 1\underline{/0^\circ}$ mA, is applied at *B*-*G*; (*b*) sketch $|\mathbf{V}_{BG}|$ versus $\omega$ if the current source is applied at *A*-*G*.

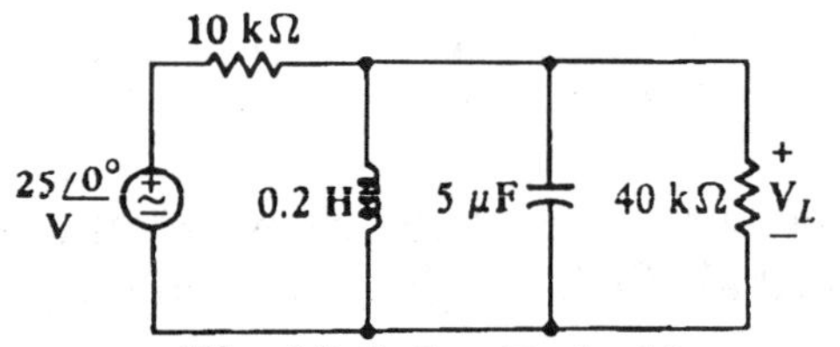


**Fig. 13-5:** See Prob. 11.

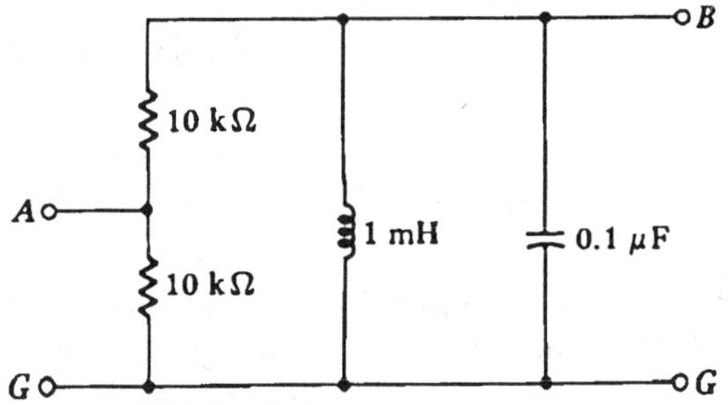


**Fig. 13-6:** See Prob. 14.

**15** Use accurate methods on the network of Fig. 13-7 to find: (*a*) $\omega_0$; (b) $\mathbf{Y}_{in,0}$; (*c*) $\omega_1$; (*d*) $\omega_2$; (*e*) $\mathcal{B}$; (*f*) $\omega_0/\mathcal{B}$.

## Section 13-4 Series resonance
## Section 13-5 Other resonant forms

**16** (*a*) Calculate values for $\omega_0$, $Q_0$, and $\mathcal{B}$ for the circuit of Fig. 13-8. (*b*) Determine $I_m$ at $\omega = \omega_0 \pm 10n$ rad/s, where $n = 0, 1, 2$, and 3, and plot the results. Use both the approximate and exact formulas.

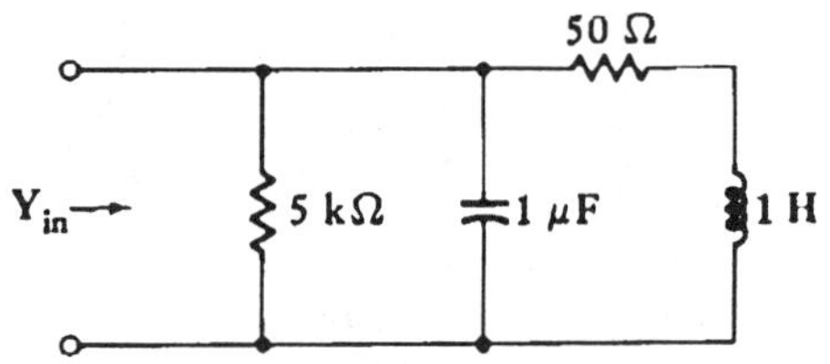

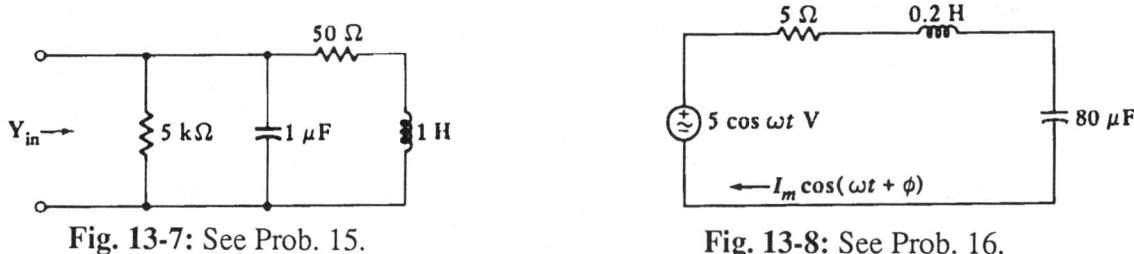

**Fig. 13-7:** See Prob. 15.

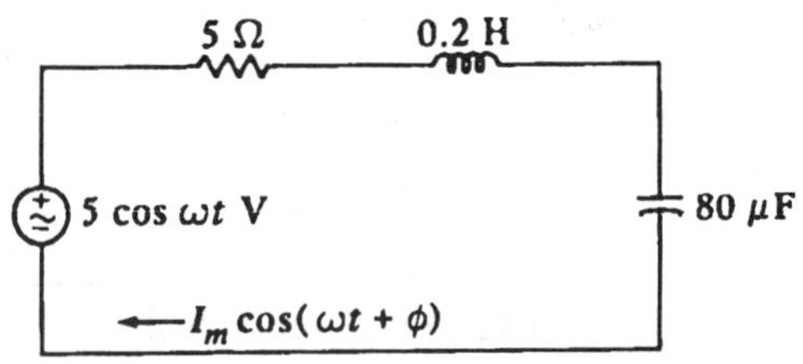


**Fig. 13-8:** See Prob. 16.

**17** A 5-mV sinusoidal signal is applied to a series *RLC* circuit containing 2 Ω, 20 μH, and a capacitor *C*. (*a*) Find the value of *C* that will cause $|V_C|$, the magnitude of the capacitor voltage, to be a maximum if the source is operating at 50 Mrad/s. What are the values of $Q_0$ and $|V_C|_{max}$? (*b*) Find the frequency at which the source should operate to cause $|V_C|$ to be a maximum if $C = 10$ pF. Find $Q_0$ and $|V_C|_{max}$. (*c*) Find the value of $|V_C|$ if $\omega = 33$ Mrad/s and $C = 45$ pF.

**18** (*a*) Find $\omega_0$ and $Q_0$ for the series resonant circuit of Fig. 13-9. (*b*) Sketch $|V_C|$ versus $\omega$.

**19** If the circuit shown in Fig. 13-10 were constructed in the lab, would you be willing to put your hands across the capacitor? Plot $|V_C|$ versus $\omega$ to help you decide.

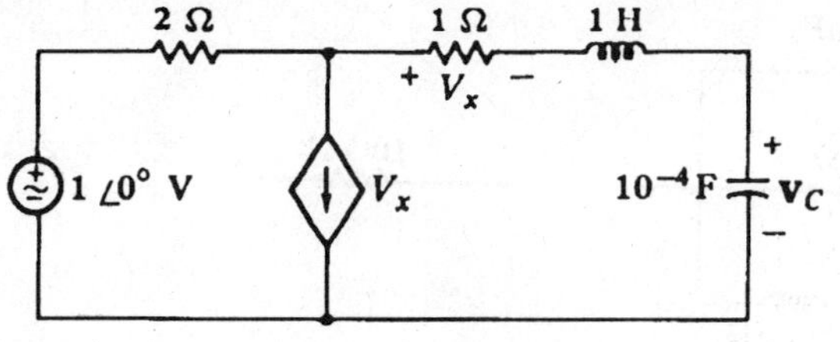


**Fig. 13-9:** See Prob. 18.

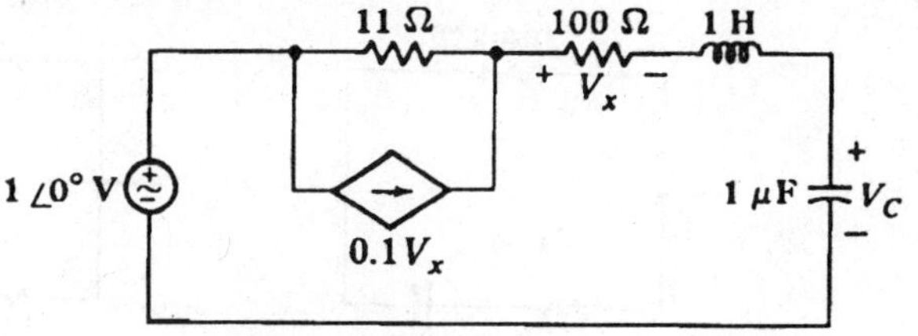


**Fig. 13-10:** See Prob. 19.

**20** (*a*) Calculate element values for a series resonant circuit having $f_0 = 5$ kHz, $\mathcal{B} = 180$ Hz, and a capacitive reactance at resonance of 600 Ω. (*b*) Apply a source voltage of 1-V amplitude and calculate the amplitude of the inductor voltage at 4.8, 5, and 5.2 kHz using exact methods.

**21** The input impedance $\mathbf{Z}_{in}(s)$ of a certain network has a pole at the origin, zeros at $-300 \pm j1000$ s$^{-1}$, and a pole at infinity. Specify the simplest configuration of the network completely if it is also true that: (*a*) $\mathbf{Z}_{in}(-200) = -5 + j0$ Ω; (*b*) $\mathbf{Z}_{in}(\mathbf{s}) \rightarrow \mathbf{s}/50$ as $\mathbf{s} \rightarrow \infty$; (*c*) $\mathrm{Re}[\mathbf{Z}_{in}(-100 + j500)| = 2$ Ω.

**22** In the circuit of Fig. 13-11, use good approximations to determine: (*a*) the resonant frequency; (*b*) the quality factor at resonance. (*c*) Sketch $|\mathbf{I}|$ versus $\omega$, showing the approximate behavior near resonance and the exact behavior as $\omega \rightarrow 0$ and $\omega \rightarrow \infty$.

**23** Use appropriate approximations on the circuit of Fig. 13-12 to find: (*a*) $Q_0$; (*b*) $\mathcal{B}$; (*c*) $v_2(t)$ if $v_s(t) = 12.5\cos 13\,000t$ V.

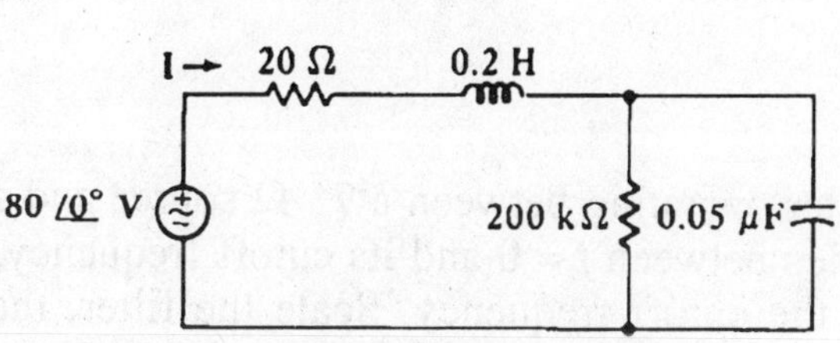


**Fig. 13-11:** See Prob. 22.

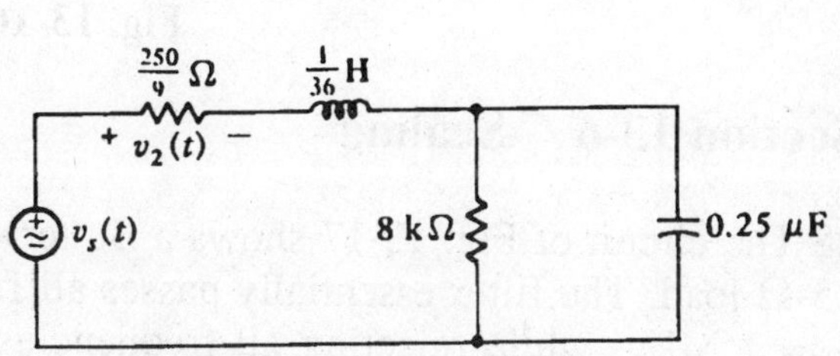


**Fig. 13-12:** See Prob. 23.

**24** For the circuit of Fig. 13-13, make a sketch of: (*a*) $|\mathbf{V}_{out}|$ versus $\omega$; (*b*) ang $\mathbf{V}_{out}$ versus $\omega$.

**25** (*a*) Show whether the circuit of Fig. 13-14 is similar to a parallel or to a series resonant circuit. (*b*) Find $\omega_0$ and $Q_0$.

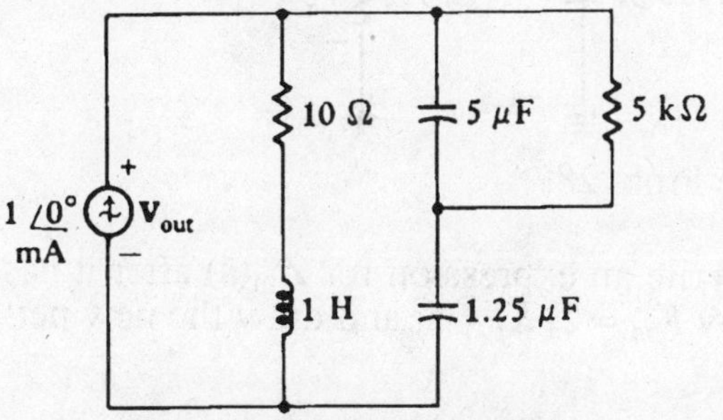


**Fig. 13-13:** See Prob. 24.

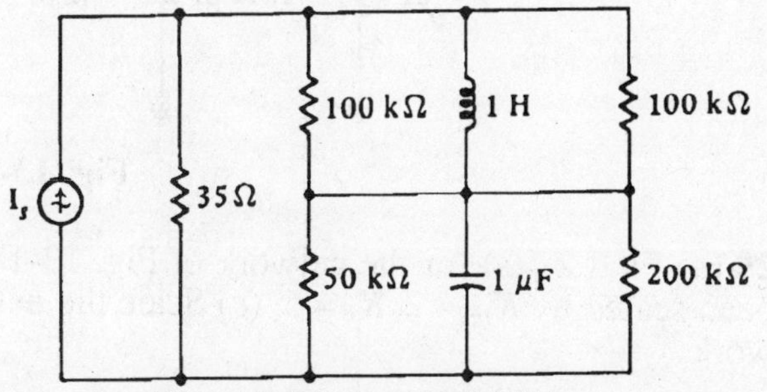


**Fig. 13-14:** See Prob. 25.

**26** A 2-nF capacitor, modeled by the network of Fig. 13-15*a*, a 100-$\mu$H inductor, modeled by *b*, and a 10-k$\Omega$ resistor, modeled by *c*, are all in parallel with a 1-mA current source. Plot the magnitude of the voltage across the 10-k$\Omega$ resistor versus $\omega$.

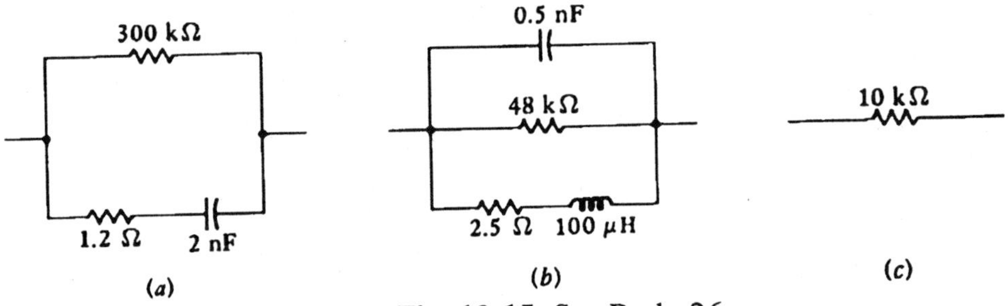


**Fig. 13-15:** See Prob. 26.

**27** Use approximate methods to find $\omega_0$ and $Q_0$ for each of the networks shown in Fig. 13-16.

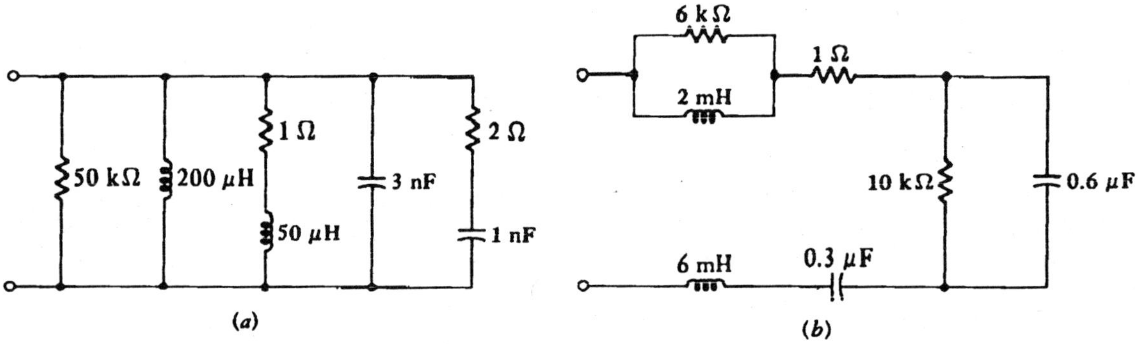


**Fig. 13-16:** See Prob. 27.

## Section 13-6 Scaling

**28** The circuit of Fig. 13-17 shows a Butterworth filter operating between a 75-$\Omega$ source and a 75-$\Omega$ load. The filter essentially passes all frequencies between $f = 0$ and its cutoff frequency, $f_0 = 1$ MHz, while rejecting all frequencies above the cutoff frequency. Scale the filter, the source, and the load so that the circuit operates between a 300-$\Omega$ source and load, and has a cutoff frequency of 60 MHz. Draw the scaled circuit.

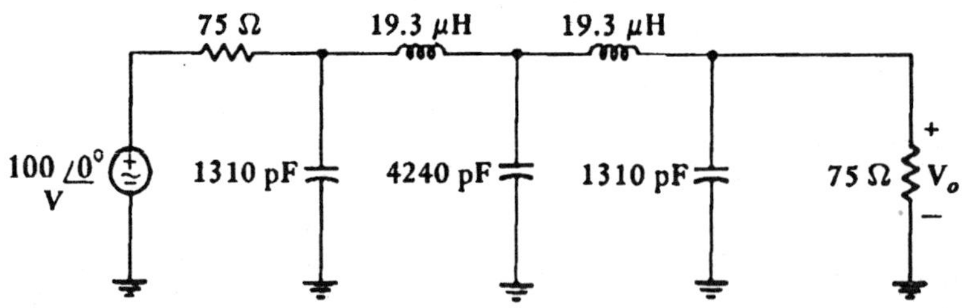


**Fig. 13-17:** See Prob. 28.

**29** (*a*) Find $\mathbf{Z}_{in}(s)$ for the network of Fig. 13-18. (*b*) Write an expression for $\mathbf{Z}_{in}(s)$ after it has been scaled by $K_m = 2$, $K_f = 5$. (*c*) Scale the network by $K_m = 2$, $K_f = 5$, and draw the new network.

**30** (*a*) Sketch |**I**| versus $\omega$ for the circuit shown in Fig. 13-19. Suitable approximations are encouraged. (*b*) Scale the circuit so that all critical frequencies are increased by a factor of 1000, while the current amplitude is increased by a factor of 20. (*c*) Sketch |**I**| versus $\omega$ for the scaled circuit.

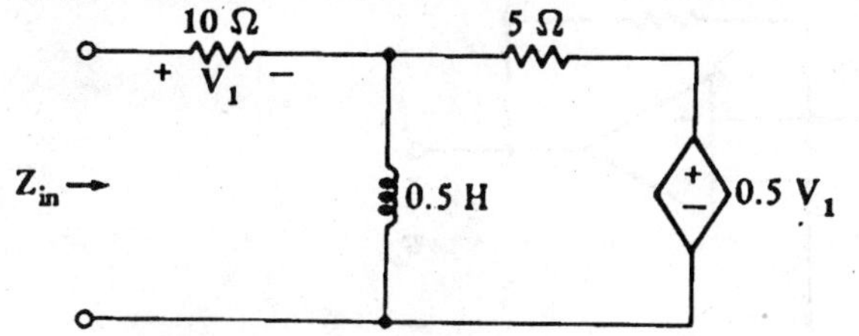

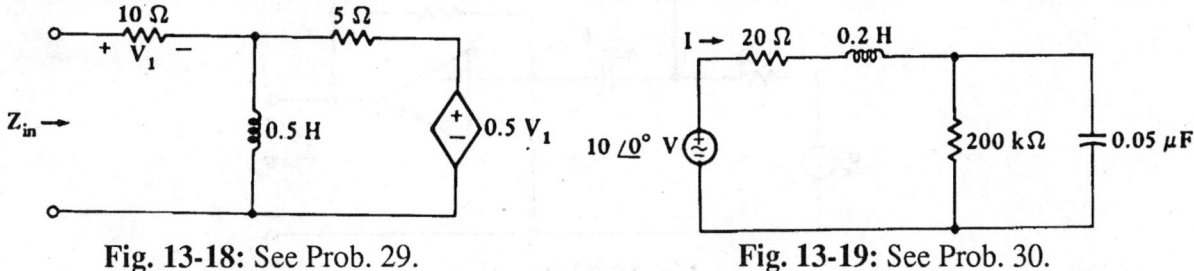

**Fig. 13-18:** See Prob. 29.

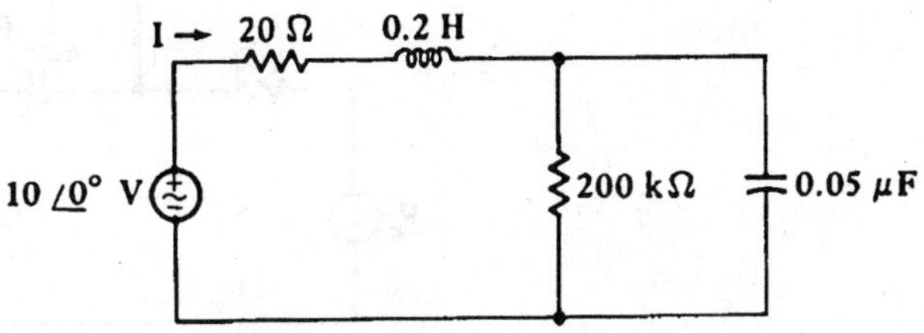


**Fig. 13-19:** See Prob. 30.

**31** A network composed entirely of ideal *R*'s. *L*'s, and *C*'s has a pair of input terminals to which a sinusoidal current source $\mathbf{I}_s$ is connected, and a pair of open-circuited output terminals at which a voltage $\mathbf{V}_{out}$ is defined. If $\mathbf{I}_s = 1\underline{/0°}$ A at $\omega = 100$ rad/s, then $\mathbf{V}_{out} = 60\underline{/40°}$ V. Specify $\mathbf{V}_{out}$ for each condition described below. If it is impossible to determine the value of $\mathbf{V}_{out}$, write OTSK.[1] (*a*) $\mathbf{I}_s = 1\underline{/0°}$ A, $\omega = 1000$ rad/s; (*b*) $\mathbf{I}_s = 1\underline{/60°}$ A, $\omega = 100$ rad/s; (*c*) $\mathbf{I}_s = 10\underline{/0°}$ A, $\omega = 100$ rad/s; (*d*) the network is scaled by $K_m = 20$, $\mathbf{I}_s = 10\underline{/0°}$ A, $\omega = 100$ rad/s; (*e*) $K_m = 20$, $K_f = 10$, $\mathbf{I}_s = 10\underline{/0°}$ A, $\omega = 1000$ rad/s.

**32** Scale the network of Fig. 13-20 and draw the result if: (*a*) the circuit is scaled in magnitude so that the new *L* is 5 mH; (*b*) the circuit is scaled in frequency so that the new *L* is 5 mH; (*c*) the circuit is scaled in magnitude and frequency so that the new *L* is 5 mH and the new *C* is 0.2 $\mu$F.

**33** (*a*) Show that the transfer function for the ideal-op-amp circuit of Fig. 13-21 is $\mathbf{H}(s) = \mathbf{V}_o(s)/\mathbf{V}_g(s) = -600s/(s^2 + 3s + 900)$. (*b*) Sketch $|\mathbf{H}(j\omega)|$ versus $\omega$, $27 \leq \omega \leq 33$ rad/s. (*c*) Find $\omega_{max}$ where $|\mathbf{H}(j\omega_{max})|_{max}$ occurs. (*d*) Determine the half-power bandwidth of the response $|\mathbf{H}(j\omega)|$. (*e*) Scale the network so that the source has an output resistance of 75 Ω and $|\mathbf{H}(j\omega)|_{max}$ occurs at 10 krad/s.

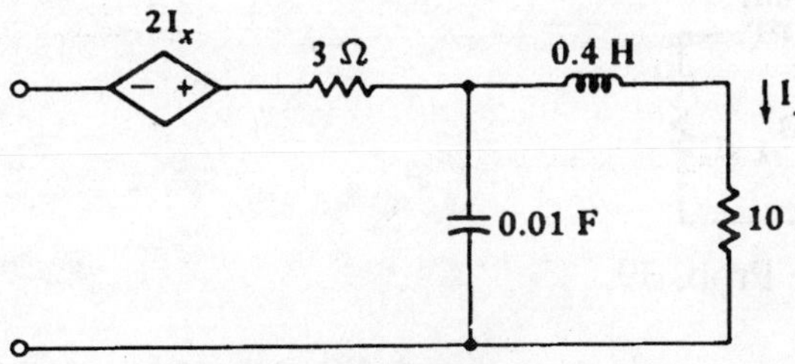

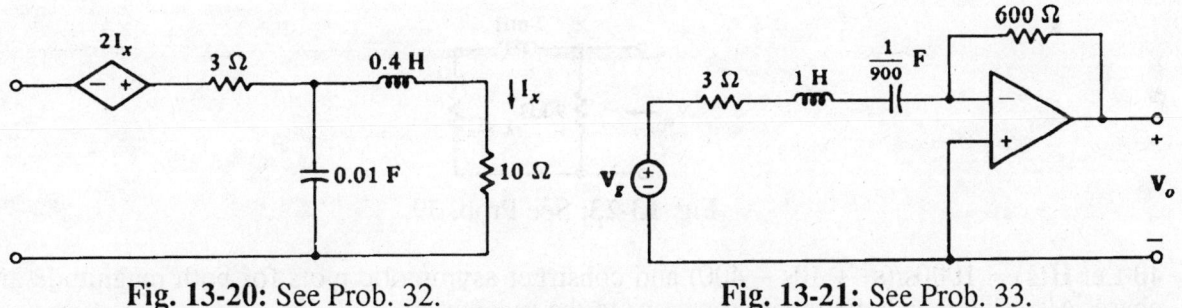

**Fig. 13-20:** See Prob. 32.

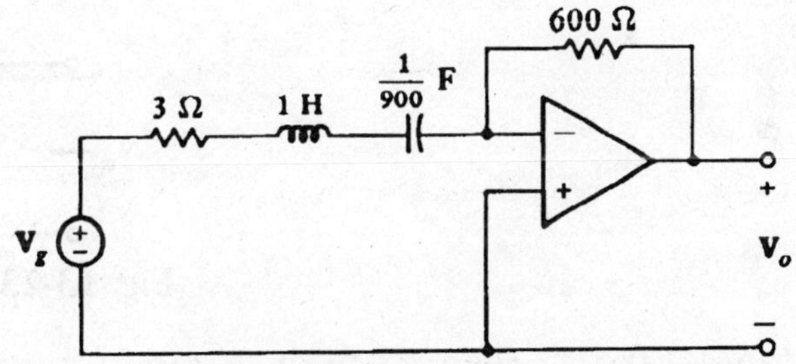


**Fig. 13-21:** See Prob. 33.

**34** The ideal-op-amp circuit of Fig. 13-22 contains no inductors, and it is interesting to compare Probs. 33 and 34. (*a*) Show that the transfer function for the ideal op-amp circuit of Fig. 13-22 is $\mathbf{H}(s) = \mathbf{V}_o(s)/\mathbf{V}_g(s) = -600s/(s^2 + 3s + 900)$. (*b*) Sketch $|\mathbf{H}(j\omega)|$ versus $\omega$, $27 \leq \omega \leq 33$ rad/s. (*c*) Find $\omega_{max}$ where $|\mathbf{H}(j\omega_{max})|_{max}$ occurs. (*d*) Determine the half-power bandwidth of the response $|\mathbf{H}(j\omega)|$. (*e*) Scale the network so that the source has an output resistance of 75 Ω and $|\mathbf{H}(j\omega)|_{max}$ occurs at 10 krad/s.

---

[1] Only The Shadow Knows.

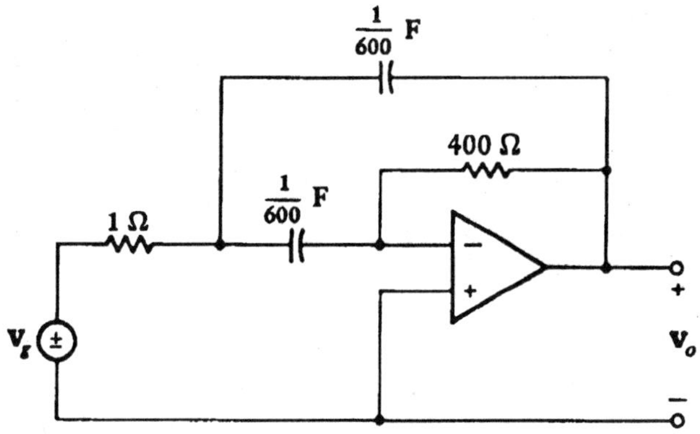


Fig. 13-22: See Prob. 34.

## Section 13-7 Bode diagrams

**35** If $\mathbf{G}(s) = 10^4 s/[(s + 1)(s + 10\,000)]$: (*a*) prepare a Bode amplitude plot and use the asymptotes to approximate the values of $\omega$ at which $|\mathbf{G}(j\omega)| = |\mathbf{G}(j100)|/10$; (*b*) calculate the exact values of $\omega$ at which $|\mathbf{G}(j\omega)| = |\mathbf{G}(j100)|/10$.

**36** Construct asymptotic magnitude and phase plots for $\mathbf{H}(s)$ =: (*a*) $5000(s + 10)/(s^2 + 100s)$; (*b*) $0.2(s + 100)^2/s$.

**37** Draw the asymptotic amplitude and asymptotic phase plots for the transfer function $\mathbf{H}(s) = (s + 1)(s + 1000)^2/[(s + 10)^2(s + 10\,000)]$.

**38** Let $\mathbf{H}(s) = 40s^2/(s + 5)$, and construct magnitude and phase diagrams, giving numerical coordinates for all corners and intercepts on the magnitude plot if the origin is at $H_{dB} = 0$, $\omega = 1$ rad/s.

**39** Draw the Bode diagram for the input admittance of the network shown in Fig. 13-23. Give both magnitude and phase.

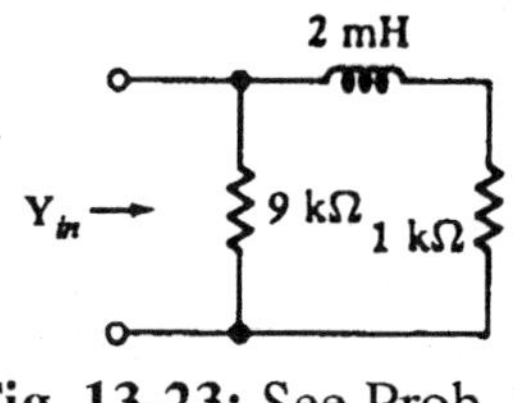


Fig. 13-23: See Prob. 39.

**40** Let $\mathbf{H}(s) = 1000s/(s^2 + 10s + 400)$ and construct asymptotic plots for both magnitude and phase. Also sketch in better approximations to the true curves.

**41** Express $\mathbf{H}(s)$ as the ratio of two polynomials in s if the Bode magnitude plot is that shown in Fig. 13-24.

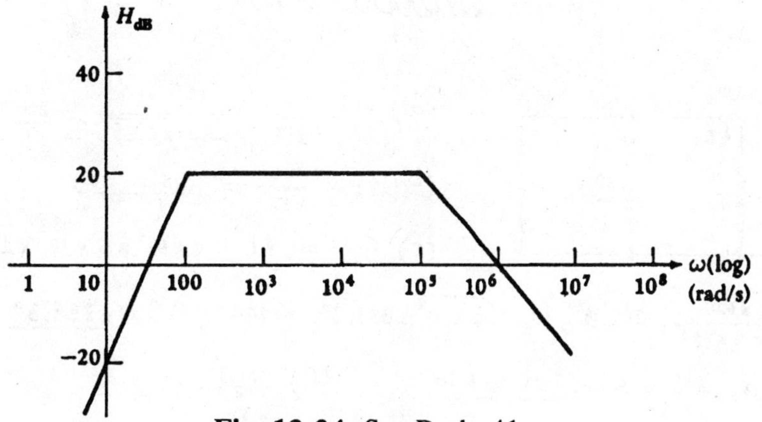


**Fig. 13-24:** See Prob. 41.

**42** Draw the pole-zero plot and construct the asymptotic magnitude plot for $|\mathbf{I}_R/\mathbf{I}_s|$ in the circuit of Fig. 13-25 if $R$ =: (*a*) 500 Ω; (*b*) 400 Ω; (*c*) 200 Ω.

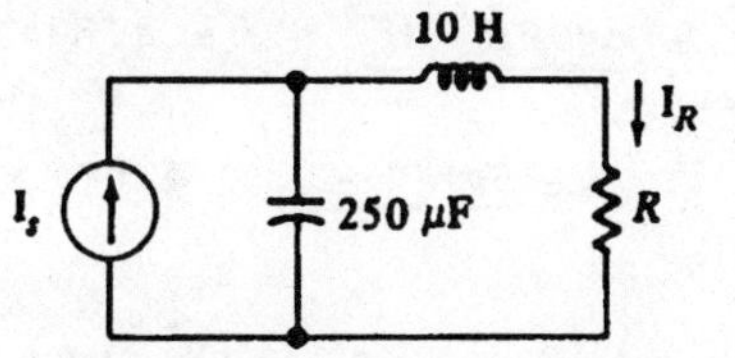


**Fig. 13-25:** See Prob. 42.

**43** (SPICE) Use SPICE on the network of Fig. 13-26 to calculate the magnitude of the input impedance at the frequencies 10, 11, 12, . . . , 22, 23, 24 Hz.

**44** (SPICE) Use a SPICE program to determine the magnitude of the output voltage in Fig. 13-27 at sixteen frequencies spaced at equal logarithmic intervals from 100 to 1000 Hz, inclusive.

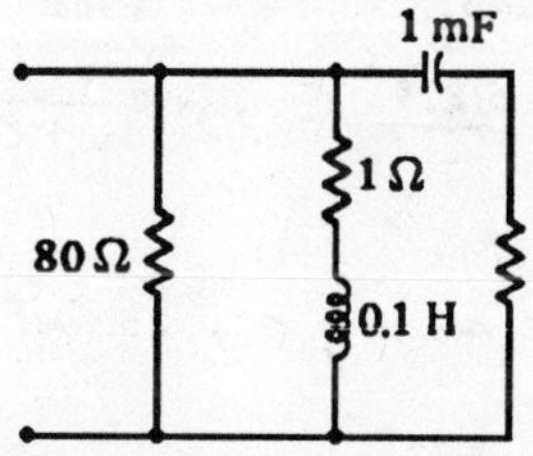


**Fig. 13-26:** See Prob. 43.

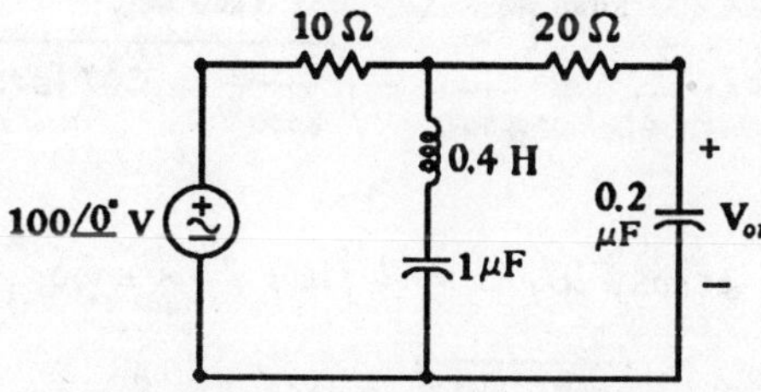


**Fig. 13-27:** See Prob. 44.

# Chapter 13

**13-1**

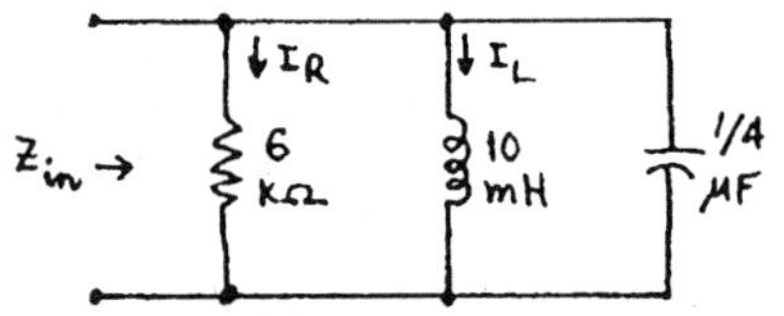


(a) $\omega_o = \dfrac{1}{\sqrt{LC}} = \dfrac{10^3}{\sqrt{0.01 \times 0.25}} = \underline{2 \times 10^4 \text{ rad/s}}$

(b) $f_o = \dfrac{\omega_o}{2\pi} = \underline{3183 \text{ Hz}}$

(c) $Q_o = \omega_o RC = 2 \times 10^4 \times 6 \times 10^3 \times \frac{1}{4} \times 10^{-6} = \underline{30}$

(d) $\alpha = \dfrac{4 \times 10^6}{2 \times 6 \times 10^3} = 333.3$, $\omega_d = \sqrt{4 \times 10^8 - 333.3^2} = 19997$; Poles: $\underline{-333.3 \pm j19{,}997 \text{ s}^{-1}}$; Zeros: $\underline{0, \infty}$

(e) $|I_L|_o = Q_o I_s = 30 \times 5 = \underline{150 \text{ mA}}$ (f) $|I_R| = \underline{5 \text{ mA}}$

**13-2**

$Y = K \dfrac{(s + 20 + j160)(s + 20 - j160)}{s} = K \dfrac{s^2 + 40s + 26{,}000}{s}$ $\therefore 2 \times 10^{-3} = K \left| \dfrac{-25{,}600 + j6400 + 26000}{160} \right|$

$\therefore K = 4.990 \times 10^{-5} = C = \underline{49.90 \ \mu F}$; $Y = 4.990 \times 10^{-5} \left(40 + s + \dfrac{26{,}000}{s}\right)$

$\dfrac{1}{R} = 40 \times 4.990 \times 10^{-5}$ $\therefore$ $R = \underline{501.0 \ \Omega}$; $4.990 \times 10^{-5} \times 26{,}000 = \dfrac{1}{L}$ $\therefore L = \underline{0.7707 \text{ H}}$

**13-3** (a)

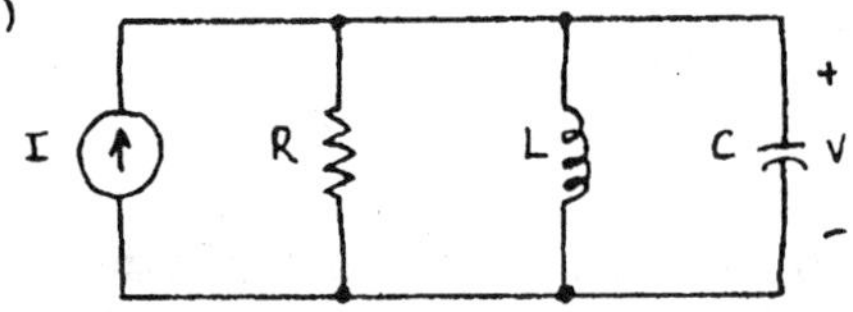


$I = 2\angle 0^\circ \ \mu A$, $|V|_{max} = 0.125$ V, $Q_o = 40$, $f_o = 1$ MHz

$R = \dfrac{|V|_{max}}{I} = \dfrac{125 \times 10^{-3}}{2 \times 10^{-6}} = \underline{62.5 \text{ k}\Omega}$

$40 = 2\pi 10^6 C \times 62.5 \times 10^3$ $\therefore C = \underline{101.86 \text{ pF}}$; $L = \dfrac{1}{\omega_o^2 C} = \underline{0.249 \text{ mH}}$

(b) $Z_{in,o} = 8 + j0 \text{ k}\Omega$ at $\omega_o = 2000$; $Z_{in} = 4 \times 10^3 \angle\theta$ at $\omega = 1800$ $\therefore R = \underline{8 \text{ k}\Omega}$

$Y = \dfrac{1}{8000} + j\left(1800C - \dfrac{1}{1800L}\right)$, $\left. LC = \dfrac{1}{4 \times 10^6} \right\}$ $\therefore Y = \dfrac{1}{8000} + j\left(1800 - \dfrac{4 \times 10^6}{1800}\right)C$

$\therefore \dfrac{1}{Z_{in}} = \dfrac{1}{4000} = \sqrt{\dfrac{1}{8000^2} + C^2\left(1800 - \dfrac{4 \times 10^6}{1800}\right)^2}$ $\therefore C = \underline{512.8 \text{ nF}}$; $L = \dfrac{1}{\omega_o^2 C} = \underline{0.4875 \text{ H}}$

**13-4** Zeros at $-100 \pm j1600$ $\therefore \alpha = 100$, $\omega_d = 1600$; Pole at $s = 0$; $C = 20 \ \mu F$

$\omega_o = \sqrt{100^2 + 1600^2} = 1603.1$, $R = \dfrac{1}{2\alpha C} = \dfrac{10^6}{2 \times 100 \times 0.02} = 250 \ \Omega$

(a) $Q_o = \omega_o RC = \underline{8.016}$ (b) $\omega_o = \underline{1603.1 \text{ rad/s}}$ (c) $\alpha = \underline{100 \text{ Np/s}}$

(d) $\zeta = \dfrac{1}{2Q_o} = \underline{0.06238}$ (e) $R = \underline{250 \ \Omega}$ (f) $L = \dfrac{1}{\omega_o^2 C} = \underline{19.455 \text{ mH}}$

(g) $Y_{in} = \dfrac{1}{250}\left[1 + j8.016\left(\dfrac{\omega}{1603.1} - \dfrac{1603.1}{\omega}\right)\right]$, $|Y_{in}| = \dfrac{2}{250}$ $\therefore 8.016\left(\dfrac{\omega}{1603.1} - \dfrac{1603.1}{\omega}\right) = \sqrt{3}$

$\therefore \omega = \underline{1785.6}$ and $\underline{1439.3 \text{ rad/s}}$

**13-5** (a)

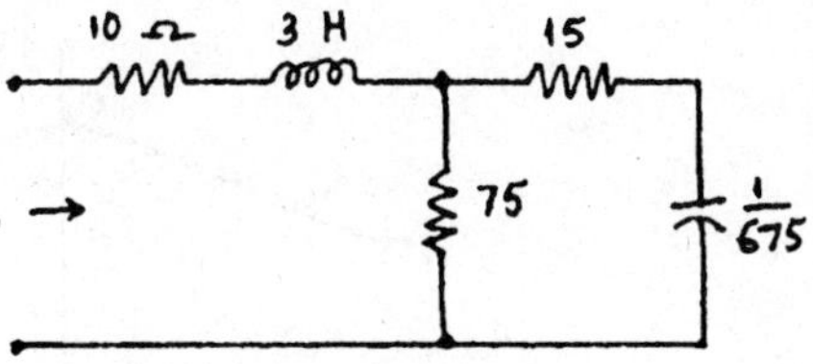


$$Z_1 = 75 // (15 + 675/s) = \frac{75(15+675/s)}{90+675/s} = \frac{75(1+45/s)}{6+45/s}$$

$$Z_1 = \frac{25(s+45)}{2s+15} \therefore Z_1(j\omega) = \frac{25(45+j\omega)}{15+j2\omega} \times \frac{15-j2\omega}{15-j2\omega}$$

$$Z_1 = \frac{25(675+2\omega^2 - j75\omega)}{225+\omega^2} \therefore 3\omega_0 = \frac{25(75\omega_0)}{225+4\omega_0^2}$$

$$\therefore 225 + 4\omega_0^2 = 625, \quad \omega_0 = \underline{10 \text{ rad/s}}$$

(b) $$Z_{in,0} = 10 + \frac{25(675+2\times10^2)}{225+400} = 10 + \frac{875}{25} = \underline{45\ \Omega}$$

**13-6** $535 < f_0 < 1605$ KHZ ; $Q_0 \le 45$ ; $R = 25$ kΩ $\therefore C_{min} L = \frac{1}{4\pi^2 \times 1.605^2 \times 10^{12}}$

$$C_{max} L = \frac{1}{4\pi^2 \times 0.535^2 \times 10^{12}} \quad \therefore \frac{C_{max}}{C_{min}} = 9 \ ; Q_0 = 2\pi f_0 C \times 25\times10^3 = \frac{\omega_0 \times 25 \times 10^3}{\omega_0^2 L} = 45$$

$$Q_0 = \frac{12{,}500}{\pi f_0 L} = \frac{12{,}500}{\pi \times 0.535\times10^6 L} = 45 \therefore L = \underline{165.27\ \mu H}, \quad C_{min} = \underline{59.50\ pF}$$

$$C_{max} = \underline{535.5\ pF}$$

**13-7**

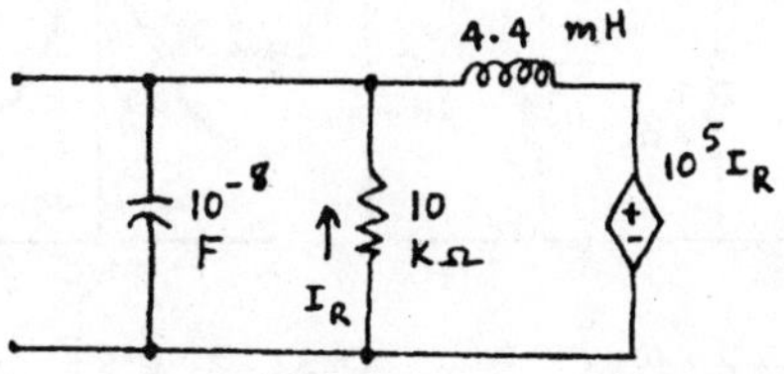


Apply 1 V $\therefore I_R = -10^{-4}$ A

$$\therefore I_{in} = 10^{-8} s + 10^{-4} + \frac{1+10^5 10^{-4}}{0.0044 s} = 10^{-4} + 10^{-8} s + \frac{2500}{s} = Y_{in}$$

$$Y_{in} = 10^{-4} + j\left(10^{-8}\omega - \frac{2500}{\omega}\right) \therefore 10^{-8}\omega_0 = \frac{2500}{\omega_0}$$

$$\therefore \omega_0 = \underline{5\times10^5 \text{ rad/s}}, \quad Q_0 = \omega_0 RC = 5\times10^5 \times 10^4 \times 10^{-8} = \underline{50}$$

**13-8** $\omega_0 = 10^4$, $Q_0 = 8$, $Z_{in,0} = 500\ \Omega$

(a) $\mathcal{B} = \frac{\omega_0}{Q_0} = \frac{10^4}{8} = \underline{1250 \text{ rad/s}}$ (b) $N = \frac{9200 - 10{,}000}{625} = \underline{-1.28}$

(c) $Z_{in} \doteq \frac{500}{1 - j1.28} = \underline{307.8 \angle 52.00^\circ\ \Omega}$ (d) $Z_{in} = \frac{500}{1 + j8(9.2/10 - 10/9.2)} = \underline{299.7 \angle 53.18^\circ\ \Omega}$

(e) $|Z_{in}|$ : % error $= 100 \frac{\text{Approx} - \text{True}}{\text{True}} = 100 \frac{307.8 - 299.7}{299.7} = \underline{2.722\ \%}$

$\angle Z_{in}$ : % error $= 100 \frac{52.00^\circ - 53.18^\circ}{53.18^\circ} = \underline{-2.213\ \%}$

**13-9** (a) $C = 50\ \mu F$ ; $\omega_0 = 100$, $\omega_2 = 120$ ; $\omega_0 = \sqrt{\omega_1 \omega_2}$ $\therefore \omega_1 = \frac{10^4}{120} = \frac{250}{3}$ ; $\mathcal{B} = \omega_2 - \omega_1 = \frac{110}{3}$

$\therefore \mathcal{B} = \underline{36.67 \text{ rad/s}}$, $Q_0 = \frac{\omega_0}{\mathcal{B}} = \frac{300}{110} = \underline{2.727}$, $L = \frac{10^6}{10^4 \times 50} = \underline{2\ H}$, $R = \frac{Q_0}{\omega_0 C} = \underline{545.5\ \Omega}$

(b) $C = 50\ \mu F$ ; $\omega_0 = 100$, $\omega_1 = 80$ $\therefore \omega_2 = \frac{10^4}{80} = 125$ ; $\mathcal{B} = 125 - 80 = \underline{45 \text{ rad/s}}$

$Q_0 = \frac{100}{45} = \underline{2.222}$ ; $L = \underline{2\ H}$ ; $R = \frac{20\times10^6}{9\times100\times50} = \underline{444.4\ \Omega}$

(c) $C = 50\ \mu F$ ; $\omega_1 = 80$, $\omega_2 = 120$ $\therefore \mathcal{B} = \underline{40 \text{ rad/s}}$ ; $\omega_0 = \sqrt{80\times120} = \underline{97.98 \text{ rad/s}}$

$L = \frac{10^6}{9600\times50} = \underline{2.083\ H}$ ; $Q_0 = \frac{97.98}{40} = \underline{2.449}$ ; $R = \frac{2.449\times10^6}{97.98\times50} = \underline{500\ \Omega}$

**13-10** (a)

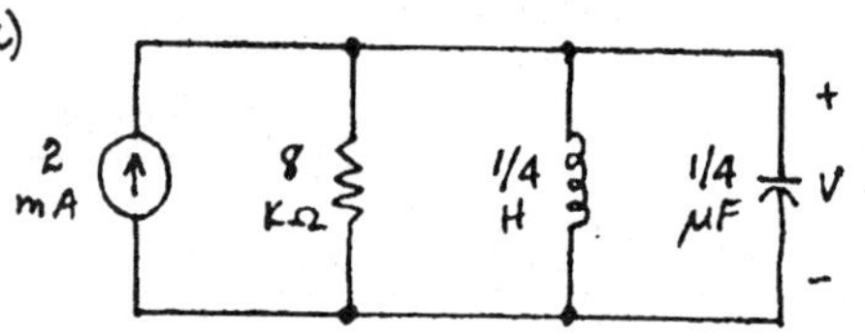


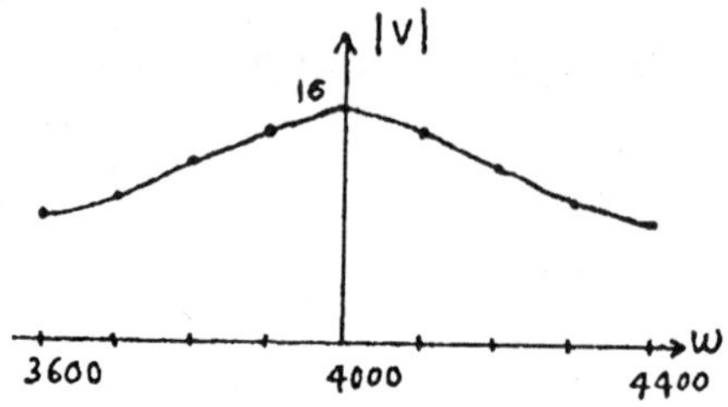


$\omega_o = \frac{1}{\sqrt{(1/4)^2 \times 10^{-6}}} = 4000$ rad/s ; $Q_o = 4000 \times \frac{1}{4} \times 10^{-6+3} \times 8 = 8$ ; $\mathcal{B} = \frac{4000}{8} = 500$ , $\frac{1}{2}\mathcal{B} = 250$

$|V_o| = 16$ , $|V| = 16 \Big/ \sqrt{1 + \left(\frac{\omega - 4000}{250}\right)^2}$

(b) $V \doteq 16 \Big/ \left(1 + j\frac{3250 - 4000}{250}\right) = \underline{5.060 \angle 71.57^\circ \text{ V}}$

(c) $V = \frac{I}{Y} = \frac{2 \times 10^{-3}}{1/8000 + j(3250 \times 1/4 \times 10^{-6} - 4/3250)} = \underline{4.581 \angle 73.36^\circ} \text{ V}$

**13-11** (a)

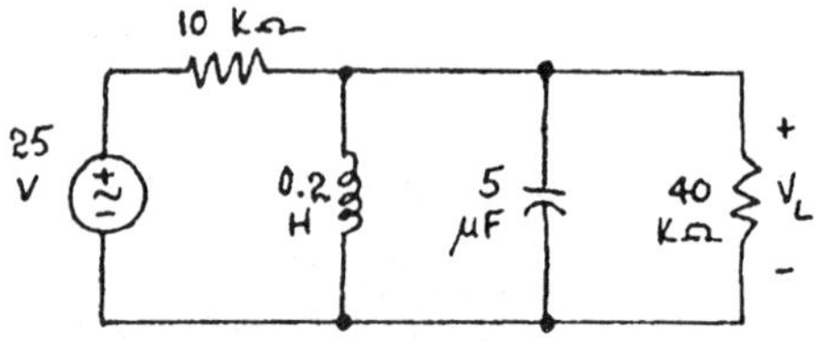


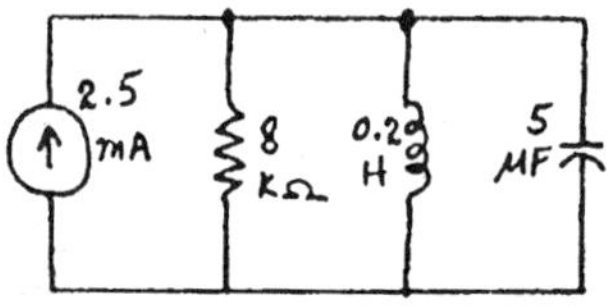


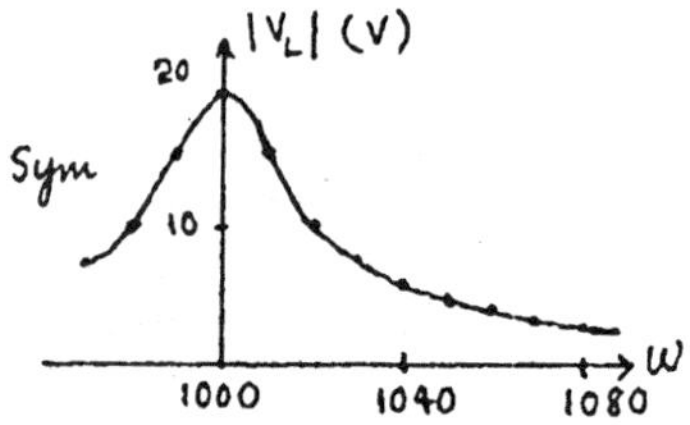


$\omega_o = \frac{1000}{\sqrt{5 \times 0.2}} = 1000$ , $Q_o = \omega_o RC = 1000 \times 8000 \times 5 \times 10^{-6} = 40$ , $\mathcal{B} = \frac{\omega_o}{Q_o} = 25$ $\therefore \frac{1}{2}\mathcal{B} = 12.5$

$|V_L|_{max} = 20$ V , $|V_L| = 20 \Big/ \sqrt{1 + \left(\frac{\omega - 10^3}{12.5}\right)^2}$ $\left(\omega = 1050 \;\therefore |V_L|_{approx} = 4.851 \text{ V}\right)$

(b) $V_{L,exact} = 2.5 \times 10^{-3} \Big/ \left[\frac{1}{8000} + j(5 \times 10^{-6}\omega - 1/0.2\omega)\right]$

$\omega = 1050$ ; $V_{L,exact} = 2.5 \times 10^{-3} \Big/ \left[\frac{1}{8000} + j(5.25 \times 10^{-3} - 1/210)\right] = \underline{4.962 \angle -75.64^\circ \text{ V}}$

**13-12** $Q_o = 20$ ; $\omega_o = 10^4$ ; $Z_{in,o} = 5000$ ; $\mathcal{B} = \frac{10^4}{20} = 500$ $\therefore \frac{1}{2}\mathcal{B} = 250$

By approx : $3 = \frac{5}{\sqrt{1 + N^2}}$ $\therefore N = \pm 4/3 = \frac{\omega - \omega_o}{1/2\,\mathcal{B}}$ $\therefore \omega = 10^4 \pm \frac{4}{3} \times 250$ $\therefore \Delta\omega = \frac{4}{3} \times 250 = \underline{666.7 \text{ rad/s}}$

By exact : $\frac{4}{3} = 20\left|\frac{\omega}{10{,}000} - \frac{10{,}000}{\omega}\right|$ $\therefore \omega = 10{,}338.89$ and $9672.22$ $\therefore \Delta\omega = \underline{666.7 \text{ rad/s}}$ again

**13-13** (a) $\mathcal{B} = \frac{500}{25} = 20$ Hz , $\frac{1}{2}\mathcal{B} = 10$ Hz , $|V|_{495} = 100 \Big/ \sqrt{1 + \left(\frac{495 - 500}{10}\right)^2} = \underline{89.44 \text{ V}}$

(b) $50 = 100 \Big/ \sqrt{1 + \left(\frac{f - 500}{10}\right)^2}$ $\therefore f = \underline{517.3}$ and $\underline{482.7}$ Hz

(c) ang $Y = -50^\circ \doteq \tan^{-1} N$ $\therefore N = \tan(-50^\circ) = \frac{f - 500}{10}$ $\therefore f = 500 - 10 \tan 50^\circ = \underline{488.08 \text{ Hz}}$

**13-14** (a)

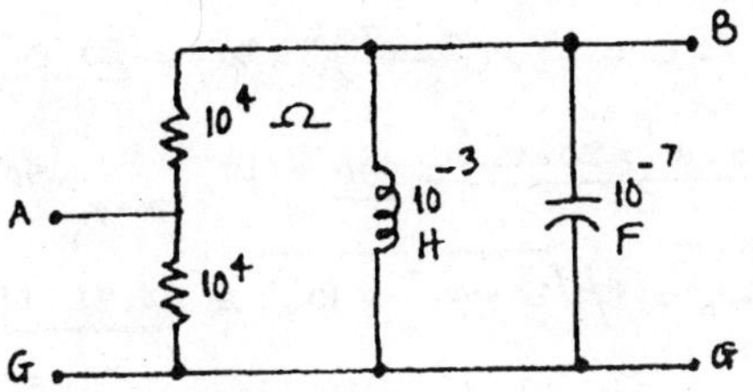


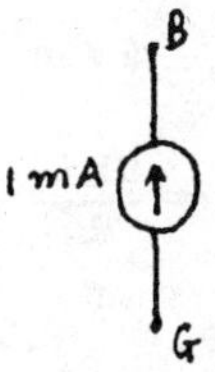


$\omega_0 = 10^5$, $Q_0 = 10^5 \times 2\times10^4 \times 10^{-7} = 200$

$\mathcal{B} = 10^5/200 = 500$, $1/2\,\mathcal{B} = 250$

$V_{AG,0} = 10$ V, $V_{AG} = 10\Big/\sqrt{1+\left(\dfrac{\omega-10^5}{250}\right)^2}$

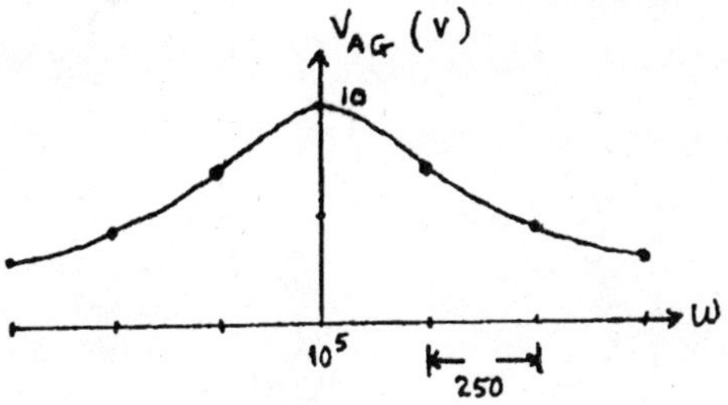


(b)

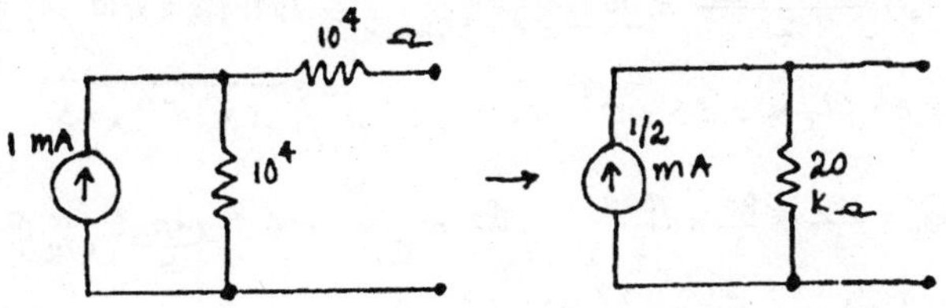


$\omega_0 = 10^5$, $Q_0 = 200$, $\mathcal{B} = 500$, $1/2\,\mathcal{B} = 250$, $V_{BG,0} = 10$ V

∴ Same curve (this is reciprocity)

**13-15**

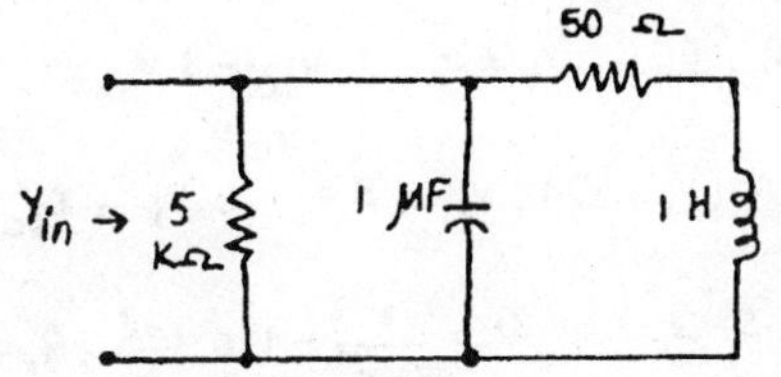


(a) $Y_{in} = \dfrac{1}{5000} + j10^{-6}\omega + \dfrac{1}{50+j\omega} \times \dfrac{50-j\omega}{50-j\omega}$, $I_m(Y_{in}) = 0$

$\therefore 10^{-6}\omega_0 = \dfrac{\omega_0}{\omega_0^2 + 2500}$ $\therefore \omega_0 = \underline{998.75 \text{ rad/s}}$

(b) $Y_{in,0} = 1/5000 + 50/10^6 = \underline{250\ \mu S}$ ($Z_{in,0} = 4$ kΩ)

(c) and (d) $|Y_{in}| = 250\sqrt{2} \times 10^{-6} = \left| 2000\times10^{-6} + j10^{-6}\omega + \dfrac{50}{\omega^2+2500} - j\dfrac{\omega}{\omega^2+2500} \right|$

$\therefore 125{,}000 = \left(200 + \dfrac{50\times10^6}{\omega^2+2500}\right)^2 + \left(\omega - \dfrac{10^6\omega}{\omega^2+2500}\right)^2$ $\therefore \omega_2 = \underline{1138.00}$, $\omega_1 = \underline{887.38}$ rad/s

(e) $\mathcal{B} = \omega_2 - \omega_1 = 1138.00 - 887.38 = \underline{250.62 \text{ rad/s}}$

(f) $\omega_0/\mathcal{B} = 998.75/250.62 = \underline{3.985^+}$

**13-16** (a)

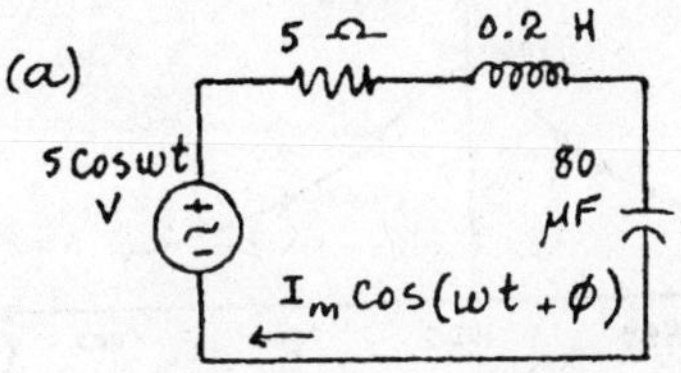


$\omega_0 = \dfrac{1}{\sqrt{LC}} = 250$ rad/s; $Q_0 = \dfrac{\omega_0 L}{R} = \underline{10}$; $\mathcal{B} = \dfrac{\omega_0}{Q_0} = \underline{25 \text{ rad/s}}$

$I_{m,exact} = 1\Big/\sqrt{1+(\omega/25 - 2500/\omega)^2}$

$I_{m,approx} \doteq 1\Big/\sqrt{1+[(\omega-250)/12.5]^2}$

(b)

| ω | $I_{m,exact}$ | $I_{m,approx}$ |
|---|---|---|
| 250 | 1.000 | 1.000 |
| 260 | 0.787 | 0.781 |
| 270 | 0.544 | 0.530 |
| 280 | 0.403 | 0.385 |
| 240 | 0.775 | 0.781 |
| 230 | 0.514 | 0.530 |
| 220 | 0.363 | 0.385 |

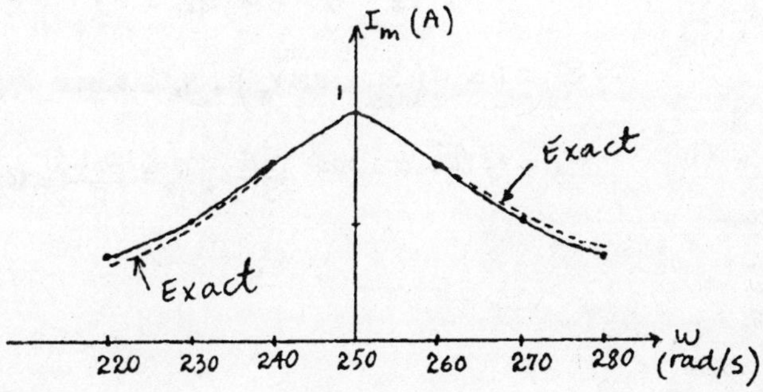

13-17

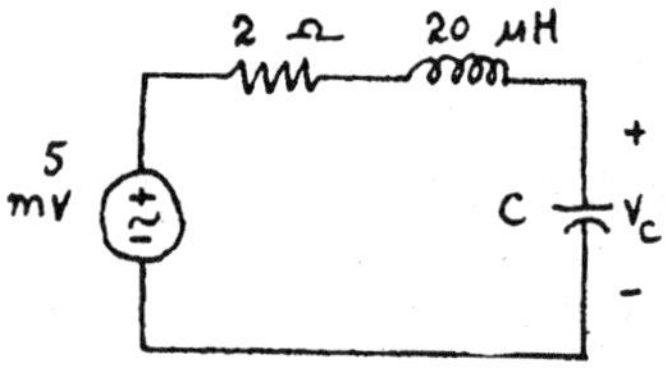


(a) $\omega_0 = 5\times 10^7 \quad \therefore C = 10^6/(5\times10^7)^2\times 20 = \underline{20\ \text{pF}}$

$Q_0 = \dfrac{\omega_0 L}{R} = \dfrac{5\times10^7\times 20\times10^{-6}}{2} = \underline{500}\ ; \ |V_C|_{max} = 500\times5\times10^{-3} = \underline{2.5\ \text{V}}$

(b) $C = 10^{-11} \quad \therefore \omega_0 = 1/\sqrt{20\times10^{-6}\times10^{-11}} = \underline{70.71\ \text{Mrad/s}}$

$Q_0 = \dfrac{70.71\times 20}{2} = \underline{707.1}\ ; \ |V_C|_{max} = 707.1\times5\times10^{-3} = \underline{3.536\ \text{V}}$

(c) $C = 45\ \text{pF},\ \omega = 33\times10^6 \quad \therefore \omega_0 = 10^6/\sqrt{20\times10^{-6}\times45} = \dfrac{100}{3}\ \text{Mrad/s}$

$Q_0 = \dfrac{100}{3}\times\dfrac{20}{2} = \dfrac{1000}{3}, \quad B = \dfrac{\omega_0}{Q_0} = 0.1\ \dfrac{\text{Mrad}}{\text{s}},\ \tfrac{1}{2}B = 0.05\ \dfrac{\text{Mrad}}{\text{s}}$

$\therefore |V_C| \doteq \dfrac{5\times10^{-3}(1000/3)}{\sqrt{1+[(33-100/3)/0.05]^2}} = \underline{0.2472\ \text{V}}$

(Exact: $|V_C| = \underline{0.2485\ \text{V}}$)

13-18 (a)

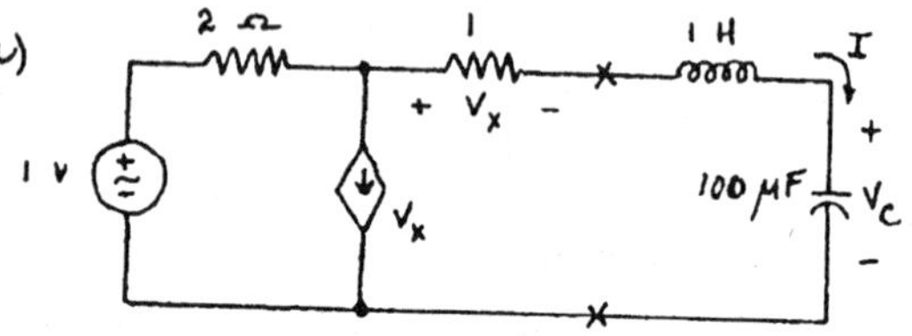


At x - x, OC: $V_x = 0 \quad \therefore V_{oc} = 1\ \text{V}$

↓SC: $1 = 2(V_x + V_x) + V_x = 5V_x \quad \therefore V_x = I_{sc}\downarrow = 1/5$

$\therefore Z_{Th} = 5\ \Omega, \quad \omega_0 = \dfrac{1}{\sqrt{LC}} = \underline{100\ \dfrac{\text{rad}}{\text{s}}},\ Q_0 = \dfrac{\omega_0 L}{Z_{Th}} = \underline{20}$

(b) $|V_C|_0 = 20\times1 = 20\ \text{V}$

$I = \dfrac{1}{5 + j(\omega - 10^4/\omega)}$

$|V_C| = \dfrac{10^4}{\omega}\ \dfrac{1}{|5+j(\omega-10^4/\omega)|}$

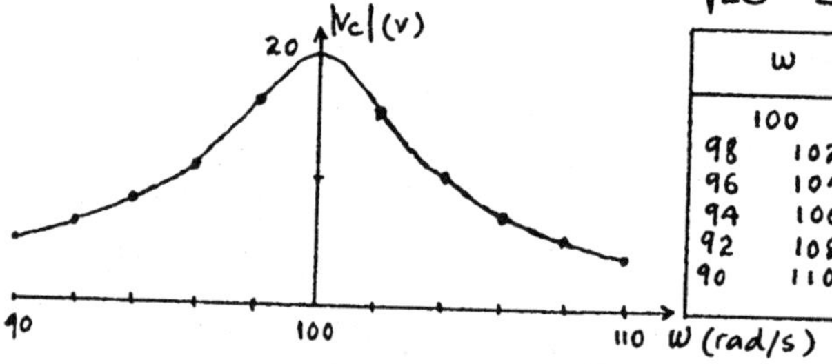


| ω | | $\lvert V_C \rvert$ | |
|---|---|---|---|
| 100 | | 20 | |
| 98 | 102 | 15.87 | 15.37 |
| 96 | 104 | 10.88 | 10.33 |
| 94 | 106 | 7.97 | 7.44 |
| 92 | 108 | 6.24 | 5.72 |
| 90 | 110 | 5.12 | 4.61 |

13-19

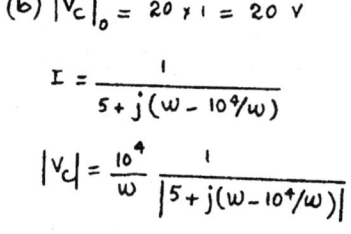

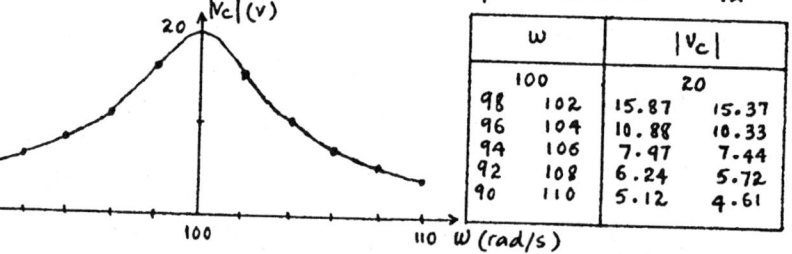

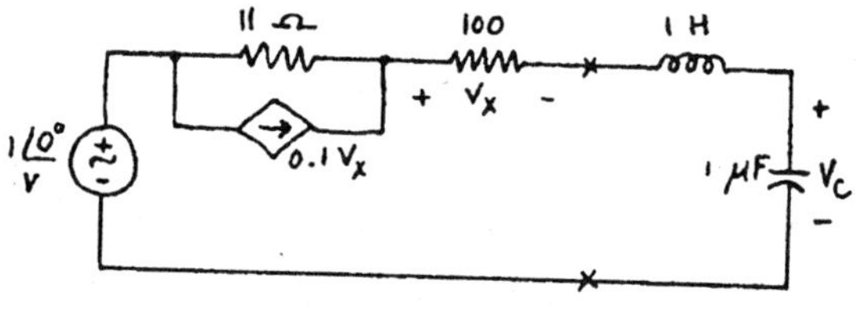


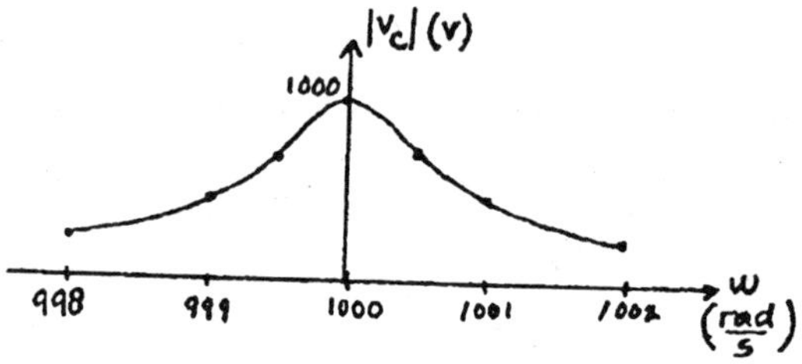


At x - x, OC: $V_x = 0 \quad \therefore V_{oc} = 1\ \text{V}$

↓SC: $1 = 11\left(\dfrac{V_x}{100} - 0.1V_x\right) + V_x = 0.01V_x \quad \therefore V_x = 100, \quad I_{sc} = 100/100 = 1\ \text{A} \quad \therefore Z_{Th} = 1\ \Omega$

$\omega_0 = 1/\sqrt{LC} = 1000\ \dfrac{\text{rad}}{\text{s}}, \quad Q_0 = \dfrac{\omega_0 L}{R} = 1000, \quad B = \dfrac{\omega_0}{Q_0} = 1, \quad \tfrac{1}{2}B = \tfrac{1}{2} \quad \therefore |V_C|_0 = \underline{1000\ \text{V}}$

**13-20** (a) $f_0 = 5000$ Hz, $\mathcal{B} = 180$ Hz, $X_{C,0} = 600\ \Omega$ $\therefore 600 = \dfrac{1}{2\pi \times 5000\,C}$ $\therefore C = \underline{53.05^+ \text{ nF}}$

$Q_0 = \dfrac{\omega_0}{\mathcal{B}} = \dfrac{5000}{180} = \underline{27.78}$, $L = \dfrac{10^9}{(2\pi \times 5000)^2 \times 53.05} = \underline{19.099 \text{ mH}}$; $Q_0 = 27.78 = \dfrac{\omega_0 L}{R}$ $\therefore R = \underline{21.60\ \Omega}$

(b) $I = \dfrac{1}{21.60 + j(0.019099\omega - 10^9/53.05\omega)}$ $\therefore |V_L| = |j\omega 0.019099 I|$

$\omega = 2\pi(5000) \therefore |V_L| = \underline{27.778 \text{ V}}$; $\omega = 2\pi(4800) \therefore |V_L| = \underline{10.756 \text{ V}}$; $\omega = 2\pi(5200) \therefore |V_L| = \underline{12.047 \text{ V}}$

**13-21** (a) $Z_{in}(s) = \dfrac{K(s + 300 + j1000)(s + 300 - j1000)}{s} = K\left(s + 600 + \dfrac{1{,}090{,}000}{s}\right)$

$Z_{in}(-200) = -5 = K\left(-200 + 600 - \dfrac{1{,}090{,}000}{200}\right) \therefore K = \dfrac{1}{1010}$ $\therefore R = \dfrac{600}{1010} = \underline{0.5941\ \Omega}$; $L = \dfrac{1}{1010} = \underline{0.9901 \text{ mH}}$

$C = 1010/1{,}090{,}000 = \underline{926.6\ \mu\text{F}}$. Series RLC

(b) $K = 1/50$ $\therefore R = \underline{12\ \Omega}$, $L = \underline{20 \text{ mH}}$, $C = 50/1{,}090{,}000 = \underline{45.87\ \mu\text{F}}$, series RLC

(c) $Z(-100 + j500) = K[500 + j500 - (10{,}900 + j54{,}500)/26]$ $\therefore 2 = K(500 - 10900/26)$ $\therefore K = 52/2100$

$R = \dfrac{600 \times 52}{2100} = \underline{14.857\ \Omega}$, $L = \dfrac{52}{2100} = \underline{24.76 \text{ mH}}$, $C = \dfrac{2100/52}{1{,}090{,}000} = \underline{37.05^+\ \mu\text{F}}$, series RLC

**13-22** (a)

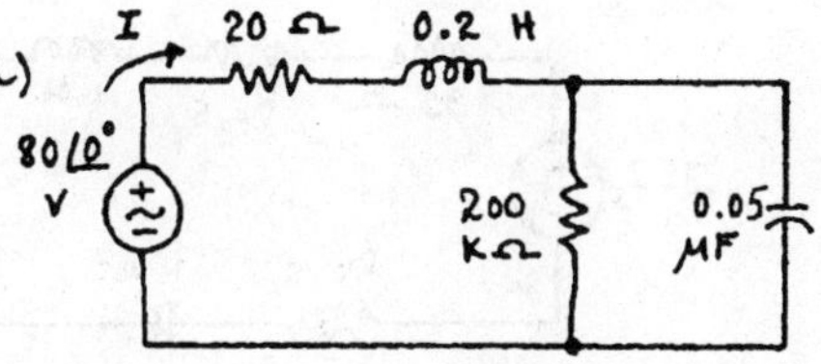


$\omega_0 \doteq \dfrac{1}{\sqrt{LC}} = \dfrac{10^3}{\sqrt{0.2 \times 0.05}} = \underline{10^4 \text{ rad/s}}$

$Q_C = 10^4 \times 0.05 \times 10^{-6} \times 2 \times 10^5 = 100$ $\therefore R_{C,S} = \dfrac{2 \times 10^5}{100^2} = 20\ \Omega$

(b) $Q_0 = \dfrac{\omega_0 L}{R_{S,tot}} = \dfrac{10^4 \times 0.2}{20 + 20} = \underline{50}$

(c) $\mathcal{B} = \dfrac{\omega_0}{Q_0} = \dfrac{10^4}{50} = 200$, $\tfrac{1}{2}\mathcal{B} = 100$

$|I|_0 \doteq \dfrac{80}{20 + 20} = 2$ A

As $\omega \to \infty$, $|I| \to 0$; $\omega \to 0$, $|I| \to 0.4$ mA

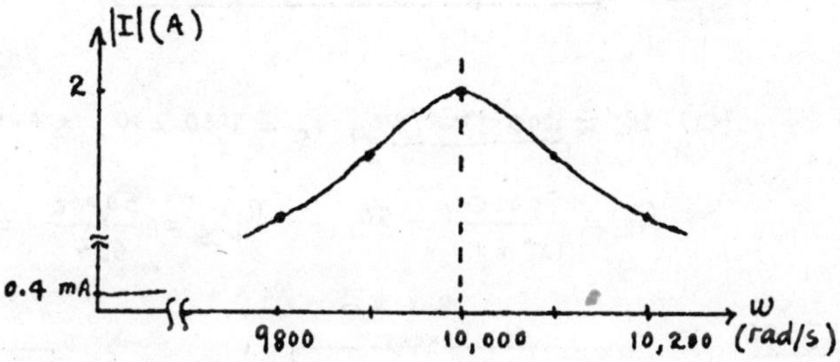


**13-23** (a)

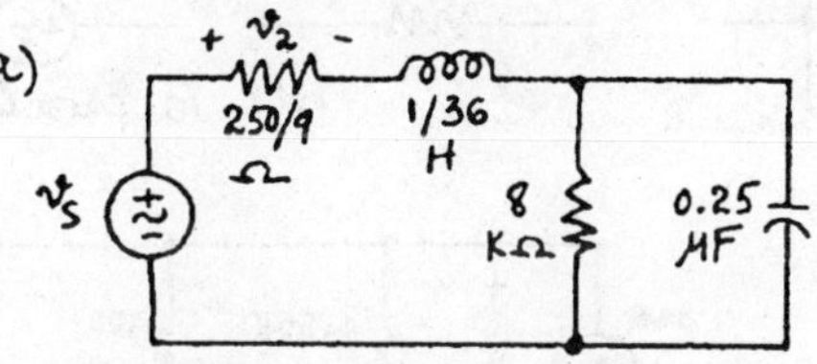


$v_S = 12.5 \cos 13{,}000t$ V

$\omega_0 \doteq \dfrac{1}{\sqrt{LC}} = \dfrac{1000}{\sqrt{(1/36)(1/4)}} = 12{,}000$ rad/s

$Q_C = 12{,}000 \times 1/4 \times 10^{-6} \times 8000 = 24$; $R_{C,S} = \dfrac{R_{C,P}}{Q_C^2}$

$R_{C,S} = \dfrac{8000}{24^2} = 13.889\ \Omega$; $R_{S,tot} = 13.889 + \dfrac{250}{9} = 41.67\ \Omega$

$Q_0 = \dfrac{\omega_0 L}{R_{S,tot}} = \dfrac{12{,}000 \times 1/36}{41.67} = \underline{8}$

(b) $\mathcal{B} = \dfrac{\omega_0}{Q_0} = \underline{1500 \text{ rad/s}}$, $\tfrac{1}{2}\mathcal{B} = 750$ rad/s

(c) $N = \dfrac{\omega - \omega_0}{\tfrac{1}{2}\mathcal{B}} = \dfrac{13{,}000 - 12{,}000}{750} = 4/3$; $I = \dfrac{12.5}{41.67(1 + j4/3)} = 0.180\angle{-53.13^\circ}$; $V_2 = 250/9\, I$

$\therefore V_2 = 5\angle{-53.13^\circ}$, $v_2(t) = \underline{5.00\cos(13{,}000t - 53.1^\circ) \text{ V}}$ [Exact: $v_2(t) = \underline{5.193\cos(13{,}000t - 53.69^\circ)}$ V]

**13-24** (a)

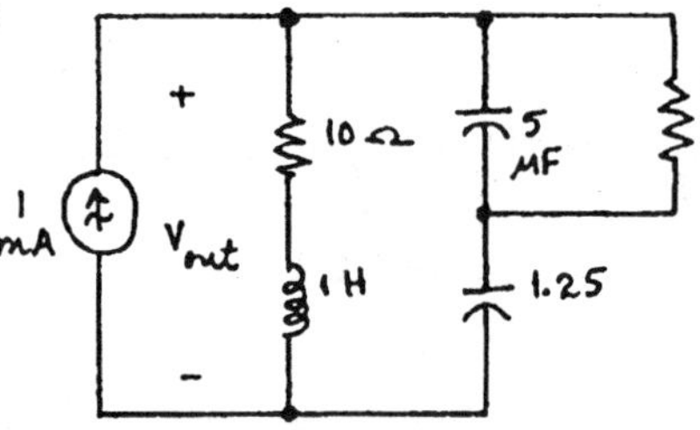


$L_{eq} \doteq 1$ H , $C_{eq} \doteq 1/(1/5 + 1/1.25) = 1$ μF

$\omega_0 \doteq \frac{1}{\sqrt{LC}} = 1000 \frac{rad}{s}$; $Q_{c,1} = 10^3 \times 5 \times 10^{-6} \times 5 \times 10^3 = 25$

$\therefore R_{c,s} = \frac{5000}{25^2} = 8$ Ω

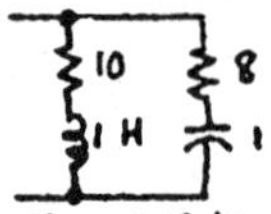


$Q_{c2} = \frac{1}{10^3\ 10^{-6}\ 8} = 125 \quad \therefore R_{c,p} = 125^2 \times 8 = 125$ kΩ

$Q_L = \frac{10^3 \times 1}{10} = 100 \quad \therefore R_{L,p} = 100^2 \times 10 = 100$ kΩ

100//125 = 55.56 kΩ

$\therefore |V_{out}|_0 = 10^{-3} \times 55.56 \times 10^3 = 55.56$ V ; $Q_0 = \omega_0 RC = 55.56$ ; $\mathcal{B} = \omega_0/Q_0 = 18$ , $\frac{1}{2}\mathcal{B} = 9$

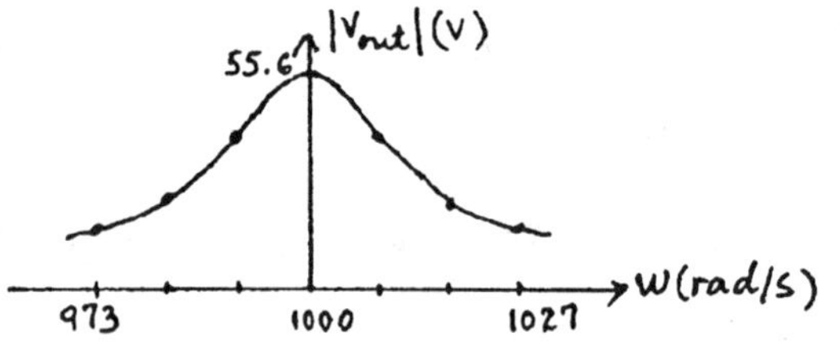


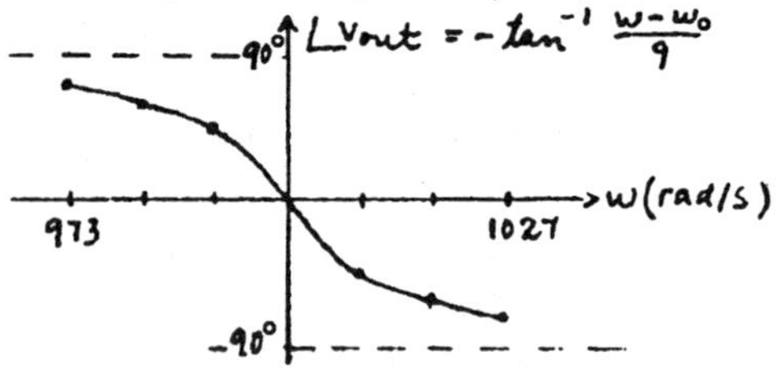


**13-25** (a)

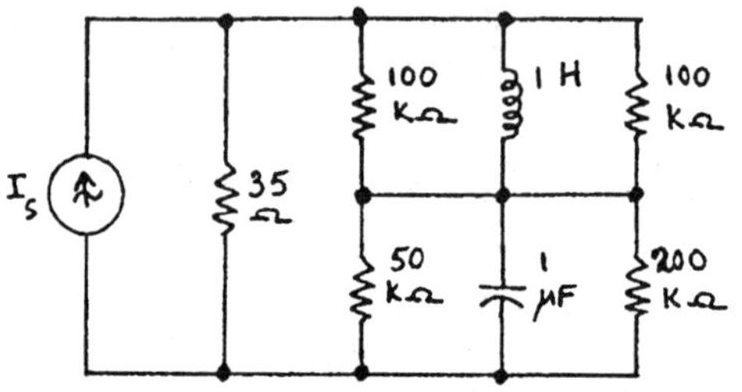


Transform source, change ∥RL and ∥RC to series, get:

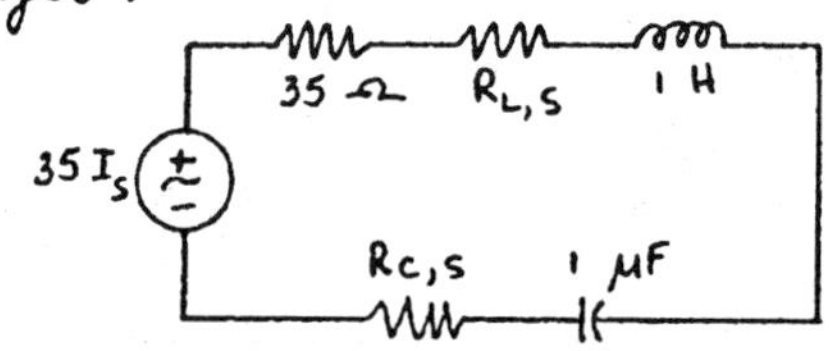


Series RLC

(b) $\omega_0 \doteq 1000$ rad/s , $Q_c = 1000 \times 10^{-6} \times 40000 = 40 \quad \therefore R_{c,s} = 40{,}000/40^2 = 25$ Ω

$Q_L = \frac{50000}{10^3 \times 1} = 50 \quad \therefore R_{L,s} = \frac{50{,}000}{50^2} = 20$ Ω $\quad \therefore Q_0 = \frac{\omega_0 L}{R} = \frac{10^3 \times 1}{35 + 25 + 20} = 12.5$

**13-26**

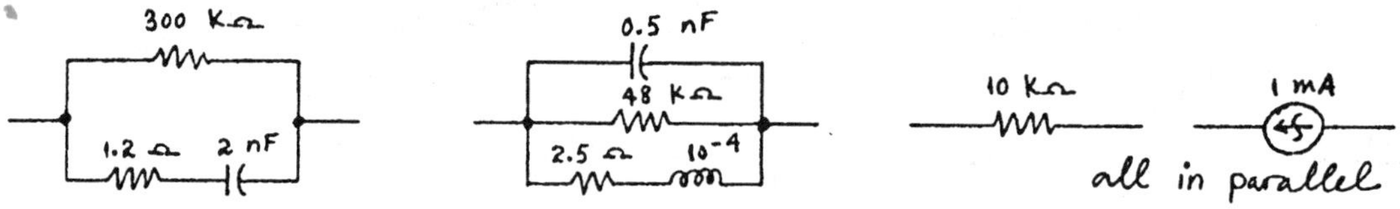


all in parallel

$\omega_0 \doteq \frac{1}{\sqrt{LC_{eq}}} = \frac{1}{\sqrt{10^{-4}(2 + 0.5)10^{-9}}} = 2 \times 10^6$ rad/s

$Q_{RC} = \frac{10^9}{2 \times 10^6 \times 2 \times 1.2} = 208.3 \quad \therefore R_{c,p} = 1.2 \times 208.3^2 = 52.083$ kΩ

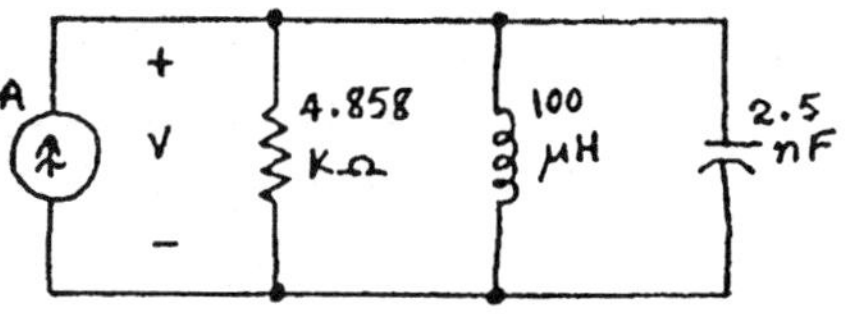


$Q_{RL} = \frac{2 \times 10^6 \times 10^{-4}}{2.5} = 80 \quad \therefore R_{L,p} = 80^2 \times 2.5 = 16$ kΩ

52.08//300//48//16//10 = 4.858 kΩ = $R_{eq}$

$Q_0 = \omega_0 R_{eq} C_{eq} = 2 \times 10^6 \times 4.858 \times 10^3 \times 2.5 \times 10^{-9}$

$\therefore Q_0 = 24.29$ , $\mathcal{B} = \frac{2 \times 10^6}{24.29} = 82.34$ , $\frac{1}{2}\mathcal{B} = 41.17 \frac{Krad}{s}$

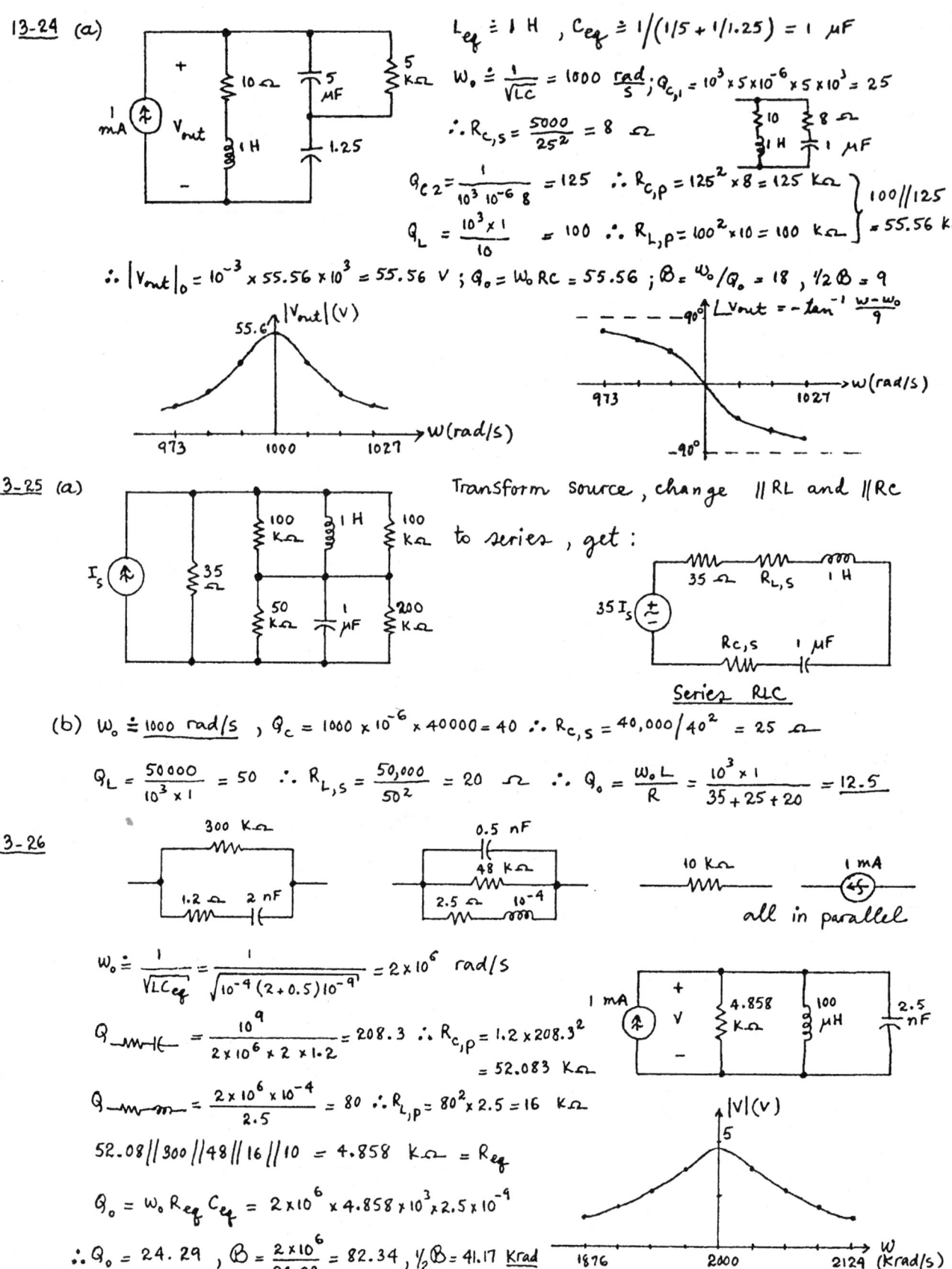

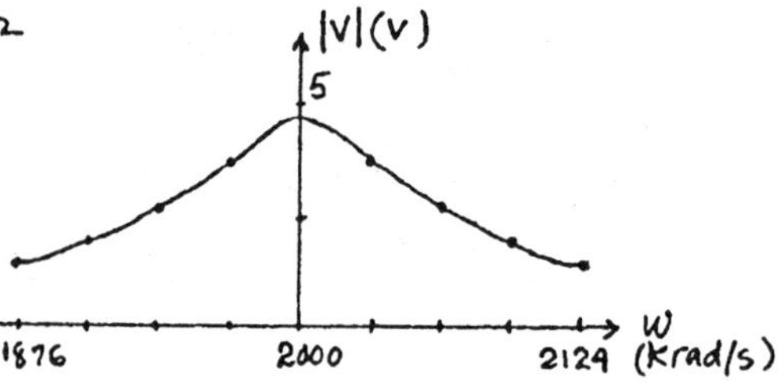

**13-27** (a)

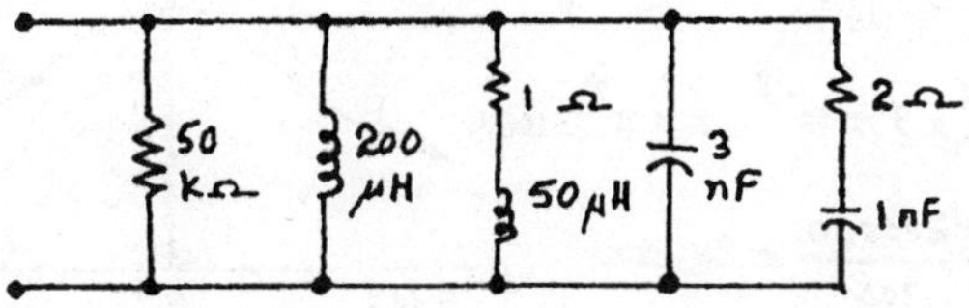


$L_{eq} = 200 \| 50 = 40\ \mu H,\ C_{eq} = 3 + 1 = 4\ nF$

$\omega_0 = \frac{1}{\sqrt{LC}} = \frac{1}{\sqrt{40 \times 4 \times 10^{-15}}} = \underline{2.5 \times 10^6\ rad/s}$

$Q_{s,RL} = \frac{50 \times 2.5}{1} = 125\ \therefore R_{P,RL} = 125^2 \times 1 = 15.625\ k\Omega$

$Q_{s,RC} = \frac{10^9}{2.5 \times 10^6 \times 2 \times 1} = 200\ \therefore R_{P,RC} = 200^2 \times 2 = 80\ k\Omega;\ R_{eq} = 15.625 \| 80 \| 50 = 10.363\ k\Omega$

$Q_0 = \omega_0 R_{eq} C_{eq} = 2.5 \times 10^6 \times 10.363 \times 10^3 \times 4 \times 10^{-9} = \underline{103.6}$

(b)

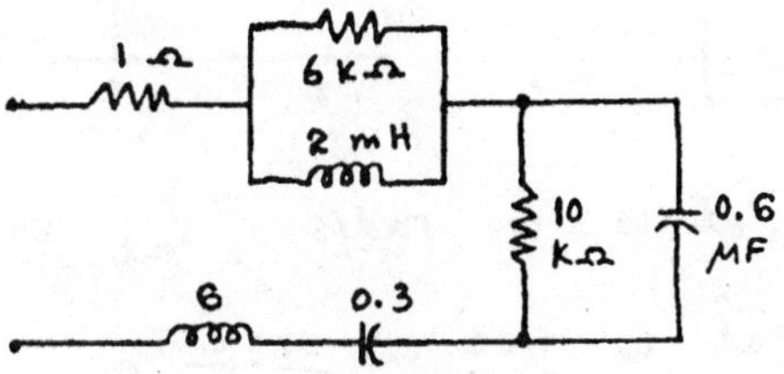


$L_{eq} = 2 + 6 = 8\ mH,\ C_{eq} = \frac{0.6 \times 0.3}{0.6 + 0.3} = 0.2\ \mu F$

$\omega_0 = \frac{1}{\sqrt{8 \times 0.2 \times 10^{-9}}} = \underline{25{,}000\ rad/s}$

$Q_{P,RL} = \frac{6000}{25 \times 10^3 \times 2 \times 10^{-3}} = 120;\ R_{s,RL} = \frac{6000}{120^2} = \frac{5}{12}\ \Omega$

$Q_{P,RC} = 25 \times 10^3 \times 10^4 \times 0.6 \times 10^{-6} = 150;\ R_{s,RC} = \frac{10{,}000}{150^2} = \frac{4}{9}\ \Omega$

$R_{s,eq} = \frac{5}{12} + \frac{4}{9} + 1 = \frac{67}{36}\ \Omega;\ Q_0 = \frac{\omega_0 L_{eq}}{R} = \frac{25 \times 10^3 \times 8 \times 10^{-3}}{67/36} = \underline{107.46}$

**13-28**

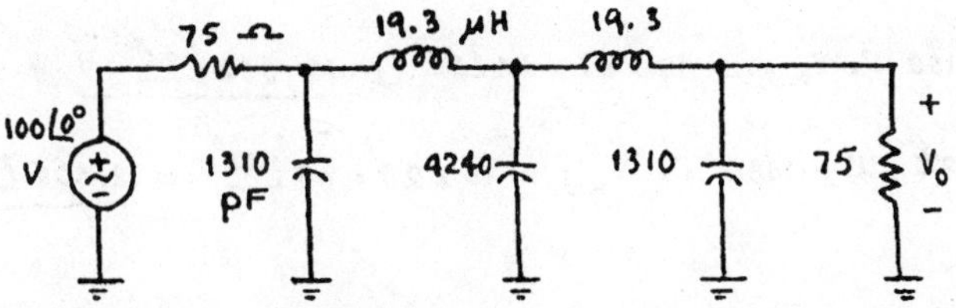


$f_0 = 1\ MHz$, scale to 300 Ω, 60 MHz

$\therefore K_m = \frac{300}{75} = 4,\ K_f = \frac{60}{1} = 60$

$\therefore R_{new} = 300\ \Omega;\ L_{new} = \frac{4 \times 19.3}{60} = 1.2867\ \mu H$

$C_{new,1} = \frac{1310}{K_m K_f} = \frac{1310}{4 \times 60} = 5.4858\ pF$

$C_{new,2} = \frac{4240}{4 \times 60} = 17.667\ pF$

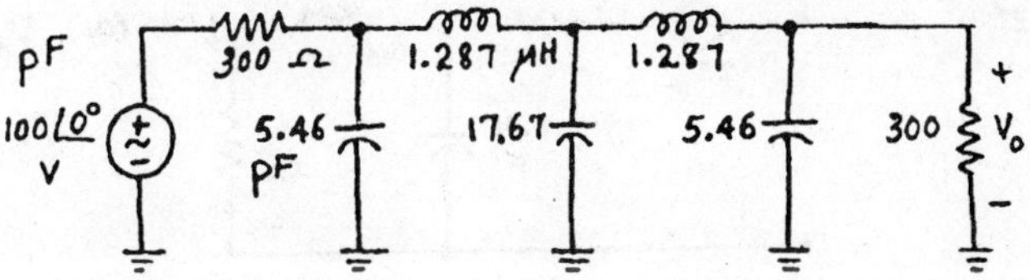


**13-29** (a)

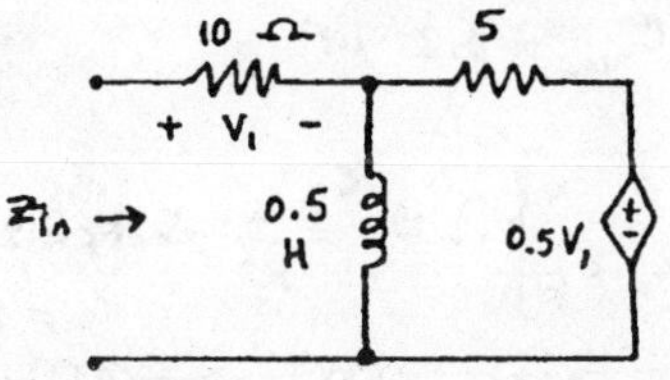


Let $I_{in} \to 1\ A\ \therefore V_1 = 10\ V,\ 0.5 V_1 = 5\ V$

Nodal: $-1 + \frac{V_{in} - 10}{0.5 s} + \frac{V_{in} - 10 - 5}{5} = 0$

$\therefore V_{in}\left(\frac{2}{s} + \frac{1}{5}\right) = 4 + \frac{20}{s},\ V_{in} = Z_{in} = \frac{20s + 100}{s + 10}$

(b) $K_m = 2,\ K_f = 5\ \therefore Z_{in,new} = 2\,\frac{20(s/5) + 100}{(s/5) + 10} = \frac{40s + 1000}{s + 50}$

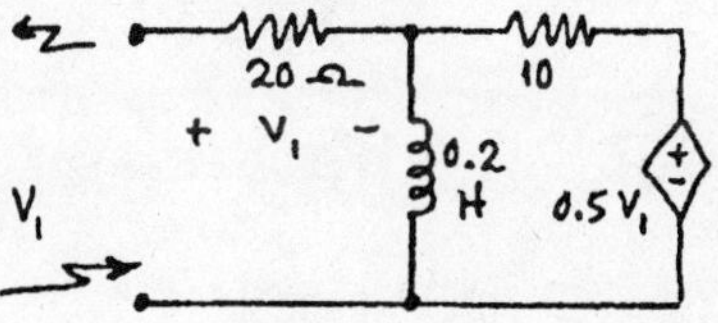


(c) 10 Ω → 20 Ω; 5 Ω → 10 Ω; 0.5 H → 0.2 H; $0.5 V_1 \to 0.5 V_1$

13-30 (a)

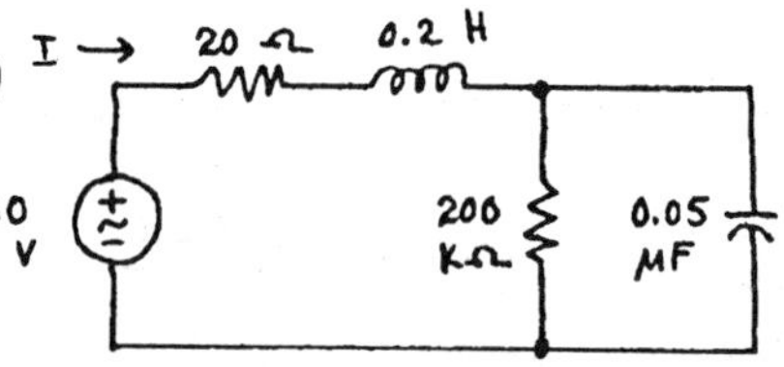


$\omega_0 \doteq \frac{1}{\sqrt{LC}} = \frac{1000}{\sqrt{0.01}} = 10^4 \ \frac{rad}{s}$

$Q_C = 10^4 \times 5 \times 10^{-8} \times 2 \times 10^5 = 100$

$R_{C,S} = \frac{200000}{100^2} = 20 \ \Omega$

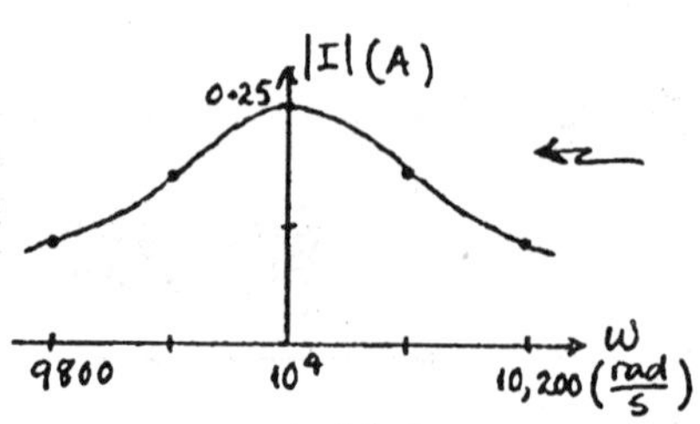


$Q_o = \frac{\omega_o L}{R_{s,eq}} = \frac{10^4 \times 0.2}{20 + 20} = 50$ ; $B = \frac{\omega_o}{Q_o} = 200$, $\frac{1}{2} B = 100$, $|I|_o = .25$ A

(b) $K_f = 1000$, $K_m = 1/20$ ∴

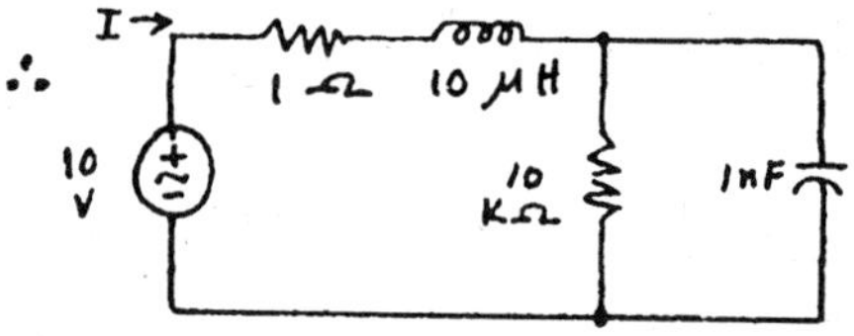


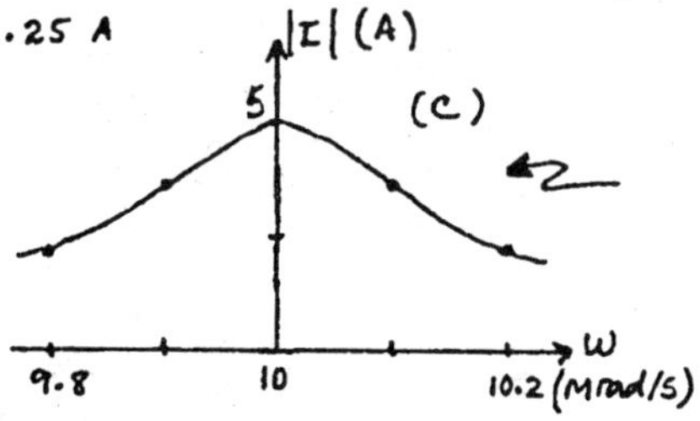


13-31

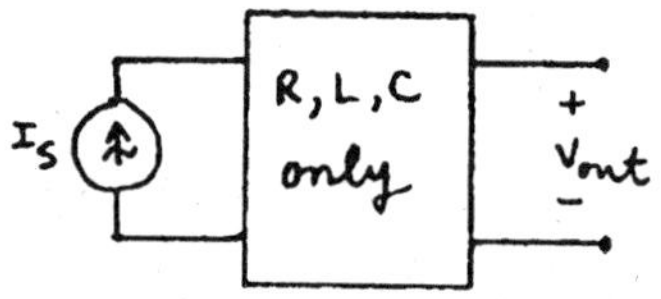


If $I_s = 1\angle 0°$ A at $\omega = 100$ rad/s, $V_{out} = 60\angle 40°$ V

(a) $I_s = 1\angle 0°$ at $\omega = 1000$ ∴ OTSK

(b) $I_s = 1\angle 60°$ at $\omega = 100$ ∴ $V_{out} = 60\angle 100°$ V

(c) $I_s = 10\angle 0°$ A at $\omega = 100$ ∴ $V_{out} = 600\angle 40°$ V

(d) $K_m = 20$, $I_s = 10\angle 0°$ A at $\omega = 100$ ∴ $V_{out} = 10 \times 20 \times 60\angle 40° = 12{,}000\angle 40°$ V

(e) $K_m = 20$, $K_f = 10$, $I_s = 10\angle 0°$ A at $\omega = 1000$ ∴ $V_{out} = 10 \times 20 \times 60\angle 40° = 12{,}000\angle 40°$ V

13-32

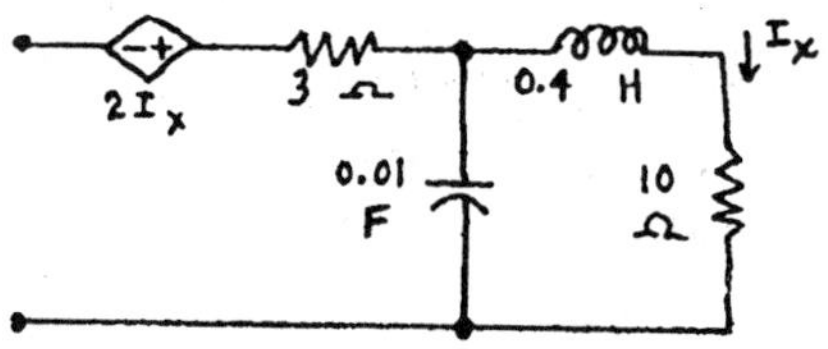


(a) $L_{new} = 5$ mH

∴ $K_m = \frac{5}{400} = \frac{1}{80}$

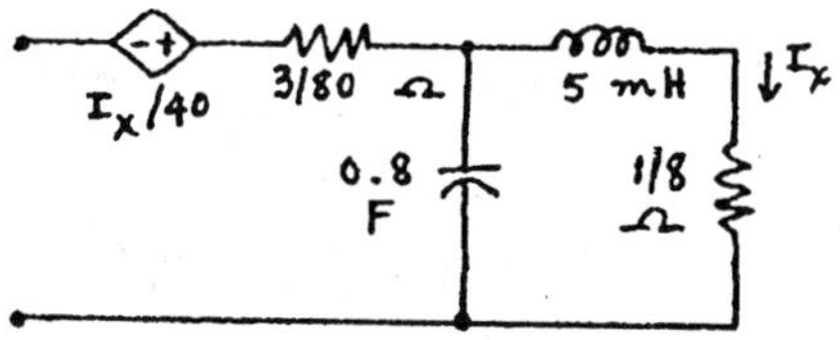


(b) $L_{new} = 5$ mH ∴ $K_f = \frac{400}{5} = 80$

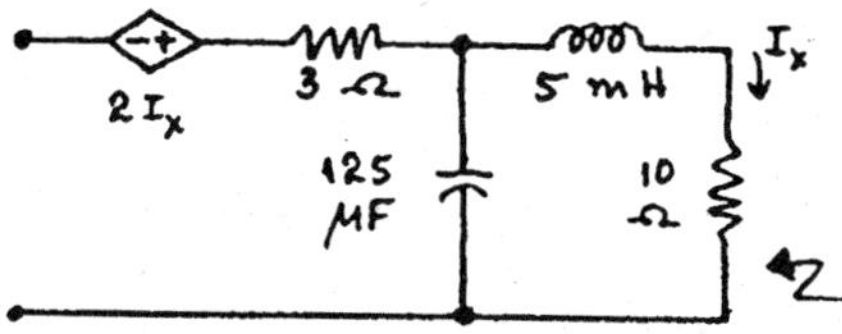


(c) $L_{new} = 5$ mH, $C_{new} = 0.2$ μF

$5 = K_m / K_f \times 400$

$2 \times 10^{-7} = 10^{-2} / K_m K_f$

∴ $10^{-6} = \frac{4}{K_f^2}$ ∴ $K_f = 2000$, $K_m = 25$

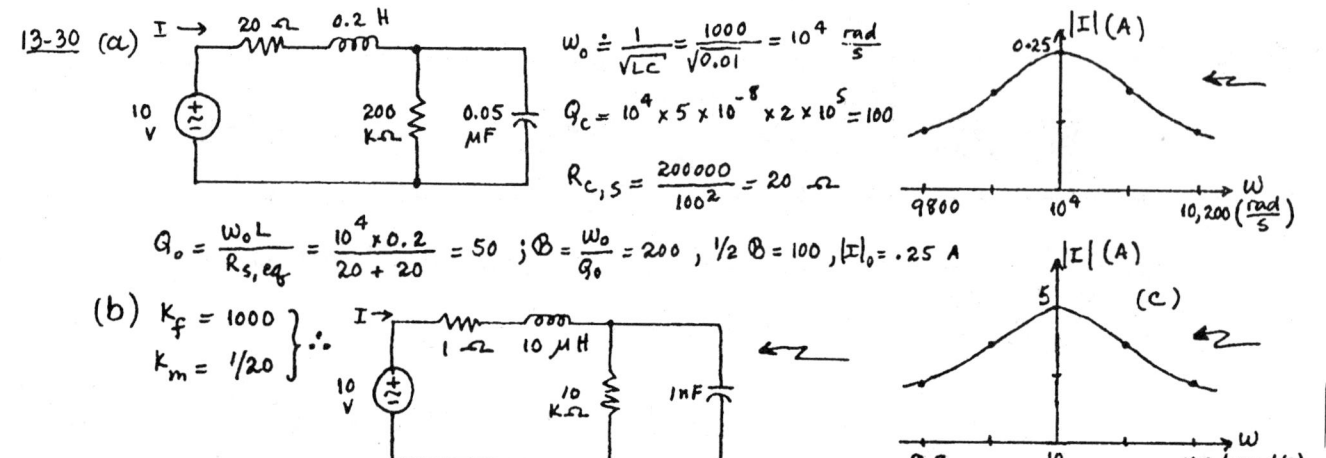

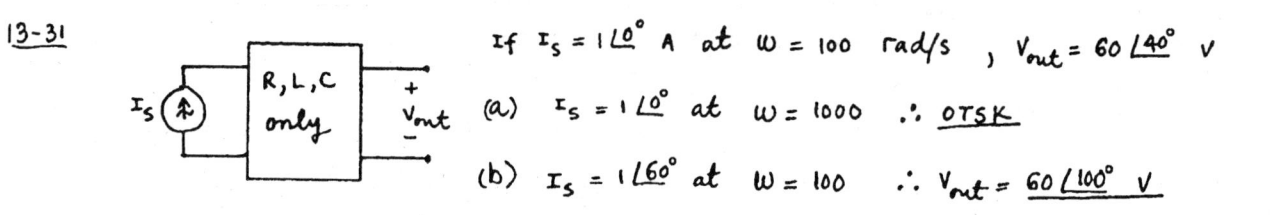

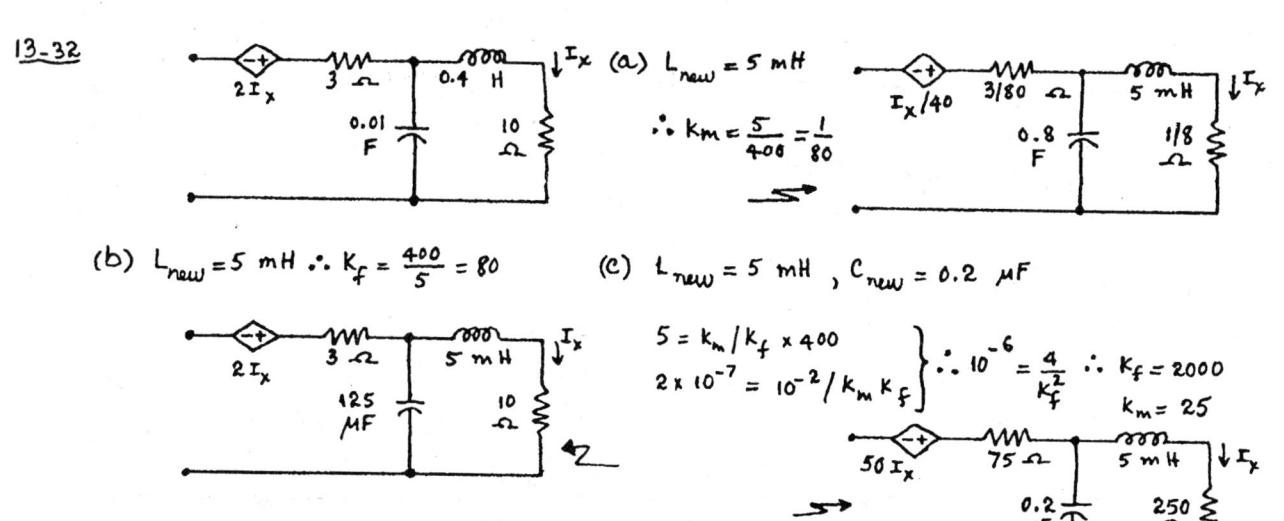

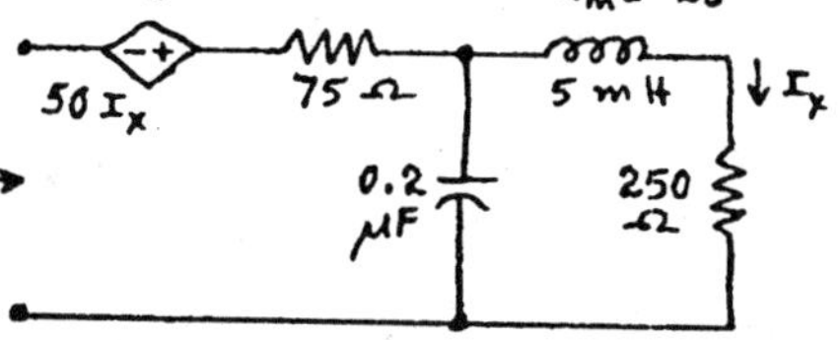

**13-33** (a)

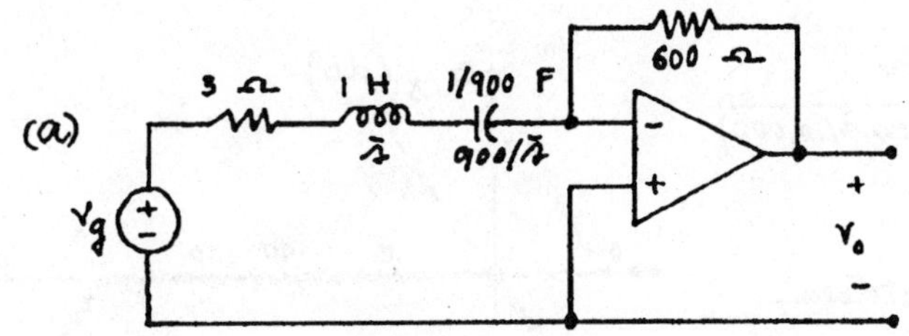


$$\frac{V_g}{3+s+900/s}+\frac{V_o}{600}=0 \quad \therefore \frac{V_o}{V_g}=\frac{-600}{3+s+900/s}$$

$$\frac{V_o}{V_g}=H(s)=\frac{-600s}{s^2+3s+900}$$

(b) $$|H(j\omega)|=\frac{600\,\omega}{\sqrt{(900-\omega^2)^2+9\omega^2}}$$

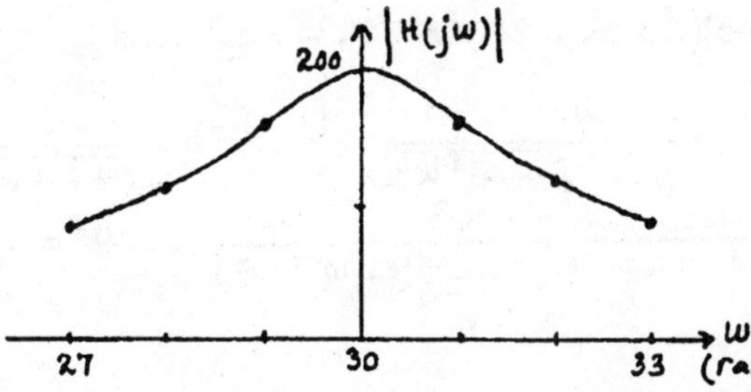


| ω | \|H(jω)\| |
|---|---|
| 30 | 200 |
| 31 | 167.2 |
| 32 | 122.4 |
| 33 | 92.8 |
| 29 | 165.5 |
| 28 | 117.3 |
| 27 | 85.6 |

(c) $$\frac{600^2}{|H(j\omega)|^2}=\frac{\omega^4-1791\,\omega^2+810{,}000}{\omega^2}$$

$$\frac{d(600^2/|H|^2)}{d\omega}=0=2\omega-\frac{2\times 810{,}000}{\omega^3}$$

$\therefore \omega^2=900$ , $\omega_{max}=$ 30 rad/s

(d) $$H(j\omega)=\frac{-600j\omega}{900-\omega^2+j3\omega}=\frac{-200}{1+j(\omega/3-300/\omega)}=\frac{-200}{1+j10(\omega/30-30/\omega)} \quad \therefore \omega_o=30,\ Q_o=10,\ \beta=3 \text{ rad/s}$$

(e) $K_m=\frac{75}{3}=25$ , $K_f=\frac{10{,}000}{30}=\frac{1000}{3}$

$\therefore$ 3 Ω → 75 Ω , 600 Ω → 15 KΩ

$$1\text{ H} \rightarrow \frac{25}{1000/3}\times 1 = 75 \text{ mH}$$

$$1/900\text{ F} \rightarrow \frac{1}{900\times 25\times 1000/3}=133.33 \text{ nF}$$

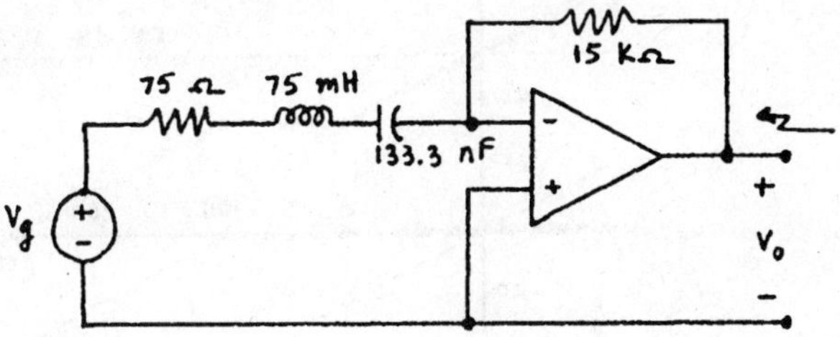


**13-34** (a)

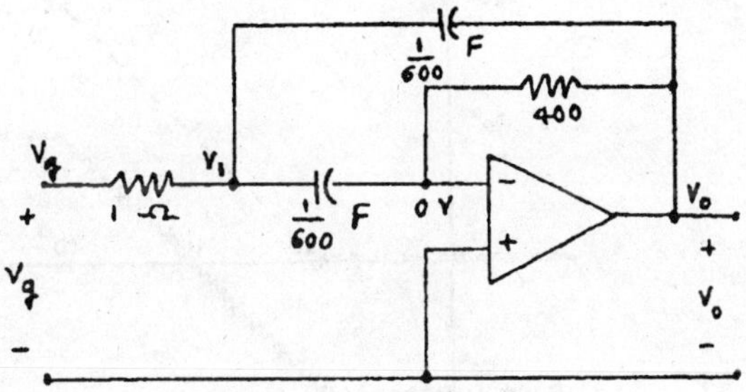


$$\begin{cases}\dfrac{V_1-V_g}{1}+\dfrac{V_1 s}{600}+\dfrac{(V_1-V_o)s}{600}=0\\[2mm] \dfrac{V_o}{400}+\dfrac{sV_1}{600}=0 \quad \therefore V_1=\dfrac{-1.5}{s}V_o\end{cases}$$

$$\therefore V_o\left(-\frac{1.5}{s}\right)\left(1+\frac{s}{300}\right)-\frac{sV_o}{600}=V_g=V_o\left(-\frac{1.5}{s}-\frac{1}{200}-\frac{s}{600}\right)$$

$$\therefore \frac{V_g}{V_o}=-\frac{s^2+3s+900}{600s} \quad \therefore H(s)=\frac{V_o}{V_g}=\frac{-600s}{s^2+3s+900}$$

(b) (c) (d) same as Prob 33 above

(e) $K_m=75$ , $K_f=\frac{10{,}000}{30}=\frac{1000}{3}$

$\therefore$ 1 Ω → 75 Ω , 400 Ω → 30 KΩ

$$1/600 \rightarrow (1/600)(1/75)(3/1000)=66.67 \text{ nF}$$

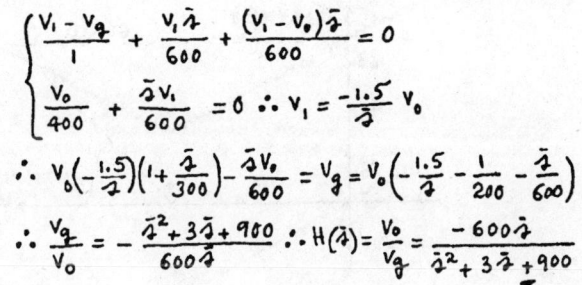

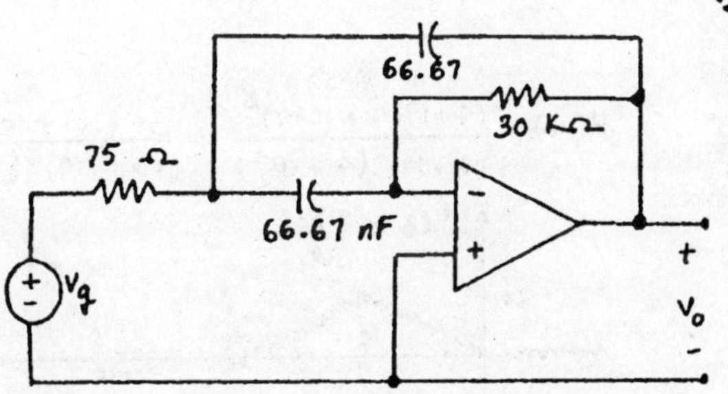

**13-35** $G(s) = \dfrac{10{,}000\, s}{(s+1)(s+10{,}000)} = \dfrac{s}{(1+s)(1+s/10{,}000)}$

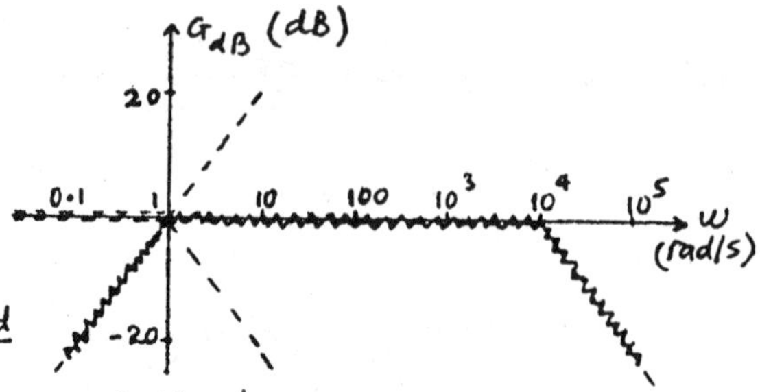


(a) $G(j100) = \dfrac{j100}{(1+j100)(1+j0.01)}$

$|G(j100)| \to 0$ dB (using asymptotes)

$|G(j100)|/10 \to -20$ dB $\therefore\ \omega = \underline{0.1}$ and $\underline{10^5}$ $\frac{\text{rad}}{\text{s}}$

(b) $|G(j\omega)| = \dfrac{\omega}{\sqrt{1+\omega^2}\sqrt{1+10^{-8}\omega^2}}$ ; $|G(j100)| = \dfrac{100}{\sqrt{10{,}001}\sqrt{1.0001}}$ ; $\dfrac{|G(j100)|}{10} = \dfrac{10}{\sqrt{10{,}001}\sqrt{1.001}} = \dfrac{\omega}{\sqrt{1+\omega^2}\sqrt{1+10^{-8}\omega^2}}$

$\therefore \dfrac{100}{10{,}001 \times 1.0001} = \dfrac{\omega^2}{(1+\omega^2)(1+10^{-8}\omega^2)}$ $\therefore\ \omega^2 = 0.010099$ and $9.90200 \times 10^9$

$\therefore\ \omega = \underline{0.100494}$ and $\underline{99{,}508.8}$ rad/s

**13-36** (a) $H(s) = \dfrac{5000(s+10)}{s(s+100)} = \dfrac{500(1+s/10)}{s(1+s/100)}$ $\qquad 500 \to 54$ dB , $\dfrac{1}{j\omega} \longrightarrow -90°$

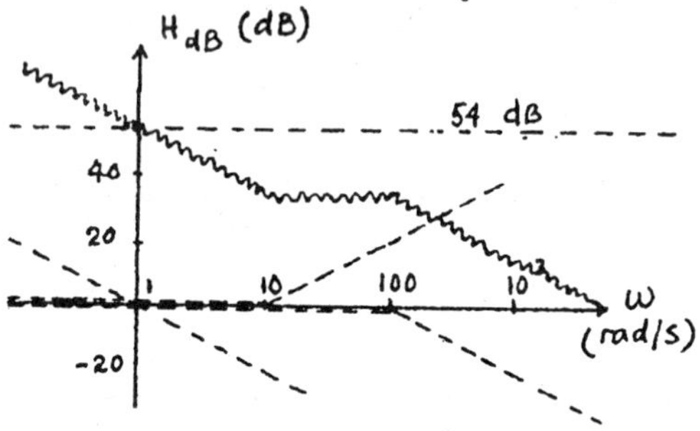


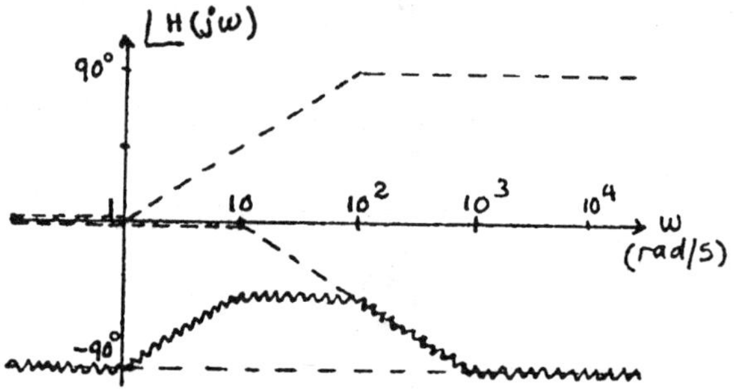


(b) $H(s) = \dfrac{0.2(s+100)^2}{s} = \dfrac{2000(1+s/100)^2}{s}$ $\qquad 2000 \longrightarrow 66$ dB , $1/j\omega = -90°$

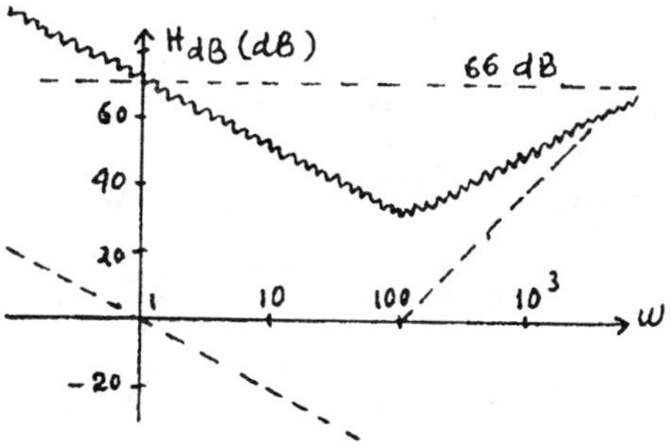


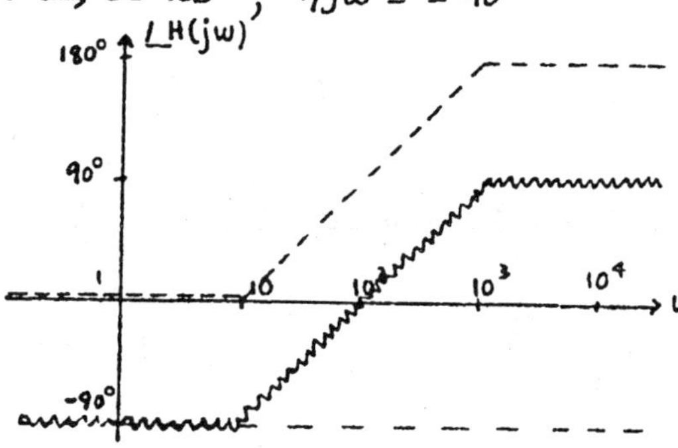


**13-37** $H(s) = \dfrac{(s+1)(s+1000)^2}{(s+10)^2(s+10^4)} = \dfrac{(1+s)(1+s/1000)^2}{(1+s/10)^2(1+s/10^4)}$

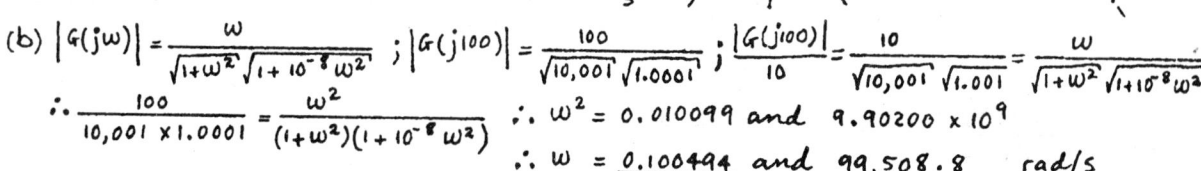

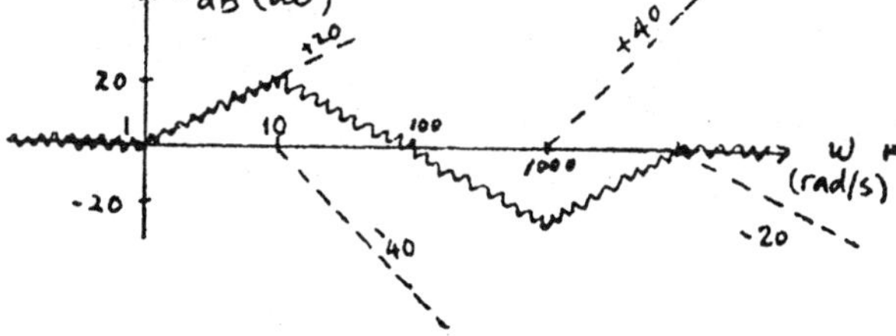


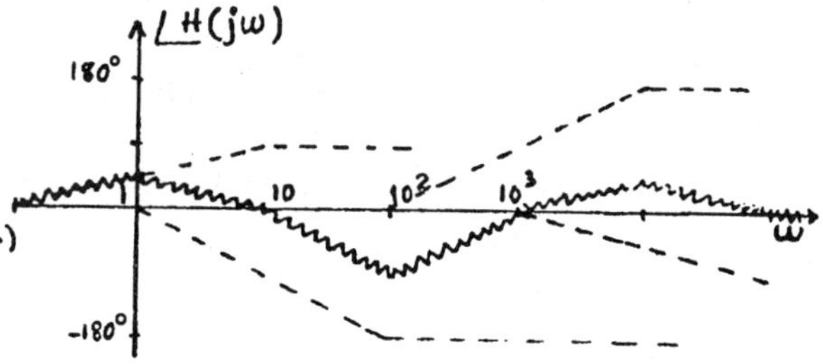

**13-38** $H(\bar{s}) = \dfrac{40\bar{s}^2}{\bar{s}+5} = \dfrac{8\bar{s}^2}{1+\bar{s}/5}$ $\quad 8 \to 18$ dB , $(j\omega)^2 \to 180°$

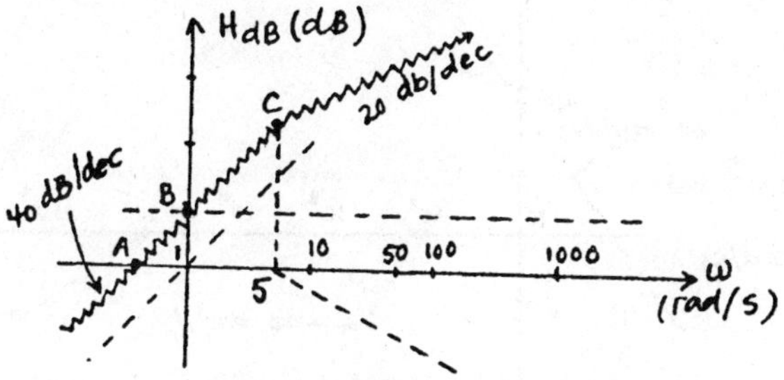


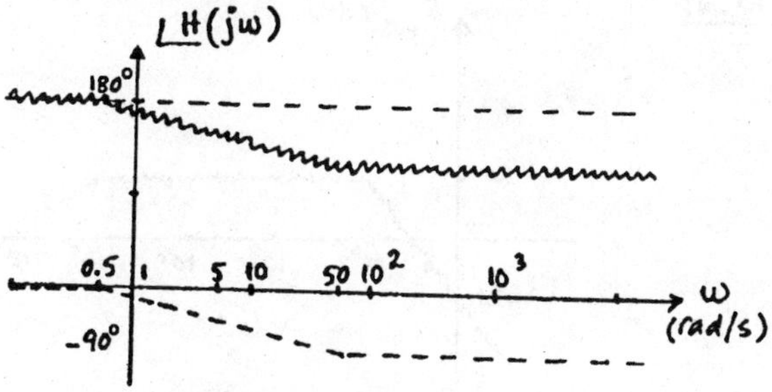


A : $H_{dB} = 0$ : $\log 8\omega^2 = 0 \;\therefore\; \omega = 1/\sqrt{8} = \underline{0.354 \text{ rad/s}}$ (intercept)

B : $\omega = \underline{1}$ : $H_{dB} = 20 \log 8 = \underline{18.06 \text{ dB}}$ (intercept)

C : $\omega = \underline{5}$ : $H_{dB} = 20 \log(8 \times 5^2) = \underline{46.02 \text{ dB}}$ (corner); (43.01 dB exact)

**13-39**

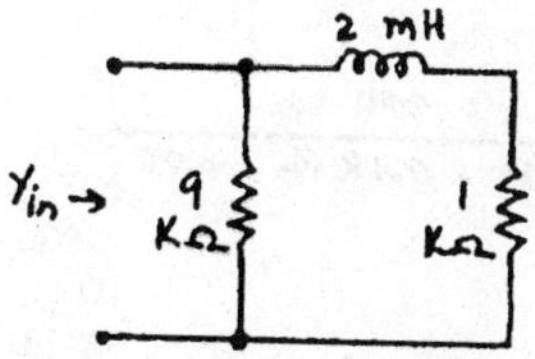


$Y_{in} = \dfrac{1}{9000} + \dfrac{1}{\bar{s}/500 + 1000} = \dfrac{\bar{s}/500 + 10^4}{18\bar{s} + 9\times 10^6} = \dfrac{1}{900}\;\dfrac{1+\bar{s}/5\times 10^6}{1+\bar{s}/5\times 10^5}$

$\dfrac{1}{900} \to -59.1$ dB , $j\omega \to 90°$

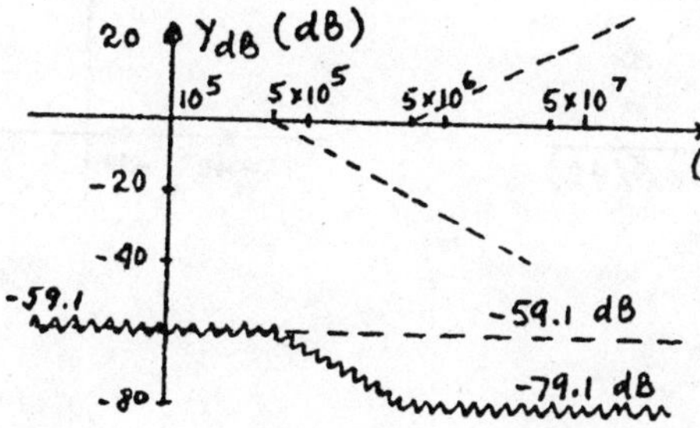


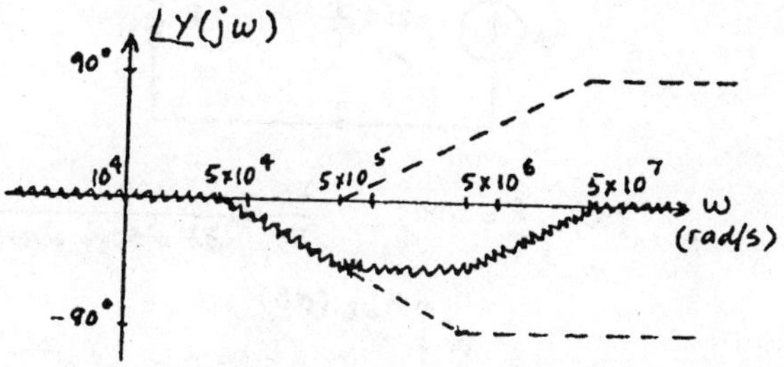


**13-40** $H(\bar{s}) = \dfrac{1000\bar{s}}{\bar{s}^2 + 10\bar{s} + 400} = \dfrac{2.5\bar{s}}{1 + 2(\frac{1}{4})(\frac{\bar{s}}{20}) + (\bar{s}/20)^2} \;\therefore\; \omega_o = 20$ , $\zeta = \frac{1}{4}$ (+6 dB correction)

$2.5 \to 8$ dB , $\angle 2.5j\omega = 90°$

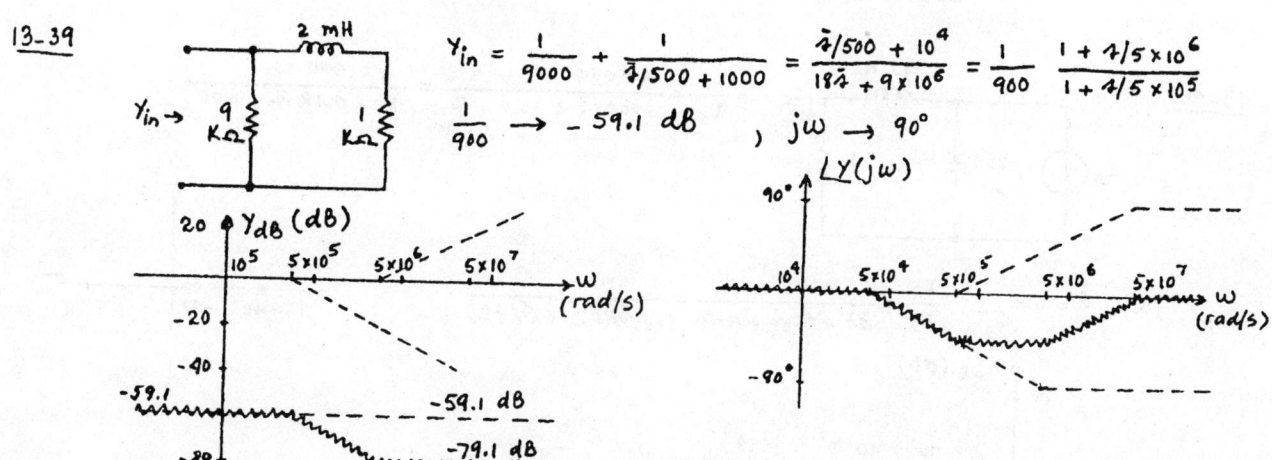

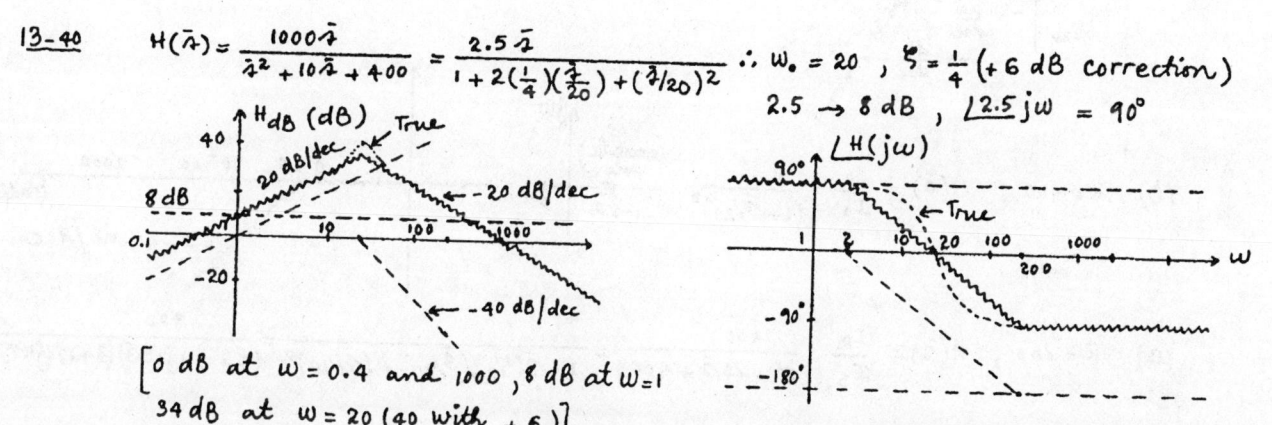

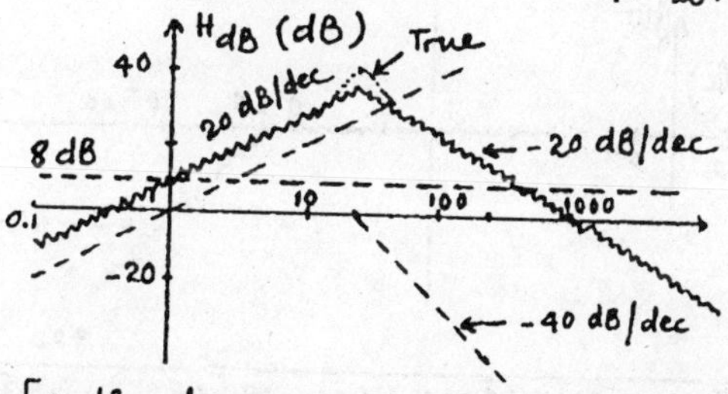


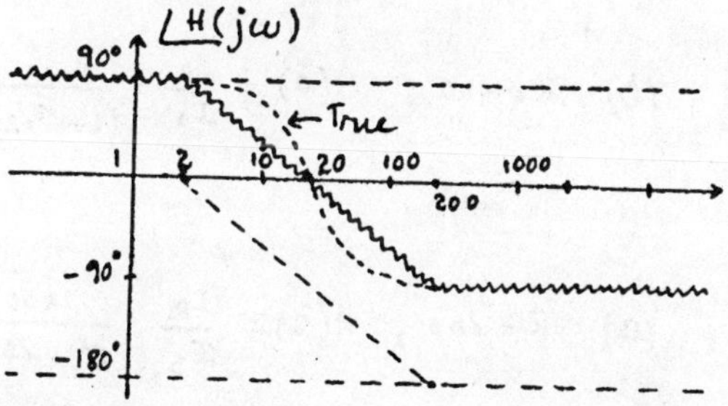


[0 dB at ω = 0.4 and 1000 , 8 dB at ω = 1
34 dB at ω = 20 (40 with + 6)]

13-41

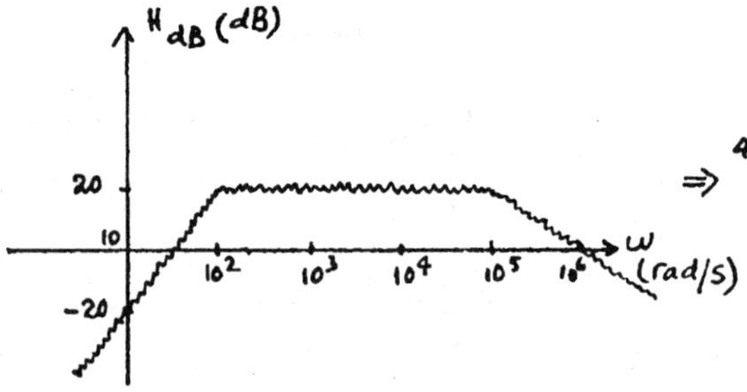


⇒

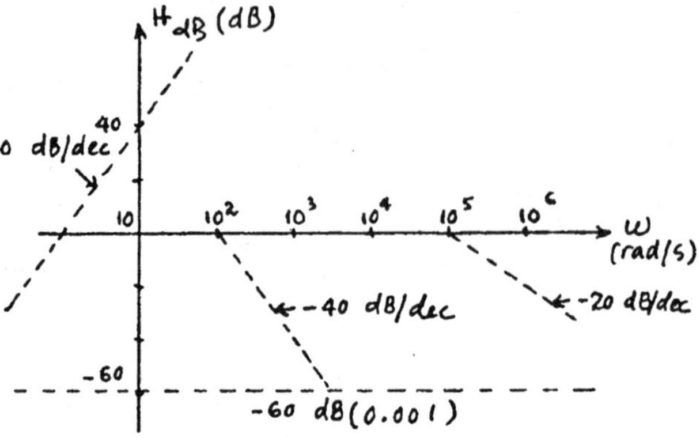


$$\therefore H(s) = \frac{0.001 s^2}{\left(1+\frac{s}{100}\right)^2\left(1+\frac{s}{10^5}\right)} = \frac{10^6 s^2}{(s+100)^2(s+10^5)} = \frac{10^6 s^2}{(s^3 + 100{,}200 s^2 + 20{,}010{,}000 s + 10^9)}$$

13-42

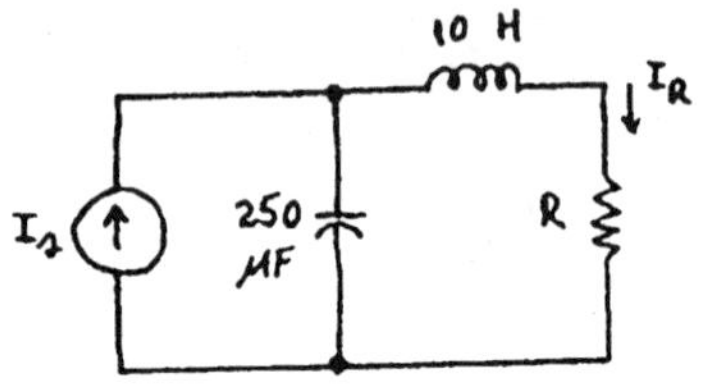


$$\frac{I_R}{I_s} = \frac{4000/s}{4000/s + 10s + R} = \frac{400}{s^2 + 0.1Rs + 400}$$

(a) $R = 500$; $\frac{I_R}{I_s} = \frac{400}{s^2 + 50s + 400} = \frac{1}{(1+s/10)(1+s/40)}$

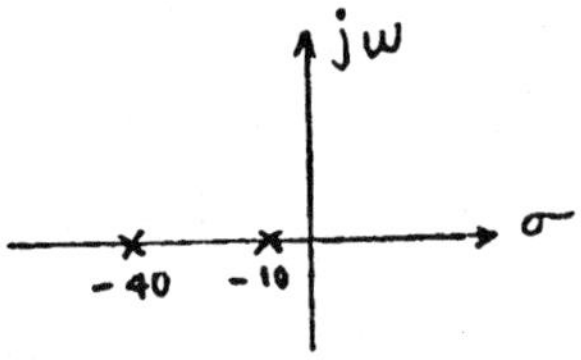


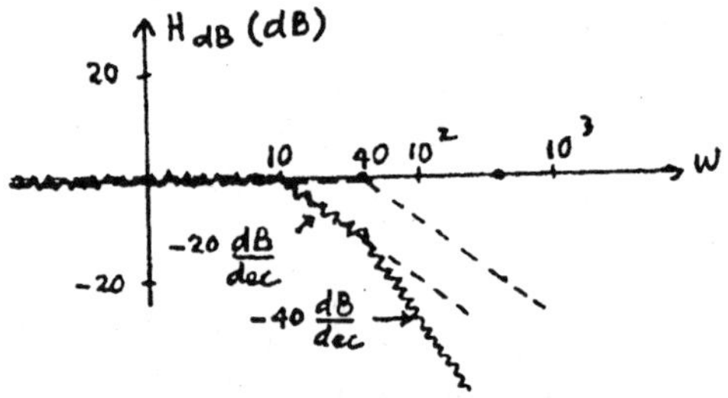


(b) $R = 400$; $H(s) = \frac{I_R}{I_s} = \frac{1}{(1+s/20)^2}$

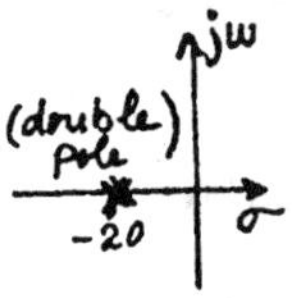


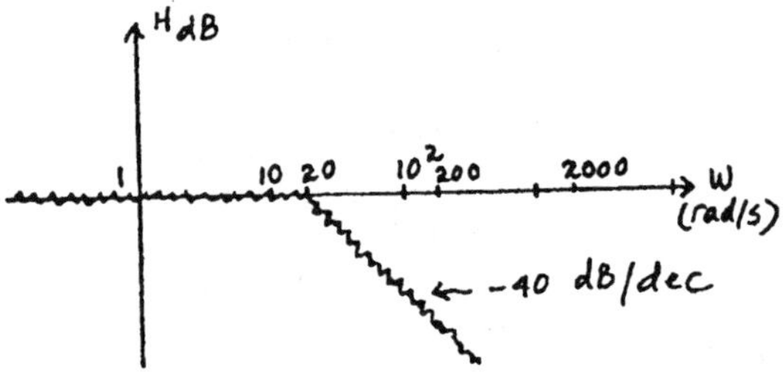


(c) $R = 200$, $H(s) = \frac{I_R}{I_s} = \frac{400}{s^2 + 20s + 400} = \frac{1}{1 + 2(1/2)(s/20) + (s/20)^2} = \frac{400}{(s+10+j10\sqrt{3})(s+10-j10\sqrt{3})}$

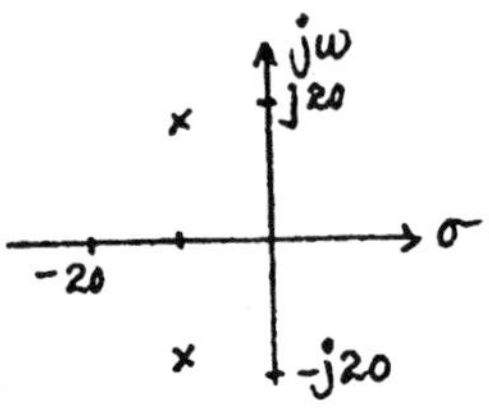


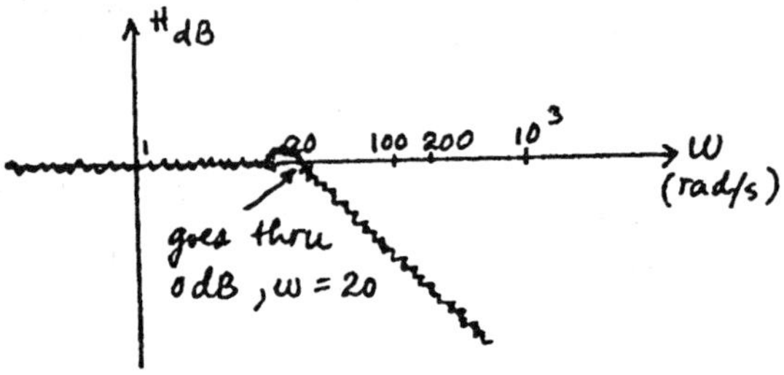

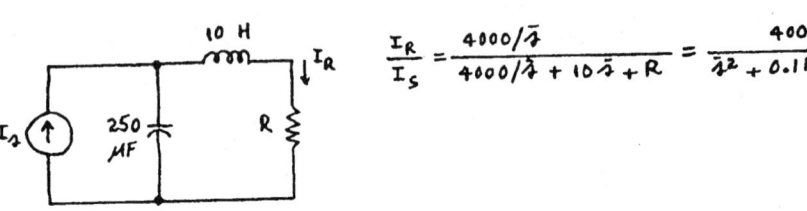

## 13-43

Applying 1 V at the input terminals, the input impedance is then numerically equal to the voltage across the input terminals, VM(1) which is listed below.

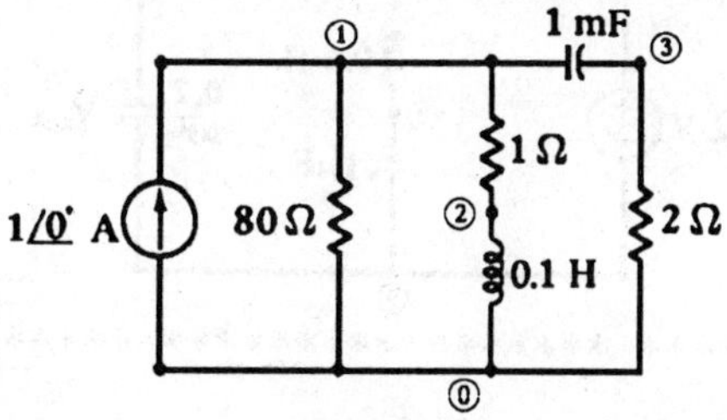


************************************************************************

**PROGRAM:**

```
I01 0 1 AC 1
R10 1 0 80
R12 1 2 1
L20 2 0 0.1
C13 1 3 1M
R30 3 0 2
.AC LIN 15 10 24
.PRINT AC VM(1)
.END
```

************************************************************************

| FREQ | VM(1) |
|---|---|
| 1.000E+01 | 9.649E+00 |
| 1.100E+01 | 1.174E+01 |
| 1.200E+01 | 1.432E+01 |
| 1.300E+01 | 1.739E+01 |
| 1.400E+01 | 2.065E+01 |
| 1.500E+01 | 2.320E+01 |
| 1.600E+01 | 2.394E+01 |
| 1.700E+01 | 2.272E+01 |
| 1.800E+01 | 2.049E+01 |
| 1.900E+01 | 1.813E+01 |
| 2.000E+01 | 1.605E+01 |
| 2.100E+01 | 1.431E+01 |
| 2.200E+01 | 1.289E+01 |
| 2.300E+01 | 1.172E+01 |
| 2.400E+01 | 1.076E+01 |

## *13-44*

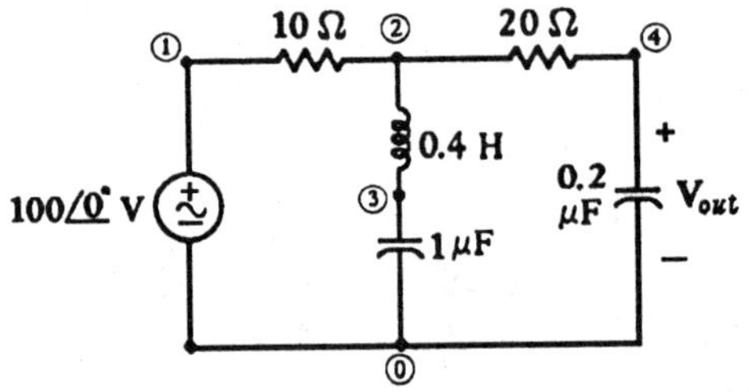


***********************************************************************

**PROGRAM:**

```
V10 1 0 AC 100
R12 1 2 10
L23 2 3 0.4
C30 3 0 1U
R24 2 4 20
C40 4 0 0.2U
.AC DEC 15 100 1000
.PRINT AC VM(4)
.END
```

***********************************************************************

| FREQ | VM(4) |
|---|---|
| 1.000E+02 | 1.000E+02 |
| 1.166E+02 | 9.999E+01 |
| 1.359E+02 | 9.999E+01 |
| 1.585E+02 | 9.998E+01 |
| 1.848E+02 | 9.996E+01 |
| 2.154E+02 | 9.985E+01 |
| 2.512E+02 | 2.241E+01 |
| 2.929E+02 | 9.988E+01 |
| 3.415E+02 | 9.997E+01 |
| 3.981E+02 | 9.998E+01 |
| 4.642E+02 | 9.998E+01 |
| 5.412E+02 | 9.998E+01 |
| 6.310E+02 | 9.997E+01 |
| 7.356E+02 | 9.997E+01 |
| 8.577E+02 | 9.995E+01 |
| 1.000E+03 | 9.993E+01 |

# CHAPTER 14: MAGNETICALLY COUPLED CIRCUITS

## Section 14-2 Mutual inductance

**1** The physical construction of three pairs of coupled coils is shown in Fig. 14-1. Show two different possible locations for the two dots on each pair of coils.

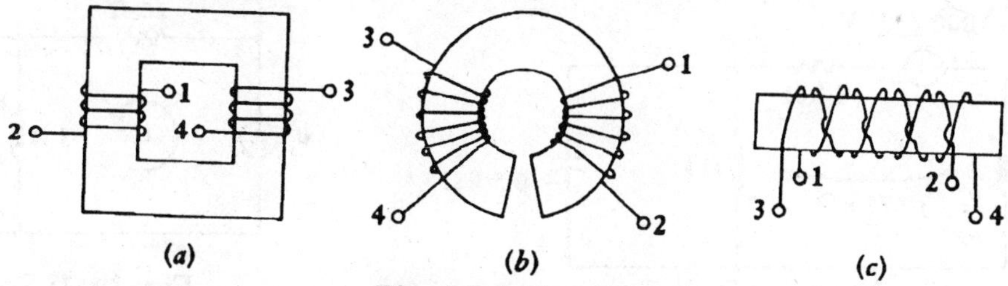


**Fig. 14-1:** See Prob. 1.

**2** In the circuit shown in Fig. 14-2: (*a*) find $v_1(t)$ and $v_2(t)$ if $i_1 = 4\cos 3t$ A and $i_2 = 0$; (*b*) find $v_1(t)$ and $v_2(t)$ if $i_1 = 4\cos 3t$ A and $i_2 = 3\cos 3t$ A; (*c*) find $\mathbf{V}_1$ and $\mathbf{V}_2$ in polar form if $\mathbf{I}_1 = 4\underline{/0°}$ A, $\mathbf{I}_2 = 5\underline{/-90°}$ A, and $\omega = 3$ rad/s.

**3** For the circuit shown in Fig. 14-2, let $\mathbf{V}_1 = 40\underline{/0°}$ V, $\mathbf{V}_2 = 60\underline{/90°}$ V, $\mathbf{s} = -2 + j3$ s$^{-1}$, and find $\mathbf{I}_1$ and $\mathbf{I}_2$ in polar form.

**4** (*a*) Refer to Fig. 14-3 and find $\mathbf{H}(\mathbf{s}) = \mathbf{I}_2(\mathbf{s})/\mathbf{V}_s(\mathbf{s})$. (*b*) At what value of $\omega$ is $|\mathbf{H}(j\omega)|$ a maximum?

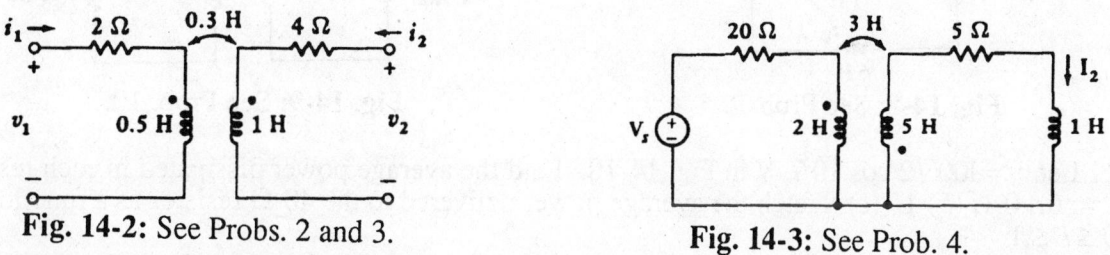

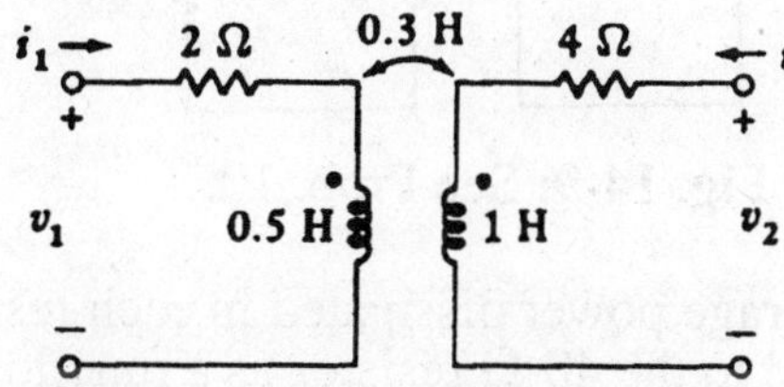


**Fig. 14-2:** See Probs. 2 and 3.

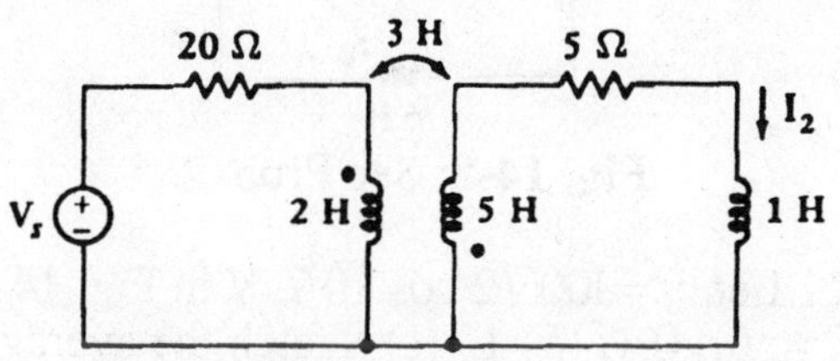


**Fig. 14-3:** See Prob. 4.

**5** Let $\mathbf{s} = j10$ s$^{-1}$ for the arrangement of coupled coils shown in Fig. 14-4 and find $\mathbf{Z}_{in}$ if the missing dot is at: (*a*) *A*; (*b*) *B*.

**6** In Fig. 14-5, the source current $i_1(t) = 10\sqrt{2}\cos 2t$ A. (*a*) Find $i_2(t)$. (b) Find the phasor-voltage ratio, $\mathbf{V}_2/\mathbf{V}_1$.

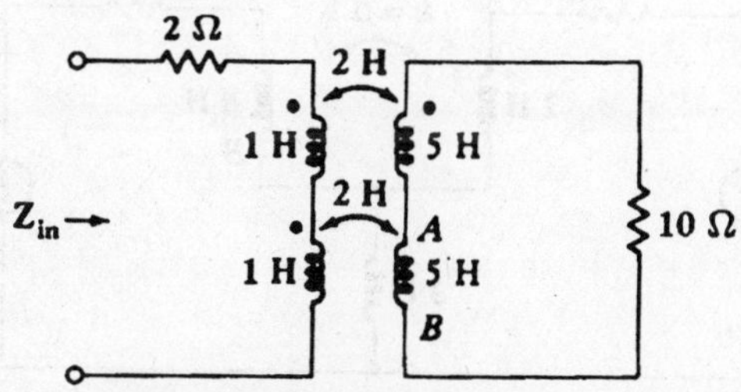

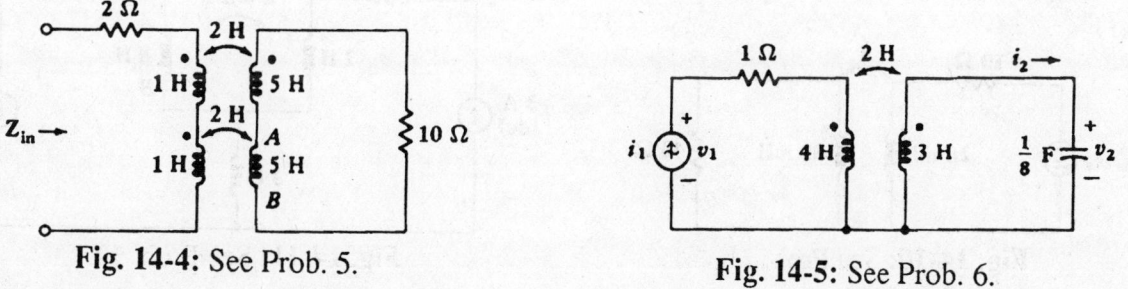

**Fig. 14-4:** See Prob. 5.

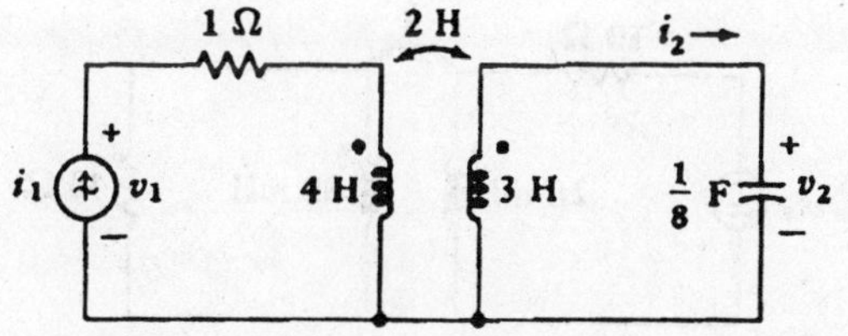


**Fig. 14-5:** See Prob. 6.

**7** For the circuit shown in Fig. 14-6, find: (*a*) the power factor at which the source is operating, specifying whether it is leading or lagging; (*b*) the average power delivered to the 1-kΩ resistor.

**8** Using the mesh currents assigned in Fig. 14-7, let $\omega = 10$ rad/s, and: (*a*) write the three mesh equations in standard form; (*b*) find the average power being supplied by $\mathbf{V}_s$, if $\mathbf{V}_s = 100\angle 0^\circ$ V.

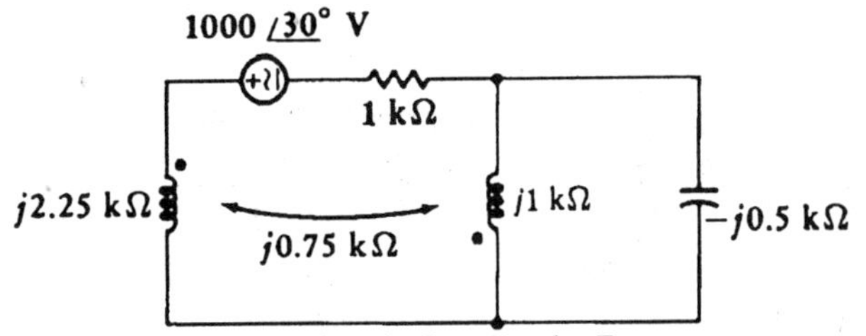


Fig. 14-6: See Prob. 7.

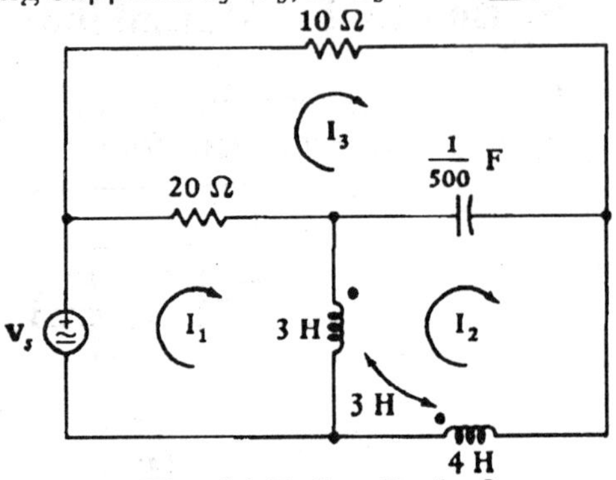


Fig. 14-7: See Prob. 8.

**9** It is possible to arrange three coils physically in such a way that there is mutual coupling between coils $A$ and $B$ and between $B$ and $C$, but not between $A$ and $C$. Such an arrangement is shown in Fig. 14-8. Find $v(t)$.

**10** With reference to Fig. 14-9, what value of $L_2$ will cause exactly 5 W average power to be delivered to the 8-Ω speaker at $\omega = 1$ krad/s?

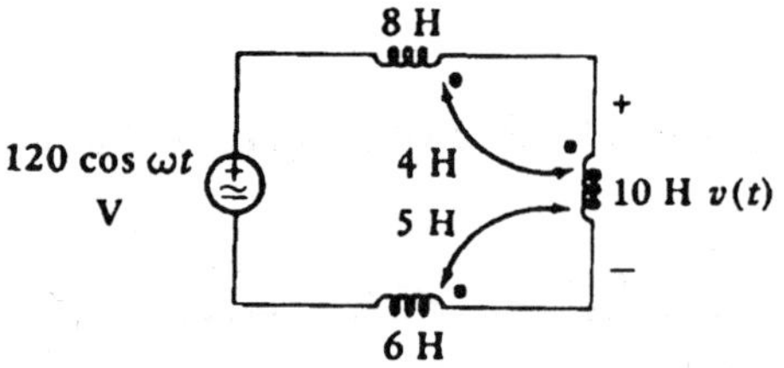


Fig. 14-8: See Prob. 9.

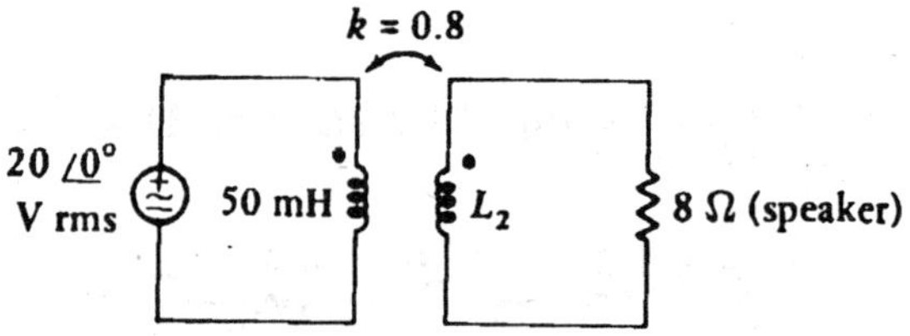


Fig. 14-9: See Prob. 10.

**11** Let $v_s = 100\sqrt{2}\cos 10^3 t$ V in Fig. 14-10. Find the average power dissipated in each resistor if $k$ =: (*a*) 0.1; (*b*) 1. (*c*) Sketch the average power delivered to the 40-Ω resistor as a function of $k$, $0 \le k \le 1$.

## Section 14-3 Energy considerations

**12** Determine the total energy stored in the system of Fig. 14-11 if the missing dot is at: (*a*) $A$, and $I_2 = 2$ A; (*b*) $B$, and $I_2 = 2$ A; (*c*) $A$, and $I_2 = -2$ A; (*d*) $B$, and $I_2 = -2$ A.

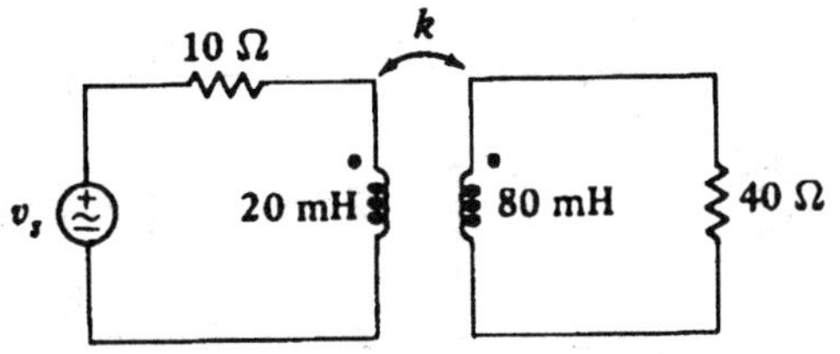


Fig. 14-10: See Prob. 11.

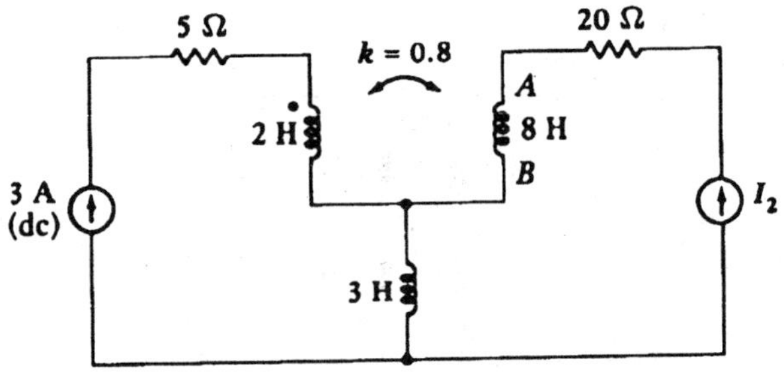


Fig. 14-11: See Prob. 12.

## Section 14-4 The linear transformer

**13** (a) Find $i_x(t)$ and $v(t)$ in Fig. 14-12 if $i_1 = 5\cos 40t$ A and $i_2 = 2\sin 40t$ A. (*b*) Is the direction of average positive power flow to the left or to the right in the center of the circuit?

**14** If $i_1 = 2\cos 10t$ A in Fig. 14-13, find the total energy stored in the network at $t = 0$ if: (*a*) *a-b* is open-circuited; (*b*) *a-b* is short-circuited.

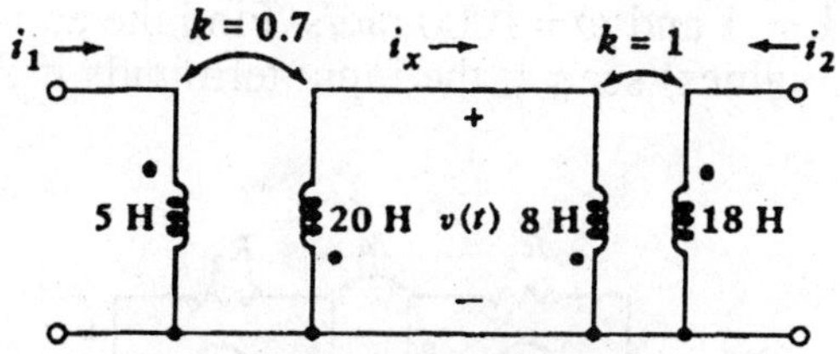


**Fig. 14-12:** See Prob. 13.

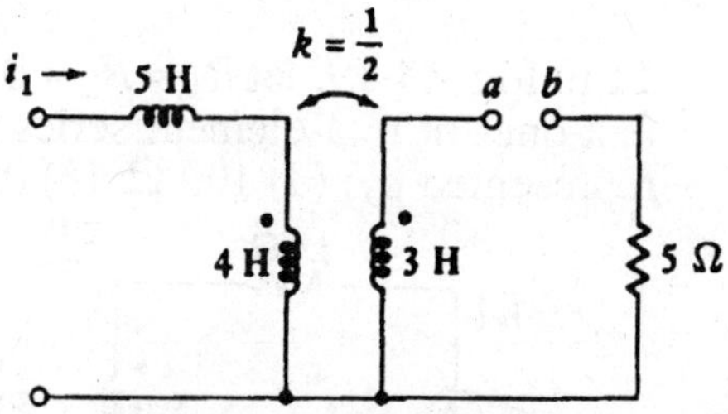


**Fig. 14-13:** See Prob. 14.

**15** After having been open for a long time, the switch appearing in the circuit of Fig. 14-14 is closed at $t = 0$. (*a*) Find the energy stored in the system of inductors at $t = 0^-$. (*b*) At what time has half this energy been dissipated in the resistor?

**16** Find two frequencies other than 0 or 10 rad/s at which $\mathbf{Z}_{in} = R_{in} + j0$ for the network of Fig. 14-15.

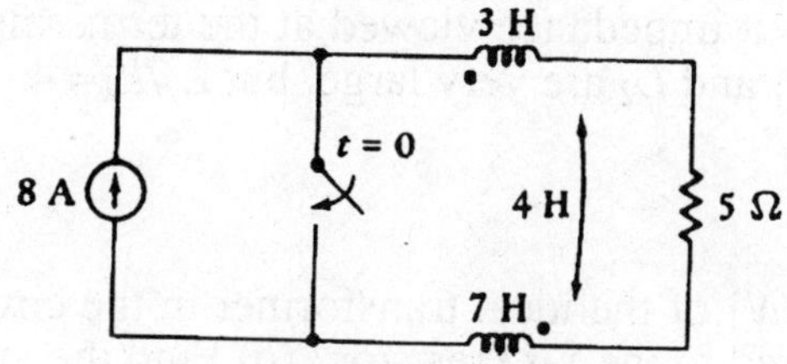


**Fig. 14-14:** See Prob. 15.

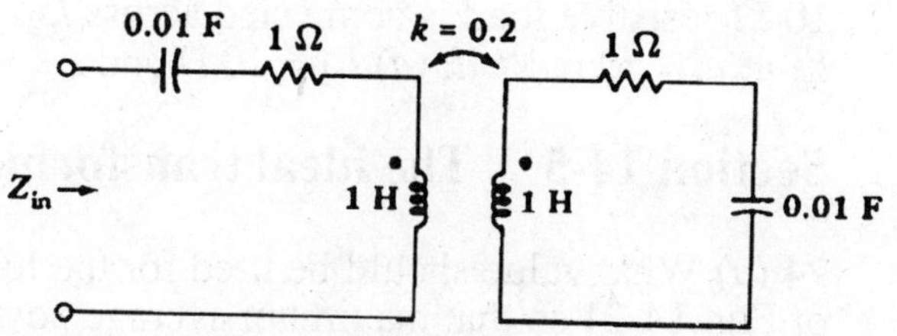


**Fig. 14-15:** See Prob. 16.

**17** Calculate the inductance presented at the specified terminals in Fig. 14-16 if these terminals are connected together: (*a*) $L_{AD}$, *B* to *C*; (*b*) $L_{AC}$, *B* to *D*; (*c*) $L_{AB}$, *C* to *D*; (*d*) $L_{CD}$, *A* to *B*; (*e*) $L_{AB}$, *A* to *C*, and *B* to *D*; (*f*) $L_{AB}$ , no other connections; (*g*) $L_{CD}$, no other connections. (*h*) Find another connection that will provide a different value of *L*.

**18** Find the equivalent inductance seen at terminals 1-2 in Fig. 14-17 if the following terminals are connected together: (*a*) none; (*b*) *A* to *B*; (*c*) *B* to *C*; (*d*) *A* to *C*.

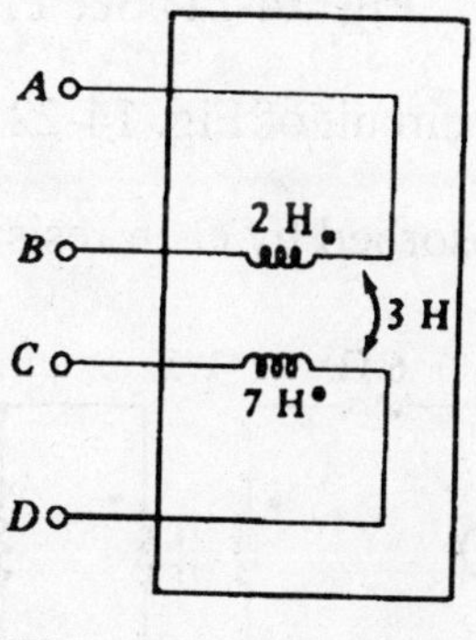


**Fig. 14-16:** See Prob. 17.

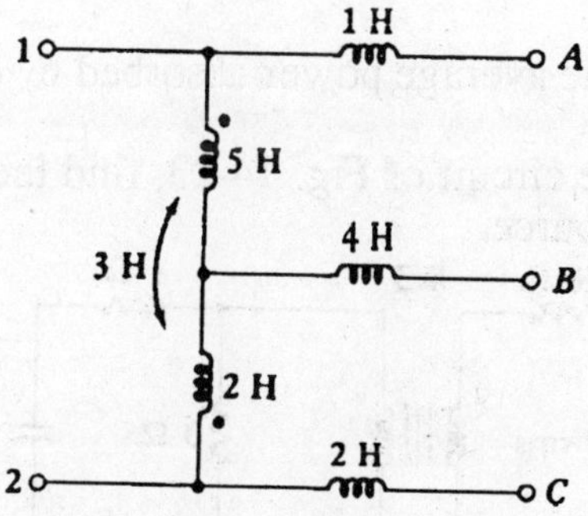


**Fig. 14-17:** See Prob. 18.

**19** For the circuit of Fig. 14-18: (*a*) find the transfer function, $\mathbf{H}(s) = \mathbf{I}_L(s)/\mathbf{V}_s(s)$; (*b*) specify the form of the natural response, $i_{Ln}(t)$; (*c*) let $v_s = 100u(t)$ V and write two differential equations in terms of the variables $i_s$ and $i_L$; (*d*) specify values for $i_L(0^+)$ and $i_L'(0^+)$; (*e*) find $i_L(t)$.

**20** In Fig. 14-19, let $R_1 = 2\ \Omega$, $R_2 = 0$, and $\mathbf{Z}_L = 40 + j0\ \Omega$. If $\omega = 1$ krad/s, determine $\mathbf{V}_L/\mathbf{V}_s$ if: (*a*) $L_1 = 1$ mH, $L_2 = 25$ mH, and $k = 1$; (*b*) $L_1 = 1$ H, $L_2 = 25$ H, and $k = 0.99$; (*c*) $L_1 = 1$H, $L_2 = 25$ H, and $k = 1$.

**21** In Fig. 14-19, let $R_1 = R_2 = 0$, $L_1 = 1$ H, $L_2 = 4$ H, $k = 1$ and $\omega = 1000$ rad/s. Find the equivalent one- or two-element series network ($R$, $L$, and $C$ values) seen at the input terminals if $\mathbf{Z}_L$ is represented by: (*a*) 100 Ω; (*b*) 0.1 H; (*c*) 10 μF.

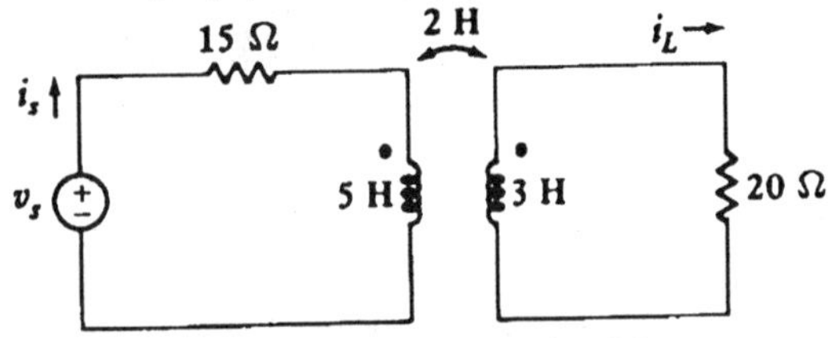


**Fig. 14-18:** See Prob. 19.

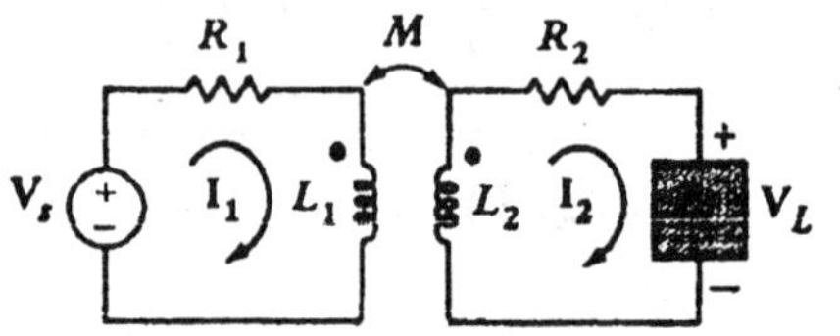


**Fig. 14-19:** See Probs. 20 and 21.

**22** At $\omega = 10$ krad/s, find $\mathbf{Z}_{\text{in}}$ for the network shown in Fig. 14-20.

**23** A transformer has a primary inductance $L_1$, a secondary inductance $L_2$, and unity coupling. A 10-Ω resistive load is connected across $L_2$. Find the input impedance viewed at the terminals of $L_1$ at $\omega = 50$ rad/s if: (*a*) $L_1 = 10$ H and $L_2 = 40$ H; (*b*) $L_1$ and $L_2$ are very large, but $L_2/L_1 = 4$.

## Section 14-5 The ideal transformer

**24** (*a*) What value should be used for the turns ratio $N_2/N_1$ of the ideal transformer in the circuit of Fig. 14-21 so that maximum average power is delivered to the 8-Ω resistor? (*b*) Find the value of $P_{8,max}$. (*c*) What value of $N_2/N_1$ will result in maximum dissipation in the 300-Ω resistor?

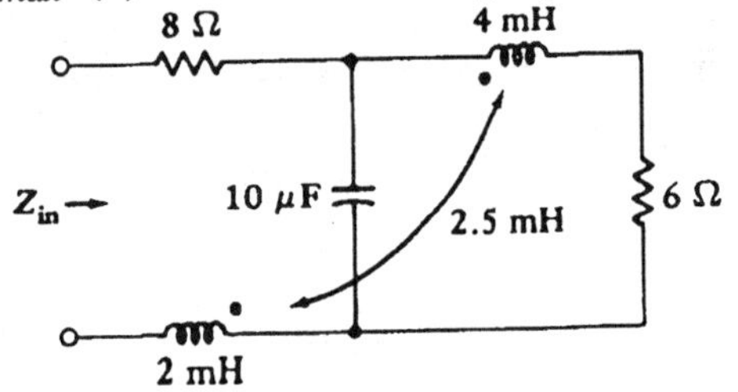

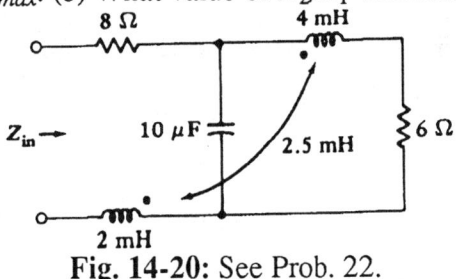

**Fig. 14-20:** See Prob. 22.

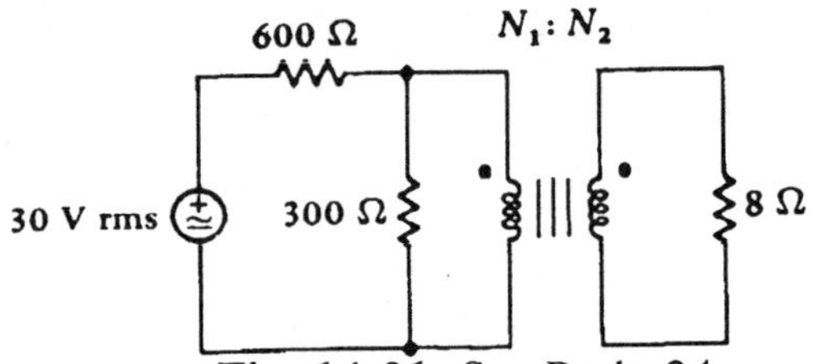


**Fig. 14-21:** See Prob. 24.

**25** Find the average power absorbed by each resistor in the circuit of Fig. 14-22.

**26** For the circuit of Fig. 14-23, find the average power absorbed by each resistor and generated by each source.

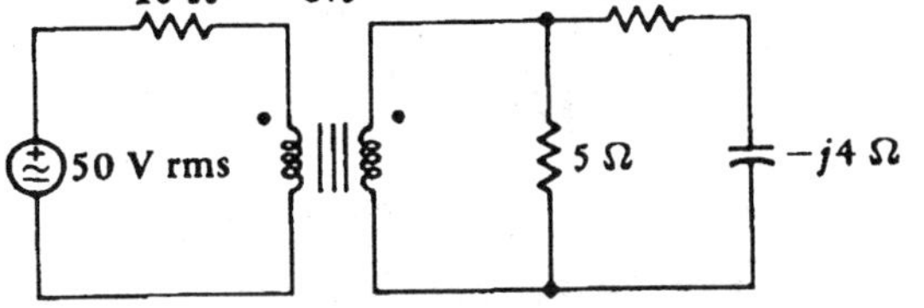


**Fig. 14-22:** See Prob. 25.

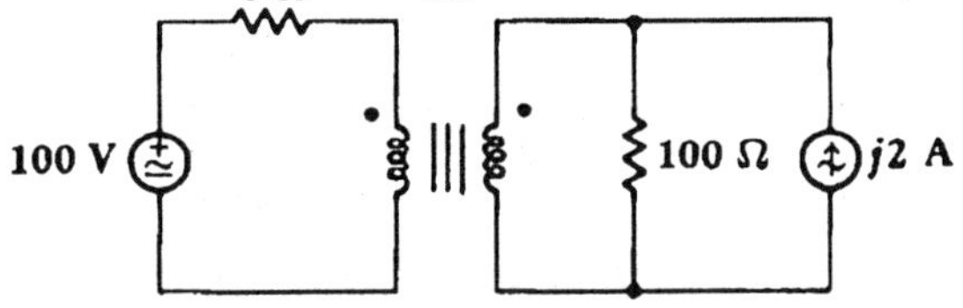


**Fig. 14-23:** See Prob. 26.

**27** (*a*) Find the average power delivered to each 10-Ω resistor in Fig. 14-24. (*b*) Repeat after connecting *A* to *C* and *B* to *D*.

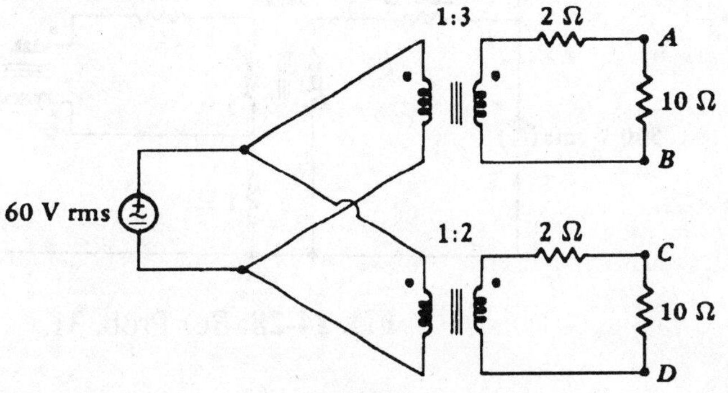


**Fig. 14-24:** See Prob. 27.

**28** Find the voltages $\mathbf{V}_8$ and $\mathbf{V}_{20}$ in the ciicuit shown in Fig. 14-25.

**29** (*a*) Find the Thévenin equivalent of the network in Fig. 14-26. (*b*) Repeat if one dot is moved to the other end of its winding.

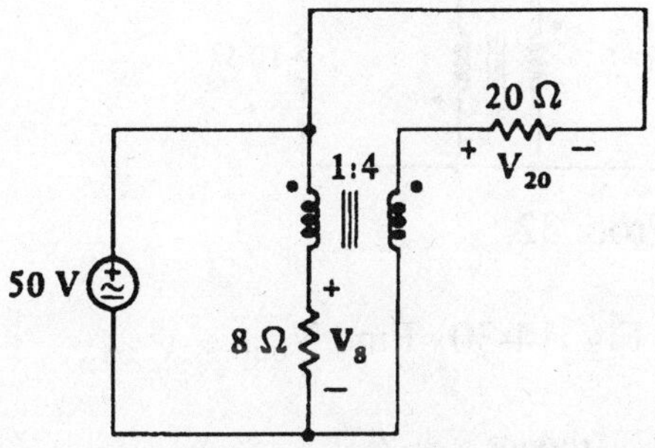

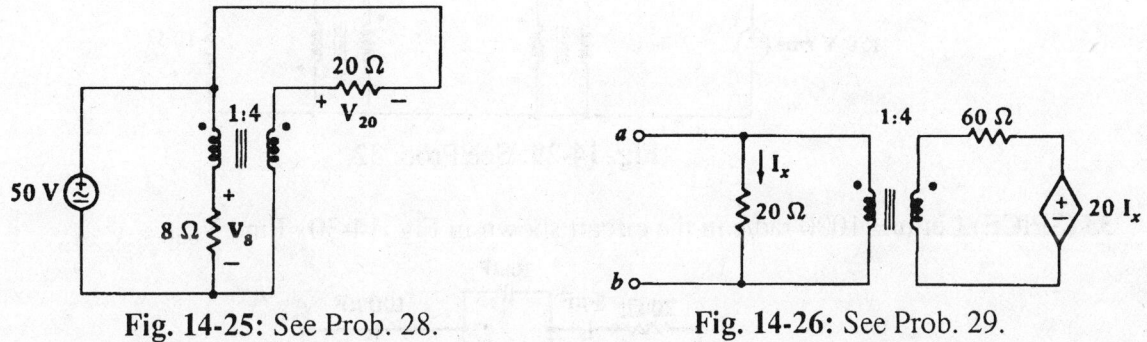

**Fig. 14-25:** See Prob. 28.

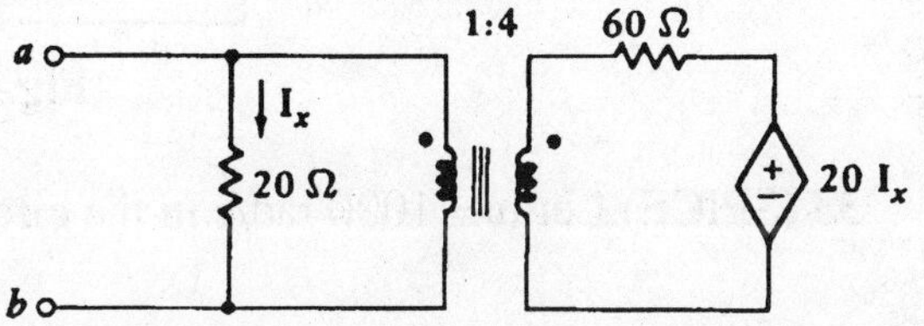


**Fig. 14-26:** See Prob. 29.

**30** Select values for the turns ratios *a* and *b* in Fig. 14-27 so that the ideal source supplies 1 kW, half of which is delivered to the 100-Ω load.

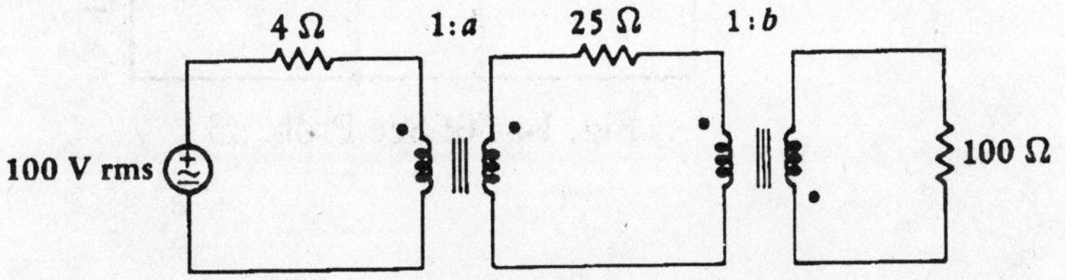


**Fig. 14-27:** See Prob. 30.

**31** Find the power being generated by the ideal source in the circuit of Fig. 14-28.

**32** In the circuit of Fig. 14-29, find the power being dissipated in each resistor and being generated by the source.

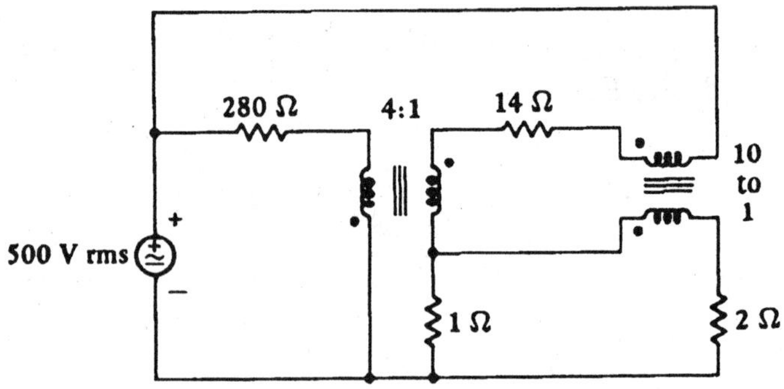


**Fig. 14-28:** See Prob. 31.

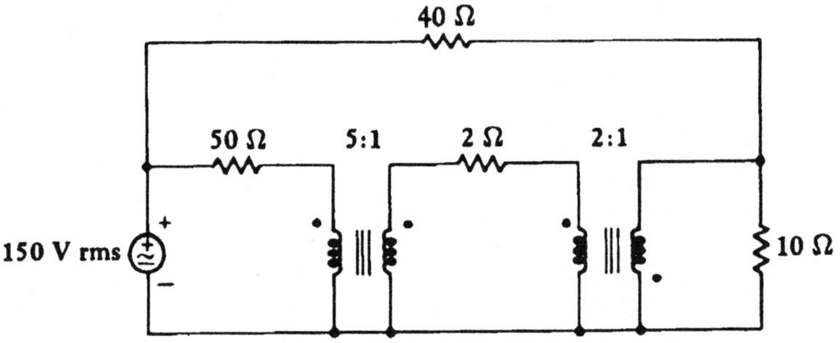


**Fig. 14-29:** See Prob. 32.

**33** (SPICE) Let $\omega = 1000$ rad/s in the circuit shown in Fig. 14-30. Find $\mathbf{V}_{out}$.

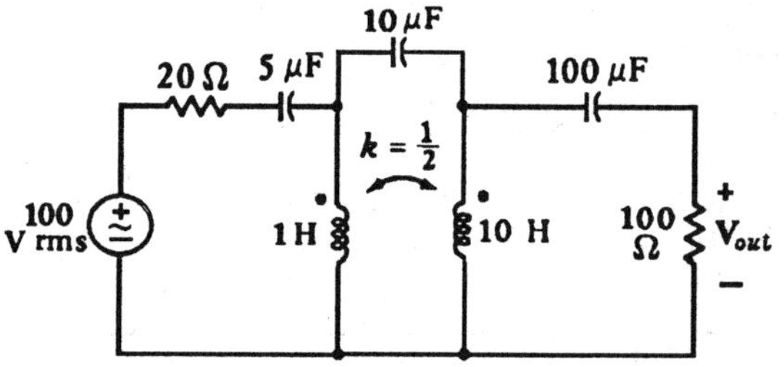


**Fig. 14-30:** See Prob. 33.

# Chapter 14

14-1 (a)

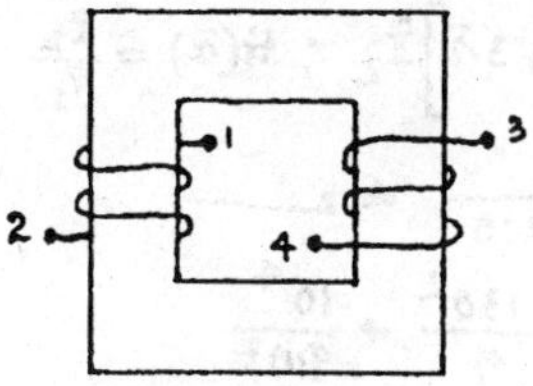


In at 1, CW
In at 3, CW
∴ 1 and 3
or 2 and 4

(b)

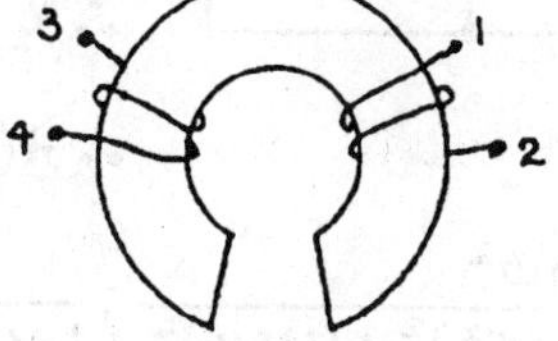


In at 1, CW
In at 4, CW
∴ 1 and 4
or 2 and 3

(c)

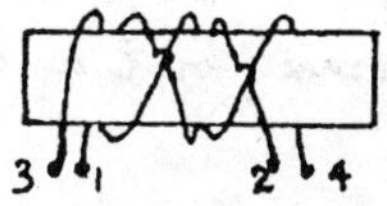


In at 1, —◇ ; In at 4 —◇

∴ 1 and 4 or 2 and 3

14-2 (a)

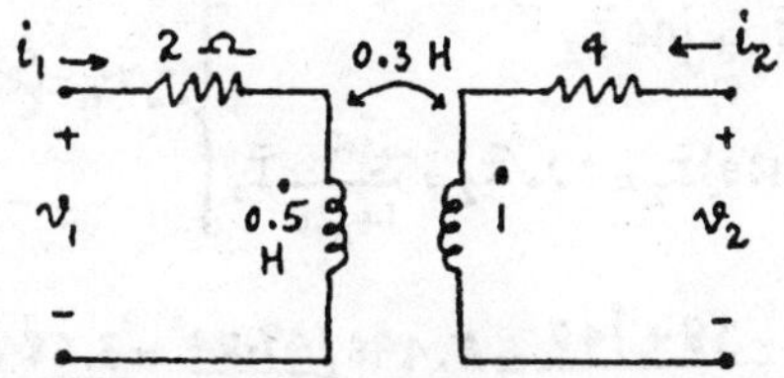


$i_1 = 4\cos 3t$ A, $i_2 = 0$ ∴ $v_1 = 2i_1 + 0.5i_1' + 0.3i_2'$

∴ $v_1 = \underline{8\cos 3t - 6\sin 3t} = \underline{10\cos(3t + 36.87°)\ \text{V}}$

$v_2 = 0.3i_1' = \underline{-3.6\sin 3t\ \text{V}}$

(b) $i_1 = 4\cos 3t$, $i_2 = 3\cos 3t$ ∴ $v_1 = 8\cos 3t - 6\sin 3t + 0.3(-3)3\sin 3t = \underline{8\cos 3t - 8.7\sin 3t\ \text{V}}$

∴ $v_1 = \underline{11.819\cos(3t + 47.40°)}$; $v_2 = -3.6\sin 3t + 12\cos 3t - 9\sin 3t = \underline{12\cos 3t - 12.6\sin 3t} = \underline{17.4\cos(3t + 46.4°)\text{V}}$

(c)

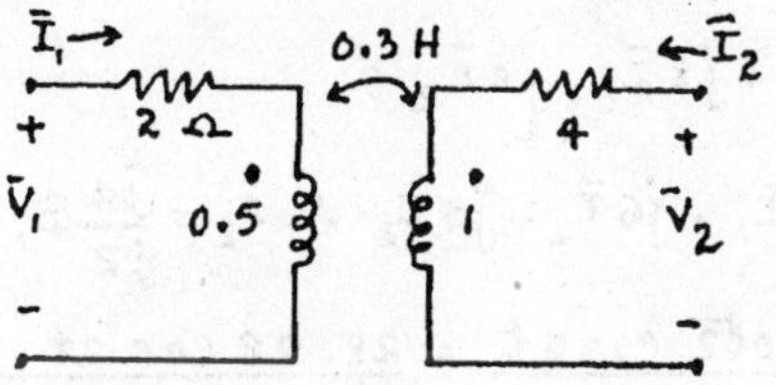


$\bar{I}_1 = 4\angle 0°$ A, $\bar{I}_2 = 5\angle -90°$ A, $\omega = 3$ rad/s

$\bar{V}_1 = 2\bar{I}_1 + j1.5\bar{I}_1 + j0.9\bar{I}_2 = (2 + j1.5)4 + j0.9(-j5)$

∴ $\bar{V}_1 = \underline{13.865^+ \angle 25.64°\ \text{V}}$

$\bar{V}_2 = 4\bar{I}_2 + j3\bar{I}_2 + j0.9\bar{I}_1 = (4 + j3)(-j5) + j3.6 = \underline{22.23\angle -47.55°\ \text{V}}$

14-3

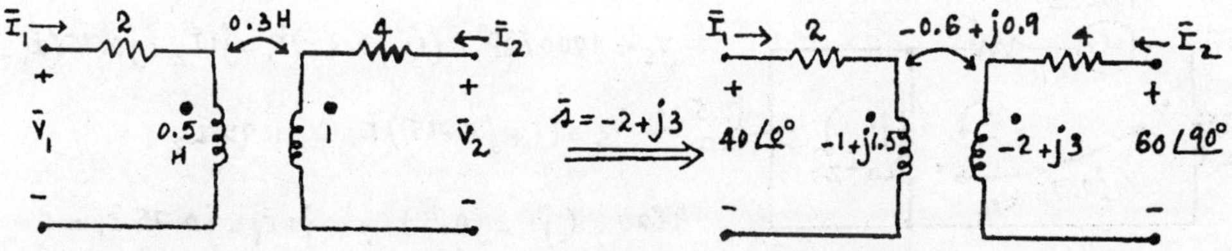


$$\left.\begin{aligned} 40 &= (2 - 1 + j1.5)\bar{I}_1 + (-0.6 + j0.9)\bar{I}_2 \\ j60 &= (4 - 2 + j3)\bar{I}_2 + (-0.6 + j0.9)\bar{I}_1 \end{aligned}\right\} \therefore \bar{I}_1 = \frac{\begin{vmatrix} 40 & -0.6 + j0.9 \\ j60 & 2 + j3 \end{vmatrix}}{\begin{vmatrix} 1 + j1.5 & -0.6 + j0.9 \\ -0.6 + j0.9 & 2 + j3 \end{vmatrix}} = \frac{134 + j156}{-2.05 + j7.08} = \underline{27.90\angle -56.81°\ \text{A}}$$

$$\bar{I}_2 = \frac{\begin{vmatrix} 1 + j1.5 & 40 \\ -0.6 + j0.9 & j60 \end{vmatrix}}{-2.05 + j7.08} = \frac{-66 + j24}{-2.05 + j7.08} = \underline{9.528\angle 53.87°\ \text{A}}$$

**14-4** (a)

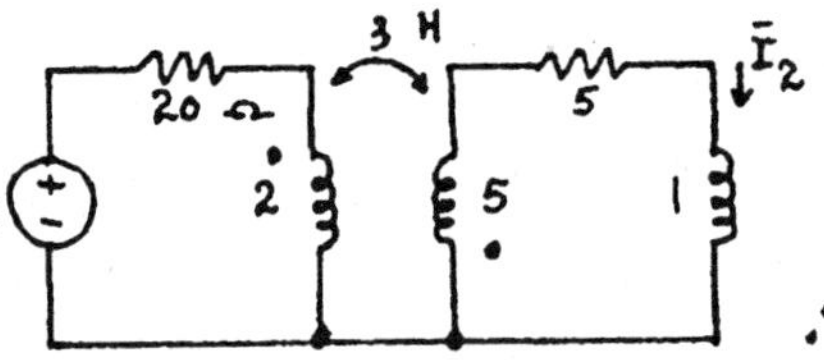


$$\begin{cases}\bar V_s = (20+2s)I_1 + 3s\bar I_2 \\ 3s\bar I_1 + (5+6s)\bar I_2 = 0 \therefore \bar I_1 = -\dfrac{5+6s}{3s}\bar I_2\end{cases}$$

$$\therefore \bar V_s = \left[\frac{(20+2s)(5+6s)}{-3s} + 3s\right]\bar I_2 \;;\; H(s) = \frac{\bar I_2}{\bar V_s}$$

$$\therefore H(s) = \frac{-3s}{3s^2 + 130s + 100}$$

(b) $$|H(j\omega)|^2 = \frac{9\omega^2}{(100-3\omega^2)^2 + (130\omega)^2} \;;\; \frac{1}{|H(j\omega)|^2} = \omega^2 - \frac{600}{9} + \frac{130^2}{9} + \frac{10^4}{9\omega^2}$$

$$\frac{d(1/|H(j\omega)|^2)}{d\omega} = 2\omega - \frac{2\times 10^4}{9\omega^3} = 0 \quad \therefore \omega^4 = 10^4/9,\ \omega^2 = \frac{100}{3},\ \omega = 10/\sqrt{3} = \underline{5.774 \text{ rad/s}}$$

(it is a max, not a min)

**14-5** (a)

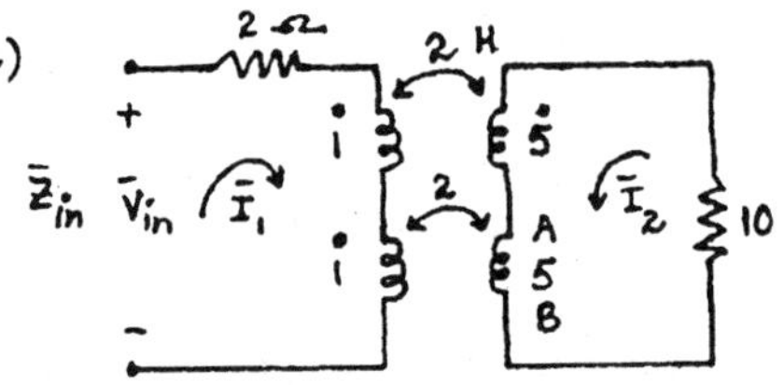


$s = j10$, Dot at A:

$$\left.\begin{aligned}&\bar V_{in} = (2+j20)\bar I_1 + j40\bar I_2 \\ &j40\bar I_1 + (10+j100)\bar I_2 = 0 \therefore \bar I_2 = \frac{-j4}{1+j10}\bar I_1\end{aligned}\right\} \therefore \bar V_{in} = \left(2+j20+\frac{160}{1+j10}\right)\bar I_1 = -\frac{38+j40}{1+j10}\bar I_1$$

$$\therefore \bar Z_{in} = \frac{\bar V_{in}}{\bar I_1} = \frac{-38+j40}{1+j10} = \underline{5.490\angle 49.24^\circ} = \underline{3.584 + j4.158\ \Omega}$$

(b) Dot at B: $\bar V_{in} = (2+j20)\bar I_1 \quad \therefore \bar Z_{in} = \underline{2+j20} = \underline{20.10\angle 84.29^\circ\ \Omega}$

**14-6** (a)

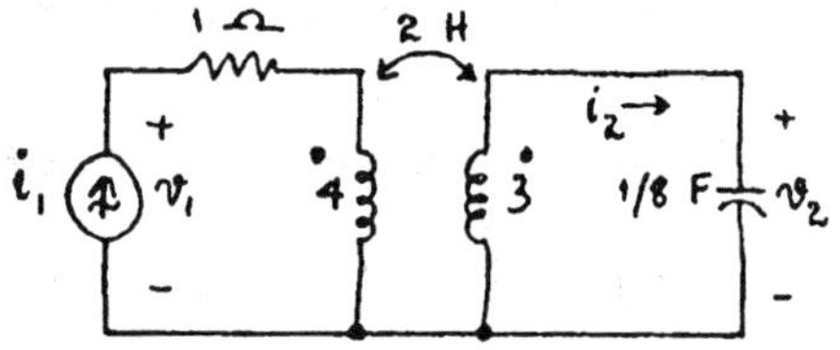


$i_1 = 10\sqrt{2}\cos 2t$ A

$\bar V_1 = \bar I_1 + j8\bar I_1 - j4\bar I_2$

$0 = -j4\bar I_1 + j6\bar I_2 - j4\bar I_2 \quad \therefore \bar I_2 = \dfrac{j4}{j2}\bar I_1 = 2\bar I_1$

$\therefore i_2 = \underline{20\sqrt{2}\cos 2t} = \underline{28.28\cos 2t \text{ A}}$

(b) $\bar V_1 = (1+j8)\bar I_1 - j8\bar I_1 = \bar I_1$, $\bar V_2 = -j4\bar I_2 = -j8\bar I_1 \quad \therefore \bar V_2/\bar V_1 = \underline{-j8} = \underline{8\angle -90^\circ}$

**14-7** (a)

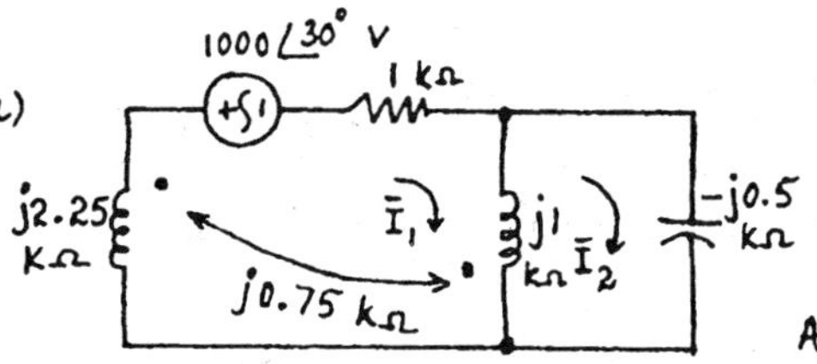


$\bar V_s = 1000\angle 30^\circ = (1+j3.25)\bar I_1 - j1\bar I_2 + j0.75(\bar I_1 - \bar I_2) + j0.75\bar I_1$

$\therefore \bar V_s = (1+j4.75)\bar I_1 - j1.75\bar I_2$

Also, $(j1 - j0.5)\bar I_2 - j1\bar I_1 - j0.75\bar I_1 = 0$

$\therefore -j1.75\bar I_1 + j0.5\bar I_2 = 0 \quad \therefore \bar I_2 = \dfrac{j1.75}{j0.5}\bar I_1 = 3.5\bar I_1$

$\therefore \bar V_s = (1+j4.75-j6.125)\bar I_1 \quad \therefore \bar Z_{in} = 1 - j1.375 = 1.7002\underline{\angle -53.97^\circ}$; PF $= \cos 53.97^\circ = \underline{0.5882 \text{ (lead)}}$

(b) $$P = \frac{1}{2}\left|\frac{1000}{1000 - j1375}\right|^2 \times 1000 = \underline{172.97 \text{ W}}$$

**14-8** (a)

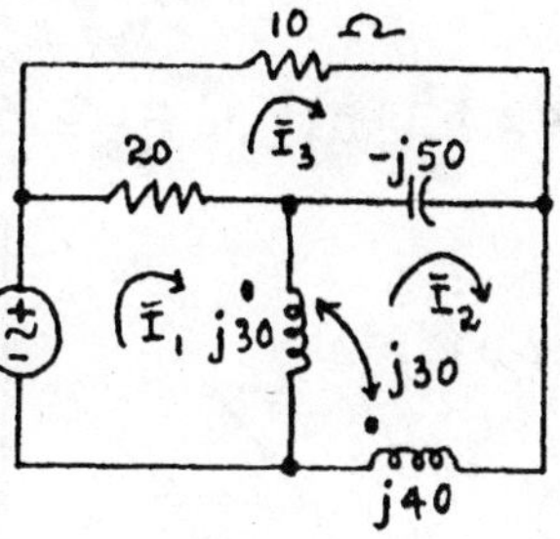


$\omega = 10$ ; $\bar{V}_s = 20(\bar{I}_1 - \bar{I}_3) + j30(\bar{I}_1 - \bar{I}_2) - j30\bar{I}_2$

$\therefore \bar{V}_s = (20 + j30)\bar{I}_1 - j60\bar{I}_2 - 20\bar{I}_3 = 100$ ①

$0 = -j50(\bar{I}_2 - \bar{I}_3) + j40\bar{I}_2 + j30(\bar{I}_2 - \bar{I}_1) + j30(\bar{I}_2 - \bar{I}_1) + j30\bar{I}_2$

$\therefore 0 = -j60\bar{I}_1 + j80\bar{I}_2 + j50\bar{I}_3$ ②

$0 = -20\bar{I}_1 + j50\bar{I}_2 + (30 - j50)\bar{I}_3$ ③

(b)

$$\bar{I}_1 = \frac{\begin{vmatrix} 100 & -j60 & -20 \\ 0 & j80 & j50 \\ 0 & j50 & 30 - j50 \end{vmatrix}}{\begin{vmatrix} 20 + j30 & -j60 & -20 \\ -j60 & j80 & j50 \\ -20 & j50 & 30 - j50 \end{vmatrix}} = \frac{10(65 + j24)}{(2 + j3)(65 + j24) + j6(-30 - j8) - 2(30 + j16)}$$

$$= \frac{650 + j240}{46 + j31} = 12.491 \angle -13.71^\circ \text{ A}$$

$\therefore P = \frac{1}{2} \times 100 \times 12.491 \cos 13.71^\circ = \underline{606.8 \text{ W}}$

**14-9**

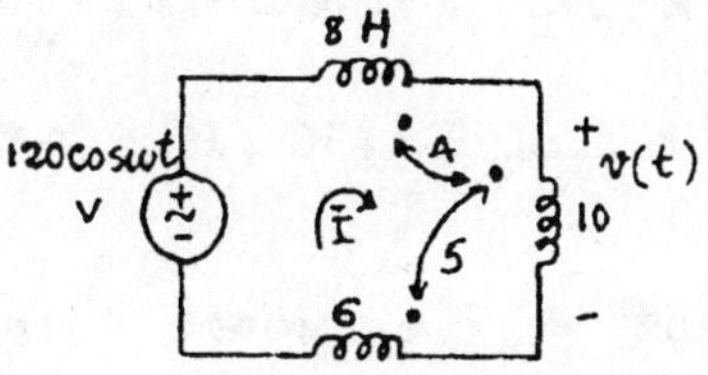


$120 = j8\omega\bar{I} - j4\omega\bar{I} + j10\omega\bar{I} - j4\omega\bar{I} + j5\omega\bar{I} + j6\omega\bar{I} + j5\omega\bar{I} = j26\omega\bar{I}$

$\therefore \bar{I} = 120/j26\omega$ , $\bar{V} = (j10\omega - j4\omega + j5\omega)\dfrac{120}{j26\omega} = 50.77$ V

$\therefore v(t) = \underline{50.77\cos\omega t \quad \text{V}}$

**14-10**

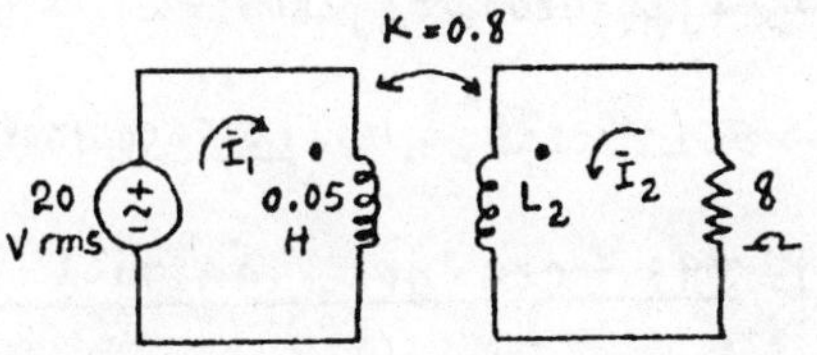


$\omega = 1000$ rad/s

$20 = j50\bar{I}_1 + j1000 \times 0.8\sqrt{0.05L_2}\,\bar{I}_2$ ; let $\ell_2 = 1000L_2$

$\therefore 20 = j50\bar{I}_1 + j0.8\sqrt{50\ell_2}\,\bar{I}_2$

$0 = j0.8\sqrt{50\ell_2}\,\bar{I}_1 + (8 + j\ell_2)\bar{I}_2 \quad \therefore \bar{I}_1 = \dfrac{-(8 + j\ell_2)}{j0.8\sqrt{50\ell_2}}\bar{I}_2$

$\therefore 20 = \bar{I}_2\left[\dfrac{-62.5(8 + j\ell_2)}{\sqrt{50\ell_2}} + j0.8\sqrt{50\ell_2}\right]$ ; $P = |\bar{I}_2|^2 8 = 5$

$\therefore \bar{I}_2 = \dfrac{20\sqrt{50\ell_2}}{-500 - j62.5\ell_2 + j40\ell_2} = \dfrac{20\sqrt{50\ell_2}}{-500 - j22.5\ell_2} \quad \therefore |\bar{I}_2|^2 = \dfrac{5}{8} = \dfrac{400 \times 50\ell_2}{250{,}000 + 22.5^2\ell_2^2}$

$\therefore \ell_2 = 9.1317$ and $54.08$ , $L_2 = \underline{9.1317 \text{ mH}}$ and $\underline{54.08 \text{ mH}}$

**14-11**

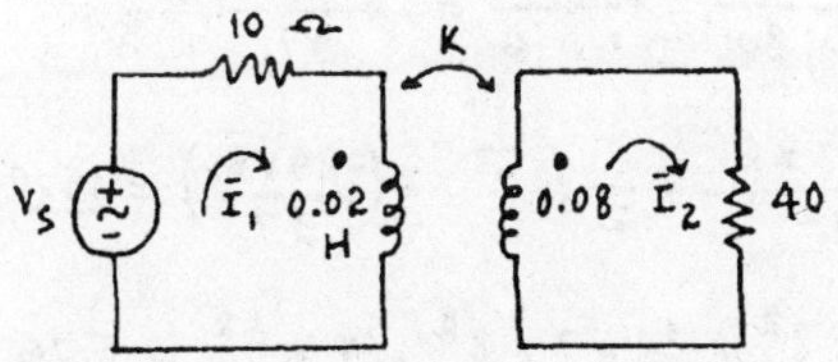


$v_s = 100\sqrt{2}\cos 10^3 t$ V

(rms) $100 = (10 + j20)\bar{I}_1 - j40K\bar{I}_2$

$0 = -j40K\bar{I}_1 + (40 + j80)\bar{I}_2 \quad \therefore \bar{I}_2 = \dfrac{jK\bar{I}_1}{1 + j2}$

$\therefore 100 = \bar{I}_1\left[10 + j20 + \dfrac{40K^2}{1 + j2}\right] = \bar{I}_1\left(\dfrac{-30 + 40K^2 + j40}{1 + j2}\right)$

$\therefore \bar{I}_1 = \dfrac{100(1 + j2)}{-30 + 40K^2 + j40}$ , $|\bar{I}_1|^2 = \dfrac{500}{25 - 24K^2 + 16K^4}$

(cont. on next page)

14.11 (cont.) (a) $K = 0.1 \therefore P_{10} = \dfrac{500 \times 10}{25 - 0.24 + 0.0016} = \underline{201.9 \text{ W}}$ ; $|\bar{I}_2|^2 = \dfrac{K^2}{5}|\bar{I}_1|^2 \therefore P_{40} = \dfrac{4000K^2}{25 - 24K^2 + 16K^4}$

$\therefore P_{40} = \dfrac{40}{25 - 0.24 + 0.0016} = \underline{1.6154 \text{ W}}$

(b) $K = 1 \quad \therefore P_{10} = \dfrac{5000}{17} = \underline{294.1 \text{ W}}$

$P_{40} = \dfrac{4000}{17} = \underline{235.3 \text{ W}}$

(c) $P_{40} = \dfrac{4000K^2}{25 - 24K^2 + 16K^4}$

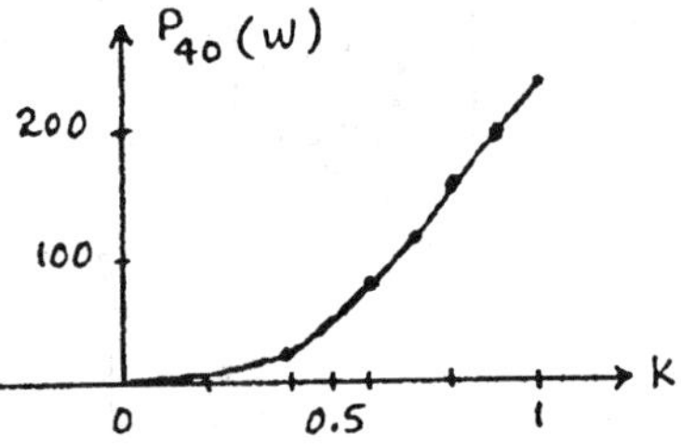


| K | $P_{40}$(W) |
|---|---|
| 0 | 0 |
| 0.2 | 6.65 |
| 0.4 | 29.7 |
| 0.5 | 50.0 |
| 0.6 | 78.1 |
| 0.7 | 114.7 |
| 0.8 | 158.1 |
| 0.9 | 201.8 |
| 1 | 235.3 |

14.12

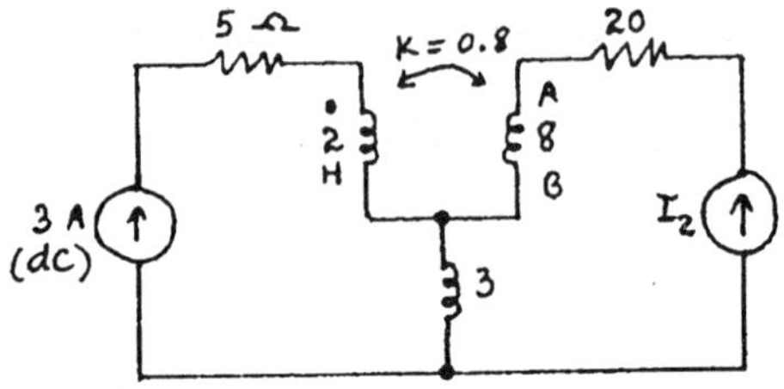


(a) $I_2 = 2$, dot at A $\therefore w_L = \dfrac{2 \times 9}{2} + \dfrac{8 \times 4}{2} + 3.2 \times 3 \times 2 + \dfrac{3 \times 25}{2}$

$\therefore w_L = \underline{81.7 \text{ J}}$

(b) $I_2 = 2$, dot at B $\therefore w_L = 9 + 16 - 19.2 + 37.5 = \underline{43.3 \text{ J}}$

(c) $I_2 = -2$, dot at A $\therefore w_L = 9 + 16 - 19.2 + \dfrac{3 \times 1}{2} = \underline{7.3 \text{ J}}$

(d) $I_2 = -2$, dot at B $\therefore w_L = 9 + 16 + 19.2 + 1.5 = \underline{45.7 \text{ J}}$

14.13 (a)

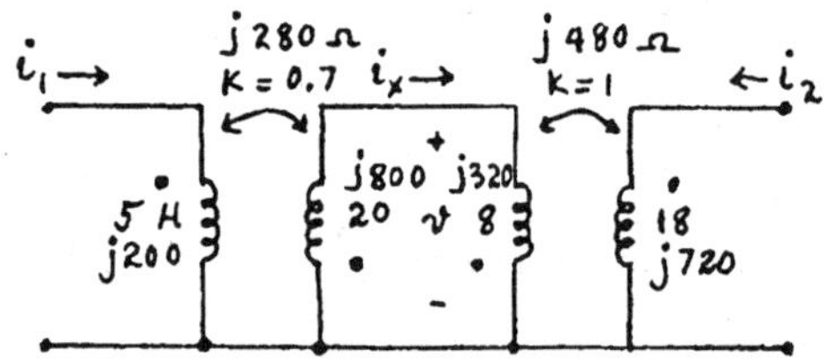


$i_1 = 5\cos 40t$ A, $i_2 = 2\sin 40t$ A $\therefore \bar{I}_1 = 5, \bar{I}_2 = -j2$

$\therefore j1120\bar{I}_x + j2(j480) + 5(j280) = 0$

$\therefore \bar{I}_x = 1.5156\angle{-145.56^\circ}$, $i_x(t) = \underline{1.5156\cos(40t - 145.56^\circ) \text{ A}}$

$\bar{V} = j320\bar{I}_x + j480(j2) = 793.9\angle{-149.74^\circ}$ V $\therefore v(t) = \underline{793.9\cos(40t - 149.74^\circ) \text{ V}}$

(b) $P_{av \text{ to right}} = \dfrac{1}{2} \times 793.9 \times 1.5156\cos 4.18^\circ = \underline{600.0 \text{ W}}$ $\therefore$ Power flow is to the $\underline{\text{right}}$

14.14

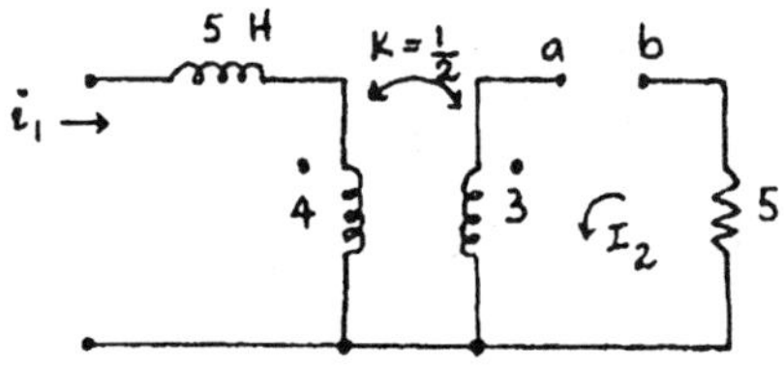


$i_1 = 2\cos 10t$ A

(a) a-b oc : $i_1(0) = 2 \therefore w_L(0) = \dfrac{1}{2}(5+4)2^2 = \underline{18 \text{ J}}$

(b) a-b sc : $\bar{I}_1 = 2 \therefore (5 + j30)\bar{I}_2 + j10 \times \dfrac{1}{2}\sqrt{12} \times 2 = 0$

$\therefore \bar{I}_2 = \dfrac{-j20\sqrt{3}}{5 + j30} = \dfrac{-j4\sqrt{3}}{1 + j6} = \dfrac{4\sqrt{3}(-6 - j1)}{37}$

$\therefore w_L(0) = 18 + \dfrac{1}{2} \times 3 \times \dfrac{16 \times 3 \times 6}{37^2} + \dfrac{1}{2}\sqrt{12} \times 2\left(\dfrac{-24\sqrt{3}}{37}\right) = \underline{16.001 \text{ J}}$

14.15

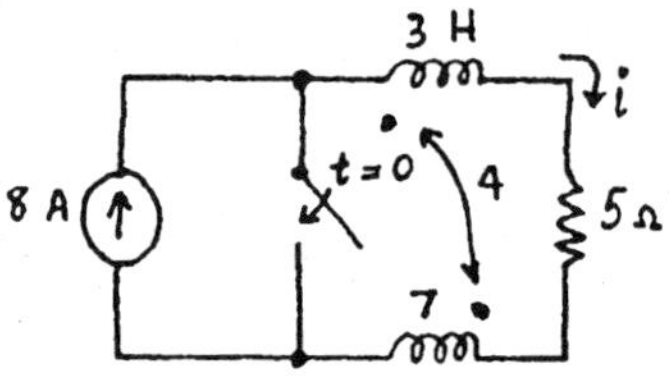


(a) $w(0^-) = \dfrac{1}{2} \times 3 \times 8^2 + \dfrac{1}{2} \times 7 \times 8^2 + 4 \times 8^2 = \underline{576 \text{ J}}$

(b) $3i' + 7i' + 4i' + 5i + 4i' = 0 \therefore 18i' + 5i = 0$, $L_{eq} = 18$ H

$\tau = \dfrac{18}{5} = 3.6$ s $\therefore i = 8e^{-t/3.6} \therefore \dfrac{8}{\sqrt{2}} = 8e^{-t_1/3.6}$

$\therefore t_1 = \underline{1.2477 \text{ s}}$

**14-16**

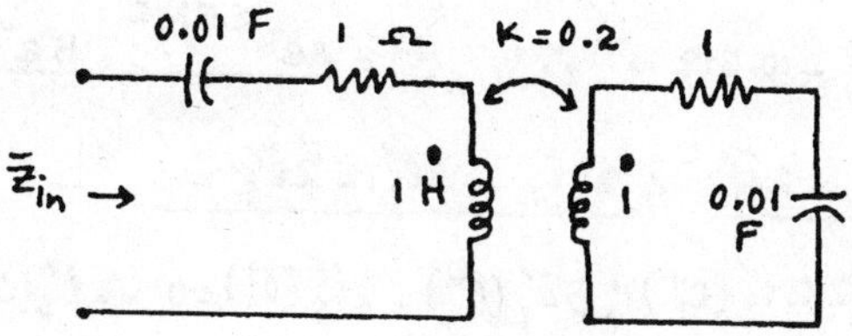


$$\bar{Z}_{in} = \bar{Z}_{11} + \frac{\omega^2 M^2}{\bar{Z}_{22}} = 1 + j\left(\omega - \frac{100}{\omega}\right) + \frac{0.04\,\omega^2}{1 + j(\omega - 100/\omega)}$$

$$\therefore \bar{Z}_{in} = 1 + j\left(\omega - \frac{100}{\omega}\right) + \frac{0.04\,\omega^2 - j0.04\,\omega^2(\omega - \frac{100}{\omega})}{1 + (\omega - 100/\omega)^2}$$

$$X_{in} = 0 \therefore \omega - \frac{100}{\omega} = \frac{0.04\,\omega^2(\omega - 100/\omega)}{1 + (\omega - 100/\omega)^2}$$

[Cancel $(\omega - \frac{100}{\omega})$ ∴ $\omega = 10$ is a solution ✓]

$$\therefore 1 + \omega^2 - 200 + 10^4/\omega^2 = 0.04\,\omega^2 \therefore \omega^2 = \frac{199 \pm \sqrt{1201}}{1.92}, \quad \omega = \underline{9.252} \text{ and } \underline{11.032} \ \frac{\text{rad}}{\text{s}}$$

**14-17**

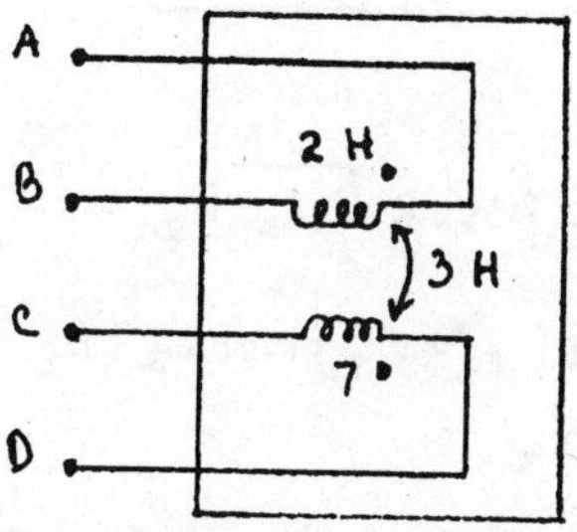


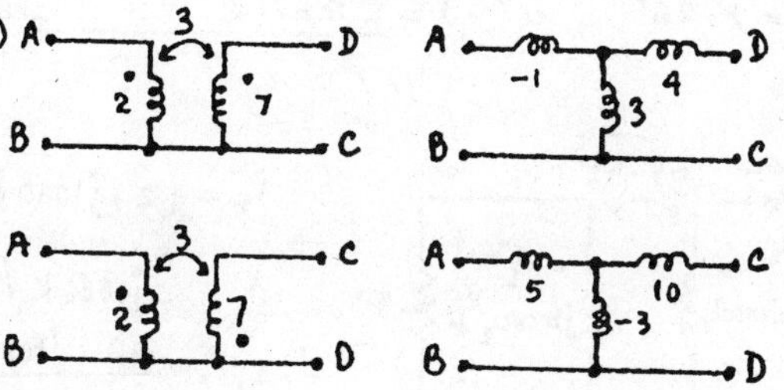


(a) $L_{AD}$, B to C ∴ $L_{AD} = \underline{3\ H}$

(b) $L_{AC}$, B to D ∴ $L_{AC} = \underline{15\ H}$

(c) $L_{AB}$, C to D : $L_{AB} = -1 + 4\|3 = \underline{0.7143\ H}$

(d) $L_{CD}$, A to B : $L_{CD} = 4 + 3\|(-1) = \underline{2.5\ H}$ (e) $L_{AB}$, A to C and B to D : $L_{AB} = -3 + 5\|10 = \underline{0.333\ H}$

(f) $L_{AB}$, no other connections : $L_{AB} = \underline{2\ H}$ (g) $L_{CD}$, no other connections : $L_{CD} = \underline{7\ H}$

(h) $L_{AB}$, A to D and B to C : $L_{AB} = 3 + 4\|(-1) = 1.6667\ H$

**14-18**

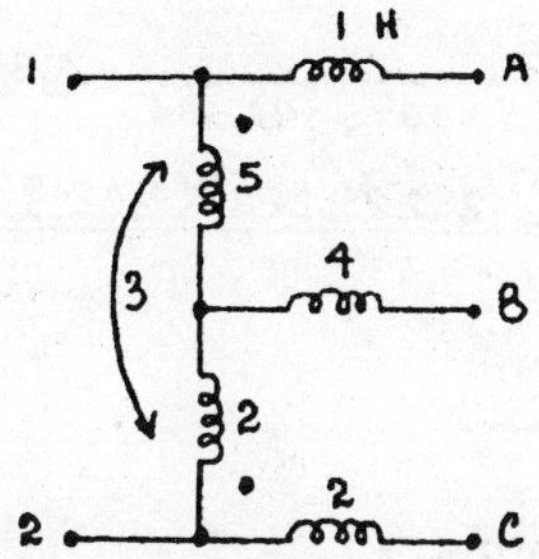


(a) None: $L_{12} = 5 + 2 - 2 \times 3 = \underline{1\ H}$

(b) A to B:

$L_{12} = -1 + 2\|8$

$= \underline{0.6\ H}$

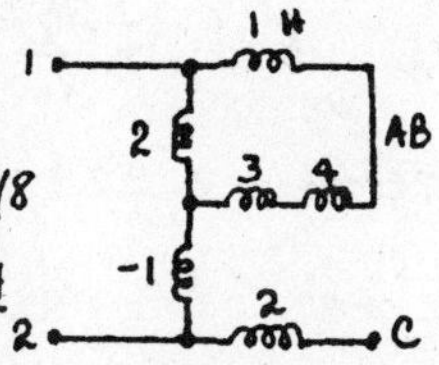


(c) B to C:

$L_{12} = 2 + 9\|(-1)$

$= \underline{0.875\ H}$

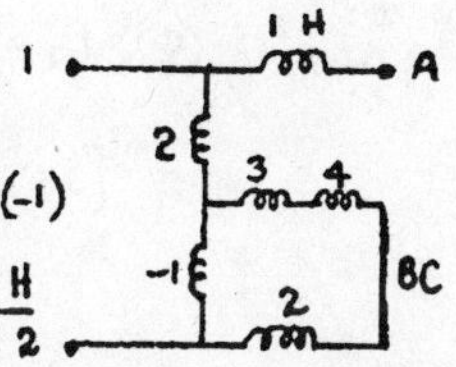


(d) A to C:

$L_{12} = (2-1)\|(1+2)$

$= \underline{0.750\ H}$

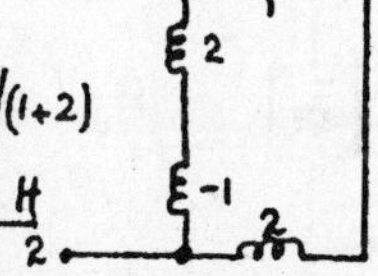


**14-19** (a)

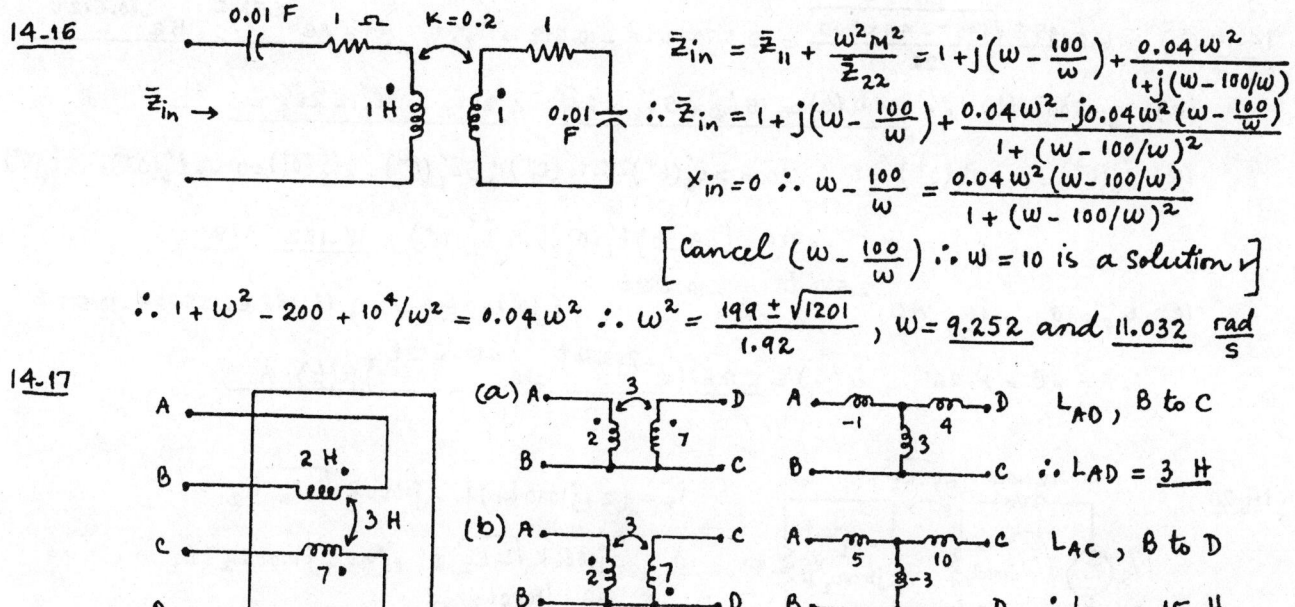

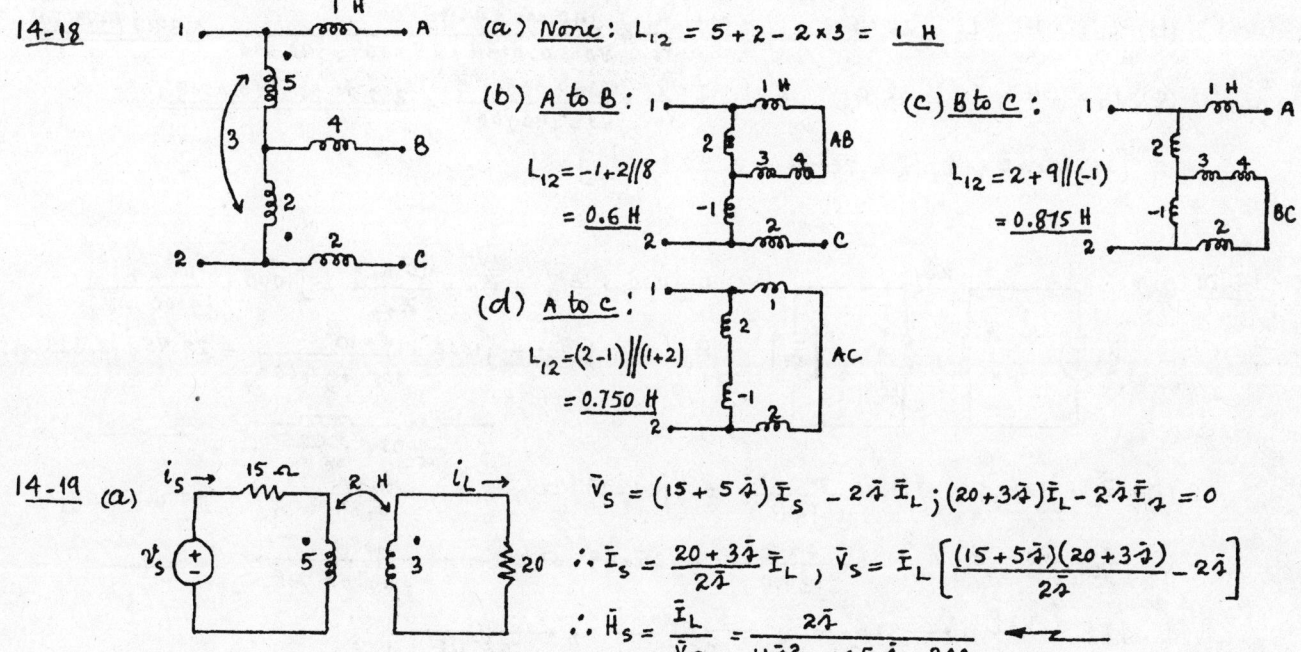

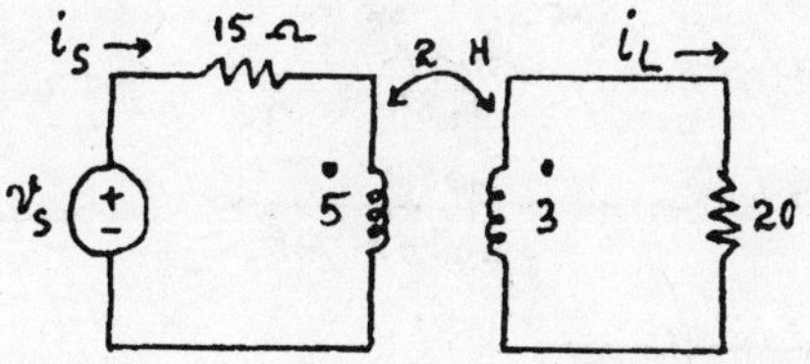


$$\bar{V}_s = (15 + 5s)\bar{I}_s - 2s\,\bar{I}_L;\ (20 + 3s)\bar{I}_L - 2s\,\bar{I}_s = 0$$

$$\therefore \bar{I}_s = \frac{20 + 3s}{2s}\bar{I}_L, \quad \bar{V}_s = \bar{I}_L\left[\frac{(15 + 5s)(20 + 3s)}{2s} - 2s\right]$$

$$\therefore \bar{H}_s = \frac{\bar{I}_L}{\bar{V}_s} = \frac{2s}{11s^2 + 145s + 300} \longleftarrow$$

(cont. on next page)

14-19 (cont.)

(b) $s = \dfrac{-145 \pm \sqrt{145^2 - 44 \times 300}}{22} = -2.570$ and $-10.612\ s^{-1}$ $\therefore i_{L,n}(t) = \underline{Ae^{-2.570t} + Be^{-10.612t}}$

(c) $v_s = 100\,u(t)$ $\therefore$ $\underline{100\,u(t) = 15 i_s + 5 i_s' - 2 i_L'}$ ; $\underline{20 i_L + 3 i_L' - 2 i_s' = 0}$

(d) $i_s(0^+) = 0$ , $i_L(0^+) = 0$ $\therefore 100 = 5 i_s'(0^+) - 2 i_L'(0^+)$ ; $3 i_L'(0^+) - 2 i_s'(0^+) = 0$ $\therefore i_s'(0^+) = 1.5\, i_L'(0^+)$

$\therefore 100 = (7.5 - 2)\, i_L'(0^+)$ $\therefore i_L'(0^+) = \underline{18.182\ \text{A/s}}$

(e) $i_{Lf} = 0$ $\therefore i_L = Ae^{-2.570t} + Be^{-10.612t}$ ; $i_L(0^+) = 0 = A + B$ , $18.182 = -2.570A - 10.612B$

$\therefore A = -B = 2.261$ , $i_L(t) = \underline{2.261(e^{-2.570t} - e^{-10.612t})\,u(t)\ \text{A}}$

14-20

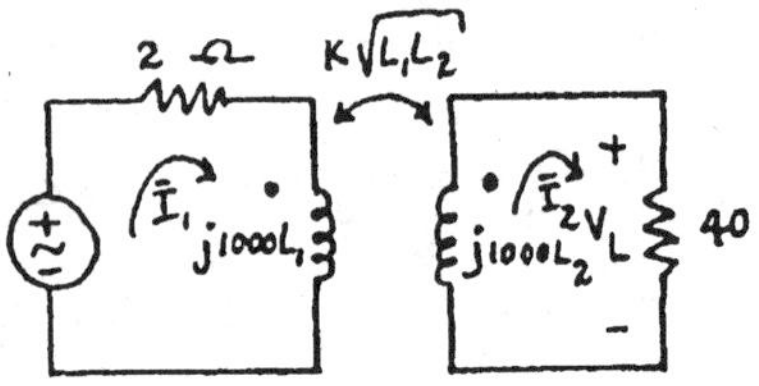


$\bar{V}_s = (2 + j1000L_1)\bar{I}_1 - j1000K\sqrt{L_1L_2}\,\bar{I}_2$

$0 = -j1000K\sqrt{L_1L_2}\,\bar{I}_1 + (40 + j1000L_2)\bar{I}_2$

$\therefore \bar{I}_1 = \dfrac{40 + j1000L_2}{j1000K\sqrt{L_1L_2}}\,\bar{I}_2$

$\therefore \bar{V}_s = \left[\dfrac{(2 + j10{,}000L_1)(40 + j1000L_2)}{j1000K\sqrt{L_1L_2}} - j1000K\sqrt{L_1L_2}\right]\bar{I}_2$

$$\bar{V}_s = \frac{80 - 10^6 L_1L_2 + j1000(40L_1 + 2L_2) + 10^6K^2L_1L_2}{j1000K\sqrt{L_1L_2}}\,\bar{I}_2$$

$$\therefore \frac{\bar{V}_L}{\bar{V}_s} = \frac{j40{,}000K\sqrt{L_1L_2}}{80 - (1 - K^2)10^6L_1L_2 + j1000(40L_1 + 2L_2)}$$

(a) $L_1 = 1$ mH, $L_2 = 25$ mH , $K = 1$ : $\dfrac{\bar{V}_L}{\bar{V}_s} = \dfrac{j40 \times 5}{80 + j90} = \underline{1.6609\,\angle 41.63°}$

(b) $L_1 = 1$ H , $L_2 = 25$ H , $K = 0.99$: $\dfrac{\bar{V}_L}{\bar{V}_s} = \dfrac{j40{,}000 \times 4.95}{80 - 0.0199 \times 25 \times 10^6 + j90{,}000} = \underline{0.3917\,\angle -79.74°}$

(c) $L_1 = 1$ H , $L_2 = 25$ H , $K = 1$ : $\dfrac{\bar{V}_L}{\bar{V}_s} = \dfrac{j40{,}000 \times 5}{80 + j90{,}000} = \underline{2.222221\,\angle 0.0509°}$

[If ideal, $\bar{V}_L/\bar{V}_s = 200/9 = 2.222\ldots$ ]

14-21 (a)

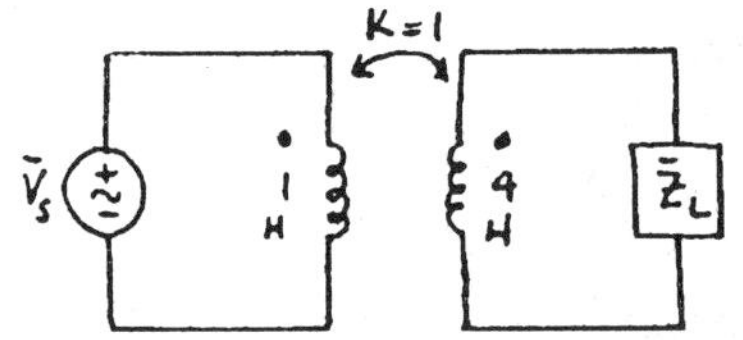


$\omega = 1000$ : $\bar{Z}_{in} = \bar{Z}_{11} + \dfrac{\omega^2M^2}{\bar{Z}_{22}} = j1000 + \dfrac{10^6 \times 4}{j4000 + \bar{Z}_L}$

$\bar{Z}_L = 100$ ; $\bar{Z}_{in} = j1000 + \dfrac{4 \times 10^6}{100 + j4000} = 24.98 + j0.625\ \Omega$

24.98 Ω, 0.625 mH

(b) $\bar{Z}_L = j100$ , $Z_{in} = j1000 + \dfrac{4 \times 10^6}{j4100} = j24.39\ \Omega$ — 24.39 mH

(c) $\bar{Z}_L = -j100$ , $Z_{in} = j1000 + \dfrac{4 \times 10^6}{j3900} = -j25.64\ \Omega$ — 39.00 μF

[If ideal , 25 Ω , 25 mH , 40 μF ]

**14-22**

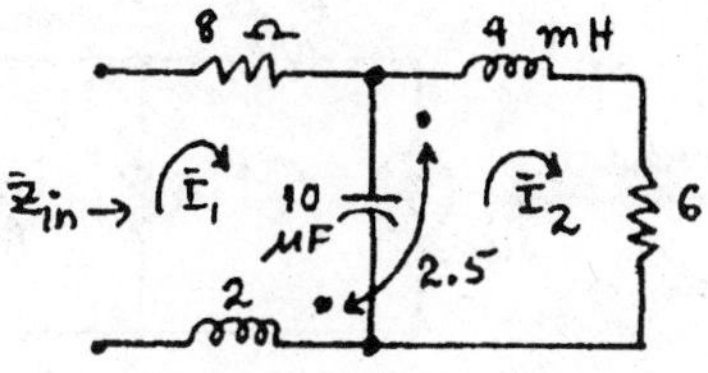


$\omega = 10^4$ rad/s

$$\bar{V}_{in} = (8 + j20 - j10)\bar{I}_1 + j10\bar{I}_2 + j25\bar{I}_2$$

$$0 = (j25 + j10)\bar{I}_1 + (6 + j40 - j10)\bar{I}_2 \therefore \bar{I}_2 = \frac{-j35}{6 + j30}\bar{I}_1$$

$$\bar{Z}_{in} = \frac{\bar{V}_{in}}{\bar{I}_1} = 8 + j10 + j35 \times \frac{-j35}{6 + j30} = \underline{33.28\angle -61.55^\circ\ \Omega}$$

**14-23**

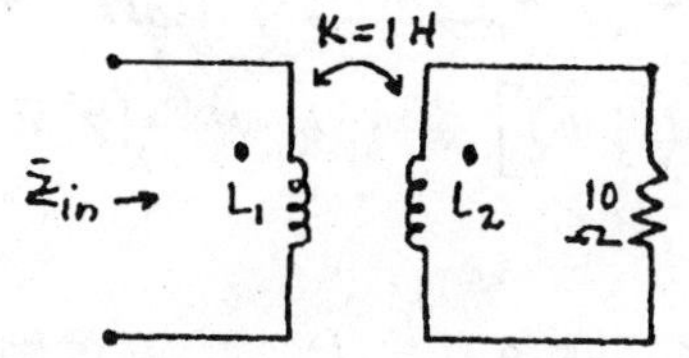


$\omega = 50$ rad/s

(a) $$\bar{Z}_{in} = j\omega L_1 + \frac{\omega^2 L_1 L_2}{10 + j\omega L_2}\ ;\ L_1 = 10\text{ H},\ L_2 = 40\text{ H}$$

$$\therefore \bar{Z}_{in} = j500 + \frac{10^6}{10 + j2000} = \frac{j5000}{10 + j2000}$$

$$= \underline{2.49994 + j0.01250\ \Omega}$$

(b) $\frac{L_2}{L_1} = 4 = a^2$, $a = 2$ ∴ ideal transformer

$$\bar{Z}_{in} = \frac{\bar{Z}_L}{a^2} = \frac{10}{4} = \underline{2.5 + j0\ \Omega}$$

**14-24**

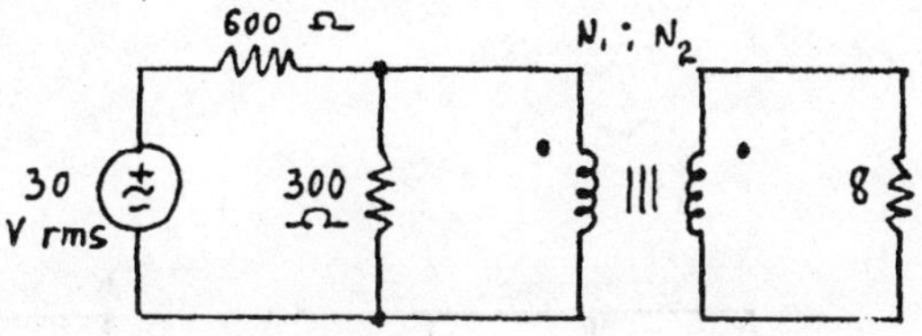


(a) $R_{Th} = 600 \| 300 = 200$ ; $a = \frac{N_2}{N_1} = \sqrt{\frac{8}{200}} = \frac{1}{5}$ ←

(b) $\bar{V}_{Th} = 10$ Vrms ∴ $P_8 = \left(\frac{10}{200 + 200}\right)^2 \times 200$

$P_8 = \underline{125\text{ mW}}$

(c)

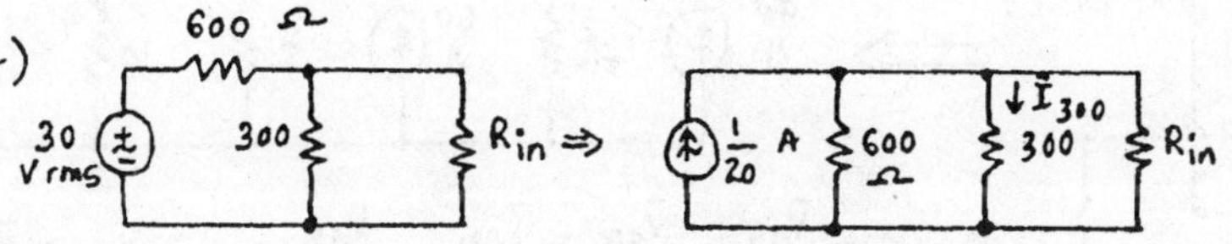


$P_{300}$ is max when $I_{300}$ is max

∴ $R_{in} = \infty$, $\frac{N_1}{N_2} = \infty$ ∴ $\frac{N_2}{N_1} = \underline{0}$

**14-25**

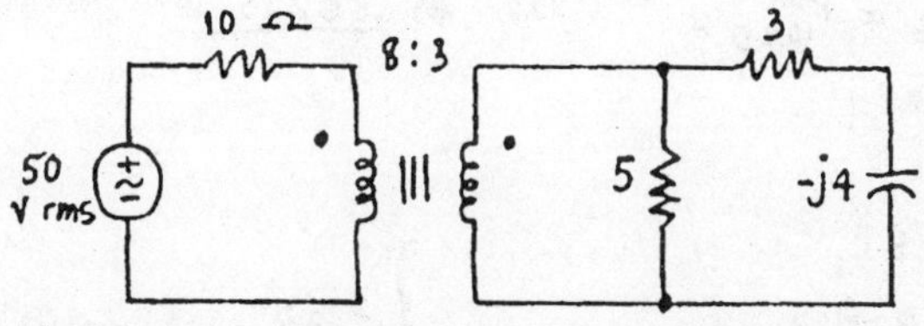


Refer to secondary →

$50 \times \frac{3}{8} = \frac{75}{4}$ V, $10 \times \frac{9}{64} = \frac{45}{32}$ Ω

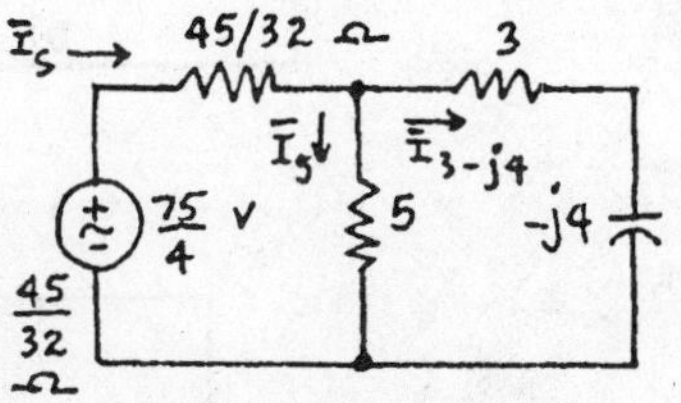


$$\bar{I}_S = \frac{75/4}{45/32 + \frac{5(3 - j4)}{8 - j4}} = \frac{75(2 - j1)}{\frac{45}{8}(2 - j1) + 15 - j20} = \frac{120(2 - j1)}{42 - j41} = 4.5716\angle 17.7447^\circ\text{ A}$$

$$\therefore P_{10} = 4.5716^2 \times \frac{45}{32} = \underline{29.39\text{ W}}\ ;\ \bar{I}_5 = \frac{120(2 - j1)}{42 - j41} \times \frac{3 - j4}{8 - j4} = \frac{30(3 - j4)}{42 - j41},$$

$$\bar{I}_{3-j4} = \frac{120(2 - j1)}{42 - j41} \times \frac{5}{8 - j4} = \frac{30 \times 5}{42 - j41}\ ;\ P_5 = \frac{900 \times 25}{41^2 + 42^2} \times 5 = \underline{32.656\text{ W}}\ ;\ P_3 = \frac{900 \times 25}{42^2 + 41^2} \times 3 = \underline{19.594\text{ W}}$$

**14-26**

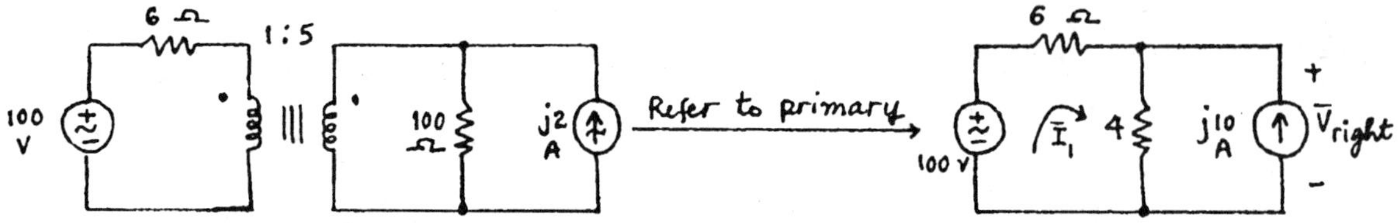


$100 = 6\bar{I}_1 + 4(\bar{I}_1 + j10)$ $\therefore$ $\bar{I}_1 = 10 - j4$ , $P_{100} = \frac{1}{2} \times 100 \times 10 = \underline{500\ W}$

$P_6 = \frac{1}{2} \times 6(10^2 + 4^2) = \underline{348\ W}$ , $P_4 = \frac{1}{2} \times 4(10^2 + 6^2) = \underline{272\ W}$ ; $\bar{V}_{right} = 4(10 + j6)$

$\therefore$ $\bar{V}_{right} = 40 + j24$ , $P_{j10} = \text{Re}\left[\frac{1}{2}(40 + j24)(-j10)\right] = \underline{120\ W}$ ; ($\Sigma P$ ok)

**14-27**

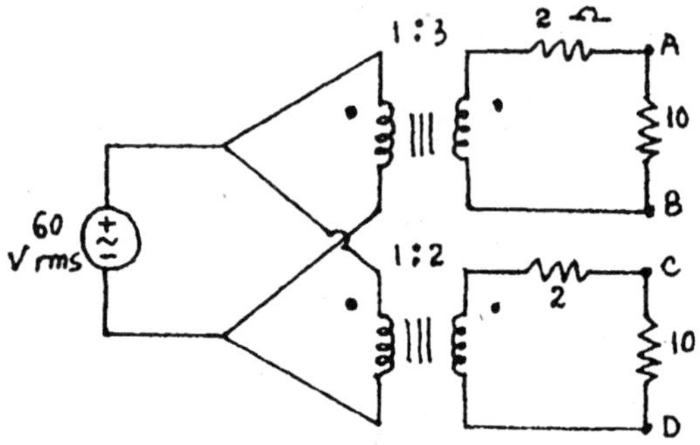

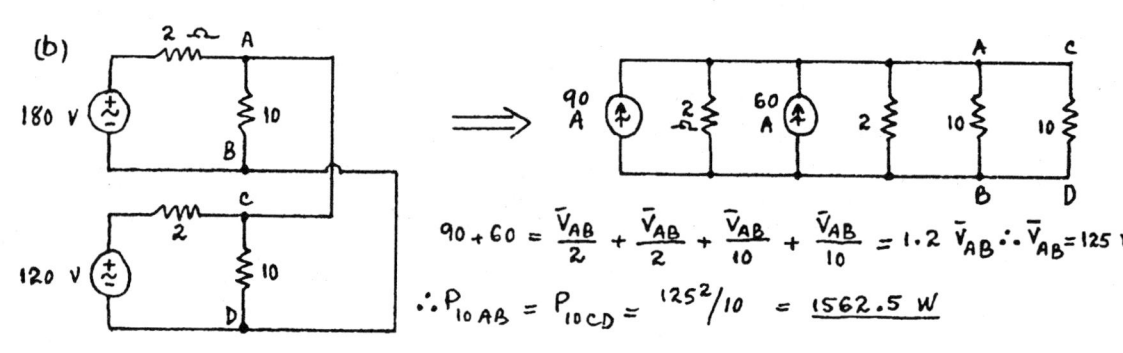

(a) $\frac{60 \times 3}{2 + 10} = 15\ A$ $\therefore$ $P_{10AB} = 15^2 \times 10 = \underline{2250\ W}$

$\frac{60 \times 2}{2 + 10} = 10\ A$ $\therefore$ $P_{10CD} = 10^2 \times 10 = \underline{1000\ W}$

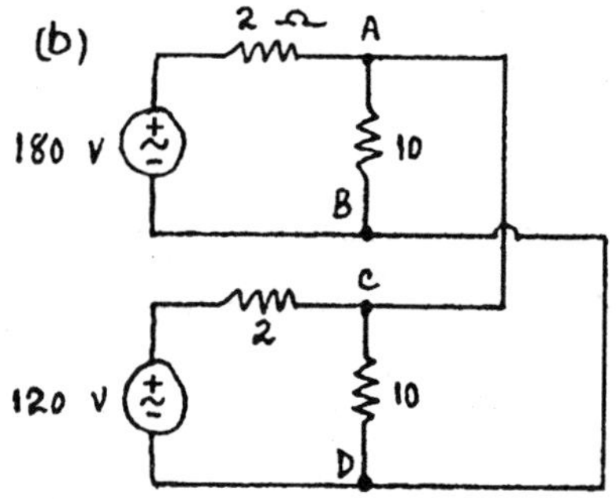


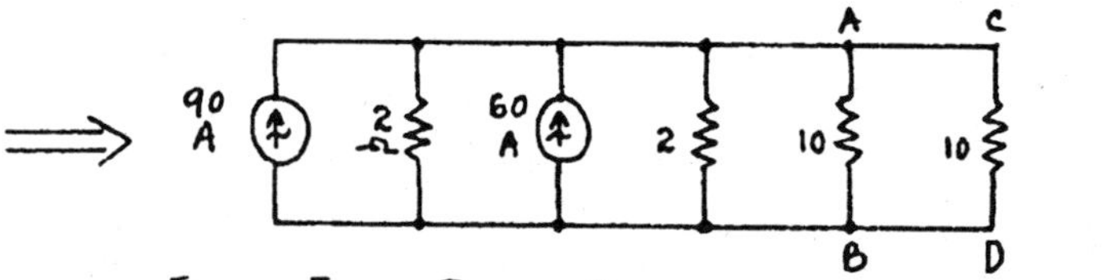


$90 + 60 = \frac{\bar{V}_{AB}}{2} + \frac{\bar{V}_{AB}}{2} + \frac{\bar{V}_{AB}}{10} + \frac{\bar{V}_{AB}}{10} = 1.2\,\bar{V}_{AB}$ $\therefore$ $\bar{V}_{AB} = 125\ V$

$\therefore$ $P_{10AB} = P_{10CD} = 125^2/10 = \underline{1562.5\ W}$

**14-28**

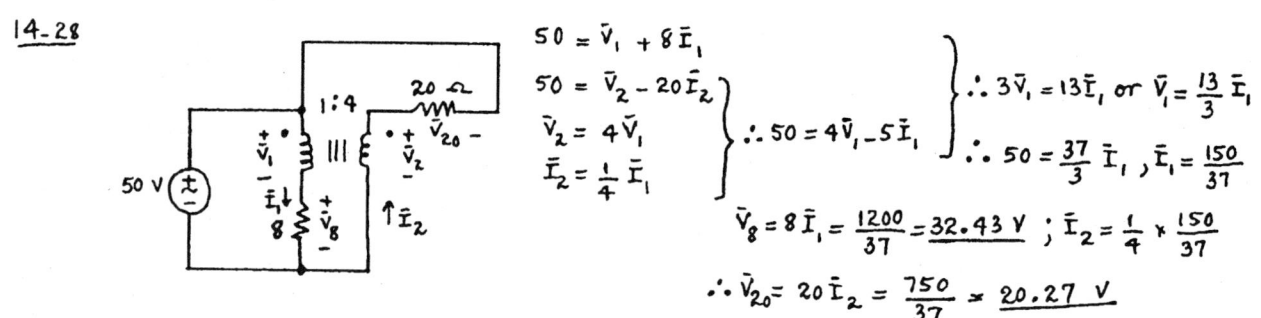

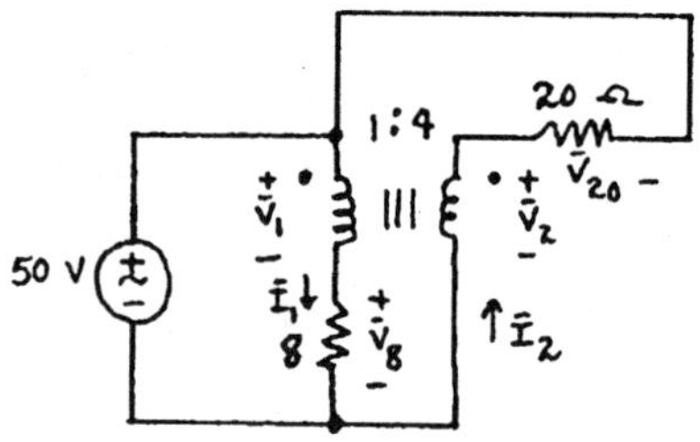


$50 = \bar{V}_1 + 8\bar{I}_1$

$50 = \bar{V}_2 - 20\bar{I}_2$

$\bar{V}_2 = 4\bar{V}_1$

$\bar{I}_2 = \frac{1}{4}\bar{I}_1$

$\therefore$ $50 = 4\bar{V}_1 - 5\bar{I}_1$

$\therefore$ $3\bar{V}_1 = 13\bar{I}_1$ or $\bar{V}_1 = \frac{13}{3}\bar{I}_1$

$\therefore$ $50 = \frac{37}{3}\bar{I}_1$ , $\bar{I}_1 = \frac{150}{37}$

$\bar{V}_8 = 8\bar{I}_1 = \frac{1200}{37} = \underline{32.43\ V}$ ; $\bar{I}_2 = \frac{1}{4} \times \frac{150}{37}$

$\therefore$ $\bar{V}_{20} = 20\bar{I}_2 = \frac{750}{37} = \underline{20.27\ V}$

14-29 (a)

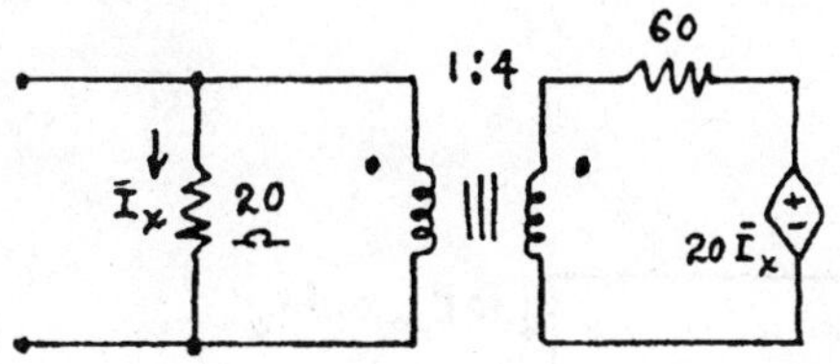


(a)

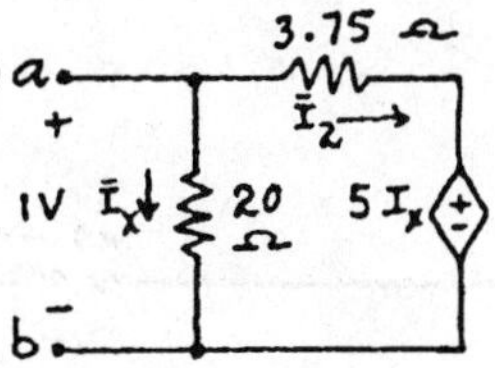


$I_x = \frac{1}{20}$, $I_2 = \frac{1 - 5(1/20)}{3.75}$

$I_2 = 0.2 \therefore I_{in} = 0.2 + 0.05$

$I_{in} = 0.25\ A \therefore R_{Th} = \underline{4\ \Omega}$

(b)

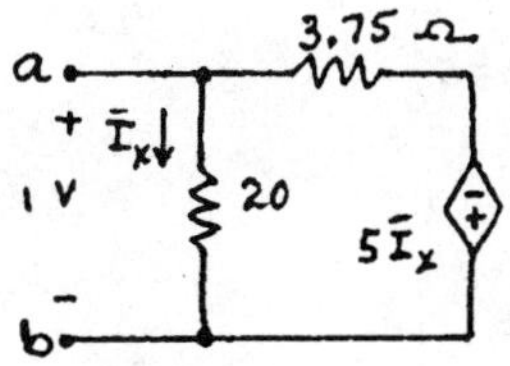


$I_x = \frac{1}{20}$, $I_2 = \frac{1 + 5(1/20)}{3.75} = 0.3333$

$I_{in} = 0.3333 + 0.05 = 0.3833 \therefore R_{Th} = \frac{1}{0.3833} = \underline{2.609\ \Omega}$

14-30

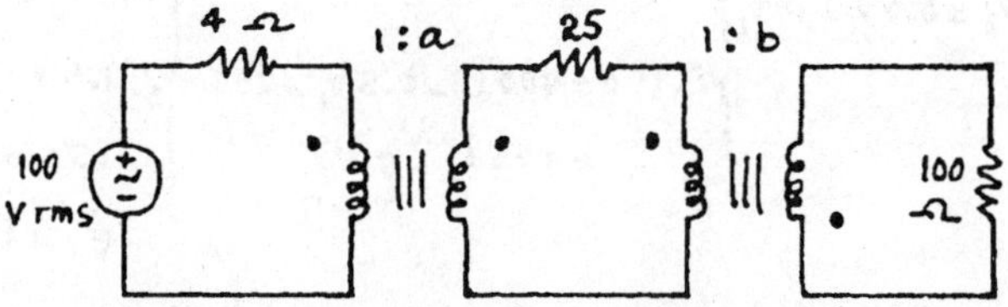


$1000 = 100^2/R_{in} \therefore R_{in} = 10\ \Omega$

$\therefore 10 = 4 + \frac{25 + (100/b^2)}{a^2} \therefore 6a^2 = \frac{100}{b^2} + 25$

Also, $\frac{100}{b^2} = 25 + 4a^2$ (equal P)

$\therefore 6a^2 = 25 + 4a^2 + 25 \therefore 2a^2 = 50$

$a = \underline{5}$ (or $\underline{-5}$)

$\therefore \frac{100}{b^2} = 25 + 100$, $b = \pm\sqrt{0.8} = \underline{\pm 0.8944}$

14-31

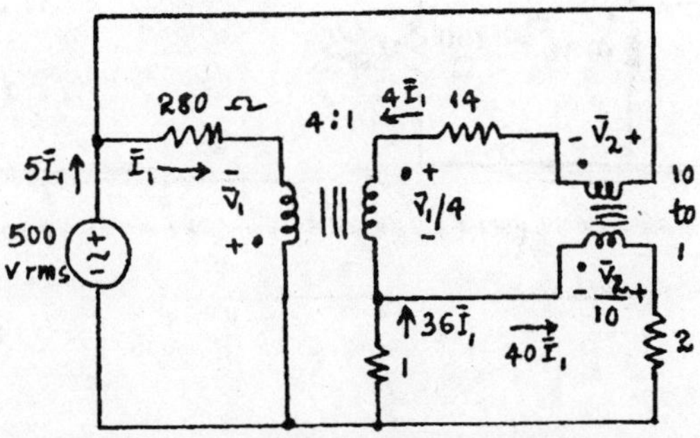


$500 = 280 I_1 - V_1$

$500 = V_2 + 56 I_1 + 0.25 V_1 - 36 I_1$

$0 = 36 I_1 - 0.1 V_2 + 80 I_1 \therefore V_2 = 1160 I_1$

$\therefore 500 = 1180 I_1 + 0.25 V_1 = 280 I_1 - V_1 \therefore V_1 = -\frac{900}{1.25} I_1$

$V_1 = -720 I_1 \therefore 500 = 280 I_1 + 720 I_1 \therefore I_1 = 0.5\ A$

$P_s = 5 I_1 \times 500 = \underline{1250\ W}$ ($\Sigma P$ ok)

14-32

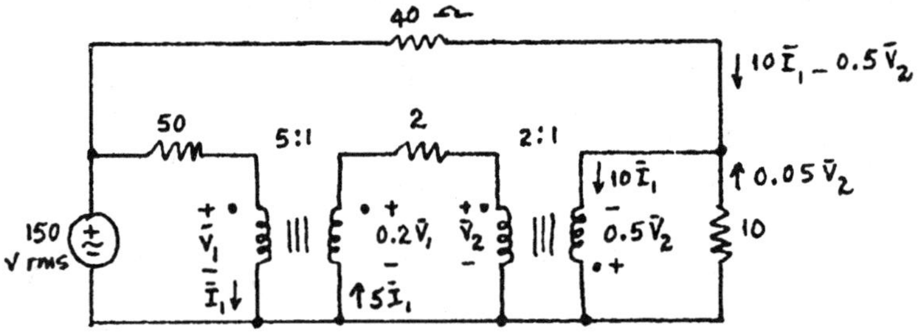


$$150 = 50\bar{I}_1 + \bar{V}_1$$

$$0 = -0.2\bar{V}_1 + 10\bar{I}_1 + \bar{V}_2 \therefore \bar{V}_2 = 0.2\bar{V}_1 - 10\bar{I}_1$$

$$150 = 400\bar{I}_1 - 2\bar{V}_2 - 0.5\bar{V}_2$$

$$\therefore 150 = 400\bar{I}_1 - 0.5\bar{V}_1 + 25\bar{I}_1 = 425\bar{I}_1 - 0.5\bar{V}_1$$

$$\therefore 1.5\bar{V}_1 = 375\bar{I}_1 , \bar{V}_1 = 2$$

$$\therefore 150 = 300\bar{I}_1 , \bar{I}_1 = 0$$

$$\bar{V}_1 = 125 \text{ V} , \bar{V}_2 = 25$$

$$\therefore P_{150} = 150(0.5 + 5 - 1) = \underline{675 \text{ W}} \qquad P_{10} = 1^2 \times 10 = \underline{10 \text{ W}}$$

$$P_{50} = 0.25(50) = \underline{12.5 \text{ W}} \qquad P_{40} = 4^2 \times 40 = \underline{640 \text{ W}}$$

$$P_2 = 2.5^2 \times 2 = \underline{12.5 \text{ W}}$$

14-33

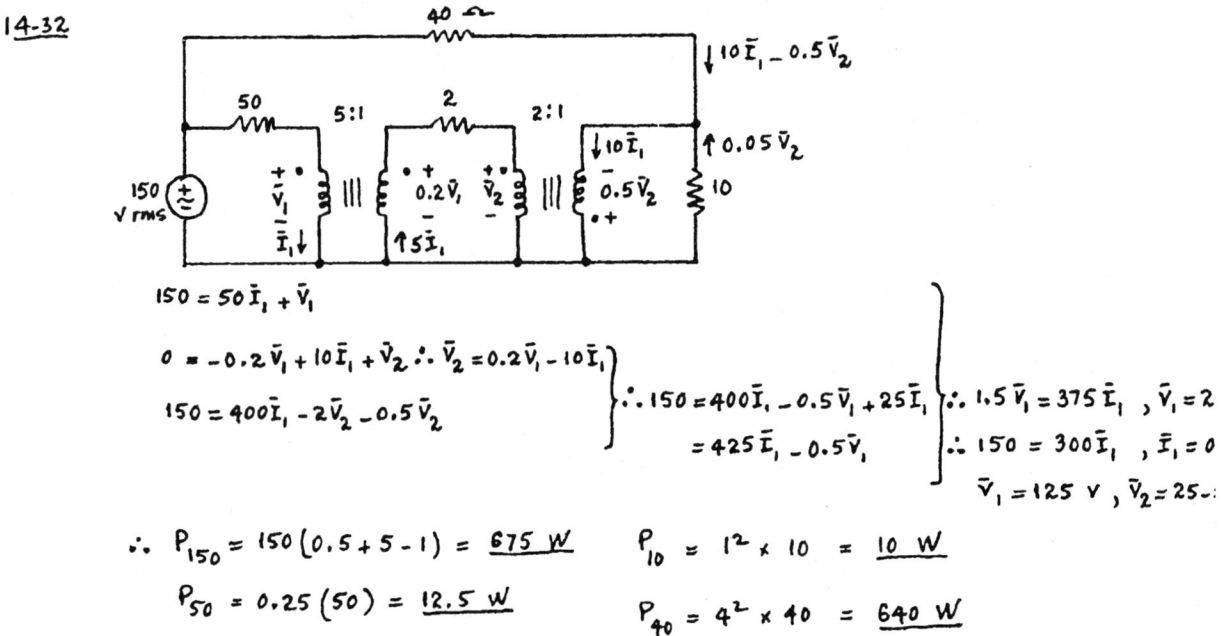

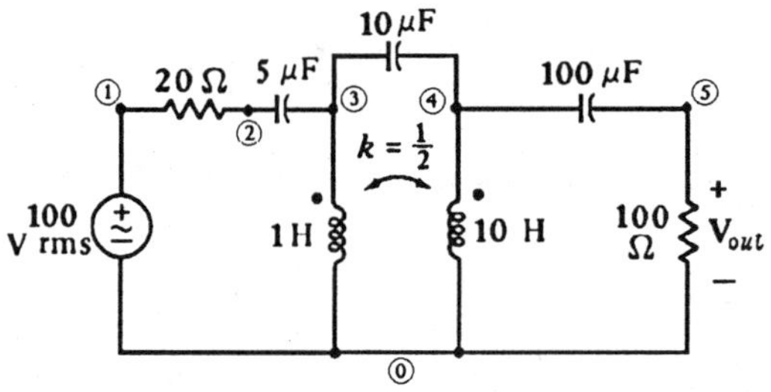


************************************************************************

**PROGRAM:**

```
V10 1 0 AC 100
R12 1 2 20
C23 2 3 5U
L30 3 0 1
L40 4 0 10
K34 L30 L40 0.5
C34 3 4 10U
C45 4 5 100U
R50 5 0 100
.AC LIN 1 159.1549 159.1549
.PRINT AC VM(5) VP(5)
.END
```

************************************************************************

| FREQ | VM(5) | VP(5) |
|---|---|---|
| 1.592E+02 | 3.277E+01 | 7.143E+01 |

Hence, $V_{out} = 32.77 \underline{/71.43^\circ}$ V rms

# CHAPTER 15: TWO-PORT NETWORKS

## Section 15-2 One-port networks

**1** Using nodal analysis and the knowledge that $\mathbf{Y}_{in} = 1/\mathbf{Z}_{in} = \Delta_Y/\Delta_{11}$, as expressed in Eq. (4), find the input admittance of the network shown in Fig. 15-1 if the arrow in the dependent current source is reversed in direction.

**2** Find $\mathbf{Z}_{in}$ for the one port of Fig. 15-2 by: (*a*) determining $\Delta_Z$; (*b*) finding $\Delta_Y$ and $\mathbf{Y}_{in}$ first, and then $\mathbf{Z}_{in}$.

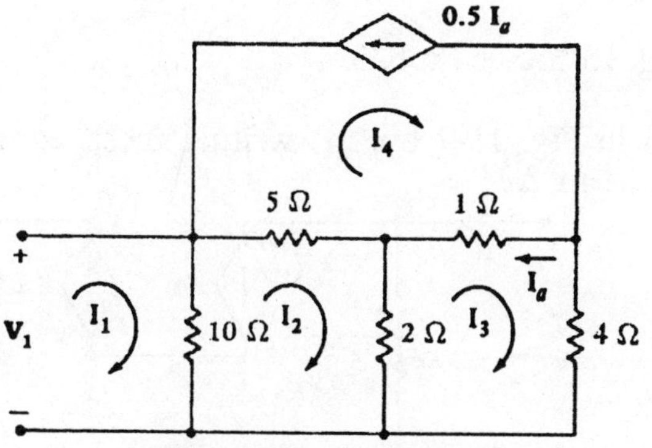


**Fig. 15-1:** See Prob. 1.

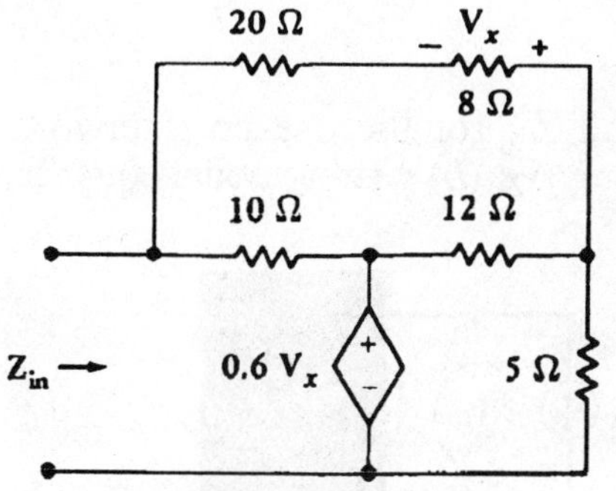


**Fig. 15-2:** See Prob. 2.

**3** Find the output impedance of the network shown in Fig. 15-3 as a function of **s**.

**4** Assume that the op-amp shown in Fig. 15-4 is ideal and find $R_{in}$.

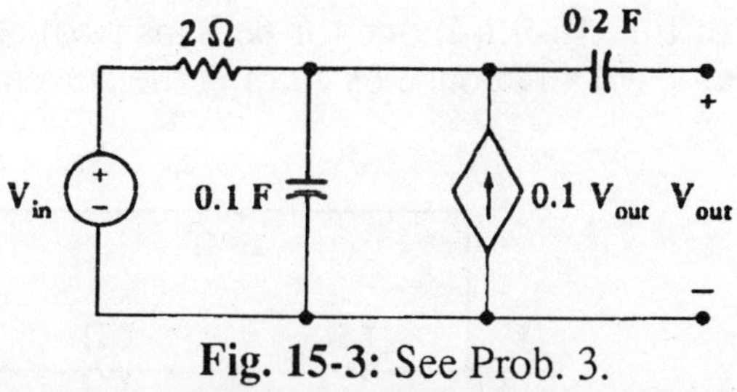


**Fig. 15-3:** See Prob. 3.

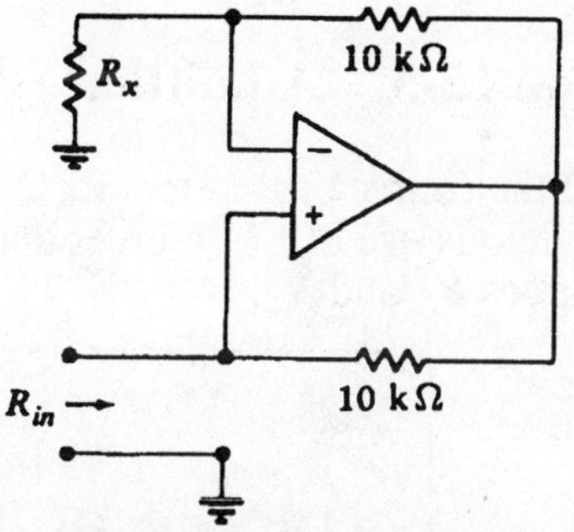


**Fig. 15-4:** See Prob. 4.

**5** (*a*) Assume that both op-amps appearing in the circuit of Fig. 15-5 are ideal and find $\mathbf{Z}_{in}$. (*b*) If $R_1 = 1\ \text{k}\Omega$, $R_2 = 2\ \text{k}\Omega$, $R_3 = 3\ \text{k}\Omega$, $R_4 = 4\ \text{k}\Omega$, and $C = 100$ pF, show that $\mathbf{Z}_{in} = j\omega L_{in}$, where $L_{in} =$ 0.6 mH.

**6** The impedance determinant of the one-port network shown in Fig. 15-6 is $\Delta_Z = \begin{vmatrix} 8 & -1 & -3 \\ -1 & 10 & -2 \\ -3 & -2 & 6 \end{vmatrix}$

where all units are ohms. (*a*) Determine $\mathbf{Z}_{in}$; (*b*) What will $\mathbf{Z}_{in}$ become if a 3-Ω resistor is inserted in series with the branch common to meshes 2 and 3?

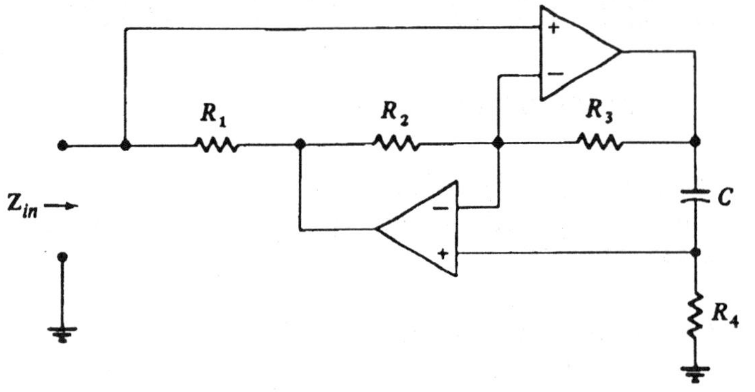


**Fig. 15-5:** See Prob. 5.

**7** Find $\mathbf{Z}_{in}$ for the one-port network shown in Fig. 15-7 by: (*a*) writing mesh equations and finding $\Delta_Z$; (*b*) writing nodal equations and finding $\Delta_Y$.

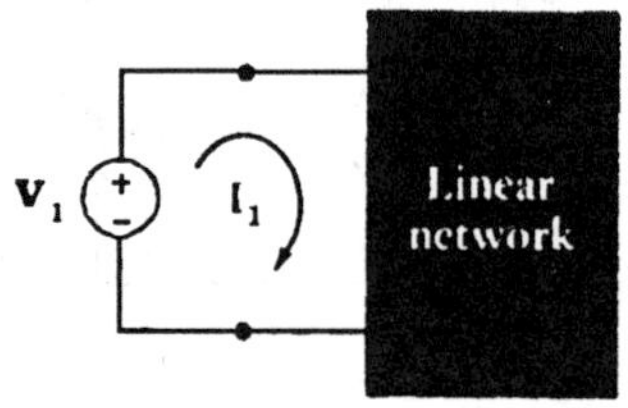


**Fig. 15-6:** See Prob. 6.

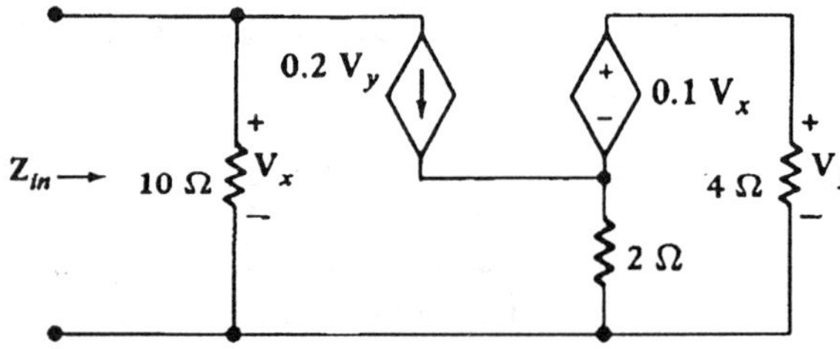


**Fig. 15-7:** See Prob. 7.

**8** Find all the critical frequencies of the input impedance $\mathbf{Z}_{in}(s)$ for the one-port that is shown in Fig. 15-8.

## Section 15-3 Admittance parameters

**9** Find the four **y** parameters for the two-port of Fig. 15-9: (*a*) one (or perhaps two) at a time by the judicious use of 1-V sources and short circuits; (*b*) all at once by finding the currents supplied by sources $\mathbf{V}_1$ and $\mathbf{V}_2$.

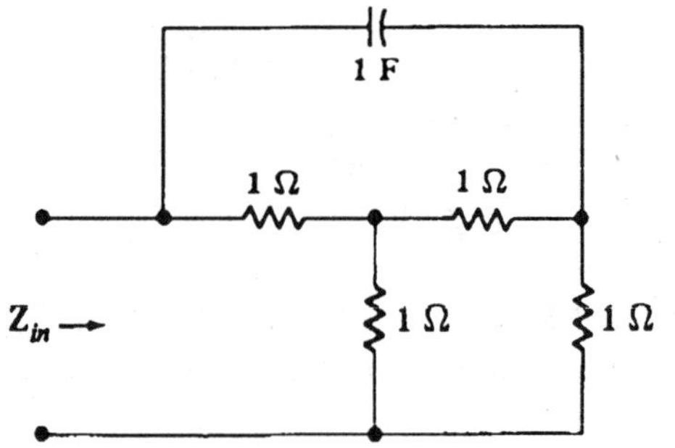


**Fig. 15-8:** See Prob. 8.

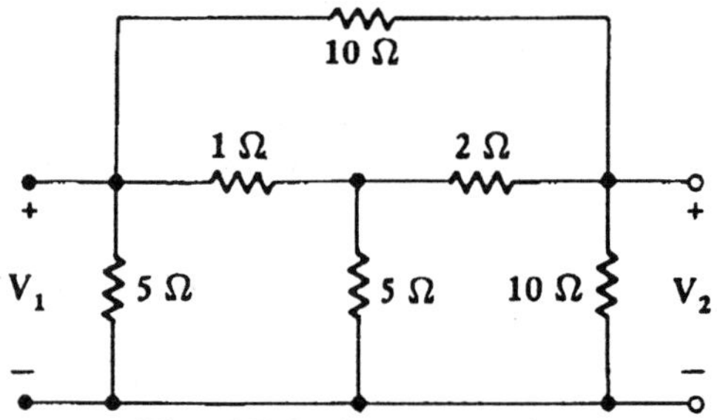


**Fig. 15-9:** See Prob. 9.

**10** (*a*) Find $\mathbf{y}_{12}$ and $\mathbf{y}_{21}$ for the network of Fig. 15-10. (*b*) To what value should the factor 10 in $10\mathbf{I}_1$ be changed to provide a reciprocal network?

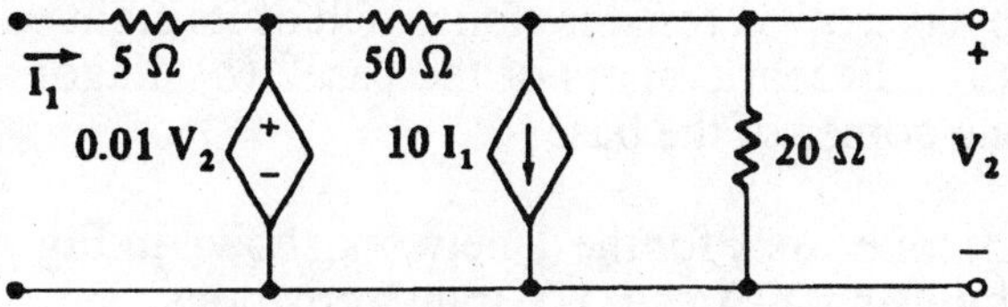


**Fig. 15-10:** See Prob. 10.

**11** (*a*) What value of $k$ in the two-port of Fig. 15-11 will produce a reciprocal network? (*b*) Let $k = 0.8$, connect a 5-Ω load at the right and a 50-V source at the left, and determine the power delivered to the load.

**12** Find [**y**(**s**)] for the two-port of Fig. 15-12.

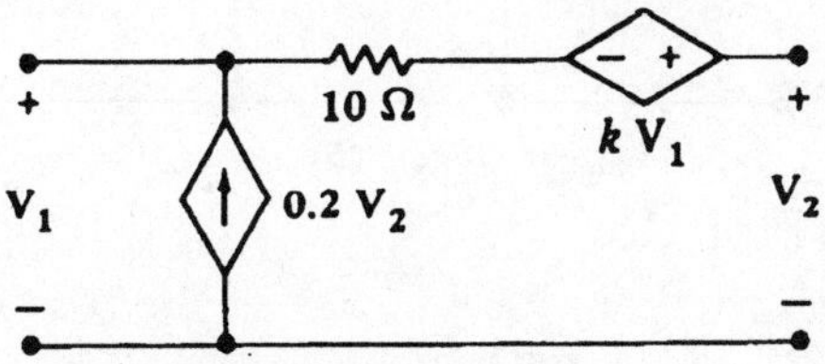

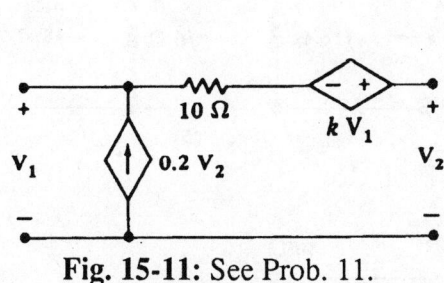

**Fig. 15-11:** See Prob. 11.

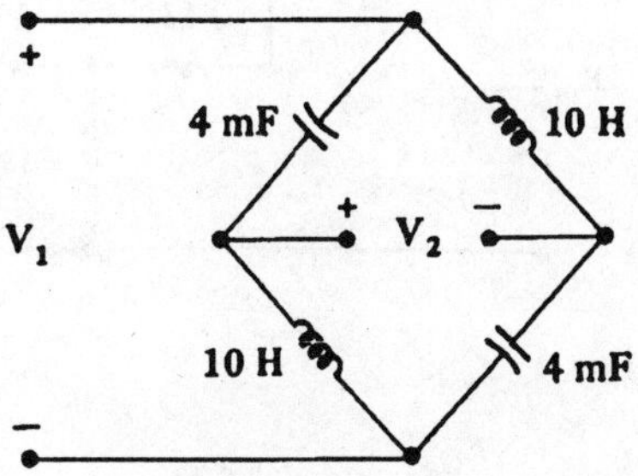


**Fig. 15-12:** See Prob. 12.

**13** A two-port network has the **y** parameters: $y_{11} = 10$, $y_{12} = -5$, $y_{21} = -20$, and $y_{22} = 2$, all in mS. Find the new [**y**] if a 100-Ω resistor is connected: (*a*) across the input; (*b*) across the output; (*c*) between the upper input and output terminals inside the two-port.

**14** Complete the table given as part of Fig. 15-13 and also give values for the **y** parameters.

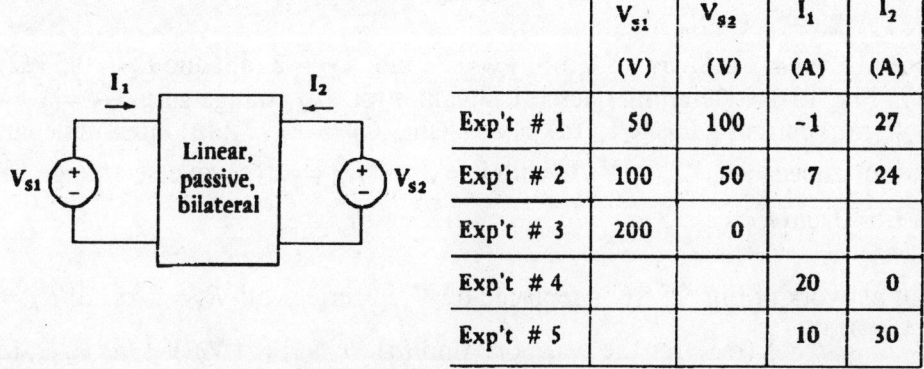

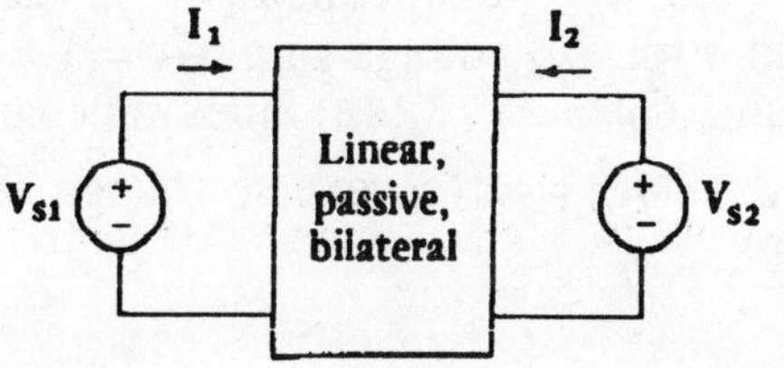


| | $V_{s1}$ (V) | $V_{s2}$ (V) | $I_1$ (A) | $I_2$ (A) |
|---|---|---|---|---|
| Exp't # 1 | 50 | 100 | -1 | 27 |
| Exp't # 2 | 100 | 50 | 7 | 24 |
| Exp't # 3 | 200 | 0 | | |
| Exp't # 4 | | | 20 | 0 |
| Exp't # 5 | | | 10 | 30 |

**Fig. 15-13:** See Prob. 14.

**15** A two-port is characterized by the equations, $12v_1 - 3i_1 - 6i_2 + 4v_2 = 0$ and $5i_1 = 6v_1 - 2i_2 + 0.1v_2$. (*a)* Find [**y**]. (*b*) Find $[\mathbf{y}]_{new}$ after a 5-Ω resistor is added in series with the input.

## Section 15-4 Some equivalent networks

**16** (*a*) Find $R_{in}$ in Fig. 15-14*a* by making Y-Δ or Δ-Y transformations and using resistance combinations. (*b*) Find $\mathbf{Z}_{in}$ for Fig. 15-14*b* by using T-π and π-T transformations and impedance combinations.

**17** A four-sided pyramid has a 60-Ω resistor along each of its eight edges. What resistance would be measured between: (*a*) adjacent corners of the base? (*b*) diagonally opposite corners of the base? (*c*) the apex and one corner of the base?

**18** (*a*) Find the equivalent $\pi$ network for the T network shown in Fig. 15-15*a*. (*b*) Find $R_{\text{in}}$ for the network of Fig. 15-15*b* by using Y-Δ and Δ-Y transformations.

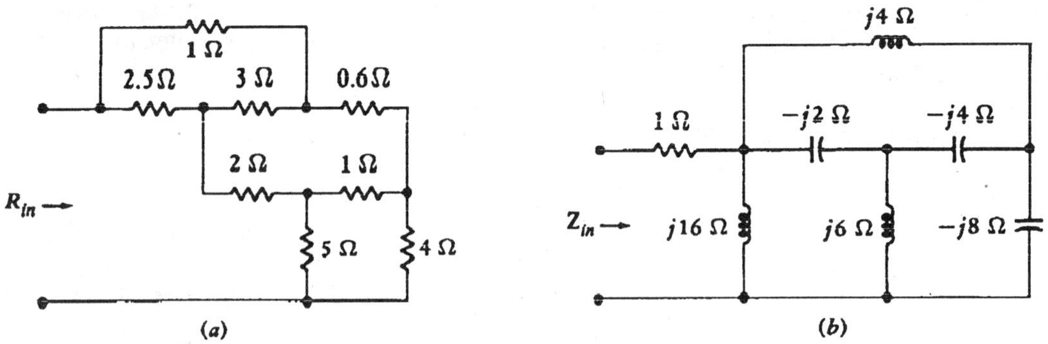


**Fig- 15-14:** See Prob. 16.

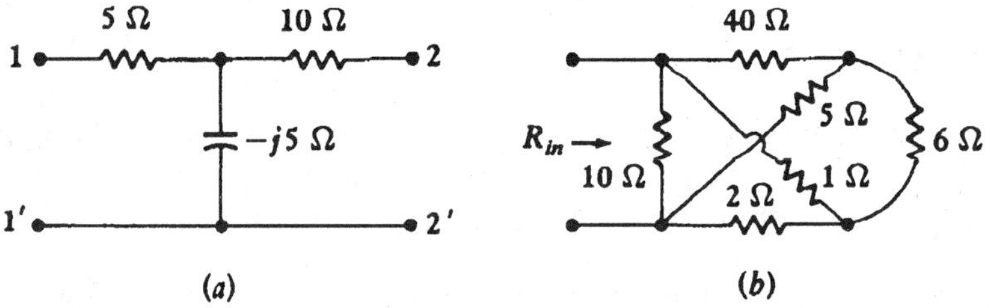


**Fig. 15-15:** See Prob. 18.

**19** Let $R_s = 2$ kΩ, $y_{11} = 1$ mS, $y_{12} = -1$ mS, $y_{21} = 20$ mS, $y_{22} = 2$ mS, and $R_L = 0.5$ kΩ for the circuit shown in Fig. 15-16. Determine numerical values for: (*a*) voltage gain, $\mathbf{G}_V = \mathbf{V}_2/\mathbf{V}_1$; (*b*) current gain, $\mathbf{I}_2/\mathbf{I}_1$; (*c*) current gain, $\mathbf{I}_2/\mathbf{I}_s$; (*d*) power gain, $G_P = P_L/P_s$; (*e*) input impedance, $\mathbf{Z}_{\text{in}} = \mathbf{V}_1/\mathbf{I}_1$; (*f*) output impedance, $\mathbf{Z}_{\text{out}} = \mathbf{V}_2/\mathbf{I}_2$ with $\mathbf{I}_s = 0$ and $R_L = \infty$; (*g*) reverse voltage gain, $\mathbf{G}_{VR} = \mathbf{V}_1/\mathbf{V}_2$ with $\mathbf{I}_s = 0$ and $R_L = \infty$.

**20** The input network of Fig. 15-16 is replaced by $\mathbf{V}_s$ in series with $R_s = 2$ kΩ. If $R_L = 20$ kΩ and $[\mathbf{y}] = \begin{bmatrix} 0.4 & -0.002 \\ -5 & 0.04 \end{bmatrix}$ (mS) for the two-port, find (*a*) $\mathbf{V}_2/\mathbf{V}_1$; (*b*) $\mathbf{V}_2/\mathbf{V}_s$; (*c*) $\mathbf{I}_2/\mathbf{I}_1$; (*d*) $\mathbf{V}_1/\mathbf{I}_1$; (*e*) $R_{\text{out}}$.

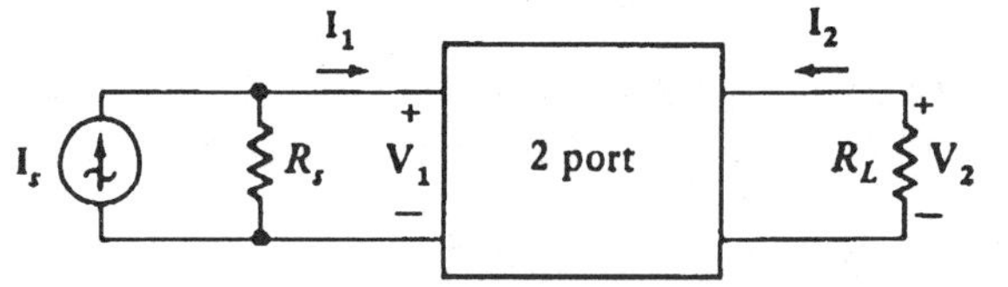


**Fig. 15-16:** See Probs. 19 and 20.

**21** (*a*) Find $[\mathbf{y}]_a$ for the two-port of Fig. 15-17*a*. (*b*) Find $[\mathbf{y}]_b$ for Fig. 15-17*b*. (*c*) Draw the network that is obtained when these two-ports are connected in parallel, and show that [y] for this network is equal to $[\mathbf{y}]_a + [\mathbf{y}]_b$.

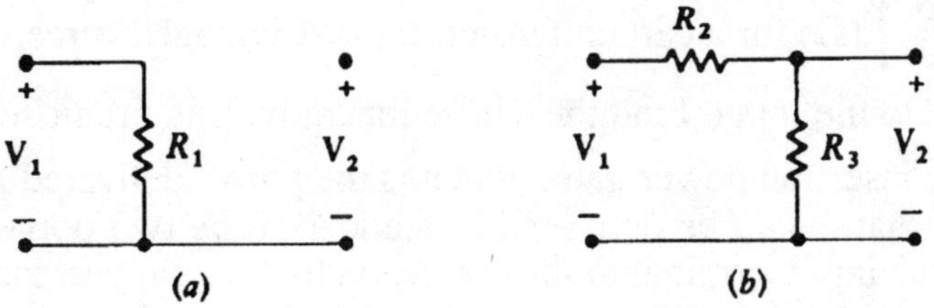


**Fig. 15-17:** See Prob. 21.

**22** (*a*) Draw an equivalent circuit in the form of Fig. 15-18 for which $[\mathbf{y}]_A = \begin{bmatrix} 1.5 & -1 \\ 4 & 3 \end{bmatrix}$ (mS). (*b*) If two of these two-ports are connected in parallel, draw the new equivalent circuit and show that $[\mathbf{y}]_{\text{new}} = 2[\mathbf{y}]_A$.

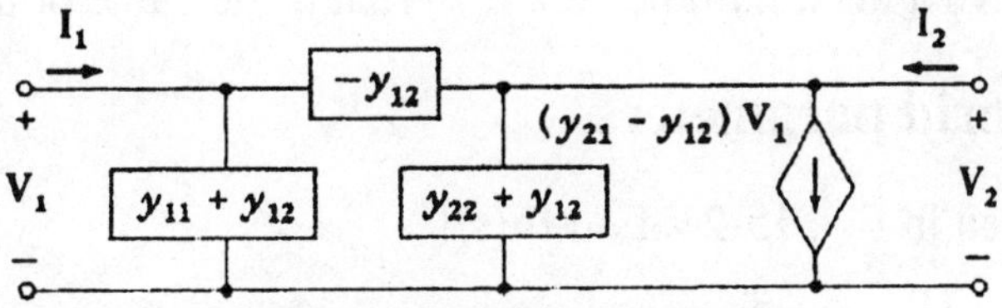


**Fig. 15-18:** See Prob. 22.

## Section 15-5 Impedance parameters

**23** (*a*) Find the four **z**-parameters at $\omega = 10^8$ rad/s for the transistor, high-frequency equivalent circuit shown in Fig. 15-19. (*b*) Which of these four parameter values are functions of frequency?

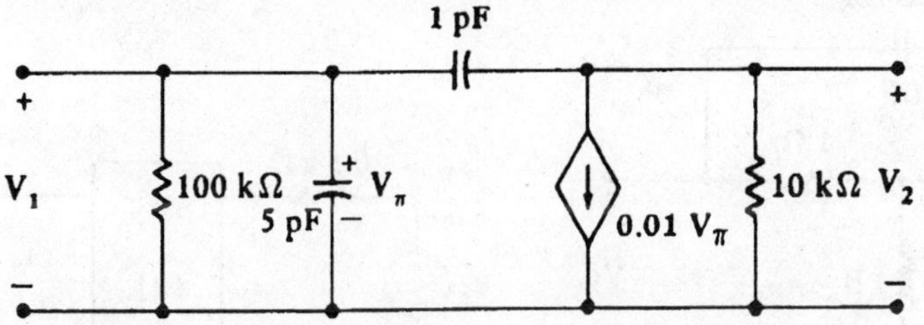


**Fig. 15-19:** See Prob. 23.

**24** A sinusoidal source, 100 V rms in series with 50 Ω, is connected to the input of a two-port network for which $[\mathbf{z}] = \begin{bmatrix} 200 & 20 \\ 400 & 100 \end{bmatrix}$ (Ω), and a load, $R_L = 250\ \Omega$, is placed at the output. Find the Thévenin equivalent circuit seen at the: (*a*) input terminals of the two-port; (*b*) output terminals of the two-port.

**25** For the two-port shown in Fig. 15-20: (*a*) find [**z**]; (*b*) place $\mathbf{V}_s$ in series with $R_s = 200\ \Omega$ at the input and $R_L = 4\ \text{k}\Omega$ at the output, and determine $\mathbf{G}_V = \mathbf{V}_2/\mathbf{V}_1$ and $R_{in} = \mathbf{V}_1/\mathbf{I}_1$.

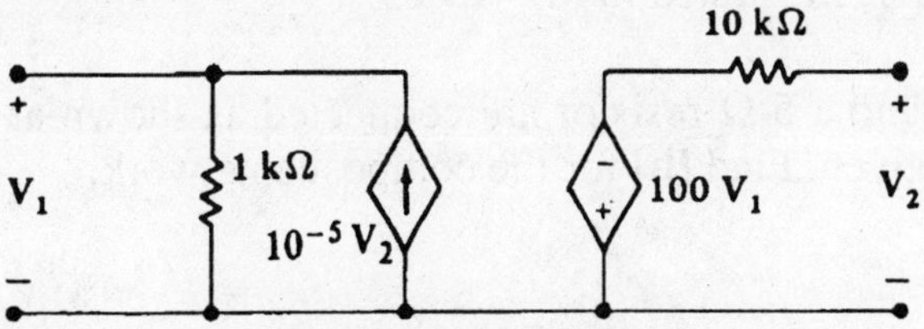


**Fig. 15-20:** See Prob. 25.

**26** Let $[\mathbf{z}] = \begin{bmatrix} 10 & -2 \\ 40 & 20 \end{bmatrix}$ ($\Omega$) for a certain two-port. (*a*) A voltage source, $\mathbf{V}_s = 1\angle 0^\circ$ V, in series with 2 $\Omega$ is connected to the input. Find the Thévenin equivalent circuit seen at the output terminals. (*b*) Calculate the insertion power gain, which is the power delivered to the load, $R_L = 25\ \Omega$, divided by the power that *would* be delivered to the load if the two-port were replaced by short circuits connecting each input terminal to the corresponding output terminal. (*c*) Calculate the reverse voltage gain, $\mathbf{G}_{VR} = \mathbf{V}_1/\mathbf{V}_2$, if an independent voltage source $\mathbf{V}_2$ is applied to the output and $\mathbf{V}_s$ is set equal to zero.

**27** A two-port network for which $[\mathbf{z}] = \begin{bmatrix} 10 & 2 \\ 40 & 5 \end{bmatrix}$ ($\Omega$) is terminated in $\mathbf{Z}_L = \mathbf{z}_{22}$ and driven by a 20-V rms source in series with $\mathbf{Z}_s = \mathbf{z}_{11}$. (*a*) How much power is delivered to $\mathbf{Z}_L$? (*b*) What value should $\mathbf{Z}_L$ have to receive a maximum power? (*c*) What is the value of this maximum power?

## Section 15-6 Hybrid parameters

**28** For the network given in Fig. 15-21, find [$\mathbf{h}$(s)].

**29** A certain two-port is characterized by the following pair of equations: $2\mathbf{V}_1 + 4\mathbf{I}_2 = \mathbf{I}_1$ and $8\mathbf{I}_2 = \mathbf{V}_2 + 6\mathbf{V}_1$. Find the value of: (*a*) $\mathbf{y}_{11}$; (*b*) $\mathbf{z}_{21}$; (*c*) $\mathbf{h}_{21}$.

**30** The two identical stages of the amplifier which is shown in block form in Fig. 15-22 each have $[\mathbf{h}] = \begin{bmatrix} 1\ \text{k}\Omega & 0.001 \\ 100 & 0.1\ \text{mS} \end{bmatrix}$. What amplitude of the sinusoidal input voltage $\mathbf{V}_s$ is required to obtain a 25-V amplitude for $\mathbf{V}_o$?

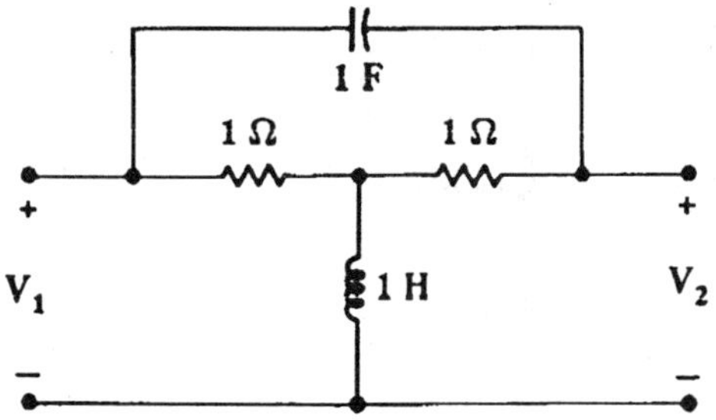

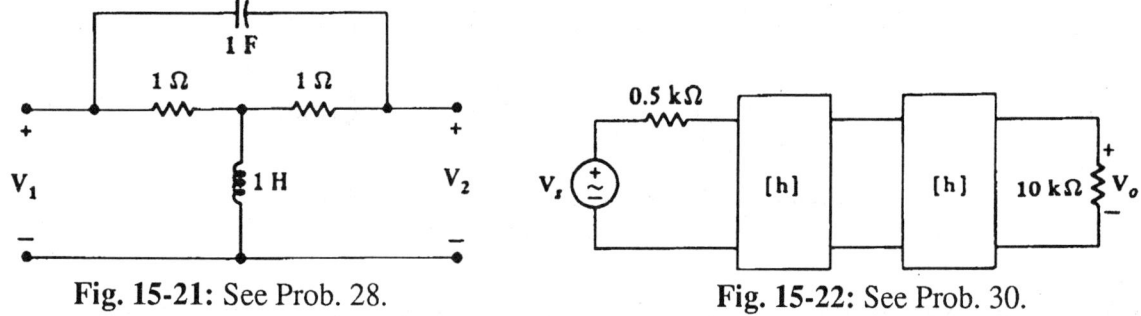

**Fig. 15-21:** See Prob. 28.

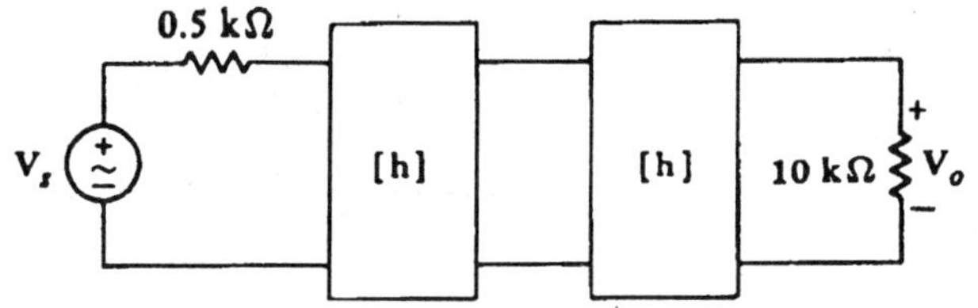

**Fig. 15-22:** See Prob. 30.

**31** A two-port for which $[\mathbf{h}] = \begin{bmatrix} 10\ \Omega & -1 \\ 4 & 0.05\ \text{S} \end{bmatrix}$ is terminated by $R_L = 10\ \Omega$ and driven by an ideal sinusoidal voltage source, $\mathbf{V}_s = 100$ V rms, in series with $R_s = 2\ \Omega$. Find the average power delivered to: (*a*) $R_s = 2\ \Omega$; (*b*) $R_L = 10\ \Omega$; (*c*) the entire two-port network.

**32** Find [$\mathbf{h}$] for the two-port illustrated in Fig. 15-23.

**33** In Fig. 15-24, a 2-$\Omega$ and a 5-$\Omega$ resistor are connected as shown at the input of the two-port whose $\mathbf{h}$ parameters are given. Find [$\mathbf{h}$] for the composite network.

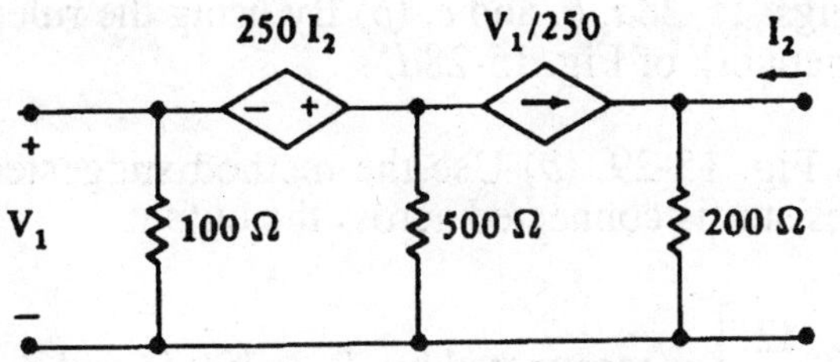


Fig. 15-23: See Prob. 32.

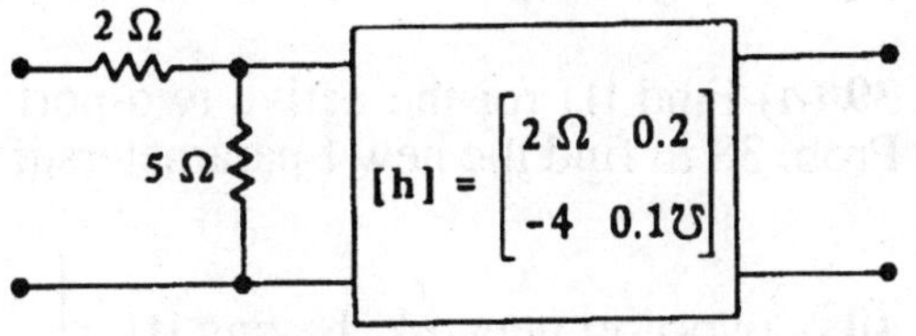


Fig. 15-24: See Prob. 33.

## Section 15-7 Transmission parameters

**34** Find the **t** parameters for: (*a*) the two-port shown in Fig. 15-25; (*b*) a two-port for which $[\mathbf{h}] = \begin{bmatrix} 20\ \Omega & 0.3 \\ -4 & 0.1\ \text{S} \end{bmatrix}$.

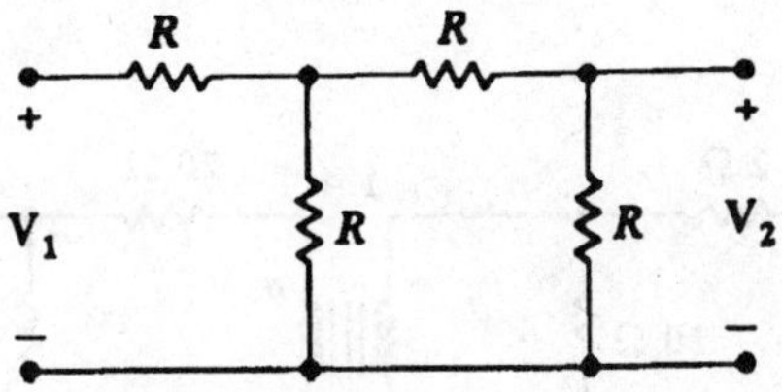

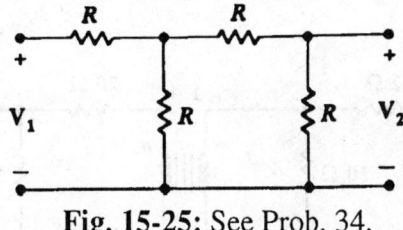

Fig. 15-25: See Prob. 34.

**35** (*a*) Find $[\mathbf{t}]_a$ for the two-port of Fig. 15-26*a*. (*b*) Find $[\mathbf{t}]_b$ for Fig. 15-26*b*. (*c*) Find $[\mathbf{t}]_a[\mathbf{t}]_b$ and show that [**t**] for the cascaded combination is the same. (*d*) Find $[\mathbf{t}]_b[\mathbf{t}]_a$.

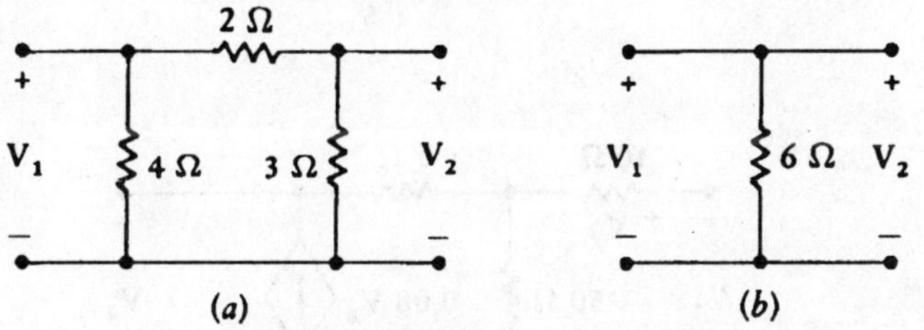


Fig. 15-26: See Prob. 35.

**36** A two-port contains a single 1-Ω resistor connected from the upper input terminal to the upper output terminal; the lower terminals are connected directly to each other. (*a*) Find $[\mathbf{t}]_a$ for this network. (*b*) Show that the **t** parameters for a similarly connected 5-Ω resistor can be obtained by $([\mathbf{t}]_a)^5$.

**37** Find [**t**] for the two-port shown in Fig. 15-27.

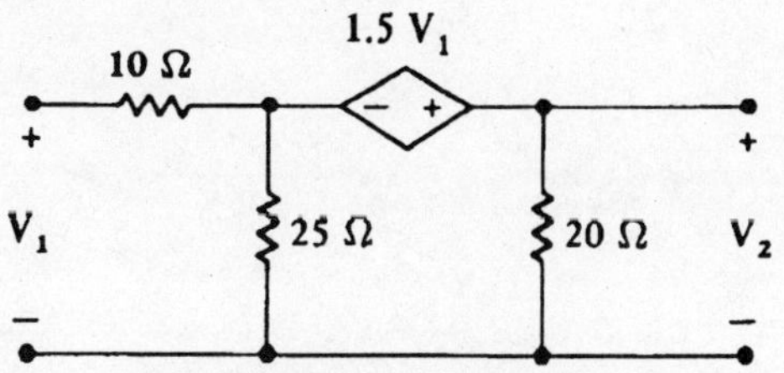


Fig. 15-27: See Prob. 37.

**38** (*a*) Determine [**t**] for the networks shown in Figs. 15-28*a*, *b*, and *c*. (*b*) By using the rules for connecting two-ports in cascade, find [**t**] for the network of Fig. 15-28*d*.

**39** (*a*) Find [**t**] for the active two-port shown in Fig. 15-29. (*b*) Use the method suggested by Prob. 38 to find the new **t** parameters if a 20-Ω resistor is connected across the output.

**40** A two-port network, having $[\mathbf{t}] = \begin{bmatrix} 10^{-2} & 10^{-1}\,\Omega \\ 10^{-5}\,\text{S} & 10^{-4} \end{bmatrix}$, is terminated by $R_L = 100\ \Omega$ and driven by a sinusoidal source $\mathbf{V}_s$ in series with $R_s = 100\ \Omega$. Find $\mathbf{G}_V$, $\mathbf{G}_I$, $G_P$, $\mathbf{Z}_{\text{in}}$, and $\mathbf{Z}_{\text{out}}$. Is the two-port reciprocal ?

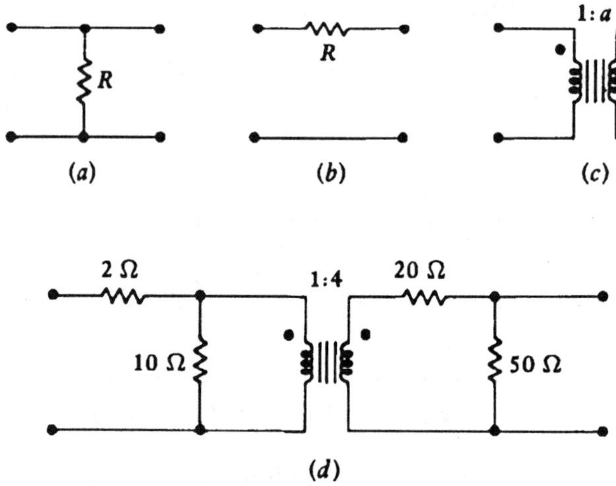


**Fig. 15-28:** See Prob. 38.

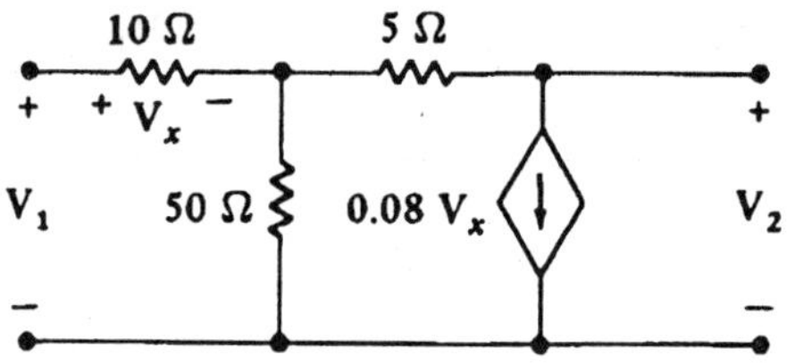


**Fig. 15-29:** See Prob. 39.

# Chapter 15

**15-1**

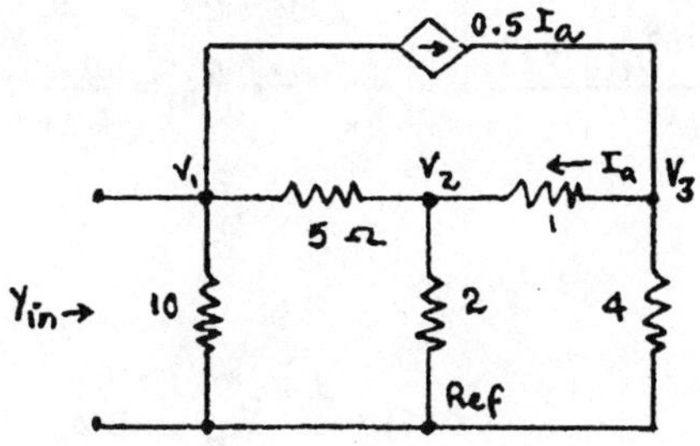


$$I_a = V_3 - V_2$$

$$\left.\begin{aligned} &0.1V_1 + 0.2(V_1 - V_2) + 0.5(V_3 - V_2) = 0 \\ &-0.2V_1 + 1.7V_2 - V_3 = 0 \\ &0.25V_3 + V_3 - V_2 - 0.5(V_3 - V_2) = 0 \end{aligned}\right\} \therefore \Delta_Y = \begin{vmatrix} 0.3 & -0.7 & 0.5 \\ -0.2 & 1.7 & -1 \\ 0 & -0.5 & 0.75 \end{vmatrix}$$

$$\therefore \Delta_Y = 0.3(0.775) + 0.2(-0.275) = 0.1775 \text{ S}^3 \; ; \; Y_{in} = \frac{\Delta_Y}{\Delta_{11}} = \frac{0.1775}{0.775} = \underline{0.2290 \text{ S}}$$

**15-2** (a)

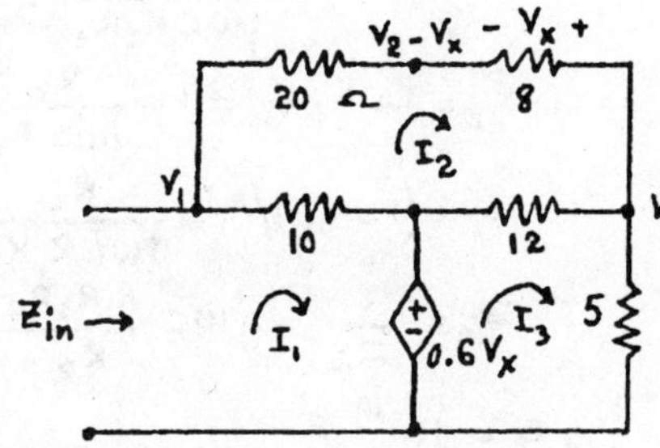


$$\left.\begin{aligned} &V_{in} = 10(I_1 - I_2) + 0.6(-8I_2) \\ &0 = -10I_1 + 50I_2 - 12I_3 \\ &0 = 0.6(8I_2) - 12I_2 + 17I_3 \end{aligned}\right\} \therefore \Delta_Z = \begin{vmatrix} 10 & -14.8 & 0 \\ -10 & 50 & -12 \\ 0 & -7.2 & 17 \end{vmatrix}$$

$$\Delta_Z = 5120$$

$$\therefore Z_{in} = \frac{\Delta Z}{\Delta_{11}} = \frac{5120}{763.6} = \underline{6.705^+ \ \Omega}$$

(b)

$$\left.\begin{aligned} &I_{in} = \frac{V_1 - V_2 + V_x}{20} + \frac{V_1 - 0.6V_x}{10} \\ &0 = \frac{V_2 - V_x - V_1}{20} - \frac{V_x}{8} \\ &0 = \frac{V_x}{8} + \frac{V_2 - 0.6V_x}{12} + \frac{V_2}{5} \end{aligned}\right\} \therefore \Delta_Y = \begin{vmatrix} 0.15 & -0.05 & -0.01 \\ -0.05 & 0.05 & -0.175 \\ 0 & 0.283 & 0.075 \end{vmatrix} \begin{aligned} &= 0.15(0.0533) + 0.05(-0.00092) \\ &= 0.007954 \end{aligned}$$

$$Y_{in} = \frac{\Delta_Y}{\Delta_{11}} = \frac{0.007954}{0.0533} = 0.14914 \quad \therefore Z_{in} = \underline{6.705^+ \ \Omega}$$

**15-3**

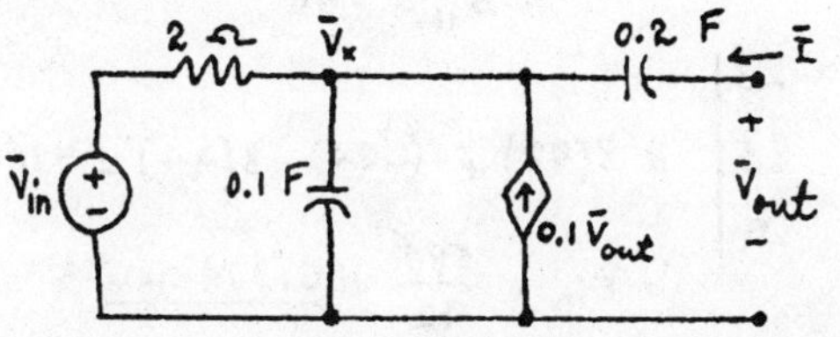


let $\bar{V}_s = 0$, $\bar{V}_{out} = 1$ V

$$\therefore \frac{\bar{V}_x}{2} + \frac{\bar{s}\bar{V}_x}{10} - 0.1 \times 1 + \frac{\bar{s}}{5}(\bar{V}_x - 1) = 0$$

$$\therefore \bar{V}_x(0.5 + 0.3\bar{s}) = 0.1 + 0.2\bar{s}, \quad \bar{V}_x = \frac{0.2\bar{s} + 1}{0.3\bar{s} + 0.5}$$

$$\bar{I} = (1 - \bar{V}_x)0.2\bar{s} = \left(1 - \frac{0.2\bar{s} + 1}{0.3\bar{s} + 0.5}\right)0.2\bar{s} = 0.2\bar{s} \times \frac{0.1\bar{s} + 0.4}{0.3\bar{s} + 0.5} = \frac{\bar{s}(0.1\bar{s} + 0.4)}{1.5\bar{s} + 2.5} = \bar{Y}_{out}$$

$$\bar{Z}_{out} = \frac{1}{\bar{Y}_{out}} = \frac{5(3\bar{s} + 5)}{\bar{s}(\bar{s} + 4)} \longleftarrow$$

15-4

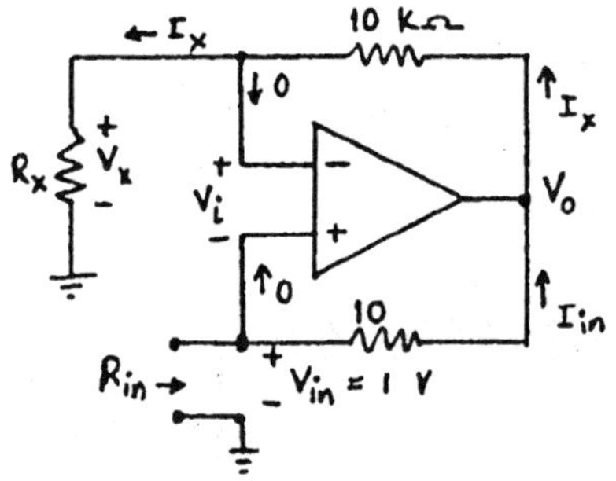


Assume $V_{in} = 1$ V $\therefore V_i = 0$ and $V_x = 1$ V

$\therefore I_x = \dfrac{1}{R_x}$ $\quad \therefore V_o = I_x 10^4 + 1 = \dfrac{10^4}{R_x} + 1$

$I_{in} = \dfrac{1 - (10^4/R_x + 1)}{10^4} = -\dfrac{1}{R_x}$ $\quad \therefore R_{in} = \dfrac{1}{I_{in}} = \underline{-R_x}$

15-5 (a)

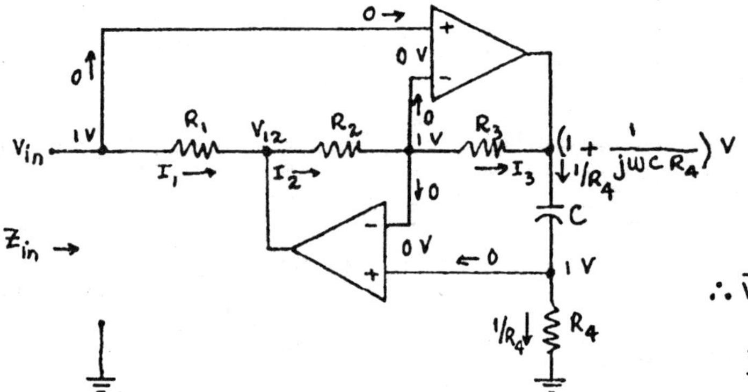


Assume $\bar{V}_{in} = 1$ V

$\therefore \bar{I}_3 = \left(1 - 1 - \dfrac{1}{j\omega C R_4}\right)/R_3$

$= -\dfrac{1}{j\omega C R_4 R_3} = I_2$

$\therefore \bar{V}_{12} = 1 + R_2 I_2 = 1 - \dfrac{R_2}{j\omega C R_4 R_3}$

$\bar{I}_1 = (1 - V_{12})/R_1 = \dfrac{R_2}{j\omega C R_1 R_3 R_4} = I_{in}$

$\bar{Z}_{in} = \dfrac{1}{I_{in}} = j\omega C \dfrac{R_1 R_3 R_4}{R_2}$ ←

(b) $R_1 = 10^3$, $R_2 = 2\times 10^3$, $R_3 = 3\times 10^3$, $R_4 = 4\times 10^3$, $C = 10^{-10}$ F

$\therefore \bar{Z}_{in} = j\omega 10^{-10} \dfrac{10^3 \times 3\times 10^3 \times 4 \times 10^3}{2\times 10^3} = j\omega 6\times 10^{-4} = j\omega L_{in}$ $\quad \therefore L_{in} = \underline{6\times 10^{-4} \text{ H}}$

15-6 (a)

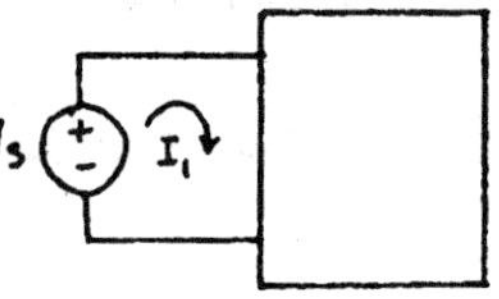


$$\Delta_Z = \begin{vmatrix} 8 & -1 & -3 \\ -1 & 10 & -2 \\ -3 & -2 & 6 \end{vmatrix} (\Omega) = 8(56) + 1(-12) - 3(32) = 340\ \Omega^3$$

$\therefore Z_{in} = \dfrac{\Delta_Z}{\Delta_{11}} = \dfrac{340}{56} = \underline{6.071\ \Omega}$

(b)

$$\Delta_Z = \begin{vmatrix} 8 & -1 & -3 \\ -1 & 10+3 & -2-3 \\ -3 & -2-3 & 6+3 \end{vmatrix} = \begin{vmatrix} 8 & -1 & -3 \\ -1 & 13 & -5 \\ -3 & -5 & 9 \end{vmatrix} = 8(92) + 1(-24) - 3(44) = 580\ \Omega^3$$

$\therefore Z_{in} = \dfrac{580}{92} = \underline{6.304\ \Omega}$

15-7 (a)

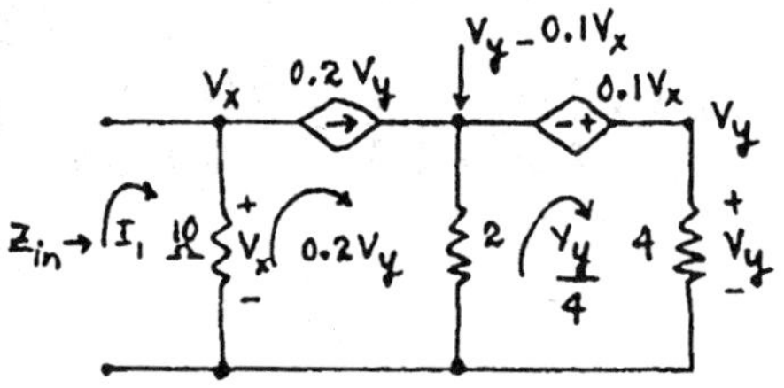


$V_{in} = 10(I_1 - 0.2V_y) = 10I_1 - 2V_y$ ①

$2(0.25V_y - 0.2V_y) - 0.1V_x + V_y = 0 \therefore 1.1V_y - 0.1(10I_1 - 2V_y) = 0$ ②

$\therefore \Delta_Z = \begin{vmatrix} 10 & -2 \\ -1 & 1.3 \end{vmatrix} = 11$, $Z_{in} = \dfrac{11}{1.3} = \underline{8.462\ \Omega}$

(b) $I_{in} = 0.1V_x + 0.2V_y$ ①

$-0.2V_y + (V_y - 0.1V_x)/2 + 0.25V_y = 0 \therefore 0.55V_y - 0.05V_x = 0$ ②

$\therefore \Delta_Y = \begin{vmatrix} 0.1 & 0.2 \\ -0.05 & 0.55 \end{vmatrix} = 0.065$

$\therefore Y_{in} = \dfrac{0.065}{0.55}$ $\quad \therefore Z_{in} = \dfrac{0.55}{0.065} = \underline{8.462\ \Omega}$

15-8

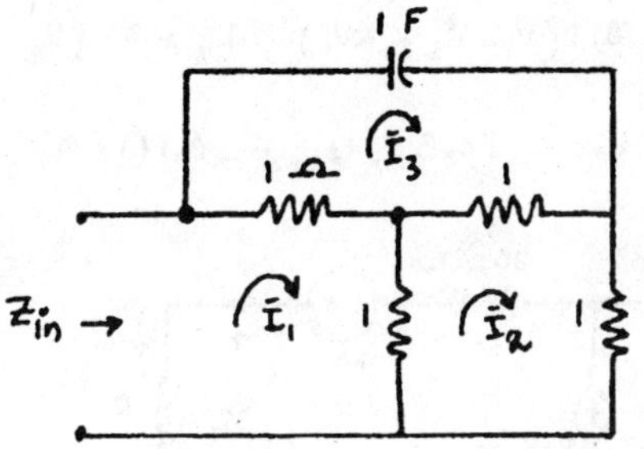


$$\bar{\Delta}_Z = \begin{vmatrix} 2 & -1 & -1 \\ -1 & 3 & -1 \\ -1 & -1 & 2+1/\bar{s} \end{vmatrix} = 2\left(5+\frac{3}{\bar{s}}\right) + 1\left(-3-\frac{1}{\bar{s}}\right) - 4 = 3 + 5/\bar{s}$$

$$\therefore \bar{Z}_{in}(\bar{s}) = \frac{3+5/\bar{s}}{5+3/\bar{s}} = \frac{3\bar{s}+5}{5\bar{s}+3}$$

$\therefore$ Pole at $\bar{s} = \underline{-3/5}$, zero at $\bar{s} = \underline{-5/3}$

15-9 (a)

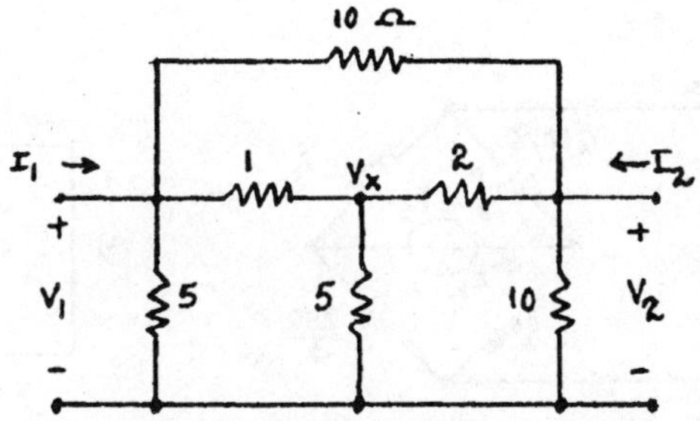


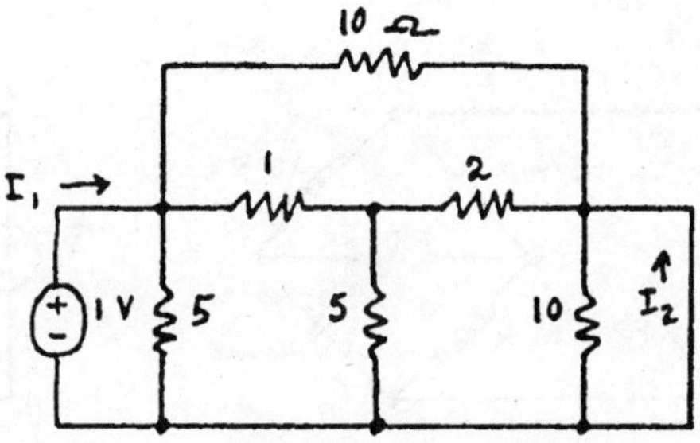


$$I_1 = 1/\left((1+2\|5)\|5\|10\right) = \frac{121}{170} \therefore y_{11} = \underline{0.7118\ S}\ ;\ I_2 = -\left(\frac{1}{10} + \frac{1}{17/7} \times \frac{5}{7}\right) = -\frac{67}{170} \therefore y_{21} = \underline{-0.3941\ S}$$

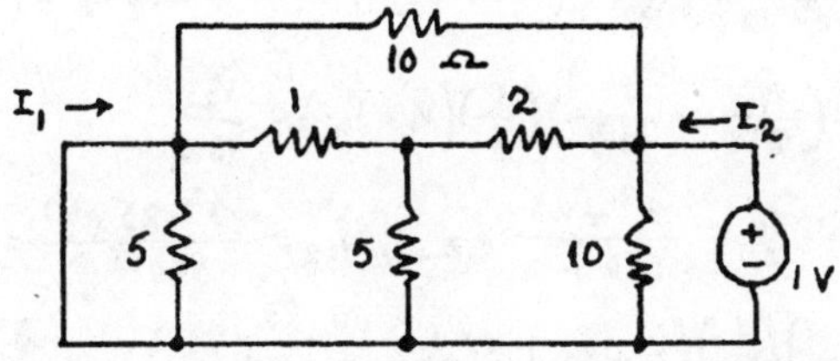


$$I_2 = \frac{1}{(2+1\|5)\|10\|10} = \frac{47}{85} = y_{22} = \underline{0.5529}\ S$$

$$I_1 = -\left(\frac{1}{17/6} \times \frac{5}{6} + \frac{1}{10}\right) = -\frac{67}{170} = \underline{-0.3941\ S} = y_{12}$$

(b) Center node: $\frac{V_x - V_1}{1} + \frac{V_x}{5} + \frac{V_x - V_2}{2} = 0 \therefore V_x = \frac{10}{17}V_1 + \frac{5}{17}V_2$

$$I_1 = \frac{V_1}{5} + (V_1 - V_x) + \frac{1}{10}(V_1 - V_2) = \frac{V_1}{5} + \left(V_1 - \frac{10}{17}V_1 - \frac{5}{17}V_2\right) + \frac{1}{10}(V_1 - V_2) = \underline{\frac{121}{170}}V_1 - \underline{\frac{67}{170}}V_2$$

$$I_2 = \frac{V_2}{10} + \frac{1}{2}(V_2 - V_x) + \frac{1}{10}(V_2 - V_1) = \frac{V_2}{10} + \frac{1}{2}\left(V_2 - \frac{10}{17}V_1 - \frac{5}{17}V_2\right) + \frac{1}{10}(V_2 - V_1) = -\underline{\frac{67}{170}}V_1 - \underline{\frac{47}{85}}V_2$$

15-10 (a)

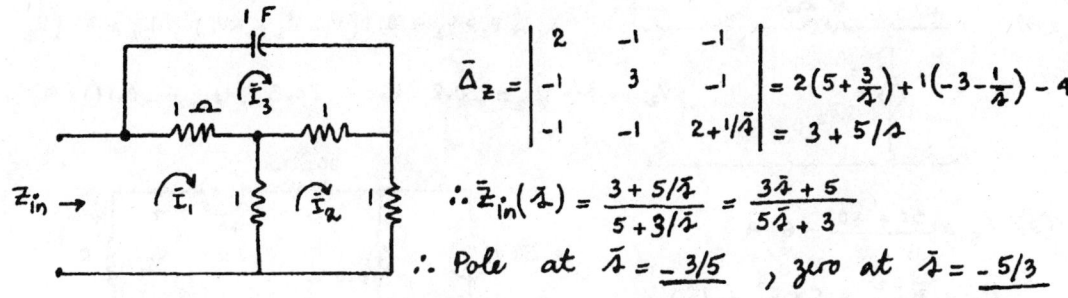

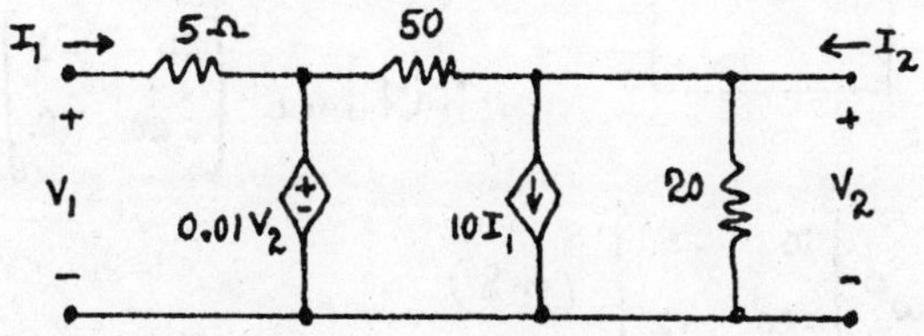


$V_2 = 0$, $V_1 = 1$ $\therefore 0.01V_2 = 0$, $1 = 5I_1 + 0$

$\therefore I_1 = 0.2$ ; $I_2 = 10I_1 = 10(0.2) = \underline{2} = y_{21}$

$V_1 = 0$, $V_2 = 1$ $\therefore 0.01V_2 = 0.01$

$\therefore I_1 = -\frac{0.01}{5} = \underline{-0.002} = y_{12}$

(b) $V_2 = 0$, $V_1 = 1$ $\therefore 1 = 5I_1$, $I_1 = 0.2$, $I_2 = kI_1 = \frac{k}{5} = y_{21}$ ; $y_{12} = -0.002$

$y_{21} = y_{12} \therefore \frac{k}{5} = -0.002$, $k = \underline{-0.01}$

**15-11** (a)

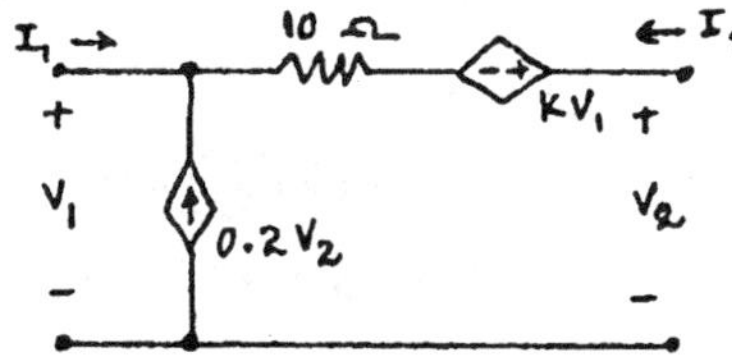


$I_1 + 0.2V_2 = 0.1(V_1 - V_2 + KV_1)$ ; $I_2 = 0.1(V_2 - KV_1 - V_1)$

$\therefore y_{12} = -0.2 - 0.1 = -0.3$ , $y_{21} = -0.1(1+K) = -0.3$

$\therefore K = \underline{2}$

(b) $I_5 = \frac{50+40}{10+5} = 6$ A

$P_5 = RI_5^2 = 5 \times 6^2 = \underline{180 \text{ W}}$

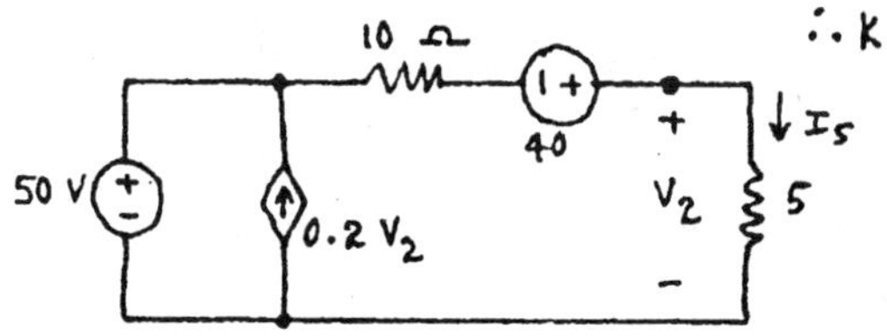


**15-12**

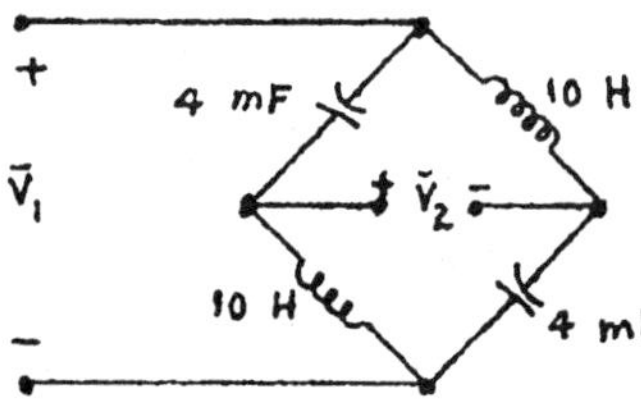


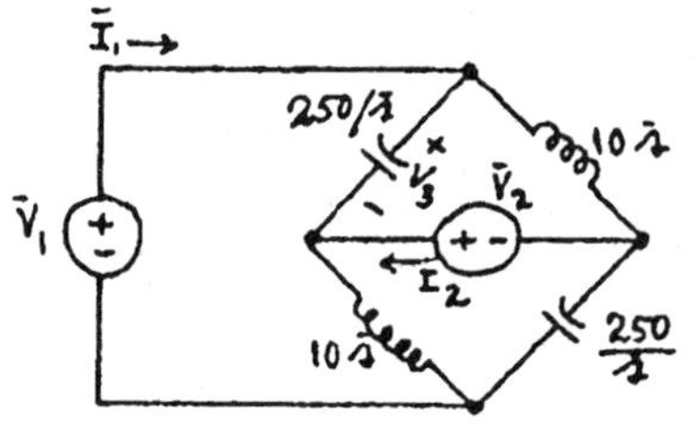


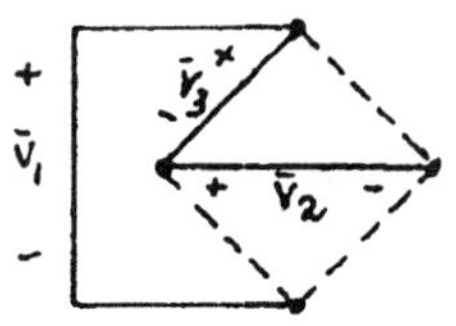


Left and right : $-\frac{s\bar V_3}{250} + \frac{-\bar V_3+\bar V_1}{10s} + \frac{-\bar V_2-\bar V_3}{10s} + \frac{(-\bar V_2-\bar V_3+\bar V_1)s}{250} = 0$

$\therefore \left(\frac{s}{250}+\frac{1}{10s}\right)\bar V_1 + \left(-\frac{1}{10s}-\frac{s}{250}\right)\bar V_2 = \left(\frac{2s}{250}+\frac{2}{10s}\right)\bar V_3 \therefore (s^2+25)\bar V_1 - (s^2+25)\bar V_2 = (2s^2+50)\bar V_3$

$\therefore \bar V_3 = \frac{1}{2}(\bar V_1 - \bar V_2)$ ; $\bar I_1 = \frac{s}{250}\bar V_3 + \frac{\bar V_3+\bar V_2}{10s} = \left(\frac{s}{250}+\frac{1}{10s}\right)\left(\frac{1}{2}\right)(\bar V_1-\bar V_2) + \frac{\bar V_2}{10s}$

$\therefore \bar I_1 = \frac{s^2+25}{250s\times 2}\bar V_1 - \frac{s^2+25}{500s}\bar V_2 + \frac{1}{10s}\bar V_2 \therefore \bar y_{11} = \frac{s^2+25}{500s}$ ; $\bar y_{12} = \frac{-s^2-25+50}{500s} = \frac{25-s^2}{500s}$

$\bar I_2 = \frac{-\bar V_3+\bar V_1}{10s} - \frac{s}{250}\bar V_3 = \frac{\bar V_1}{10s} - \left(\frac{1}{10s}+\frac{s}{250}\right)\left(\frac{1}{2}\right)(\bar V_1-\bar V_2) = \left(\frac{1}{10s} - \frac{s^2+25}{500s}\right)\bar V_1 + \frac{s^2+25}{500s}\bar V_2$

$\therefore \bar y_{21} = \frac{25-s^2}{500s} = \bar y_{12}$ $\quad \bar y_{22} = \frac{s^2+25}{500s}$ (= $y_{11}$ by sym)

**15-13** (a)

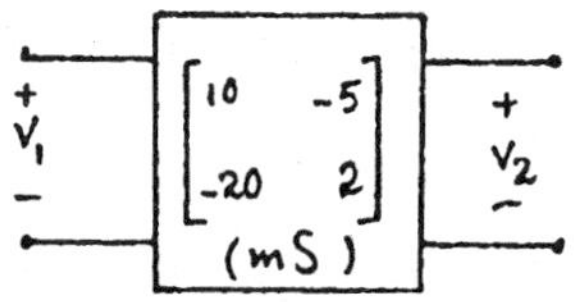


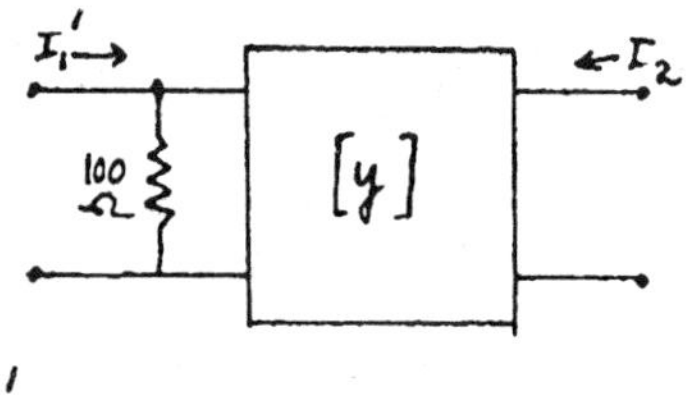


$I_1' = (10+10)V_1 - 5V_2$

$I_2 = -20V_1 + 2V_2$

$\therefore [y]_{new} = \begin{bmatrix} 20 & -5 \\ -20 & 2 \end{bmatrix}$ (mS)

(b)

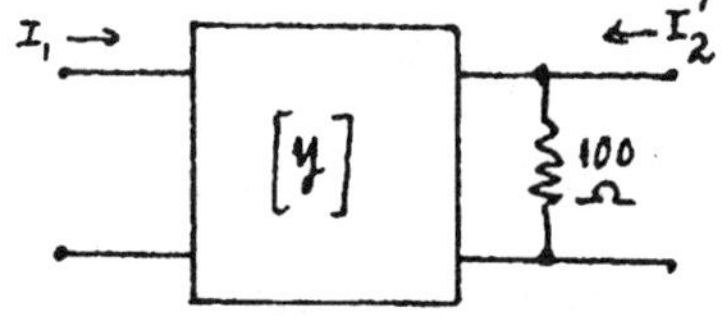


$[y]_{new} = \begin{bmatrix} 10 & -5 \\ -20 & 12 \end{bmatrix}$ (mS)

(c)

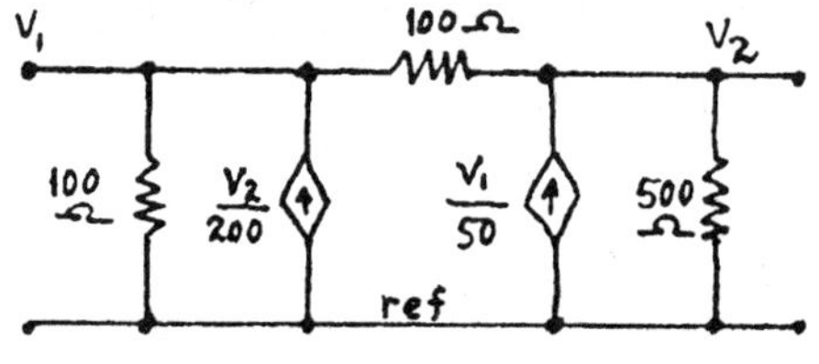


$I_1 = \frac{V_1}{100} - \frac{V_2}{200} + \frac{V_1 - V_2}{100} = \frac{V_1}{50} - \frac{3V_2}{200}$

$I_2 = \frac{V_2}{500} - \frac{V_1}{50} + \frac{V_2 - V_1}{100} = -\frac{3V_1}{100} + \frac{6V_2}{500}$

$\therefore [y]_{new} = \begin{bmatrix} 20 & -15 \\ -30 & 12 \end{bmatrix}$ (mS)

15-14

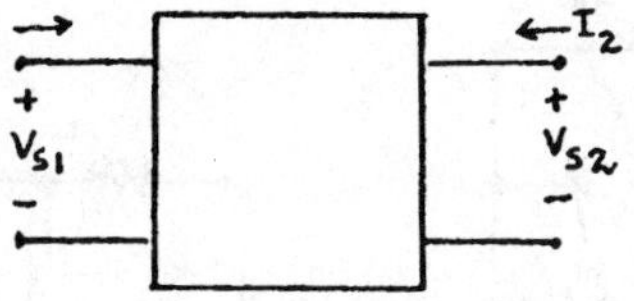


| | $V_{s1}$ | $V_{s2}$ | $I_1$ | $I_2$ |
|---|---|---|---|---|
| ① | 50 | 100 | −1 | 27 |
| ② | 100 | 50 | 7 | 24 |
| ③ | 200 | 0 | | |
| ④ | | | 20 | 0 |
| ⑤ | | | 10 | 30 |

Expt 1 + Expt 2

$-1 = y_{11}50 + y_{12}100$

$7 = y_{11}100 + y_{12}50$

$\therefore y_{11} = \underline{0.1\ S}$ , $y_{12} = \underline{-0.06\ S}$

Also $27 = 50y_{21} + 100y_{22}$ , $24 = 100y_{21} + 50 + y_{22}$ $\therefore y_{21} = \underline{0.14\ S}$ , $y_{22} = \underline{0.2\ S}$

Expt 3: $I_1 = 0.1\times 200 + 0 = \underline{20\ A}$ , $I_2 = 0.14\times 200 + 0 = \underline{28\ A}$

Expt 4: $20 = 0.1V_{s1} - 0.06V_{s2}$ , $0 = 0.14V_{s1} + 0.2V_{s2}$ $\therefore V_{s1} = \underline{140.85\ V}$ , $V_{s2} = \underline{-98.59\ V}$

Expt 5: $10 = 0.1V_{s1} - 0.06V_{s2}$ , $30 = 0.14V_{s1} + 0.2V_{s2}$ $\therefore V_{s1} = \underline{133.80\ V}$ , $V_{s2} = \underline{56.34\ V}$

15-15 (a) $12v_1 - 3i_1 - 6i_2 + 4v_2 = 0 \therefore i_1 = 4v_1 + \frac{4}{3}v_2 - 2i_2$

$\therefore 5i_1 = 6v_1 - 2i_2 + 0.1v_2 = 20v_1 + 20/3\, v_2 - 10i_2 \therefore 14v_1 + \frac{197}{30}v_2 = 8i_2$

$\therefore i_2 = \frac{7}{4}v_1 + \frac{197}{240}v_2$ ← ; $i_1 = 4v_1 + \frac{4}{3}v_2 - \frac{7}{2}v_1 - \frac{197}{120}v_2 = \frac{1}{2}v_1 - \frac{37}{120}v_2$ ←

$$[y] = \begin{bmatrix} 0.5 & -0.3083 \\ 1.75 & 0.8208 \end{bmatrix} (S) \leftarrow$$

(b)

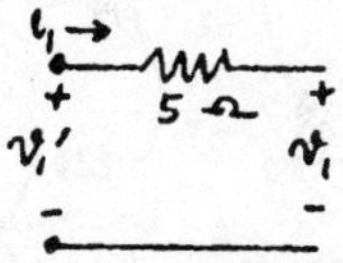


$v_1' = 5i_1 + v_1 \therefore v_1 = v_1' - 5i_1$ , $i_1 = \frac{1}{2}v_1 - \frac{37}{120}v_2 = \frac{1}{2}v_1' - \frac{5}{2}i_1 - \frac{37}{120}v_2$

$\therefore i_1 = \frac{1}{7}v_1' - \frac{37}{420}v_2$ ← , $i_2 = \frac{7}{4}v_1 + \frac{197}{240}v_2$

$\therefore i_2 = \frac{7}{4}v_1' - \frac{35}{4}\left(\frac{1}{7}v_1' - \frac{37}{420}v_2\right) + \frac{197}{240}v_2 \therefore i_2 = \frac{1}{2}v_1' + \frac{382}{240}v_2$ ←

$$[y]_{new} = \begin{bmatrix} 0.14286 & 0.08810 \\ 0.500 & 1.5917 \end{bmatrix} (S) \leftarrow$$

15-16 (a)

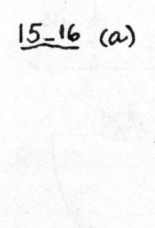

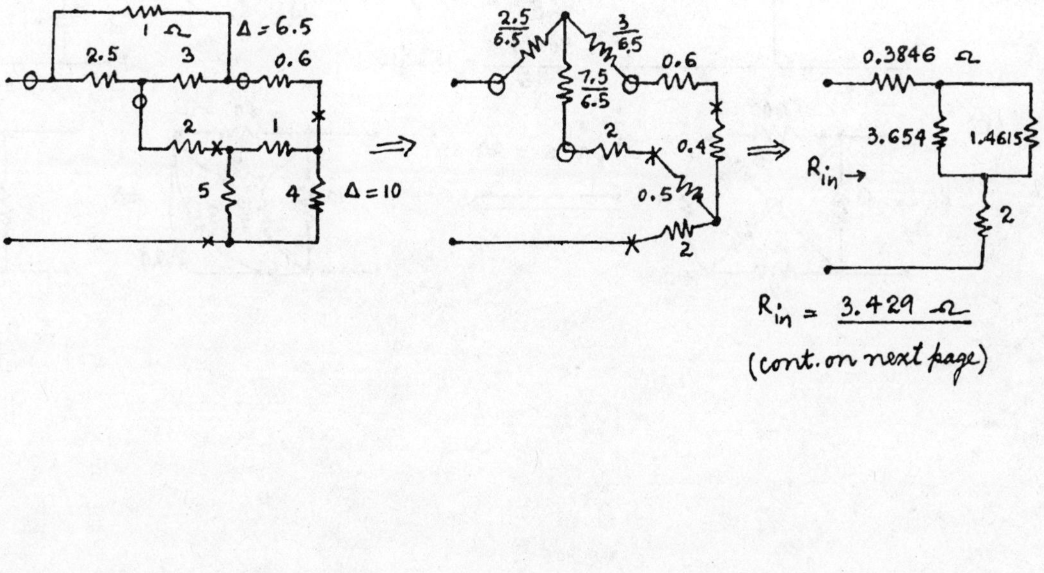


$R_{in} = \underline{3.429\ \Omega}$

(cont. on next page)

15-16 (cont) (b)

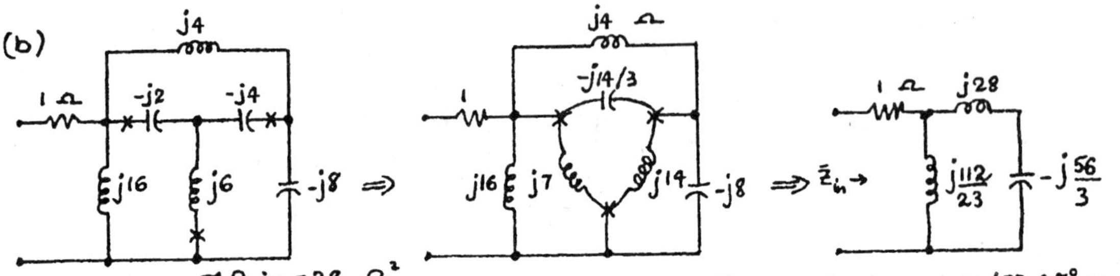


$\Sigma$ Pair = 28 $\Omega^2$

$\therefore\ \bar{Z}_{in} = \underline{1 + j3.2} = \underline{3.353\angle 72.65^\circ\ \Omega}$

15-17 (a)

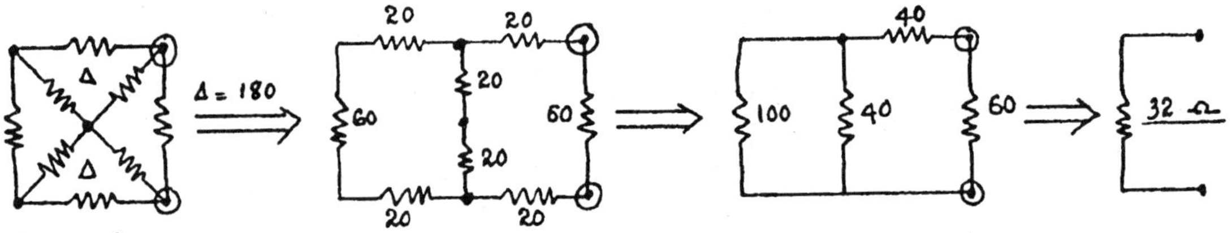


60 each

(b)

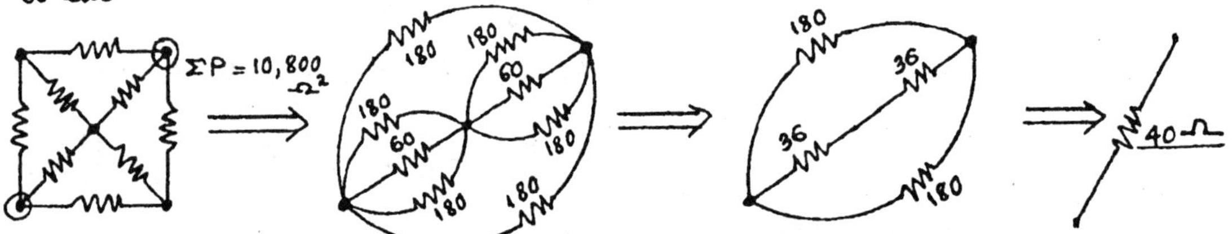


60 each

(c)

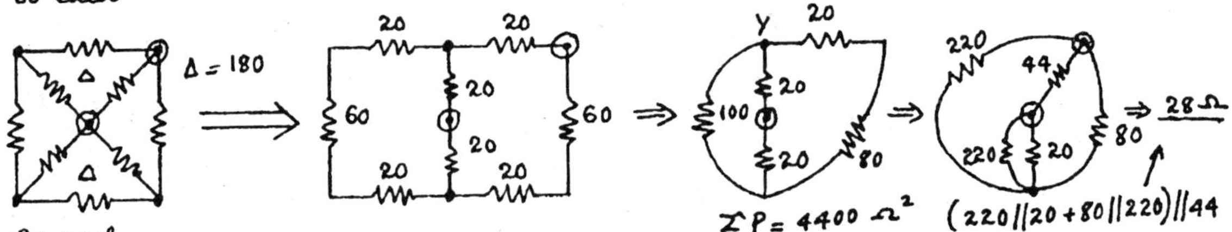


60 each

$\Sigma P = 4400\ \Omega^2$

$(220 \| 20 + 80 \| 220) \| 44$

15-18 (a)

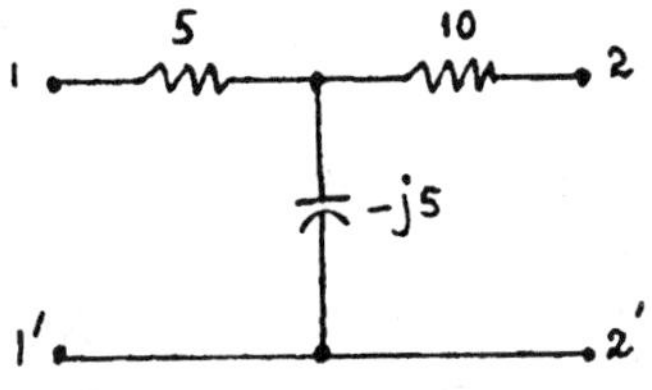


$\Sigma$ Pairs $= 50 - j25 - j50 = 50 - j75$

$\frac{50 - j75}{5} = 10 - j15\ ;\ \frac{50 - j75}{10} = 5 - j7.5$

$\frac{50 - j75}{-j5} = 15 + j10$

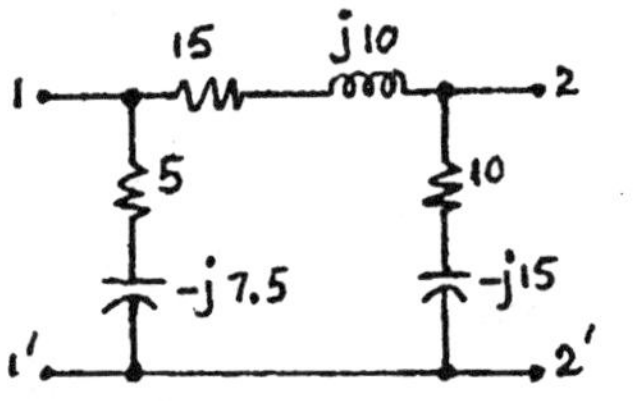


(b)

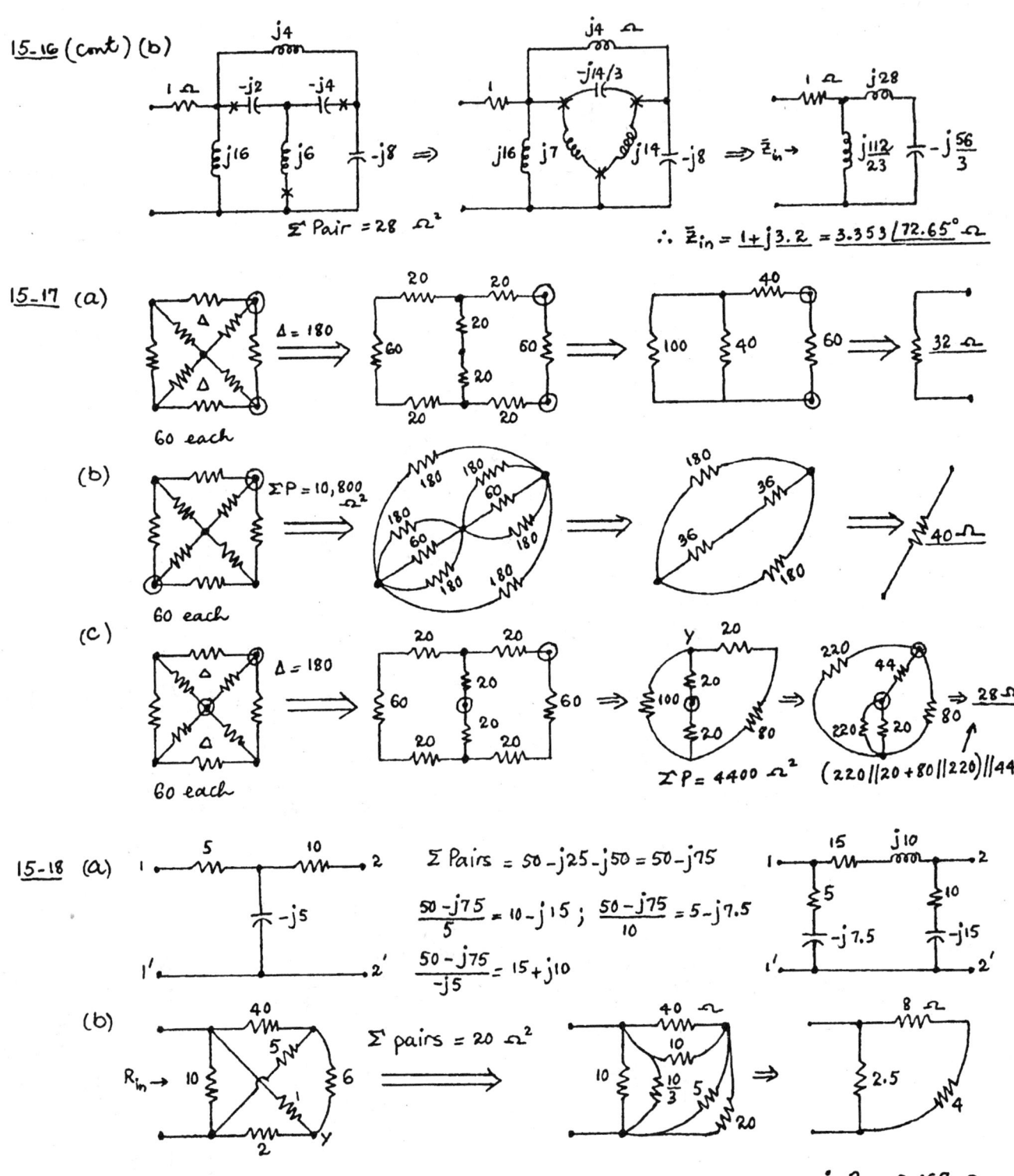

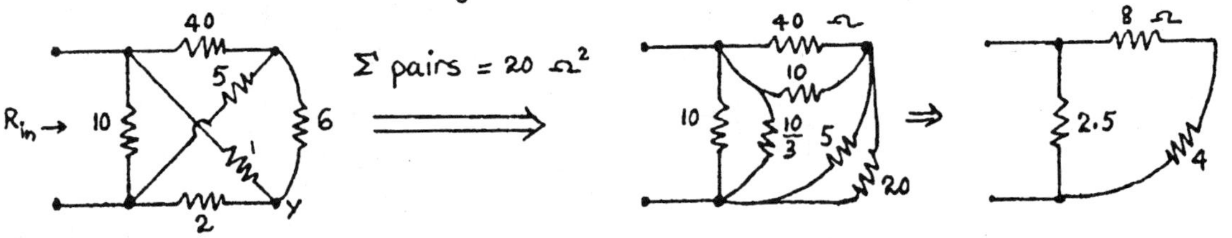


$\therefore\ R_{in} = \underline{2.069\ \Omega}$

**15-19**

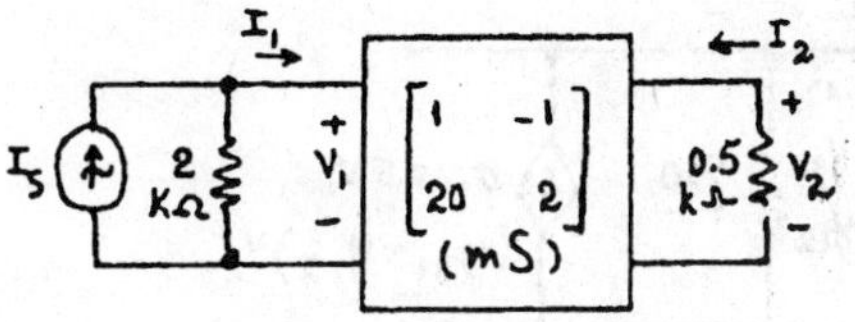


$I_s = I_1 + 0.5V_1$ ; $I_1 = V_1 - V_2$ ; $I_2 = 20V_1 + 2V_2$ ; $I_2 = -2V_2$

(a) $I_2 = -2V_2 = 20V_1 + 2V_2 \therefore G_V = \frac{V_2}{V_1} = \underline{-5}$

(b) $G_I = \frac{I_2}{I_1} = \frac{20 + 2(-5)}{1 - 1(-5)} = \underline{1.6667}$

(c) $I_1 = V_1 - V_2 = V_1 + 5V_1 = 6V_1 \therefore I_s = I_1 + 0.5V_1 = I_1\left(1 + \frac{1}{12}\right) = \frac{13}{12}I_1$ ; $\frac{I_2}{I_s} = \frac{I_2}{I_1}\frac{I_1}{I_s} = \underline{1.5385}$

(d) $\frac{P_L}{P_S} = -\frac{V_2 I_2}{V_1 I_s} = 5 \times 1.5385 = \underline{7.692}$

(e) $Z_{in} = \frac{V_1}{I_1} = \frac{1}{6} = \underline{166.67\ \Omega}$

(f) $I_s = 0 \therefore I_1 = -0.5V_1 = V_1 - V_2 \therefore V_1 = \frac{2}{3}V_2$ , $I_2 = \left(\frac{40}{3} + 2\right)V_2 = \frac{46}{3}V_2 \therefore Z_{out} = \frac{3}{46} = \underline{65.22\ \Omega}$

(g) $\left.\frac{V_1}{V_2}\right|_{rev} = G_{V,rev} = \frac{2}{3} = \underline{0.6667}$

**15-20** (a)

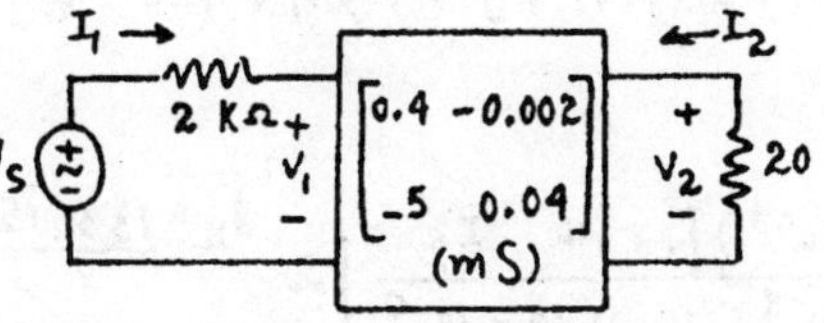


$V_s = 2I_1 + V_1$ , $I_1 = 0.4V_1 - 0.002V_2$ , $I_2 = -5V_1 + 0.04V_2$

and $V_2 = -20I_2 \therefore -0.05V_2 = -5V_1 + 0.04V_2$

$\therefore G_V = \frac{V_2}{V_1} = \frac{5}{0.09} = \underline{55.56}$

(b) $I_1 = 0.4V_1 - \frac{1}{9}V_1 = \frac{2.6}{9}V_1 \therefore V_s = \frac{5.2}{9}V_1 + V_1 = \frac{14.2}{9}V_1$ ; $\frac{V_2}{V_s} = \frac{V_2}{V_1}\frac{V_1}{V_s} = \frac{500}{9}\frac{9}{14.2} = \underline{35.21}$

(c) $\frac{I_2}{I_1} = \frac{-5 + 20/9}{0.4 - 1/9} = \frac{-25}{2.6} = \underline{-9.615}$

(d) $\frac{V_1}{I_1} = \frac{9}{2.6} = \underline{3.462\ k\Omega}$

(e) $I_1 = -0.5V_1 = 0.4V_1 - 0.002V_2 \therefore 0.9V_1 = 0.002V_2$ , $V_1 = \frac{V_2}{450}$ , $I_2 = -\frac{1}{90}V_2 + \frac{1}{25}V_2$

$\therefore R_{out} = \left.\frac{V_2}{I_2}\right|_{V_s=0} = \underline{34.62\ k\Omega}$

**15-21** (a)

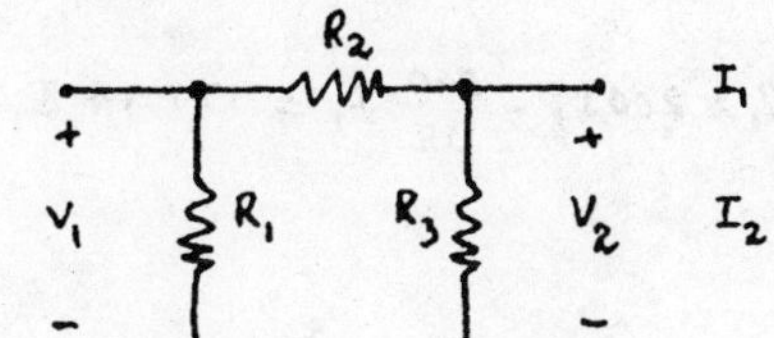


$I_1 = \frac{V_1}{R_1}$, $I_2 = 0$ $\therefore [y]_a = \begin{bmatrix} 1/R_1 & 0 \\ 0 & 0 \end{bmatrix}$

(b)

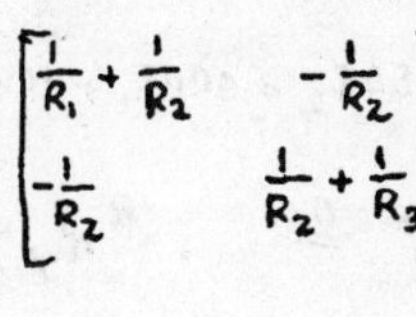


$I_1 = \frac{V_1 - V_2}{R_2}$

$I_2 = \frac{V_2}{R_3} + \frac{V_2 - V_1}{R_2}$

$[y]_b = \begin{bmatrix} 1/R_2 & -1/R_2 \\ -\frac{1}{R_2} & \frac{1}{R_2} + \frac{1}{R_3} \end{bmatrix}$

(c)




$I_1 = \frac{V_1}{R_1} + \frac{V_1 - V_2}{R_2}$

$I_2 = \frac{V_2}{R_3} + \frac{V_2 - V_1}{R_2}$

$\therefore [y] = \begin{bmatrix} \frac{1}{R_1} + \frac{1}{R_2} & -\frac{1}{R_2} \\ -\frac{1}{R_2} & \frac{1}{R_2} + \frac{1}{R_3} \end{bmatrix} = [y]_a + [y]_b$ ✓

**15-22** (a) $[\bar{y}]_A = \begin{bmatrix} 1.5 & -1 \\ 4 & 3 \end{bmatrix}$ (mS)

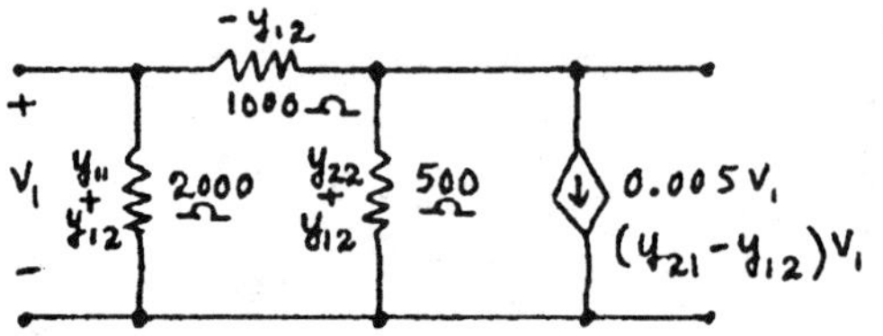


(b)

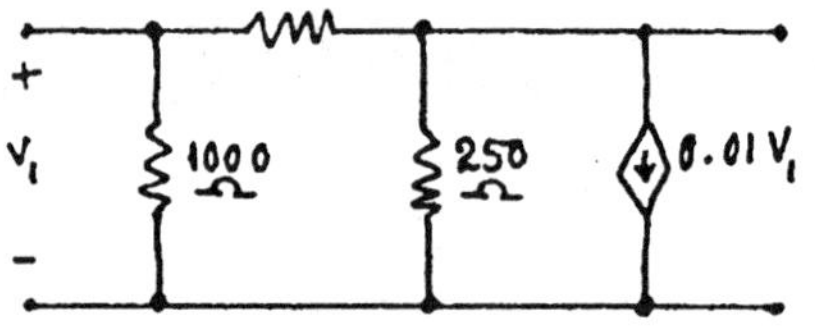


$$\left.\begin{aligned} I_1 &= V_1 + 2(V_1 - V_2) = 3V_1 - 2V_2 \\ I_2 &= 10V_1 + 4V_2 + 2V_2 - 2V_1 \\ &= 8V_1 + 6V_2 \end{aligned}\right\} \therefore [\bar{y}] = \begin{bmatrix} 3 & -2 \\ 8 & 6 \end{bmatrix} = 2[\bar{y}]_A \quad \text{(mS)}$$

**15-23** (a)

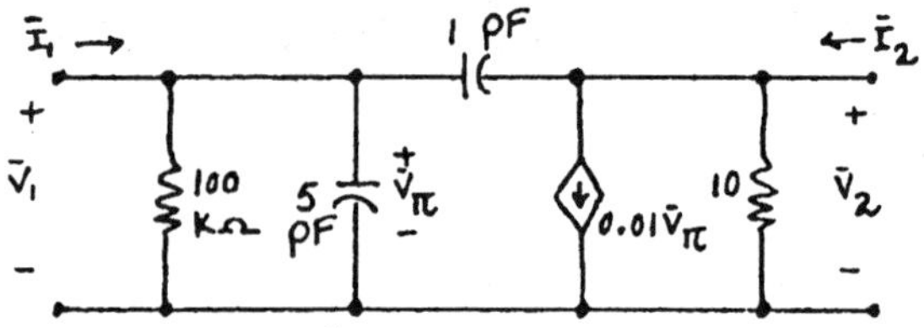


$\omega = 10^8$, $\bar{V}_\pi = \bar{V}_1$

$$\begin{aligned} \bar{I}_1 &= 10^{-5}\bar{V}_1 + j5\times10^{-4}\bar{V}_1 + j10^{-4}(\bar{V}_1 - \bar{V}_2) \\ &= (10^{-5} + j6\times10^{-4})\bar{V}_1 - j10^{-4}\bar{V}_2 \end{aligned}$$

$$\begin{aligned} \bar{I}_2 &= 10^{-4}\bar{V}_2 + 0.01\bar{V}_1 + j10^{-4}(\bar{V}_2 - \bar{V}_1) \\ &= (0.01 + j10^{-4})\bar{V}_1 + (10^{-4} + j10^{-4})\bar{V}_2 \end{aligned}$$

$$\bar{V}_1 = \frac{\begin{vmatrix} \bar{I}_1 & -j10^{-4} \\ \bar{I}_2 & 10^{-4}+j10^{-4} \end{vmatrix}}{\begin{vmatrix} 10^{-5}+j6\times10^{-4} & -j10^{-4} \\ 10^{-2}-j10^{-4} & 10^{-4}+j10^{-4} \end{vmatrix}} = \frac{(10^{-4}+j10^{-4})\bar{I}_1 + j10^{-4}\bar{I}_2}{-4.9\times10^{-8} + j1.0610\times10^{-6}} \quad \therefore \; \begin{aligned} \bar{z}_{11} &= \underline{133.15\angle -47.64^\circ\ \Omega} \\ \bar{z}_{12} &= \underline{94.15^+\angle -2.642^\circ\ \Omega} \end{aligned}$$

$$\bar{V}_2 = \frac{\begin{vmatrix} 10^{-5}+j6\times10^{-4} & \bar{I}_1 \\ 10^{-2}-j10^{-4} & \bar{I}_2 \end{vmatrix}}{-4.9\times10^{-8} + j1.0610\times10^{-6}} = \frac{-(10^{-2}-j10^{-4})\bar{I}_1 + (10^{-5}+j6\times10^{-4})\bar{I}_2}{-4.9\times10^{-8} + j1.0610\times10^{-6}} \quad \therefore \; \begin{aligned} \bar{z}_{21} &= \underline{9416\angle 186.78^\circ\ \Omega} \\ \bar{z}_{22} &= \underline{565.0\angle -3.60^\circ\ \Omega} \end{aligned}$$

(b) All four

**15-24** (b)

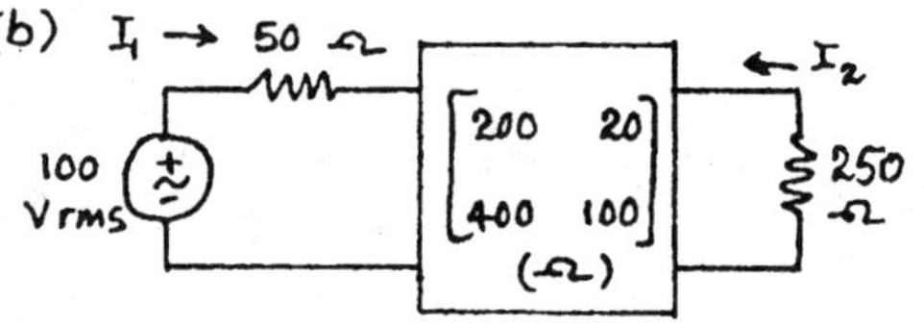


$$\left.\begin{aligned} 100 &= 50I_1 + V_1 \\ V_1 &= 200I_1 + 20I_2 \end{aligned}\right\} \therefore 100 = 250I_1 + 20I_2$$

$$V_2 = 400I_1 + 100I_2 \;\therefore\; I_1 = \frac{V_2}{400} - \frac{I_2}{4}$$

$\therefore V_2 = 160 + 68I_2$

$\therefore V_{2,oc} = \underline{160\text{ V rms}}$

$R_{Th,2} = \underline{68\ \Omega} = R_{out}$

(a) $V_2 = -250I_2 = 400I_1 + 100I_2 \;\therefore\; I_2 = -\frac{400}{350}I_1 \;\therefore\; V_1 = 200I_1 - \frac{800}{35}I_1 = 177.14\,I_1$

$\therefore V_{1,oc} = \underline{0}$, $R_{Th,1} = \underline{177.14\ \Omega} = R_{in}$

**15-25** (a)

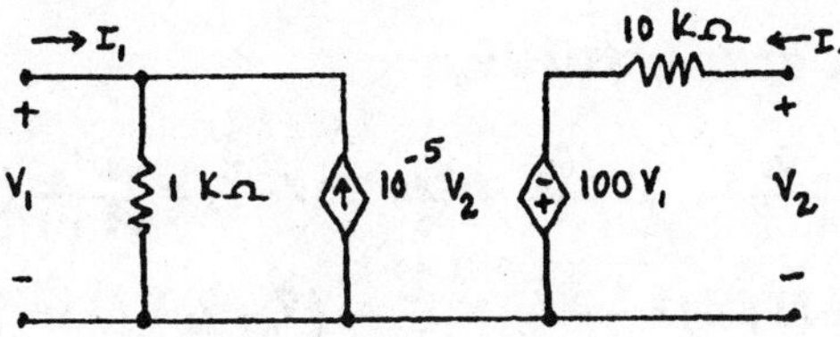


$$\left.\begin{aligned} I_1 &= 10^{-3} V_1 - 10^{-5} V_2 \\ V_2 &= 10^4 I_2 - 100 V_1 \end{aligned}\right\} \therefore V_1 = \underline{500 I_1 + 50 I_2}$$

$$\therefore V_2 = 10^4 I_2 - 100(500 I_1 + 50 I_2) = \underline{-5\times10^4 I_1 + 5000 I_2}$$

$$\therefore [z] = \begin{bmatrix} 500 & 50 \\ -50{,}000 & 5000 \end{bmatrix} \ (\Omega)$$

(b)

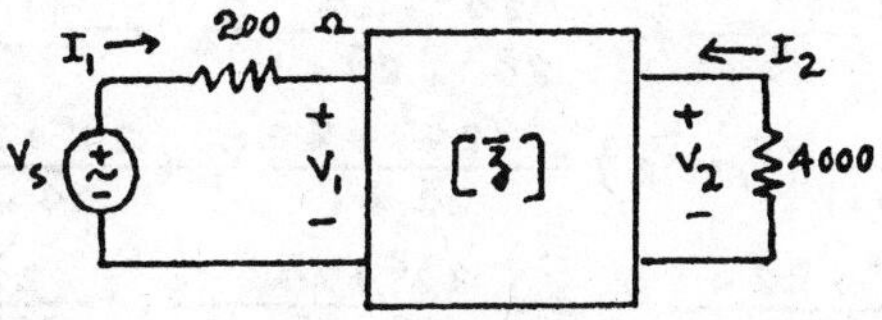


$$V_2 = -4000 I_2 = -50{,}000 I_1 + 5000 I_2 \ \therefore I_2 = \frac{50}{9} I_1$$

$$\therefore V_1 = 500 I_1 + 50\left(\frac{50}{9} I_1\right) = 777.8 I_1,\ R_{in} = \frac{V_1}{I_1} = \underline{777.8\ \Omega}$$

$$V_2 = -50000\left(\frac{V_1}{777.8}\right) + 5000\left(\frac{-V_2}{4000}\right) \therefore \frac{V_2}{V_1} = \underline{-28.57}$$

**15-26** (a)

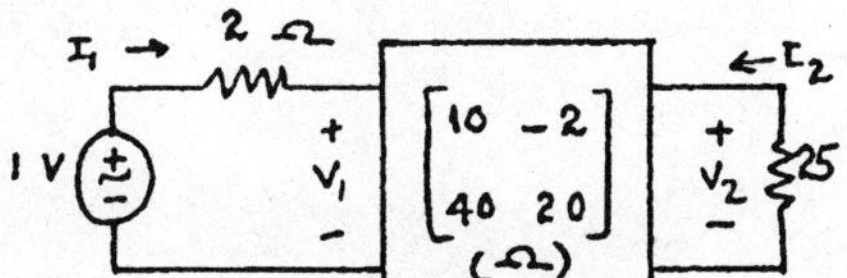


$$\left.\begin{aligned} 1 &= 2I_1 + V_1 \\ V_1 &= 10 I_1 - 2 I_2 \end{aligned}\right\} \therefore I_1 = \frac{1}{12} + \frac{1}{6} I_2$$

$$V_2 = 40 I_1 + 20 I_2$$

$$\therefore V_2 = \frac{10}{3} + \frac{80}{3} I_2$$

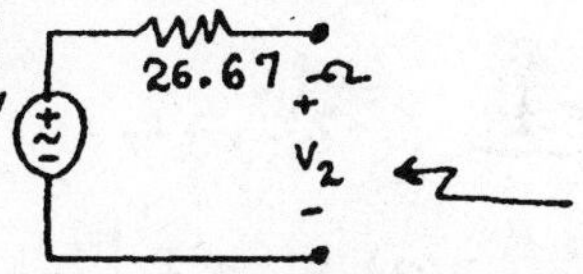


(b) $G_{ins} = \dfrac{P_{L,with}}{P_{L,without}}$ ; $P_{L,with} = \dfrac{1}{2}\left(\dfrac{3.333}{26.67+25}\right)^2 (25)$

$\therefore P_{L,with} = 52.03$ mW ; $P_{L,without} = \dfrac{1}{2}\left(\dfrac{1}{2+25}\right)^2 25 = 17.147$ mW $\therefore G_{ins} = \dfrac{52.03}{17.147} = \underline{3.034}$

(c) $V_1 = -2I_1 \ \therefore V_1 = 10\left(-\dfrac{V_1}{2}\right) - 2I_2$ or $V_1 = -\dfrac{1}{3} I_2$ ; $V_2 = 40\left(-\dfrac{V_1}{2}\right) + 20 I_2 = \dfrac{20}{3} I_2 + 20 I_2$

$\therefore G_{V,rev} = \dfrac{V_1}{V_2} = -\dfrac{1}{3} I_2 \Big/ (80 I_2/3) = -\dfrac{1}{80} = \underline{-0.0125}$

**15-27** (a)

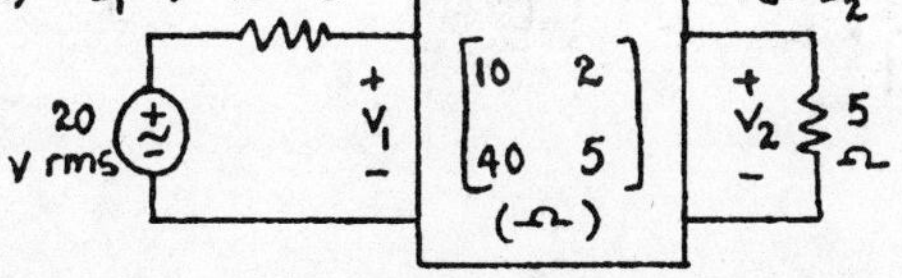


$$20 = 10 I_1 + V_1,\ V_1 = 10 I_1 + 2 I_2,\ V_2 = 40 I_1 + 5 I_2 = -5 I_2$$

$$\therefore I_1 = -0.25 I_2$$

$$20 = -2.5 I_2 + (-2.5 I_2 + 2 I_2) = -3 I_2 \ \therefore I_2 = -\frac{20}{3}$$

$$P_L = \left(\frac{20}{3}\right)^2 5 = \underline{222.2\ W}$$

(b) $V_1 = -10 I_1 = 10 I_1 + 2 I_2 \ \therefore I_1 = -0.1 I_2$, $V_2 = -4 I_2 + 5 I_2 = I_2$ , $Z_{th} = \dfrac{V_2}{I_2} = \underline{1\ \Omega}$

(c) $V_2 = -I_2 = 40 I_1 + 5 I_2 \ \therefore I_1 = -0.15 I_2$, $20 = -1.5 I_2 + (-1.5 I_2 + 2 I_2) = -I_2 \ \therefore I_2 = -20$ A

$P_L = 20^2 \times 1 = \underline{400\ W}$

**15-28**

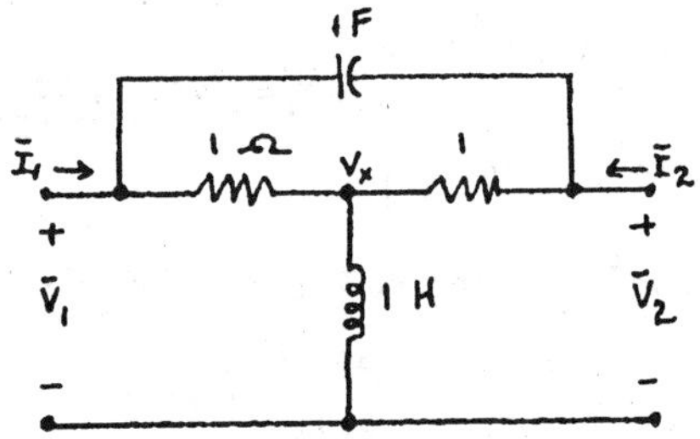

$$\bar{I}_1 = \frac{\bar{V}_1 - \bar{V}_x}{1} + \bar{\lambda}(\bar{V}_1 - \bar{V}_2)$$

$$\frac{\bar{V}_x - \bar{V}_1}{1} + \frac{\bar{V}_x - \bar{V}_2}{1} + \frac{\bar{V}_x}{\bar{\lambda}} = 0 \therefore \bar{V}_x = \frac{\bar{\lambda}}{2\bar{\lambda}+1}(\bar{V}_1 + \bar{V}_2)$$

$$\therefore \bar{I}_1 = (1+\bar{\lambda})\bar{V}_1 - \bar{\lambda}\bar{V}_2 - \frac{\bar{\lambda}}{2\bar{\lambda}+1}(\bar{V}_1 + \bar{V}_2)$$

$$\bar{I}_1 = \frac{2\bar{\lambda}^2 + 2\bar{\lambda} + 1}{2\bar{\lambda}+1}\bar{V}_1 - \frac{2\bar{\lambda}^2 + 2\bar{\lambda}}{2\bar{\lambda}+1}\bar{V}_2$$

$$\therefore \bar{V}_1 = \frac{2\bar{\lambda}+1}{2\bar{\lambda}^2 + 2\bar{\lambda} + 1}\bar{I}_1 + \frac{2\bar{\lambda}^2 + 2\bar{\lambda}}{2\bar{\lambda}^2 + 2\bar{\lambda} + 1}\bar{V}_2 \leftarrow$$

$$\bar{I}_2 = \frac{\bar{V}_2 - \bar{V}_x}{1} + \bar{\lambda}(\bar{V}_2 - \bar{V}_1) = -\bar{\lambda}\bar{V}_1 + (\bar{\lambda}+1)\bar{V}_2 - \frac{\bar{\lambda}}{2\bar{\lambda}+1}(\bar{V}_1 + \bar{V}_2) = -\frac{2\bar{\lambda}^2+2\bar{\lambda}}{2\bar{\lambda}+1}\bar{V}_1 + \frac{2\bar{\lambda}^2+2\bar{\lambda}+1}{2\bar{\lambda}+1}\bar{V}_2$$

$$\bar{I}_2 = -\left(\frac{2\bar{\lambda}^2+2\bar{\lambda}}{2\bar{\lambda}+1}\right)\left[\frac{(2\bar{\lambda}+1)\bar{I}_1 + (2\bar{\lambda}^2+2)\bar{V}_2}{2\bar{\lambda}^2+2\bar{\lambda}+1}\right] + \frac{2\bar{\lambda}^2+2\bar{\lambda}+1}{2\bar{\lambda}+1}\bar{V}_2 = -\frac{2\bar{\lambda}^2+2\bar{\lambda}}{2\bar{\lambda}^2+2\bar{\lambda}+1}\bar{I}_1 + \frac{2\bar{\lambda}+1}{2\bar{\lambda}^2+2\bar{\lambda}+1}\bar{V}_2$$

$$[\bar{h}] = \begin{bmatrix} \dfrac{2\bar{\lambda}+1}{2\bar{\lambda}^2+2\bar{\lambda}+1} & \dfrac{2\bar{\lambda}^2+2\bar{\lambda}}{2\bar{\lambda}^2+2\bar{\lambda}+1} \\ -\dfrac{2\bar{\lambda}^2+2\bar{\lambda}}{2\bar{\lambda}^2+2\bar{\lambda}+1} & \dfrac{2\bar{\lambda}^2+1}{2\bar{\lambda}^2+2\bar{\lambda}+1} \end{bmatrix} \leftarrow$$

**15-29** $\quad 2V_1 + 4I_2 = I_1,\quad 8I_2 = V_2 + 6V_1$

(a) $y_{11} = I_1\big|_{V_2=0, V_1=1} \therefore I_2 = \frac{6}{8},\ I_1 = 2\times 1 + 4\left(\frac{6}{8}\right) = \underline{5\ S} = y_{11}$

(b) $z_{21} = V_2\big|_{I_2=0, I_1=1} \therefore V_1 = \frac{1}{2},\ 0 = V_2 + 3 \therefore V_2 = \underline{-3\ \Omega} = z_{21}$

(c) $h_{21} = I_2\big|_{V_2=0, I_1=1} \therefore 8I_2 = 6V_1 = 6\left(\frac{1-4I_2}{2}\right) = 3 - 12I_2 \therefore I_2 = \underline{\frac{3}{20}} = h_{21}$

**15-30**

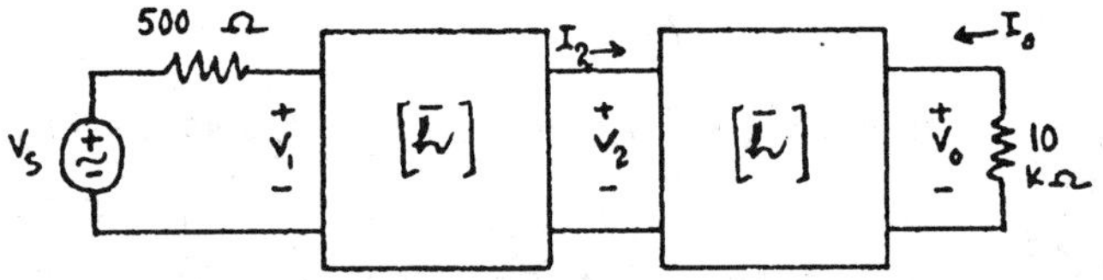

$$[\bar{h}] = \begin{bmatrix} 1\ K\Omega & 0.001 \\ 100 & 0.1\ mS \end{bmatrix}$$

2nd stage: $V_2 = I_2 + 10^{-3}V_0$
$I_0 = 100 I_2 + 0.1 V_0$
$V_0 = -10 I_0$ $\Big\} \therefore I_0 = 50 I_2 \therefore V_2 = I_2 + 10^{-3}(-10)(50)I_2,\ Z_{in,2} = 0.5\ K\Omega$
$\qquad\qquad = 0.5 I_2$

1rst stage: $V_S = 0.5 I_1 + V_1$
$V_1 = I_1 + 0.001 V_2$ $\Big\} \therefore V_S = 1.5 I_1 + 10^{-3} V_2$
$V_2 = 0.5 I_2$
$-I_2 = 100 I_1 + 0.1 V_2$ $\Big\} \therefore I_1 = -0.0105 I_2 = -\frac{0.0105}{0.5}V_2$

2nd stage + 1rst stage
$\therefore V_2 = -\frac{V_S}{0.0305} = I_2 + 10^{-3}V_0$

Also, $I_0 = -0.1 V_0 = 100 I_2 + 0.1 V_0 \therefore I_2 = -0.002 V_0 \therefore -\frac{V_S}{0.0305} = -0.002 V_0 + 0.001 V_0$

$\therefore V_S = -0.0305(-0.002 V_0 + 0.001 V_0),\ V_0 = 25\ V \therefore V_S = \underline{0.7625\ mV}$

**15-31** $[\bar{h}] = \begin{bmatrix} 10\ \Omega & -1 \\ 4 & 0.05\ \text{S} \end{bmatrix}$, $R_L = 10\ \Omega$, $V_s = 100$ V rms, $R_s = 2\ \Omega$

(a) $V_s = 100 = 2I_1 + V_1 = 2I_1 + 10I_1 - V_2 = 12I_1 - V_2$ ; $I_2 = 4I_1 + 0.05V_2 = -0.1V_2$

$\therefore 4I_1 = -0.15V_2$ or $V_2 = -\frac{80}{3}I_1$ $\therefore 100 = 12I_1 + \frac{80}{3}I_1 = \frac{116}{3}I_1$ $\therefore I_1 = \frac{300}{116}$ A rms

$P_{2\Omega} = \left(\frac{300}{116}\right)^2 2 = \underline{13.377\ \text{W}}$

(b) $V_2 = -\frac{80}{3}I_1 = -\frac{80}{3}\left(\frac{300}{116}\right) = -68.97$ V $\therefore P_{10\Omega} = 68.97^2 \times 10 = \underline{475.6\ \text{W}}$

(c) $P_{s,gen} = \frac{300}{116} \times 100 = 258.6$ W, $P_{2\,port} = 258.6 - 13.4 - 475.6 = \underline{-230.4\ \text{W}}$

**15-32**

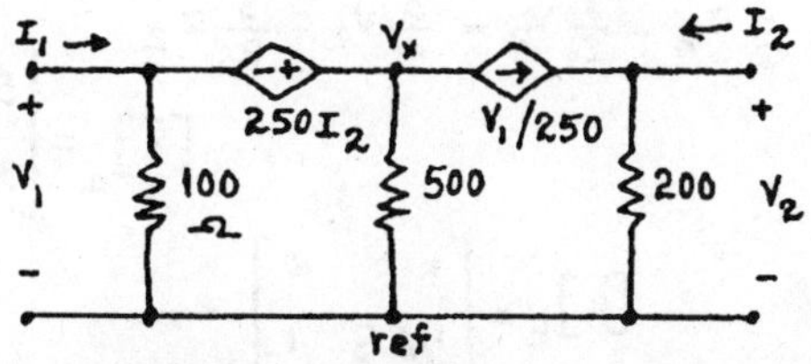


$I_1 = 0.01V_1 + 0.002V_x + 0.004V_1 = 0.014V_1 + 0.002V_x$

$I_2 = 0.005V_2 - 0.004V_1$, $V_x = 250I_2 + V_1$

$\therefore I_1 = 0.014V_1 + 0.002(250I_2 + V_1) = 0.016V_1 + 0.5I_2$

$\therefore I_1 = 0.016V_1 + 0.5(0.005V_2 - 0.004V_1) = 0.014V_1 + 0.0025V_2$ $\therefore V_1 = \frac{1}{0.014}I_1 - \frac{0.0025}{0.014}V_2$

$\therefore h_{11} = \underline{71.43\ \Omega}$, $h_{12} = \underline{-0.17857}$ ; $I_2 = 0.005V_2 - 0.004\left(\frac{1}{0.014}I_1 - \frac{0.0025}{0.014}V_2\right)$

$= -\frac{2}{7}I_1 + \frac{0.04}{7}V_2$

$\therefore h_{21} = \underline{-0.2857}$, $h_{22} = \underline{0.00571\ \text{S}}$

**15-33**

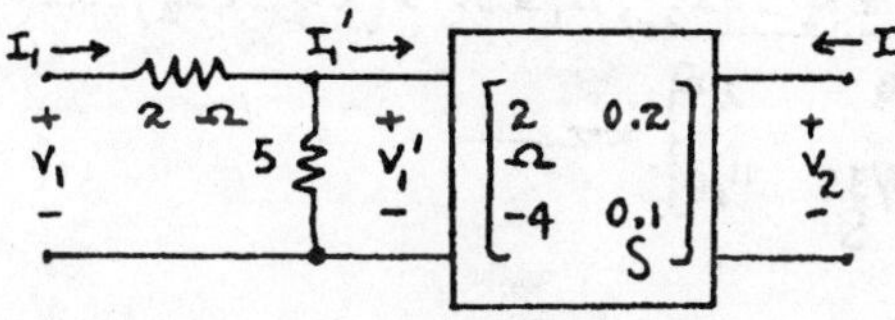


$V_1' = 2I_1' + 0.2V_2$, $I_2 = -4I_1' + 0.1V_2$

$V_1 = 2I_1 + V_1'$ $\therefore V_1' = V_1 - 2I_1$
$I_1 = 0.2V_1' + I_1'$
$\Big\}$ $\therefore I_1' = 1.4I_1 - 0.2V_1$

$V_1' = V_1 - 2I_1 = 2(1.4I_1 - 0.2V_1) + 0.2V_2$

$\therefore V_1 = \frac{4.8}{1.4}I_1 + \frac{0.2}{1.4}V_2$ $\therefore h_{11} = \underline{3.429\ \Omega}$, $h_{12} = \underline{0.14286}$

$I_2 = -4(1.4I_1) + 0.8\left(\frac{4.8}{1.4}I_1 + \frac{0.2}{1.4}V_2\right) + 0.1V_2 = -\frac{4}{1.4}I_1 + \frac{0.3}{1.4}V_2$

$\therefore h_{21} = \underline{-2.857}$, $h_{22} = \underline{0.2143\ \text{S}}$ $\therefore [\bar{h}] = \begin{bmatrix} 3.43\ (\Omega) & 0.1429 \\ -2.86 & 0.214\ (\text{S}) \end{bmatrix}$

**15-34** (a)

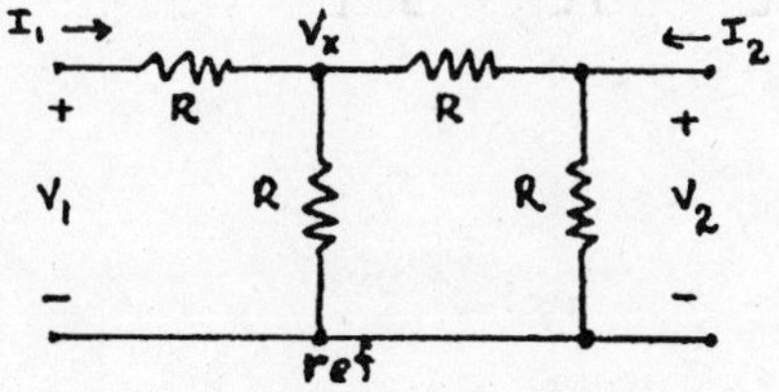


$\frac{V_x - V_1}{R} + \frac{V_x}{R} + \frac{V_x - V_2}{R} = 0$ $\therefore V_x = \frac{1}{3}V_1 + \frac{1}{3}V_2$

$I_1 = \frac{V_1 - V_x}{R} = \frac{1}{R}\left(V_1 - \frac{1}{3}V_1 - \frac{1}{3}V_2\right) = \frac{2}{3R}V_1 - \frac{1}{3R}V_2$

$I_2 = \frac{1}{R}(V_2 + V_2 - V_x) = \frac{1}{R}\left(2V_2 - \frac{1}{3}V_1 - \frac{1}{3}V_2\right)$

(cont. on next page)

**15-34 (cont)** $\therefore I_2 = -\frac{1}{3R}V_1 + \frac{5}{3R}V_2 \therefore V_1 = \underline{5V_2 - 3RI_2}$ ; $I_1 = \frac{2}{3R}V_1 - \frac{1}{3R}V_2 = \frac{2}{3R}(5V_2 - 3RI_2) - \frac{1}{3R}V_2$

$\therefore I_1 = \underline{\frac{3}{R}V_2 - 2I_2}$ $\therefore [\bar{t}] = \begin{bmatrix} 5 & 3R \\ 3/R & 2 \end{bmatrix}$ ←

(b) $[\bar{h}] = \begin{bmatrix} 20 & 0.3 \\ -4 & 0.1 \end{bmatrix} \therefore V_1 = 20I_1 + 0.3V_2$ , $I_2 = -4I_1 + 0.1V_2 \therefore I_1 = 0.025V_2 - 0.25I_2$

$V_1 = 20(0.025V_2 - 0.25I_2) + 0.3V_2 = 0.8V_2 - 5I_2 \therefore [\bar{t}] = \begin{bmatrix} 0.8 & 5\,\Omega \\ 0.025\text{ S} & 0.25 \end{bmatrix}$ ←

**15-35** (a)

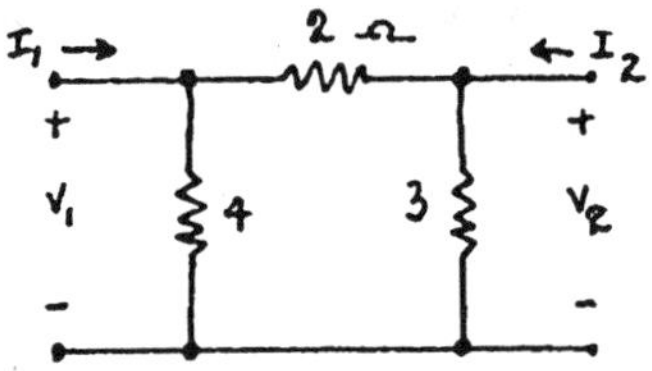


$I_1 = 0.25V_1 + 0.5(V_1 - V_2) = 0.75V_1 - 0.5V_2$

$I_2 = \frac{1}{3}V_2 + \frac{1}{2}(V_2 - V_1) = -\frac{1}{2}V_1 + \frac{5}{6}V_2 \therefore V_1 = \underline{\frac{5}{3}V_2 - 2I_2}$

$I_1 = \frac{5}{4}V_2 - \frac{3}{2}I_2 - \frac{1}{2}V_2 = \underline{\frac{3}{4}V_2 - \frac{3}{2}I_2}$

$\therefore [\bar{t}]_a = \begin{bmatrix} 5/3 & 2\,\Omega \\ 3/4\text{ S} & 3/2 \end{bmatrix}$ ←

(b)

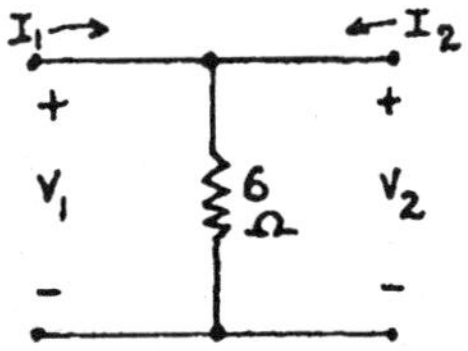


$V_1 = V_2$

$I_1 = \frac{1}{6}V_2 - I_2$ $\therefore [\bar{t}]_b = \begin{bmatrix} 1 & 0 \\ 1/6\text{ S} & 1 \end{bmatrix}$ ←

(c) $[\bar{t}]_a[\bar{t}]_b = \begin{bmatrix} 5/3 & 2 \\ 3/4 & 3/2 \end{bmatrix}\begin{bmatrix} 1 & 0 \\ 1/6 & 1 \end{bmatrix} = \begin{bmatrix} 2 & 2\,\Omega \\ 1\text{ S} & 1.5 \end{bmatrix}$ ←

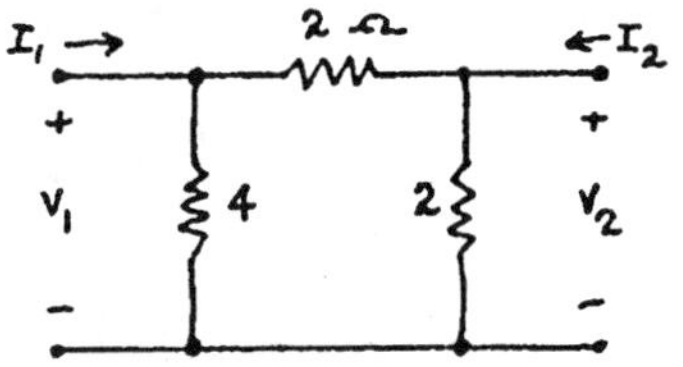


$I_1 = 0.25V_1 + 0.5(V_1 - V_2) = \underline{0.75V_1 - 0.5V_2}$

$I_2 = 0.5V_2 + 0.5(V_2 - V_1) = V_2 - 0.5V_1 \therefore V_1 = \underline{2V_2 - 2I_2}$ ; $I_1 = 0.75(2V_2 - 2I_2) - 0.5V_2 = \underline{V_2 - 1.5I_2}$

(d) $[\bar{t}]_b[\bar{t}]_a = \begin{bmatrix} 1 & 0 \\ 1/6 & 1 \end{bmatrix}\begin{bmatrix} 5/3 & 2 \\ 3/4 & 3/2 \end{bmatrix} = \begin{bmatrix} 5/3 & 2\,\Omega \\ 37/36\text{ S} & 11/6 \end{bmatrix}$ ←

**15-36** (a)

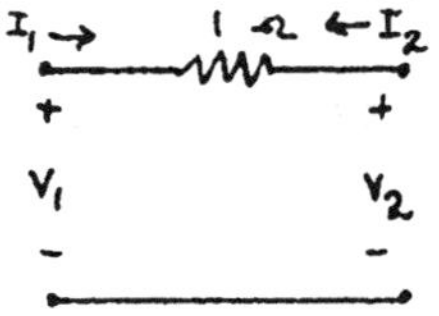


$V_1 = V_2 - I_2$

$I_1 = -I_2$

$[\bar{t}]_a = \begin{bmatrix} 1 & 1 \\ 0 & 1 \end{bmatrix}$

(b)

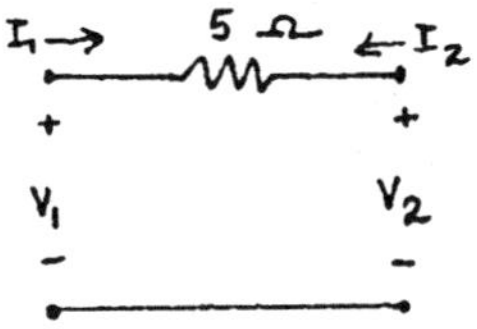


$V_1 = V_2 - 5I_2$

$I_1 = -I_2$

$[\bar{t}]_b = \begin{bmatrix} 1 & 5 \\ 0 & 1 \end{bmatrix}$

(b) $\begin{bmatrix} 1 & 1 \\ 0 & 1 \end{bmatrix}^2 = \begin{bmatrix} 1 & 1 \\ 0 & 1 \end{bmatrix}\begin{bmatrix} 1 & 1 \\ 0 & 1 \end{bmatrix} = \begin{bmatrix} 1 & 2 \\ 0 & 1 \end{bmatrix}$ ; $\begin{bmatrix} 1 & 1 \\ 0 & 1 \end{bmatrix}^4 = \begin{bmatrix} 1 & 2 \\ 0 & 1 \end{bmatrix}\begin{bmatrix} 1 & 2 \\ 0 & 1 \end{bmatrix} = \begin{bmatrix} 1 & 4 \\ 0 & 1 \end{bmatrix}$

$\begin{bmatrix} 1 & 1 \\ 0 & 1 \end{bmatrix}^5 = \begin{bmatrix} 1 & 4 \\ 0 & 1 \end{bmatrix}\begin{bmatrix} 1 & 1 \\ 0 & 1 \end{bmatrix} = \begin{bmatrix} 1 & 5 \\ 0 & 1 \end{bmatrix}$ ←

**15-37**

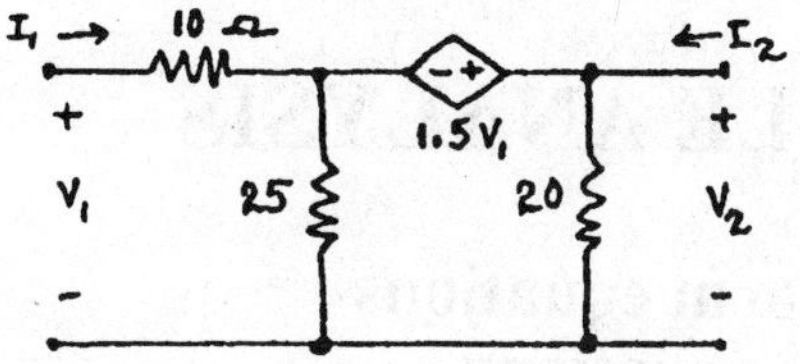


$V_1 = 10I_1 - 1.5V_1 + V_2 \therefore 10I_1 = 2.5V_1 - V_2$

$\frac{V_2 - 1.5V_1 - V_1}{10} + \frac{V_2 - 1.5V_1}{25} + \frac{V_2}{20} = I_2$

$\therefore 0.19V_2 - 0.31V_1 = I_2,\ V_1 = 0.6129V_2 - 3.226I_2$

$\therefore I_1 = \frac{2.5}{10}(0.6129V_2 - 3.226I_2) - \frac{V_2}{10} = 0.05323V_2 - 0.8065I_2 \therefore [t] = \begin{bmatrix} 0.613 & 3.23\ \Omega \\ 0.0532\ \text{S} & 0.806 \end{bmatrix}$ ←

**15-38** (a)

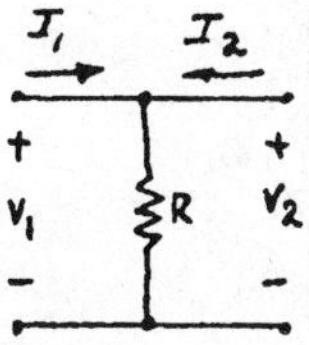


$V_1 = V_2$

$I_1 = \frac{V_2}{R} - I_2$

$[t] = \begin{bmatrix} 1 & 0 \\ 1/R & 1 \end{bmatrix}$ ←

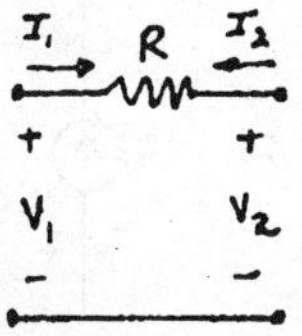


$V_1 = V_2 - RI_2$

$I_1 = -I_2$

$[t] = \begin{bmatrix} 1 & R \\ 0 & 1 \end{bmatrix}$ ←

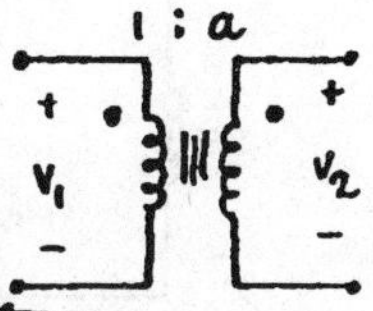


$V_1 = V_2/a$

$I_1 = -aI_2$

$[t] = \begin{bmatrix} 1/a & 0 \\ 0 & a \end{bmatrix}$ ←

(b)

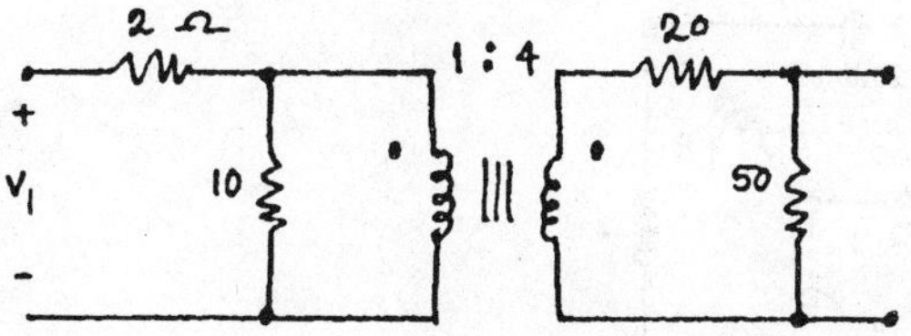


$[t] = \begin{bmatrix} 1 & 2 \\ 0 & 1 \end{bmatrix}\begin{bmatrix} 1 & 0 \\ 0.1 & 1 \end{bmatrix}\begin{bmatrix} 0.25 & 0 \\ 0 & 4 \end{bmatrix}\begin{bmatrix} 1 & 20 \\ 0 & 1 \end{bmatrix}\begin{bmatrix} 1 & 0 \\ 0.02 & 1 \end{bmatrix}$

$= \begin{bmatrix} 1.2 & 2 \\ 0.1 & 1 \end{bmatrix}\begin{bmatrix} 0.25 & 5 \\ 0 & 4 \end{bmatrix}\begin{bmatrix} 1 & 0 \\ 0.02 & 1 \end{bmatrix}$

$[t] = \begin{bmatrix} 0.3 & 14 \\ 0.025 & 4.5 \end{bmatrix}\begin{bmatrix} 1 & 0 \\ 0.02 & 1 \end{bmatrix} = \begin{bmatrix} 0.58 & 14\ \Omega \\ 0.115\ \text{S} & 4.5 \end{bmatrix}$ ←

**15-39** (a)

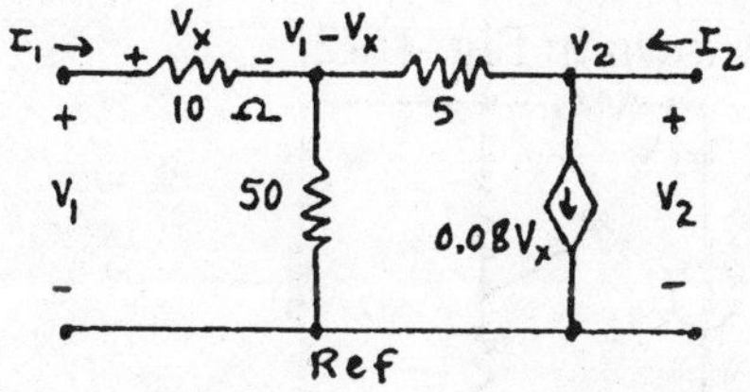


$I_1 = 0.1V_x$

$-0.1V_x + 0.02(V_1 - V_x) + 0.2(V_1 - V_x - V_2) = 0$

$\therefore 0.32V_x = 0.22V_1 - 0.2V_2,\ V_x = \frac{11}{16}V_1 - \frac{5}{8}V_2 \therefore I_1 = \frac{11}{160}V_1 - \frac{1}{16}V_2$

$I_2 = 0.08V_x + 0.2(V_2 - V_1 + V_x)$

$\therefore I_2 = 0.28\left(\frac{11}{16}V_1 - \frac{5}{8}V_2\right) + 0.2V_2 - 0.2V_1 \therefore I_2 = -\frac{3}{400}V_1 + \frac{1}{40}V_2 \therefore V_1 = \frac{10}{3}V_2 - \frac{400}{3}I_2$

$I_1 = \frac{11}{160}\left(\frac{10}{3}V_2 - \frac{400}{3}I_2\right) - \frac{1}{16}V_2 = \frac{1}{6}V_2 - \frac{55}{6}I_2 \quad \therefore [t] = \begin{bmatrix} 3.333 & 133.33\ \Omega \\ 0.16667\ \text{S} & 9.167 \end{bmatrix}$ ←

(b) $[t] = \begin{bmatrix} 10/3 & 400/3 \\ 1/6 & 55/6 \end{bmatrix}\begin{bmatrix} 1 & 0 \\ 1/20 & 1 \end{bmatrix} = \begin{bmatrix} 10 & 133.33\ \Omega \\ 0.625\ \text{S} & 9.167 \end{bmatrix}$ ←

**15-40**

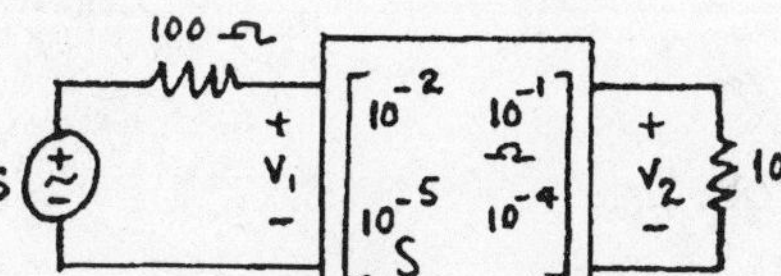


$V_1 = 10^{-2}V_2 - 10^{-1}I_2,\ I_1 = 10^{-5}V_2 - 10^{-4}I_2 \therefore V_1 = 10^3 I_1$

$\therefore Z_{in} = \frac{V_1}{I_1} = \underline{1\ k\Omega}$; $V_2 = -100I_2$

$\therefore V_1 = 10^{-2}V_2 - 10^{-1}(-10^{-2}V_2) = 0.011V_2,\ G_V = \frac{V_2}{V_1} = \underline{90.91}$

$I_1 = 10^{-5}(-10^2 I_2) - 10^{-4}I_2 = -0.0011I_2 \therefore G_I = \frac{I_2}{I_1} = -\frac{1}{0.0011} = \underline{-90.91}$; $G_p = -G_V G_I = \underline{82640}$

For $Z_{out}$: $V_1 = -100I_1 \therefore -100(10^{-5}V_2 - 10^{-4}I_2) = 10^{-2}V_2 - 10^{-1}I_2 \therefore 0.11I_2 = 0.011V_2$

$Z_{out} = \frac{V_2}{I_2} = \underline{10\ \Omega}$; $\Delta_t = 10^{-6} - 10^{-6} = 0 \therefore$ Not reciprocal

# CHAPTER 16: STATE-VARIABLE ANALYSIS

## Section 16-2 State variables and normal-form equations
## Section 16-3 Writing a set of normal-form equations

**1** Draw a normal tree for the circuit shown in: (*a*) Fig. 16-1*a*; (*b*) Fig. 16-1*b*; (*c*) Fig. 16-1*c*.

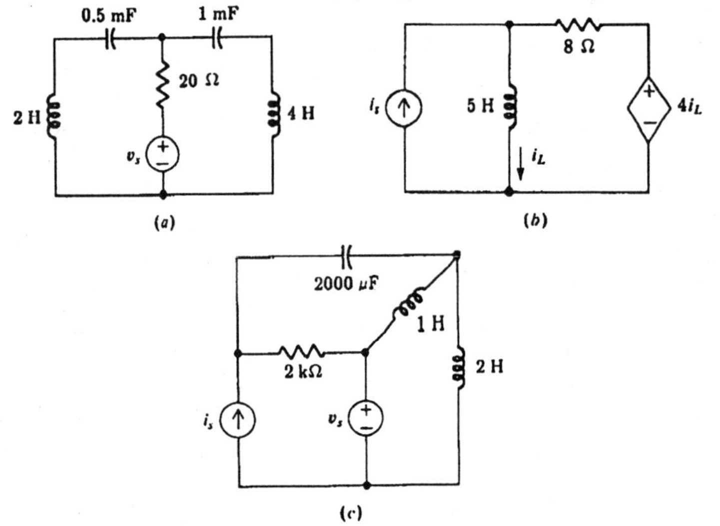


**Fig. 16-1:** See Probs. 1 and 3.

**2** Construct all possible normal trees for each network shown in Fig. 16-2.

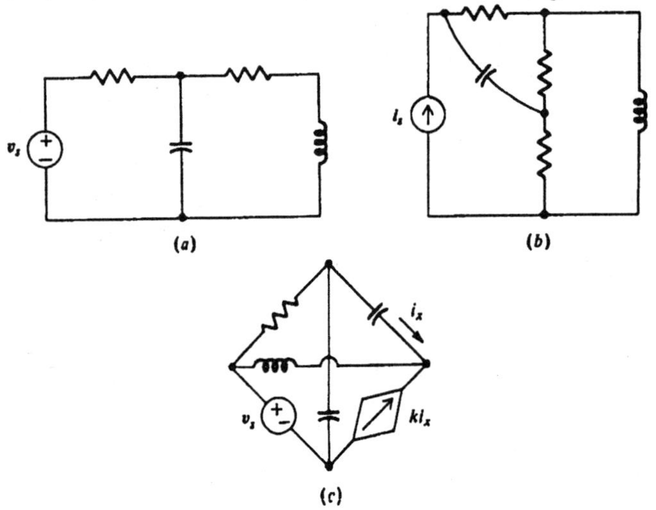


**Fig. 16-2:** See Prob. 2.

**3** Write a set of normal-form equations for the circuit shown in: (*a*) Fig. 16-1*a*; (b) Fig. 16-1*b*; (*c*) Fig. 16-1*c*.

**4** Write each of the following sets of equations in normal form using the variable order indicated: (*a*) order: $i$, $v$; $2i - 3v = 4i' - 5v' + f(t)$, $3 - v' + i' = 2i - v + g(t)$; (*b*) order: $x_1, x_2, x_3$; $x_1' = x_2' - x_3 + \sin 2t$, $x_3' - x_2' = x_1' + x_2$, $x_1 + x_2' + 3x_3' = 0$.

**5** (*a*) Write the following three equations in normal form using the order $x$, $y$, $z$: $x' = 2x - y + z + t^2 + 1$, $y' = x' - 2y$, $z' = x + 2z - 2y'$. (*b*) If $x(0) = 1$, $y(0) = 2$, and $z(0) = 3$, find $x'(0)$, $y'(0)$, and $z'(0)$. (*c*) Knowing $x'(0)$, $y'(0)$, and $z'(0)$, estimate $x(0.1)$, $y(0.1)$, and $z(0.1)$.

**6** Write a set of normal-form equations for each circuit shown in Fig. 16-3. Identify your variables on a circuit diagram and specify the order in which the state variables are used.

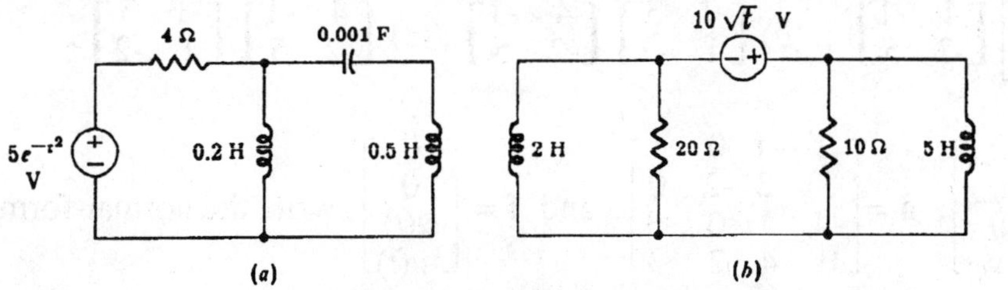


**Fig. 16-3:** See Prob. 6.

**7** Find a system of normal-form equations for the circuit shown in Fig. 16-4. Let the order of the state variables be $v$ and $i$.

**8** Determine a suitable tree for the circuit shown in Fig. 16-5, select a state variable, and write the corresponding normal-form equation.

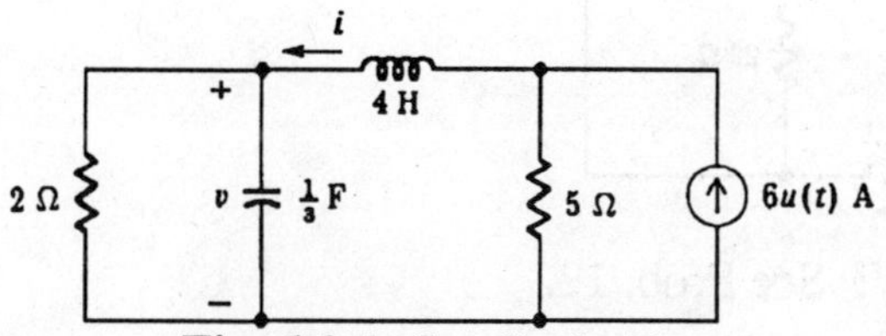


**Fig. 16-4:** See Prob. 7.

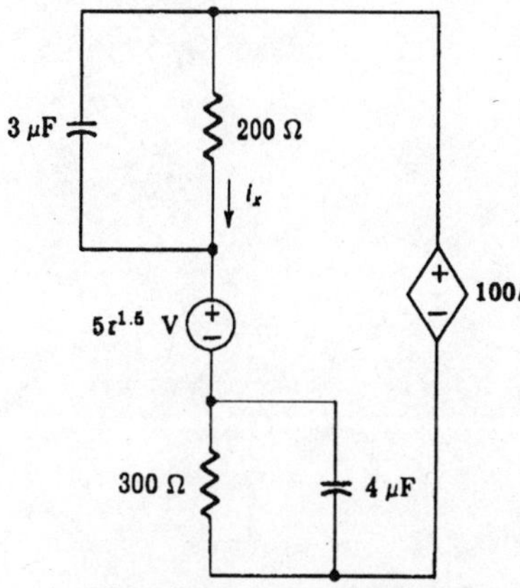


**Fig. 16-5:** See Prob. 8.

**9** For the circuit illustrated in Fig. 16-6: (*a*) draw a normal tree in which the 5-Ω resistor is part of the tree; (*b*) assign state variables $v_C$ and $i_L$ and resistor variables $v_R$ and $i_R$; (*c*) determine the normal-form equations with the state variables ordered $v_C$, $i_L$.

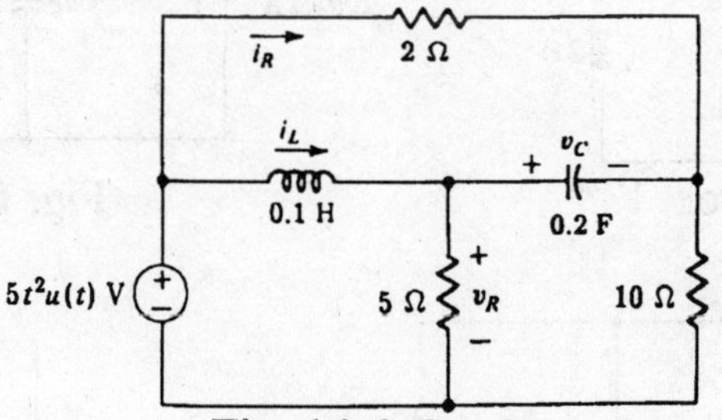


**Fig. 16-6:** See Prob. 9.

## Section 16-4 The use of matrix notation
## Section 16-5 Solution of the first-order equation

**10** Find the matrix that is equal to the indicated product:

$$(a)\begin{bmatrix} 2 & 0 \\ -1 & 3 \\ 1 & -2 \end{bmatrix}\begin{bmatrix} 4 & -1 \\ -2 & 5 \end{bmatrix}\ ;\ (b)\begin{bmatrix} -1 & 3 \\ 1 & -2 \end{bmatrix}\begin{bmatrix} 4 & -1 \\ -2 & 5 \end{bmatrix};\ (c)\begin{bmatrix} 4 & -1 \\ -2 & 5 \end{bmatrix}\begin{bmatrix} -1 & 3 \\ 1 & -2 \end{bmatrix}.$$

**11** If $\mathbf{q} = \begin{bmatrix} i_{L1} \\ i_{L2} \\ i_{L3} \\ v_C \end{bmatrix}$, $\mathbf{a} = \begin{bmatrix} 2 & -1 & 0 & 1 \\ 3 & 2 & -2 & 0 \\ -4 & 1 & 0 & -3 \\ 0 & 4 & 2 & 3 \end{bmatrix}$, and $\mathbf{f} = \begin{bmatrix} 0 \\ 0 \\ u(t) \\ -u(t) \end{bmatrix}$, write the normal-form equations.

**12** Using the state vector, $\mathbf{q} = \begin{bmatrix} v_C \\ i_{L1} \\ i_{L2} \end{bmatrix}$, write the system matrix **a** and the forcing-function vector **f** for the circuit shown in Fig. 16-7 if the unknown element is a: (*a*) 5-Ω resistor; (*b*) 15-V battery, positive terminal on the right.

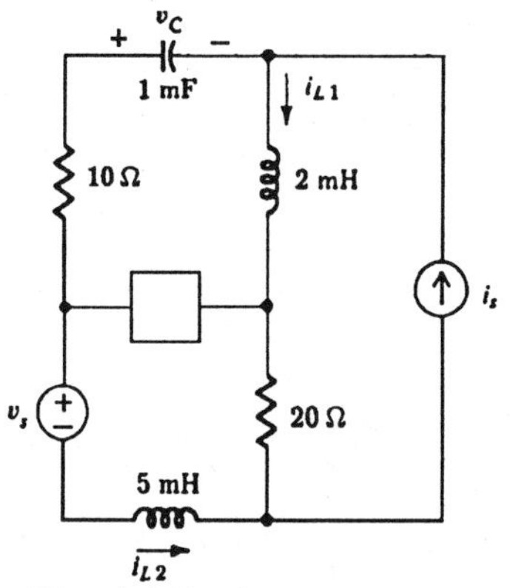


**Fig. 16-7:** See Prob. 12.

**13** Using the indicated state vector, determine **a** and **f** for the circuit illustrated in: (*a*) Fig. 16-8, $q_1 = i_L$ (down); (*b*) Fig. 16-9, $q_1 = v_C$; (*c*) Fig. 16-10, $q_1 = i_L$, $q_2 = v_C$ (+ at top); (*d*) Fig. 16-11, $q_1 = v_C$, $q_2 = i_L$ (to the right).

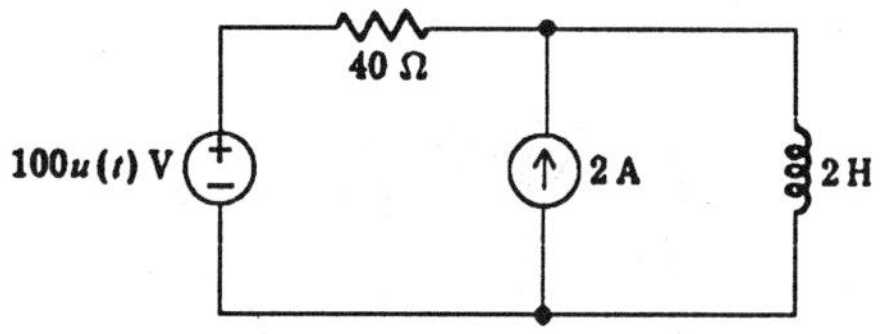


**Fig. 16-8:** See Prob. 13.

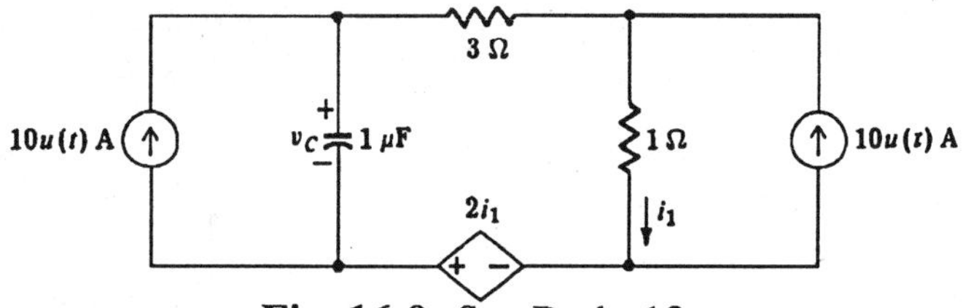


**Fig. 16-9:** See Prob. 13.

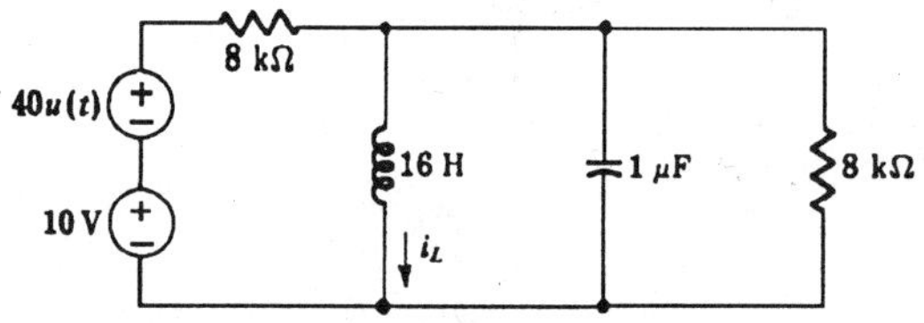


**Fig. 16-10:** See Prob. 13.

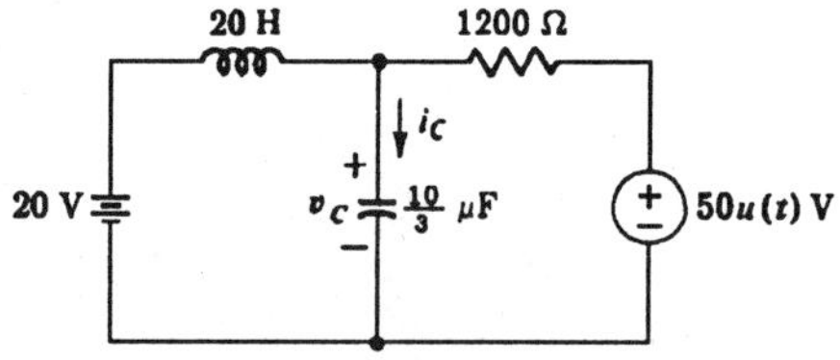


**Fig. 16-11:** See Prob. 13.

**14** Find the inductor current $i_L(t)$ in Fig. 16-12 if the sources are: (*a*) $v_s = 60e^{-10t}[u(t) - u(t-1)]$ V and $i_s = 0$; (*b*) $v_s = 60$ V and $i_s = -6u(t)$A; (*c*) $v_s = 0$ and $i_s = 4\cos 5t\, u(t)$ A.

**15** Find $i(t)$ in the circuit of Fig. 16-13 by state-variable techniques.

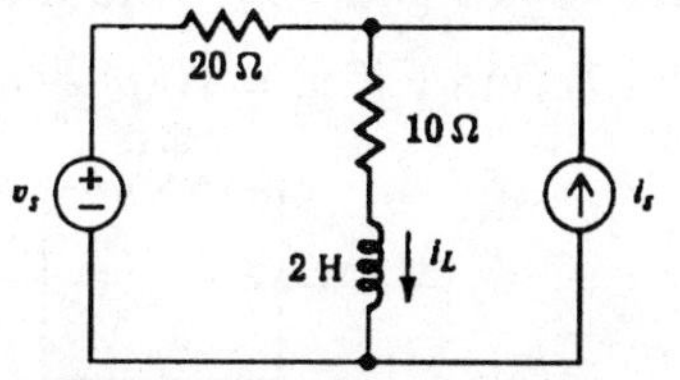


**Fig. 16-12:** See Prob. 14.

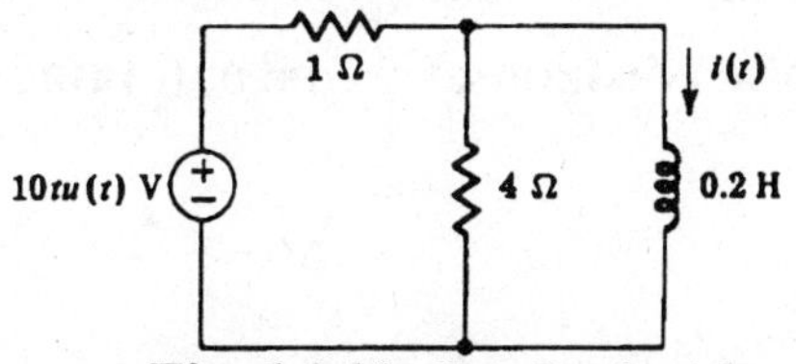


**Fig. 16-13:** See Prob. 15.

**16** Write the normal-form equation for the voltage $v(t)$ in the circuit shown in Fig. 16-14 and solve it if $v_s(t)$ =: (*a*) $27u(t)$ V; (*b*) $27u(-t)$ V.

**17** Use state-variable techniques to find $v(t)$ in the circuit of Fig. 16-15.

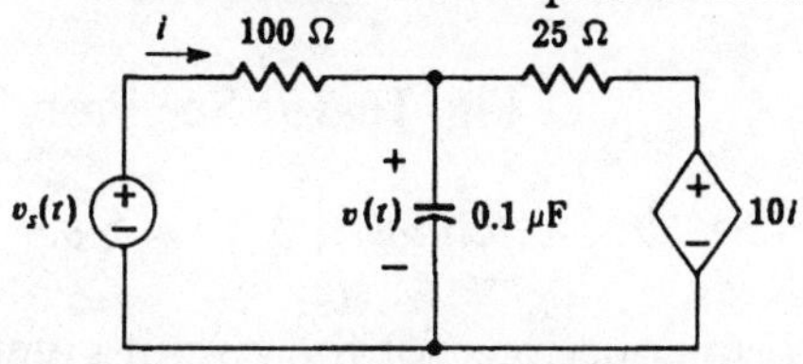

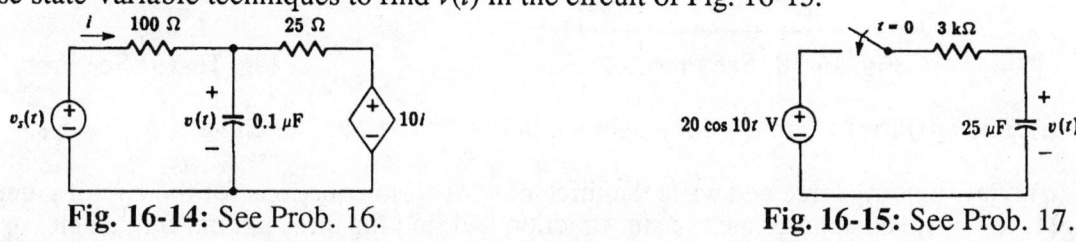

**Fig. 16-14:** See Prob. 16.

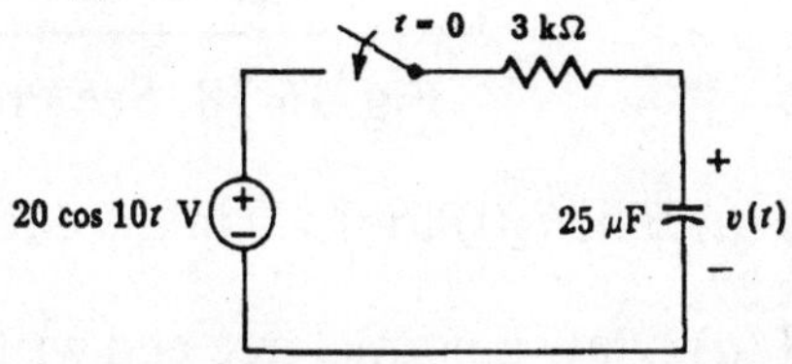


**Fig. 16-15:** See Prob. 17.

## Section 16-6 The solution of the matrix equation

**18** Write the normal-form equation for the circuit of Fig. 16-16, identify $a_{11}$ and $f_1(t)$, and find $i_L(t)$ for all time.

**19** Determine $i_1(t)$ for $t > 0$ in the circuit shown in Fig. 16-17.

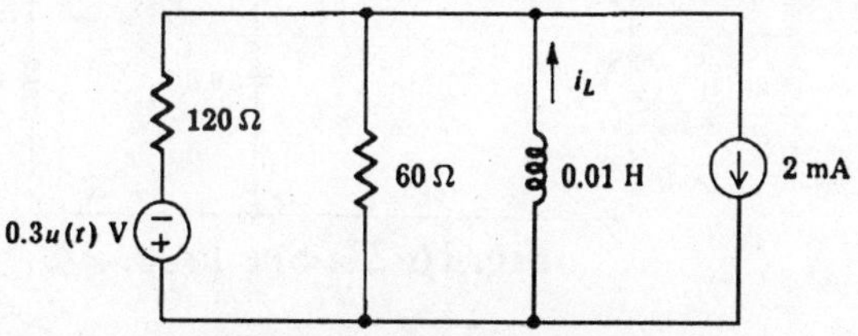


**Fig. 16-16:** See Prob. 18.

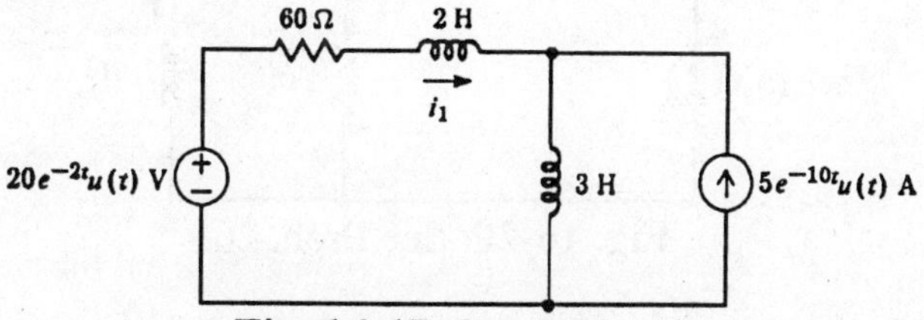


**Fig. 16-17:** See Prob. 19.

**20** If $\mathbf{b} = \begin{bmatrix} -0.10 & 0.12 \\ 0.20 & -0.05 \end{bmatrix}$, use the power series expansion for $e^x$ to find $e^{\mathbf{b}}$ and $e^{-\mathbf{b}}$, and then show that their product is **I**.

**21** Find the eigenvalues of the matrix: (*a*) $\begin{bmatrix} -2 & 1 \\ 3 & -4 \end{bmatrix}$; (*b*) $\begin{bmatrix} -2 & 1 \\ -2 & -4 \end{bmatrix}$; (*c*) $\begin{bmatrix} 0 & 2 \\ -8 & 0 \end{bmatrix}$; (*d*) $\begin{bmatrix} -2 & 1 & -3 \\ 0 & -2 & 0 \\ -1 & 0 & -4 \end{bmatrix}$

## Section 16-7 A further look at the state-transition matrix

**22** Determine the elements of $e^{t\mathbf{a}}$ in closed form if **a** =: (*a*) $\begin{bmatrix} -1 & -1 \\ 2 & -4 \end{bmatrix}$; (*b*) $\begin{bmatrix} 0 & -4 \\ 1 & 0 \end{bmatrix}$

**23** Show that the evaluation of $e^{t\mathbf{a}}$ at $t = 0$ leads to the identity matrix.

**24** Use state-variable methods to find the capacitor voltage in Fig. 16-18 as a function of time for $t > 0$.

**25** For the circuit shown in Fig. 16-19, use $\begin{bmatrix} v \\ i \end{bmatrix}$ as the state vector and: (*a*) write the set of normal-form equations; (*b*) find **a**; (*c*) find $v$ if $v_s = 100u(t)$ V; (*d*) find $v$ if $v_s = 100e^{-20t}u(t)$ V.

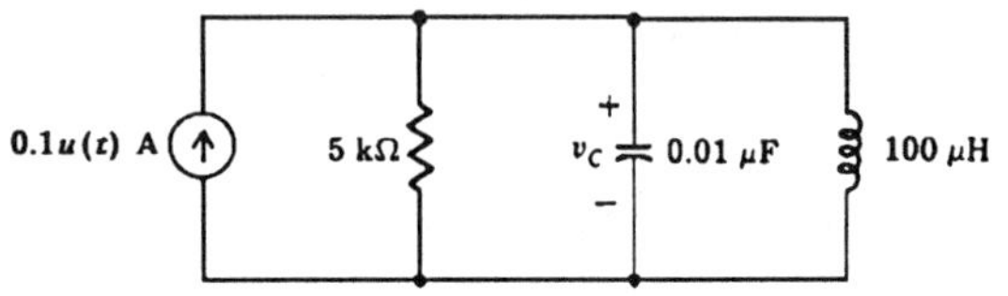


**Fig. 16-18:** See Prob. 24.

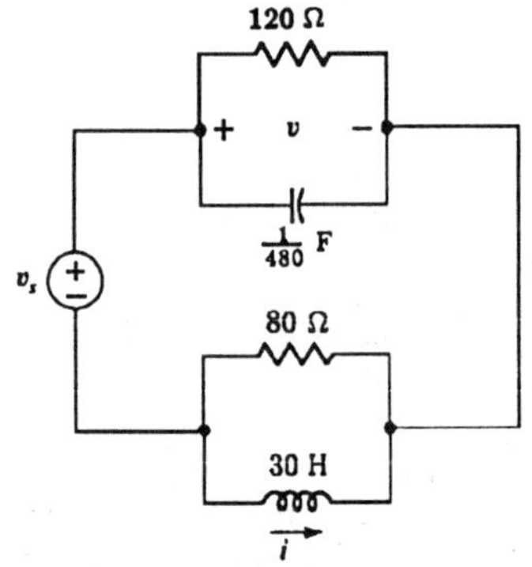


**Fig. 16-19:** See Prob. 25.

**26** (*a*) Find $i_2(t)$ for $t > 0$ in the circuit shown in Fig. 16-20. (*b*) Sketch $i_2(t)$, $0 \leq t \leq 5$ s.

**27** (*a*) Draw a normal tree and write the three normal-form equations for the circuit given in Fig. 16-21. (*b*) Write an appropriate transfer function and find the three natural frequencies associated with the response $i_1(t)$. (*c*) If $i_1(t)$ is represented as $i_{1n}(t) + i_{1f}(t)$, find $i_{1f}(t)$ by working in the frequency domain. (*d*) Determine the initial values of the three state variables. (*e*) Determine the initial value of the derivative $i_1'(0^+)$ from the normal-form equations. (*f*) By taking the derivative of each normal-form equation, find the value of $i_1''(0^+)$. (*g*) Evaluate the constants appearing in $i_{1n}(t)$ and determine the complete response $i_1(t)$.

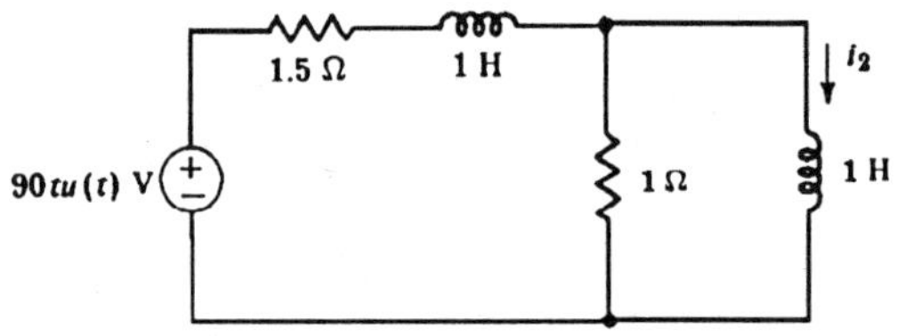


**Fig. 16-20:** See Prob. 26.

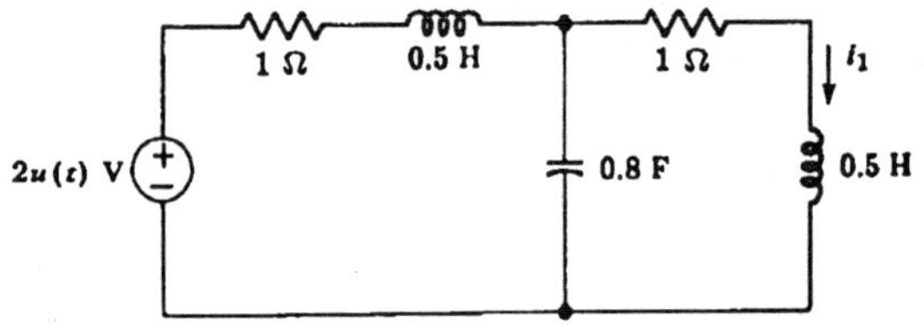


**Fig. 16-21:** See Prob. 27.

# Chapter 16

16-1 (a)

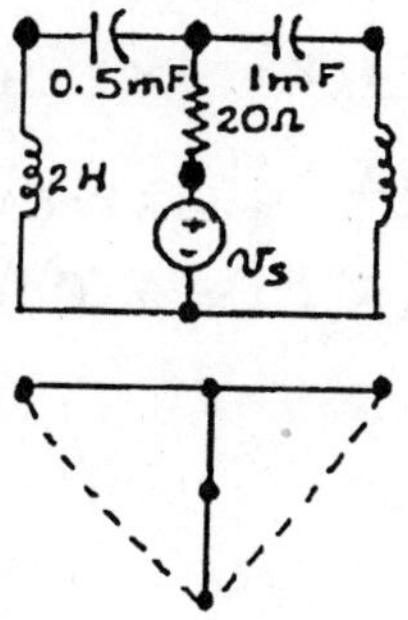


(b)

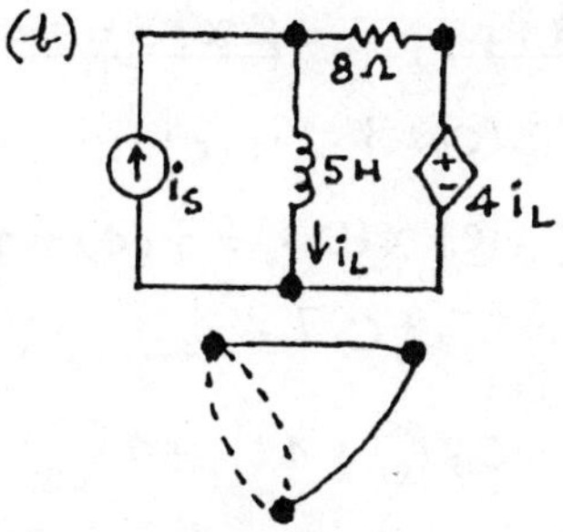


(c)

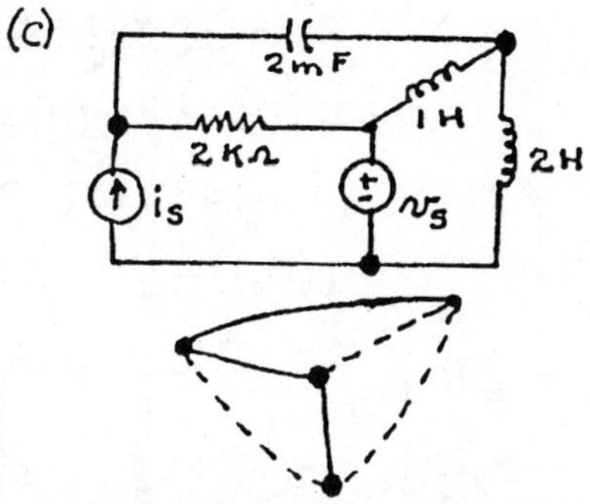


16-2 (a)

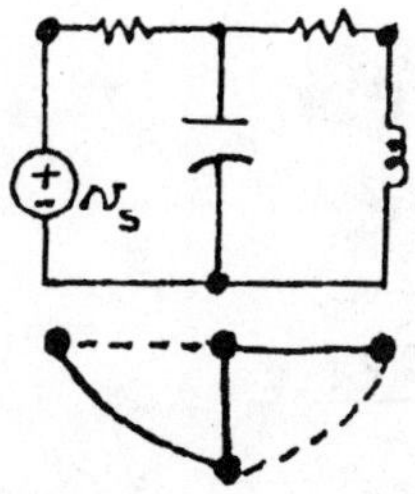


(b)

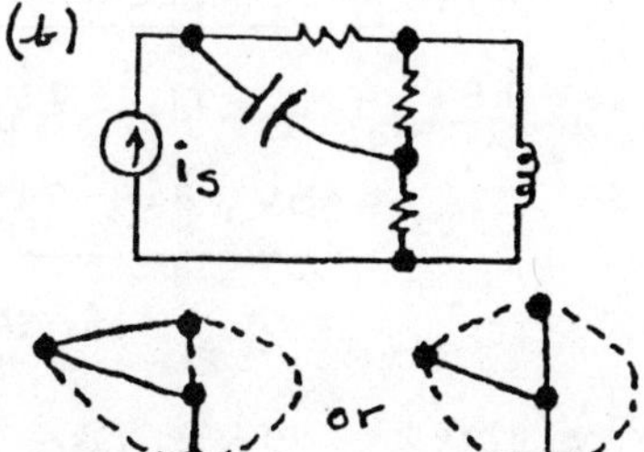


(c)

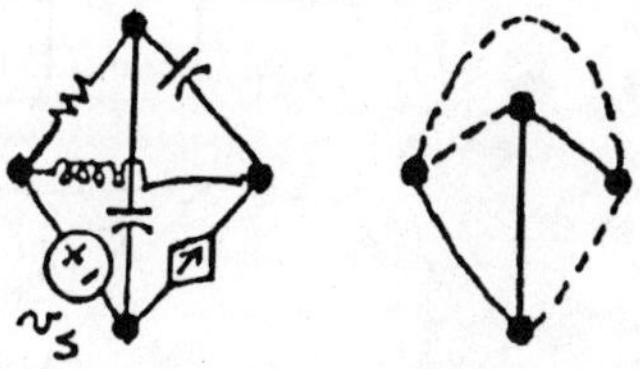


16-3 (a)

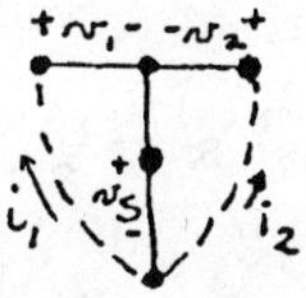


$0.5\times 10^{-3} v_1' = i_1$, $10^{-3} v_2' = i_2$, $2i_1' = -v_1 - 20(i_1 + i_2) - v_s$,

$4i_2' = -v_2 - 20(i_1 + i_2) - v_s$,

$v_1' = \underline{2000 i_1}$, $v_2' = \underline{1000 i_2}$, $i_1' = \underline{-\frac{1}{2}v_1 - 10i_1 - 10i_2 - \frac{1}{2}v_s}$,

$i_2' = \underline{-\frac{1}{4}v_2 - 5i_1 - 5i_2 - \frac{1}{4}v_s}$

(b)

$5i_L' = 8(i_s - i_L) + 4i_L$ ∴ $i_L' = \underline{-0.8 i_L + 16 i_s}$

(c)

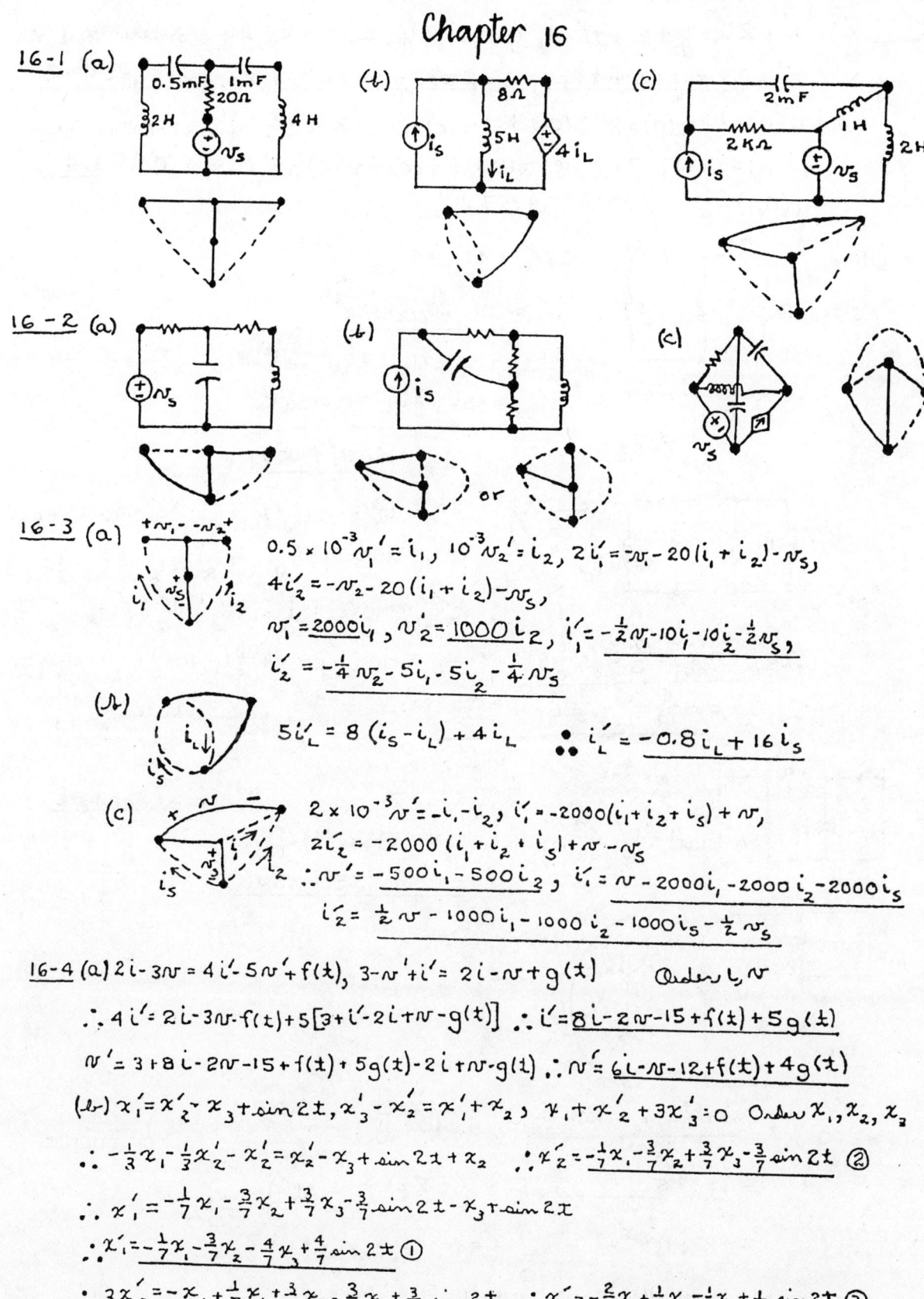

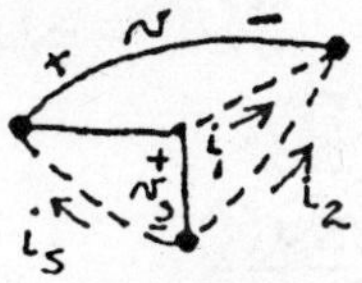


$2\times 10^{-3} v' = -i_1 - i_2$, $i_1' = -2000(i_1 + i_2 + i_s) + v$,

$2i_2' = -2000(i_1 + i_2 + i_s) + v - v_s$

∴ $v' = \underline{-500 i_1 - 500 i_2}$, $i_1' = \underline{v - 2000 i_1 - 2000 i_2 - 2000 i_s}$

$i_2' = \underline{\frac{1}{2}v - 1000 i_1 - 1000 i_2 - 1000 i_s - \frac{1}{2}v_s}$

16-4 (a) $2i - 3v = 4i' - 5v' + f(t)$, $3 - v' + i' = 2i - v + g(t)$ Order $i, v$

∴ $4i' = 2i - 3v - f(t) + 5[3 + i' - 2i + v - g(t)]$ ∴ $i' = \underline{8i - 2v - 15 + f(t) + 5g(t)}$

$v' = 3 + 8i - 2v - 15 + f(t) + 5g(t) - 2i + v - g(t)$ ∴ $v' = \underline{6i - v - 12 + f(t) + 4g(t)}$

(b) $x_1' = x_2' - x_3 + \sin 2t$, $x_3' - x_2' = x_1' + x_2$, $x_1 + x_2' + 3x_3' = 0$ Order $x_1, x_2, x_3$

∴ $-\frac{1}{3}x_1 - \frac{1}{3}x_2' - x_2' = x_2' - x_3 + \sin 2t + x_2$ ∴ $x_2' = \underline{-\frac{1}{7}x_1 - \frac{3}{7}x_2 + \frac{3}{7}x_3 - \frac{3}{7}\sin 2t}$ ②

∴ $x_1' = -\frac{1}{7}x_1 - \frac{3}{7}x_2 + \frac{3}{7}x_3 - \frac{3}{7}\sin 2t - x_3 + \sin 2t$

∴ $x_1' = \underline{-\frac{1}{7}x_1 - \frac{3}{7}x_2 - \frac{4}{7}x_3 + \frac{4}{7}\sin 2t}$ ①

∴ $3x_3' = -x_1 + \frac{1}{7}x_1 + \frac{3}{7}x_2 - \frac{3}{7}x_3 + \frac{3}{7}\sin 2t$ ∴ $x_3' = \underline{-\frac{2}{7}x_1 + \frac{1}{7}x_2 - \frac{1}{7}x_3 + \frac{1}{7}\sin 2t}$ ③

**16-5** (a) $x' = 2x - y + z + t^2 + 1$, $y' = x' - 2y$, $z' = x + 2z - 2y'$ order: $x, y, z$

$\therefore x' = \underline{2x - y + z + t^2 + 1}$, $y' = \underline{2x - 3y + z + t^2 + 1}$, $z' = \underline{-3x + 6y - 2t^2 - 2}$

(b) $x(0) = 1$ $y(0) = 2$ $z(0) = 3$ $\therefore x'(0) = 2 - 2 + 3 + 1 = \underline{4}$, $y'(0) = 2 - 6 + 3 + 1 = \underline{0}$

$z'(0) = -3 + 12 - 2 = \underline{7}$ (c) $x(0.1) \doteq x(0) + x'(0)\Delta t = 1 + 4(0.1) = \underline{1.4}$

$y(0.1) \doteq \underline{2}$ $z(0.1) \doteq 3 + 0.7 = \underline{3.7}$

**16-6** (a)

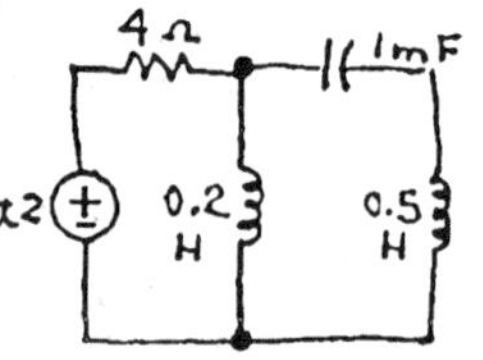


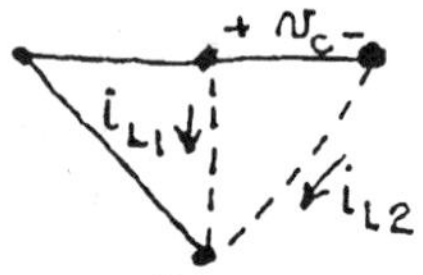


$0.2 i_{L1}' + 4 i_{L1} + 4 i_{L2} = 5e^{-t^2}$

$\therefore i_{L1}' = \underline{-20 i_{L1} - 20 i_{L2} + 25 e^{-t^2}}$

$0.5 i_{L2}' + v_C + 4 i_{L1} + 4 i_{L2} = 5e^{-t^2}$

$\therefore i_{L2}' = \underline{-8 i_{L1} - 8 i_{L2} - 2 v_c + 10 e^{-t^2}}$

$10^{-3} v_C' = i_{L2}$ $\therefore v_c' = \underline{1000\, i_{L2}}$

(b)

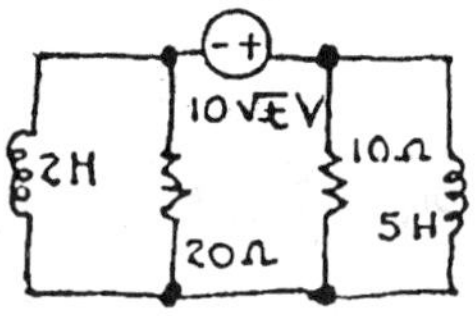


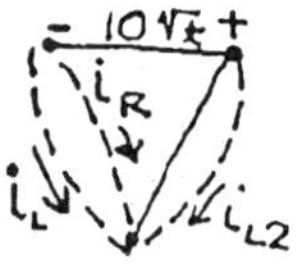


$2 i_{L1}' = -10\sqrt{t} - 10(i_{L1} + i_{L2} + i_R)$, $5 i_{L2}' = -10(i_{L1} + i_{L2} + i_R)$

$20 i_R = -10\sqrt{t} - 10(i_{L1} + i_{L2} + i_R)$ $\therefore i_R = -\frac{1}{3} i_{L1} - \frac{1}{3} i_{L2} - \frac{1}{3}\sqrt{t}$

$\therefore 2 i_{L1}' = -10\sqrt{t} - 10 i_{L1} - 10 i_{L2} + \frac{10}{3} i_{L1} + \frac{10}{3} i_{L2} + \frac{10}{3}\sqrt{t}$

$\therefore i_{L1}' = \underline{-\frac{10}{3} i_{L1} - \frac{10}{3} i_{L2} - \frac{10}{3}\sqrt{t}}$ $\therefore 5 i_{L2}' = -10 i_{L1} - 10 i_{L2} + \frac{10}{3} i_{L1} + \frac{10}{3} i_{L2} + \frac{10}{3}\sqrt{t}$

$\therefore i_{L2}' = \underline{-\frac{4}{3} i_{L1} - \frac{4}{3} i_{L2} + \frac{2}{3}\sqrt{t}}$ Order: $i_{L1}, i_{L2}$

**16-7**

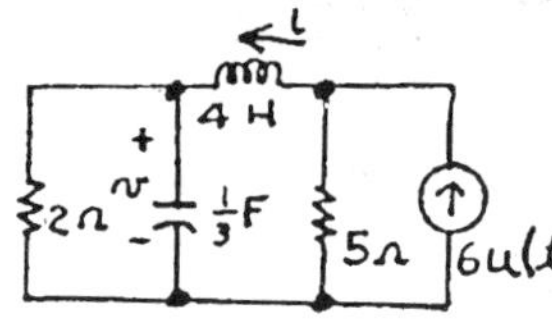


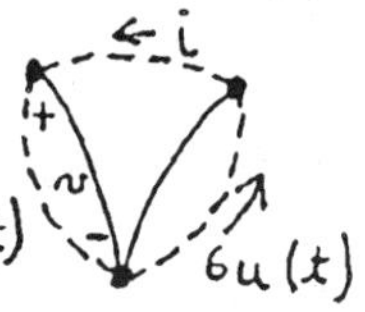


$\frac{1}{3} v' = i - \frac{v}{2}$ $\therefore v' = \underline{-1.5v + 3i}$

$4i' = -v + 5[6u(t) - i]$

$\therefore i' = \underline{-0.25v - 1.25i + 7.5u(t)}$

**16-8**

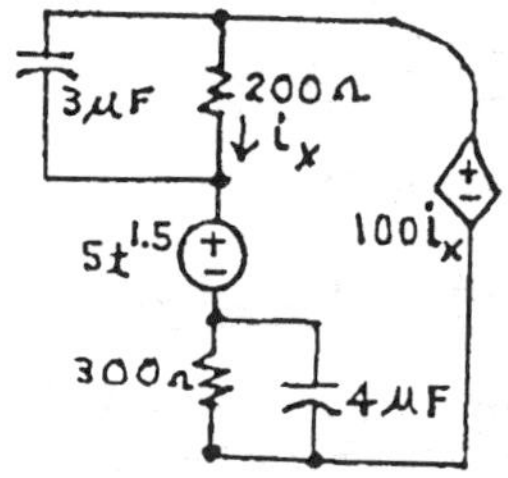


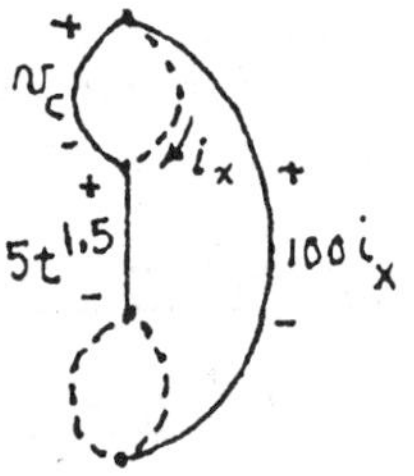


$$3 \times 10^{-6} v_c' = \frac{-5t^{1.5} - v_c + 100\frac{v_c}{200}}{300} + 4 \times 10^{-6} \frac{d}{dt}\left(-5t^{1.5} - v_c + \frac{1}{2}v_c\right) - \frac{v_c}{200}$$

$$= -\frac{1}{60}t^{1.5} - \frac{v_c}{600} + 4 \times 10^{-6}\left(-7.5t^{0.5} - \frac{1}{2}v_c'\right) - \frac{v_c}{200}$$

$$= -2 \times 10^{-6} v_c' - \frac{v_c}{150} - 5t^{1.5} - 30 \times 10^{-6} t^{0.5}$$

$$\therefore v_c' = \underline{-\frac{4000}{3} v_c - \frac{10\,000}{3} t^{1.5} - 6t^{0.5}}$$

16-9

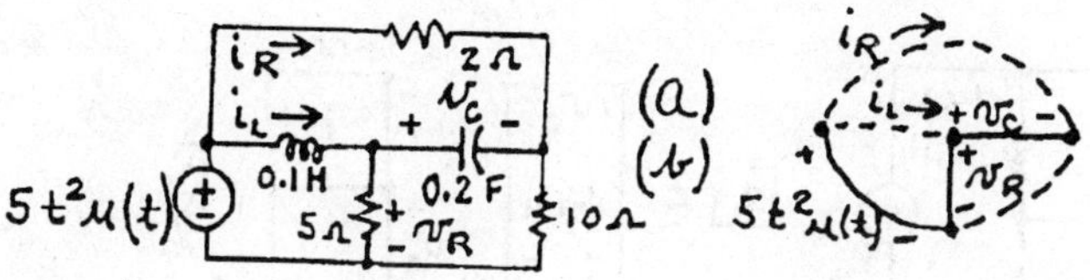


(c) $0.2v_C' = -i_R + \dfrac{v_R - v_C}{10}$, $0.1 i_L' = 5t^2u(t) - v_R$

$$0.2v_R = i_L + i_R - \frac{v_R - v_C}{10}, \quad 2i_R = 5t^2u(t) - v_R + v_C$$

$\therefore 0.4v_R - 2i_L + 0.2v_R - 0.2v_C = 5t^2u(t) - v_R + v_C$

$\therefore v_R = 0.75v_C + 1.25i_L + 3.125t^2u(t)$,

$2i_R = 5t^2u(t) + v_C - 0.75v_C - 1.25i_L - 3.125t^2u(t)$

$\therefore i_R = 0.125v_C - 0.625i_L + 0.9375t^2u(t)$

$\therefore 0.2v_C' = -0.225v_C + 0.625i_L - 0.9375t^2u(t) + 0.075v_C$
$+ 0.125i_L + 0.3125t^2u(t)$

$\therefore v_C' = \underline{-0.75v_C + 3.75i_L - 3.125t^2u(t)}$, $i_L' = \underline{-7.5v_C - 12.5i_L + 18.75t^2u(t)}$

16-10 (a) $\begin{bmatrix} 2 & 0 \\ -1 & 3 \\ 1 & -2 \end{bmatrix}\begin{bmatrix} 4 & -1 \\ -2 & 5 \end{bmatrix} = \begin{bmatrix} 8 & -2 \\ -10 & 16 \\ 8 & -11 \end{bmatrix}$ (b) $\begin{bmatrix} -1 & 3 \\ 1 & -2 \end{bmatrix}\begin{bmatrix} 4 & -1 \\ -2 & 5 \end{bmatrix} = \begin{bmatrix} -10 & 16 \\ 8 & -11 \end{bmatrix}$

(c) $\begin{bmatrix} 4 & -1 \\ -2 & 5 \end{bmatrix}\begin{bmatrix} -1 & 3 \\ 1 & -2 \end{bmatrix} = \begin{bmatrix} -5 & 14 \\ 7 & -16 \end{bmatrix}$

16-11 $[q] = \begin{bmatrix} i_{L1} \\ i_{L2} \\ i_{L3} \\ v_C \end{bmatrix}$, $[a] = \begin{bmatrix} 2 & -1 & 0 & 1 \\ 3 & 2 & -2 & 0 \\ -4 & 1 & 0 & -3 \\ 0 & 4 & 2 & 3 \end{bmatrix}$, $[f] = \begin{bmatrix} 0 \\ 0 \\ u(t) \\ -u(t) \end{bmatrix}$

$\therefore i_{L1}' = 2i_{L1} - i_{L2} + v_C$
$i_{L2}' = 3i_{L1} + 2i_{L2} - 2i_{L3}$
$i_{L3}' = -4i_{L1} + i_{L2} - 3v_C + u(t)$
$v_C' = 4i_{L2} + 2i_{L3} + 3v_C - u(t)$

16-12 (a) 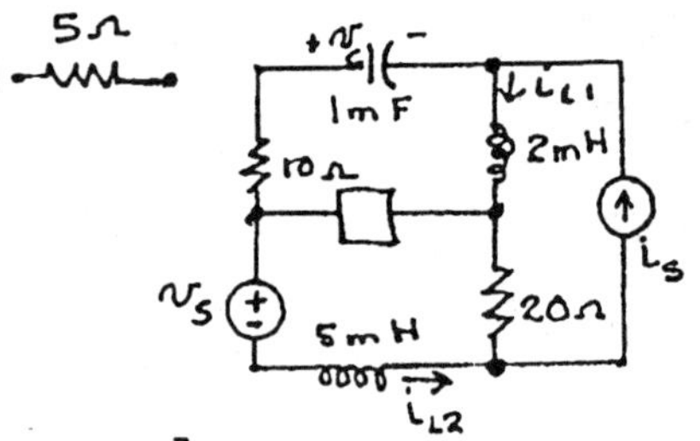


$$[q] = \begin{bmatrix} v_c \\ i_{L1} \\ i_{L2} \end{bmatrix}$$

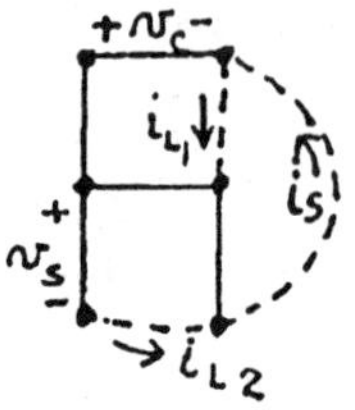


$$10^{-3} v_c' = i_{L1} - i_s \quad \therefore v_c' = 1000 i_{L1} - 1000 i_s \leftarrow$$

$$2\times10^{-3} i_{L1}' = -v_c + 10 i_s - 10 i_{L1} + 5 i_s - 5 i_{L1} - 5 i_{L2}$$

$$\therefore i_{L1}' = -500 v_c - 7500 i_{L1} - 2500 i_{L2} + 7500 i_s \leftarrow$$

$$5\times10^{-3} i_{L2}' = -v_s + 5 i_s - 5 i_{L1} - 5 i_{L2} - 20 i_{L2} + 20 i_s$$

$$\therefore i_{L2}' = -1000 i_{L1} - 5000 i_{L2} - 200 v_s + 5000 i_s \leftarrow$$

$$[a] = \begin{bmatrix} 0 & 1000 & 0 \\ -500 & -7500 & -2500 \\ 0 & -1000 & -5000 \end{bmatrix}, \quad [f] = \begin{bmatrix} -1000 i_s \\ 7500 i_s \\ -200 v_s + 5000 i_s \end{bmatrix}$$

(b) 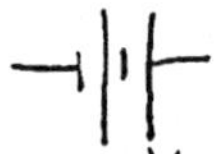


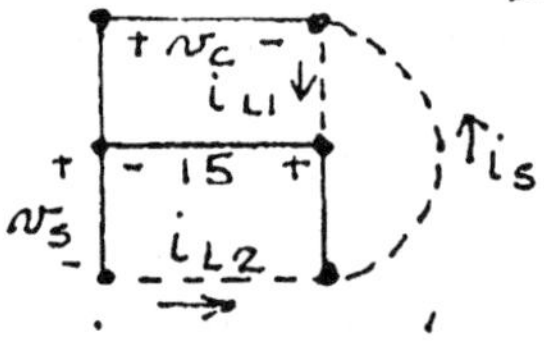


$$10^{-3} v_c' = i_{L1} - i_s \quad \therefore v_c' = 1000 i_{L1} - 1000 i_s \leftarrow$$

$$2\times10^{-3} i_{L1}' = -v_c + 10 i_s - 10 i_{L1} - 15$$

$$\therefore i_{L1}' = -500 v_c - 5000 i_{L1} - 7500 + 5000 i_s \leftarrow$$

$$5\times10^{-3} i_{L2}' = -v_s - 15 - 20 i_{L2} + 20 i_s$$

$$\therefore i_{L2}' = -4000 i_{L2} - 3000 + 4000 i_s - 200 v_s \leftarrow$$

$$[a] = \begin{bmatrix} 0 & 1000 & 0 \\ -500 & -5000 & 0 \\ 0 & 0 & -4000 \end{bmatrix}, \quad [f] = \begin{bmatrix} -1000 i_s \\ -7500 + 5000 i_s \\ -3000 + 4000 i_s - 200 v_s \end{bmatrix}$$

16-13

(a) 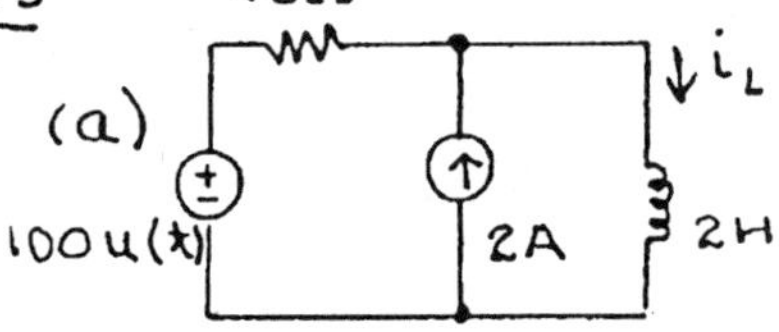


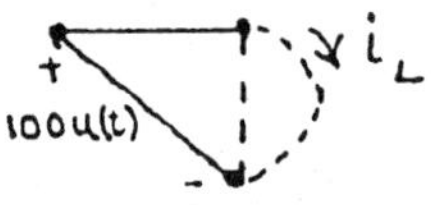


$$2 i_L' = 100u(t) + 40(2 - i_L)$$

$$\therefore i_L' = -20 i_L + 40 + 50u(t)$$

$$[a] = [20], \ [f] = [40 + 50u(t)]$$

(b) 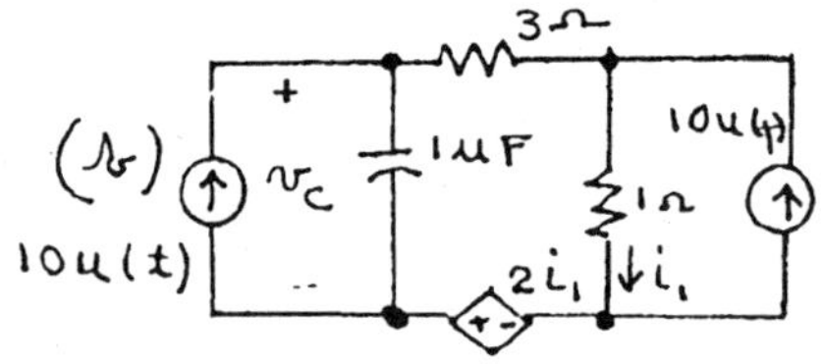

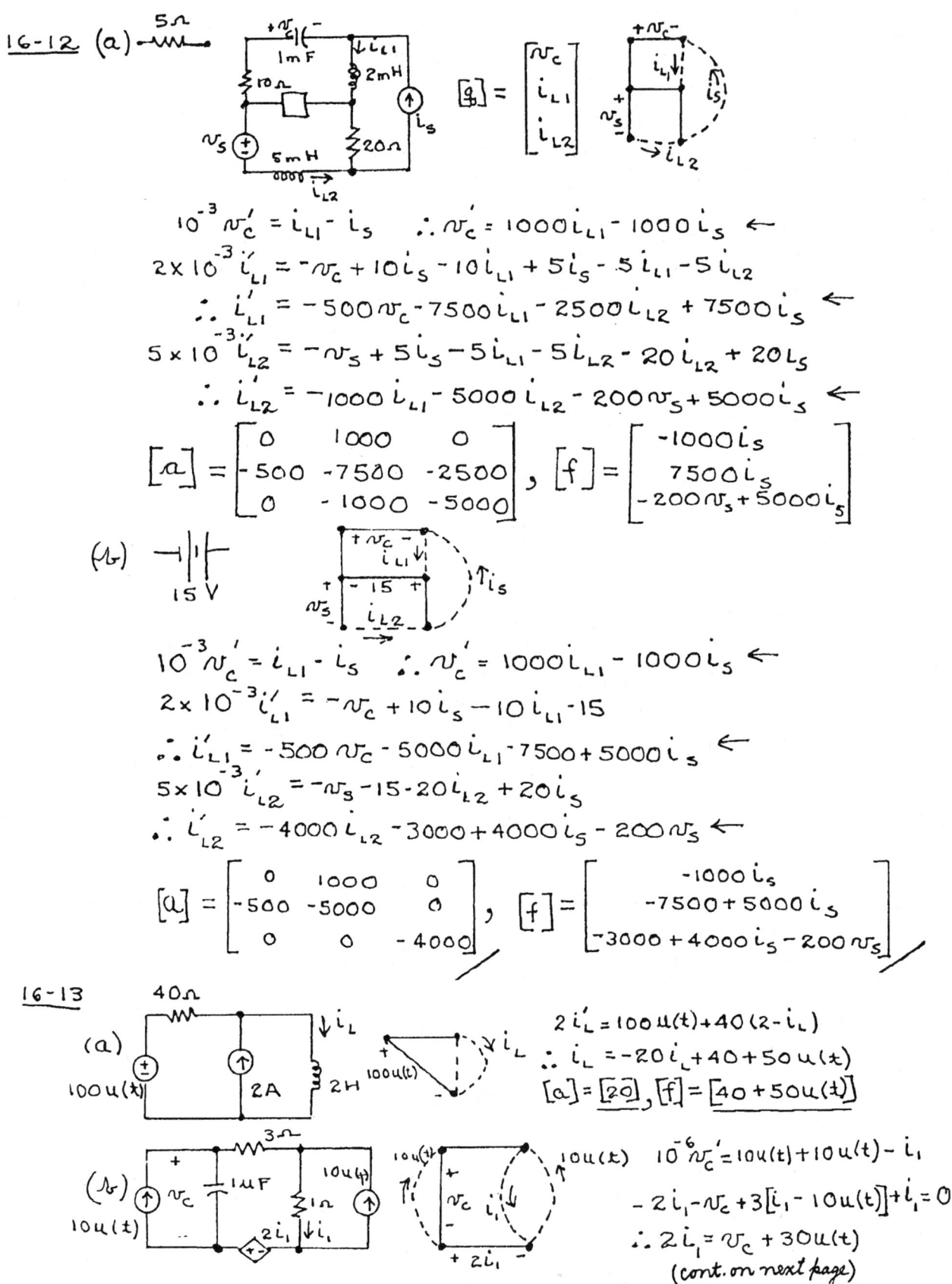

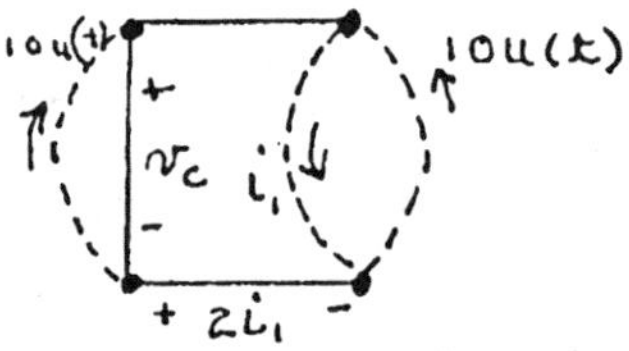


$$10^{-6} v_c' = 10u(t) + 10u(t) - i_1$$

$$-2i_1 - v_c + 3[i_1 - 10u(t)] + i_1 = 0$$

$$\therefore 2i_1 = v_c + 30u(t)$$

(cont. on next page)

**16-13 (cont.)** $\therefore 10^{-6} v_c' = 20u(t) - 0.5 v_c - 15u(t)$

$v_c' = -0.5\times10^6 v_c + 5\times10^6 u(t)$ $\qquad [a] = [-0.5\times10^6],\ [f] = [5\times10^6 u(t)]$

(c) 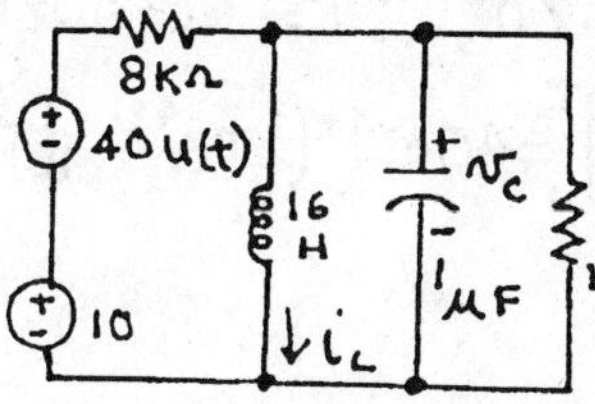


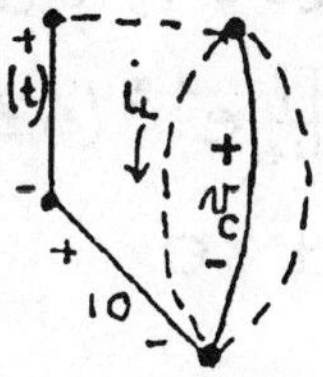


$[q] = \begin{bmatrix} i_L \\ v_c \end{bmatrix}$

$[a] = \begin{bmatrix} 0 & \frac{1}{16} \\ -10^6 & -250 \end{bmatrix},\ [f] = \begin{bmatrix} 0 \\ 1250 + 5000u(t) \end{bmatrix}$

$16 i_L' = v_c,\quad 10^{-6} v_c' = -i_L - \dfrac{v_c}{8000} + \dfrac{10 + 40u(t) - v_c}{8000}$

$\therefore i_L' = \frac{1}{16} v_c,\quad v_c' = -10^6 i_L - 250 v_c + 1250 + 5000u(t)$

(d) 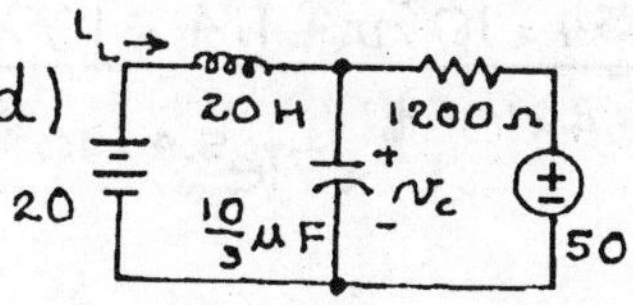


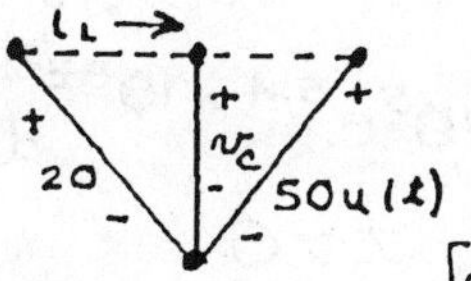


$[q] = \begin{bmatrix} v_c \\ i_L \end{bmatrix}$

$[a] = \begin{bmatrix} -250 & 3\times10^5 \\ -0.05 & 0 \end{bmatrix},\ [f] = \begin{bmatrix} 12500u(t) \\ 1 \end{bmatrix}$

$\frac{10}{3}\times10^{-6} v_c' = -\dfrac{v_c}{1200} + \dfrac{50u(t)}{1200} + i_L,\quad 20 i_L' = 20 - v_c$

$\therefore v_c' = -250 v_c + 3\times10^5 i_L + 12500u(t),\quad i_L' = -0.05 v_c + 1$

**16-14** 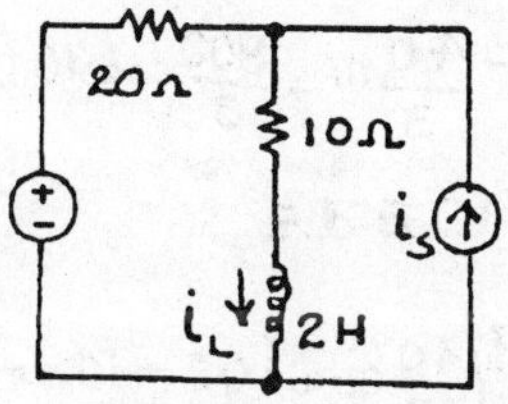


(a) $i_s = 0,\ v_s = 60e^{-10t}[u(t) - u(t-1)]$

$\therefore 2i_L' = v_s - 30 i_L,\quad i_L' = -15 i_L + 0.5 v_s$

$\therefore i_L = e^{-15t}\int_0^t e^{15z}\, 30 e^{-10z}\, dz = \underline{e^{-15t}\, 6(e^{5t} - 1)},\ 0<t<1$

and $i_L = e^{-15t}\int_0^1 e^{-15z}\, 30 e^{-10z}\, dz = \underline{e^{-15t}\, 6(e^5 - 1)},\ t>1$

(b) $v_s = 60,\ i_s = -6u(t)$

$\therefore 2i_L' = v_s - 10 i_L + 20(i_s - i_L) \quad \therefore i_L' = -15 i_L + 30 - 60u(t)$

$\therefore i_L = e^{-15t}\int_{-\infty}^t e^{15z}[30 - 60u(z)]\,dz = e^{-15t}[2e^{15t} - 4(e^{15t} - 1)]$

$\therefore i_L = \underline{-2 + 4e^{-15t}}$ A, $t>0$

(c) $v_s = 0\quad i_s = 4\cos 5t\, u(t)\quad \therefore 2i_L' = -10 i_L + 20(i_s - i_L)$

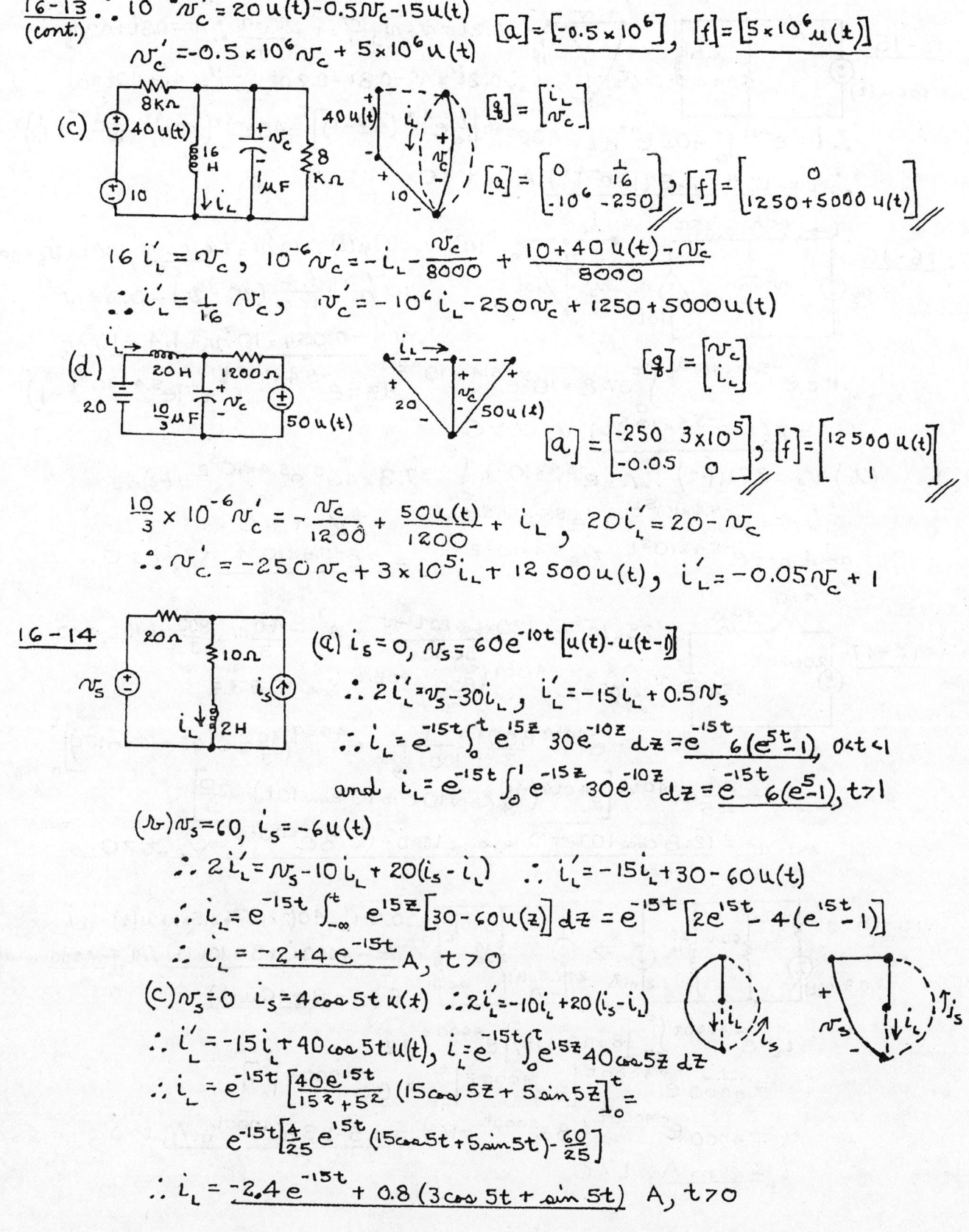

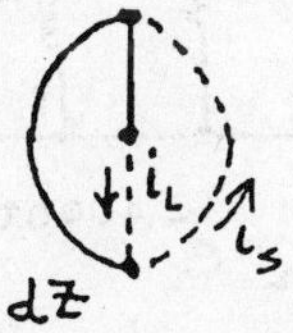

$\therefore i_L' = -15 i_L + 40\cos 5t\, u(t),\quad i_L = e^{-15t}\int_0^t e^{15z}\, 40\cos 5z\, dz$

$\therefore i_L = e^{-15t}\left[\dfrac{40e^{15t}}{15^2 + 5^2}(15\cos 5z + 5\sin 5z)\right]_0^t =$

$e^{-15t}\left[\frac{4}{25} e^{15t}(15\cos 5t + 5\sin 5t) - \frac{60}{25}\right]$

$\therefore i_L = \underline{-2.4e^{-15t} + 0.8(3\cos 5t + \sin 5t)}$ A, $t>0$

**16-15**

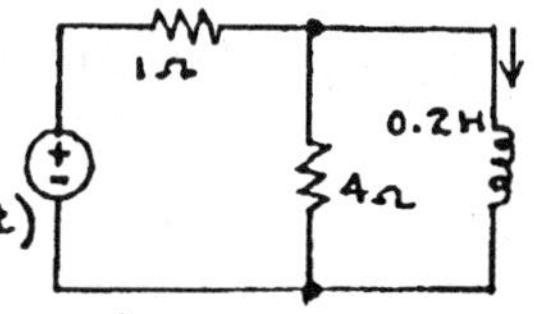


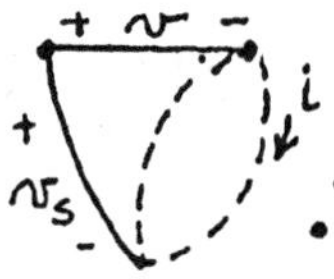


$0.2i' = v_s - v$, $\frac{v}{1} = i + \frac{v_s - v}{4}$ ∴ $v = 0.8i + 0.2v_s$

∴ $0.2i' = v_s - 0.8i - 0.2v_s$ ∴ $i' = -4i + 40t\,u(t)$

∴ $i = e^{-4t}\int_0^t 40z e^{4z}\,dz = 40e^{-4t}\left[\frac{1}{16}e^{4z}(4z-1)\right]_0^t = 40e^{-4t}\left(\frac{1}{4}te^{4t} - \frac{1}{16}e^{4t} + \frac{1}{16}\right)$

∴ $i = \underline{10t - 2.5(1-e^{-4t})}$ A, $t>0$

**16-16**

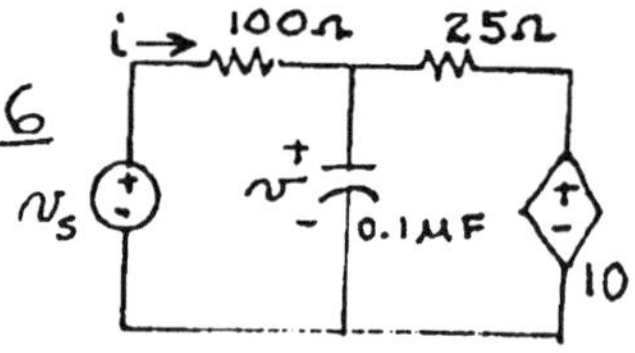


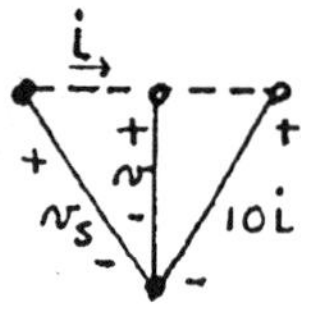


(a) $v_s = 27u(t)$ ∴ $10^{-7}v' = i + \frac{10i - v}{25}$, $100i = v_s - v$

∴ $10^{-7}v' = \frac{1.4}{100}(v_s - v) - 0.04v$

∴ $v' = \underline{-0.054\times10^{7}v + 1.4\times10^{5}v_s}$

∴ $v = e^{-5.4\times10^5 t}\int_0^t 37.8\times10^5 e^{5.4\times10^5 z}\,dz = e^{-5.4\times10^5 t}\times 7(e^{5.4\times10^5 t} - 1)$

$= \underline{7(1-e^{-5.4\times10^5 t})}$ V, $t>0$

(b) $v_s = 27u(-t)$ ∴ $v = e^{-5.4\times10^5 t}\int_{-\infty}^t 37.8\times10^5 e^{5.4\times10^5 z}u(-z)\,dz$

∴ $v = e^{-5.4\times10^5 t}\,7(e^{5.4\times10^5 z})_{-\infty}^{t} = \underline{7V}$, $t<0$

and $v = e^{-5.4\times10^5 t}\,7(e^{5.4\times10^5 z})_{-\infty}^{0} = \underline{7e^{-5.4\times10^5 t}}$ V, $t>0$

**16-17**

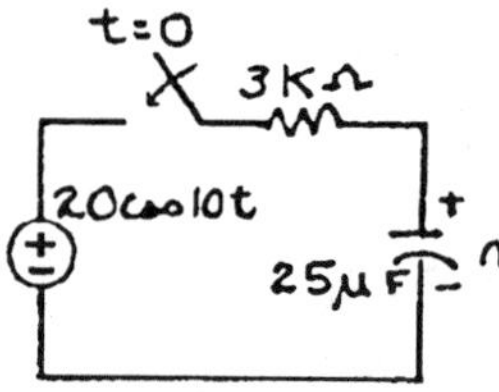


$25\times10^{-6}v' = \frac{20\cos10t - v}{3000}$ ∴ $v' = -\frac{40}{3}v + \frac{800}{3}\cos10t$, $t>0$

∴ $v = e^{-40t/3}\int_0^t \frac{800}{3}e^{40z/3}\cos10z\,dz$

$= e^{-40t/3}\frac{800}{3}\left[\frac{1}{100+\frac{1600}{9}}e^{40z/3}\left(\frac{40}{3}\cos10z + 10\sin10z\right)\right]_0^t$

∴ $v = \frac{24}{25}e^{-40t/3}\left[e^{40t/3}\left(\frac{40}{3}\cos10t + 10\sin10t\right) - \frac{40}{3}\right]$

or, $v = \underline{12.8\cos10t + 9.6\sin10t - 12.8e^{-40t/3}}$ V, $t>0$

**16-18**

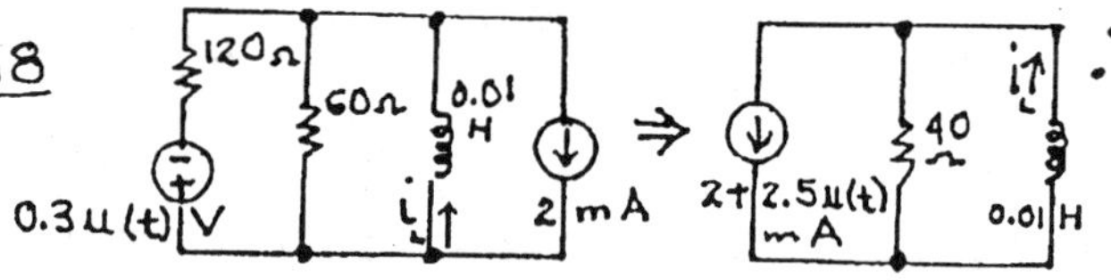


∴ $0.01i_L' = 40(2\times10^{-3} + 2.5\times10^{-3}u(t) - i_L)$

$i_L' = \underline{-4000i_L + 8 + 10u(t)}$ ∴ $a_{11} = \underline{-4000}$

$f_1(t) = \underline{8 + 10u(t)}$

∴ $i_L = e^{-4000t}\int_{-\infty}^t [8 + 10u(z)]e^{4000z}\,dz$

$= \frac{1}{4000}e^{-4000t}\left[8e^{4000z}\Big|_{-\infty}^t + 10e^{4000z}\Big|_0^t\right]$

$= \frac{1}{4000}e^{-4000t}(18e^{4000t} - 10) = \underline{4.5 - 2.5e^{-4000t}}$ mA, $t>0$;

$i_L = \underline{2\text{ mA}}$, $t<0$

**16-19**

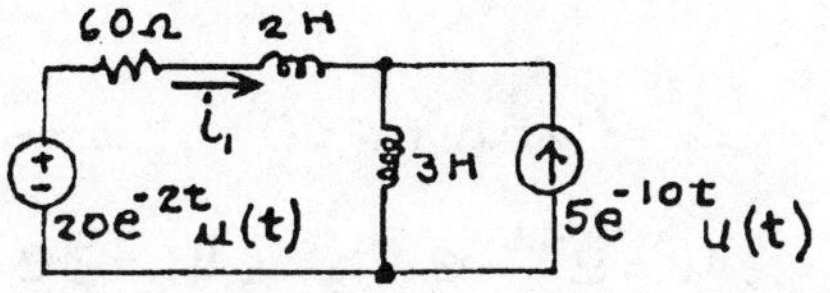


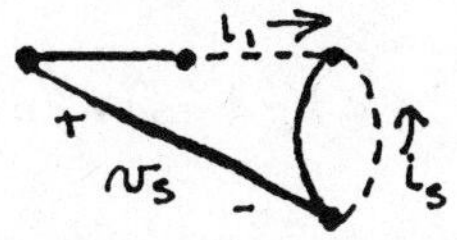


$$2i_1' = -60i_1 + v_s - 3\frac{d}{dt}(i_1 + i_s)$$

$$\therefore 2i_1' = -60i_1 + 20e^{-2t}u(t) + 150e^{-10t}u(t) - 3i_1' - 15u'(t)$$

note: $u'(t) = 0$ since $t > 0$

$$i_1' = -12i_1 + 4e^{-2t}u(t) + 30e^{-10t}u(t) - 3u'(t)$$

$$\therefore i_1 = e^{-12t}\int_0^t e^{12z}\left[4e^{-2z} + 30e^{-10z} - 3u'(z)\right]dz$$

$$\therefore i_1 = e^{-12t}\int_0^t (4e^{10z} + 30e^{2z} - 3u'(z))dz$$

$$= e^{-12t}(0.4e^{10t} - 0.4 + 15e^{2t} - 15)$$

$$\therefore i_1 = \underline{-15.4e^{-12t} + 0.4e^{-2t} + 15e^{-10t}\ \text{A}},\quad t > 0$$

**16-20** $[b] = \begin{bmatrix} -0.10 & 0.12 \\ 0.20 & -0.05 \end{bmatrix}$

$$\therefore e^{[b]} = \begin{bmatrix} 1 & 0 \\ 0 & 1 \end{bmatrix} + \begin{bmatrix} -0.1 & 0.12 \\ 0.2 & -0.05 \end{bmatrix} + \frac{1}{2}\begin{bmatrix} 0.034 & -0.018 \\ -0.03 & 0.0265 \end{bmatrix} + \frac{1}{6}\begin{bmatrix} -0.007 & 0.00498 \\ 0.0083 & -0.004925 \end{bmatrix}$$

$$+ \frac{1}{24}\begin{bmatrix} 0.001696 & -0.001089 \\ -0.001815 & 0.001242 \end{bmatrix}$$

$$= \begin{bmatrix} 0.9 + 0.017 - 0.001167 + 0.000071 & 0.12 - 0.009 + 0.00083 - 0.000045 \\ 0.2 - 0.015 + 0.001383 - 0.000076 & 0.95 + 0.01325 - 0.000821 + 0.000052 \end{bmatrix}$$

$$= \underline{\underline{\begin{bmatrix} 0.915904 & 0.111785 \\ 0.186307 & 0.962481 \end{bmatrix}}}$$

$$\therefore e^{-[b]} = \begin{bmatrix} 1.1 + 0.017 + 0.001167 + 0.000071 & -0.12 - 0.009 - 0.00083 - 0.000045 \\ -0.2 - 0.015 - 0.001383 - 0.000076 & 1.05 + 0.01325 + 0.000821 + 0.000052 \end{bmatrix}$$

$$\therefore e^{-[b]} = \underline{\underline{\begin{bmatrix} 1.118238 & -0.129875 \\ -0.216459 & 1.064123 \end{bmatrix}}} \qquad e^{[b]}e^{-[b]} = \underline{\underline{\begin{bmatrix} 1.000002 & 0.000000 \\ -0.000002 & 1.000002 \end{bmatrix}}}$$

**16-21** (a) $\begin{bmatrix} -2 & 1 \\ 3 & -4 \end{bmatrix}$ $\therefore (-2-s)(-4-s) - 3 = 0$, $s^2 + 6s + 5 = 0$ $\therefore s = \underline{-1}$ and $\underline{-5}$

(b) $\begin{bmatrix} -2 & 1 \\ -2 & -4 \end{bmatrix}$ $\therefore (-2-s)(-4-s) + 2 = 0$,

$s^2 + 6s + 10 = 0$ $\quad s = \dfrac{-6 \pm \sqrt{36-40}}{2} = \underline{-3 \pm j1}$

(c) $\begin{bmatrix} 0 & 2 \\ -8 & 0 \end{bmatrix}$ $\therefore s^2 + 16 = 0$ $\quad s = \underline{\pm j4}$

(d) $\begin{bmatrix} -2 & 1 & -3 \\ 0 & -2 & 0 \\ -1 & 0 & -4 \end{bmatrix}$ $\therefore (-2-s)(-2-s)(-4-s) - 1(-3)(s+2) = 0$ $\therefore s^2 + 6s + 8 - 3 = 0$

$\therefore s = \underline{-1}, \underline{-2}$, and $\underline{-5}$

**16-22** (a) $[a]=\begin{bmatrix}-1 & -1\\ 2 & -4\end{bmatrix}$ $\therefore (-1-s)(-4-s)+2=0$ $\quad\therefore s^2+5s+6=0$ $\quad\therefore s=-2$ and $-3$

$\therefore e^{-2t}=u_0-2u_1$ and $e^{-3t}=u_0-3u_1$ $\quad\therefore u_1=e^{-2t}-e^{-3t}$ and $u_0=3e^{-2t}-2e^{-3t}$

$$e^{t[a]}=\begin{bmatrix}(3e^{-2t}-2e^{-3t}) & \\ 0 & (3e^{-2t}-2e^{-3t})\end{bmatrix}+\begin{bmatrix}(-e^{-2t}+e^{-3t}) & (-e^{-2t}+e^{-3t})\\ (2e^{-2t}-2e^{-3t}) & (-4e^{-2t}+4e^{-3t})\end{bmatrix}$$

$$=\begin{bmatrix}(2e^{-2t}-e^{-3t}) & (-e^{-2t}+e^{-3t})\\ (2e^{-2t}-2e^{-3t}) & (-e^{-2t}+2e^{-3t})\end{bmatrix}$$

(b) $[a]=\begin{bmatrix}0 & -4\\ 1 & 0\end{bmatrix}$ $\therefore s^2+4=0$ $\quad s=\pm j2$ $\therefore e^{j2t}=u_0+j2u_1,\ e^{-j2t}=u_0-j2u_1$

$\therefore u_0=\cos 2t,\quad \cos 2t+j\sin 2t=\cos 2t+j2u_1$ $\quad\therefore u_1=\frac{1}{2}\sin 2t$

$$\therefore e^{t[a]}=\begin{bmatrix}\cos 2t & 0\\ 0 & \cos 2t\end{bmatrix}+\begin{bmatrix}0 & -2\sin 2t\\ \frac{1}{2}\sin 2t & 0\end{bmatrix}=\begin{bmatrix}\cos 2t & -2\sin 2t\\ \frac{1}{2}\sin 2t & \cos 2t\end{bmatrix}$$

**16-23** $e^{t[a]}=[I]+t[a]+\frac{1}{2!}t^2[a]^2+\ldots=[I]$ at $t=0$

**16-24**

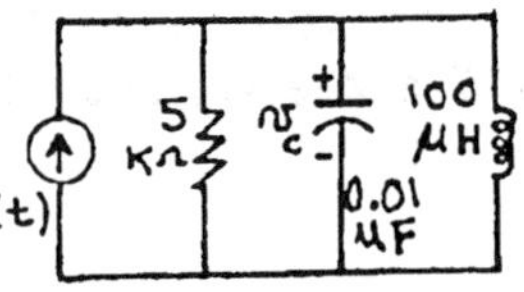


$\therefore 10^{-8}v_c'=0.1u(t)-\frac{v_c}{5000}-i,$

$\therefore v_c'=-2\times 10^4 v_c-10^8 i+10^7 u(t)$ and $i'=10^4 v_c$

$[a]=\begin{bmatrix}-2\times 10^4 & -10^8\\ 10^4 & 0\end{bmatrix}$, $\det[a]=10^{12}$, $s_{1,2}=-10^4\pm\sqrt{10^8-10^{12}}\doteq -10^4\pm j10^6$

$\therefore K_{1,2}=\frac{-10^4\pm j10^6+2\times 10^4}{-10^8}=-10^{-4}\mp j10^{-2},\quad K_2-K_1=j2\times 10^{-2}$

$$v_c=\frac{e^{(-10^4+j10^6)t}}{j0.02}\int_0^t(-10^{-4}+j0.02)10^7 e^{(10^4-j10^6)z}dz$$

$$+\frac{e^{(-10^4-j10^6)t}}{j0.02}\int_0^t(10^{-4}+j0.02)10^7 e^{(10^4+j10^6)z}dz$$

$$=\frac{e^{(-10^4+j10^6)t}}{j0.02}\ \cancel{\frac{-10^3+j10^5}{10^4-j10^6}}^{\,-0.1}\left[e^{(-10^4-j10^6)t}-1\right]$$

$$+\frac{e^{(-10^4-j10^6)t}}{j0.02}\ \cancel{\frac{10^3+j10^5}{10^4+j10^6}}^{\,0.1}\left[e^{(-10^4+j10^6)t}-1\right]$$

$=j5-j5e^{(-10^4+j10^6)t}-j5+j5e^{(-10^4-j10^6)t}$

$=\underline{10e^{-10^4 t}\sin 10^6 t}\ \text{V},\quad t>0$

**16-25**

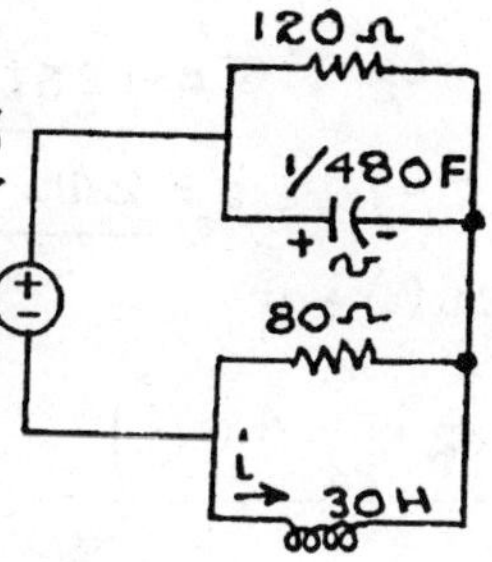


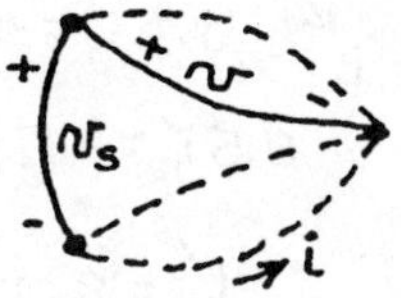

(a) $\frac{1}{480}v' = \frac{-v}{120} - i + \frac{v_s - v}{80}$

$\therefore v' = -10v - 480i + 6v_s$

$30i' = v - v_s \quad \therefore i' = \frac{1}{30}v - \frac{1}{30}v_s$

(b) $[a] = \begin{bmatrix} -10 & -480 \\ 1/30 & 0 \end{bmatrix}$

(c) $v_s = 100u(t)\,\text{V} \quad s_{1,2} = -5 \pm \sqrt{25-16} = -2 \text{ and } -8$

$k_{1,2} = \frac{(-2,-8)+10}{-480} = -\frac{1}{60}, -\frac{1}{240} \quad \therefore k_2 - k_1 = \frac{1}{80}$

$v = 80e^{-2t}\int_0^t \left[-\frac{1}{240}\times 600 + \frac{100}{30}\right]e^{2z}dz + 80e^{-8t}\int_0^t\left[-\frac{100}{30} + \frac{1}{60}\times 600\right]e^{8z}dz$

$= \left(-200 + \frac{800}{3}\right)e^{-2t}\frac{1}{2}(e^{2t}-1) + \left(-\frac{800}{3} + 800\right)e^{-8t}\frac{1}{8}(e^{8t}-1)$

$= \left(100 - \frac{100}{3}e^{-2t} - \frac{200}{3}e^{-8t}\right)\text{V}, \; t > 0$

(d) $v_s = 100e^{-20t}u(t) \therefore v = \frac{200}{3}e^{-2t}\int_0^t e^{-20z}e^{2z}dz + \frac{1600}{3}e^{-8t}\int_0^t e^{-20z}e^{8z}dz$

$\therefore v = -\frac{200}{54}e^{-2t}(e^{-18t}-1) - \frac{400}{9}e^{-8t}(e^{-12t}-1)$

$= -\frac{1300}{27}e^{-20t} + \frac{100}{27}e^{-2t} + \frac{400}{9}e^{-8t} \;\text{V}, \; t > 0$

**16-26**

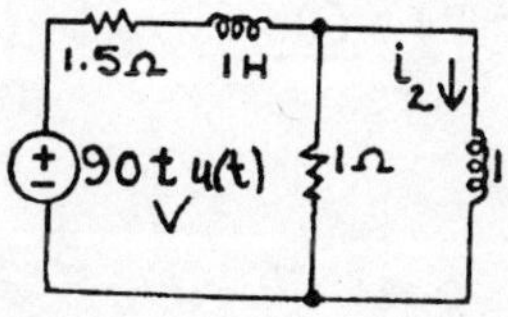


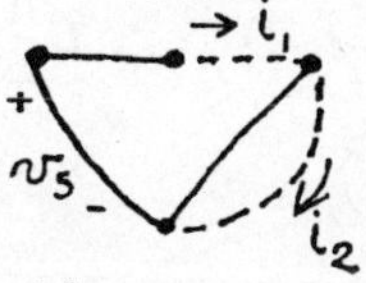

(a) $i_1' = -1.5i_1 + v_s + i_2 - i_1$

$\therefore i_1' = -2.5i_1 + i_2 + v_s$

$i_2' = 1(i_1 - i_2) \quad \therefore i_2' = i_1 - i_2$

$[a] = \begin{bmatrix} -2.5 & 1 \\ 1 & -1 \end{bmatrix}, \; [f] = \begin{bmatrix} 90t\,u(t) \\ 0 \end{bmatrix}, \; s_{1,2} = -\frac{7}{4} \pm \sqrt{\frac{49}{16} - \frac{24}{16}} = -\frac{7}{4} \pm \frac{5}{4} = -\frac{1}{2}, -3$

$k_{1,2} = \frac{(-\frac{1}{2}, -3) + \frac{5}{2}}{1} = 2, -\frac{1}{2} \quad \therefore k_2 - k_1 = -\frac{5}{2}$

$i_2 = 2\left(-\frac{2}{5}\right)e^{-t/2}\int_0^t\left[-\frac{1}{2}90z\right]e^{z/2}dz - \frac{1}{2}\left(-\frac{2}{5}\right)e^{-3t}\int_0^t\left[-2\times 90z\right]e^{3z}dz$

$= 36e^{-t/2}\left[4e^{z/2}\left(\frac{1}{2}z - 1\right)\right]_0^t - 36e^{-3t}\left[\frac{1}{9}e^{3z}(3z-1)\right]_0^t$

$= -140 + 60t + 144e^{-t/2} - 4e^{-3t} \;\text{A}, \; t > 0$

(b)

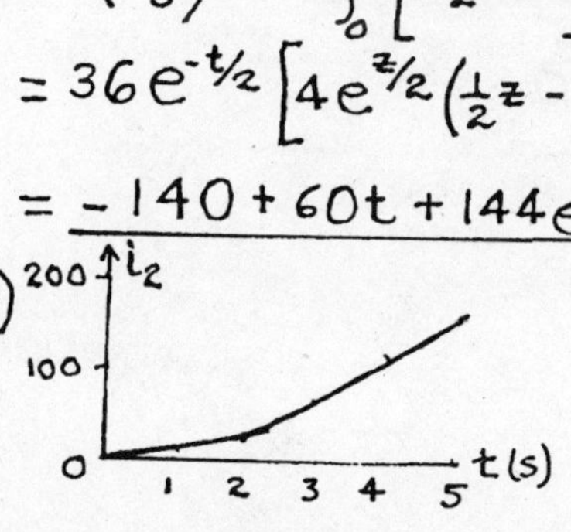


(b)

| t | $i_2$ |
|---|---|
| 0 | 0 |
| 0.5 | 1.255 |
| 1 | 7.14 |
| 1.5 | 18.0 |
| 2 | 33.0 |
| 2.5 | 51.25 |
| 3 | 72.1 |
| 4 | 119.5 |
| 5 | 171.8 |

**16-27**

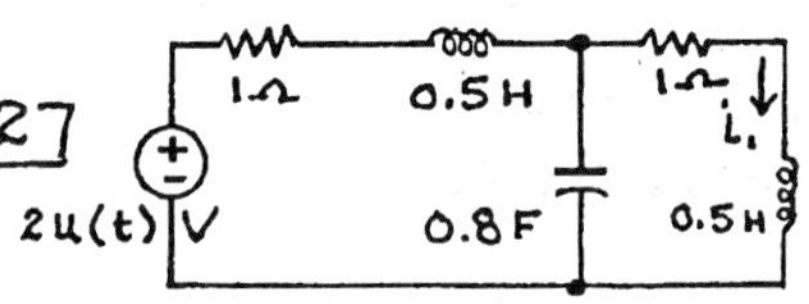


(a)

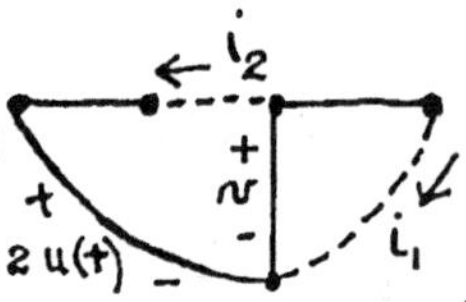


$0.8v' = -i_1 - i_2 \quad \therefore v' = \underline{-1.25 i_1 - 1.25 i_2},$

$0.5 i_1' = v - i_1, \quad \therefore i_1' = \underline{2v - 2i_1}$

$0.5 i_2' = -i_2 - 2u(t) + v$

$\therefore i_2' = \underline{2v - 2i_2 - 4u(t)}$

(b)

$$\frac{I_1}{V_s} = \frac{1}{1 + 0.5s + \dfrac{\frac{1.25}{s}(1+0.5s)}{\frac{1.25}{s} + 1 + 0.5s}} \times \frac{\frac{1.25}{s}}{\frac{1.25}{s} + 1 + 0.5s}$$

$$= \frac{1}{1 + 0.5s + \dfrac{1+0.5s}{1+0.8s+0.4s^2}} \qquad \frac{1}{1 + 0.8s + 0.4s^2}$$

$$= \frac{1}{1 + 0.8s + 0.4s^2 + 0.5s + 0.4s^2 + 0.2s^3 + 1 + 0.5s}$$

$$= \frac{1}{2 + 1.8s + 0.8s^2 + 0.2s^3} = \frac{5}{s^3 + 4s^2 + 9s + 10} = \frac{5}{(s+2)(s^2+2s+5)}$$

$\therefore s = \dfrac{-2 \pm \sqrt{4-20}}{2} = -1 \pm j2 \quad \therefore s_1 = \underline{-2}, \; s_2 = \underline{-1+j2}, \; s_3 = \underline{-1-j2}$

(c) $i_{1F}(t) = \underline{1\text{ A}}$

(d) $v(0) = i_1(0) = i_2(0) = \underline{0}$ (e) $i_1'(0^+) = \underline{0}$

(f) $i_1'' = 2v' - 2i_1' = -2.5 i_1 - 2.5 i_2 - 2i_1' \qquad \therefore i_1''(0^+) = \underline{0}$

(g) $i_1 = 1 + Ae^{-2t} + e^{-t}(B\cos 2t + C\sin 2t)$

$\therefore 0 = 1 + A + B, \quad i_1'(0^+) = -2A + 2C - B = 0$

$i_1' = -2Ae^{-2t} + e^{-t}(-2B\sin 2t + 2C\cos 2t - B\cos 2t - C\sin 2t)$

$= -2Ae^{-2t} + e^{-t}\left[(2C - B)\cos 2t - (2B + C)\sin 2t\right]$

$\therefore i_1''(0^+) = 4A - 4B - 2C - 2C + B = 0 \quad \therefore 4A - 3B - 4C = 0$

But, $-4A - 2B + 4C = 0 \quad \therefore B = 0, \quad A = -1, \quad C = -1$

$\therefore i_1(t) = \underline{1 - e^{-2t} - e^{-t}\sin 2t}$ A, $\quad t > 0$

# CHAPTER 17: FOURIER ANALYSIS

## Section 17-2 Trigonometric form of the Fourier series

**1** If $f(t) = 2 + 3\cos(10\pi t + 30°) + 4\cos(20\pi t + 60°) + \cos(30\pi t + 90°)$, find: (*a*) the average value of $f(t)$; (*b*) the effective value of $f(t)$; (*c*) the period of $f(t)$; (*d*) the value of $f(t)$ at $t = 0.05$ s.

**2** (*a*) Sketch $f(t) = 2\cos 2\pi t - \sin 4\pi t + 2\sin 6\pi t$ over the interval $0 < t < T$. (*b*) What is the maximum (positive) value of $f(t)$ in this interval?

**3** The function shown in Fig. 17-1 is periodic with $T = 10$ s. Find: *(a)* the average value; *(b)* the effective value; (c) the amplitude of the third harmonic.

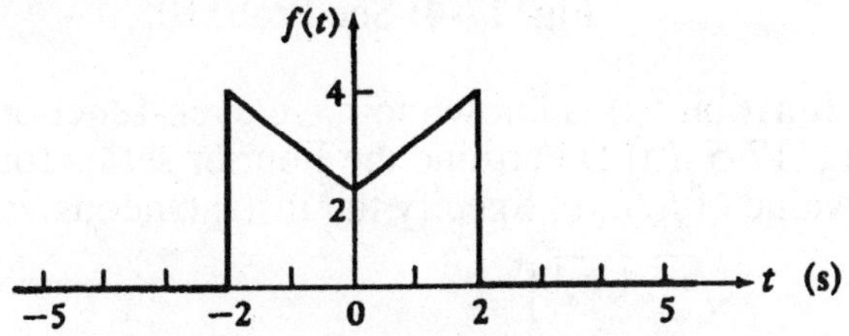

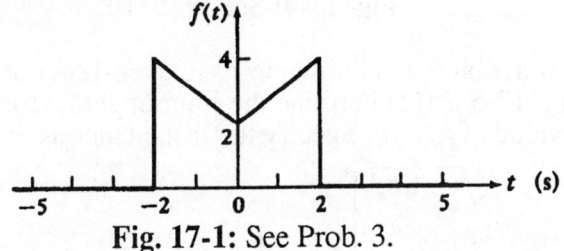

**Fig. 17-1:** See Prob. 3.

**4** For the periodic voltage shown in Fig. 17-2, find: (*a*) $T$; (*b*) $f_0$; (*c*) $\omega_0$; (*d*) $a_0$; (*e*) $b_1$.

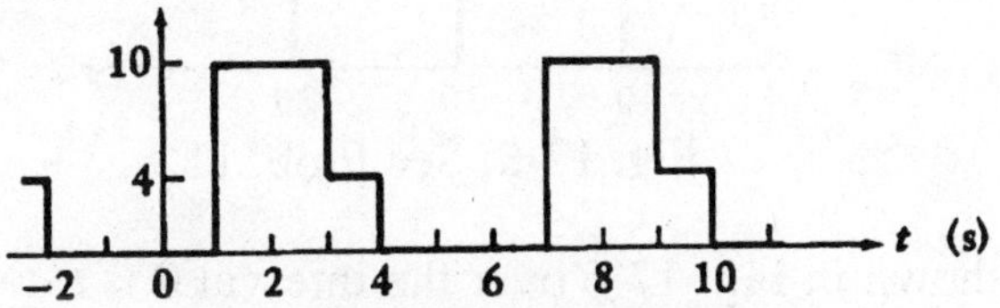


**Fig. 17-2:** See Prob. 4.

**5** Find $a_3$, $b_3$, and $\sqrt{a_3{}^2 + b_3{}^2}$ for the periodic waveform sketched in Fig. 17-3.

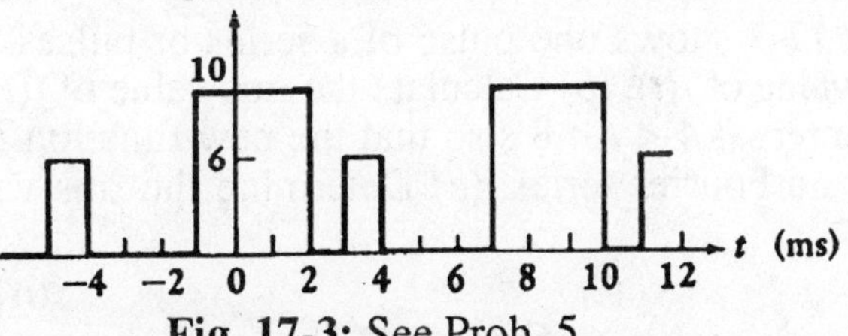


**Fig. 17-3:** See Prob. 5.

**6** Obtain the trigonometric form of the Fourier series, give the value of $T$, and determine the average value for each of these periodic functions of time: (*a*) $2.7\sin^2 90\pi t$; (*b*) $2.7\sin^3 90\pi t$; (*c*) $2.7\cos 89\pi t - 1.2\sin 90\pi t$.

**7** Given the periodic function, $v(t) = -50t$, $-0.02 < t < 0$ s, and $v(t) = 100t$, $0 < t < 0.01$ s, with $T = 0.03$ s, find: (a) $a_0$; (*b*) $a_n$.

**8** Let $f(t) = 20 + 8\cos 5t - 6\cos 10t + 3\cos 15t - 5\sin 5t + 6\sin 10t - 4\sin 15t + 2\sin 20t$. Find: (*a*) the period of $f(t)$; (*b*) the bandwidth in hertz required for the signal; (*c*) the average value of $f(t)$; (*d*) the effective value of $f(t)$; (*e*) the discrete amplitude and phase spectra of the signal.

**9** A periodic function is defined as $g(t) = 10\sin 20t$, $0 < t < \pi/60$ s, with $T = \pi/60$ s. Determine values for: (*a*) $a_0$; (*b*) $a_3$; (*c*) $b_3$.

## Section 17-3 The use of symmetry

**10** (*a*) Identify the symmetry of the waveform shown in Fig. 17-4. (*b*) Which of the $a_n$, $b_n$, or $a_0$ are zero? (*c*) Find $a_0$, $a_1$, $b_1$, $a_2$, and $b_2$.

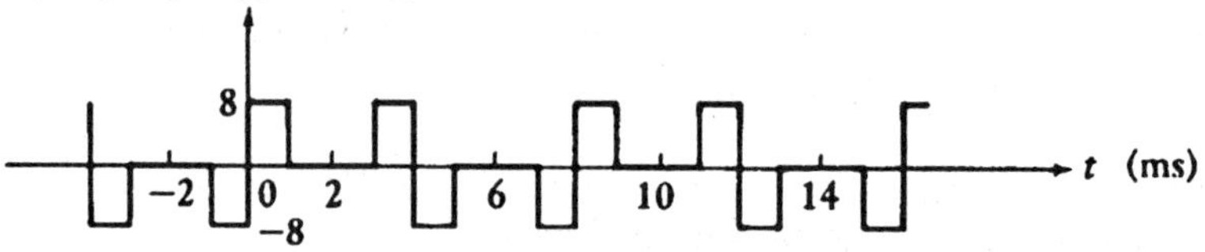


**Fig. 17-4:** See Prob. 10.

**11** A certain periodic function $f(t)$ is known to have even-function symmetry and the amplitude spectrum shown in Fig. 17-5. (*a*) Determine the Fourier series for $f(t)$ if $a_n$ and $b_n \geq 0$ for all $n$. (*b*) Find the effective value of $f(t)$. (*c*) Specify the instantaneous value of $f(t)$ at $t = 0.04$ s.

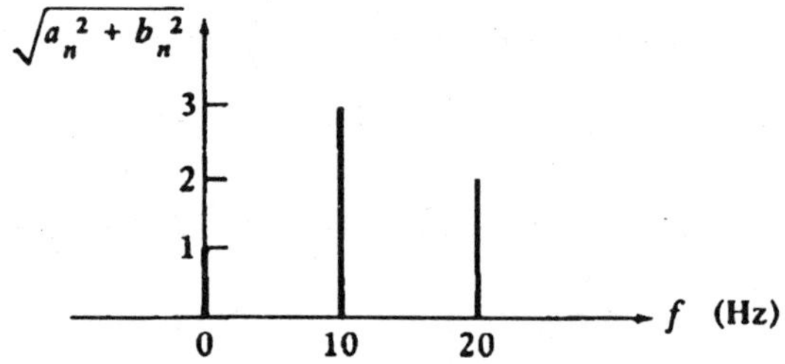

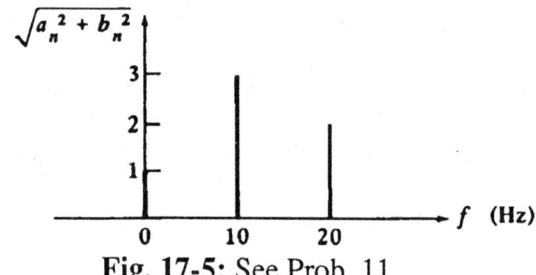

**Fig. 17-5:** See Prob. 11.

**12** A function $f(t)$ is shown in Fig. 17-6 over the interval $0 < t < 4$. Sketch a function $g(t)$ over $0 < t < 20$ that equals $f(t)$ for $0 < t < 4$ and also has: (*a*) $T = 8$ and even symmetry; (*b*) $T = 8$ and odd symmetry; (*c*) $T = 16$, even, and half-wave symmetry; (*d*) $T = 16$, odd, and half-wave symmetry.

**13** Assume that Fig. 17-7 shows one pulse of a series of pulses that repeats every 4 s. (*a*) Determine the average value of $f(t)$. (*b*) Calculate the rms value of $f(t)$. (*c*) Define a continuation of this function in the interval $4 < t < 8$ s so that the new function $f_{\text{new}}(t)$ is odd and has $T = 8$ s. (*d*) Express $f_{\text{new}}(t)$ as a Fourier series. (*e*) Determine the rms value of the third harmonic of $f_{\text{new}}(t)$.

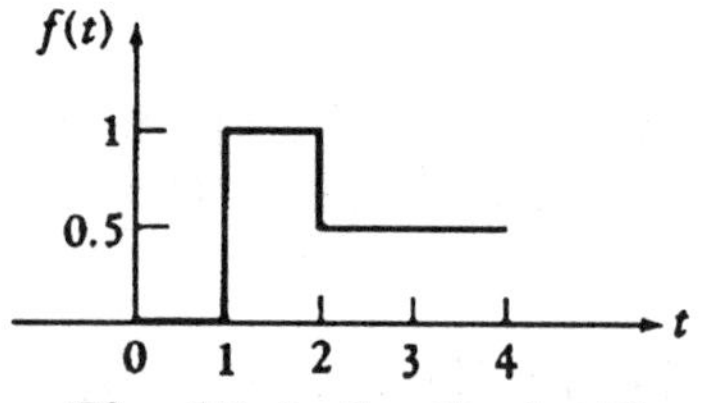


**Fig. 17-6:** See Prob. 12.

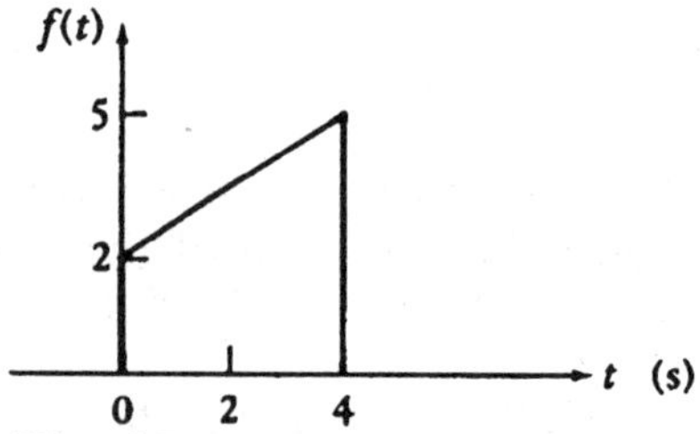


**Fig. 17-7:** See Prob. 13.

**14** Make use of symmetry as much as possible to obtain numerical values of $a_0$, $a_n$, and $b_n$, $1 \leq n \leq 10$, for the waveform shown in Fig. 17-8.

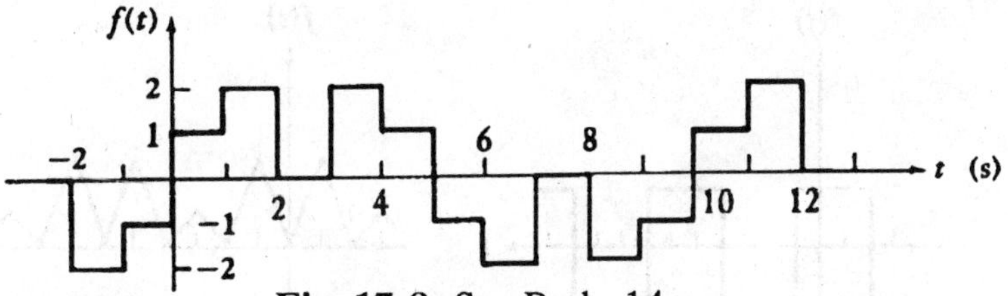


**Fig. 17-8:** See Prob. 14.

**15** In the interval $2 < t < 5$ ms, a function is given as $10\sin 1000\pi t$ for $2 < t < 3$ ms and zero for $0 < t < 2$ ms. The function has a period of 12 ms and possesses both even and half-wave symmetry. Find numerical values for $a_0$, $a_1$, $a_2$, $a_3$, $a_4$, and $a_5$.

**16** After investigating the symmetry shown by the waveform in Fig. 17-9, write its trigonometric Fourier series, giving complete numerical values for all harmonics up to the eighth.

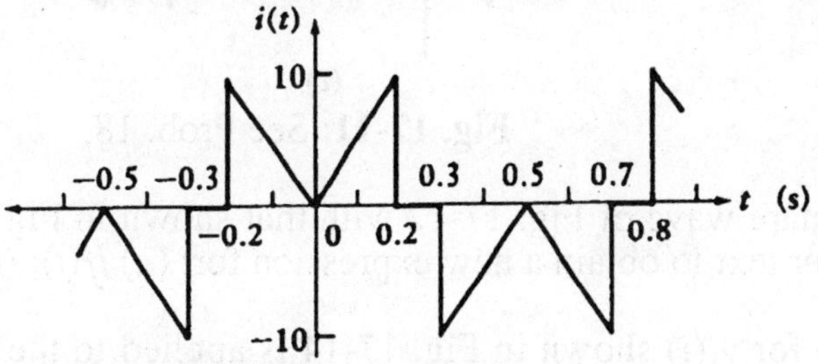


**Fig. 17-9:** See Prob. 16.

**17** For each of the waveforms sketched in Fig. 17-10, state whether or not it has even symmetry, odd symmetry, or half-wave symmetry, or some combination of these symmetries .

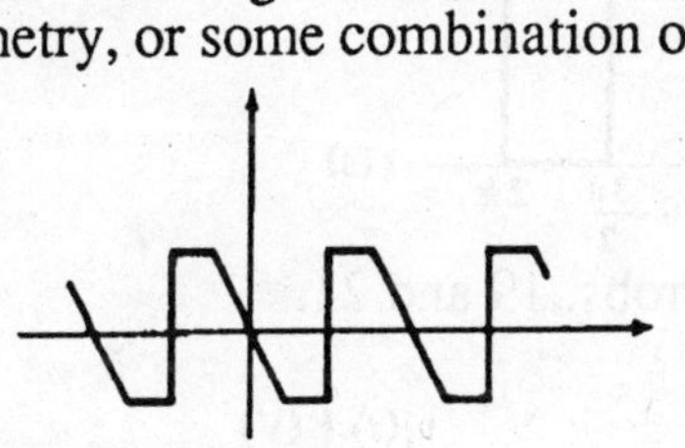

(*a*)

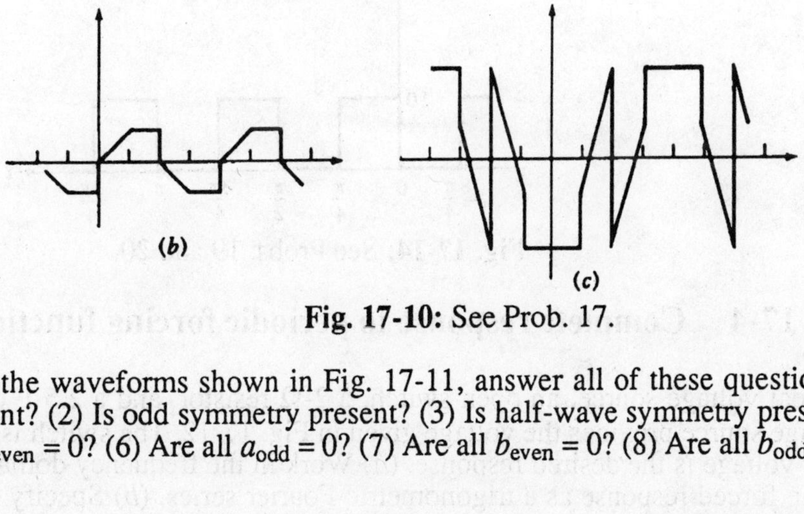

(*b*) (*c*)

**Fig. 17-10:** See Prob. 17.

**18** For each of the waveforms shown in Fig. 17-11, answer all of these questions: (1) Is even symmetry present? (2) Is odd symmetry present? (3) Is half-wave symmetry present? (4) Is $a_0 = 0$? (5) Are all $a_{\text{even}} = 0$? (6) Are all $a_{\text{odd}} = 0$? (7) Are all $b_{\text{even}} = 0$? (8) Are all $b_{\text{odd}} = 0$?

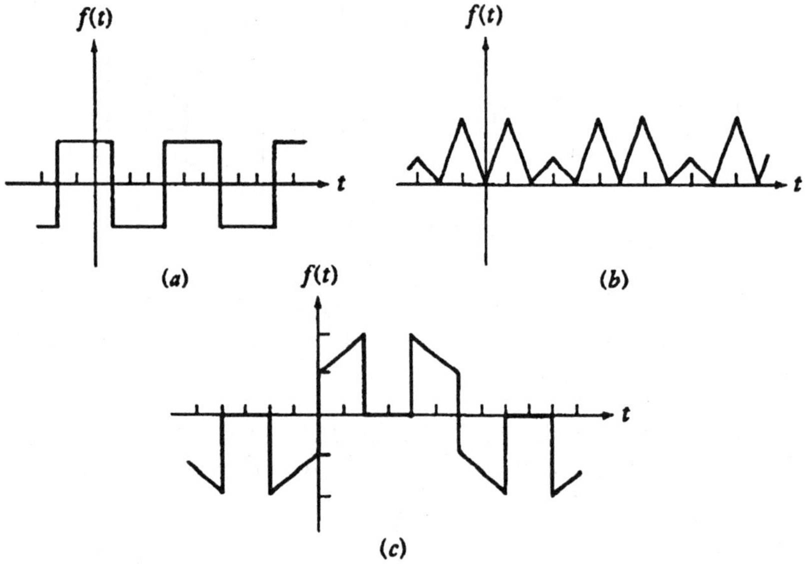


**Fig. 17-11:** See Prob. 18.

**19** Replace the square wave of Fig. 17-12 with that shown in Fig. 17-14 and repeat the analysis of Sec. 17-4 in your text to obtain a new expression for: (*a*) $i_f(t)$; (*b*) $i(t)$.

**20** The waveform for $v_s(t)$ shown in Fig. 17-14 is applied to the circuit of Fig. 17-13. Use the standard methods of transient analysis to calculate $i(t)$ at $t =$: (*a*) $\pi/4$ s; (*b*) $\pi/2$ s; (*c*) $3\pi/4$ s.

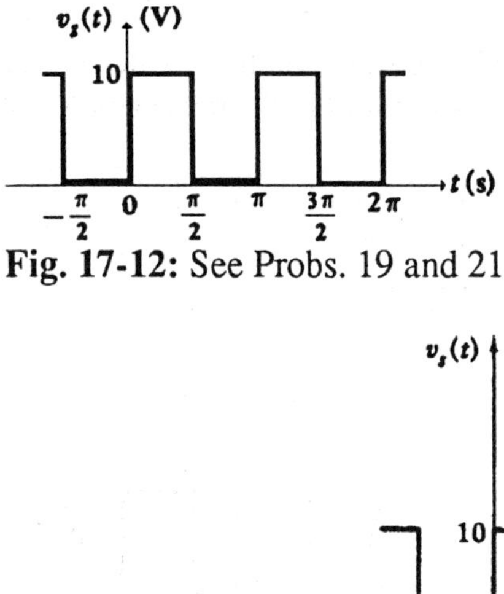


**Fig. 17-12:** See Probs. 19 and 21.

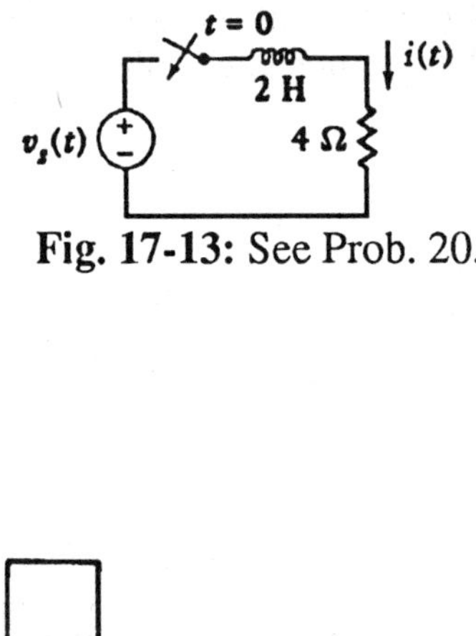


**Fig. 17-13:** See Prob. 20.

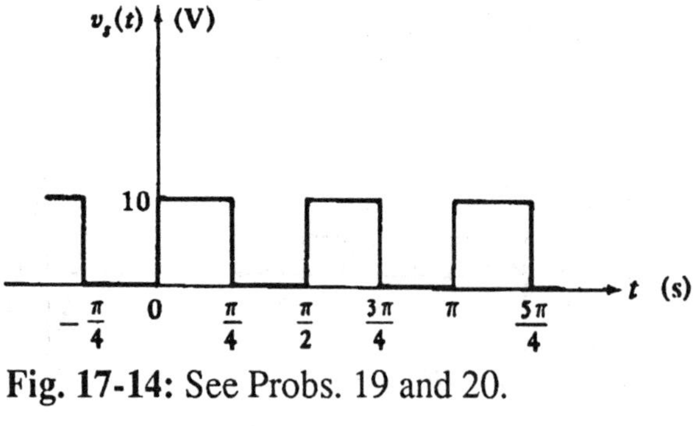


**Fig. 17-14:** See Probs. 19 and 20.

## Section 17-4 Complete response to periodic forcing functions

**21** An ideal voltage source, an open switch, a 2-Ω resistor, and a 2.5-F capacitor are in series. The voltage source provides the voltage given in Fig. 17-12. The switch is closed at $t = 0$ and the capacitor voltage is the desired response. (*a*) Work in the frequency domain of the *n*th harmonic to find the forced response as a trigonometric Fourier series. (*b*) Specify the functional form of the natural response. (*c*) Determine the complete response.

## Section 17-5 Complex form of the Fourier series

**22** Determine $\mathbf{c}_3$, $\mathbf{c}_{-3}$, $|\mathbf{c}_3|$, $a_3$, $b_3$, and $\sqrt{a_3^2 + b_3^2}$ for the waveform of Fig. 17-15 knowing that $T = 6$ ms.

**23** Find the complex Fourier series for the periodic waveform shown in Fig. 17-16. Give numerical values for $\mathbf{c}_n$, $n = 0, \pm 1, \pm 2, \pm 3$.

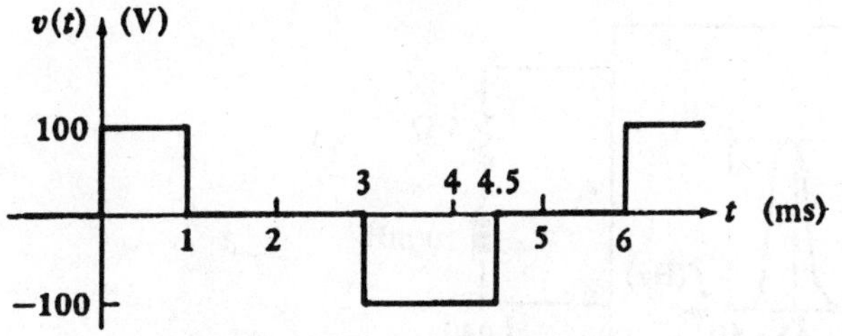

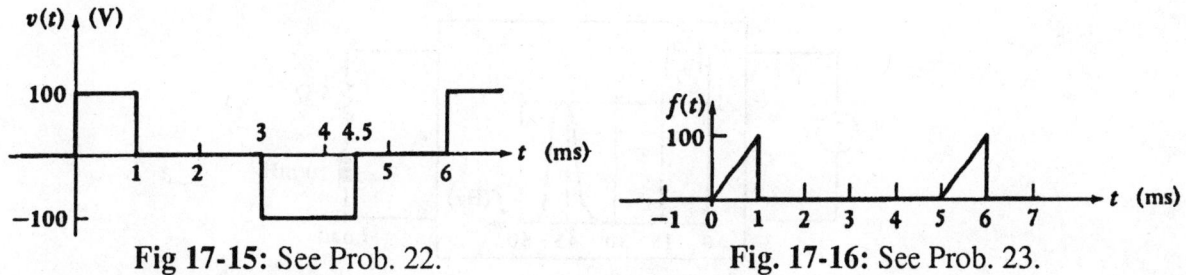

Fig 17-15: See Prob. 22.

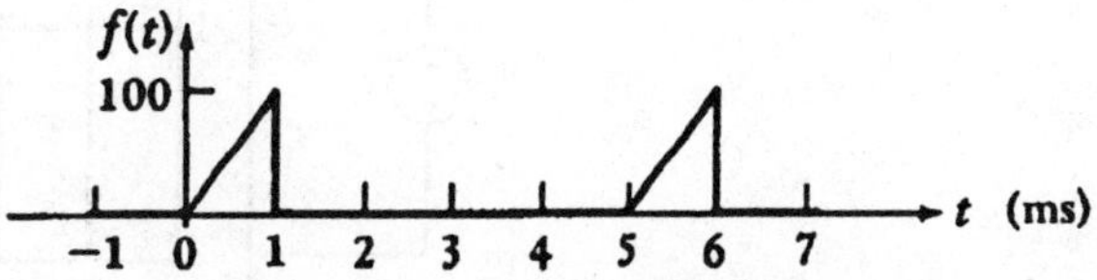


**Fig. 17-16:** See Prob. 23.

**24** A periodic sequence of positive pulses is composed of pulses having an amplitude of 20 V, a duration of 0.4 $\mu$s, and a repetition rate of 3000 pulses per second. (*a*) At what frequency is the envelope of the frequency spectrum equal to zero? (*b*) What is the frequency separation of the spectral lines? (*c*) What is the value of $|\mathbf{c}_n|$ for the frequency component nearest to 10 kHz? (*d*) . . . nearest to 1 MHz? (*e*) What nominal bandwidth is required in an amplifier to transmit this pulse train with reasonable fidelity? (*f*) How many spectral components lie between 1 and 1.1 Mrad/s? (*g*) Find $|\mathbf{c}_{227}|$ and the frequency to which it corresponds.

**25** The complex Fourier coefficients for a periodic voltage are: $\mathbf{c}_0 = 10$, $\mathbf{c}_1 = 6 - j6$, $\mathbf{c}_2 = 4 + j2$, $\mathbf{c}_3 = -2 - j3$, and $\mathbf{c}_n = 0$ for $|n| \geq 4$. Let $T = 10$ ms. (*a*) Evaluate $v(0.002 \text{ s})$. (*b*) What is the effective value of this voltage?

**26** A periodic wave with $T = 20$ ms has complex Fourier coefficients given by $\mathbf{c}_0 = 1$ A and $\mathbf{c}_n = (1/n^2) - j(2/n^3)$ A, for $|n| \geq 1$. (*a*) Sketch a line spectrum for $|\mathbf{c}_n|$. (*b*) What average power does this current deliver to a 5-$\Omega$ resistor?

**27** Let $f(t) = 1$ for $-3 < t < -2$ $\mu$s and $2 < t < 3$ $\mu$s, and $f(t) = 0$ elsewhere in the period interval, $-12.5 < t < 12.5$ $\mu$s; $T = 25$ $\mu$s. This series of pulses might represent the decimal number 3 being transmitted in binary form by a digital computer. After determining the complex Fourier series for $f(t)$: (*a*) give the value of $\mathbf{c}_4$; (*b*) find $\mathbf{c}_0$; (*c*) find $|\mathbf{c}_n|_{max}$; (*d*) find $N$ so that $|\mathbf{c}_n| \leq 0.01|\mathbf{c}_n|_{max}$ for all $n > N$. What bandwidth is required to transmit this portion of the spectrum?

**28** The signal shown in Fig. 17-17 is periodic with $T = 1$ ms. (*a*) Find $\mathbf{c}_n$. (*b*) Evaluate $\mathbf{c}_5$.

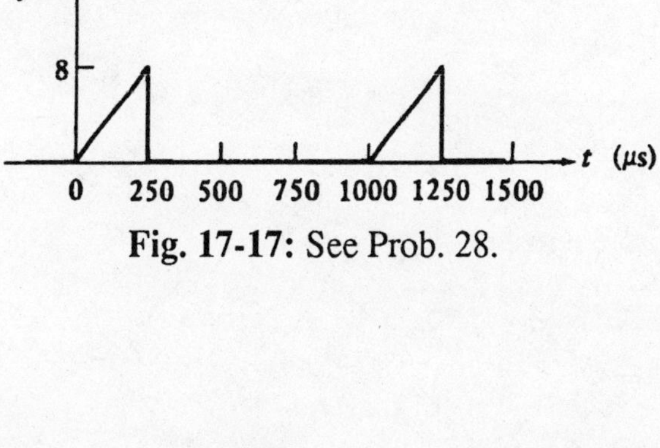


**Fig. 17-17:** See Prob. 28.

**29** A certain periodic signal, having a period of 9 s, is described as follows: $f(t) = 2t + 4$, $-2 < t < -1$ s; $f(t) = 2$, $-1 < t < 1$ s; $f(t) = 4 - 2t$, $1 < t < 2$ s; $f(t) = 0$ elsewhere. (*a*) Find the general coefficient $\mathbf{c}_n$ of the complex Fourier series. (*b*) Evaluate $\mathbf{c}_0$ and $\mathbf{c}_1$.

**30** A voltage waveform $v_s(t)$ has a period of $\frac{1}{12}$ s and is defined by: $v_s = 60$ V, $0 < t < \frac{1}{96}$ s; $v_s = 0$, $\frac{1}{96} < t < \frac{1}{12}$ s. (*a*) Find $\mathbf{c}_4$. (*b*) The voltage $v_s$, is applied as the source in the circuit shown in Fig. 17-18. What average power is delivered to the load?

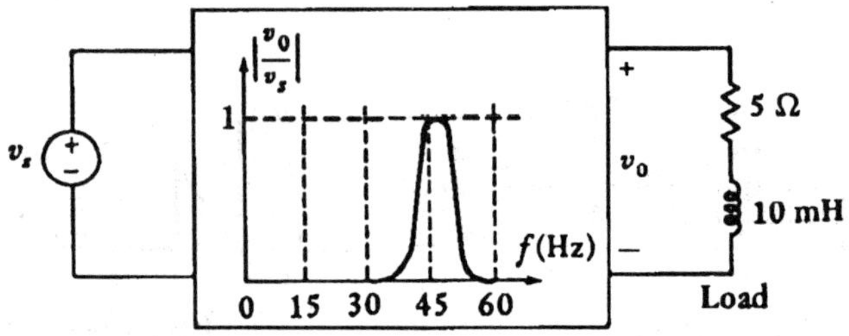


**Fig. 17-18:** See Prob. 30.

# Chapter 17

**17-1** $f(t) = 2 + 3\cos(10\pi t + 30^\circ) + 4\cos(20\pi t + 60^\circ) + \cos(30\pi t + 90^\circ)$

(a) $F_{av} = \underline{2}$ (b) $F_{eff} = \sqrt{2^2 + \frac{1}{2}\times 3^2 + \frac{1}{2}\times 4^2 + \frac{1}{2}\times 1^2} = \sqrt{17} = \underline{4.123}$

(c) $T = \frac{2\pi}{\omega_0} = \frac{2\pi}{10\pi} = \underline{0.2\ s}$

(d) $f(0.05) = 2 + 3\cos(90^\circ + 30^\circ) + 4\cos(180^\circ + 60^\circ) + \cos(270^\circ + 90^\circ) = 2 - 1.5 - 2 + 1 = \underline{-0.500}$

**17-2** (a)

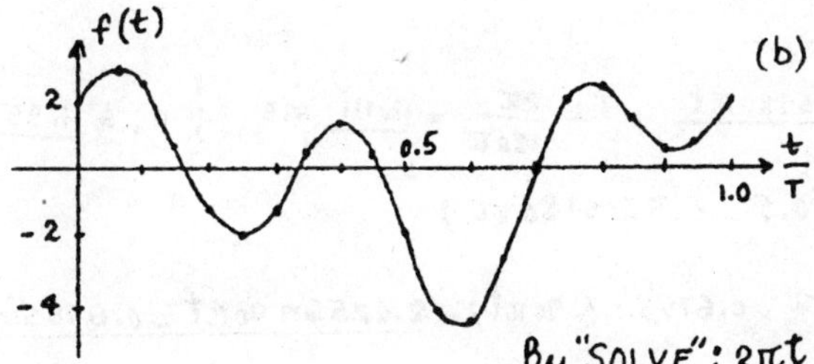


(b) $f(t) = 2\cos 2\pi t - \sin 4\pi t + 2\sin 6\pi t$

$f'(t) = -4\pi\sin 2\pi t - 4\pi\cos 4\pi t + 12\pi\cos 6\pi t = 0$

$\therefore 3\cos 6\pi t = \sin 2\pi t + \cos 4\pi t$

By "SOLVE": $2\pi t = 0.400120406 \therefore \frac{t}{T} = 0.063681140$

$f_{max} = \underline{2.988844184}$

**17-3**

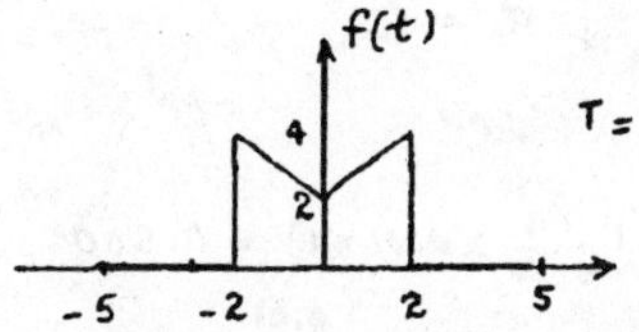


$T = 10\ s$ (a) $F_{av} = \frac{1}{10}(2\times 2 + 2\times 1 + 2\times 2 + 2\times 1) = \underline{1.2}$

(b) $F_{eff}^2 = \frac{1}{10}\left[2\times\int_0^2 (2+t)^2\,dt\right] = 0.2\int_0^2 (4 + 4t + t^2)\,dt$

$= 0.2\left[4t + 2t^2 + \frac{1}{3}t^3\right]_0^2 = 0.2\left(8 + 8 + \frac{8}{3}\right) = 3.733$

$\therefore F_{eff} = \underline{1.9322}$

(c) $a_3 = 2\times\frac{2}{10}\int_0^2 (2+t)\cos 0.6\pi t\,dt = 0.8\int_0^2 \cos 0.6\pi t\,dt + 0.4\int_0^2 t\cos 0.6\pi t\,dt$

$= \frac{4}{3\pi}\sin 0.6\pi t\Big]_0^2 + 0.4\frac{1}{0.36\pi^2}\cos 0.6\pi t\Big]_0^2 + \frac{2t}{3\pi}\sin 0.6\pi t\Big]_0^2$

$= \frac{4}{3\pi}\sin 1.2\pi + \frac{10}{9\pi^2}(\cos 1.2\pi - 1) + \frac{4}{3\pi}\sin 1.2\pi = -0.7026 \quad \therefore |a_3| = \underline{0.7026}$

**17-4**

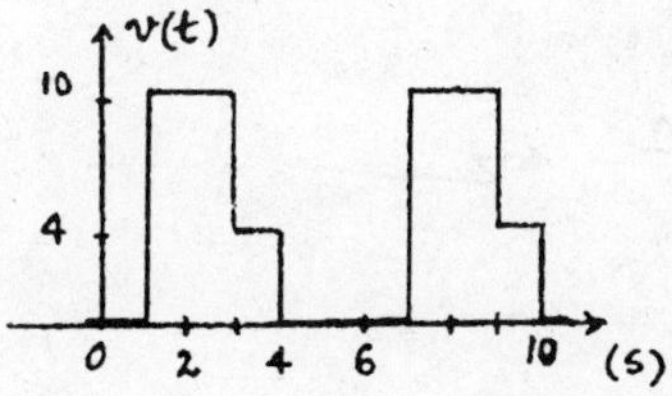


(a) $T = \underline{6\ s}$ (b) $f_0 = \underline{1/6\ Hz}$ (c) $\omega_0 = \underline{\pi/3}\ \frac{rad}{s}$

(d) $a_0 = \frac{1}{6}(2\times 10 + 1\times 4) = \underline{4}$

(e) $b_1 = \frac{2}{6}\left[\int_1^3 10\sin\frac{\pi t}{3}\,dt + \int_3^4 4\sin\frac{\pi t}{3}\,dt\right]$

$\therefore b_1 = \frac{10}{3}\left(-\frac{3}{\pi}\right)\left(\cos\pi - \cos\frac{\pi}{3}\right) + \frac{4}{3}\left(-\frac{3}{\pi}\right)\left(\cos\frac{4\pi}{3} - \cos\pi\right) = -\frac{10}{\pi}(-1 - 0.5) - \frac{4}{\pi}(-0.5 + 1) = \frac{13}{\pi} = \underline{4.138}$

17-5

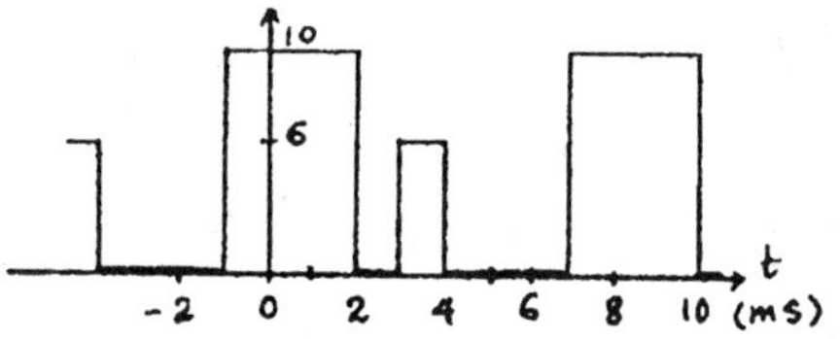


$T = 8\text{ ms}$, $f_0 = \frac{1}{T} = 125$, $\omega_0 = 2\pi T = 250\pi$

$$a_3 = \frac{2}{0.008}\left[\int_{-0.001}^{0.002} 10\cos 750\pi t\,dt + \int_{0.003}^{0.004} 6\cos 750\pi t\,dt\right]$$

$$= 250\left(\frac{1}{75\pi}\right)\sin 750\pi t\Big|_{-0.001}^{0.002} + 250\left(\frac{1}{125\pi}\right)\sin 750\pi t\Big|_{0.003}^{0.004}$$

$$a_3 = \frac{1}{0.3\pi}(\sin 1.5\pi + \sin 0.75\pi) + \frac{2}{\pi}(\sin 3\pi - \sin 2.25\pi) = \frac{1}{0.3\pi}\left(-1 + \frac{\sqrt{2}}{2}\right) + \frac{2}{\pi}\left(0 - \frac{\sqrt{2}}{2}\right) = \underline{-0.7609}$$

$$b_3 = 250\left[\int_{-0.001}^{0.002} 10\sin 750\pi t\,dt + \int_{0.003}^{0.004} 6\sin 750\pi t\,dt\right] = -\frac{10}{3\pi}(\cos 1.5\pi - \cos 0.75\pi) - \frac{2}{\pi}(\cos 3\pi - \cos 2.25\pi)$$

$$= \frac{10}{-3\pi}\left(0 + \frac{\sqrt{2}}{2}\right) - \frac{2}{\pi}\left(-1 - \frac{\sqrt{2}}{2}\right) = \underline{0.3365}^{+}\ ;\ \sqrt{a_3^2 + b_3^2} = \underline{0.8320}$$

17-6 (a) $f(t) = 2.7\sin^2 90\pi t = \underline{1.35 - 1.35\cos 180\pi t}$, $T = \frac{2\pi}{180\pi} = \underline{11.111\text{ ms}}$ ; $a_0 = \underline{1.35}$

(b) $f(t) = 2.7\sin^3 90\pi t = 2.7\sin 90\pi t\,(0.5 - 0.5\cos 180\pi t)$

$$= 1.35\sin 90\pi t - 0.675\sin 270\pi t - 0.675\sin(-90\pi t) = \underline{2.025\sin 90\pi t - 0.675\sin 270\pi t}$$

$T = \frac{2\pi}{90\pi} = \underline{22.22\text{ ms}}$, $a_0 = \underline{0}$

(c) $f(t) = \underline{2.7\cos 89\pi t - 1.2\sin 90\pi t}$, $T = \frac{2\pi}{\pi} = \underline{2\text{ s}}$, $a_0 = \underline{0}$

17-7

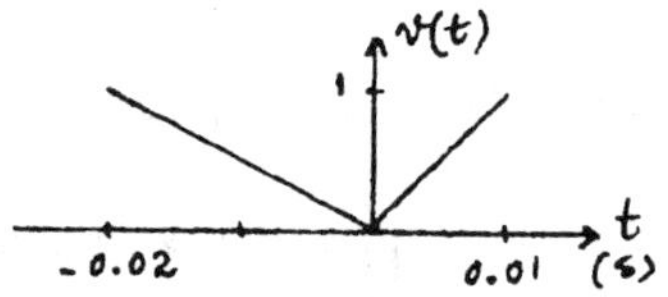


$T = 0.03$, $\omega_0 = \frac{2\pi}{0.03}$

(a) $a_0 = \frac{1}{0.03}\left(\frac{1}{2}\times 0.02\times 1 + \frac{1}{2}\times 0.01\times 1\right) = \underline{0.500}$

(b) $$a_n = \frac{2}{0.03}\left[\int_{-0.02}^{0} -50t\cos\frac{2\pi n t}{0.03}dt + \int_{0}^{0.01} 100t\cos\frac{2\pi n}{0.03}t\,dt\right]$$

$$= -\frac{100}{0.03}\left(\frac{0.03^2}{4\pi^2 n^2}\cos\frac{2\pi n t}{0.03}\Big|_{-0.02}^{0} + \frac{0.03t}{2\pi n}\sin\frac{2\pi n t}{0.03}\Big|_{-0.02}^{0}\right) +$$

$$+ \frac{200}{0.03}\left(\frac{0.03^2}{4\pi^2 n^2}\cos\frac{2\pi n t}{0.03}\Big|_{0}^{0.01} + \frac{0.03t}{2\pi n}\sin\frac{2\pi n t}{0.03}\Big|_{0}^{0.01}\right)$$

(b) $$a_n = -\frac{3}{4\pi^2 n^2}\left(1 - \cos\frac{4\pi n}{3}\right) - \frac{50}{n\pi}\left(-0.02\sin\frac{4\pi n}{3}\right) + \frac{6}{4\pi^2 n^2}\left(\cos\frac{2\pi n}{3} - 1\right) + \frac{100}{n\pi}\left(0.01\sin\frac{2\pi n}{3}\right)$$

$$= -\frac{9}{4\pi^2 n^2} + \frac{3}{4\pi^2 n^2}\cos\frac{4\pi n}{3} + \frac{6}{4\pi^2 n^2}\cos\frac{2\pi n}{3} + \frac{1}{n\pi}\sin\frac{4\pi n}{3} + \frac{1}{n\pi}\sin\frac{2\pi n}{3} \leftarrow$$

But $\sin\frac{4\pi n}{3} = -\sin\frac{2\pi n}{3}$ for all $n$

$$\therefore a_n = -\frac{9}{4\pi^2 n^2} + \frac{3}{4\pi^2 n^2}\cos\frac{4\pi n}{3} + \frac{6}{4\pi^2 n^2}\cos\frac{2\pi n}{3} \leftarrow$$

17-8 $f(t) = 20 + 8\cos 5t - 6\cos 10t + 3\cos 15t - 5\sin 5t + 6\sin 10t - 4\sin 15t + 2\sin 20t$

(a) $\omega_0 = 5$ , $T = \underline{0.4\pi}$ s

(b) $f_0 = \frac{\omega_0}{2\pi} = 5/2\pi$ , $B = 4f_0 = 10/\pi = \underline{3.183}$ Hz (c) $F_{av} = \underline{20}$

(d) $F_{eff} = \sqrt{20^2 + \frac{1}{2}(8^2 + 6^2 + 3^2 + 5^2 + 6^2 + 4^2 + 2^2)} = \sqrt{495} = \underline{22.25}$

(e)

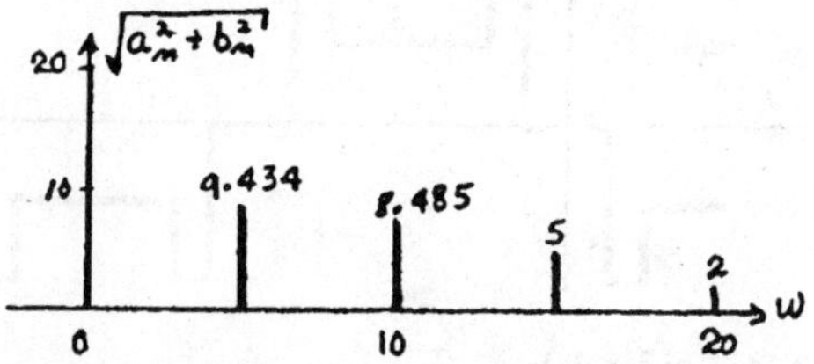


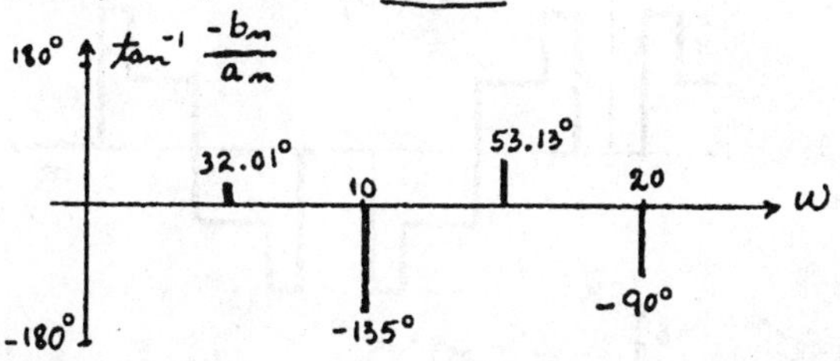


17-9

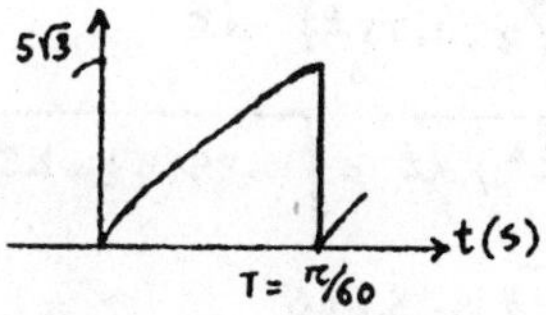


(a) $a_0 = \frac{60}{\pi}\int_0^{\pi/60} 10\sin 20t\,dt = \frac{30}{\pi}(-\cos 20t)\Big|_0^{\pi/60} = \frac{30}{\pi}\left(1 - \frac{1}{2}\right) = \underline{4.775}$

(b) $a_3 = \frac{120}{\pi}\int_0^{\pi/60} 10\sin 20t \cos\frac{6\pi t}{\pi/60}\,dt = \frac{1200}{\pi}\int_0^{\pi/60} \sin 20t\cos 360t\,dt$

$= \frac{600}{\pi}\int_0^{\pi/60}(\sin 380t - \sin 340t)\,dt = \frac{600}{\pi}\left(-\frac{1}{380}\cos 380t + \frac{1}{340}\cos 340t\right]_0^{\pi/60}$

$\therefore a_3 = \frac{600}{\pi}\left(-\frac{1}{380}\cos\frac{19\pi}{3} + \frac{1}{340}\cos\frac{17\pi}{3} + \frac{1}{380} - \frac{1}{340}\right) = \frac{600}{\pi}\left(\frac{1}{760} - \frac{1}{680}\right) = \underline{-0.02956}$ ←

(c) $b_3 = \frac{1200}{\pi}\int_0^{\pi/60}\sin 20t\sin 360t\,dt = \frac{600}{\pi}\int_0^{\pi/60}(-\cos 380t + \cos 340t)\,dt = \frac{600}{\pi}\left(-\frac{\sin 380t}{380} + \frac{\sin 340t}{340}\right)\Big|_0^{\pi/60}$

$= \frac{600}{\pi}\left(-\frac{1}{380}\sin\frac{19\pi}{3} + \frac{1}{340}\sin\frac{17\pi}{3}\right) = \frac{600}{\pi}\left(-\frac{\sqrt{3}}{2}\frac{1}{380} - \frac{\sqrt{3}}{2}\frac{1}{340}\right) = \underline{-0.9217}$ ←

17-10

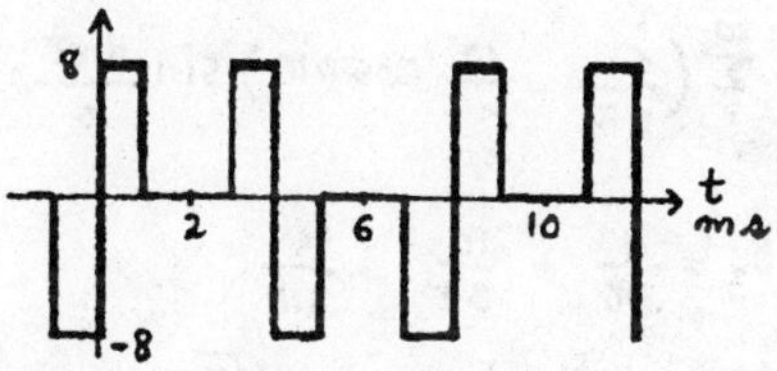


(a) Odd , half-wave

(b) $a_0 = \underline{0}$ , $a_n = \underline{0}$ , $b_{even} = \underline{0}$

(c) $b_1 = \frac{8}{T}\int_0^{T/4} f(t)\sin\omega_0 t\,dt = \frac{8}{0.008}\int_0^{0.001} 8\sin\frac{2\pi t}{0.008}\,t\,dt$

$= 8\left(-\frac{4}{\pi}\right)\cos\frac{\pi t}{0.004}\Big|_0^{0.001} = -\frac{32}{\pi}\left(\frac{\sqrt{2}}{2} - 1\right) = \underline{2.983}$

$a_0 = a_1 = a_2 = b_2 = \underline{0}$

17-11

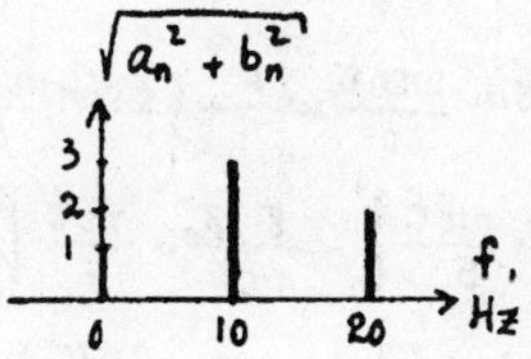


(a) Even sym. $\therefore f(t) = \underline{1 + 3\cos 20\pi t + 2\cos 40\pi t}$

(b) $F_{eff} = \sqrt{1^2 + \frac{1}{2}(3^2 + 2^2)} = \sqrt{7.5} = \underline{2.739}$

(c) $f(0.04) = 1 + 3\cos 0.8\pi + 2\cos 1.6\pi = \underline{-0.8090}$

17-12

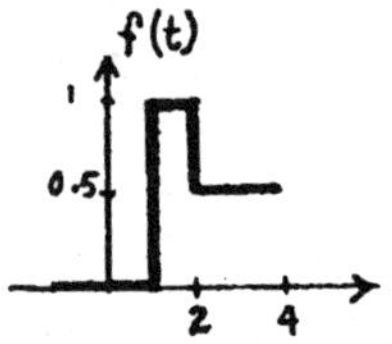


(a) even

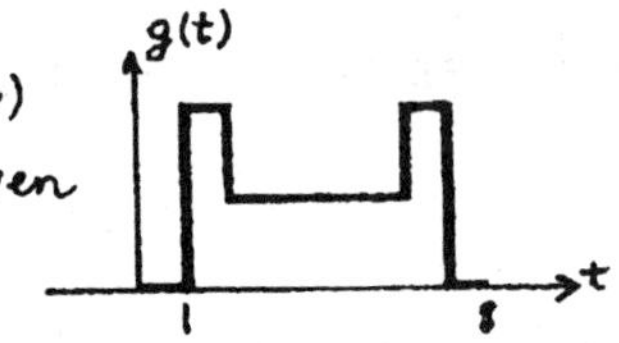


(b) odd

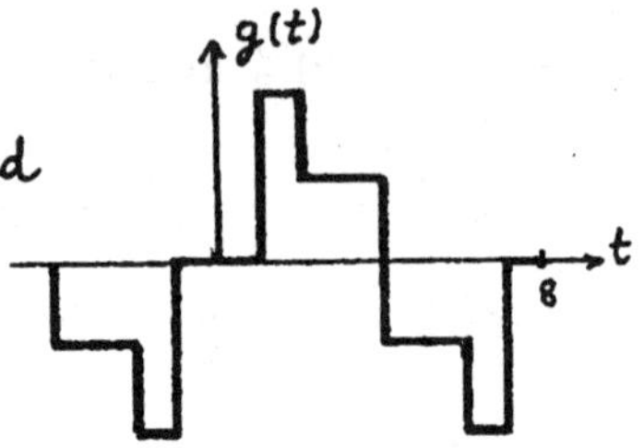


(c) even, HW

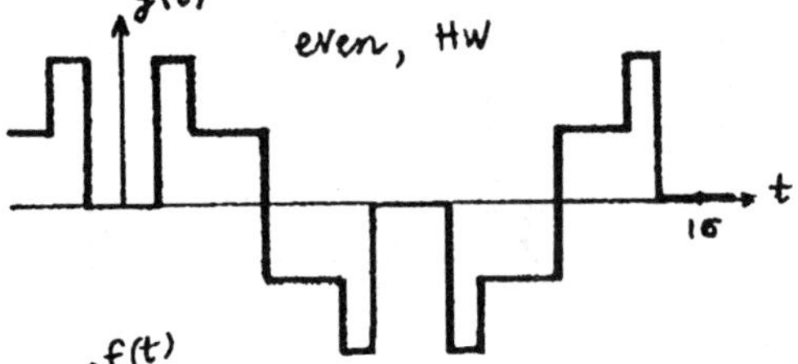


(d) odd, HW

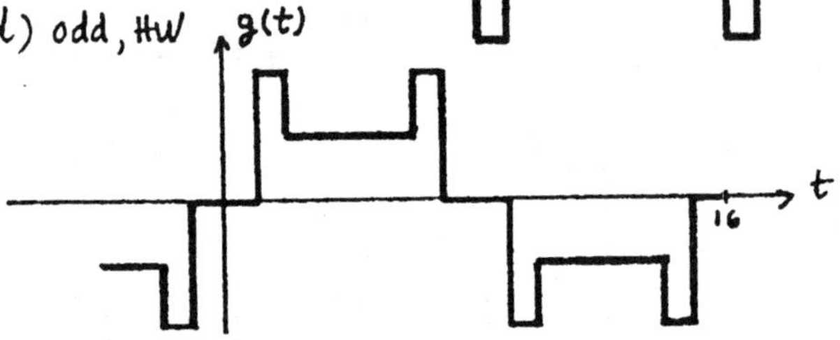


17-13

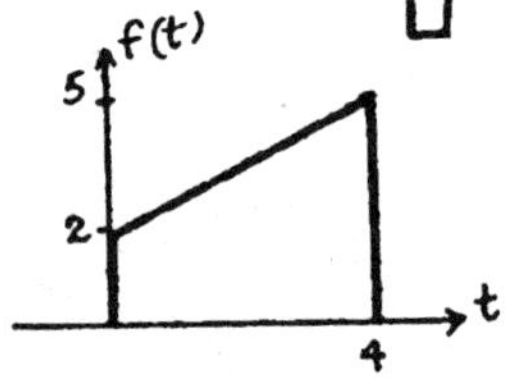


(a) $F_{av} = \frac{1}{4}\left(4\times 2 + \frac{1}{2}\times 3\times 4\right) = \underline{3.5}$

(b) $F_{rms} = \sqrt{\frac{1}{T}\int_0^T [f(t)]^2\, dt} = \sqrt{\frac{1}{4}\int_0^4 (2+0.75t)^2\, dt}$

$= \sqrt{0.25\int_0^4 (4+3t+0.5625t^2)\, dt} = \sqrt{0.25\left(4t+1.5t^2+0.1875t^3\right)\Big|_0^4}$

$= \sqrt{0.25(16+24+12)} = \sqrt{13} = \underline{3.606}$

(c)

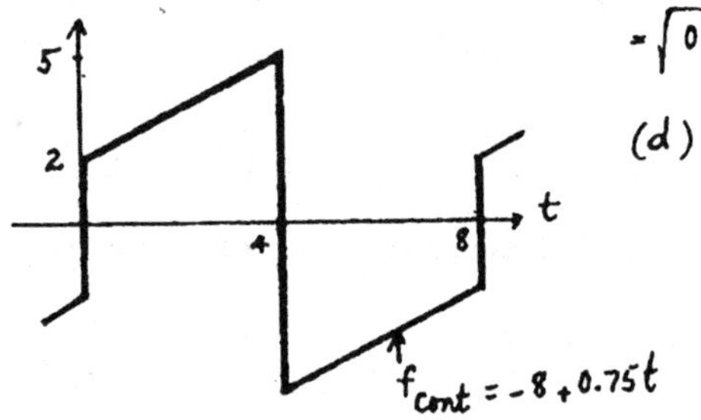


(d) $b_n = \frac{4}{8}\int_0^4 (2+0.75t)\sin\frac{2\pi n}{8}t\, dt$

$= -\frac{4}{n\pi}\cos\frac{n\pi t}{4}\Big|_0^4 + \frac{3}{8}\int_0^4 t\sin\frac{n\pi t}{4}\, dt$

$= \frac{4}{n\pi}(1-\cos n\pi) + \frac{3}{8}\left(\frac{16}{n^2\pi^2}\sin\frac{n\pi t}{4} - \frac{4t}{n\pi}\cos\frac{n\pi t}{4}\right)\Big|_0^4$

$= \frac{4}{n\pi}(1-\cos n\pi) - \frac{3}{8}\times\frac{4}{n\pi}\times 4\cos n\pi = \frac{4}{n\pi} - \frac{10}{n\pi}\cos n\pi$

$\therefore f_{new}(t) = \sum_1^\infty \left(\frac{4}{n\pi} - \frac{10}{n\pi}\cos n\pi\right)\sin\frac{n\pi t}{4}$ ←

(e) $n = 3 \quad \therefore b_3 = \frac{4}{3\pi} + \frac{10}{3\pi} = \frac{14}{3\pi}$

$\therefore b_{3,rms} = \frac{14}{3\pi\sqrt{2}} = \frac{7\sqrt{2}}{3\pi} = \underline{1.0504}$

17-14

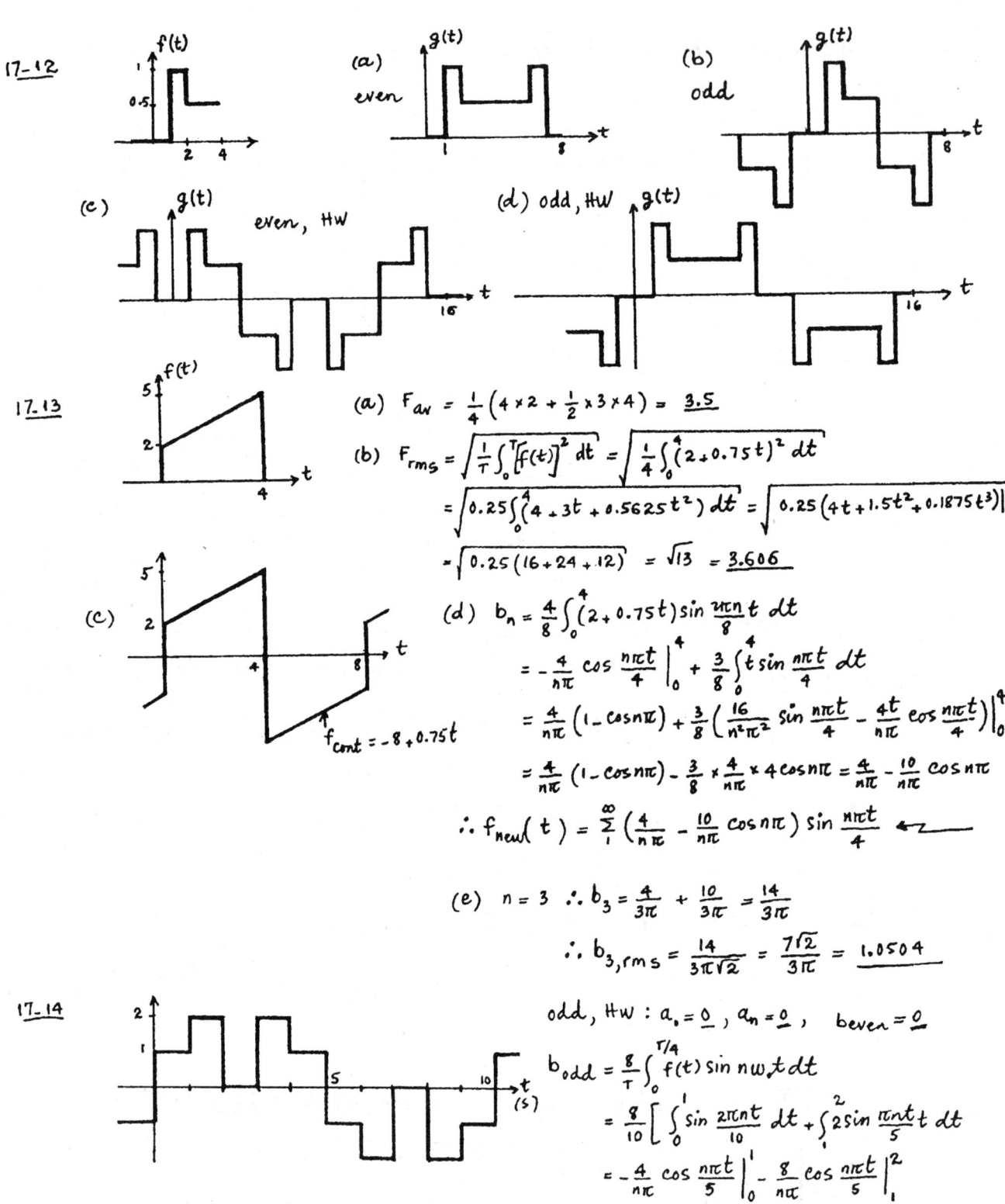

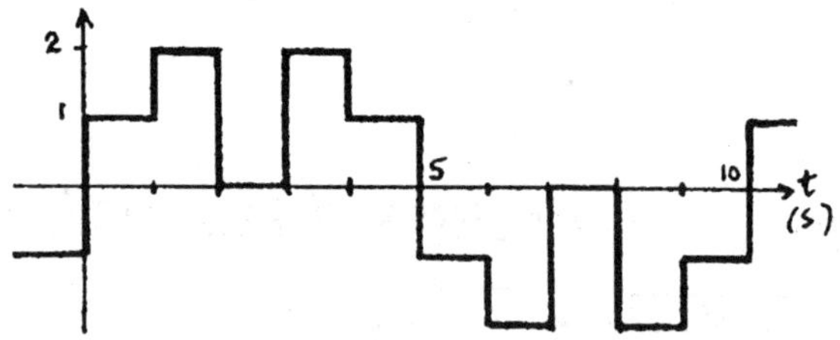


odd, HW : $a_0 = \underline{0}$, $a_n = \underline{0}$, $b_{even} = \underline{0}$

$b_{odd} = \frac{8}{T}\int_0^{T/4} f(t)\sin n\omega_0 t\, dt$

$= \frac{8}{10}\left[\int_0^1 \sin\frac{2\pi n t}{10}\, dt + \int_1^2 2\sin\frac{\pi n t}{5}\, dt\right]$

$= -\frac{4}{n\pi}\cos\frac{n\pi t}{5}\Big|_0^1 - \frac{8}{n\pi}\cos\frac{n\pi t}{5}\Big|_1^2$

$\therefore b_{odd} = -\frac{4}{n\pi}\cos\frac{n\pi}{5} + \frac{4}{n\pi} - \frac{8}{n\pi}\cos\frac{2n\pi}{5} + \frac{8}{n\pi}\cos\frac{n\pi}{5} = \frac{4}{n\pi}\left(1+\cos\frac{n\pi}{5} - 2\cos\frac{2n\pi}{5}\right)$

$\therefore b_1 = \underline{1.5164}$, $b_3 = \underline{0.9800}$, $b_5 = \underline{-0.5093}$, $b_7 = \underline{0.4200}$, $b_9 = \underline{0.16849}$

17-15

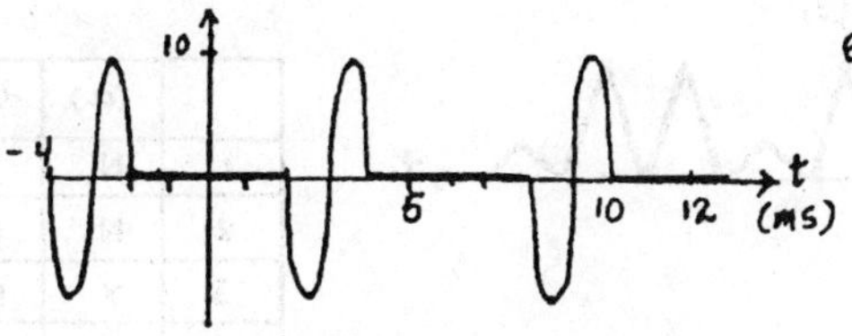


even, HW $\therefore a_0 = a_2 = a_4 = 0$ ←

$$a_n = \frac{8}{T}\int_0^{T/4} f(t)\cos n\omega_0 t\,dt$$

$$= \frac{8\times 10}{0.012}\int_{0.002}^{0.003} \sin\frac{6000\pi t}{6}\cos\frac{1000 n\pi t}{6}\,dt$$

$$\therefore a_n = \frac{40}{0.012}\int_{0.002}^{0.003}\left[\sin\frac{1000\pi(n+6)t}{6} + \sin\frac{1000\pi(6-n)t}{6}\right]dt$$

$$= -\frac{40}{12}\left[\frac{6}{\pi(6+n)}\cos\frac{1000\pi(n+6)t}{6} + \frac{6}{\pi(6-n)}\cos\frac{1000\pi(6-n)t}{6}\right]_{0.002}^{0.003}$$

$$= -\frac{20}{\pi}\left[\frac{1}{6+n}\cos 1000\pi(n+6)t + \frac{1}{6-n}\cos 1000\pi(6-n)t\right]_{0.002}^{0.003}$$

$$= -\frac{20}{\pi}\left[\frac{\cos(3\pi + n\pi/2)}{6+n} - \frac{\cos(2\pi + n\pi/3)}{6+n} + \frac{\cos(3\pi - n\pi/2)}{6-n} - \frac{\cos(2\pi + n\pi/3)}{6-n}\right]$$

$$a_n = -\frac{20}{\pi}\left[0 - \frac{\cos n\pi/3}{6+n} + 0 - \frac{\cos n\pi/3}{6-n}\right] = \frac{20}{\pi}\left(\frac{1}{6-n} + \frac{1}{6+n}\right)\cos\frac{n\pi}{3} = \frac{20}{\pi}\times\frac{12}{36-n^2}\cos\frac{n\pi}{3}$$

(only for n odd)

$\therefore a_1 = \underline{1.0913}$, $a_3 = \underline{-2.829}$, $a_5 = \underline{3.472}$

17-16

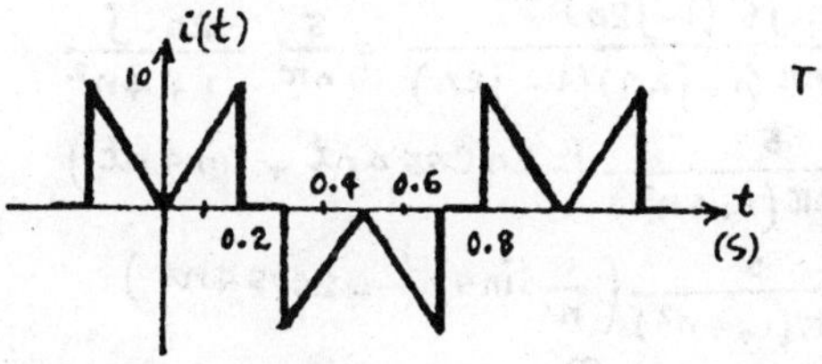


$T = 1\,s$, even, HW $\therefore a_0 = a_{even} = b_n = 0$

$$a_{odd} = \frac{8}{T}\int_0^{T/4} f(t)\cos n\omega_0 t\,dt$$

$$= \frac{8}{1}\int_0^{0.2} 50t\cos 2\pi nt\,dt$$

$$\therefore a_{odd} = 400\left(\frac{1}{4\pi^2 n^2}\cos 2\pi nt + \frac{t}{2\pi n}\sin 2\pi nt\right)\Big|_0^{0.2} = \frac{100}{\pi^2 n^2}(\cos 0.4\pi n - 1) + \frac{40}{n\pi}\sin 0.4n\pi$$

$\therefore a_1 = \underline{5.108}$, $a_3 = \underline{-4.531}$, $a_5 = 0$, $a_7 = \underline{0.6951}$

$$i(t) = \sum_{n=1,\,odd}^{\infty}\left[\frac{100}{\pi^2 n^2}(\cos 0.4\pi n - 1) + \frac{40}{\pi n}\sin 0.4\pi n\right]\cos 2\pi nt$$ ←

17-17 (a)

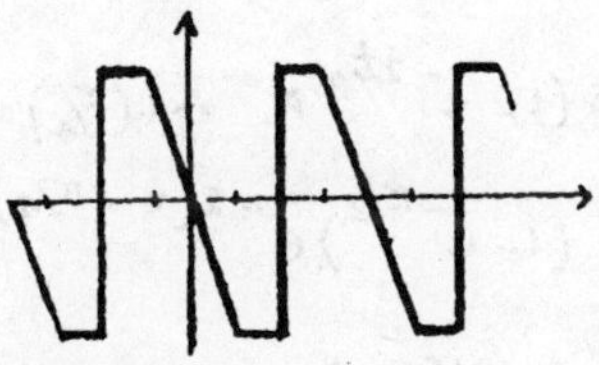

Odd only

(b)

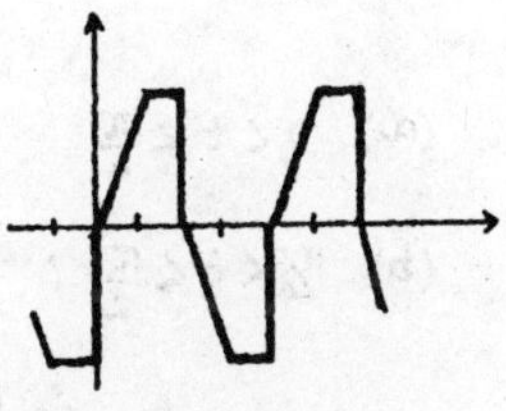

Half-wave only

(c)

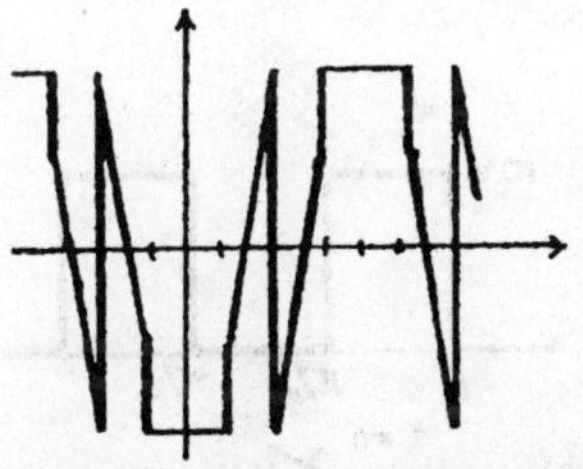

Even and half-wave

**17-18** (a)

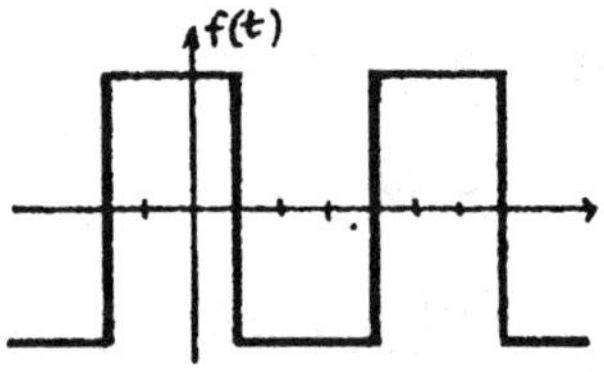


(b)

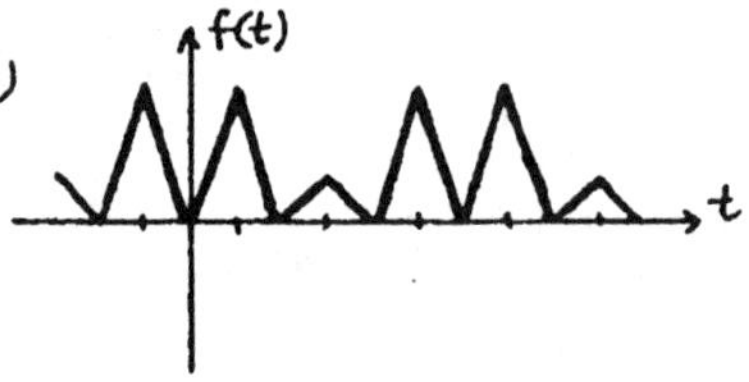


(c)

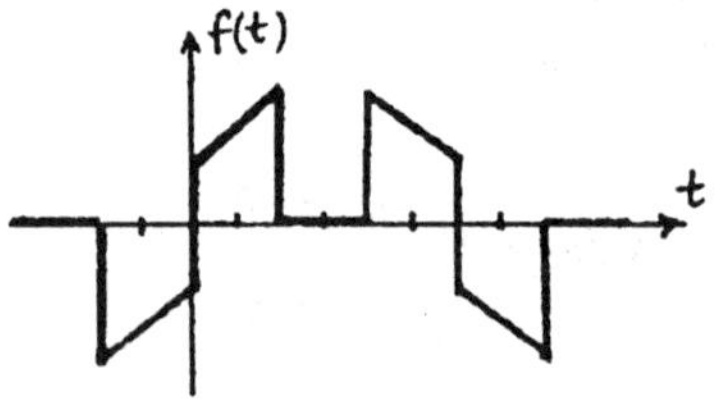


| | (a) | (b) | (c) |
|---|---|---|---|
| 1 | N | Y | N |
| 2 | N | N | Y |
| 3 | Y | N | Y |
| 4 | Y | N | Y |
| 5 | Y | N | Y |
| 6 | N | N | Y |
| 7 | Y | Y | Y |
| 8 | N | Y | N |

**17-19**

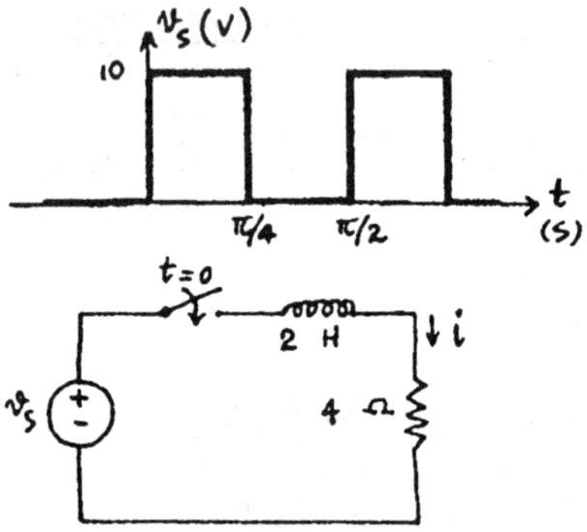


(a) $v_s = 5 + \frac{20}{\pi}\sum_{1,odd}^{\infty}\frac{1}{n}\sin\frac{2\pi nt}{\pi/2}$

$v_{sn} = \frac{20}{n\pi}\sin 4nt \quad \therefore \bar{V}_{sn} = -j\frac{20}{n\pi}, \quad \bar{s}_n = j4n$

$\bar{Z}_n = 4 + j2(4n) = 4 + j8n; \quad \bar{I}_{fn} = \frac{\bar{V}_{sn}}{\bar{Z}_n} = -\frac{j20}{n\pi}\frac{1}{4(1+j2n)}$

$\bar{I}_{fn} = \frac{-j5(1-j2n)}{n\pi(1+j2n)(1-j2n)} = \frac{5}{n\pi}\frac{-2n-j}{1+4n^2}$

$\therefore i_{fn} = \frac{5}{n\pi(1+4n^2)}(-2n\cos 4nt + \sin 4nt)$

$= \frac{5}{\pi(1+4n^2)}\left(\frac{1}{n}\sin 4nt - 2\cos 4nt\right)$

$\therefore i_f = 1.25 + \frac{5}{\pi}\sum_{1,odd}^{\infty}\frac{1}{1+4n^2}\left(\frac{1}{n}\sin 4nt - 2\cos 4nt\right)$ ←

(b) $i_n = Ae^{-2t}; \quad i = i_f + i_n, \quad i(0) = 0, \quad i_f(0) = 1.25 - \frac{10}{\pi}\sum_{1,odd}^{\infty}\frac{1}{1+4n^2} = 1.25 - \frac{2.5}{\pi}\sum_{1,odd}^{\infty}\frac{1}{n^2+0.25}$

$\therefore i_f(0) = 1.25 - \frac{2.5}{\pi}\times\frac{\pi}{2}\tanh\frac{\pi}{4} = 0.43026 \quad \therefore A = -0.43026$

$\therefore i = -0.43026e^{-2t} + 1.25 + \frac{5}{\pi}\sum_{1,odd}^{\infty}\frac{1}{1+4n^2}\left(\frac{1}{n}\sin 4nt - 2\cos 4nt\right)$ ←

**17-20**

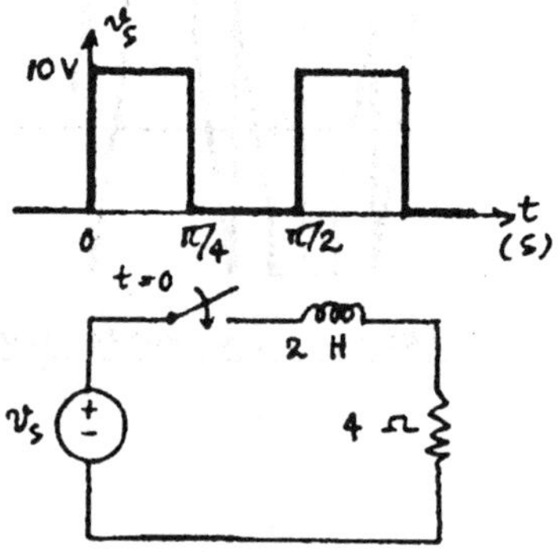


(a) $0 < t < \frac{\pi}{4}: \quad i = 2.5(1 - e^{-2t})$ A $\quad \therefore i(\pi/4) = \underline{1.9803 \text{ A}}$

(b) $\pi/4 < t < \frac{\pi}{2}: \quad i = 2.5(1 - e^{-\pi/2})e^{-2(t-\pi/4)}$ A

$\therefore i(\pi/2) = \underline{0.41166 \text{ A}}$

(c) $\frac{\pi}{2} < t < \frac{3\pi}{4}: \quad i = 2.5 - (2.5 - 0.41166)e^{-2(t-\pi/2)}$ A

$\therefore i(3\pi/4) = \underline{2.06588 \text{ A}}$

**17-21**

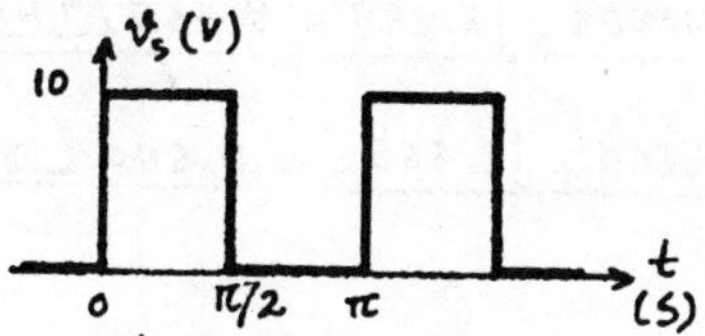


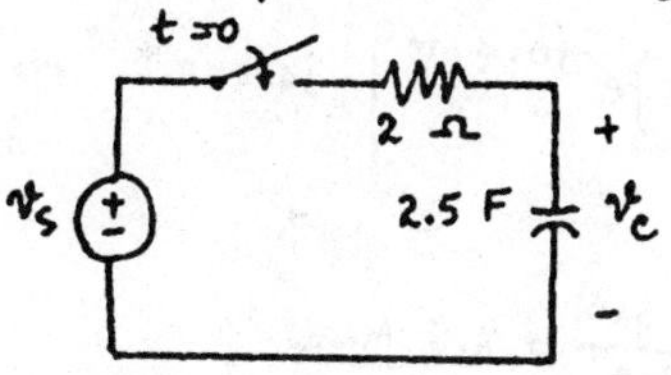


(a) $v_s = 5 + \frac{20}{\pi}\sum_{1,odd}^{\infty} \frac{1}{n}\sin 2nt$ ; $v_{sn} = \frac{20}{n\pi}\sin 2nt$

$\therefore \bar{V}_{sn} = -j\frac{20}{n\pi}$ , $\bar{A}_n = j2n$

$\bar{Z}_n = 2 + \frac{1}{j(2.5)(2n)} = 2 + \frac{1}{j5n}$ $\therefore \bar{V}_{cn} = \frac{1/j5n}{2 + 1/j5n}\ \frac{-j20}{n\pi}$

$\bar{V}_{cn} = \frac{-j20}{(1+j10n)n\pi}\frac{(1-j10n)}{(1-j10n)} = \frac{-j20(1-j10n)}{n\pi(1+100n^2)} = \frac{-200n - j20}{n\pi(1+100n^2)}$

$v_{cn} = \frac{1}{n\pi(1+100n^2)}(-200n\cos 2nt + 20\sin 2nt)$

$\therefore v_{cf} = 5 + \frac{20}{\pi}\sum_{1,odd}^{\infty}\frac{1}{1+100n^2}\left(\frac{1}{n}\sin 2nt - 10\cos 2nt\right)$ ←

(b) $v_n = \underline{Ae^{-t/5}}$

(c) $v_c(0) = A + 5 + \frac{20}{\pi}\sum_{1,odd}^{\infty} -\frac{10}{1+100n^2} = A + 5 - \frac{2}{\pi}\sum_{1,odd}^{\infty}\frac{1}{n^2+0.01} = A + 5 - \frac{2}{\pi}\left(\frac{\pi}{0.4}\tanh\frac{\pi}{20}\right)$

$\therefore A = -5 + \frac{2}{\pi}\times 1.223653 = -4.220998$

$\therefore v_c(t) = -4.221e^{-t/5} + 5 + \frac{20}{\pi}\sum_{1,odd}^{\infty}\frac{1}{1+100n^2}\left(\frac{1}{n}\sin 2nt - 10\cos 2nt\right)$ ←

**17-22**

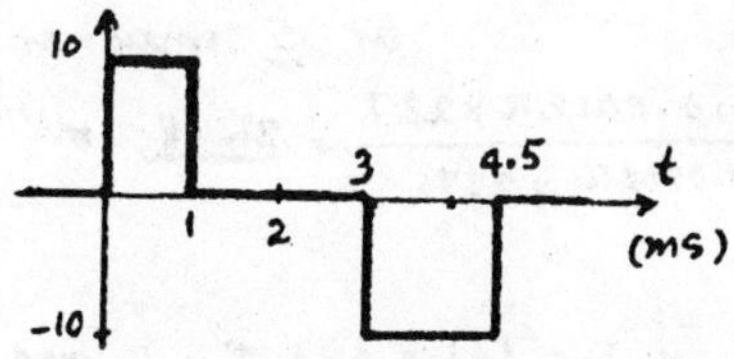


$T = 6\text{ ms}$ , $f = \frac{1000}{6}$ , $\omega_o = 2\pi f = \frac{1000\pi}{3}$

$C_3 = \frac{1000}{6}\left[\int_0^{0.001} 100e^{-j3\times 1000\pi t/3}dt - 100\int_{0.003}^{0.0045} e^{-j1000\pi t}dt\right]$

$= \frac{1000}{6(-j1000\pi)}\left[100e^{-j1000\pi t}\Big|_0^{0.001} - 100e^{-j1000\pi t}\Big|_{0.003}^{0.0045}\right]$

$C_3 = \frac{j}{6\pi}\left[100(e^{-j\pi} - 1) - 100(e^{-j4.5\pi} - e^{-j3\pi})\right] = \frac{j}{6\pi}(-100 - 100 + j100 - 100) = \underline{-5.305^+ - j15.915^+}$

or $C_3 = \underline{16.776\angle -108.44°}$ ; $C_{-3} = \underline{-5.305^+ + j15.915^+} = \underline{16.776\angle 108.44°}$ , $|C_3| = \underline{16.776}$

$C_3 = \frac{1}{2}(a_3 - jb_3)$ $\therefore a_3 = \underline{-10.610}$ , $b_3 = \underline{31.83}$ , $\sqrt{a_3^2 + b_3^2} = 2|C_3| = \underline{33.55^+}$

**17-23**

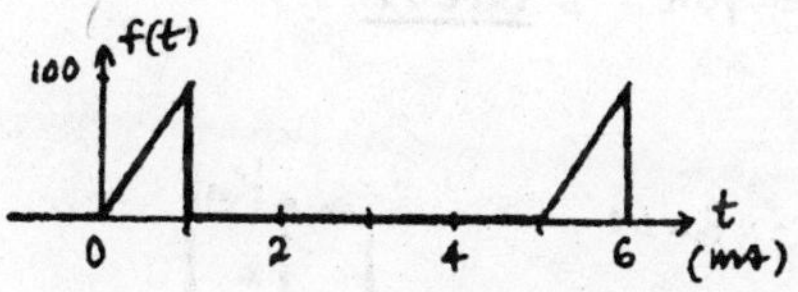


$T = 0.005\text{ s}$ , $f_o = 200$ , $\omega_o = 400\pi$

$C_n = 200\int_0^{0.001} 10^5 t\, e^{-jn400\pi t}dt$

$= 2\times 10^7\left[e^{-j400\pi nt}\left(\frac{t}{-j400\pi n} + \frac{1}{400^2\pi^2 n^2}\right)\right]_0^{0.001}$

$\therefore C_n = 2\times 10^7\left[e^{-j0.4\pi n}\left(\frac{10^{-6}}{-j0.4\pi n} + \frac{10^{-6}}{0.4^2\pi^2 n^2}\right) - \frac{10^{-6}}{0.4^2\pi^2 n^2}\right]$

$= -\frac{125}{n^2\pi^2} + \left(\frac{125}{n^2\pi^2} + j\frac{50}{n\pi}\right)e^{-j0.4n\pi}$

$\therefore C_1 = \underline{6.385^+ - j7.127} = \underline{9.569\angle -48.14°}$ , $C_{-1} = \underline{6.385^+ + j7.127} = \underline{9.569\angle 48.14°}$

(cont. on next page)

**17-23 (cont.)**

$c_2 = -1.0504 - j8.299 = 8.365\angle -97.21^\circ$ , $c_{-2} = -1.0504 + j8.299 = 8.365\angle 97.21^\circ$

$c_3 = -5.664 - j3.465 = 6.640\angle -148.54^\circ$ , $c_{-3} = -5.664 + j3.465 = 6.640\angle 148.54^\circ$

By inspection : $c_0 = \frac{1}{2} \times 100 \times 1 \times \frac{1}{5} = \underline{10}$

Series is $f(t) = 10 + \sum_{n=\pm 1}^{\pm\infty} \left[ -\frac{125}{n^2\pi^2} + \left( \frac{125}{n^2\pi^2} + j\frac{50}{n\pi} \right) e^{-j0.4n\pi} \right] e^{j400n\pi t}$ ←

**17-24**

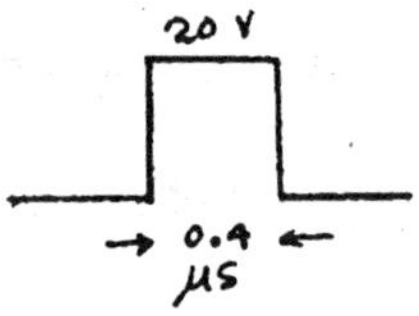


(a) $T = 1/3000$ , $f_0 = 3000$ , $\frac{1}{\tau} = \frac{10^6}{0.4} = \underline{2.5 \text{ MHz}}$

(b) $1/T = \underline{3000 \text{ Hz}}$

(c) $9000 = 3 \times 3000$ , $|c_n| = \frac{V_0 \tau}{T} \left| \frac{\sin(\frac{1}{2} n\omega_0 \tau)}{\frac{1}{2} n\omega_0 \tau} \right|$

$\therefore |c_3| = \frac{20 \times 0.4 \times 10^{-6}}{1/3000} \left| \frac{\sin(\frac{1}{2} \times 3 \times 6000 \times \pi \times 0.4 \times 10^{-6})}{0.0036\pi} \right| = \underline{24.00 \text{ mV}}$

(d) $\frac{999{,}000}{3000} = 333$ $\therefore |c_{333}| = 0.024 \frac{\sin(0.0012\pi \times 333)}{0.0012\pi \times 333} = \underline{18.175 \text{ mV}}$

(e) $B = \frac{1}{\tau} = \underline{2.5 \text{ MHz}}$

(f) $\frac{10^6}{2\pi} = 159{,}150$ Hz , $\frac{1.1 \times 10^6}{2\pi} = 175{,}070$ Hz $\therefore$ components at 162, 165, 168, 171, 174 kHz or $\underline{5}$ components

(g) $f = 3 \times 10^3 \times 227 = \underline{681 \text{ kHz}}$ ; $|c_{227}| = 0.024 \frac{\sin 0.0012\pi \times 227}{0.0012\pi \times 227} = \underline{21.18}$ mV

**17-25**

$c_0 = 10$ , $c_1 = 6 - j6$ , $c_2 = 4 + j2$ , $c_3 = -2 - j3$ , $c_n = 0$ for $|n| \geq 4$ ; $T = 10$ ms

(a) $2c_n = a_n - jb_n$ $\therefore a_0 = c_0 = 10$ ; $a_1 + jb_1 = 12 + j12$ ; $a_2 + jb_2 = 8 - j4$ ; $a_3 + jb_3 = -4 + j6$

$\therefore v(t) = 10 + 12\cos 200\pi t + 8\cos 400\pi t - 4\cos 600\pi t + 12\sin 200\pi t - 4\sin 400\pi t + 6\sin 600\pi t$

$v(0.002) = 10 + 12\cos 72^\circ + 8\cos 144^\circ - 4\cos 216^\circ + 12\sin 72^\circ - 4\sin 144^\circ + 6\sin 216^\circ = \underline{16.007}$

(b) $V_{rms} = \sqrt{10^2 + \frac{1}{2}(144 + 64 + 16 + 144 + 16 + 36)} = \sqrt{310} = \underline{17.607}$ V

**17-26**

$T = 0.02$ , $c_0 = 1$ A , $c_n = \frac{1}{n^2} - j\frac{2}{n^3}$ A , $|n| \geq 1$ $\therefore f_0 = 50$

(a) $|c_1| = |1 - j2| = 2.236$ , $|c_2| = \left|\frac{1}{4} - j\frac{1}{4}\right| = 0.354$

$|c_3| = 0.134$ , $|c_4| = 0.070$ , $|c_5| = 0.043$ , $|c_6| = 0.029$

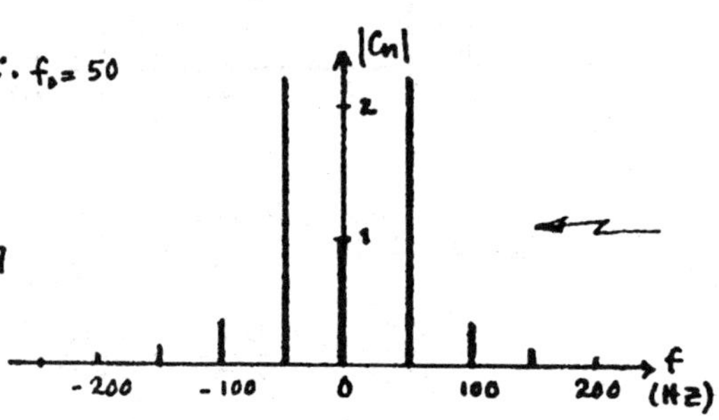


(b) Amplitude $= 2|c_n| = \sqrt{a_n^2 + b_n^2}$

$P_{5\Omega} = 1^2 \times 5 + \frac{1}{2} \times 5 \times 2^2 (2.236^2 + 0.354^2 + 0.134^2 + 0.070^2 + \ldots) = \underline{56.517 \text{ W}}$

17-27 (a)

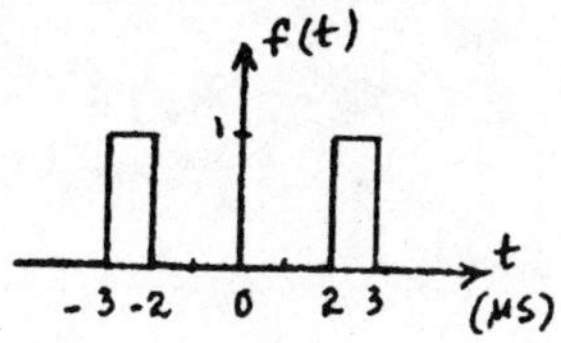


$T = 25\times10^{-6}$, $\omega_0 = 8\pi\times10^4$, $C_n = \frac{1}{T}\int_{-T/2}^{T/2} f(t)e^{-jn\omega_0 t}\,dt$

$\therefore C_n = \frac{10^6}{25}\int_{-T/2}^{T/2} f(t)\,e^{-j8\pi n10^4 t}\,dt$

$= \frac{10^6}{25}\left[\int_{-3\times10^{-6}}^{-2\times10^{-6}} e^{-j8\pi n10^4 t}\,dt + \int_{2\times10^{-6}}^{3\times10^{-6}} e^{-j8\pi n10^4 t}\,dt\right]$

Let $q = -t$ in $\int$ #1 $\therefore$ $C_n = \frac{10^6}{25}\left[\int_{2\times10^{-6}}^{3\times10^{-6}} e^{j8\pi n10^4 q}\,dq + \int_{2\times10^{-6}}^{3\times10^{-6}} e^{-j8\pi n\times10^4 q}\,dq\right]$

$\therefore C_n = \frac{2\times10^6}{25}\int_{2\times10^{-6}}^{3\times10^{-6}} \cos 8\pi n10^4 q\,dq = \frac{2\times10^6}{25\times8n\pi10^4}(\sin 0.24n\pi - \sin 0.16n\pi)$

$= \frac{1}{n\pi}(\sin 0.24n\pi - \sin 0.16n\pi)$ $\therefore C_4 = \frac{1}{4\pi}(\sin 0.96\pi - \sin 0.64\pi) = \underline{-0.06203}$

(b) $C_0 = \lim_{n\to0}\left[\frac{1}{n\pi}(\sin 0.24n\pi - \sin 0.16n\pi)\right] = 0.24 - 0.16 = \underline{0.08}$

(c) A little testing shows that $|C_0|$ is max $\therefore |C_n|_{max} = \underline{0.08}$

(d) $0.01|C_n|_{max} = 0.8\times10^{-3}$ ; $|\sin 0.24n\pi - \sin 0.16n\pi|_{max} \doteq 2$

$\therefore \frac{1}{n\pi}(2) \doteq 0.8\times10^{-3}$ $\therefore n \doteq 796$. Checking a few values, we find

$|C_{760}| = 0.797\times10^{-3}$, $|C_{740}| = 0.818\times10^{-3}$, $|C_{741}| = 0.629\times10^{-3}$, others $< 0.8\times10^{-3}$

$\therefore N = \underline{740}$, $B = 740 f_0 = 740\,\frac{1}{25\times10^{-6}} = \underline{29.6}$ MHz or $\underline{186.0}$ Mrad/s

17-28 (a)

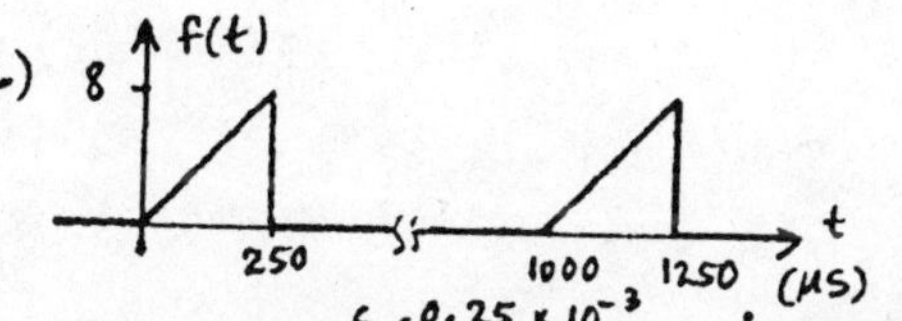


$T = 0.001$ s, $\omega_0 = 2000\pi$, $f(t) = 32{,}000t$ $0 < t < 250$ μs

$C_n = \frac{1}{T}\int_{-T/2}^{T/2} f(t)e^{-jn\omega_0 t}\,dt = 1000\int_0^{0.25\times10^{-3}} 32\times10^3 t\,e^{-j2000n\pi t}\,dt$

$\therefore C_n = 32\times10^6\int_0^{0.25\times10^{-3}} t e^{-j2000n\pi t}\,dt = 32\times10^6\left(\frac{t}{-j2000n\pi} - \frac{1}{-4\times10^6n^2\pi^2}\right)e^{-j2000n\pi t}\Big|_0^{0.25\times10^{-3}}$

$C_n = 32\left(\frac{j1}{8n\pi} + \frac{1}{4n^2\pi^2}\right)e^{-jn\pi/2} - 32\times\frac{1}{4n^2\pi^2} = \frac{j4}{n\pi}e^{-jn\pi/2} - \frac{8}{n^2\pi^2}(1 - e^{-jn\pi/2})$ ←

(b) $C_5 = \frac{j4}{5\pi}(-j) - \frac{8}{25\pi^2}(1 + j1) = \frac{4}{5\pi} - \frac{8}{25\pi^2} - j\frac{8}{25\pi^2} = 0.2246\,\underline{/-8.301^\circ}$ ←

17-29 (a)

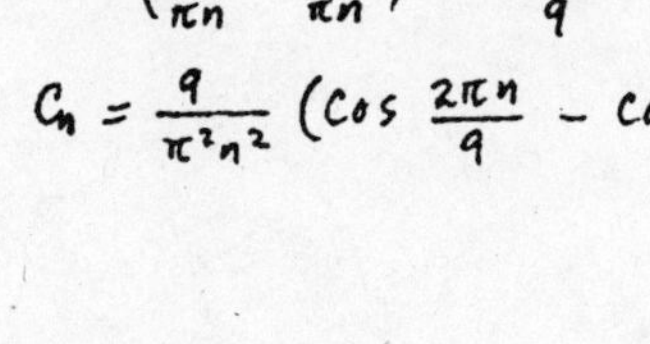


$\omega_0 = \frac{2\pi}{9}$, even sym $\therefore b_n = 0$

$C_n = \frac{1}{2}a_n = \frac{1}{2}\times\frac{2}{9}\times2\left[\int_0^1 2\cos\frac{2\pi nt}{9}\,dt + \int_1^2(4-2t)\cos\frac{2\pi nt}{9}\,dt\right]$

$= \frac{4}{9}\left[\frac{9}{2\pi n}\sin\frac{2\pi nt}{9}\Big|_0^1 + 2\times\frac{9}{2\pi n}\sin\frac{2\pi nt}{9}\Big|_1^2 - \int_1^2 t\cos\frac{2\pi nt}{9}\,dt\right]$

$\therefore C_n = \frac{4}{9}\left[\frac{9}{2\pi n}\sin\frac{2\pi n}{9} + \frac{18}{2\pi n}\sin\frac{4\pi n}{9} - \frac{18}{2\pi n}\sin\frac{2\pi n}{9} - \left(\frac{81}{4\pi^2n^2}\cos\frac{2\pi nt}{9} + \frac{9t}{2\pi n}\sin\frac{2\pi t}{9}\right)\Big|_1^2\right]$

$= \left(\frac{2}{\pi n} - \frac{4}{\pi n}\right)\sin\frac{2\pi n}{9} + \frac{4}{\pi n}\sin\frac{4\pi n}{9} - \frac{9}{\pi^2n^2}\cos\frac{4\pi n}{9} + \frac{9}{\pi^2n^2}\cos\frac{2\pi n}{9} - \frac{4}{\pi n}\sin\frac{4\pi n}{9} + \frac{2}{\pi n}\sin\frac{2\pi n}{9}$

$C_n = \frac{9}{\pi^2n^2}\left(\cos\frac{2\pi n}{9} - \cos\frac{4\pi n}{9}\right)$ ← (cont. on next page)

17-29 (b) (cont.)

$$c_0 = \frac{\text{Area}}{9} = \frac{1\times 2+4}{9} = \underline{2/3}$$

$$c_1 = \frac{9}{\pi^2}\left(\cos\frac{2\pi}{9} - \cos\frac{4\pi}{9}\right) = \underline{0.5402}$$

17-30

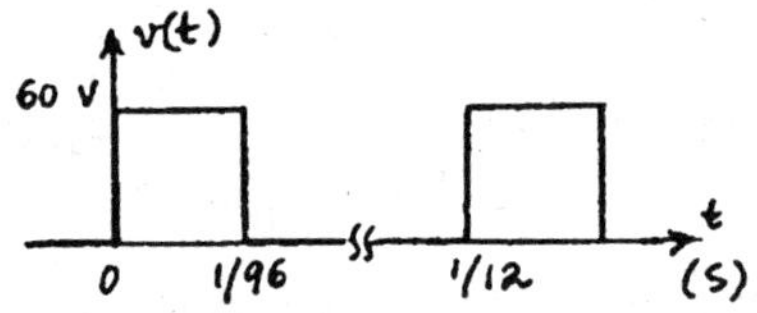


$T = 1/12$, $f_0 = 12$, $\omega_0 = 24\pi$

(a) $$c_4 = 12\int_0^{1/96} 60e^{-j96\pi t}\,dt = \frac{720}{-j96\pi} e^{-j96\pi t}\Big|_0^{1/96}$$

$$= j\frac{7.5}{\pi}\left(e^{-j\pi} - 1\right) = -j\frac{15}{\pi} = \underline{-j4.775^-}$$

(b) $4f_0 = 48$ Hz goes thru filter

$$\bar{V}_{4,out} = 2(-j4.775^-) = -j9.549 \text{ V}$$

$$\therefore \bar{I}_{4,out} = \frac{-j9.549}{5 + j0.01\times 96\pi} = 1.6354\,\underline{/-99.81^\circ}\text{ A}$$

$$\therefore P_{av} = \frac{1}{2}\times 1.6354^2 \times 5 = \underline{6.686\text{ W}}$$

# CHAPTER 18: FOURIER TRANSFORMS

## Section 18-2 Definition of the Fourier transform
## Section 18-3 Some properties of the Fourier transform

**1** Use Eq. (8*a*) in your text to determine the Fourier transform of the following: (*a*) $f(t) = e^{-at}u(t - t_0)$, $a > 0$; (*b*) $g(t) = te^{-at}u(t)$, $a > 0$; (*c*) $v(t) = (t - t_0)e^{-a(t - t_0)}u(t - t_0)$.

**2** Let $f(t) = 100u(t + 2) + 100u(t + 1) - 100u(t - 1) - 100u(t - 2)$. (*a*) Sketch $f(t)$. (*b*) Find $\mathbf{F}(j\omega)$ by using the defining equations for the Fourier transform.

**3** Find the Fourier transform of the single sawtooth pulse sketched in Fig. 18-1.

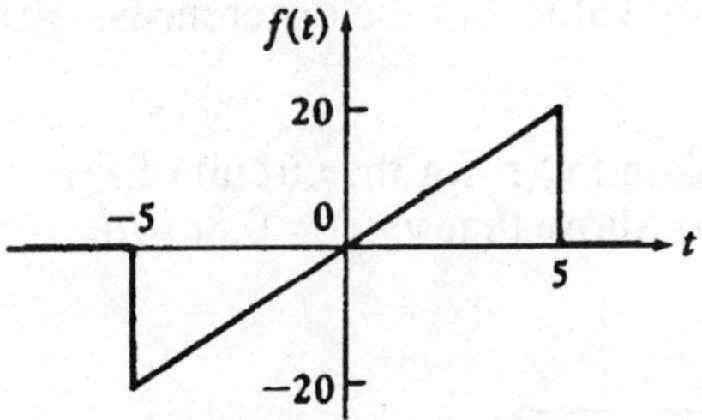


**Fig. 18-1:** See Prob. 3.

**4** Find the Fourier transforms for the waveforms illustrated in Fig. 18-2.

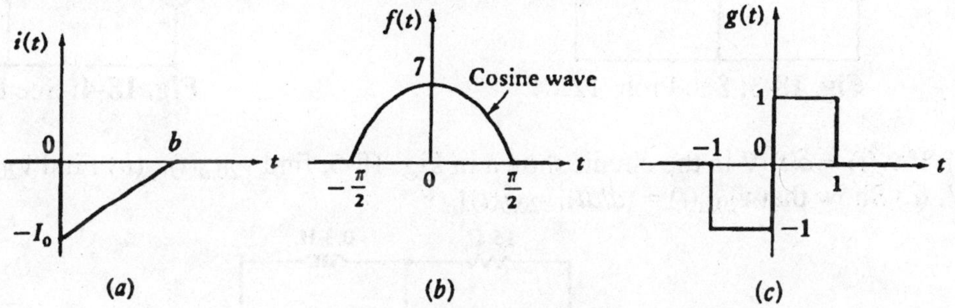


**Fig. 18-2:** See Prob. 4.

**5** Determine the time function that corresponds to each of these continuous frequency spectra and then evaluate it at $t = 0.5$: (*a*) $\mathbf{F}(j\omega) = 1$, $1 < |\omega| < 2$; $\mathbf{F}(j\omega) = 0$, $|\omega| < 1$ and $|\omega| > 2$; (*b*) $\mathbf{F}(j\omega) = e^{-|\omega|}$; (*c*) $\mathbf{F}(j\omega) = j\sin\omega$, $|\omega| < \pi$; $\mathbf{F}(j\omega) = 0$, $|\omega| > \pi$.

**6** A single pulse is defined by $f(t) = 12\cos 200\pi t$ V for $-2.5 \le t \le 2.5$ ms, and $f(t) = 0$ for $t < -2.5$ ms and $t > 2.5$ ms. Evaluate $\mathbf{F}(j\omega)$ for $\omega =$: (*a*) 0; (*b*) $100\pi$ rad/s; (*c*) $-100\pi$ rad/s; (*d*) $200\pi$ rad/s.

**7** The voltage $v_5(t)$ is present across a 5-Ω resistor and has a Fourier transform whose magnitude is $|\mathbf{F}_5(j\omega)| = 9 - \omega^2$ V/(rad/s) for $|\omega| < 3$ and 0 for $|\omega| > 3$ rad/s. (*a*) What is the total energy present in the signal? (*b*) In what range, $|\omega| < \omega_1$, is 50 percent of this energy located?

**8** Let $\mathbf{F}(j\omega) = \pi/\sqrt{|\omega|}$ be the Fourier transform of $f(t)$. (*a*) Given the definite integral, $\int_0^\infty [(\cos mx)/\sqrt{x}]dx = \sqrt{\pi/2m}$, find $f(t)$. (*b*) Calculate the 1-Ω energy in $f(t)$ that is found in the frequency band, $1 < |\omega| < 3$ rad/s. (*c*) Calculate the 1-Ω energy in $f(t)$ that is found in the time interval, $1 < t < 3$ s.

**9** Let $i(t) = 10te^{-5t}u(t)$ A be the current in a 1-Ω resistor. Calculate: (*a*) the total energy absorbed by the resistor over all time, and (*b*) the percentage of the energy that lies in the frequency range, $|\omega| < 5$ rad/s.

**10** If $f(t) = 2e^{-8t}u(t)$, find: (*a*) the 1-Ω energy associated with this signal; (*b*) the amplitude spectrum of $\mathbf{F}(j\omega)$; and (*c*) the frequency range $|\omega| < \omega_1$ in which 90 percent of the 1-Ω energy lies.

## Section 18-4 The unit-impulse function

**11** Find the strength of the impulse that is defined by the limit as $\varepsilon \to 0$ of: (*a*) $(1/\varepsilon)[\cos^2(\pi t/2\varepsilon)][u(t+\varepsilon) - u(t-\varepsilon)]$; (*b*) $(1/\varepsilon)[1 - (t/\varepsilon)^2][u(t+\varepsilon) - u(t-\varepsilon)]$.

**12** (*a*) Find $i_R(t)$, $i_C(t)$, and $i_s(t)$ for the circuit of Fig. 18-3. Express all currents in terms of singularity functions. (*b*) Specify values for the differences, $i_R(0^+) - i_R(0^-)$, $i_C(0^+) - i_C(0^-)$, and $q_C(0^+) - q_C(0^-)$.

**13** (*a*) Find $v_L(t)$, $i_x(t)$, $v_R(t)$, and $v_s(t)$ for the circuit of Fig. 18-4. Express each quantity in terms of singularity functions. (*b*) Show that $v_R(t) = 0.8di_x/dt$. Remember that $(d/dt)(uv) = u\,dv/dt + v\,du/dt$.

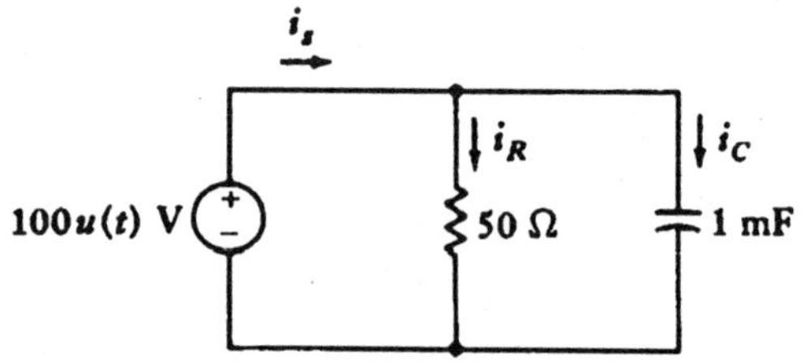

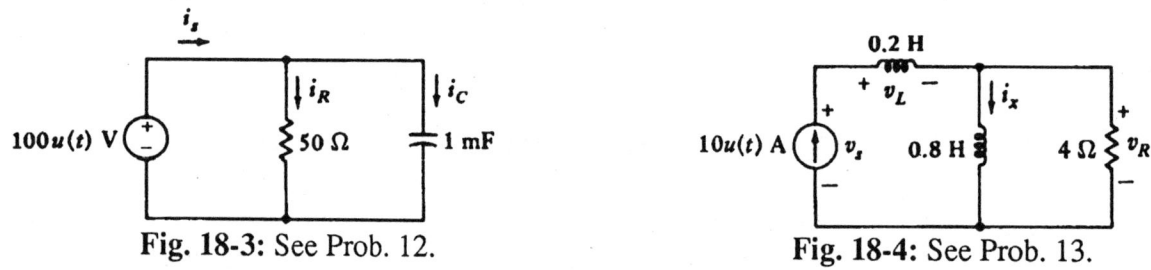

**Fig. 18-3:** See Prob. 12.

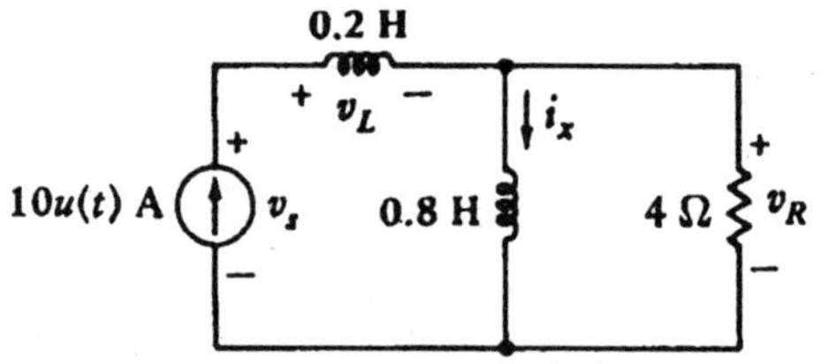


**Fig. 18-4:** See Prob. 13.

**14** (*a*) If $v_s(t) = \delta(t)$ V in the circuit shown in Fig. 18-5, find $v_{20,a}(t)$. (*b*) Find $v_{20,b}(t)$ if $v_s(t) = u(t)$ V. (*c*) Show that $v_{20,a}(t) = (d/dt)[v_{20,b}(t)]$.

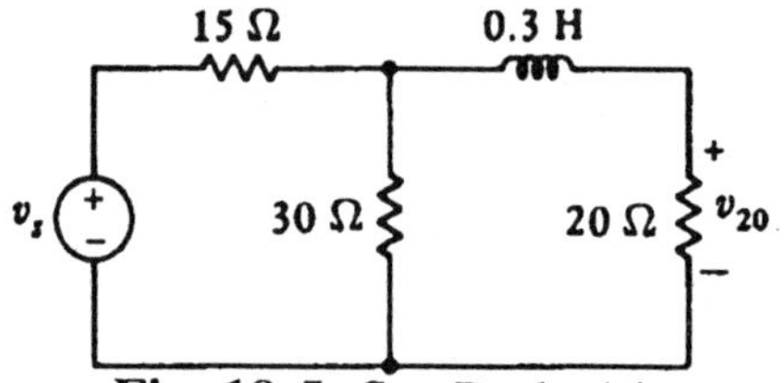


**Fig. 18-5:** See Prob. 14.

**15** The voltage $v(t) = (20/a)[u(t) - u(t-a)]$ V is applied to the series combination of a 10-kΩ resistor and a 5-μF capacitor. Find the magnitude of the capacitor voltage at $t = 30$ ms if: (*a*) $a = 40$ ms; (*b*) $a = 20$ ms; (*c*) $a = 2$ ms; (*d*) $a \to 0$, as a limit.

## Section 18-5 Fourier transform pairs for some simple time functions

**16** Use the definition of the Fourier transform to prove the following results, where $\mathcal{F}\{f(t)\} = \mathbf{F}(j\omega)$: (*a*) $\mathcal{F}\{f(t-t_0)\} = e^{-j\omega t_0}\mathcal{F}\{f(t)\}$; (*b*) $\mathcal{F}\{df(t)/dt\} = j\omega\mathcal{F}\{f(t)\}$; (*c*) $\mathcal{F}\{f(kt)\} = (1/|k|)\mathbf{F}(j\omega/k)$; (*d*) $\mathcal{F}\{f(-t)\} = \mathbf{F}(-j\omega)$; (*e*) $\mathcal{F}\{tf(t)\} = jd[\mathbf{F}(j\omega)]/d\omega$.

**17** Find the Fourier transform of $f(t) =$: (*a*) $5[\text{sgn}(t-1)]\,\delta(t)$; (*b*) $5[\text{sgn}(t-1)]\,\delta(t-2)$; (*c*) $5\sin 2t\,u(t)$; (*d*) $5\sin 2t[u(t+\pi/2) - u(t-\pi/2)]$.

**18** Find $\mathbf{F}(j\omega)$ if $f(t)$ =: (*a*) $A \sin(\omega_0 t + \theta)$; (*b*) $2\delta(t) - 4\text{sgn}(t) + 3u(t)$; (*c*) $\cosh 5t\, u(t)$.

**19** Find $f(t)$ at $t = 2$ if $\mathbf{F}(j\omega)$ = : (*a*) $2u(\omega + 2) - 2u(\omega - 2)$; (*b*) $2u(-2 - \omega) + 2u(-2 + \omega)$; (*c*) $\delta(\omega) + 2u(-2 - \omega) + 2u(-2 + \omega)$.

**20** Find the function of time that corresponds to each of the following Fourier transforms: (*a*) $2 - 2\delta(\omega - 2) + (j2/\omega) + 2/(2 + j\omega)$; (*b*) $(2 + j\omega)/[(2 + j\omega)^2 + 2^2]$; (*c*) $(\sin\ 2\omega)/(2\omega)$.

## Section 18-6 The Fourier transform of a general periodic time function

**21** Find the Fourier transform of the periodic waveform shown in Fig. 18-6.

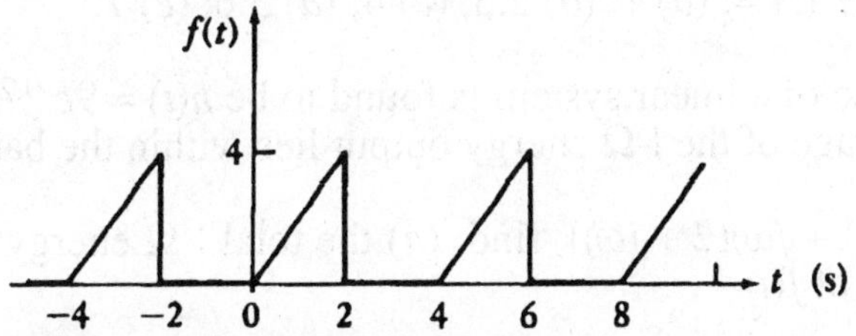

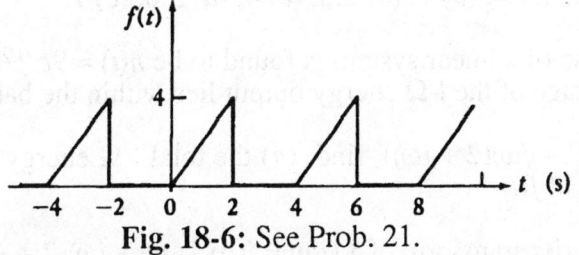

**Fig. 18-6:** See Prob. 21.

**22** A periodic waveform $f(t)$ is described over the period $0 < t < 5$ by the expression $f_1(t) = 2u(t) + 3u(t - 2) - 5u(t - 4)$. Find $\mathbf{F}(j\omega)$.

**23** Let $\mathbf{F}(j\omega) = 100 \sum_{n=-\infty}^{\infty} \frac{1}{|n|!}\delta(\omega - 0.1n)$. Find the value of the corresponding $f(t)$ at $t = 2$.

## Section 18-7 Convolution and circuit response in the time domain

**24** Given $x(t) = h(t) = u(t) - u(t - 1)$: (*a*) use convolution in the time domain to find $y(t) = x(t) * h(t)$ and (*b*) sketch the result.

**25** Let $x(t) = 2[u(t) - u(t - 2)]$ and $h(t) = u(t)$. Find $y(t) = x(t) * h(t)$ by using: (*a*) the first integral in Eq. (43) in your text; (*b*) the second integral in (43).

**26** Use convolution in the time domain to find $f(t) = f_1(t) * f_2(t)$, if $f_1(t) = (1 + 2e^{-t})u(t)$ and $f_2(t) = 4[u(t + 1) - u(t - 2)]$.

**27** The unit-impulse response and the input to a certain linear system are shown in Fig. 18-7. (*a*) Obtain an integral expression for the output that is valid in the range, $4 < t < 6$, and does not contain any singularity functions. (*b*) Evaluate the output at $t = 5$.

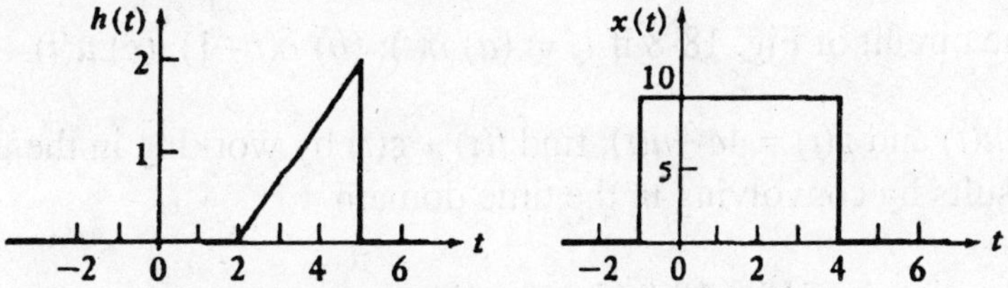

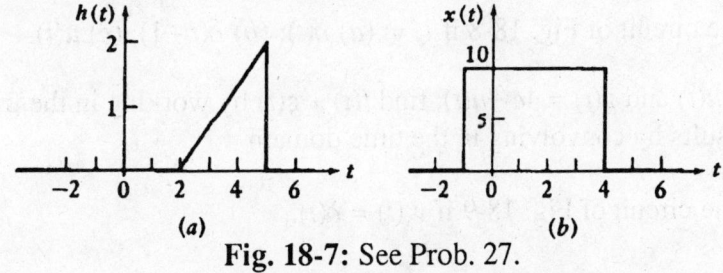

**Fig. 18-7:** See Prob. 27.

**28** The input signal for a certain linear system is $x(t) = 4e^{-(t-1)}u(t-1)$, while the impulse response is $h(t) = (16 - 2t)[u(t-5) - u(t-8)]$. Find the output signal value at: (*a*) $t = 5$; (*b*) $t = 8$.

**29** If the input to a linear system is $x_1(t) = \delta(t)$, then the output is $y_1(t) = 8 - 2t$ for $0 < t < 4$, and zero elsewhere. A different input signal, $x_2(t) = 10\delta(t-2) + 6u(t)$, is now applied. Calculate values for the output $y_2(t)$ at $t$ =: (*a*) 1; (*b*) $2^-$; (*c*) $2^+$; (*d*) 3; (*e*) 5; (*f*) 7.

**30** If $y(t) = x(t) * h(t)$, where $x(t) = 5\delta(t-4) + 6[u(t) - u(t-4)]$ and $h(t) = 9e^{-0.2t}u(t)$: (*a*) sketch $y(t)$ for $0 < t < 8$, and (*b*) find values for $y(3)$ and $y(5)$.

**31** If $h(t) = u(t) - u(t-1)$ and $x(t) = 6[u(t-2) - u(t-5)]$, use convolution in the time domain to evaluate $y(t) = x(t) * h(t)$ at $t$ =: (*a*) 1; (*b*) 2.5; (*c*) 4; (*d*) 5.6; (*e*) 7.

**32** The impulse response of a linear system is found to be $h(t) = 9e^{-0.2t}u(t)$. When a unit impulse is applied, what percentage of the 1-Ω energy output lies within the band, $-0.02 < f < 0.02$ Hz?

**33** If $\mathbf{F}(j\omega) = (2 - j\omega)/[(1 + j\omega)(2 + j\omega)]$, find: (*a*) the total 1-Ω energy present in the signal, and (*b*) the maximum value of $f(t)$.

**34** Find the inverse Fourier transform of $\mathbf{F}(j\omega)$ =: (*a*) $1/[(1 + j\omega)(2 + j\omega))(3 + j\omega)]$; (*b*) $j\omega/[(1 + j\omega)(2 + j\omega)(3 + j\omega)]$; (*c*) $(j\omega)^2/[(1 + j\omega)(2 + j\omega))(3 + j\omega)]$; (*d*) $(j\omega)^3/[(1 + j\omega)(2 + j\omega))(3 + j\omega)]$.

## Section 18-8 The system function and response in the frequency domain
## Section 18-9 The physical significance of the system function

**35** In Sec. 18-7 of your text, the impulse response, $h(t) = 2e^{-t}u(t)$, was hypothesized. In order to develop one network which has such a response: (*a*) determine $\mathbf{H}(j\omega) = \mathbf{V}_o(j\omega)/\mathbf{V}_i(j\omega)$. (*b*) Either by inspecting $h(t)$ or $\mathbf{H}(j\omega)$, note that the network has a single energy storage element. Arbitrarily selecting an $RC$ circuit with $R = 1\ \Omega$, $C = 1$ F, to provide the necessary time constant, determine the form of the circuit to give $\frac{1}{2}h(t)$ or $\frac{1}{2}\mathbf{H}(j\omega)$. (*c*) Place an ideal voltage amplifier in cascade with the network to provide the proper multiplicative constant. What is the gain of the amplifier?

**36** A linear system is characterized by $h(t) = 3e^{-4(t-2)}u(t-2)$. If a unit impulse, $x(t) = \delta(t)$, is applied to the system input, use the statements of Prob. 16 (if necessary) to help find: (*a*) the 1-Ω energy density of the system output, and (*b*) the total energy in the output.

**37** Consider a linear system for which $\mathbf{H}(j\omega) = 2/(3 + j\omega)$. (*a*) Find the unit-impulse response of the system. (*b*) If the input, $x(t) = 5\cos 4t$, is applied to the system, find the output $y(t)$.

**38** Find $i_x(t)$ for the circuit of Fig. 18-8 if $i_s$ =: (*a*) $\delta(t)$; (*b*) $\delta(t-1)$; (*c*) $u(t)$.

**39** (*a*) If $f(t) = 3e^{-t}u(t)$ and $g(t) = 4e^{-2t}u(t)$, find $f(t) * g(t)$ by working in the frequency domain. (*b*) Check your results by convolving in the time domain.

**40** Find $v_o(t)$ for the circuit of Fig. 18-9 if $v_s(t) = \delta(t)$.

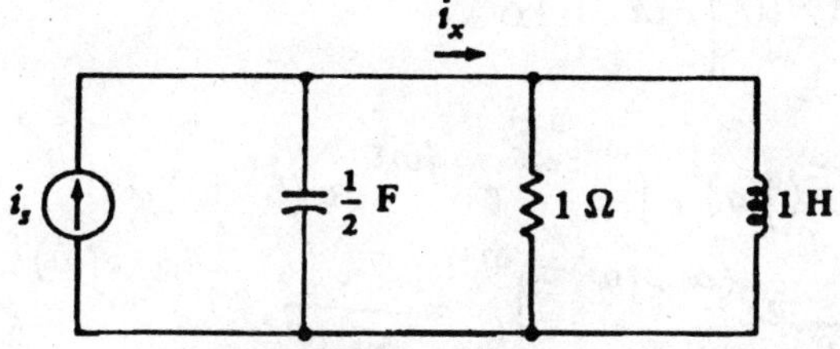

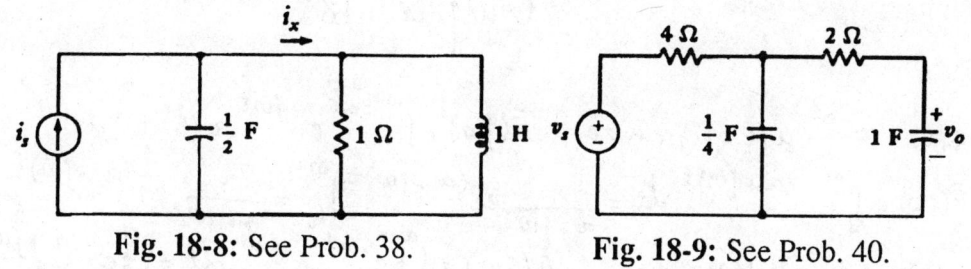

Fig. 18-8: See Prob. 38.

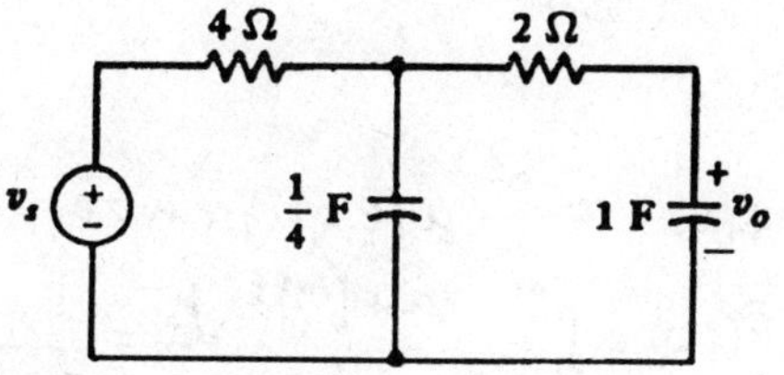


Fig. 18-9: See Prob. 40.

# Chapter 18

**18-1** (a) $f(t) = e^{-at}\, u(t - t_0)$, $a > 0$ $\therefore\ \bar{F}(j\omega) = \int_{-\infty}^{\infty} e^{-at}\, e^{-j\omega t}\, u(t - t_0)\, dt$

$$\bar{F}(j\omega) = \int_{t_0}^{\infty} e^{-(a+j\omega)t}\, dt = \frac{-1}{a+j\omega}\, e^{-(a+j\omega)t}\Big|_{t_0}^{\infty} = \frac{1}{a+j\omega}\, e^{-(a+j\omega)t_0} \leftarrow$$

(b) $g(t) = t e^{-at}\, u(t)$, $a > 0$ $\therefore\ \bar{G}(j\omega) = \int_{0}^{\infty} t e^{-at}\, e^{-j\omega t}\, dt = \int_{0}^{\infty} t e^{-(a+j\omega)t}\, dt$

$$\therefore\ \bar{G}(j\omega) = e^{-(a+j\omega)t}\left[\frac{-t}{a+j\omega} - \frac{1}{(a+j\omega)^2}\right]_0^{\infty} = \frac{1}{(a+j\omega)^2} \leftarrow$$

(c) $v(t) = (t - t_0)e^{-a(t-t_0)}\, u(t - t_0)$, $a > 0$ $\therefore\ \bar{V}(j\omega) = \int_{-\infty}^{\infty} (t - t_0) e^{-a(t-t_0)}\, e^{-j\omega t}\, u(t - t_0)\, dt$

Let $x = t - t_0$ $\therefore\ \bar{V}(j\omega) = \int_{-\infty}^{\infty} x e^{-ax}\, e^{-j\omega t_0}\, e^{-j\omega x}\, u(x)\, dx$

$$= e^{-j\omega t_0}\int_0^{\infty} x e^{-(a+j\omega)x}\, dx = e^{-j\omega t_0}\, e^{-(a+j\omega)x}\left[\frac{-x}{a+j\omega} - \frac{1}{(a+j\omega)^2}\right]_0^{\infty}$$

$$\bar{V}(j\omega) = e^{-j\omega t_0}\, \frac{1}{(a+j\omega)^2} \leftarrow$$

**18-2** $f(t) = 100\, u(t+2) + 100\, u(t+1) - 100\, u(t-1) - 100\, u(t-2)$

(a)

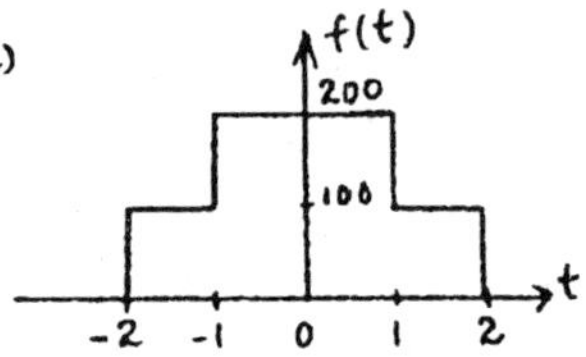


(b) $\bar{F}(j\omega) = \int_{-2}^{2} 100\, e^{-j\omega t}\, dt + \int_{-1}^{1} 100\, e^{-j\omega t}\, dt$

$$= \frac{100}{-j\omega}\left(e^{-j2\omega} - e^{j2\omega}\right) + \frac{100}{-j\omega}\left(e^{-j\omega} - e^{j\omega}\right)$$

$$= \frac{100}{-j\omega}(-j2\sin 2\omega) + \frac{100}{-j\omega}(-j2\sin\omega) = \underline{\frac{200}{\omega}(\sin\omega + \sin 2\omega)} \leftarrow$$

**18-3**

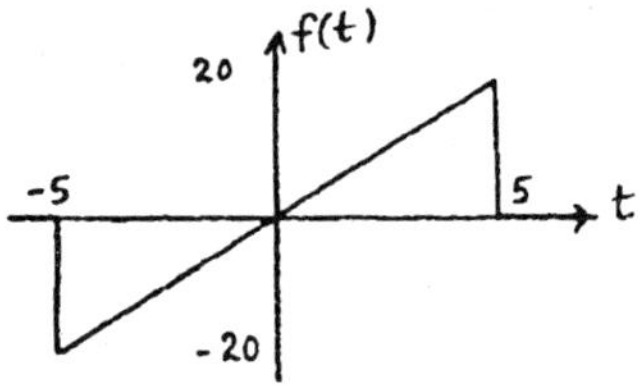


$$\bar{F}(j\omega) = \int_{-5}^{5} 4t e^{-j\omega t}\, dt = 4e^{-j\omega t}\left(\frac{-t}{j\omega} - \frac{1}{-\omega^2}\right)\Big|_{-5}^{5}$$

$$= 4\left[e^{-j5\omega}\left(\frac{1}{\omega^2} - \frac{5}{j\omega}\right) - e^{j5\omega}\left(\frac{1}{\omega^2} + \frac{5}{j\omega}\right)\right]$$

$$= 4\left(-j2 \times \frac{1}{\omega^2}\sin 5\omega\right) - \frac{4 \times 5}{j\omega} \times 2\cos 5\omega$$

$$\bar{F}(j\omega) = \underline{-j\frac{8}{\omega^2}\sin 5\omega + j\frac{40}{\omega}\cos 5\omega} \leftarrow$$

**18-4** (a)

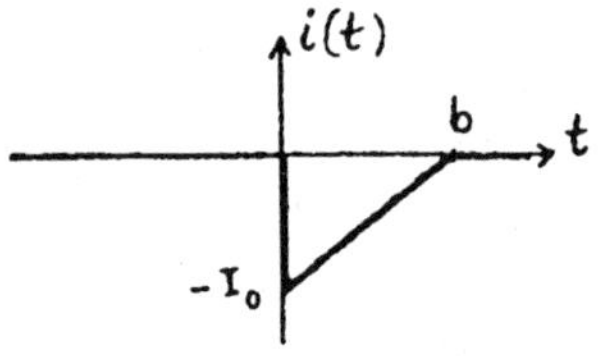


$$\bar{I}(j\omega) = \int_0^{b} -I_0(1 - t/b)\, e^{-j\omega t}\, dt$$

$$= \frac{-I_0}{-j\omega}\left(e^{-j\omega b} - 1\right) - \frac{I_0}{b}\, e^{-j\omega t}\left(\frac{t}{j\omega} + \frac{1}{-\omega^2}\right)\Big|_0^{b}$$

$$= -j\frac{I_0}{\omega}\left(e^{-j\omega b} - 1\right) - \frac{I_0}{b}\left[e^{-j\omega b}\left(\frac{b}{j\omega} - \frac{1}{\omega^2}\right) - \left(-\frac{1}{\omega^2}\right)\right]$$

$$= j\frac{I_0}{\omega} - \frac{I_0}{b\omega^2} + e^{-j\omega b}\frac{I}{b\omega^2} = \underline{-\frac{I_0}{b\omega^2}\left(1 - e^{-j\omega b}\right) + j\frac{I_0}{\omega}} \leftarrow$$

18-4 (cont.)

(b)

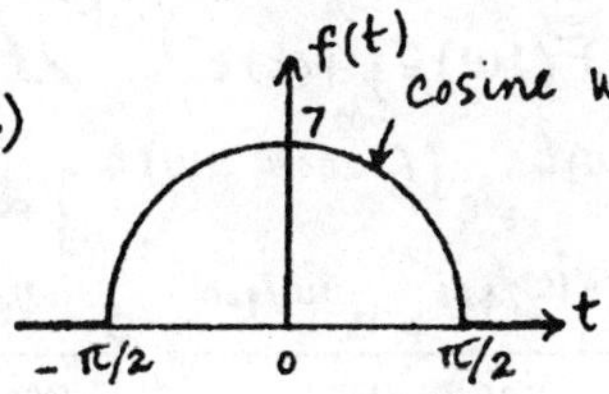


$$\bar{F}(j\omega) = 7\int_{-\pi/2}^{\pi/2} \cos t \, e^{-j\omega t}\, dt = 3.5\int_{-\pi/2}^{\pi/2} (e^{jt} + e^{-jt})\, e^{-j\omega t}\, dt$$

$$= \frac{3.5}{j1 - j\omega} e^{(j-j\omega)t}\Big|_{-\pi/2}^{\pi/2} + \frac{3.5}{-j1-j\omega} e^{-(j+j\omega)t}\Big|_{-\pi/2}^{\pi/2}$$

$$= \frac{3.5}{1-\omega}\left(e^{-j\omega\pi/2} + e^{j\omega\pi/2}\right) - \frac{3.5}{1+\omega}\left(-e^{j\omega\pi/2} - e^{-j\omega\pi/2}\right)$$

$$= \frac{7}{1-\omega}\cos\frac{\omega\pi}{2} + \frac{7}{1+\omega}\cos\frac{\omega\pi}{2} = \underline{\frac{14}{1-\omega^2}\cos\frac{\omega\pi}{2}} \leftarrow$$

(c)

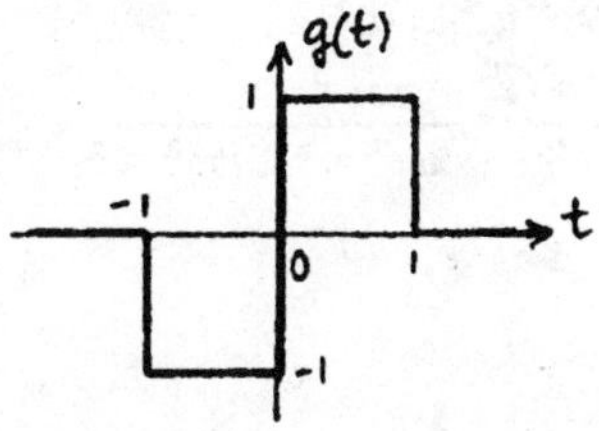


$$\bar{G}(j\omega) = \int_{-1}^{0} -e^{-j\omega t}\, dt + \int_0^1 e^{-j\omega t}\, dt$$

$$= \int_{-1}^{0} (-\cos\omega t + j\sin\omega t)\, dt + \int_0^1 (\cos\omega t - j\sin\omega t)\, dt$$

$$= \frac{1}{\omega}\left[(-\sin\omega t - j\cos\omega t)\Big|_{-1}^{0} + (\sin\omega t + j\cos\omega t)\Big|_0^1\right]$$

$$= \frac{1}{\omega}\left[-j1 - \sin\omega + j\cos\omega + \sin\omega + j\cos\omega - j1\right]$$

$$= \underline{\frac{j2}{\omega}(\cos\omega - 1)} \leftarrow$$

18-5 (a)

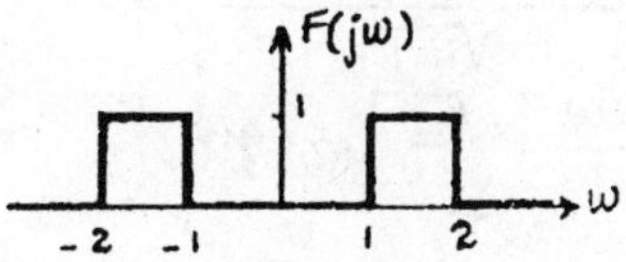


$$f(t) = \frac{1}{2\pi}\int_{-2}^{-1} e^{j\omega t}\, d\omega + \frac{1}{2\pi}\int_1^2 e^{j\omega t}\, d\omega$$

$$= \frac{1}{2\pi}\left[\frac{1}{jt}(e^{-jt} - e^{-j2t}) + \frac{1}{jt}(e^{j2t} - e^{jt})\right]$$

$$= \frac{1}{\pi t}(\sin 2t - \sin t) \quad \therefore f(\tfrac{1}{2}) = \underline{0.2305}$$

(b)

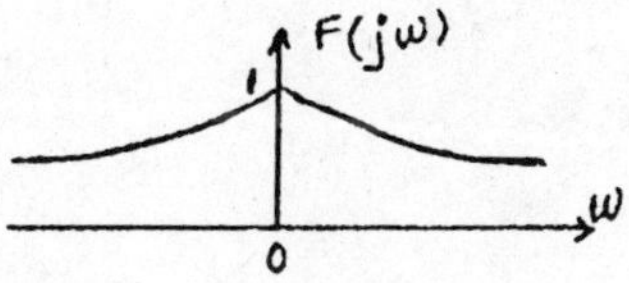


$$f(t) = \frac{1}{2\pi}\int_{-\infty}^{0} e^{\omega} e^{j\omega t}\, d\omega + \frac{1}{2\pi}\int_0^{\infty} e^{-\omega} e^{j\omega t}\, d\omega$$

$$= \frac{1}{2\pi}\left[\frac{1}{1+jt} e^{(1+jt)\omega}\Big|_{-\infty}^{0} + \frac{1}{-1+jt} e^{(-1+jt)\omega}\Big|_0^{\infty}\right]$$

$$= \frac{1}{2\pi}\,\frac{2}{(1+t^2)} \quad \therefore f(\tfrac{1}{2}) = \frac{1}{1.25\pi} = \underline{0.2546}$$

(c)

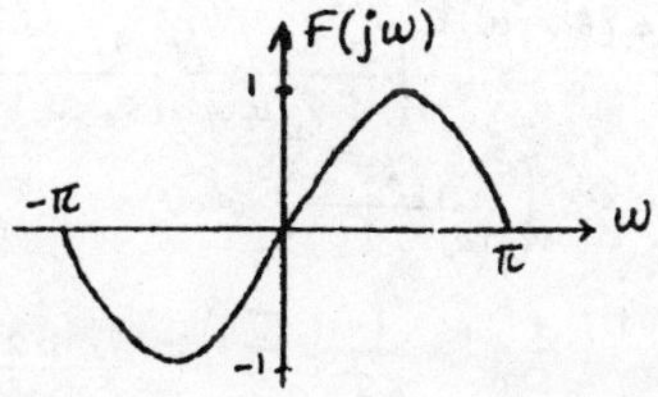


$$f(t) = \frac{1}{2\pi}\int_{-\pi}^{\pi} j\sin\omega\, e^{j\omega t}\, d\omega = \frac{1}{2\pi}\int_{-\pi}^{\pi}(\overset{\text{odd}}{j\sin\omega\cos\omega t} - \overset{\text{even}}{\sin\omega\sin\omega t})\, d\omega$$

$$= -\frac{1}{\pi}\int_0^{\pi} \sin\omega\sin\omega t\, d\omega = \frac{1}{2\pi}\int_0^{\pi}\left[\cos(\omega + \omega t) - \cos(\omega - \omega t)\right] d\omega$$

$$= \frac{1}{2\pi(1+t)}\sin(1+t)\omega\Big|_0^{\pi} - \frac{1}{2\pi(1-t)}\sin(1-t)\omega\Big|_0^{\pi}$$

$$= \frac{1}{2\pi(1+t)}(-\sin\pi t) - \frac{1}{2\pi(1-t)}\sin\pi t = -\frac{\sin\pi t}{2\pi}\left(\frac{1}{1+t} + \frac{1}{1-t}\right)$$

$$= \frac{-\sin\pi t}{\pi(1-t^2)} \quad \therefore f(1/2) = \frac{-\sin\pi/2}{\pi(1-0.25)} = \underline{-0.4244}$$

**18-6** $f(t) = 12\cos 200\pi t$ V, $-2.5 \le t \le 2.5$ ms only ; $\bar{F}(j\omega) = \int_{-\infty}^{\infty} f(t) e^{-j\omega t} dt$

$$\bar{F}(j\omega) = \int_{-0.0025}^{0.0025} 12\cos 200\pi t \times e^{-j\omega t} dt = 6\int_{-0.0025}^{0.0025} \left[e^{j(200\pi - \omega)t} + e^{j(-200\pi - \omega)t}\right] dt$$

$$= 6\left[\frac{e^{j(200\pi - \omega)t}}{j(200\pi - \omega)} + \frac{e^{-j(200\pi + \omega)t}}{-j(200\pi + \omega)}\right]_{-0.0025}^{0.0025} = 6\left[\frac{e^{-j\omega/400} + e^{j\omega/400}}{200\pi - \omega} + \frac{e^{-j\omega/400} + e^{j\omega/400}}{200\pi + \omega}\right]$$

$$= 12\cos\frac{\omega}{400}\left(\frac{1}{200\pi - \omega} + \frac{1}{200\pi + \omega}\right) = 24\cos\frac{\omega}{400} \times \frac{200\pi}{200^2\pi^2 - \omega^2}$$

(a) $\bar{F}(j0) = \frac{24}{200\pi} = \underline{38.20}$ mV/(rad/s) (b) $\bar{F}(j100\pi) = \frac{24\sqrt{2}}{2}\,\frac{200\pi}{200^2\pi^2 - 100^2\pi^2} = \underline{36.01}$ mV/(rad/s)

(c) $\bar{F}(-j100\pi) = \underline{36.01}$ mV/(rad/s) (d) $\bar{F}(j200\pi) = \underline{0}$

**18-7** (a) $W_5 = \frac{1}{5} \times \frac{1}{2\pi}\int_{-3}^{3} |\bar{F}_5(j\omega)|^2 d\omega = \frac{1}{5\pi}\int_0^3 (9 - \omega^2)^2 d\omega = \frac{1}{5\pi}\int_0^3 (81 - 18\omega^2 + \omega^4) d\omega$

$$= \frac{1}{5\pi}\left(81 \times 3 - 6 \times 27 + \frac{1}{5} \times 243\right) = \frac{25.92}{\pi} = \underline{8.251 \text{ J}}$$

(b) $\frac{1}{5\pi}\left(81\omega_1 - 6\omega_1^3 + \frac{1}{5}\omega_1^5\right) = \frac{1}{2} \times \frac{25.92}{\pi}$ By SOLVE, $\omega_1 = \underline{0.84338 \text{ rad/s}}$

**18-8** (a) $\bar{F}(j\omega) = \frac{\pi}{\sqrt{|\omega|}}$ , $f(t) = \frac{1}{2\pi}\int_{-\infty}^{\infty} \frac{\pi}{\sqrt{|\omega|}} e^{j\omega t} d\omega = \frac{1}{2}\int_{-\infty}^{\infty} \frac{\cos\omega t + j\sin\omega t}{\sqrt{|\omega|}} d\omega$

$$= \frac{1}{2}\int_{-\infty}^{\infty} \frac{\cos\omega t\, d\omega}{\sqrt{|\omega|}} = \int_0^{\infty} \frac{\cos\omega t\, d\omega}{\sqrt{|\omega|}} = \sqrt{\frac{\pi}{2t}} \leftarrow \text{(given)}$$

(b) $W_{1\Omega} = \frac{1}{2\pi} \times 2\int_1^3 \frac{\pi^2}{\omega} d\omega = \pi \ln 3 = \underline{3.451 \text{ J}}$

(c) $W_{1\Omega} = \int_1^3 \frac{\pi}{2t} dt = \frac{\pi}{2}\ln 3 = \underline{1.7257 \text{ J}}$

**18-9** (a) $W_{1\Omega} = \int_0^{\infty} 100 t^2 e^{-10t} dt = 100 e^{-10t}\left(\frac{t^2}{-10} - \frac{2t}{100} - \frac{2}{1000}\right)\Big|_0^{\infty} = \underline{0.2 \text{ J}}$

(b) $\bar{F}(j\omega) = \int_{-\infty}^{\infty} f(t) e^{-j\omega t} dt = 10\int_0^{\infty} t e^{-(5 + j\omega)t} dt = 10 e^{-(5+j\omega)t}\left[\frac{-t}{5 + j\omega} - \frac{1}{(5 + j\omega)^2}\right]_0^{\infty}$

$\therefore \bar{F}(j\omega) = \frac{10}{(5 + j\omega)^2}$ $\therefore |\bar{F}(j\omega)|^2 = \frac{100}{(\omega^2 + 25)^2}$ , $W_{1\Omega} = \frac{1}{2\pi} \times 2\int_0^5 \frac{100}{(\omega^2 + 25)^2} d\omega$

$\therefore W_{1\Omega} = \frac{100}{\pi}\left[\frac{\omega}{2 \times 25(\omega^2 + 25)} + \frac{1}{2 \times 125}\tan^{-1}\frac{\omega}{5}\right]_0^5 = \frac{100}{\pi}\left(\frac{1}{500} + \frac{1}{250}\,\frac{\pi}{4}\right) = \underline{0.16366} \text{ J}$

$\therefore \%E = \frac{0.16366}{0.2} \times 100\% = \underline{81.83\%}$

**18-10** (a) $f(t) = 2e^{-8t} u(t)$ , $W_{1\Omega} = \int_0^{\infty} 4e^{-16t} dt = \underline{0.25 \text{ J}}$

(b) $\bar{F}(j\omega) = \int_0^{\infty} 2e^{-(8 + j\omega)t} dt = \frac{-2}{8 + j\omega}(-1)$ , $|\bar{F}(j\omega)| = \frac{2}{\sqrt{\omega^2 + 64}} \leftarrow$

(c) $W_{1\Omega} = \frac{1}{2\pi} \times 2\int_0^{\omega_1} \frac{4}{\omega^2 + 64} d\omega = \frac{4}{\pi} \times \frac{1}{8}\tan^{-1}\frac{\omega}{8}\Big|_0^{\omega_1} = \frac{1}{2\pi}\tan^{-1}\frac{\omega_1}{8} = 0.9 \times 0.25$ $\therefore \omega_1 = \underline{50.51}$ rad/s

**18-11** (a) $\lim_{\epsilon\to 0}\int_{-\epsilon}^{\epsilon}\frac{1}{\epsilon}\cos^2\frac{\pi t}{2\epsilon}\,dt = \lim_{\epsilon\to 0}\frac{1}{\epsilon}\int_{-\epsilon}^{\epsilon}\left(0.5+0.5\cos\frac{\pi t}{\epsilon}\right)dt$

$= \lim_{\epsilon\to 0}\frac{1}{\epsilon}\left[0.5t+\frac{0.5\epsilon}{\pi}\sin\frac{\pi t}{\epsilon}\right]_{-\epsilon}^{\epsilon} = 1$ ←

(b) $\lim_{\epsilon\to 0}\int_{-\epsilon}^{\epsilon}\frac{1}{\epsilon}\left(1-\frac{t^2}{\epsilon^2}\right)dt = \lim_{\epsilon\to 0}\left[\frac{t}{\epsilon}-\frac{t^3}{3\epsilon^3}\right]_{-\epsilon}^{\epsilon} = \lim_{\epsilon\to 0}\left(2-\frac{2}{3}\right) = \frac{4}{3}$ ←

**18-12**

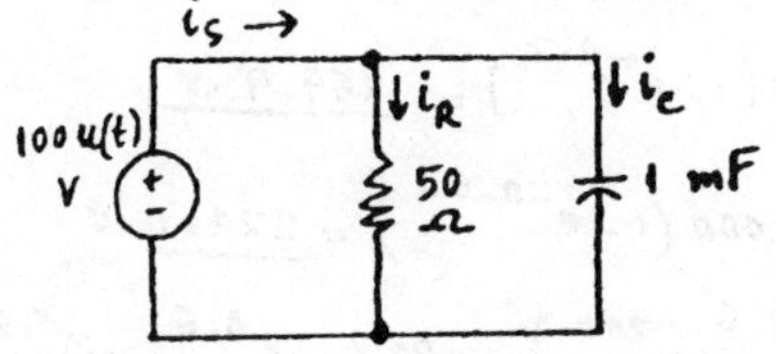


(a) $i_R(t) = \frac{100\,u(t)}{50} = \underline{2\,u(t)}$ A

$i_c(t) = 10^{-3}\frac{d}{dt}\left[100u(t)\right] = \underline{0.1\,\delta(t)}$ A

$i_s(t) = \underline{2u(t) + 0.1\delta(t)}$ A

(b) $i_R(0^+) - i_R(0^-) = 2 - 0 = \underline{2}$ A , $i_c(0^+) - i_c(0^-) = 0 - 0 = \underline{0}$

$q_c(t) = \int i_c(t)\,dt = 10^{-3}\times 100\,u(t) = 0.1\,u(t)$ C

$q_c(0^+) - q_c(0^-) = 0.1 - 0 = \underline{0.1\text{ C}}$

**18-13** (a)

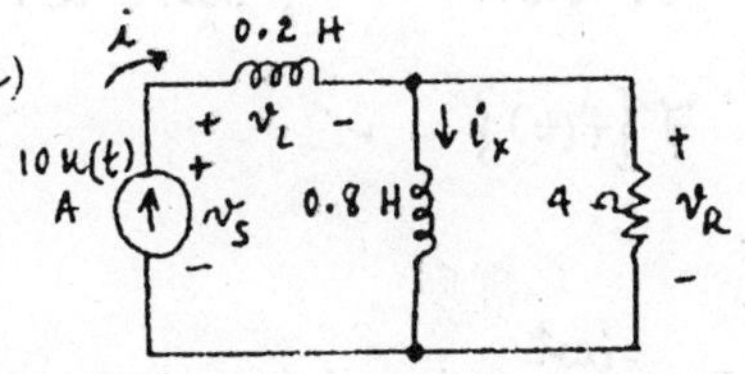


$v_L(t) = L\frac{di}{dt} = 0.2\frac{d}{dt}\left[10u(t)\right] = \underline{2\,\delta(t)}$ V

$i_x = i_{xf} + i_{xn}$ ; $i_{xf} = 10$ , $i_x(t) = \underline{(10 - 10e^{-5t})\,u(t)}$ A

$v_R = v_{Rf} + v_{Rn}$ , $v_{rf} = 0$ ; $v_R(0^+) = 10\times 4 = 40$ V

$\therefore v_R(t) = \underline{40e^{-5t}\,u(t)}$ V ; $v_s = v_L + v_R = \underline{2\delta(t) + 40e^{-5t}\,u(t)}$ V

(b) $\frac{di_x}{dt} = \frac{d}{dt}\left[(10 - 10e^{-5t})\,u(t)\right] = (-10 - 10e^{-5t})\delta(t) + 50e^{-5t}\,u(t) = 50e^{-5t}\,u(t)$

$\therefore v_R(t) = 0.8\times 50e^{-5t}\,u(t) = \underline{40e^{-5t}\,u(t)}$ V

**18-14** (a)

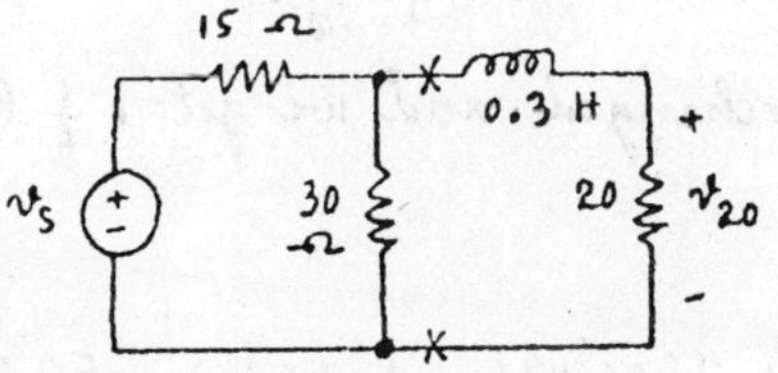


$v_s = \delta(t)$ ; use Thev. at x - x

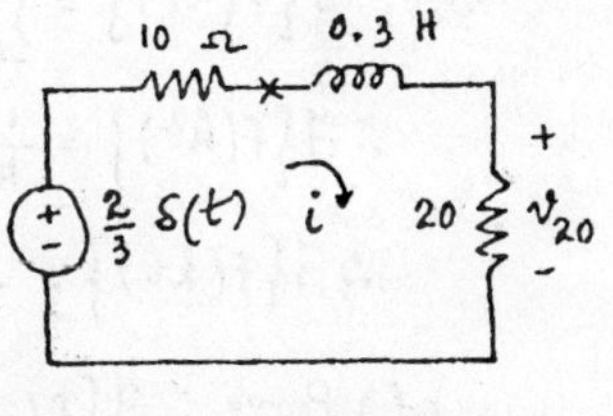


$\therefore \frac{2}{3}\delta(t) = 30i + 0.3i'$

$\int_{0^-}^{0^+}\frac{2}{3}\delta(t) = \frac{2}{3} = 0 + \int_{0^-}^{0^+}0.3\frac{di}{dt}\,dt$

$\therefore \frac{2}{3} = 0.3\left[i(0^+) - i(0^-)\right] = 0.3\,i(0^+)$ $\therefore i(0^+) = \frac{20}{9}$

$\therefore i(t) = \frac{20}{9}e^{-30t/0.3}\,u(t)$ A , $v_{20,a}(t) = 20\,i(t) = \underline{\frac{400}{9}e^{-100t}\,u(t)}$ V ←

(b) $v_s = u(t)$ ; use Thev. again, get $v_{oc} = \frac{2}{3}u(t)$ $\therefore v_{20,f} = \frac{2}{3}\times\frac{20}{30} = \frac{4}{9}$ V

$\therefore v_{20,b}(t) = \underline{\left(\frac{4}{9} - \frac{4}{9}e^{-100t}\right)u(t)}$ ←

(c) $\frac{d}{dt}\left[\left(\frac{4}{9} - \frac{4}{9}e^{-100t}\right)u(t)\right] = \frac{4}{9}(1 - e^{-100t})\delta(t) + \frac{400}{9}e^{-100t}\,u(t) = \underline{\frac{400}{9}e^{-100t}\,u(t)}$ V ←

**18-15**

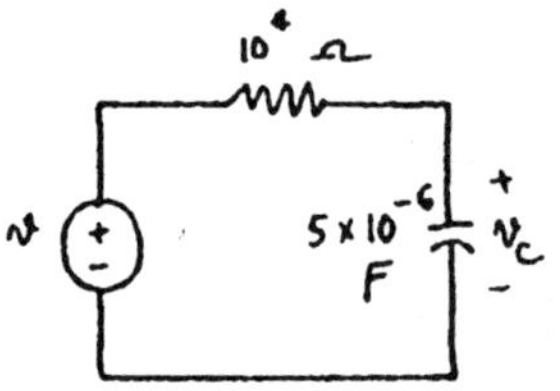


$v(t) = \frac{20}{a}\left[u(t) - u(t-a)\right]$ V

$\therefore v_c(t) = \frac{20}{a}\left[1 - e^{-20t}\right]u(t) - \frac{20}{a}\left[1 - e^{-20(t-a)}\right]u(t-a)$

$v_c(0.03) = \frac{20}{a}\left[1 - e^{-0.6}\right]u(t) - \frac{20}{a}\left[1 - e^{20a-0.6}\right]u(0.03-a)$

(a) $a = 0.04$ $\quad \therefore v_c(0.03) = 500(1 - e^{-0.6}) = \underline{225.6 \text{ V}}$

(b) $a = 0.02$ $\quad \therefore v_c(0.03) = 1000(1 - e^{-0.6}) - 1000(1 - e^{-0.2}) = \underline{269.9 \text{ V}}$

(c) $a = 0.002$ $\quad \therefore v_c(0.03) = 10{,}000(1 - e^{-0.6}) - 10{,}000(1 - e^{-0.56}) = \underline{224.0} \text{ V}$

(d) $a \to 0$ $\quad \therefore v_c(0.03) \to \frac{20}{a}(\cancel{1} - e^{-0.6} - \cancel{1} + e^{-0.6}e^{20a}) \to \frac{20}{a}(-\cancel{e^{-0.6}} + \cancel{e^{-0.6}} + 20ae^{-0.6})$

$\therefore v_c(0.03) \to 400e^{-0.6} = \underline{219.5 \text{ V}}$

**18-16** (a) Prove: $\mathcal{F}\{f(t-t_0)\} \stackrel{?}{=} e^{-j\omega t_0}\mathcal{F}\{f(t)\}$

$\mathcal{F}\{f(t-t_0)\} = \int_{-\infty}^{\infty} f(t-t_0)e^{-j\omega t}\,dt$, let $t - t_0 = \tau$ $\therefore t = \tau - t_0$, $dt = d\tau$

$\therefore \mathcal{F}\{f(t-t_0)\} = \int_{-\infty}^{\infty} f(\tau)e^{-j\omega\tau}e^{-j\omega t_0}\,d\tau = e^{-j\omega t_0}\mathcal{F}\{f(t)\}$ ✓

(b) Prove: $\mathcal{F}\{f'(t)\} \stackrel{?}{=} j\omega\mathcal{F}\{f(t)\}$

$\mathcal{F}\{f'(t)\} = \int_{-\infty}^{\infty} e^{-j\omega t}\frac{df}{dt}\,dt$, Let $u = e^{-j\omega t}$, $du = -j\omega e^{-j\omega t}$, $dv = df$, $v = f$

$\therefore \mathcal{F}\{f'(t)\} = f(t)e^{-j\omega t}\Big|_{-\infty}^{\infty} + \int_{-\infty}^{\infty} j\omega f(t)e^{-j\omega t}\,dt$, We assume $f(\pm\infty) = 0$ $\therefore \mathcal{F}\{f'(t)\} = j\omega\mathcal{F}\{f(t)\}$ ✓

(c) Prove: $\mathcal{F}\{f(kt)\} \stackrel{?}{=} \frac{1}{|k|}F\left(\frac{j\omega}{k}\right)$

$\mathcal{F}\{f(kt)\} = \int_{-\infty}^{\infty} f(kt)e^{-j\omega t}\,dt$, let $\tau = kt$, $k > 0$ $\therefore \mathcal{F}\{f(kt)\} = \int_{-\infty}^{\infty} f(\tau)e^{-j\omega\tau/k}\frac{1}{k}\,d\tau$

$\therefore \mathcal{F}\{f(kt)\} = \frac{1}{k}F\left(\frac{j\omega}{k}\right)$ If $k < 0$ limits are interchanged and we get $-\frac{1}{k}F\left(\frac{j\omega}{k}\right)$

$\therefore \mathcal{F}\{f(kt)\} = \frac{1}{|k|}F\left(\frac{j\omega}{k}\right)$ ✓

(d) Prove: $\mathcal{F}\{f(-t)\} \stackrel{?}{=} F(-j\omega)$. Let $k = -1$ in (c) $\therefore \mathcal{F}\{f(-t)\} = \frac{1}{|-1|}F(-j\omega) = F(-j\omega)$ ✓

(e) Prove: $\mathcal{F}\{tf(t)\} \stackrel{?}{=} j\frac{d}{d\omega}F(j\omega)$

$F(j\omega) = \mathcal{F}\{f(t)\} = \int_{-\infty}^{\infty} f(t)e^{-j\omega t}\,dt$ $\therefore \frac{dF(j\omega)}{d\omega} = \int_{-\infty}^{\infty} f(t)(-jt)e^{-j\omega t}\,dt = -j\mathcal{F}\{tf(t)\}$

$\therefore j\frac{dF(j\omega)}{d\omega} = \mathcal{F}\{tf(t)\}$ ✓

18-17 (a) $\mathfrak{F}\{5[\text{sgn}(t-1)]\delta(t)\} = \mathfrak{F}\{5[\text{sgn}(-1)]\delta(t)\} = \mathfrak{F}\{-5\delta(t)\} = \underline{-5}$

(b) $\mathfrak{F}\{5[\text{sgn}(t-1)]\delta(t-2)\} = \mathfrak{F}\{5[\text{sgn}(1)]\delta(t-2)\} = \mathfrak{F}\{5\delta(t-2)\} = \underline{5e^{-j2\omega}}$

(c) $\mathfrak{F}\{5\sin 2t\, u(t)\} = \mathfrak{F}\{(-j2.5e^{j2t} + j2.5e^{-j2t})u(t)\} = -j2.5\frac{1}{-j2+j\omega} + j2.5\frac{1}{j2+j\omega}$

$= \frac{-2.5}{\omega-2} + \frac{2.5}{\omega+2} = \frac{-10}{\omega^2-4} \leftarrow$

(d) $f(t) = 5\sin 2t\,[u(t+\pi/2) - u(t-\pi/2)]$

$F(j\omega) = \int_{-\pi/2}^{\pi/2} 5e^{-j\omega t}\sin 2t\,dt = \int_{-\pi/2}^{\pi/2} 5e^{-j\omega t}\left(\frac{e^{j2t} - e^{-j2t}}{j2}\right)dt$

$= \int_{-\pi/2}^{\pi/2} -j2.5\left[e^{j(2-\omega)t} - e^{-j(2+\omega)t}\right]dt = -j2.5\left[\frac{1}{j(2-\omega)}e^{j(2-\omega)t} + \frac{1}{j(2+\omega)}e^{-j(2+\omega)t}\right]_{-\pi/2}^{\pi/2}$

$= -2.5\left[\frac{1}{2-\omega}(e^{j\omega\pi/2} - e^{-j\omega\pi/2}) + \frac{1}{2+\omega}(e^{j\omega\pi/2} - e^{-j\omega\pi/2})\right]$

$= 2.5\left[\frac{-j2}{2-\omega}\sin\frac{\omega\pi}{2} - \frac{j2}{2+\omega}\sin\frac{\omega\pi}{2}\right] = j5\sin\frac{\omega\pi}{2}\left(\frac{-1}{2-\omega} - \frac{1}{2+\omega}\right) = \frac{j20}{\omega^2-4}\sin\frac{\omega\pi}{2} \leftarrow$

18-18 (a) $f(t) = A\sin(\omega_0 t + \theta) = \frac{A}{j2}\left(e^{j\omega_0 t + j\theta} - e^{-j\omega_0 t - j\theta}\right) = \frac{Ae^{j\theta}}{j2}e^{j\omega_0 t} - \frac{Ae^{-j\theta}}{j2}e^{-j\omega_0 t}$

$\bar{F}(j\omega) = \frac{Ae^{j\theta}}{j2}2\pi\delta(\omega-\omega_0) - \frac{Ae^{-j\theta}}{j2}2\pi\delta(\omega+\omega_0) = \underline{-jA\pi e^{j\theta}\delta(\omega-\omega_0) + jA\pi e^{-j\theta}\delta(\omega+\omega_0)}$

(b) $f(t) = 2\delta(t) + 3u(t) - 4\,\text{sgn}(t) \Leftrightarrow 2 + 3\pi\delta(\omega) + \frac{3}{j\omega} - \frac{8}{j\omega} = \underline{2 + 3\pi\delta(\omega) + j5/\omega}$

(c) $f(t) = \cosh 5t\, u(t) = \frac{1}{2}e^{5t}u(t) + \frac{1}{2}e^{-5t}u(t) \Leftrightarrow \frac{1/2}{-5+j\omega} + \frac{1/2}{5+j\omega} = \frac{-j\omega}{\omega^2+25} \leftarrow$

18-19 (a) $\bar{F}(j\omega) = 2u(\omega+2) - 2u(\omega-2)$

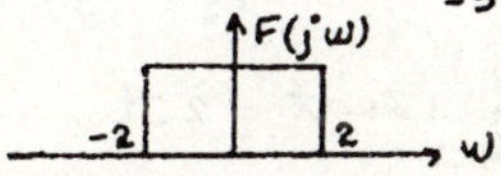


$f(t) = \frac{1}{2\pi}\int_{-\infty}^{\infty}[2u(\omega+2) - 2u(\omega-2)]e^{j\omega t}d\omega = \frac{1}{\pi}\int_{-2}^{2}e^{j\omega t}d\omega = \frac{2}{\pi}\int_0^2 \cos\omega t\,d\omega$

$= \frac{2}{\pi t}\sin\omega t\Big|_0^2 = \frac{2}{\pi t}\sin 2t \quad \therefore f(2) = \underline{-0.2409}$

(b) $\bar{F}(j\omega) = 2u(-2-\omega) + 2u(-2+\omega)$

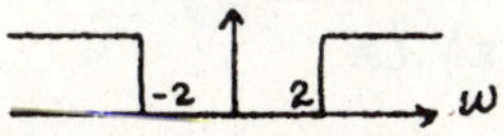


$\bar{F}_b(j\omega) = 2 - F_a(j\omega)$ (above) $\therefore f(t) = 2\delta(t) - \frac{2}{\pi t}\sin 2t \quad , f(2) = \underline{0.2409}$

(c) $\bar{F}(j\omega) = \delta(\omega) + 2u(-2-\omega) + 2u(-2+\omega)$

$f(t) = \frac{1}{2\pi} + 2\delta(t) - \frac{2}{\pi t}\sin 2t \quad \therefore f(2) = \frac{1}{2\pi} + 0.2409 = \underline{0.4001}$

18-20 (a) $\bar{F}(j\omega) = 2 - 2\delta(\omega-2) + \frac{j2}{\omega} + \frac{2}{2+j\omega} \Leftrightarrow 2\delta(t) - \frac{1}{\pi}e^{j2t} - \text{sgn}(t) + 2e^{-2t}u(t) \leftarrow$

(b) $\bar{F}(j\omega) = \frac{2+j\omega}{(2+j\omega)^2 + 2^2} \quad \therefore \alpha = 2,\ \omega_d = 2 \quad , f(t) = \underline{e^{-2t}\cos 2t\, u(t)}$

(c) $\bar{F}(j\omega) = \frac{\sin 2\omega}{2\omega}$ In table 18-1, $T=4 \quad \therefore f(t) = \underline{0.25u(t+2) - 0.25u(t-2)}$

**18-21**

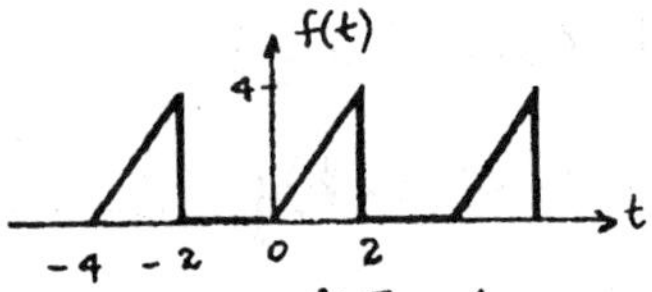


$T = 4 \;;\; C_n = \frac{1}{T}\int_{-T/2}^{T/2} f(t)\, e^{-jn\omega_o t}\, dt$

$\therefore C_n = \frac{1}{4}\int_0^2 2t\, e^{-jn\pi t/2}\, dt = \frac{1}{2}\left[e^{-jn\pi t/2}\left(\frac{t}{-jn\pi/2} - \frac{1}{-n^2\pi^2/4}\right)\right]_0^2$

$\therefore C_n = \frac{1}{2} e^{-jn\pi}\left(\frac{j4}{n\pi} + \frac{4}{n^2\pi^2}\right) - \frac{1}{2}\,\frac{4}{n^2\pi^2} = e^{-jn\pi}\left(\frac{j2}{n\pi} + \frac{2}{n^2\pi^2}\right) - \frac{2}{n^2\pi^2}$

$\bar{F}(j\omega) = 2\pi \sum_{-\infty}^{\infty} C_n\, \delta(\omega - n\omega_o) = 2\pi \sum_{-\infty}^{\infty}\left[e^{-jn\pi}\left(\frac{j2}{n\pi} + \frac{2}{n^2\pi^2}\right) - \frac{2}{n^2\pi^2}\right]\delta\left(\omega - \frac{n\pi}{2}\right)$ ←

**18-22**

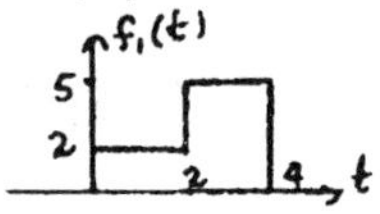


$f_1(t) = 2u(t) + 3u(t-2) - 5u(t-4)\;,\; 0 < t < 5\;,\; T = 5$

$C_n = \frac{1}{T}\int_{-T/2}^{T/2} f(t)\, e^{-jn\omega_o t}\, dt = 0.2\left[\int_0^2 2e^{-j0.4\pi nt}\, dt + \int_2^4 5e^{-j0.4\pi nt}\, dt\right]$

$= \frac{0.4}{-j0.4\pi n} e^{-j0.4\pi nt}\Big|_0^2 + \frac{1}{-j0.4\pi n} e^{-j0.4\pi nt}\Big|_2^4 = \frac{j}{n\pi}\left(e^{-j0.8\pi n} - 1\right) + \frac{j2.5}{n\pi}\left(e^{-j1.6\pi n} - e^{-j0.8\pi n}\right)$

$= \frac{j}{n\pi}\left(-1 - 1.5e^{-j0.8\pi n} + 2.5e^{-j1.6\pi n}\right) \therefore \bar{F}(j\omega) = 2\pi\sum_{-\infty}^{\infty}\left[\frac{j}{n\pi}\left(-1 - 1.5e^{-j0.8\pi n} + 2.5e^{-j1.6\pi n}\right)\right]\delta(\omega - 0.4\pi n)$

**18-23**

$\bar{F}(j\omega) = 100\sum_{n=-\infty}^{\infty}\frac{1}{|n|!}\,\delta(\omega - 0.1n) \;\therefore\; f(t) = \frac{100}{2\pi} + \frac{100}{\pi}\sum_{1}^{\infty}\frac{1}{n!}\cos 0.1nt$

$\therefore f(2) = \frac{50}{\pi} + \frac{100}{\pi}\left(\cos 0.2 + \frac{1}{2}\cos 0.4 + \frac{1}{6}\cos 0.6 + \dots\right) = \frac{50}{\pi} + \frac{100}{\pi}(1.6122) = \underline{67.23}$

**18-24**

$x(t) = h(t) = u(t) - u(t-1)$

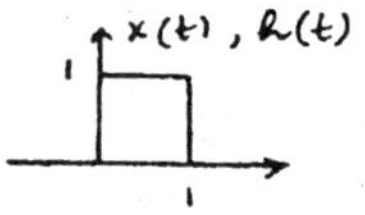


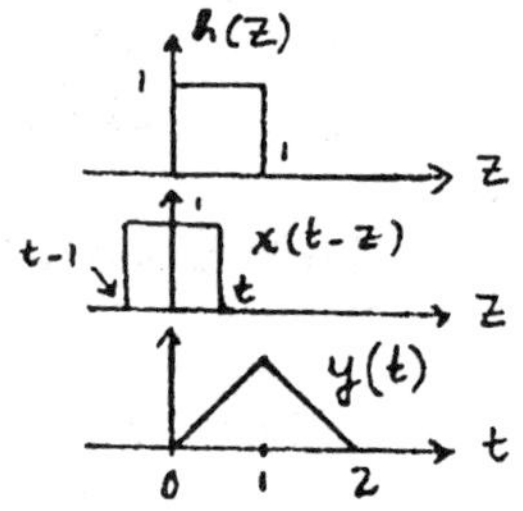


(a) $t < 0: y(t) = \underline{0}\;;\; 0 < t < 1: y(t) = \int_0^t 1\times 1\, dt = \underline{t}$

$1 < t < 2: y(t) = \int_{t-1}^{1} 1\times 1\, dt = 2 - t\;;\; t > 2: y(t) = \underline{0}$

**18-25**

$x(t) = 2[u(t) - u(t-2)]\;,\; h(t) = u(t)$

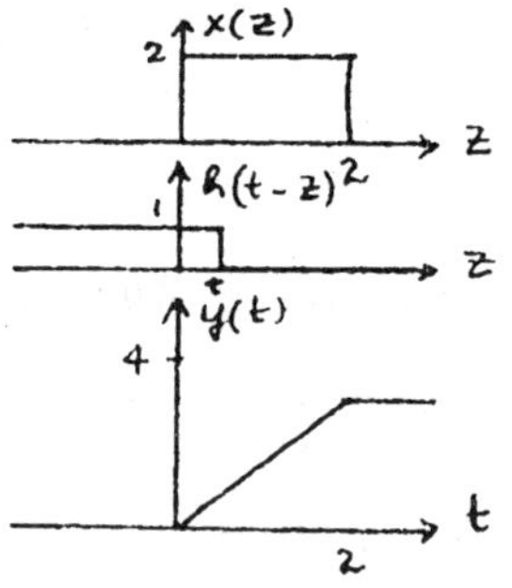


(a) $y(t) = \int_{-\infty}^{t} x(z)\,h(t-z)\,dz$

$t < 0: y(t) = \underline{0}\;;\; 0 < t < 2: y(t) = \int_0^t 2\times 1\, dt = \underline{2t}$

$t > 2: \int_0^2 2\times 1\, dt = \underline{4} = y(t)$

(b) $y(t) = \int_0^{\infty} x(t-z)\,h(z)\,dz$

$t < 0: y(t) = 0\;;\; 0 < t < 2: y(t) = \int_0^t 2\times 1\, dz = \underline{2t}$

$t > 2: y(t) = \int_{t-2}^{t} 2\times 1\, dz = \underline{4}$

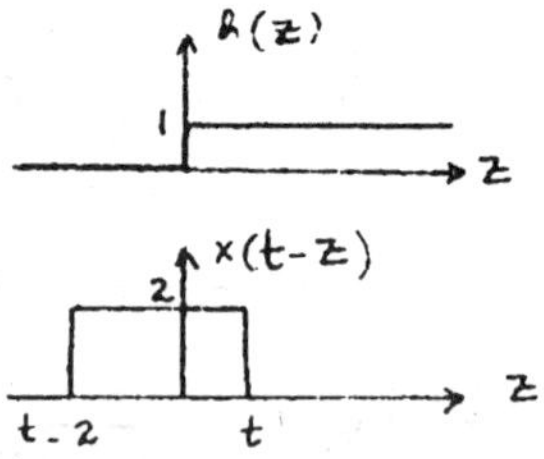

**18-26** $f_1(t) = (1+2e^{-t})u(t)$ , $f_2(t) = 4[u(t+1) - u(t-2)]$

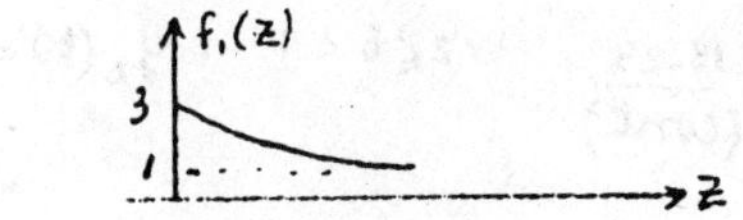

$t < -1$ : $f(t) = \underline{0}$

$-1 < t \le 2$ : $f(t) = \int_0^{t+1} (4 + 8e^{-z})\,dz = (4z - 8e^{-z})\Big|_0^{t+1}$

$= 4t + 4 - 8e^{-(t+1)} + 8 = \underline{4t + 12 - 8e^{-(t+1)}}$

$t > 2$ : $f(t) = \int_{t-2}^{t+1} (4 + 8e^{-z})\,dz = (4z - 8e^{-z})\Big|_{t-2}^{t+1}$

$= 4 + 8 - 8e^{-(t+1)} + 8e^{-(t-2)}$

$= \underline{12 + 8e^{-t}(-e^{-1} + e^{2})}$

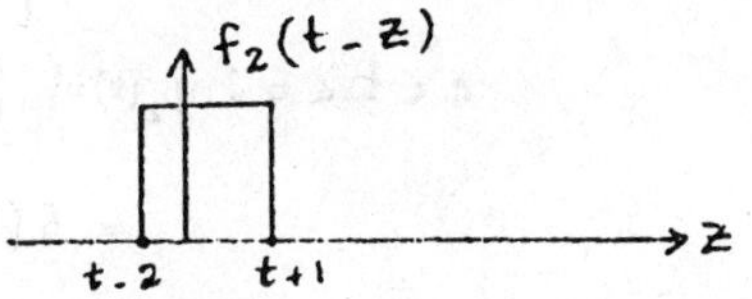

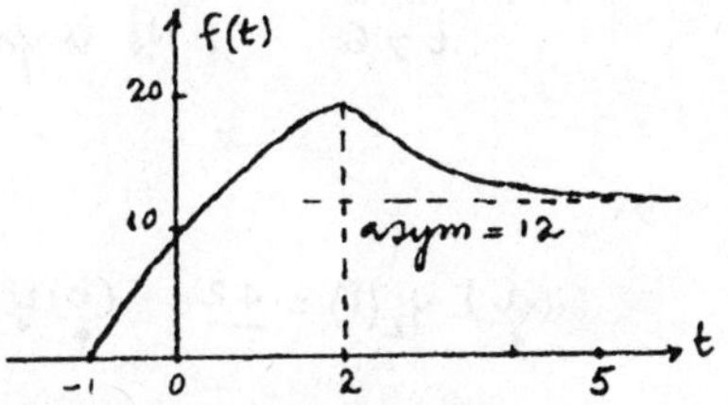


**18-27**

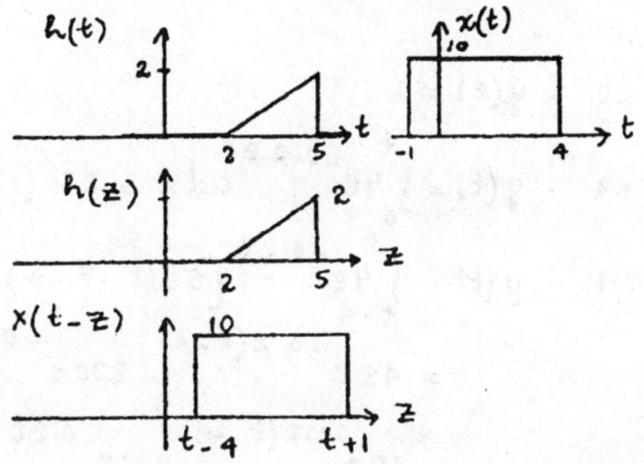

$y(t) = \int_0^{\infty} x(t-z)h(z)\,dz$

$h(t) = \frac{2}{3}(t-2)[u(t-2) - u(t-5)]$

$\therefore h(z) = \frac{2}{3}(z-2)[u(z-2) - u(z-5)]$

$x(t) = 10[u(t+1) - u(t-4)]$

$\therefore x(t-z) = 10[u(t-z+1) - u(t-z-4)]$

For $4 < t < 6$, $h(z)$ is in window, limits $z = 2$ and $z = 5$

(a) $\therefore y(t) = \underline{10 \times \frac{2}{3}\int_2^5 (z-2)\,dz} = \frac{20}{3} \times \frac{1}{2}(z-2)^2\Big|_2^5 = 30$ (b) $y(5) = \underline{30}$

**18-28**

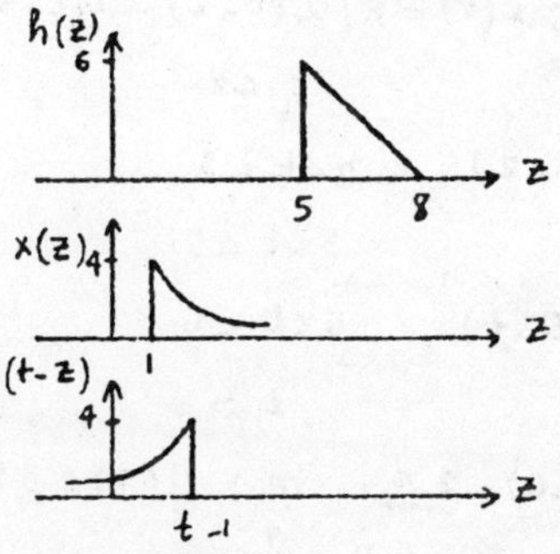

$x(t) = 4e^{-(t-1)}u(t-1)$ ; $h(t) = (16-2t)[u(t-5) - u(t-8)]$

(a) $t = 5$ ; $y(5) = \underline{0}$

(b) $t = 8$ : $y(8) = \int_5^7 4e^{-7+z}(16 - 2z)\,dz$

$= 4e^{-7}\int_5^7 (16e^{z} - 2ze^{z})\,dz$

$= 4e^{-7}\left[16(e^7 - e^5) - 2e^{z}(z-1)\Big|_5^7\right]$

$= 64(1 - e^{-2}) - 8e^{-7}(e^7 \times 6 - e^5 \times 4) = 16 - 32e^{-2}$

$y(8) = \underline{11.669}$

**18-29**

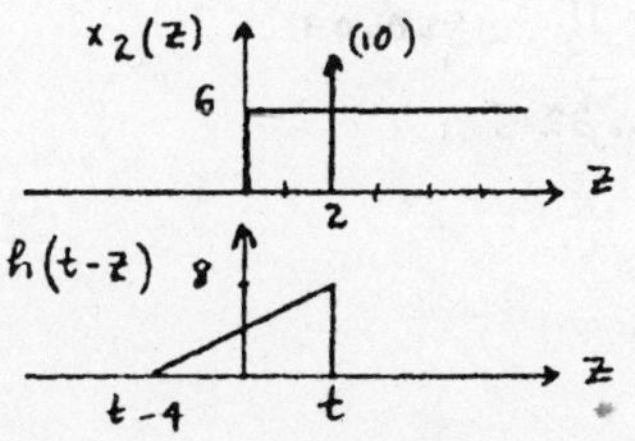

$x_1(t) = \delta(t)$ ; $y_1(t) = 8 - 2t$ ; $x_2(t) = 10\delta(t-2) + 6u(t)$

$t < 0$ : $y_2(t) = \underline{0}$

$0 < t < 2$ : $y_2(t) = \int_0^t [8 - 2(t-z)]6\,dz = 6(8-2t)t + 6t^2 = \underline{48t - 6t^2}$

(cont. on next page)

**18-29 (cont.)**

$2 \le t < 4$ : $y_2(t) = \int_0^t [8-2(t-z)][6+10\delta(z-2)]\,dz$

$= 48t - 6t^2 + [8-2(t-2)]10 = \underline{120 + 28t - 6t^2}$

$4 < t < 6$ : $y_2(t) = \int_{t-4}^{t} [8-2(t-z)][6+10\delta(z-2)]\,dz = \left[6(8-2t)z + 6z^2\right]_{t-4}^{t} + [8-2(t-2)]10$

$= 6(8-2t)4 + 6t^2 - 6(t-4)^2 + 120 - 120t = \underline{216 - 20t}$

$t > 6$ : No impulse, $y_2 = \int_{t-4}^{t} 6[8-2(t-z)]\,dz = \int_{t-4}^{t} (48 - 12t + 12z)\,dz$

$= (48-12t)z\Big|_{t-4}^{t} + 6z^2\Big|_{t-4}^{t} = (48-12t)4 + 6t^2 - 6(t-4)^2 = \underline{96}$

(a) $y_2(1) = \underline{42}$ , (b) $y_2(2^-) = \underline{72}$ (c) $y_2(2^+) = \underline{152}$ (d) $y_2(3) = \underline{150}$ (e) $y_2(5) = \underline{116}$ (f) $y_2(7) = \underline{96}$

**18-30** (a)

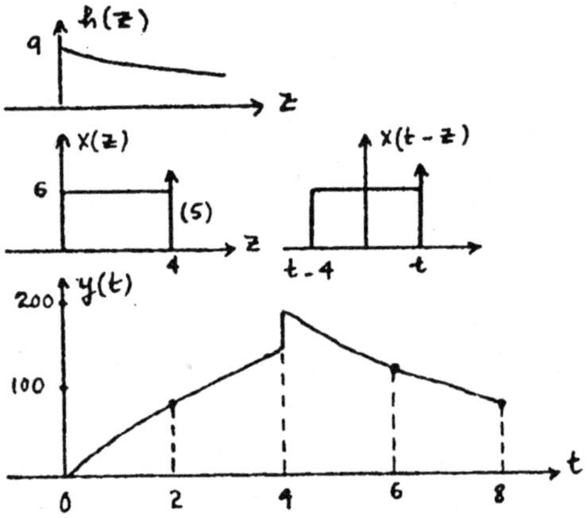


$t < 0$ : $y(t) = 0$

$0 < t < 4$ : $y(t) = \int_0^t 9e^{-0.2z}\,6\,dz = 270(1 - e^{-0.2t})$

$t > 4$ : $y(t) = \int_{t-4}^{t} 9e^{-0.2z}[5\delta(t-z-4) + 6]\,dz$

$= 45e^{-0.2(t-4)} - 270e^{-0.2z}\Big|_{t-4}^{t}$

$= 45e^{-0.2(t-4)} - 270e^{-0.2t} + 270e^{-0.2(t-4)}$

$y(t) = \underline{431.0\,e^{-0.2t}}$

(b) $y(3) = \underline{121.82}$ ; $y(5) = \underline{158.56}$

**18-31**

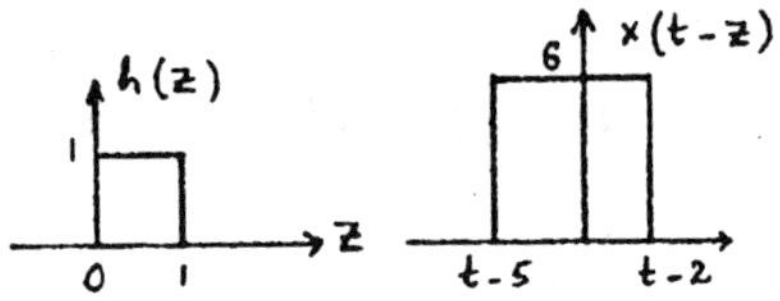


$h(t) = u(t) - u(t-1)$ ; $x(t) = 6[u(t-2) - u(t-5)]$

$$y(t) = \begin{cases} 0 & t < 2 \\ 6(t-2) & 2 < t < 3 \\ 6 & 3 < t < 5 \\ 6(6-t) & 5 < t < 6 \\ 0 & t > 6 \end{cases}$$

(a) $y(1) = \underline{0}$ (b) $y(2.5) = \underline{3}$ (c) $y(4) = \underline{6}$ (d) $y(5.6) = \underline{2.4}$ (e) $y(7) = \underline{0}$

**18-32** $h(t) = 9e^{-0.2t}$ $\therefore$ $\bar{H}(j\omega) = \frac{9}{j\omega + 0.2}$ $\therefore$ $W_{1\Omega} = \frac{1}{\pi}\int_0^{0.02\times 2\pi} \frac{81}{\omega^2 + 0.04}\,d\omega$

$\therefore$ $W_{1\Omega} = \frac{81}{\pi}\,\frac{1}{0.2}\tan^{-1}5\omega\Big|_0^{0.04\pi} = \frac{405}{\pi}\tan 0.2\pi = 72.32$ J

$\% E = \frac{72.32}{(405/\pi)(\pi/2)} = \underline{35.71\ \%}$

**18-33** $\bar{F}(j\omega)=\dfrac{2-j\omega}{(1+j\omega)(2+j\omega)}=\dfrac{3}{1+j\omega}-\dfrac{4}{2+j\omega}\quad\therefore f(t)=(3e^{-t}-4e^{-2t})u(t)$

(a) $W_{1\Omega}=\int_0^{\infty}(9e^{-2t}+16e^{-4t}-24e^{-3t})dt=\frac{9}{2}+4-8=\underline{0.500\text{ J}}$

(b) $f'(t)=-3e^{-t}+8e^{-2t}=0\quad\therefore e^{-t}=3/8\quad\therefore f_{max}=3\times 3/8-4(3/8)^2=\underline{0.5625}$

**18-34** (a) $f(t)=\mathcal{F}^{-1}\left\{\dfrac{1}{(1+j\omega)(2+j\omega)(3+j\omega)}\right\}=\mathcal{F}^{-1}\left\{\dfrac{1/2}{1+j\omega}-\dfrac{1}{2+j\omega}+\dfrac{1/2}{3+j\omega}\right\}=\underline{\left(\frac{1}{2}e^{-t}-e^{-2t}+\frac{1}{2}e^{-3t}\right)u(t)}$

(b) $\mathcal{F}^{-1}\left\{\dfrac{j\omega}{(1+j\omega)(2+j\omega)(3+j\omega)}\right\}=\mathcal{F}^{-1}\left\{\dfrac{-1/2}{1+j\omega}+\dfrac{2}{2+j\omega}-\dfrac{3/2}{3+j\omega}\right\}=\underline{\left(-\frac{1}{2}e^{-t}+2e^{-2t}-\frac{3}{2}e^{-3t}\right)u(t)}$

(c) $\mathcal{F}^{-1}\left\{\dfrac{(j\omega)^2}{(1+j\omega)(2+j\omega)(3+j\omega)}\right\}=\mathcal{F}^{-1}\left\{\dfrac{1/2}{1+j\omega}-\dfrac{4}{2+j\omega}+\dfrac{9/2}{3+j\omega}\right\}=\underline{\left(\frac{1}{2}e^{-t}-4e^{-2t}+\frac{9}{2}e^{-3t}\right)u(t)}$

(d) $\mathcal{F}^{-1}\left\{\dfrac{(j\omega)^3}{(1+j\omega)(2+j\omega)(3+j\omega)}\right\}=\mathcal{F}^{-1}\left\{\dfrac{(j\omega)^3}{(j\omega)^3+6(j\omega)^2+11(j\omega)+6}\right\}=\mathcal{F}^{-1}\left\{1-\dfrac{6(j\omega)^2+11(j\omega)+6}{(1+j\omega)(2+j\omega)(3+j\omega)}\right\}$

$=\mathcal{F}^{-1}\left\{1-\left(\dfrac{1/2}{1+j\omega}-\dfrac{8}{2+j\omega}+\dfrac{27/2}{3+j\omega}\right)\right\}$

$=\underline{\delta(t)-\left(\frac{1}{2}e^{-t}-8e^{-2t}+\frac{27}{2}e^{-3t}\right)u(t)}$

**18-35** (a) $h(t)=2e^{-t}u(t)$, $\bar{H}(j\omega)=\dfrac{\bar{V}_o(j\omega)}{\bar{V}_i(j\omega)}=2\times\dfrac{1}{1+j\omega}=\underline{\underline{\dfrac{2}{1+j\omega}}}$

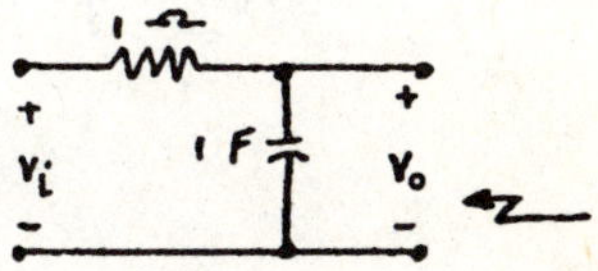


(b) $\frac{1}{2}\bar{H}(j\omega)=\dfrac{1}{1+j\omega}=\dfrac{1}{2}\dfrac{\bar{V}_o}{\bar{V}_i}=\dfrac{1/j\omega}{1+1/j\omega}$

(c) Gain $=\underline{\underline{2}}$

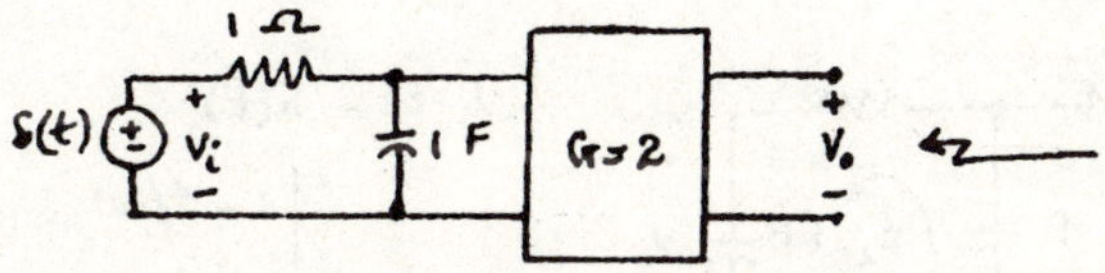


**18-36** (a) $h(t)=3e^{-4(t-2)}u(t-2)$ ; if $h_1(t)=3e^{-4t}u(t)$, then $\bar{H}_1(j\omega)=\dfrac{3}{4+j\omega}$

But $\mathcal{F}\{f(t-t_0)\}=e^{-j\omega t_0}\mathcal{F}\{f(t)\}\quad\therefore \bar{H}(j\omega)=\dfrac{3e^{-j2\omega}}{4+j\omega}$, $\bar{H}(j\omega)\bar{H}^*(j\omega)=\dfrac{9}{\omega^2+16}$

(b) $W_{1\Omega}=\dfrac{1}{2\pi}\int_{-\infty}^{\infty}\dfrac{9}{\omega^2+16}d\omega=\dfrac{9}{\pi}\int_0^{\infty}\dfrac{d\omega}{\omega^2+16}=\dfrac{9}{\pi}\times\dfrac{1}{4}\times\dfrac{\pi}{2}=\underline{1.125\text{ J}}$

**18-37** (a) $\bar{H}(j\omega)=\dfrac{2}{3+j\omega}\quad\therefore h(t)=2e^{-3t}u(t)$

(b) $x(t)=5\cos 4t \Rightarrow 5\angle 0^\circ$ at $\omega=4$. Now $\bar{H}(j4)=\dfrac{2}{3+j4}=0.4\angle -53.13^\circ$

$\therefore y(t)=5\times 0.4\cos(4t-53.13^\circ)=\underline{2\cos(4t-53.13^\circ)}$

**18-38** (a)

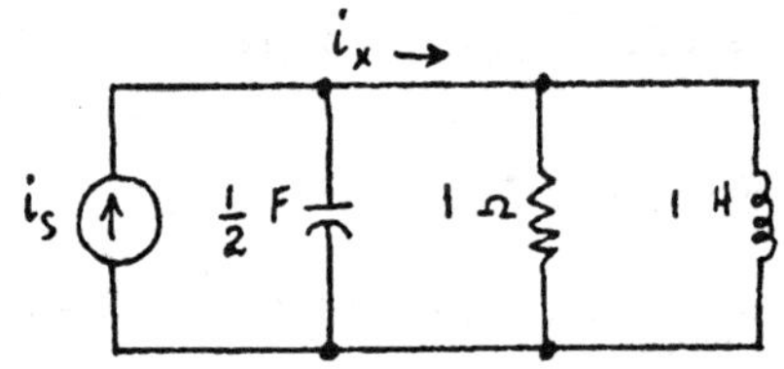


$i_s(t) = \delta(t)$ ; $\bar{H}(j\omega) = \dfrac{\bar{I}_x}{\bar{I}_s} = \dfrac{1+(1/j\omega)}{1+(1/j\omega)+(j\omega/2)}$

$\therefore \bar{H}(j\omega) = \dfrac{2(1+j\omega)}{j2\omega+2+(j\omega)^2} = \dfrac{2(1+j\omega)}{(1+j\omega)^2+1}$

$I_s = 1$ $\therefore \bar{Y}(j\omega) = \bar{H}(j\omega)$ From Table 18-1,

$y(t) = h(t) = \underline{2e^{-t}\cos t\, u(t)} = i_x(t)$

(b) $i_s = \delta(t-1)$ $\therefore i_x(t) = \left[2e^{-(t-1)}\cos(t-1)\right]u(t-1)$

(c) $i_s = u(t)$ $\therefore i_x(t) = \int_0^t 2e^{-t}\cos t\, dt = e^{-t}(-\cos t + \sin t)\,u(t)\Big|_0^t$

$i_x(t) = \left[1 + e^{-t}(\sin t - \cos t)\right]u(t)$ ←

**18-39** (a) $f(t) = 3e^{-t}u(t)$, $g(t) = 4e^{-2t}u(t)$

$\bar{F}(j\omega) = \dfrac{3}{1+j\omega}$, $\bar{G}(j\omega) = \dfrac{4}{2+j\omega}$

$[\bar{F}(j\omega)][\bar{G}(j\omega)] = \dfrac{12}{(1+j\omega)(2+j\omega)} = \dfrac{12}{1+j\omega} - \dfrac{12}{2+j\omega}$

$\therefore f(t) * g(t) = \underline{12(e^{-t} - e^{-2t})u(t)}$ ←

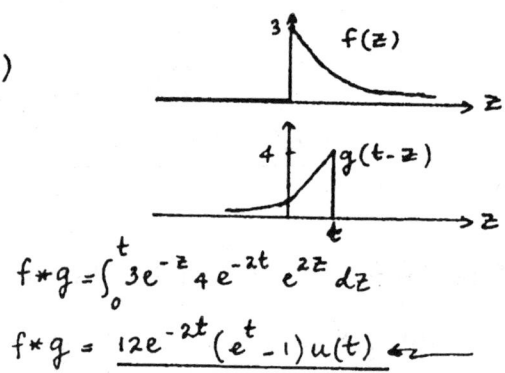

(b)

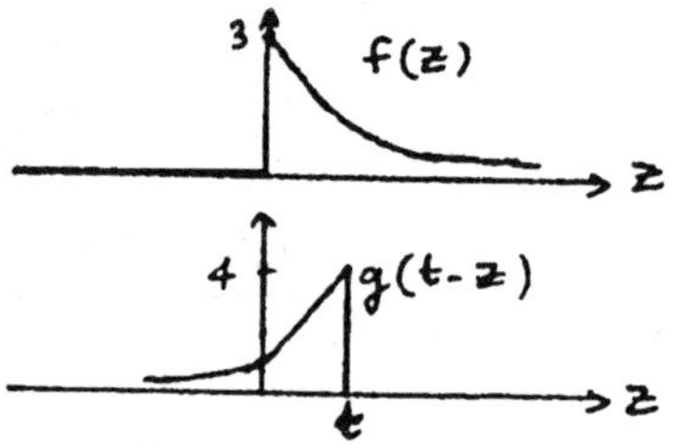


$f*g = \int_0^t 3e^{-z}\,4e^{-2t}\,e^{2z}\,dz$

$f*g = \underline{12e^{-2t}(e^{t}-1)u(t)}$ ←

**18-40**

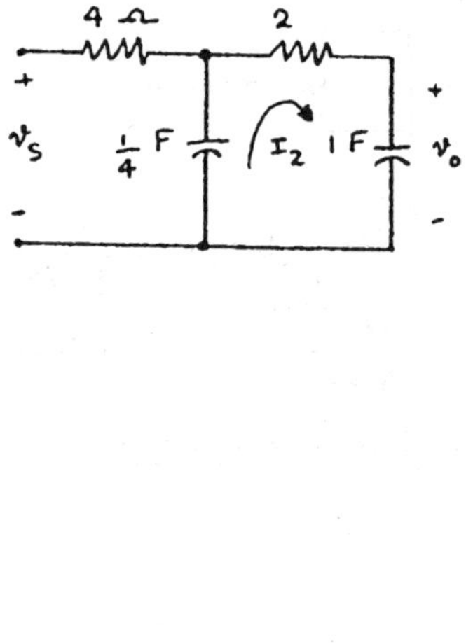


$v_s(t) = \delta(t)$

$$\bar{I}_2 = \frac{\begin{vmatrix} 4+4/j\omega & V_s \\ -4/j\omega & 0 \end{vmatrix}}{\begin{vmatrix} 4+4/j\omega & -4/j\omega \\ -4/j\omega & 2+5/j\omega \end{vmatrix}} = \frac{4\bar{V}_s/j\omega}{8+28/j\omega+20/(j\omega)^2-16/(j\omega)^2}$$

$\therefore \bar{I}_2 = \dfrac{\bar{V}_s/j\omega}{1/(j\omega)^2+7/(j\omega)+2}$

$\therefore \dfrac{\bar{V}_o}{\bar{V}_s} = \dfrac{1}{1+7j\omega+2(j\omega)^2} = \dfrac{1/2}{(j\omega)^2+3.5(j\omega)+0.5} = \bar{H}(j\omega)$

$\therefore \bar{H}(j\omega) = \dfrac{1/2}{(j\omega+3.3508)(j\omega+0.14922)} = \dfrac{-0.15617}{j\omega+3.3508} + \dfrac{0.15617}{j\omega+0.14922}$

$v_o(t) = \underline{0.15617(e^{-0.14922t} - e^{-3.3508t})\,u(t)}$

# CHAPTER 19: LAPLACE TRANSFORM TECHNIQUES

## Section 19-2 Definition of the Laplace transform

**1** Give the range of $\sigma$ over which the Laplace transform exists if $f(t)$ =: (*a*) $tu(t)$; (*b*) $t$; (*c*) $e^{-2t}u(t)$; (*d*) $e^{2t}u(t)$; (*e*) $e^{-2t}u(t - 3)$; (*f*) $e^{-2t}u(t + 3)$.

**2** Determine the Fourier transform, the two-sided Laplace transform, and the Laplace transform of $f(t)$ =: (*a*) $4e^{-2t}[u(t + 3) - u(t - 2)]$; (*b*) $4e^{2t}[u(t + 3) - u(t - 2)]$; (*c*) $4e^{-2|t|}[u(t + 3) - u(t—2)]$.

**3** Use the definition of the Laplace transform to find **F**(s) if $f(t)$ =: (*a*) $[u(t - 1)]\,[u(4 - t)]\,[u(t)]$; (*b*) $2\delta(t - 4)$; (*c*) $2(\sin \pi t)\,\delta(t - 0.2)$; (*d*) $2u(t - 4)$; (*e*) $2e^{-4t}u(t - 1)$.

**4** Determine the Laplace transform of each of the following functions, giving the range of $\sigma$ over which the transform exists (*a*) $[u(t + 3)]\,[u(t)]\,[u(5 - t)]$; (*b*) $5\delta(t - 5)$; (*c*) $5\delta(t - 0.1) \cos 2\pi t$; (*d*) $5u(t - 5)$; (*e*) $5(t - 2)u(t - 1)$.

## Section 19-3 Laplace transforms of some simple time functions
## Section 19-4 Several basic theorems for the Laplace transform

**5** Find $f(t)$ if **F**(s) =: (*a*) $1 + (1/s) + [1/(s + 1)]$; (*b*) $(s + 1)e^{-s}/s$; (*c*) $(1 + e^{-s})^2$; (*d*) $2e^{-3s} \sinh 2s$.

**6** Apply the defining expression for the Laplace transform to calculate **F**$(0.1 + j1)$ if $f(t)$ =: (*a*) $u(t - 1)$; (*b*) $tu(t - 1)$; (*c*) $(t - 1)u(t - 1)$; (*d*) $e^{-2t}u(t - 1)$; (*e*) $\delta(t - 1)u(t)$.

**7** Find $f(t)$ if **F**(s) =: (*a*) $5 - (2/s)$; (*b*) $(5/s) + [3/(s + 1)]$; (*c*) $(s + 2)/(s + 1)$; (*d*) $(s^2 + 5)/s^2$; (*e*) $6/[(s + 1)(s + 2)]$.

**8** Find the inverse Laplace transform of: (*a*) $(s + 6)/(s + 4)$; (*b*) $(2 + 3s)/(s^2 + 4)$; (*c*) $(4s + 100)/(10s^2 + 20s)$; (*d*) $60/[(s + 2)(s + 3)(s + 5)]$.

**9** (*a*) Write a single, nodal, time-domain equation in terms of $v_C(t)$ for the circuit of Fig. 19-1. (*b*) Take the Laplace transform of the equation and use it to find $v_C(t)$.

**10** Let $i_s(t) = 5 + 5u(t)$ A in the circuit shown in Fig. 19-2. Write a single differential equation in terms of $i_L(t)$, take the Laplace transform of the equation, solve for $\mathbf{I}_L(s)$, and then find $i_L(t)$.

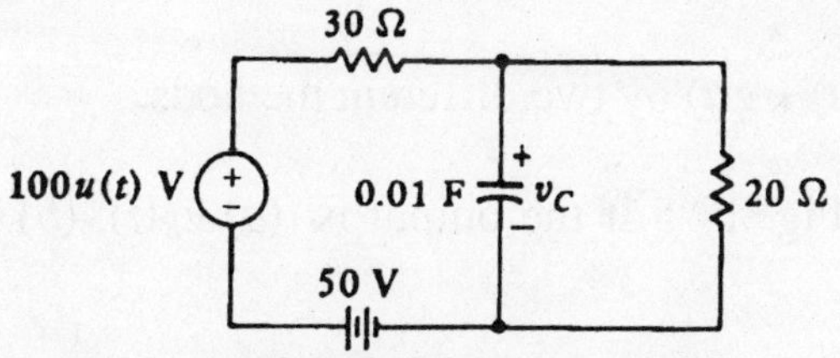


**Fig. 19-1:** See Prob. 9.

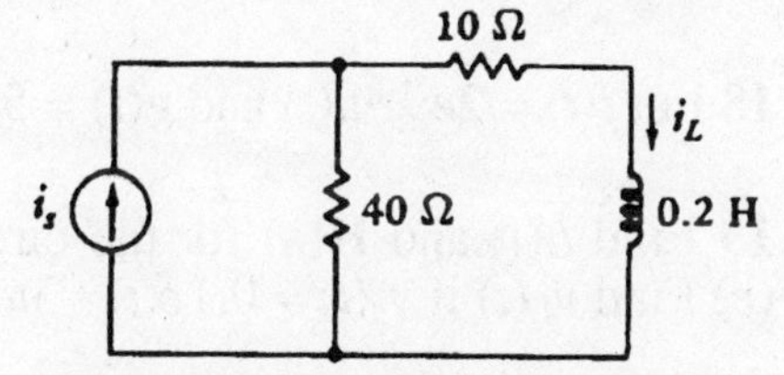


**Fig. 19-2:** See Prob. 10.

**11** If $d^3y/dt^3 + 2d^2y/dt^2 = 0$ and $y(0^-) = 0$, $y'(0^-) = 1$ and $y''(0^-) = -1$, take the Laplace transform of the equation, solve for **Y**(s), and then obtain $y(t)$.

**12** Given the equation, $\int_{-\infty}^{t} 25f(x)\,dx = 30u(t) - df/dt$, where $f(0^-) = 2$ and $\int_{-\infty}^{0^-} f(t)\,dt = 0.2$, find $f(t)$.

**13** An ideal voltage source, $v_s(t) = 100e^{-5t}u(t)$ V, a resistor $R$, an inductor $L$, and a capacitor $C$ are in series. Let $i(t)$ leave the positive terminal of $v_s$. Then $v_s = i' + 4i + 3\int_{0^-}^{t} i\,dt$. (*a*) Find $R$, $L$, and $C$. (*b*) Use Laplace-transform methods to find $i(t)$.

**14** Two signals, $x(t)$ and $y(t)$, are related by the equations, $y'(t) + 2x(t) = t$ and $y + \int_{0^-}^{t} x(z)\,dz = 2t$, where $y(0^-) = 0$ and $\int_{-\infty}^{0^-} x(t)\,dt = 0$. Find $x(t)$ and $y(t)$ for $t > 0$.

**15** Write a single integrodifferential equation in terms of $v_L(t)$ for the circuit of Fig. 19-3, take its Laplace transform, solve for $\mathbf{V}_L(s)$, and obtain $v_L(t)$ by the inverse transform .

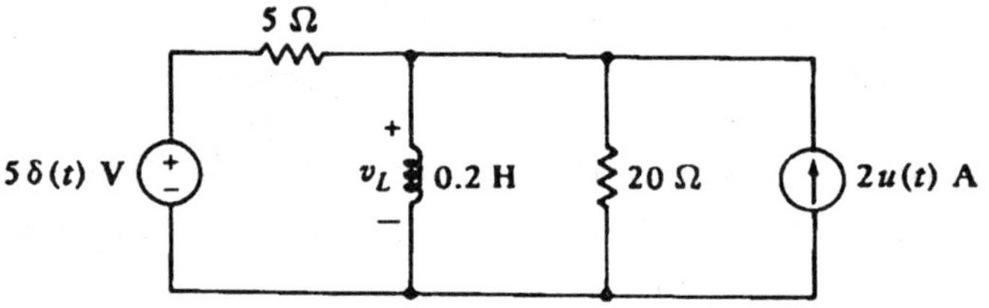


**Fig. 19-3:** See Prob. 15.

## Section 19-5 Convolution again

**16** Let $x(t) = 4tu(t)$ and $h(t) = 3u(t)$. (*a*) Find $\mathbf{X}(s)$, $\mathbf{H}(s)$, and $\mathcal{L}^{-1}\{\mathbf{X}(s)\mathbf{H}(s)\}$. (*b*) Find $x(t) * h(t)$ by working entirely in the time domain.

**17** The input $x(t)$ and the impulse response of a certain linear system are shown in Fig. 19-4. Find the output $y(t)$ as: (*a*) $x(t) * h(t)$ in the time domain; (*b*) $\mathcal{L}^{-1}\{\mathbf{X}(s)\mathbf{H}(s)\}$ in the frequency domain.

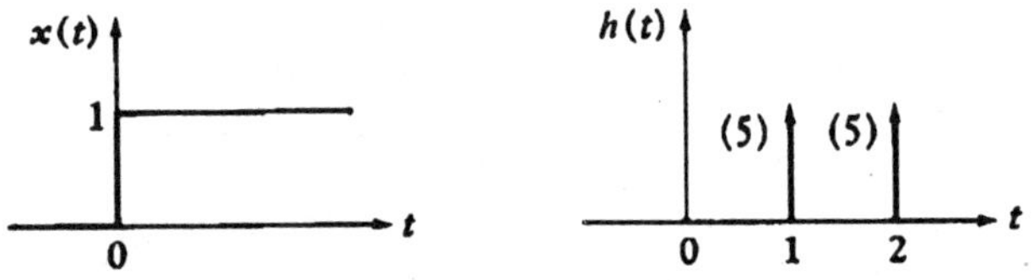


**Fig. 19-4:** See Prob. 17.

**18** Let $f(t) = 2e^{-10t}u(t)$ and $g(t) = 5e^{-3t}u(t)$. Find $f(t) * g(t)$ by two different methods.

**19** Find $h(t)$ and $\mathbf{H}(s)$ for the circuit shown in Fig. 19-5 if the output is: (*a*) $v_R(t)$; (*b*) $v_L(t)$. (*c*) Find $v_L(t)$ if $v_s(t) = 0.1\delta(t) - 3u(t)$ V.

**20** If $x(t) = 2[u(t) - u(t-1)]$ and $h(t) = u(t) - u(t-2)$: (*a*) find $\mathcal{L}^{-1}\{\mathcal{L}[x(t)]\mathcal{L}[h(t)]\}$, and (*b*) sketch the result.

**21** Given the waveforms for $x(t)$ and $h(t)$ shown in Fig. 19-6: (*a*) find $\mathbf{X}(s)$, $\mathbf{H}(s)$, and $y(t) = \mathcal{L}^{-1}\{\mathbf{X}(s)\mathbf{H}(s)\}$; (*b*) find $y(t) = x(t) * h(t)$ by working entirely in the time domain; (*c*) evaluate $y(2.5)$.

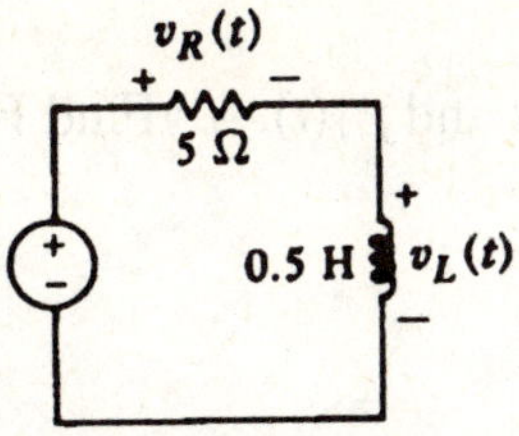

Fig. 19-5: See Prob. 19.

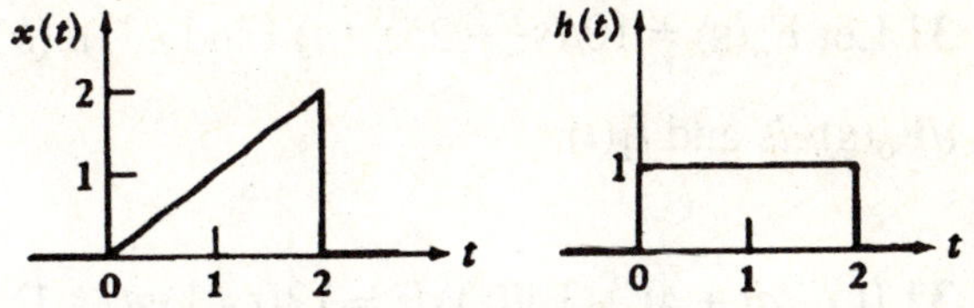

Fig. 19-6: See Prob. 21.

## Section 19-6 Time-shift and periodic functions
## Section 19-7 Shifting, differentiation, integration, and scaling in the frequency domain

**22** Use the concept of time shifting to help in finding: (*a*) **F**(s), if $f(t)$ is zero for $t < 4$, $20\sin(\pi t/2)$, for $4 < t < 5$, and 20 for $t > 5$; (*b*) $\mathcal{L}^{-1}\{4e^{-s}/[(s+1)(s+3)]\}$.

**23** Use the time-shift theorem to help determine: (*a*) **F**(s) if $f(t) = [\cos \pi t][u(t) - u(t-1)]$; (*b*) $\mathcal{L}^{-1}\{e^{-3s}/[(s+2)^2(s+1)]\}$.

**24** For the periodic waveform shown in Fig. 19-7: (*a*) find **F**(s); (*b*) evaluate **F**(0.5).

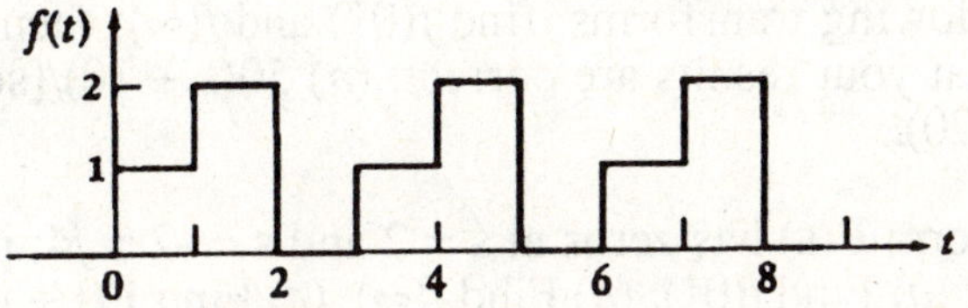

Fig. 19-7: See Prob. 24.

**25** Determine the Laplace transform of the rectified sine wave shown in Fig. 19-8.

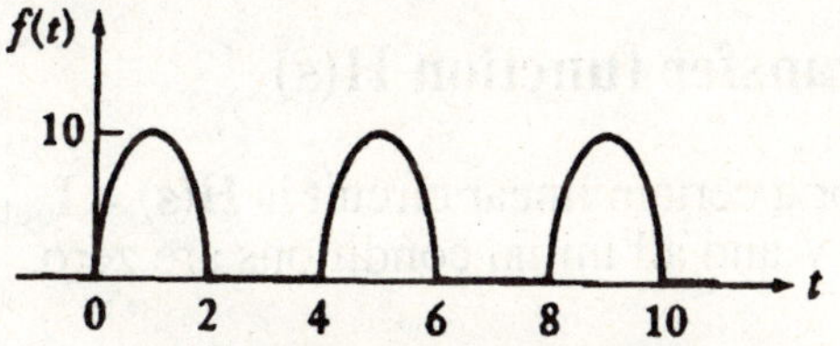

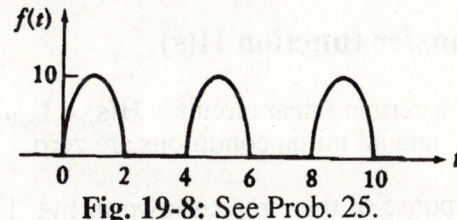

Fig. 19-8: See Prob. 25.

**26** Sketch $f(t)$ versus $t$ if **F**(s) =: (*a*) $5s/(s^2+4)$; (*b*) $5se^{-\pi s/2}/(s^2+4)$; (*c*) $5s(1+e^{-\pi s/2})/(s^2+4)$; (*d*) $5s(1+e^{-\pi s/2})/[(s^2+4)(1-e^{-1.25\pi s})]$.

**27** Find **F**(s) if $f(t)$ =: (*a*) $50\cos(10t - 0.8)u(t)$; (*b*) $(t-1)u(t-2)$; (*c*) $\sum_{n=0}^{\infty} 13(t-4n)\,[u(t-4n) - u(t-2-4n)]$.

**28** Find the Laplace transform of: (*a*) $5t^3e^{-2t}u(t)$; (*b*) $(1/t)(e^{-5t} - e^{-4t})u(t)$; (*c*) $f(2t)$ if $\mathcal{L}\{f(t)\} = s/[5 + \ln(s+1)]$.

**29** Find the inverse Laplace transform of: (*a*) $(s^3 + 6s^2 + 13s + 11)/[(s+2)^3(s+3)]$; (*b*) $(s^2-4)/(s^2+4)^2$; (*c*) **F**(2s) if $\mathcal{L}\{f(t)\} = e^{-4s}(s^2+100)$.

**30** Find $f(t)$ if **F**(s) =: (*a*) $3/(s+10)^4$; (*b*) $e^{-2(s+1)}/(s+1)^2$; (*c*) $2/[s^2(s+2)]$.

31 Let $\mathbf{F}_o(s) = 10/(s^2 + 25)$. (*a*) Find $f_o(t)$. (*b*) Find $\mathbf{F}_{-1}(s) = \int_s^{\infty} \mathbf{F}_o(s)\,ds$ and $f_{-1}(t)$. (*c*) Find $\mathbf{F}_1(s) = d\mathbf{F}_o(s)/ds$ and $f_1(t)$.

32 If $di/dt + 4i + 3 \int_{0^-}^{t} i(z)\,dz = 12(t - 1)u(t - 1)$ and $i(0^-) = 0$, find $i(t)$.

## Section 19-8 The initial-value and final-value theorems

33 Find $f(0^+)$ and $f(\infty)$ for each of the following transforms, without finding $f(t)$ first: (*a*) $(3 - e^{-2s})/[s(s^2 + 2s + 5)]$; (b) $3s/(s^2 + 2s + 5)$; (*c*) $3/(1 - e^{-2s})$.

34 (*a*) Synthesize an F(s) as the quotient of two polynomials so that $\lim_{s\to 0} [s\mathbf{F}(s)] = 10$ and $\lim_{s\to\infty} [s\mathbf{F}(s)] = 25$. (*b*) Find $f(t)$. (*c*) Find $f(0^+)$ and $f(\infty)$ by limiting operations, and also by applying the initial- and final-value theorems.

35 For each of the following transforms, find $f(0^+)$ and $f(\infty)$; then find the corresponding time functions and show that your results are correct: (*a*) $50(s + 10)/[s(s + 20)]$; (*b*) $25(1 + e^{-10s})/s$; (*c*) $(2s + 6)/(s^2 + 8s + 20)$.

36 The Laplace transform F(s) has zeros at s = 2 and $s = -2 \pm j4$; poles at s = 0, -3, -5, and $s = -1 + j1$; and F(1) = -5. (*a*) Find $f(0^+)$. (*b*) Find $f(\infty)$. (*c*) Find $\mathbf{F}(1 + j5)$.

37 Find $h(t)$ for the circuit of Fig. 19-9 by drawing the frequency-domain circuit, finding $\mathbf{V}_o(s)/\mathbf{V}_s(s)$, and then evaluating the inverse transform.

## Section 19-9 The transfer function H(s)

38 The transfer function for a certain linear circuit is $\mathbf{H}(s) = \mathbf{I}_{out}/\mathbf{V}_{in} = 2(s + 10)/[(s + 4)(s^2 + 4)]$ Find $i_{out}(t)$ if $v_{in}(t) = 8u(t)$ V and all initial conditions are zero.

39 (*a*) Find the impulse response of the circuit shown in Fig. 19-10 by working first in the frequency domain and then taking $\mathcal{L}^{-1}\{\mathbf{V}_{out}(s)\}$. (*b*) Find $v_{out}(t)$ if $v_{in}(t) = 10u(t)$.

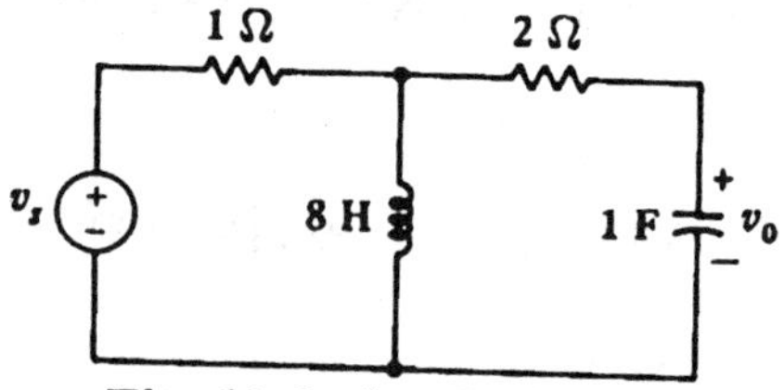


**Fig. 19-9:** See Prob. 37.

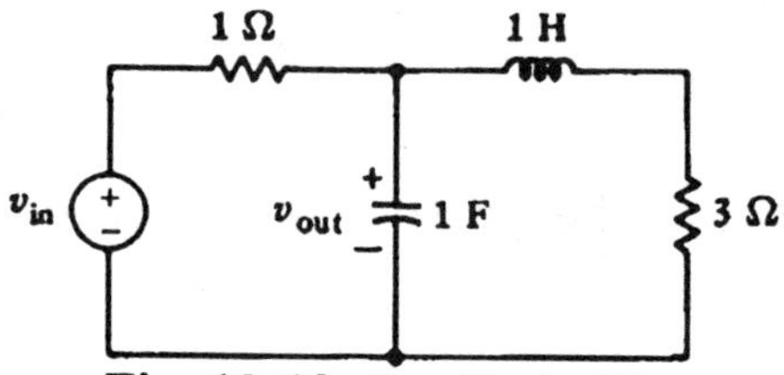


**Fig. 19-10:** See Prob. 39.

40 Find $i_s(t)$ for the circuit shown in Fig. 19-11.

41 Two identical linear systems are connected in cascade. The input signal to the first stage is $x(t)$, its output is $y(t)$, and the output of the second stage is $z(t)$. If an input $tu(t)$ is applied to a single stage, the output of that stage is $(e^{-2t} - 1)u(t)$. What is $z(t)$ if $x(t) = u(t)$?

**42** Determine $i_L(t)$ for the circuit of Fig. 19-12 if $i_s(t) = 16te^{-t}u(t)$ A.

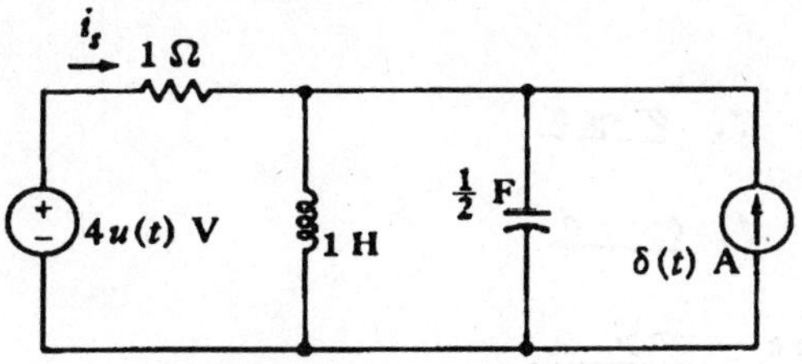


**Fig. 19-11:** See Prob. 40.

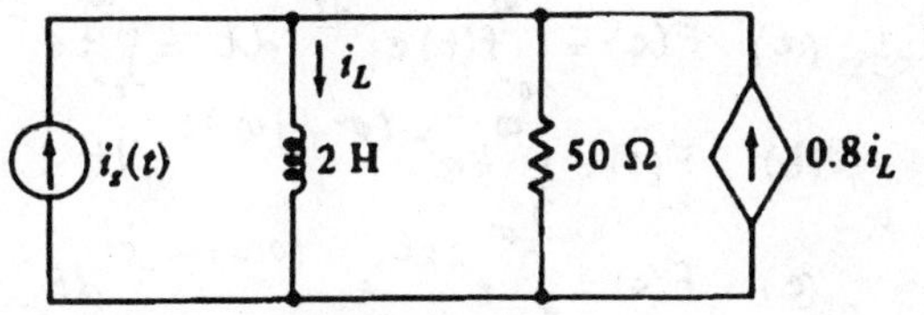


**Fig. 19-12:** See Prob. 42.

**43** (*a*) Find $\mathbf{V}_C(\mathbf{s})$ for the circuit of Fig. 19-13 by writing one or more differential equations, reducing them to obtain one second-order differential equation, and taking the Laplace transform. (*b*) Obtain $v_C(t)$.

## Section 19-10 The complete response

**44** (*a*) Draw the transform circuit for Prob. 43 and Fig. 19-13, including two initial- condition generators. (*b*) Write one or more frequency-domain equations and solve for $\mathbf{V}_C(\mathbf{s})$. (*c*) Find $v_C(t)$.

**45** The sawtooth voltage pulse, $v_s(t) = 40t[u(t) - u(t - 0.1)]$ V, is applied to the series combination of 2 Ω and 0.05 F. Find the two instants of time at which the capacitor voltage is 1 V.

**46** Sources that have been switched out of the circuit of Fig. 19-14 prior to $t = 0^-$ cause $i_L(0^-)$ to be 5 A and $v_C(0^-)$ to be 10 V. Supply two initial-condition generators for the transform circuit and find $v_C(t)$. What are $i_L(0^+)$ and $v_C(0^+)$?

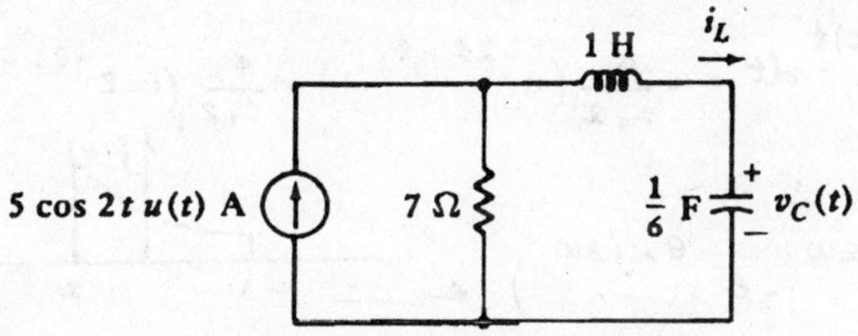


$i_L(0^-) = 4$ A, $v_C(0^-) = 7$ V

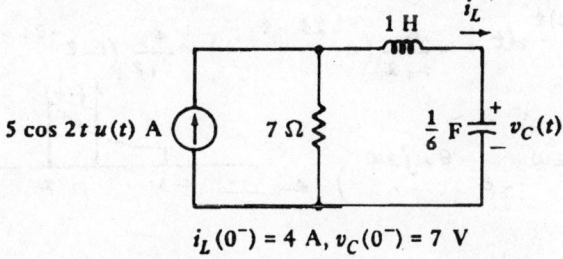

**Fig. 19-13:** See Probs. 43 and 44.

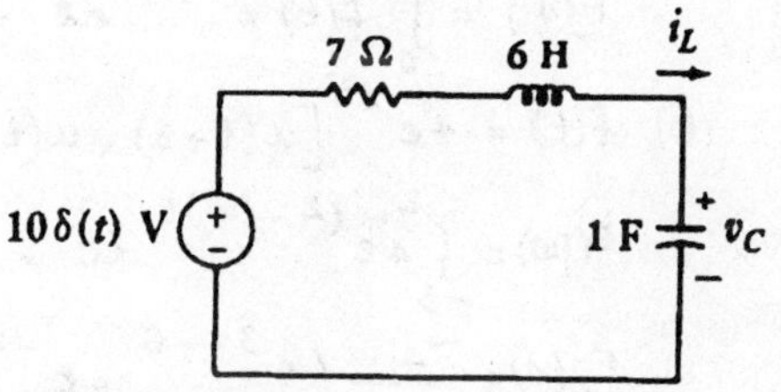


**Fig. 19-14:** See Prob. 46.

**47** Find $v_C(t)$ for all $t$ in the circuit of Fig. 19-15.

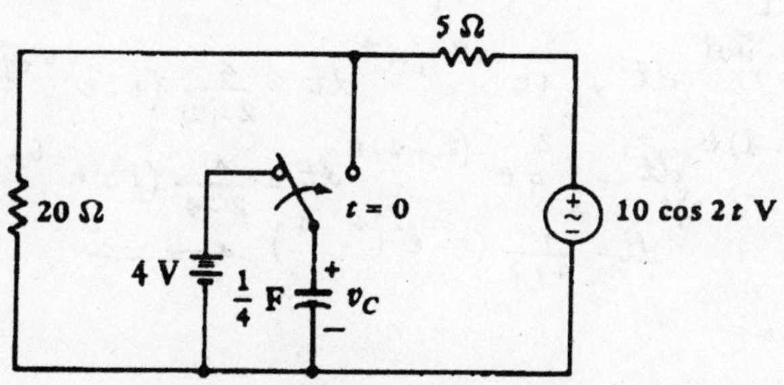


**Fig. 19-15:** See Prob. 47.

# Chapter 19

**19-1** (a) $F(s) = \int_{0^-}^{\infty} f(t)e^{-st}\,dt = \int_{0^-}^{\infty} t e^{-(\sigma+j\omega)t}\,dt \quad \therefore \ \underline{\sigma > 0}$

(b) $F(s) = \int_{0^-}^{\infty} t e^{-(\sigma+j\omega)t}\,dt \quad \therefore \ \underline{\sigma > 0}$

(c) $F(s) = \int_{0^-}^{\infty} e^{-2t} e^{-(\sigma+j\omega)t}\,dt \quad \therefore \sigma + 2 > 0, \ \underline{\sigma > -2}$

(d) $F(s) = \int_{0^-}^{\infty} e^{2t} e^{-(\sigma+j\omega)t}\,dt \quad \therefore \sigma - 2 > 0, \ \underline{\sigma > 2}$

(e) $F(s) = \int_{3}^{\infty} e^{-2t} e^{-(\sigma+j\omega)t}\,dt \quad \therefore \sigma + 2 > 0, \ \underline{\sigma > -2}$

(f) $F(s) = \int_{0^-}^{\infty} e^{-2t} e^{-(\sigma+j\omega)t}\,dt \quad \therefore \sigma + 2 > 0, \ \underline{\sigma > -2}$

**19-2** (a) $f(t) = 4e^{-2t}[u(t+3) - u(t-2)]$

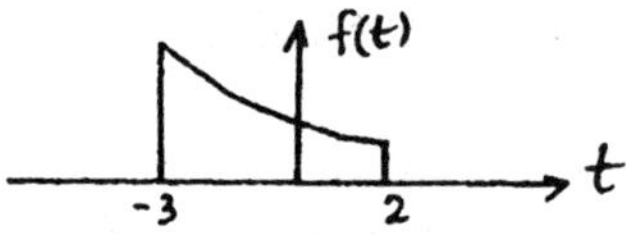


$F(j\omega) = \int_{-\infty}^{\infty} f(t)e^{-j\omega t}\,dt = \int_{-3}^{2} 4e^{(-2-j\omega)t}\,dt = \frac{-4}{2+j\omega}\left[e^{(-2-j\omega)2} - e^{(-2-j\omega)(-3)}\right]$

$= \frac{4}{2+j\omega}\left(e^{6+j3\omega} - e^{-4-j2\omega}\right) \leftarrow$

$F_2(s) = \int_{-\infty}^{\infty} f(t)e^{-st}\,dt = \int_{-3}^{2} 4e^{(-2-s)t}\,dt = \frac{-4}{s+2}\left[e^{(-2-s)2} - e^{(-2-s)(-3)}\right]$

$= \frac{4}{s+2}\left(e^{3s+6} - e^{-2s-4}\right) \leftarrow$

$F(s) = \int_{0^-}^{\infty} f(t)e^{-st}\,dt = \int_{0^-}^{2} 4e^{-(s+2)t}\,dt = -\frac{4}{s+2}\left(e^{-2s-4} - 1\right) = \frac{4}{s+2}\left(1 - e^{-2s-4}\right) \leftarrow$

(b) $f(t) = 4e^{2t}[u(t+3) - u(t-2)]$

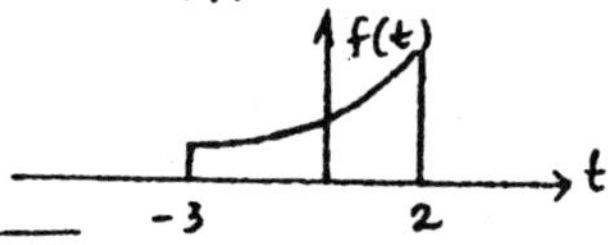


$F(j\omega) = \int_{-3}^{2} 4e^{(2-j\omega)t}\,dt = \frac{4}{2-j\omega}\left(e^{4-j2\omega} - e^{-6+j3\omega}\right) \leftarrow$

$F_2(s) = \frac{4}{s-2}\left(e^{3s-6} - e^{-2s+4}\right) \leftarrow$

$F(s) = \frac{4}{s-2}\left(1 - e^{-2s+4}\right) \leftarrow$

(c) $f(t) = 4e^{-2|t|}[u(t+3) - u(t-2)]$

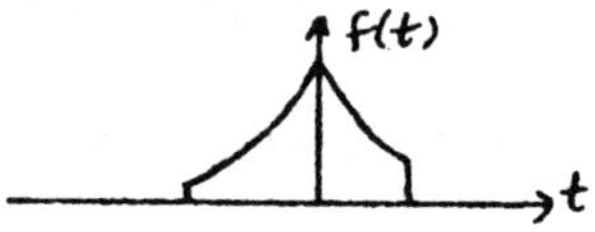


$F(j\omega) = \int_{-3}^{0} 4e^{2t-j\omega t}\,dt + \int_{0}^{2} 4e^{-2t-j\omega t}\,dt = \frac{4}{2-j\omega}\left(1 - e^{-6+j3\omega}\right) + \frac{4}{2+j\omega}\left(1 - e^{-4-j2\omega}\right) \leftarrow$

$F_2(s) = \int_{-3}^{0} 4e^{(2-s)t}\,dt + \int_{0}^{2} 4e^{-(2+s)t}\,dt = \frac{4}{2-s}\left(1 - e^{-6+3s}\right) + \frac{4}{2+s}\left(1 - e^{-4-2s}\right) \leftarrow$

$F(s) = \int_{0^-}^{2} 4e^{-(s+2)t}\,dt = \frac{4}{s+2}\left(1 - e^{-2s-4}\right) \leftarrow$

**19-3** (a) $F(s) = \int_{0^-}^{\infty} u(t-1)\,u(4-t)\,u(t)\,e^{-st}\,dt = \int_{1}^{4} e^{-st}\,dt = \underline{\frac{1}{s}\left(e^{-s} - e^{-4s}\right)}$ ←

(b) $F(s) = \int_{0^-}^{\infty} 2\delta(t-4)\,e^{-st}\,dt = \int_{4^-}^{4^+} 2\delta(t-4)\,e^{-st}\,dt = \underline{2e^{-4s}}$ ←

(c) $F(s) = \int_{0^-}^{\infty} 2\delta(t-0.2)\sin \pi t\, e^{-st}\,dt = 2\sin 0.2\pi\, e^{-0.2s} = \underline{1.1756e^{-0.2s}}$ ←

(d) $F(s) = \int_{0^-}^{\infty} 2u(t-4)\,e^{-st}\,dt = \int_{4}^{\infty} 2e^{-st}\,dt = \frac{2}{-s} e^{-st}\Big|_4^{\infty} = \underline{\frac{2}{s} e^{-4s}}$ ←

(e) $F(s) = \int_{0^-}^{\infty} 2e^{-4t}\,u(t-1)\,e^{-st}\,dt = \int_{1}^{\infty} 2e^{-(4+s)t}\,dt = \frac{2}{-(4+s)} e^{-(4+s)t}\Big|_1^{\infty} = \underline{\frac{2}{s+4} e^{-4-s}}$

**19-4** (a) $F(s) = \int_{0^-}^{\infty} u(t+3)\,u(t)\,u(5-t)\,e^{-st}\,dt = \int_{0^-}^{5} e^{-st}\,dt = \underline{\frac{1}{s}\left(1 - e^{-5s}\right)}$ $\underline{\text{all } \sigma}$

(b) $F(s) = \int_{0^-}^{\infty} 5\delta(t-5)\,e^{-st}\,dt = \underline{5e^{-5s}}$, $\underline{\text{all } \sigma}$

(c) $F(s) = \int_{0^-}^{\infty} 5\delta(t-0.1)\cos 2\pi t\, e^{-st}\,dt = 5\cos 0.2\pi\, e^{-0.1s} = \underline{4.045\, e^{-0.1s}}$ $\underline{\text{all } \sigma}$

(d) $F(s) = \int_{0^-}^{\infty} 5u(t-5)\,e^{-st}\,dt = \int_{5}^{\infty} 5e^{-st}\,dt = \underline{\frac{5}{s} e^{-5s}}$, $\underline{\sigma > 0}$

(e) $F(s) = \int_{0^-}^{\infty} 5(t-2)\,u(t-1)\,e^{-st}\,dt = \int_{1}^{\infty} 5(t-2)\,e^{-st}\,dt = 5\int_{1}^{\infty} t e^{-st}\,dt - 10\int_{1}^{\infty} e^{-st}\,dt$

$= 5e^{-st}\left(-\frac{t}{s} - \frac{1}{s^2}\right)\Big|_1^{\infty} + \frac{10}{s} e^{-st}\Big|_1^{\infty} = 5e^{-s}\left(\frac{1}{s} + \frac{1}{s^2}\right) - 10e^{-s}\,\frac{1}{s}$

$\underline{F(s) = \left(\frac{5}{s^2} - \frac{5}{s}\right)e^{-s}}$, $\underline{\sigma > 0}$

**19-5** (a) $\mathcal{L}^{-1}\left\{1 + \frac{1}{s} + \frac{1}{s+1}\right\} = \underline{\delta(t) + \left(1 + e^{-t}\right)u(t)}$

(b) $\mathcal{L}^{-1}\left\{\frac{s+1}{s} e^{-s}\right\} = \mathcal{L}^{-1}\left\{e^{-s} + \frac{1}{s} e^{-s}\right\} = \underline{\delta(t-1) + u(t-1)}$

(c) $\mathcal{L}^{-1}\left\{\left(1 + e^{-s}\right)^2\right\} = \mathcal{L}^{-1}\left\{1 + 2e^{-s} + e^{-2s}\right\} = \underline{\delta(t) + 2\delta(t-1) + \delta(t-2)}$

(d) $\mathcal{L}^{-1}\left\{2e^{-3s}\sinh 2s\right\} = \mathcal{L}^{-1}\left\{2e^{-3s} \times \frac{1}{2}\left(e^{2s} - e^{-2s}\right)\right\} = \mathcal{L}^{-1}\left\{e^{-s} - e^{-5s}\right\}$

$= \underline{\delta(t-1) - \delta(t-5)}$

**19-6** (a) $F(s) = \int_{0^-}^{\infty} u(t-1)\,e^{-st}\,dt = \int_{1}^{\infty} e^{-st}\,dt = \frac{1}{s} e^{-s}$ $\therefore F(0.1+j1) = \frac{e^{-0.1-j1}}{0.1+j1}$

$= \underline{-0.7055 - j0.5594}$

(b) $F(s) = \int_{0^-}^{\infty} t\,u(t-1)\,e^{-st}\,dt = \int_{1}^{\infty} t e^{-st}\,dt = e^{-st}\left(-\frac{t}{s} - \frac{1}{s^2}\right)\Big|_1^{\infty} = e^{-s}\left(\frac{1}{s} + \frac{1}{s^2}\right)$

$\therefore F(0.1+j1) = e^{-0.1-j1}\left[\frac{1}{0.1+j1} + \frac{1}{(0.1+j1)^2}\right] = \underline{-1.3292 + j0.08365}$

(c) $F(s) = \int_{0^-}^{\infty} (t-1)\,u(t-1)\,e^{-st}\,dt = \int_{1}^{\infty} (t-1)\,e^{-st}\,dt = \int_{1}^{\infty} t e^{-st}\,dt - \int_{1}^{\infty} e^{-st}\,dt$

$= e^{-st}\left(-\frac{t}{s} - \frac{1}{s^2}\right)\Big|_1^{\infty} + \frac{1}{s} e^{-st}\Big|_1^{\infty} = \frac{1}{s^2} e^{-s}$ $\therefore F(0.1+j1) = \frac{e^{-0.1-j1}}{(0.1+j1)^2}$

$\therefore F(0.1+j1) = \underline{-0.6237 + j0.6431}$ (cont. on next page)

19-6 (cont.)

(d) $F(s) = \int_{0^-}^{\infty} u(t-1)e^{-2t}e^{-st}\,dt = \int_{1}^{\infty} e^{-(2+s)t}\,dt = \frac{-1}{s+2}e^{-(s+2)t}\Big|_1^{\infty} = \frac{1}{s+2}e^{-2-s}$

$\therefore F(0.1+j1) = \frac{1}{2.1+j1}e^{-2.1-j1} = \underline{0.006636 - j0.05223}$

(e) $F(s) = \int_{0^-}^{\infty} \delta(t-1)\,u(t)\,e^{-st}\,dt = e^{-s} \quad \therefore F(0.1+j1) = e^{-0.1-j1} = \underline{0.4889 - j0.7614}$

19-7 (a) $\mathcal{L}^{-1}\left\{5 - \frac{2}{s}\right\} = \underline{5\delta(t) - 2u(t)}$

(b) $\mathcal{L}^{-1}\left\{\frac{5}{s} + \frac{3}{s+1}\right\} = \underline{(5+3e^{-t})\,u(t)}$

(c) $\mathcal{L}^{-1}\left\{\frac{s+2}{s+1}\right\} = \mathcal{L}^{-1}\left\{1 + \frac{1}{s+1}\right\} = \underline{\delta(t) + e^{-t}u(t)}$

(d) $\mathcal{L}^{-1}\left\{\frac{s^2+5}{s^2}\right\} = \mathcal{L}^{-1}\left\{1 + \frac{5}{s^2}\right\} = \underline{\delta(t) + 5t\,u(t)}$

(e) $\mathcal{L}^{-1}\left\{\frac{6}{(s+1)(s+2)}\right\} = \mathcal{L}^{-1}\left\{\frac{6}{s+1} - \frac{6}{s+2}\right\} = \underline{6(e^{-t} - e^{-2t})\,u(t)}$

19-8 (a) $F(s) = \frac{s+6}{s+4} = 1 + \frac{2}{s+4} \quad \therefore f(t) = \underline{\delta(t) + 2e^{-4t}u(t)}$

(b) $F(s) = \frac{2+3s}{s^2+4} = \frac{2}{s^2+4} + \frac{3s}{s^2+4} \quad \therefore f(t) = \underline{(\sin 2t + 3\cos 2t)\,u(t)}$

(c) $F(s) = \frac{4s+100}{10s^2+20s} = 0.4\,\frac{s+25}{s(s+2)} = 0.4\left(\frac{12.5}{s} - \frac{11.5}{s+2}\right) = \frac{5}{s} - \frac{4.6}{s+2}$

$\therefore f(t) = \underline{(5 - 4.6e^{-2t})\,u(t)}$

(d) $F(s) = \frac{60}{(s+2)(s+3)(s+5)} = \frac{20}{s+2} - \frac{30}{s+3} + \frac{10}{s+5} \quad \therefore f(t) = \underline{(20e^{-2t} - 30e^{-3t} + 10e^{-5t})\,u(t)}$

19-9

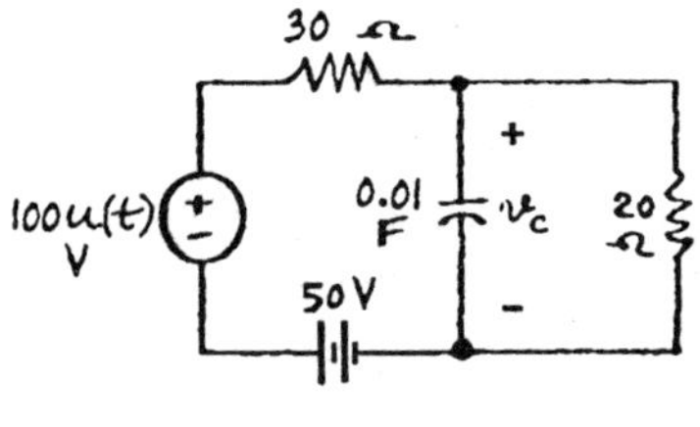


(a) $\frac{v_c - 100 - 50}{30} + 0.01\,v_c' + 0.05\,v_c = 0$

$\therefore \underline{0.3v_c' + 2.5v_c - 150 = 0} \quad (t > 0)$

(b) $0.3s\,V_c - 0.3v_c(0^-) + 2.5V_c - \frac{150}{s} = 0, \quad v_c(0^-) = 20$

$\therefore 0.3s\,V_c - 6 + 2.5V_c - \frac{150}{s} = 0, \quad V_c = \frac{6 + 150/s}{0.3s + 2.5}$

$\therefore V_c = \frac{20(s+25)}{s(s+25/3)} = \frac{60}{s} - \frac{40}{s+25/3} \quad \therefore v_c = 60 - 40e^{-25t/3}, \; t > 0 \;\therefore v_c = \underline{20u(-t) + (60 - 40e^{-\frac{25t}{3}})u(t)}$

**19-10**

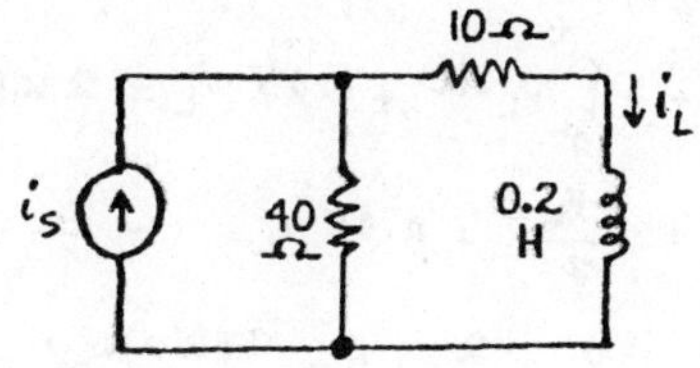


$i_s = 5 + 5u(t)$ A

$\therefore v_s = 200 + 200\,u(t)$ V

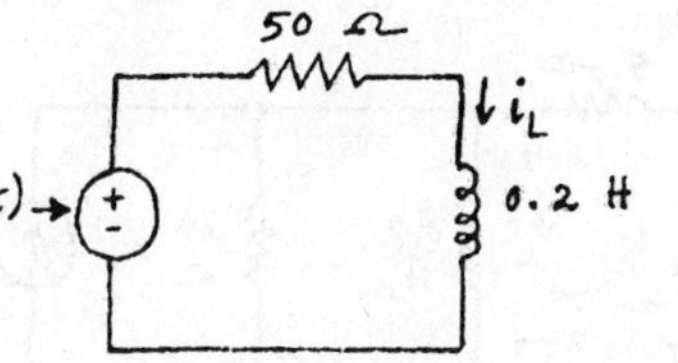


$\underline{0.2i_L' + 50\,i_L = 400\ (t>0)}$ $\therefore 0.2sI_L - 0.2\,i_L(0^-) + 50\,I_L = \frac{400}{s}$, $i_L(0^-) = 4$ A

$\therefore 0.2sI_L - 0.8 + 50I_L = \frac{400}{s}$, $I_L = \frac{0.8 + 400/s}{50 + 0.2s} = \frac{0.8s + 400}{s(0.2s + 50)} = \frac{4(s+500)}{s(s+250)}$

$\therefore I_L = \frac{8}{s} - \frac{4}{s+250}$ $\therefore \underline{i_L = 4u(-t) + (8 - 4e^{-250t})u(t)\ \text{A}}$

**19-11** $y''' + 2y'' = 0$, $y(0^-) = 0$, $y'(0^-) = 1$, $y''(0^-) = -1$

$\therefore s^3Y - s^2y(0^-) - sy'(0^-) - y''(0^-) + 2s^2Y - 2sy(0^-) - 2y'(0^-) = 0$

$\therefore s^3Y - s + 1 + 2s^2Y - 2 = 0$, $(s^3 + 2s^2)Y = s + 1$ $\therefore Y = \frac{s+1}{s^2(s+2)}$

$\therefore Y = \frac{0.5}{s^2} + \frac{A}{s} - \frac{0.25}{s+2}$, $s = 1: \frac{2}{3} = 0.5 + A - \frac{1}{12}$ $\therefore A = 0.25$

$Y = \frac{0.5}{s^2} + \frac{0.25}{s} - \frac{0.25}{s+2}$ $\therefore y(t) = \underline{(0.25 + 0.5t - 0.25e^{-2t})u(t)}$

**19-12** $\int_{-\infty}^{t} 25\,f(x)\,dx = 30\,u(t) - \frac{df}{dt}$, $f(0^-) = 2$, $\int_{-\infty}^{0^-} f(t)\,dt = 0.2$

$\therefore \int_{0^-}^{t} 25\,f(x)\,dx + f' = 30 - 5 = 25$ $\therefore \frac{25}{s}F(s) + sF(s) - 2 = \frac{25}{s}$

$\therefore (s^2 + 25)F(s) = 25 + 2s$, $F(s) = \frac{25 + 2s}{s^2 + 25} = \frac{5\times 5}{s^2 + 25} + \frac{2s}{s^2 + 25}$ $\therefore f(t) = \underline{(5\sin 5t + 2\cos 5t)\,u(t)}$

**19-13**

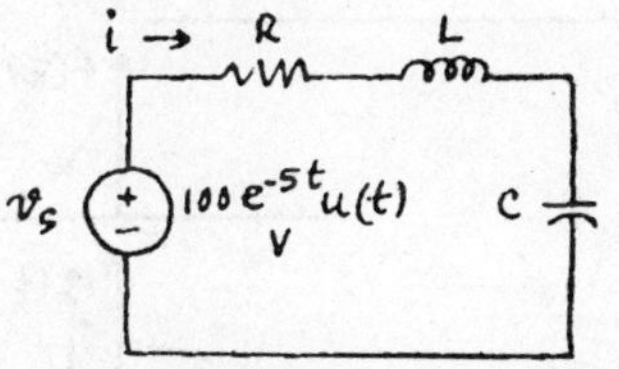


(a) $v_s = i' + 4i + 3\int_{0^-}^{t} i\,dt$ $\therefore L = \underline{1\ \text{H}}$, $R = \underline{4\ \Omega}$, $C = \underline{\frac{1}{3}}$ F

(b) $100e^{-5t} = i' + 4i + 3\int_{0^-}^{t} i\,dt$

$\therefore \frac{100}{s+5} = sI - 0 + 4I + \frac{3I}{s} = \frac{s^2 + 4s + 3}{s}I$

$\therefore I = \frac{100s}{(s+5)(s+1)(s+3)} = \frac{-12.5}{s+1} + \frac{75}{s+3} - \frac{62.5}{s+5}$ $\therefore i(t) = \underline{(75e^{-3t} - 12.5e^{-t} - 62.5e^{-5t})u(t)}$ A

**19-14** $y' + 2x = t$, $y + \int_{0^-}^{t} x(z)\,dz = 2t$, $x(0^-) = 0$, $\int_{-\infty}^{0^-} x(t)\,dt = 0$

$\therefore sY + 2X = \frac{1}{s^2}$, $Y + \frac{1}{s}X = \frac{2}{s^2}$ $\therefore s\left(\frac{2}{s^2} - \frac{X}{s}\right) + 2X = \frac{1}{s^2}$ $\therefore \frac{2}{s} + X = \frac{1}{s^2}$

$\therefore X = \frac{1}{s^2} - \frac{2}{s}$ $\therefore x(t) = \underline{(t - 2)u(t)}$

$y = 2t - \int_{0^-}^{t}(z - 2)\,dz = 2t - \frac{1}{2}t^2 + 2t$, $y(t) = \underline{(4t - \frac{1}{2}t^2)u(t)}$

**19-15**

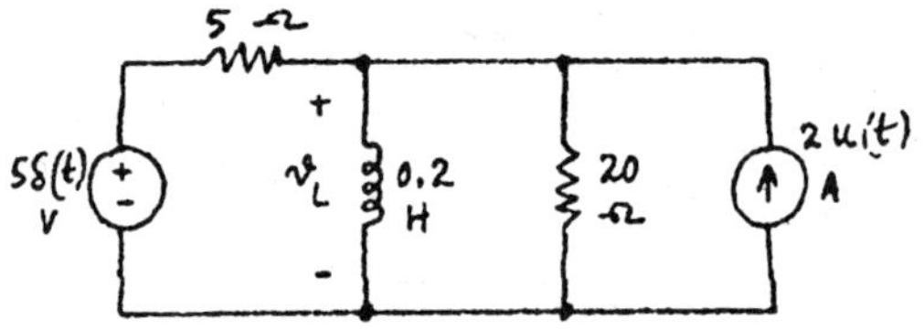


$$\frac{v_L - 5\delta(t)}{5} + 5\int_{0^-}^{t} v_L\,dt + 0.05\,v_L = 2u(t)$$

$$\therefore 0.2\,V_L - 1 + \frac{5V_L}{s} + 0.05\,V_L = \frac{2}{s}$$

$$\therefore \left(0.25 + \frac{5}{s}\right)V_L = \frac{2}{s} + 1 = \frac{s+2}{s} \quad \therefore V_L = \frac{s+2}{0.25s+5}$$

$$\therefore V_L = \frac{4s+8}{s+20} = 4 - \frac{72}{s+20} \quad \therefore v_L(t) = \underline{4\delta(t) - 72e^{-20t}\,u(t)}\ \text{V}$$

**19-16** $x(t) = 4t\,u(t)$, $h(t) = 3u(t)$

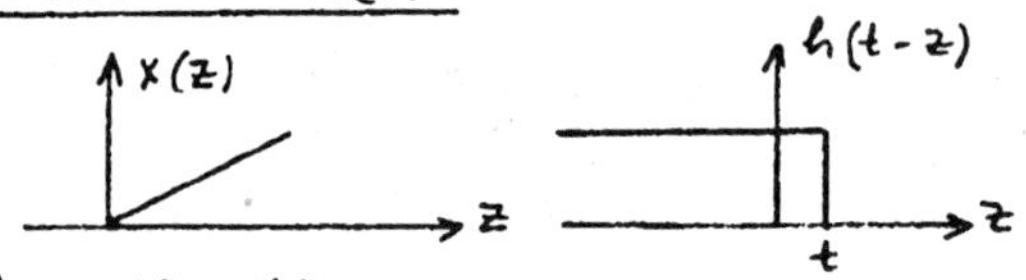


(a) $X(s) = \frac{4}{s^2}$, $H(s) = \frac{3}{s}$, $\mathcal{L}^{-1}\left\{\frac{4}{s^2}\frac{3}{s}\right\} = \mathcal{L}^{-1}\left\{\frac{12}{s^3}\right\} = \underline{6t^2 u(t)}$

(b) $x(t) * h(t) = \int_0^t 4t \times 3\,dt = \underline{6t^2 u(t)}$

**19-17**

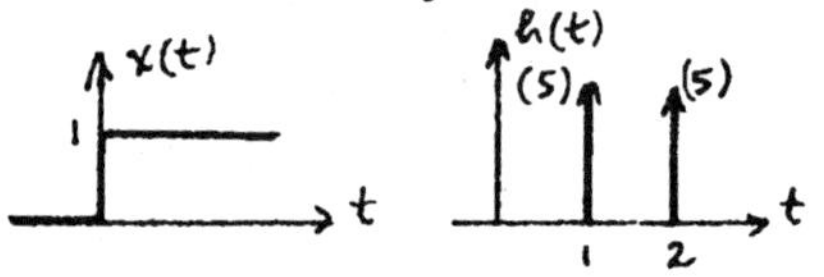


(a)

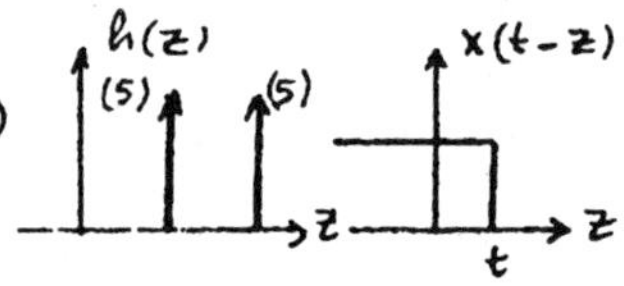


$$y(t) = \begin{cases} \underline{0} & t < 1 \\ \underline{5} & 1 < t < 2 \\ \underline{10} & t > 2 \end{cases}$$

(b) $X(s) = \frac{1}{s}$, $H(s) = 5e^{-s} + 5e^{-2s}$

$$\mathcal{L}^{-1}\left\{\frac{5}{s}e^{-s} + \frac{5}{s}e^{-2s}\right\} = \underline{5u(t-1) + 5u(t-2)}\ ✓$$

**19-18** $f(t) = 2e^{-10t}u(t)$, $g(t) = 5e^{-3t}u(t)$

(1) $F(s) = \frac{2}{s+10}$, $G(s) = \frac{5}{s+3}$, $\mathcal{L}^{-1}\left\{\frac{2}{s+10} \times \frac{5}{s+3}\right\} = \mathcal{L}^{-1}\left\{\frac{10/7}{s+3} - \frac{10/7}{s+10}\right\}$

$$= \underline{\frac{10}{7}\left(e^{-3t} - e^{-10t}\right)u(t)} = f(t) * g(t)$$

$f(t) = 2e^{-10t}u(t)$, $g(t) = 5e^{-3t}u(t)$

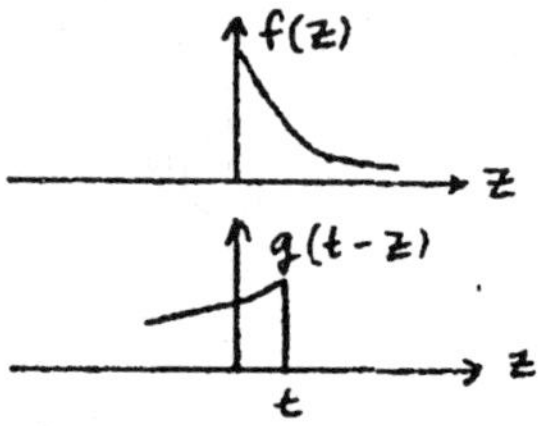


(2) $f * g = \int_0^t 2e^{-10z}\,5e^{-3(t-z)}\,dz = 10e^{-3t}\int_0^t e^{-7z}\,dz$

$$= 10e^{-3t}\left(-\frac{1}{7}\right)\left(e^{-7t} - 1\right) = \underline{\frac{10}{7}\left(e^{-3t} - e^{-10t}\right)u(t)}$$

**19-19**

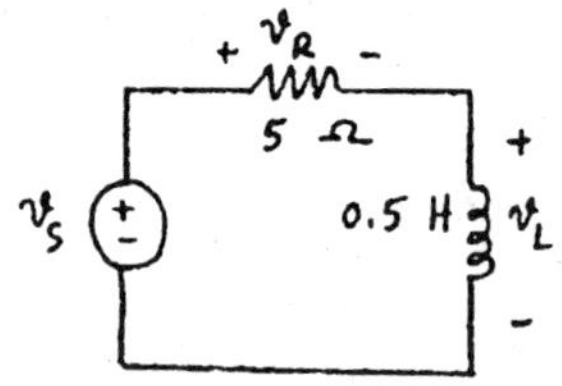


(a) let $V_s = 1$ $\therefore V_R(s) = \frac{5}{0.5s+5} = \frac{10}{s+10} = H(s)$ $\therefore h(t) = \underline{10e^{-10t}u(t)}$

(b) $V_L(s) = \frac{0.5s}{0.5s+5} = \frac{s}{s+10} = H(s) = 1 - \frac{10}{s+10}$

$$\therefore h(t) = \underline{\delta(t) - 10e^{-10t}u(t)}$$

(c) $v_s(t) = 0.1\delta(t) - 3u(t)$ V $\therefore V_s = 0.1 - \frac{3}{s}$ $\therefore V_L(s) = \frac{(0.1 - 3/s)\,0.5s}{0.5s+5}$

$$\therefore V_L(s) = \frac{0.05s - 1.5}{0.5s+5} = \frac{0.1s - 3}{s+10} = \frac{0.1s + 1 - 4}{s+10} = 0.1 - \frac{4}{s+10} \quad \therefore v_L(t) = \underline{0.1\delta(t) - 4e^{-10t}u(t)\ \text{V}}$$

**19-20** $x(t) = 2[u(t) - u(t-1)]$, $h(t) = u(t) - u(t-2)$

(a) $\mathcal{L}\{x(t)\} = \frac{2}{s}(1-e^{-s})$, $\mathcal{L}\{h(t)\} = \frac{1}{s}(1-e^{-2s})$

$$\mathcal{L}^{-1}\{\mathcal{L}\{x(t)\}\mathcal{L}\{h(t)\}\} = \mathcal{L}^{-1}\left\{\frac{2}{s^2}(1-e^{-s}-e^{-2s}+e^{-3s})\right\}$$

$$= 2[tu(t) - (t-1)u(t-1) - (t-2)u(t-2) + (t-3)u(t-3)]$$

(b)

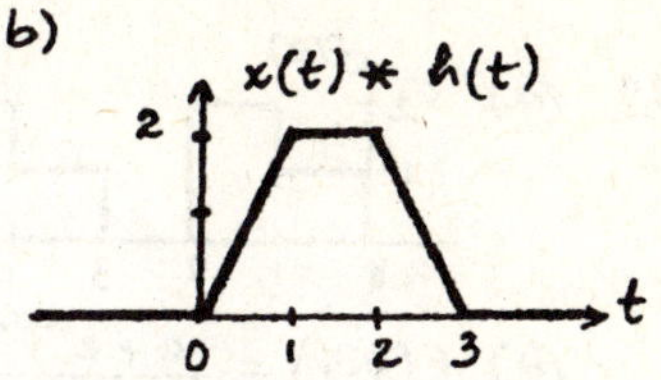


**19-21** (a)

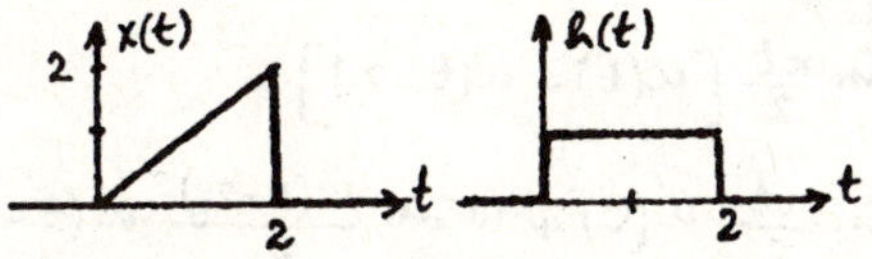


$x(t) = t[u(t) - u(t-2)] = tu(t) - (t-2)u(t-2) - 2u(t-2)$

$\therefore X(s) = \frac{1}{s^2} - \frac{1}{s^2}e^{-2s} - \frac{2}{s}e^{-2s}$

$h(t) = u(t) - u(t-2)$ $\therefore H(s) = \frac{1}{s}(1-e^{-2s})$

$$\therefore X(s)H(s) = \frac{1}{s^3} - \frac{1}{s^3}e^{-2s} - \frac{2}{s^2}e^{-2s} - \frac{1}{s^3}e^{-2s} + \frac{1}{s^3}e^{-4s} + \frac{2}{s^2}e^{-4s}$$

$$\therefore y(t) = x(t)*h(t) = \frac{1}{2}t^2u(t) - \frac{1}{2}\times 2(t-2)^2u(t-2) - 2(t-2)u(t-2) + \frac{1}{2}(t-4)^2u(t-4) + 2(t-4)u(t-4)$$

$$y(t) = \frac{1}{2}t^2u(t) + (2t-t^2)u(t-2) + \left(\frac{1}{2}t^2 - 2t\right)u(t-4) \leftarrow$$

(b)

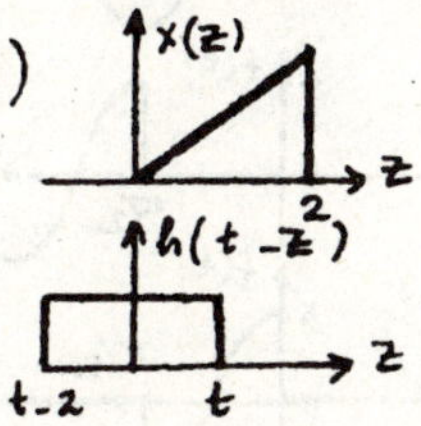


$$y(t) = \begin{cases} 0 & t<0 \text{ and } t>4 \\ \int_0^t z\,dz = \frac{1}{2}t^2 & 0<t<2 \\ \int_{t-2}^{2} z\,dz = \frac{1}{2}(4-t^2+4t-4) = 2t - \frac{1}{2}t^2 & 2<t<4 \end{cases}$$

(c) $y(2.5) = 2\times 2.5 - \frac{1}{2}\times 2.5^2 = \underline{1.875}$

**19-22** (a) $f(t) = 20\sin\frac{\pi t}{2}[u(t-4) - u(t-5)] + 20u(t-5)$

$$\therefore f(t) = 20\sin\frac{\pi(t-4)}{2}u(t-4) - 20\cos\frac{\pi(t-5)}{2}u(t-5) + 20u(t-5)$$

$$\therefore F(s) = e^{-4s}\frac{\pi/2}{s^2+\pi^2/4}\times 20 - e^{-5s}\frac{s}{s^2+\pi^2/4}\times 20 + 20\frac{e^{-5s}}{s}$$

$$F(s) = \frac{20}{s^2+\pi^2/4}\left(\frac{\pi}{2}e^{-4s} - se^{-5s}\right) + \frac{20}{s}e^{-5s} \leftarrow$$

(b) $$\mathcal{L}^{-1}\left\{\frac{4}{(s+1)(s+3)}\right\} = \mathcal{L}^{-1}\left\{\frac{2}{s+1} - \frac{2}{s+3}\right\} = (2e^{-t} - 2e^{-3t})u(t)$$

$$\therefore \mathcal{L}^{-1}\left\{\frac{4e^{-s}}{(s+1)(s+3)}\right\} = 2[e^{-(t-1)} - e^{-3(t-1)}]u(t-1) \leftarrow$$

**19-23** (a) $f(t) = \cos\pi t[u(t) - u(t-1)] = \cos\pi t\, u(t) + \cos\pi(t-1)\,u(t-1)$

$$\therefore F(s) = \frac{s}{s^2+\pi^2} + \frac{s}{s^2+\pi^2}e^{-s} = \frac{s}{s^2+\pi^2}(1+e^{-s}) \leftarrow$$

(b) $F(s) = \frac{e^{-3s}}{(s+2)^2(s+1)}$; $\frac{1}{(s+2)^2(s+1)} = \frac{1}{s+1} - \frac{1}{(s+2)^2} + \frac{A}{s+2}$, $s=0: \frac{1}{4} = 1 - \frac{1}{4} + \frac{A}{2}$ $\therefore A = -1$

$$\therefore F(s) = e^{-3s}\left[\frac{1}{s+1} - \frac{1}{(s+2)^2} - \frac{1}{s+2}\right] \therefore f(t) = \underline{[e^{-(t-3)} - (t-3)e^{-2(t-3)} - e^{-2(t-3)}]u(t-3)}$$

**19-24** (a)

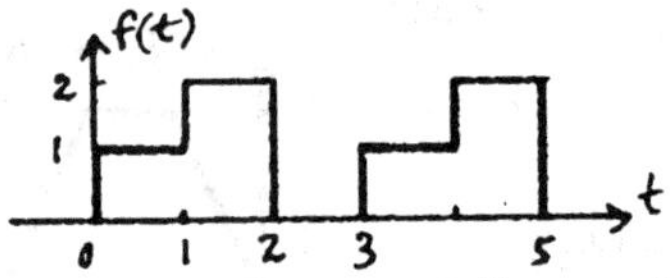


$T = 3$, $f(t) = u(t) + u(t-1) - 2u(t-2)$

$$\therefore F_1(s) = \frac{1}{s}\left(1 + e^{-s} - 2e^{-2s}\right), \therefore F(s) = \frac{1 + e^{-s} - 2e^{-2s}}{s(1 - e^{-3s})}$$

(b) $F(0.5) = \dfrac{1 + e^{-0.5} - 2e^{-1}}{0.5(1 - e^{-1.5})} = \underline{2.242}$

**19-25**

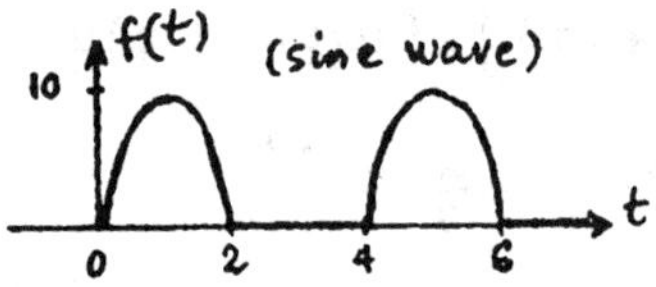


$T = 4$; $f_1(t) = 10 \sin \frac{\pi t}{2}\left[u(t) - u(t-2)\right]$

$$= 10 \sin \frac{\pi t}{2} u(t) + 10 \sin \frac{\pi(t-2)}{2} u(t-2)$$

$$\therefore F_1(s) = \frac{10(\pi/2)}{s^2 + \pi^2/4}\left(1 + e^{-2s}\right) = \frac{5\pi(1 + e^{-2s})}{s^2 + \pi^2/4}$$

$$\therefore F(s) = \frac{5\pi(1 + e^{-2s})}{(s^2 + \pi^2/4)(1 - e^{-4s})} = \frac{5\pi}{(s^2 + \pi^2/4)(1 - e^{-2s})}$$

**19-26** (a) $F(s) = \dfrac{5s}{s^2 + 4}$ $\therefore f(t) = 5 \cos 2t\, u(t)$

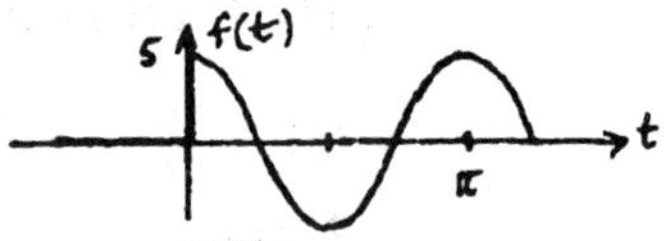


(b) $F(s) = \dfrac{5s}{s^2+4} e^{-\pi s/2}$ $\therefore f(t) = 5\cos\left[2(t - \pi/2)\right] u(t - \pi/2)$

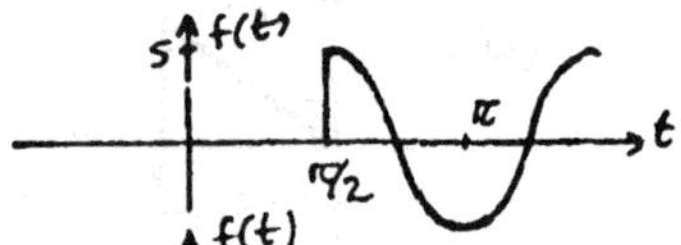


(c) $F(s) = \dfrac{5s}{s^2+4}\left(1 + e^{-\pi s/2}\right)$

$\therefore f(t) = 5\cos 2t\, u(t) + 5\cos(2t - \pi) u(t - \pi/2)$

$f(t) = 5 \cos 2t \left[u(t) - u(t - \pi/2)\right]$

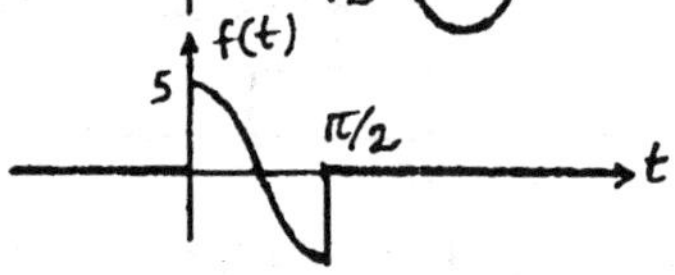


(d) $F(s) = \dfrac{5s}{s^2 + 4} \dfrac{1 + e^{-0.5\pi s}}{1 - e^{-1.25\pi s}}$

$\therefore f(t) = 5 \cos 2t \left[u(t) - u(t - \pi/2)\right]$, $T = 1.25\pi$

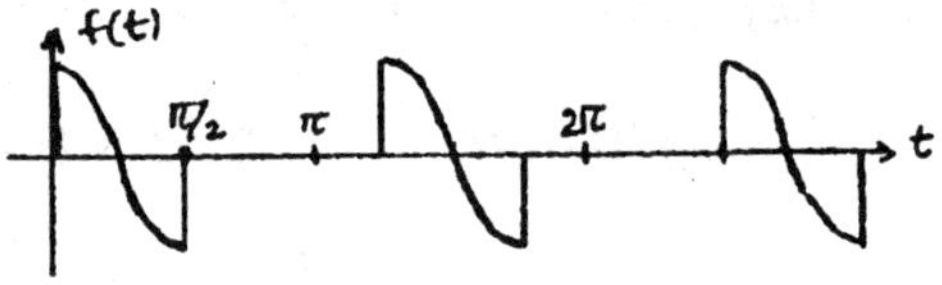


**19-27** (a) $f(t) = 50\cos(10t - 0.8) = 50\cos 0.8 \cos 10t + 50 \sin 0.8 \sin 10t$

$$\therefore F(s) = \frac{50 s \cos 0.8 + 500 \sin 0.8}{s^2 + 100} = \frac{34.84 s + 358.7}{s^2 + 100}$$

(b) $f(t) = (t-1)u(t-2) = (t-2)u(t-2) + u(t-2)$ $\therefore F(s) = \dfrac{e^{-2s}}{s^2} + \dfrac{e^{-2s}}{s} = e^{-2s}\left(\dfrac{1}{s^2} + \dfrac{1}{s}\right)$

(c) $f(t) = \sum_0^{\infty} 13(t - 4n)\left[u(t - 4n) - u(t - 2 - 4n)\right]$

$\therefore f_1(t) = 13t\left[u(t) - u(t-2)\right] = 13\left[t u(t) - (t-2)u(t-2) - 2u(t-2)\right]$, $T = 4$

$$\therefore F_1(s) = \frac{13}{s^2} - \frac{13}{s^2} e^{-2s} - \frac{26}{s} e^{-2s} \quad \therefore F(s) = \frac{13/s^2 - \frac{13}{s^2}e^{-2s} - \frac{26}{s}e^{-2s}}{1 - e^{-4s}}$$

**19-28** (a) $\mathcal{L}\{5t^3u(t)\} = \frac{5\times 6}{s^4}$ $\therefore \mathcal{L}\{5t^3e^{-2t}u(t)\} = \frac{30}{(s+2)^4}$ ←

(b) $\mathcal{L}\{e^{-5t}u(t)\} = \frac{1}{s+5}$ $\therefore \mathcal{L}\{(e^{-5t}-e^{-4t})\} = \frac{1}{s+5} - \frac{1}{s+4}$

$\therefore \mathcal{L}\{\frac{1}{t}(e^{-5t}-e^{-4t})u(t)\} = \int_s^\infty (\frac{1}{s+5} - \frac{1}{s+4})ds = \ln \frac{s+5}{s+4}\Big|_s^\infty = \ln \frac{s+4}{s+5}$ ←

(c) $\mathcal{L}\{f(t)\} = \frac{s}{5+\ln(s+1)}$, $\mathcal{L}\{f(2t)\} = \frac{1}{2}F(\frac{s}{2}) = \frac{1}{2}\frac{s/2}{5+\ln(1+s/2)} = \frac{0.25s}{5-\ln 2+\ln(s+2)}$ ←

**19-29** (a) $F(s) = \frac{s^3+6s^2+13s+11}{(s+2)^3(s+3)} = \frac{A}{s+2} + \frac{B}{(s+2)^2} + \frac{1}{(s+2)^3} + \frac{1}{s+3}$

$s=0: \frac{11}{8\times 3} = \frac{A}{2} + \frac{B}{4} + \frac{1}{8} + \frac{1}{3}$ $\therefore 11 = 12A + 6B + 11$, $B = -2A$

$s=-1: \frac{-1+6-13+11}{1\times 2} = \frac{3}{2} = A + B + 1 + \frac{1}{2}$ $\therefore B = -A$

$\therefore A = B = 0$

$\therefore f(t) = \mathcal{L}^{-1}\{\frac{1}{(s+2)^3} + \frac{1}{s+3}\} = (\frac{1}{2}t^2e^{-2t} + e^{-3t})u(t)$ ←

(b) $F_b(s) = \frac{s^2-4}{(s^2+4)^2}$ Try $\frac{d}{ds}(\frac{s}{s^2+4}) = \frac{s^2+4-2s^2}{(s^2+4)^2} = \frac{4-s^2}{(s^2+4)^2}$

$\therefore F_b(s) = \frac{d}{ds}(\frac{-s}{s^2+4})$. Now, $\mathcal{L}^{-1}\{\frac{-s}{s^2+4}\} = -\cos 2t\, u(t) = f(t)$

$\therefore \mathcal{L}^{-1}\{F_b(s)\} = \mathcal{L}^{-1}\{\frac{d}{ds}(\frac{-s}{s^2+4})\} = \mathcal{L}^{-1}\{\frac{s^2-4}{(s^2+4)^2}\} = -tf(t) = \underline{t\cos 2t\, u(t)}$

(c) $\mathcal{L}\{f(t)\} = \frac{e^{-4s}}{s^2+100} = F(s)$ $\therefore F(2s) = \frac{e^{-8s}}{4s^2+100} = \frac{0.25e^{-8s}}{s^2+25}$

$\therefore f(t/2) = 0.05\sin[5(t-8)]u(t-8)$ ←

**19-30** (a) $F(s) = \frac{3}{(s+10)^4}$ $\therefore f(t) = 3\times\frac{t^3}{6}e^{-10t}u(t) = \underline{\frac{1}{2}t^3e^{-10t}u(t)}$ (time shift)

(b) $F(s) = \frac{e^{-2(s+1)}}{(s+1)^2}$; $\mathcal{L}^{-1}\{\frac{1}{s^2}\} = tu(t)$ $\therefore \mathcal{L}^{-1}\{\frac{e^{-2s}}{s^2}\} = (t-2)u(t-2)$

$\therefore \mathcal{L}^{-1}\{\frac{e^{-2(s+1)}}{(s+1)^2}\} = (t-2)e^{-t}u(t-2)$ ← (freq. shift)

(c) $F(s) = \frac{2}{s^2(s+2)} = \frac{A}{s} + \frac{1}{s^2} + \frac{0.5}{s+2}$; $s=1: \frac{2}{3} = A + 1 + \frac{1}{6}$ $\therefore A = -\frac{1}{2}$

$\therefore f(t) = -0.5u(t) + tu(t) + 0.5e^{-2t}u(t) = (-0.5 + t + 0.5e^{-2t})u(t)$ ←

**19-31** (a) $F_0(s) = \frac{10}{s^2+25}$ $\therefore f_0(t) = \underline{2\sin 5t\, u(t)}$

(b) $F_{-1}(s) = \int_s^\infty \frac{10\,ds}{s^2+25} = \frac{10}{5}\tan^{-1}\frac{s}{5}\Big|_s^\infty = \underline{\pi - 2\tan^{-1}0.2s}$

$f_{-1}(t) = \frac{1}{t}f_0(t) = \frac{2}{t}\sin 5t\, u(t)$ ←

(c) $F_1(s) = \frac{dF_0(s)}{ds} = \frac{d}{ds}(\frac{10}{s^2+25}) = -\frac{20s}{(s^2+25)^2}$ ←; $f_1(t) = -tf_0(t) = \underline{-2t\sin 5t\, u(t)}$

**19-32** $i' + 4i + 3\int_0^t i(z)\,dz = 12(t-1)u(t-1)\ ;\ i(0^-) = 0$

$\therefore sI(s) - 0 + 4I(s) + \frac{3}{s}I(s) = \frac{12}{s^2}e^{-s} = I(s)\frac{s^2+4s+3}{s}$

$\therefore I(s) = s\left(\frac{12}{s^2}e^{-s}\right)\frac{1}{(s+1)(s+3)} = \frac{12e^{-s}}{s(s+1)(s+3)} = \frac{4}{s}e^{-s} - \frac{6}{s+1}e^{-s} + \frac{2}{s+3}e^{-s}$

$\frac{4}{s} - \frac{6}{s+1} + \frac{2}{s+3} \Leftrightarrow (4 - 6e^{-t} + 2e^{-3t})u(t)\ \therefore\ i(t) = \underline{\left[4 - 6e^{-(t-1)} + 2e^{-3(t-1)}\right]u(t-1)}$

**19-33** (a) $F(s) = \frac{3 - e^{-2s}}{s(s^2+2s+5)}\ \therefore\ f(0^+) = \lim_{s\to\infty}[sF(s)] = \lim_{s\to\infty}\frac{3-e^{-2s}}{s^2+2s+5} = \underline{\underline{0}}$

x's at $s = (-2 + \sqrt{-16})/2$ and 0 $\therefore$ OK $\therefore f(\infty) = \lim_{s\to 0}[sF(s)]$

$\therefore f(\infty) = \lim_{s\to\infty}\frac{3-e^{-2s}}{s^2+2s+5} = \underline{0.4}$

(b) $F(s) = \frac{3s}{s^2+2s+5}$, $f(0^+) = \lim_{s\to\infty}[sF(s)] = 3s$ ; x's ok $\therefore f(\infty) = \lim_{s\to 0}[sF(s)] = \underline{\underline{0}}$

(c) $F(s) = \frac{3}{1-e^{-2s}}$, $f(0^+) = \lim_{s\to\infty}\frac{3s}{1-e^{-2s}} = \underline{\infty}$

x's when $e^{-2s} = 1$ ; $s = 0$ is a simple pole since $\lim_{s\to 0}\frac{3s}{1-e^{-2s}} = 1.5$

However, for $s = j\omega$, $e^{-j2\omega} = 1$, $\omega = \pm n\pi$ $\therefore$ $j\omega$-axis poles

$\therefore f(\infty)$ is <u>indeterminate</u>.

**19-34** (a) $\lim_{s\to 0}[sF(s)] = 10\ \therefore f(\infty) = 10$ ; $\lim_{s\to\infty}[sF(s)] = 25\ \therefore f(0^+) = 25$

Let $F(s) = \frac{k(s+a)}{s(s+b)}$ (seems simplest form possible)

$\therefore k = 25$ ; $\frac{25a}{b} = 10$ let $a = 1$ (arbitrary) $\therefore b = 2.5\ \therefore F(s) = \frac{25(s+1)}{s(s+2.5)}$ ←

(b) $F(s) = \frac{10}{s} + \frac{15}{s+2.5}\ \therefore f(t) = (10 + 15e^{-2.5t})u(t)$ ←

(c) $f(0^+) = 10 + 15 = \underline{\underline{25}}$, $f(\infty) = \underline{\underline{10}}$, Initial + final value theorems are used in part (a)

**19-35** (a) $F(s) = \frac{50(s+10)}{s(s+20)}$, $f(0^+) = \lim_{s\to\infty} sF(s) = \lim_{s\to\infty}\frac{50(s+10)}{s+20} = \underline{\underline{50}}$

$f(\infty) = \lim_{s\to 0} sF(s) = \lim_{s\to 0}\frac{50(s+10)}{s+20} = 25$, x's ok ; $F(s) = \frac{25}{s} + \frac{25}{s+20}$

$\therefore f(t) = \underline{25(1 + e^{-20t})u(t)}\ \therefore f(0^+) = \underline{\underline{50}}$ and $f(\infty) = \underline{\underline{25}}$ ok.

(b) $F(s) = \frac{25(1+e^{-10s})}{s}$, $f(0^+) = \lim_{s\to\infty} 25(1+e^{-10s}) = \underline{\underline{25}}$

$f(\infty) = \lim_{s\to 0} 25(1+e^{-10s}) = \underline{\underline{50}}$, x's OK ; $F(s) = \frac{25}{s} + \frac{25}{s}e^{-10s}$

$\therefore f(t) = \underline{25u(t) + 25u(t-10)}\ \therefore f(0^+) = \underline{\underline{25}}$ and $f(\infty) = \underline{\underline{50}}$ OK

(cont. on next page)

**19-35 (cont.)** (c) $F(s) = \dfrac{2s+6}{s^2+8s+20}$, $f(0^+) = \lim_{s\to\infty} \dfrac{s(2s+6)}{s^2+8s+20} = \underline{2}$

$f(\infty) = \lim_{s\to 0} \dfrac{s(2s+6)}{s^2+8s+20} = \underline{0}$ x's OK ; $F(s) = \dfrac{2(s+4)}{(s+4)^2+4} - \dfrac{2}{(s+4)^2+4}$

$\therefore f(t) = \underline{(2e^{-4t}\cos 2t - e^{-4t}\sin 2t)\,u(t)}$ $\therefore f(0^+) = \underline{2}$ and $f(\infty) = \underline{0}$ OK

**19-36** $F(s) = K\dfrac{(s-2)(s^2+4s+20)}{s(s+3)(s+5)(s^2+2s+2)}$, $F(1) = -5 = \dfrac{K(-1)(25)}{1(4)(6)(5)}$ $\therefore K = 24$

(a) $f(0^+) = \lim_{s\to\infty} \dfrac{s(24)(s-2)(s^2+4s+20)}{s(s+3)(s+5)(s^2+2s+2)} = \underline{0}$

(b) $f(\infty) = \lim_{s\to 0} \dfrac{s(24)(s-2)(s^2+4s+20)}{s(s+3)(s+5)(s^2+2s+2)} = \dfrac{24(-2)(20)}{3(5)(2)} = \underline{-32}$ x's OK

(c) $F(1+j5) = 24(-1+j5)(1+j10-25+4+j20+20)/[(1+j5)(4+j5)(6+j5)(-24+j10+2+j10+2)]$

$F(1+j5) = \dfrac{36(5+j1)}{-206+j296} = \underline{0.5090\angle{-113.53^\circ}}$

**19-37**

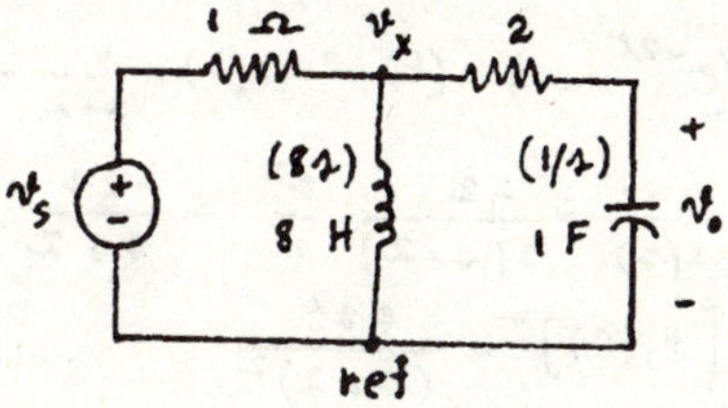


$\dfrac{V_x - V_s}{1} + \dfrac{V_x}{8s} + \dfrac{V_x - V_o}{2} = 0$ $\therefore 8s(V_x - V_s) + V_x + 4s(V_x - V_o) = 0$

$\therefore V_x = \dfrac{4sV_o + 8sV_s}{1+12s}$

Also: $\dfrac{V_o - V_x}{2} + sV_o = 0$

$\therefore \dfrac{4sV_o + 8sV_s}{1+12s} = V_o(1+2s)$

$(24s^2 + 14s + 1 - 4s)V_o = 8sV_s$

$\therefore \dfrac{V_o}{V_s} = H(s) = \dfrac{8s}{24s^2+10s+1} = \dfrac{s/3}{(s+1/4)(s+1/6)} = \dfrac{1}{s+1/4} - \dfrac{2/3}{s+1/6}$

$\therefore h(t) = (e^{-t/4} - \frac{2}{3}e^{-t/6})\,u(t)$ ←

**19-38** $H(s) = \dfrac{I_{out}}{V_{in}} = \dfrac{2(s+10)}{(s+4)(s^2+4)}$, $v_{in}(t) = 8u(t)$ $\therefore V_{in}(s) = \dfrac{8}{s}$

$\therefore I_{out}(s) = \dfrac{16(s+10)}{s(s+4)(s^2+4)} = \dfrac{10}{s} - \dfrac{1.2}{s+4} + \dfrac{As+B}{s^2+4}$

$s=1:\ \dfrac{16\times 11}{5\times 5} = 10 - 0.24 + \dfrac{A+B}{5}$ $\therefore A+B = -13.6$

$s=-1:\ \dfrac{16\times 9}{-3\times 5} = -10 - 0.4 + \dfrac{-A+B}{5}$ $\therefore -A+B = 4$

$\therefore 2B = -9.6$, $B = -4.8$, $A = -8.8$

$\therefore I_{out}(s) = \dfrac{10}{s} - \dfrac{1.2}{s+4} - \dfrac{8.8s}{s^2+4} - \dfrac{4.8}{s^2+4}$

$\therefore i_{out}(t) = (10 - 1.2e^{-4t} - 8.8\cos 2t - 2.4\sin 2t)\,u(t)$ ←

**19-39** (a)

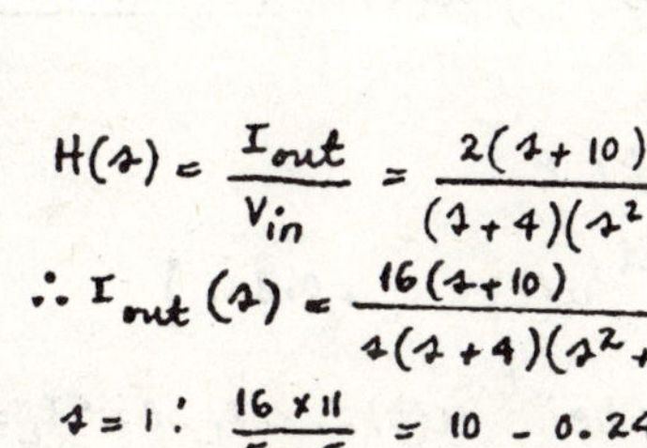


$Z_{in} = \dfrac{(s+3)/s}{s+3+\frac{1}{s}} = \dfrac{s+3}{s^2+3s+1}$ ; $V_{out} = V_{in}\dfrac{Z_{in}}{1+Z_{in}}$

$\therefore V_{out} = \dfrac{s+3}{s^2+4s+4}V_{in}$ ; $H(s) = \dfrac{V_{out}}{V_{in}} = \dfrac{s+3}{(s+2)^2} = \dfrac{1}{s+2} + \dfrac{1}{(s+2)^2}$

$\therefore h(t) = (e^{-2t} + te^{-2t})\,u(t)$ ←

(b) $v_{in}(t) = 10u(t)$ $\therefore V_{in}(s) = \dfrac{10}{s}$ $\therefore V_{out}(s) = \dfrac{10}{s}\left(\dfrac{s+3}{(s+2)^2}\right) = \dfrac{7.5}{s} + \dfrac{A}{s+2} - \dfrac{5}{(s+2)^2}$

$s=-1:\ \dfrac{10\times 2}{-1} = -7.5 + A - 5$ $\therefore A = -7.5$ $\therefore v_{out} = (7.5 - 7.5e^{-2t} - 5te^{-2t})\,u(t)$ ←

**19-40**

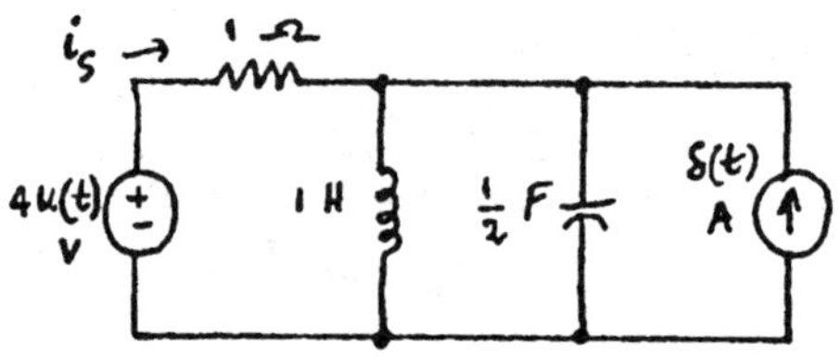


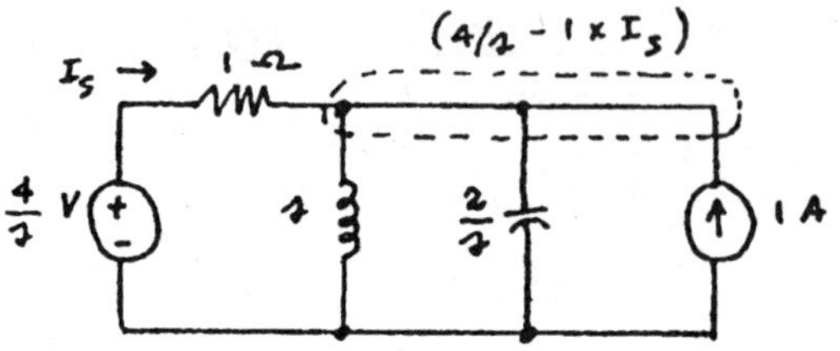


$$I_s = \frac{4/s - I_s}{s} + \frac{4/s - I_s}{2/s} - 1 \therefore I_s = \frac{4}{s^2} - \frac{1}{s} I_s + 2 - \frac{s}{2} I_s - 1 = \frac{4+s^2}{s^2} - I_s \frac{s^2+2}{2s}$$

$$\therefore I_s\left(1 + \frac{s^2+2}{2s}\right) = I_s \frac{s^2+2s+2}{2s} = \frac{s^2+4}{s^2} \therefore I_s = \frac{2(s^2+4)}{s(s^2+2s+2)} = \frac{4}{s} + \frac{As+B}{s^2+2s+2}$$

$$s=1: \frac{2\times 5}{1\times 5} = 4 + \frac{A+B}{5} \therefore A+B = -10$$

$$s=-1: \frac{2\times 5}{-1} = -4 + \frac{-A+B}{1} \therefore -A+B = -6$$

$$\therefore B=-8,\; A=-2 \quad \therefore I_s = \frac{4}{s} - \frac{2s+8}{(s+1)^2+1}$$

$$\therefore I_s = \frac{4}{s} - \frac{2(s+1)}{(s+1)^2+1} - \frac{6}{(s+1)^2+1} \therefore i_s(t) = \underline{(4 - 2e^{-t}\cos t - 6e^{-t}\sin t)\,u(t)}$$

**19-41**

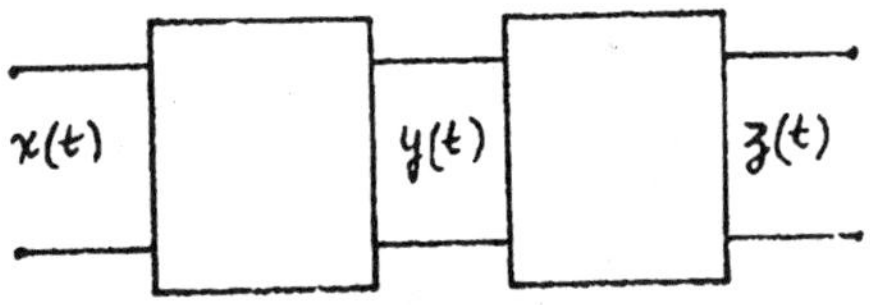


$$x(t) = t\,u(t) \therefore X(s) = \frac{1}{s^2}$$

$$\therefore y(t) = (e^{-2t} - 1)\,u(t) \therefore Y(s) = \frac{1}{s+2} - \frac{1}{s} = \frac{-2}{s(s+2)}$$

$$\therefore H_1(s) = \frac{Y(s)}{X(s)} = \frac{-2}{s(s+2)}\,\frac{s^2}{1} = \frac{-2s}{s+2}$$

$$H(s) = [H_1(s)]^2 = \frac{4s^2}{(s+2)^2}$$

Now let $x(t) = u(t) \therefore X(s) = \frac{1}{s} \therefore Z(s) = \frac{1}{s} \times \frac{4s^2}{(s+2)^2} = \frac{4s}{(s+2)^2} = \frac{4s+8-8}{(s+2)^2}$

$$\therefore Z(s) = \frac{4}{s+2} - \frac{8}{(s+2)^2} \therefore z(t) = (4e^{-2t} - 8te^{-2t})\,u(t) = \underline{4e^{-2t}(1-2t)\,u(t)}$$

**19-42**

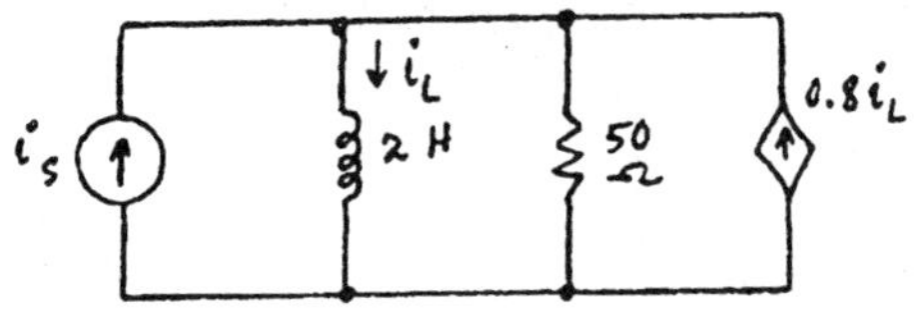


$$i_s = 16te^{-t}\,u(t) \text{ A}$$

$$2i_L' = 50\left[16te^{-t}\,u(t) - i_L + 0.8\,i_L\right]$$

$$i_L' + 5i_L = 400\,te^{-t}\,u(t) \therefore sI_L - 0 + 5I_L = \frac{400}{(s+1)^2}$$

$$\therefore I_L = \frac{400}{(s+1)^2(s+5)} = \frac{A}{s+1} + \frac{100}{(s+1)^2} + \frac{25}{s+5}\;;\; s=0: 80 = A + 100 + 5 \therefore A = -25$$

$$\therefore I_L = \frac{-25}{s+1} + \frac{100}{(s+1)^2} + \frac{25}{s+5} \therefore i_L(t) = (-25e^{-t} + 100te^{-t} + 25e^{-5t})\,u(t) \text{ A}$$

**19-43** (a)

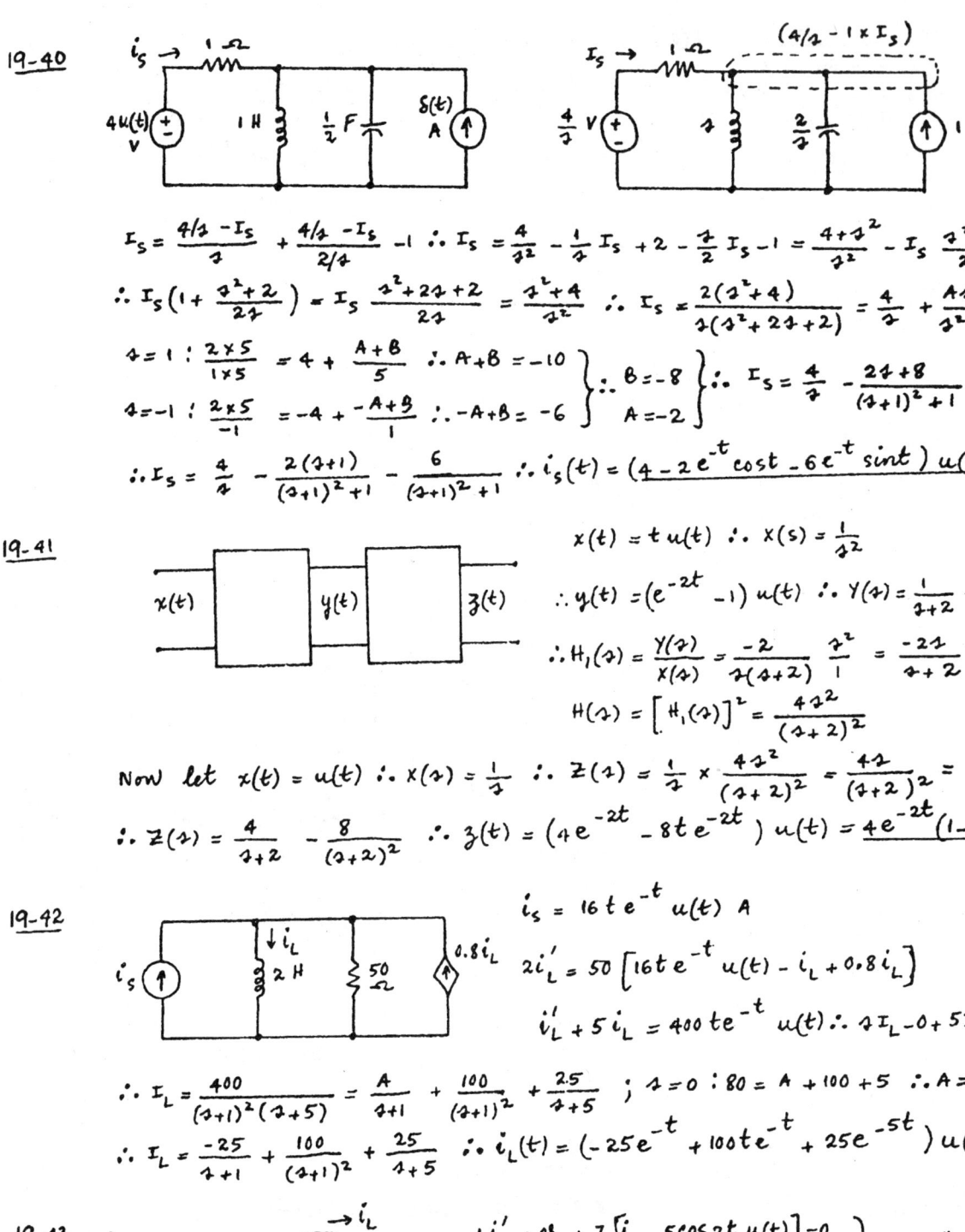

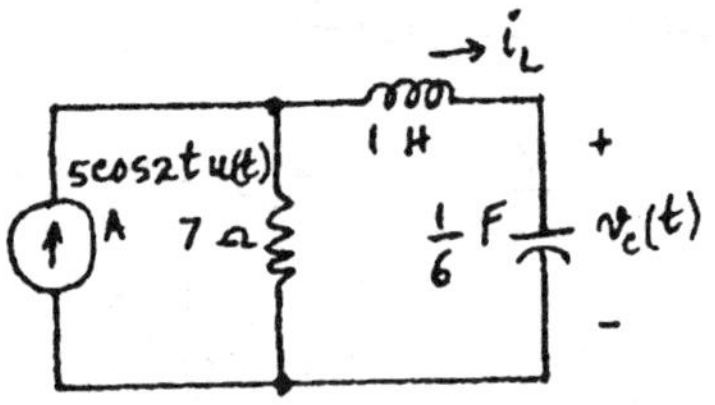


$i_L(0^-) = 4$ A, $v_c(0^-) = 7$ V

$$1i_L' + v_c + 7\left[i_L - 5\cos 2t\,u(t)\right] = 0$$

$$i_L = \frac{1}{6}v_c' \therefore v_c'(0^-) = 6\times 4 = 24 \text{ V/s}$$

$$\therefore \frac{1}{6}v_c'' + v_c + \frac{7}{6}v_c' = 35\cos 2t\,u(t)$$

$$\therefore v_c'' + 7v_c' + 6v_c = 210\cos 2t\,u(t)$$

$$\therefore s^2V_c - 7s - 24 + 7sV_c - 7\times 7 + 6V_c = \frac{210s}{s^2+4}$$

(cont. on next page)

**19-43 (cont.)**

$$\therefore V_c(s^2+7s+6) = \frac{210s}{s^2+4} + 7s + 73 \quad \therefore V_c = \frac{7s^3+73s^2+238s+292}{(s^2+4)(s+1)(s+6)} \leftarrow$$

(b) $V_c(s) = \frac{4.8}{s+1} + \frac{0.1}{s+6} + \frac{As+B}{s^2+4}$

$s=0: \frac{292}{24} = 4.8 + \frac{1}{60} + \frac{B}{4} \quad \therefore B = 29.4$

$s=1: \frac{610}{70} = 2.4 + \frac{1}{70} + \frac{A}{5} + \frac{29.4}{5} \quad \therefore A = 2.1$

$$\therefore V_c(s) = \frac{4.8}{s+1} + \frac{0.1}{s+6} + \frac{2.1s}{s^2+4} + \frac{29.4}{s^2+4}$$

$$\therefore v_c(t) = (4.8e^{-t} + 0.1e^{-6t} + 2.1\cos 2t + 14.7\sin 2t)\,u(t) \leftarrow$$

**19-44** (a)

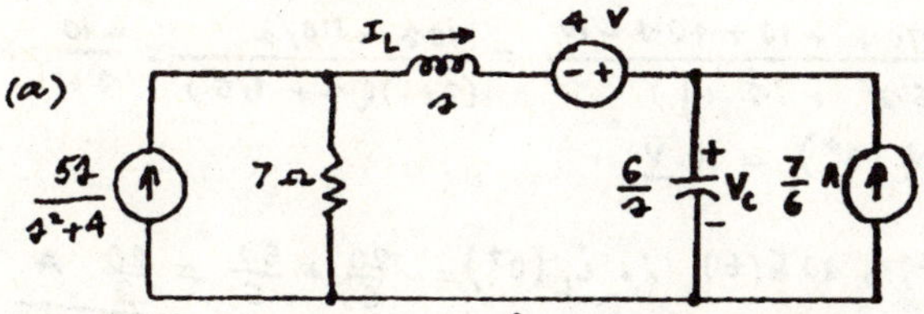


(b) Mesh around center: $7\left(I_L - \frac{5s}{s^2+4}\right) + sI_L - 4 + \frac{6}{s}\left(I_L + \frac{7}{6}\right) = 0$

$\therefore \left(7 + s + \frac{6}{s}\right)I_L = \frac{35s}{s^2+4} - \frac{7}{s} + 4$

Also $V_c = \frac{6}{s}\left(I_L + \frac{7}{6}\right) \quad \therefore I_L = \frac{sV_c - 7}{6}$

$$\therefore V_c\left(\frac{s^2+7s+6}{6}\right) = \frac{7}{6}\frac{(s^2+7s+6)}{s} + \frac{4s-7}{s} + \frac{35s}{s^2+4}$$

$$\therefore V_c(s^2+7s+6) = 49 + 7s + \frac{42}{s} + 24 - \frac{42}{s} + \frac{210s}{s^2+4}$$

$$\therefore V_c(s^2+7s+6) = 7s + 73 + \frac{210s}{s^2+4} = \frac{7s^3+73s^2+238s+292}{s^2+4}$$

$$\therefore V_c = \frac{7s^3+73s^2+238s+292}{(s+1)(s+6)(s^2+4)} \leftarrow$$

(c) Same as 19-43 above

**19-45**

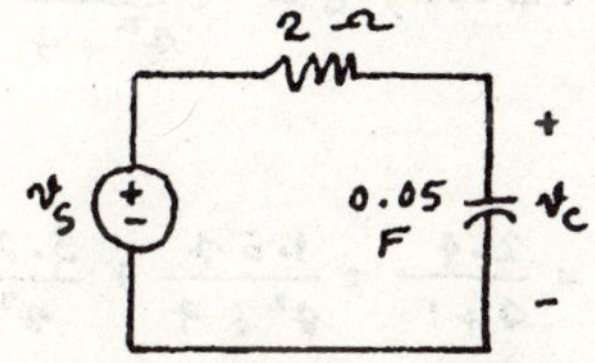


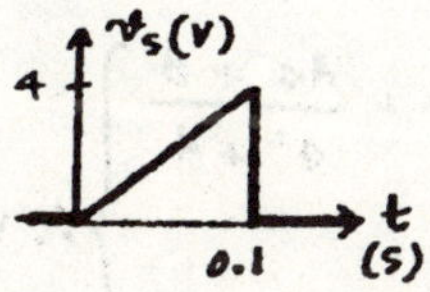


$$v_s = 40t\left[u(t) - u(t-0.1)\right] = 40t\,u(t) - 40(t-0.1)u(t-0.1) - 4u(t-0.1)$$

(s) $\therefore V_s = \frac{40}{s^2} - \frac{40}{s^2}e^{-0.1s} - \frac{4}{s}e^{-0.1s}$

$$H(s) = \frac{V_c}{V_s} = \frac{20/s}{2+20/s} = \frac{10}{s+10} \quad \therefore V_c = V_s H(s) = \frac{10}{s+10}\left(\frac{40}{s^2} - \frac{40}{s^2}e^{-0.1s} - \frac{4}{s}e^{-0.1s}\right)$$

Now, $\frac{400}{s^2(s+10)} = \frac{A}{s} + \frac{40}{s^2} + \frac{4}{s+10}$; $s = 10: 0.2 = 0.1A + 0.4 + 0.2 \quad \therefore A = -4$

Also $\frac{40}{s(s+10)} = \frac{4}{s} - \frac{4}{s+10}$

$$\therefore V_c = -\frac{4}{s} + \frac{40}{s^2} + \frac{4}{s+10} - e^{-0.1s}\left(-\frac{4}{s^2} + \frac{40}{s^2} + \frac{4}{s+10}\right) - e^{-0.1s}\left(\frac{4}{s} - \frac{4}{s+10}\right)$$

$$= -\frac{4}{s} + \frac{40}{s^2} + \frac{4}{s+10} - \frac{40}{s^2}e^{-0.1s} \quad \therefore v_c(t) = (-4 + 40t + 4e^{-10t})u(t) - 40(t-0.1)u(t-0.1)$$

$0 < t < 0.1 \quad \therefore v_c(t) = 40t - 4(1 - e^{-10t})$ V, $v_c = 1$ at $t = \underline{0.08012\ s}$

$t > 0.1 \quad v_c(t) = -4 + 40t + 4e^{-10t} - 40t + 4 \quad \therefore 4e^{-10t} = 1 \quad \therefore t = \underline{0.13863\ s}$

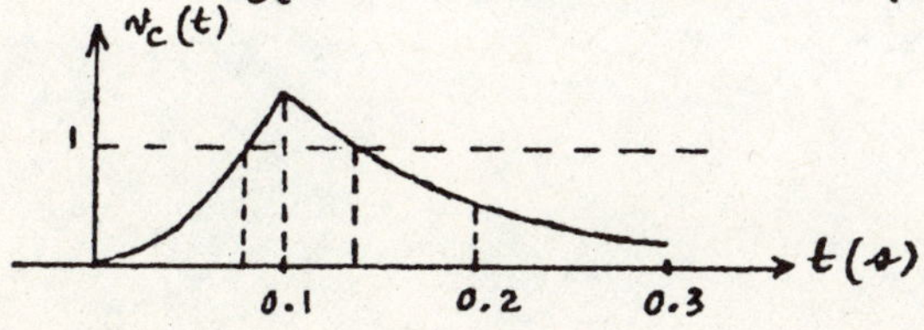

**19-46**

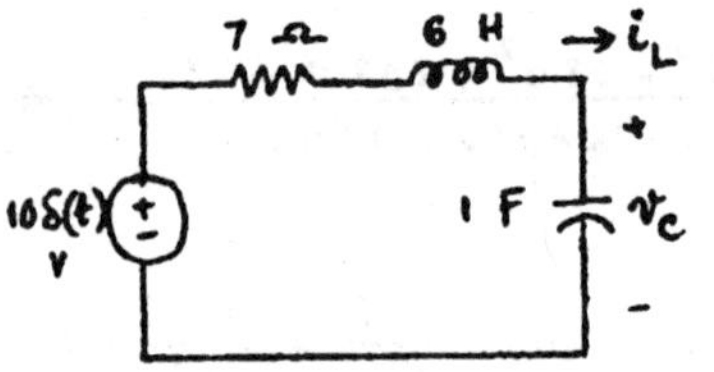


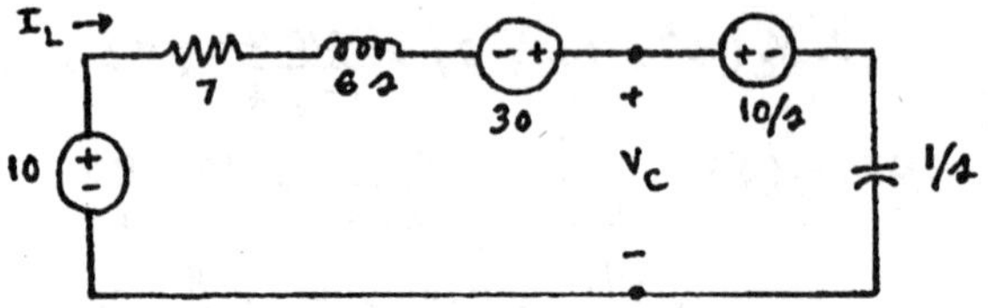


$i_L(0^-) = 5$ A , $v_C(0^-) = 10$ V

$$10 = \left(7 + 6s + \frac{1}{s}\right) I_L - 30 + \frac{10}{s}$$

$$\therefore I_L = \frac{40 - 10/s}{6s + 7 + 1/s} = \frac{40s - 10}{6s^2 + 7s + 1}$$

$$V_C = \frac{10}{s} + \frac{40s - 10}{s(6s^2 + 7s + 1)} = \frac{60s^2 + 70s + 10 + 40s - 10}{s(6s^2 + 7s + 1)} = \frac{10s + 110/6}{(s+1)(s + 1/6)} = \frac{-10}{s+1} + \frac{20}{s + 1/6}$$

$$v_C(t) = \underline{(20e^{-t/6} - 10e^{-t})u(t)} \quad \therefore v_C(0^+) = \underline{10 \text{ V}}$$

$$i_L = Cv_C' = \left(-\frac{20}{6}e^{-t/6} + 10e^{-t}\right)u(t) + 10\delta(t) \quad \therefore i_L(0^+) = -\frac{20}{6} + \frac{60}{6} = \underline{\frac{20}{3}} \text{ A}$$

**19-47**

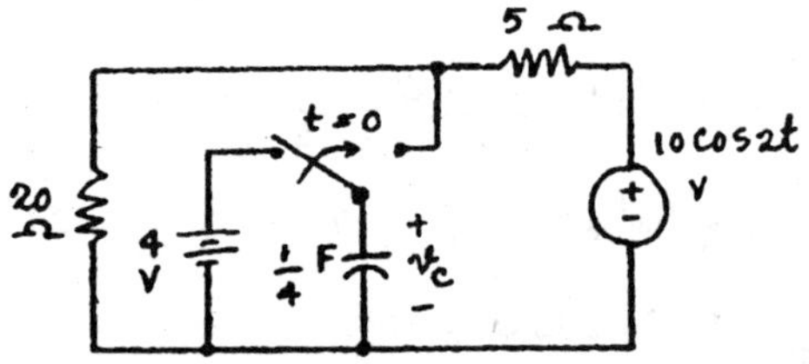


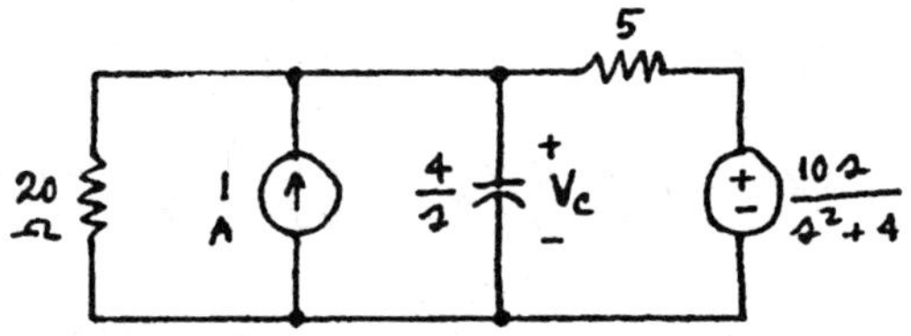


$v_C(0^+) = v_C(0^-) = 4$ V ; $Cv_C(0^-)\delta(t) = \delta(t)$ A $\rightarrow$ 1 A

$$\therefore \frac{V_C}{20} - 1 + \frac{sV_C}{4} + \frac{1}{5}\left(V_C - \frac{10s}{s^2+4}\right) = 0 , \quad (0.05 + 0.25s + 0.2)V_C = 1 + \frac{2s}{s^2+4} = \frac{s^2+2s+4}{s^2+4}$$

$$\therefore V_C = \frac{s^2 + 2s + 4}{0.25(s+1)(s^2+4)} = \frac{2.4}{s+1} + \frac{As + B}{s^2 + 4}$$

$s = 0$ : $4 = 2.4 + 0.25B$ $\therefore B = 6.4$

$s = 1$ : $\frac{7}{2.5} = 1.2 + \frac{A + 6.4}{5}$ $\therefore A = 1.6$

$$\therefore V_C = \frac{2.4}{s+1} + \frac{1.6s}{s^2+4} + \frac{3.2 \times 2}{s^2+4}$$

$\therefore v_C(t) = (2.4e^{-t} + 1.6\cos 2t + 3.2\sin 2t)$ for $t > 0$

$\therefore v_C(t) = \underline{4u(-t) + (2.4e^{-t} + 1.6\cos 2t + 3.2\sin 2t)u(t)}$ V , all $t$

NOTES

- NOTES -

NOTES

NOTES

# NOTES

- NOTES -